Mathematical Techniques of Fractional Order Systems

Mathematical Techniques of Fractional Order Systems

Edited by

Ahmad Taher Azar
Faculty of Computers and Information, Benha University, Benha, Egypt
School of Engineering and Applied Sciences, Nile University, Giza, Egypt

Ahmed G. Radwan
Engineering Mathematics and Physics Department, Faculty of Engineering, Cairo University, Giza, Egypt
Nanoelectronics Integrated Systems Center (NISC), Nile University, Cairo, Egypt

Sundarapandian Vaidyanathan
Research and Development Center, Vel Tech University, Chennai, India

ELSEVIER

Elsevier
Radarweg 29, PO Box 211, 1000 AE Amsterdam, Netherlands
The Boulevard, Langford Lane, Kidlington, Oxford OX5 1GB, United Kingdom
50 Hampshire Street, 5th Floor, Cambridge, MA 02139, United States

Notices
Knowledge and best practice in this field are constantly changing. As new research and experience broaden our understanding, changes in research methods, professional practices, or medical treatment may become necessary.

Practitioners and researchers must always rely on their own experience and knowledge in evaluating and using any information, methods, compounds, or experiments described herein. In using such information or methods they should be mindful of their own safety and the safety of others, including parties for whom they have a professional responsibility.

To the fullest extent of the law, neither the Publisher nor the authors, contributors, or editors, assume any liability for any injury and/or damage to persons or property as a matter of products liability, negligence or otherwise, or from any use or operation of any methods, products, instructions, or ideas contained in the material herein.

British Library Cataloguing-in-Publication Data
A catalogue record for this book is available from the British Library

Library of Congress Cataloging-in-Publication Data
A catalog record for this book is available from the Library of Congress

ISBN: 978-0-12-813592-1

Publisher: Mara Conner
Acquisition Editor: Sonnini R. Yura
Editorial Project Manager: Peter Jardim
Production Project Manager: Vijayaraj Purushothaman
Cover Designer: Matthew Limbert

Typeset by MPS Limited, Chennai, India

Contents

List of Contributors

Amr M. AbdelAty Engineering Mathematics and Physics Department, Faculty of Engineering, Fayoum University, El Fayoum, Egypt

Salwa K. Abd-El-Hafiz Faculty of Engineering, Cairo University, Giza, Egypt

Norelys Aguila-Camacho University of Chile, Santiago, Chile

Mohsen Alimi University of Kairouan, Kairouan, Tunisia

Dalia Allam Fayoum University, Fayoum, Egypt

Ahmad Taher Azar Faculty of Computers and Information, Benha University, Benha, Egypt; School of Engineering and Applied Sciences, Nile University, Giza, Egypt

Pagavathigounder Balasubramaniam Gandhigram Rural Institute (Deemed to be University), Dindigul, Tamil Nadu, India

Yassine Bensafia Mohand Oulhadj University of Bouira, Bouira 10000, Algeria

Muzaffar A. Bhat Jamia Millia Islamia, New Delhi, India

Y. Boukal Université de Lorraine, Cosnes et Romain, France; Université Hassan II, Casablanca, Maroc; Université de Valenciennes et du Hainaut-Cambrésis, Famars, France

Bachir Bourouba University of Sétif, Sétif 19000, Algeria

Daniele Casagrande University of Udine, Udine, Italy

M. Darouach Université de Lorraine, Cosnes et Romain, France

Subir Das Indian Institute of Technology (BHU), Varanasi, Uttar Pradesh, India

Manuel A. Duarte-Mermoud University of Chile, Santiago, Chile

Magdy Eteiba Fayoum University, Fayoum, Egypt

Ibiyinka A. Fuwape Federal University of Technology, Akure, Nigeria; Michael and Cecilia Ibru University, Ughelli, Nigeria

Gokul P.M. Lodz University of Technology, Lodz, Poland

Hany N. Hassan Benha Faculty of Engineering, Benha University, Benha, Egypt; Imam Abdulrahman Bin Faisal University, Dammam, Saudi Arabia

Samar M. Ismail Faculty of Information Engineering and Technology (IET), German University in Cairo (GUC), Cairo, Egypt

Tomasz Kapitaniak Lodz University of Technology, Lodz, Poland

Khatir Khettab Mohamed Boudiaf University of M'sila, M'sila 28000, Algeria

Wiesław Krajewski Systems Research Institute, Polish Academy of Sciences, Warsaw, Poland

Jitendra Kumar Instrumentation and Control Engineering Division, Dwarka, New Delhi, India

Vineet Kumar Instrumentation and Control Engineering Division, Dwarka, New Delhi, India

Matías G. Mayol-Suárez University of Chile, Santiago, Chile

Joana P. Neto Universidade de Lisboa, Lisbon, Portugal

Samuel T. Ogunjo Federal University of Technology, Akure, Nigeria

Kayode S. Ojo University of Lagos, Lagos, Nigeria

Adel Ouannas University of Larbi Tebessi, Tebessa, Algeria

Adel Ounnas University of Larbi Tebessi, Tebessa, Algeria

Viet–Thanh Pham Hanoi University of Science and Technology, Hanoi, Vietnam; Lodz University of Technology, Lodz, Poland

N.E. Radhy Université Hassan II, Casablanca, Maroc

Ahmed G. Radwan Faculty of Engineering, Cairo University, Egypt; Nanoelectronics Integrated Systems Center (NISC), Nile University, Cairo, Egypt

Kamal Pal Singh Rana Instrumentation and Control Engineering Division, Dwarka, New Delhi, India

Abdelwaheb Rebai University of Sfax, Sfax, Tunisia

Ahmed Rhif University of Carthage, La Marsa, Tunisia

Lobna A. Said Nano-Electronics Integrated Systems Center (NISC), Nile University, Giza, Egypt

T. Sathiyaraj Gandhigram Rural Institute (Deemed to be University), Dindigul, Tamil Nadu, India

Wafaa S. Sayed Faculty of Engineering, Cairo University, Giza, Egypt

Mourad S. Semary Benha Faculty of Engineering, Benha University, Benha, Egypt

Fernando E. Serranot Central American Technical University (UNITEC), Tegucigalpa, Honduras

Bharat B. Sharma National Institute of Technology, Hamirpur, Himachal Pradesh, India

Manoj K. Shukla National Institute of Technology, Hamirpur, Himachal Pradesh, India; Lovely Professional University, Punjab, India

Shikha Singh Jamia Millia Islamia, New Delhi, India

Mayank Srivastava Indian Institute of Technology (BHU), Varanasi, Uttar Pradesh, India

Hamed Taghavian Sharif University of Technology, Tehran, Iran

Peachimuthu Tamilalagan Amrita Vishwa Vidyapeetham (Deemed to be University), Coimbatore, Tamil Nadu, India

Mohammad Saleh Tavazoei Sharif University of Technology, Tehran, Iran

Sundarapandian Vaidyanathan Vel Tech University, Chennai, Tamil Nadu, India

Duarte Valério Universidade de Lisboa, Lisbon, Portugal

Umberto Viaro University of Udine, Udine, Italy

Susana Vinga Universidade de Lisboa, Lisbon, Portugal

Christos Volos Aristotle University of Thessaloniki, Thessaloniki, Greece

Vijay K. Yadav Indian Institute of Technology (BHU), Varanasi, Uttar Pradesh, India

Dalia Yousri Fayoum University, Fayoum, Egypt

M. Zasadzinski Université de Lorraine, Cosnes et Romain, France

Preface

Fractional calculus, as generalization of integer order integration and differentiation to its noninteger (fractional) order counterpart, has proved to be a valuable tool in the modeling of many physical phenomena and engineering problems. Fractional derivatives provide an excellent instrument for the description of memory and hereditary properties of various materials and processes.

The main reason for using integer order models was the absence of solution methods for fractional differential equations. The advantages or the real objects of the fractional order systems are that we have more degrees of freedom in the model and that a "memory" is included in the model. One of the very important areas of application of fractional calculus is the chaos theory. Chaos is a very interesting nonlinear phenomenon which has been intensively studied. It is found to be useful or has great application potential in many fields such as secure communication, data encryption, financial systems, and biomedical engineering. The research efforts have been devoted to chaos control and chaos synchronization problems in nonlinear science because of its extensive applications.

Recently, studying fractional order systems has become an active research area. The chaotic dynamics of fractional order systems began to attract much attention in recent years. It has been shown that the fractional order systems can also behave chaotically, such as the fractional order Chua's system, the fractional order Lorenz system, the fractional order Chen system, the fractional order Rössler system, the fractional order modified Duffing system, the fractional order Newton–Leipnik system, the fractional order Lotka–Volterra system, and the fractional order Liu system. Moreover, recent studies show that chaotic fractional order systems can also be synchronized. Many scientists who are interested in this field have found it to be a challenging research problem to achieve the synchronization of fractional order chaotic systems, and this research area has potential applications in secure communication and cryptography.

A wide variety of mathematical methods and techniques have been used in this book to study the dynamics, control, design, and synchronization of the fractional order control systems such as sliding mode controller, active and adaptive control methods, fuzzy logic control, nonlinear control methods, fractional order discrete maps, multiswitching synchronization, metaheuristic algorithms, backstepping control, etc.

ABOUT THE BOOK

The new Elsevier book, *Mathematical Techniques of Fractional Order Systems*, consists of 21 contributed chapters by subject experts who are specialized in the various topics addressed in this book. The special chapters have been brought out in this book after a rigorous review process in the broad areas of mathematical techniques of fractional order systems. Special importance was given to chapters offering practical solutions and novel methods for the recent research problems in the mathematical modeling and control applications of fractional order systems. This book discusses trends and applications of mathematical models of fractional order systems in engineering.

OBJECTIVES OF THE BOOK

This volume presents a selected collection of contributions on a focused treatment of mathematical techniques for fractional order systems. The book also discusses multidisciplinary applications in electrical engineering, control engineering, mechanical engineering, and computer science. Both novice and expert readers should find this book a useful reference in the field of fractional order systems.

ORGANIZATION OF THE BOOK

This well-structured book consists of 21 full chapters.

BOOK FEATURES

- The book chapters deal with the recent research problems in the areas of fractional order systems.
- The book chapters present various mathematical techniques for fractional order systems such as active control, adaptive control, backstepping control, nonlinear control, fuzzy logic control, metaheuristic algorithms, etc.
- The book chapters contain a good literature survey with a long list of references.
- The book chapters are well-written with a good exposition of the research problem, methodology, block diagrams, and mathematical techniques.
- The book chapters are lucidly illustrated with numerical examples and simulations.
- The book chapters discuss details of engineering applications and future research areas.

AUDIENCE

The book is primarily meant for researchers from academia and industry, who are working on fractional order systems in the research areas—electrical engineering, control engineering, mechanical engineering, and computer science. The book can also be used at the graduate or advanced undergraduate level as a text-book or major reference for courses such as power systems, control systems, electrical devices, scientific modeling, computational science, and many others.

ACKNOWLEDGMENTS

As the editors, we hope that the chapters in this well-structured book will stimulate further research in mathematical techniques of fractional order systems, and utilize them in real-world applications.

We hope sincerely that this book, covering so many different topics, will be very useful for all readers.

We would like to thank all the reviewers for their diligence in reviewing the chapters.

Special thanks go to Elsevier, especially the book Editorial team.

Ahmad Taher Azar

Faculty of Computers and Information, Benha University, Benha, Egypt; School of Engineering and Applied Sciences, Nile University, Giza, Egypt

Ahmed G. Radwan

Engineering Mathematics and Physics Department, Faculty of Engineering, Cairo University, Giza, Egypt; Nanoelectronics Integrated Systems Center (NISC), Nile University, Cairo, Egypt

Sundarapandian Vaidyanathan

Research and Development Center, Vel Tech University, Chennai, India

Chapter 1

Variable Order Fractional Derivatives and Bone Remodeling in the Presence of Metastases

Joana P. Neto, Duarte Valério and Susana Vinga
Universidade de Lisboa, Lisbon, Portugal

1.1 INTRODUCTION

This chapter deals with variable order fractional derivatives, which are a generalization of fractional order derivatives.

1.1.1 What Are Variable Order Fractional Derivatives?

Fractional order derivatives are, of course, themselves a generalization of usual integer order derivatives and integrals, known from Calculus. Differentiation and integration notions of orders $n \in N$ are, in Fractional Calculus, generalized to account for orders $\alpha \in R$ (Wheeler, 1997). But since all real numbers are now available as orders, it is now rather easy to conceive variable order derivatives, with an order that changes with time. Among the first papers on the subject were Samko (1995) and Lorenzo and Hartley (2002a), mostly based upon the Riemann–Liouville definition of fractional derivatives. Variable order derivatives have been further developed in Valério and Sá da Costa (2011b); Valério and Sá da Costa (2013), using not only the Riemann–Liouville definition, but also the Grünwald–Letnikoff and the Caputo definitions. Additional insights are provided in Sierociuk et al. (2015a,b) and Sierociuk and Malesza (2017), including experiments with electronic circuits implementing the relevant mathematical operators.

These variable order derivatives are useful in practice, just as fractional derivatives of constant order are. In fact, being more than 300 years old,

Mathematical Techniques of Fractional Order Systems. DOI: https://doi.org/10.1016/B978-0-12-813592-1.00001-5

fractional calculus has seen a rapid growth in its applications. As it can be said that real objects are generally fractional, even if that *fractionality* is very low (Petrás, 2009), fractional calculus is particularly well suited to describe the nonlinear relationship with time of anomalous diffusion (Margin, 2006). Indeed, noninteger derivatives occur most frequently, and naturally, in physical problems where the essential mechanisms, reactions, or interactions are governed by diffusion processes. Of course, seeing that many biological processes often present anomalous diffusion, fractional calculus is an eligible and powerful tool to describe these phenomena (e.g., subthreshold nerve conduction, viscoelasticity, bioelectrodes) (Magin, 2006). But diffusion, being one of the most relevant applications of fractional calculus, is far from being the only one. Fractional derivatives are used to formulate and solve different physical models allowing a continuous transition from relaxation to oscillation phenomena; to predict the nonlinear survival and growth curves of food-borne pathogens; to adapt the viscoelasticity equations (Hooke's Law and the Newtonian fluids Law) (Rahimy, 2010; Petrás, 2009); fractional control plays an important role in general physics, thermodynamics, electrical circuits theory and fractances, mechatronic systems, signal processing, chemical mixing, chaos theory, and many others (Petrás, 2009). And there are many possible applications of fractional derivatives in control (Valério and Sá da Costa, 2006, 2011a, 2012; Azar et al., 2017).

Many physical processes, however, also appear to exhibit a fractional order behavior that varies with time or space (Lorenzo and Hartley, 2002b). In what concerns the field of viscoelasticity of certain materials, the temperature effect in small amplitude strains is known to induce changes from an elastic to viscoelastic/viscous behavior, where real applications may require a time varying temperature to be analyzed. The relaxation processes and reaction kinetics of proteins, which are described by fractional differential equations, have been found to have an order with a temperature dependence. The behavior of some diffusion processes in response to temperature changes can be better described using variable order elements rather than time-varying coefficients, among other cases (Lorenzo and Hartley, 2002b). These are natural applications where variable order operators as a function of time (t) or some other variable (x) can be introduced with profit and in a very natural way (Lorenzo and Hartley, 2002b; Valério and Sá da Costa, 2013)

This chapter presents one such application of variable order derivatives: the development of simpler mathematical models of bone remodeling, both for healthy bone, and for bone tissue affected by cancer. There are several models published in the literature, but the novelty which is the use of variable order derivatives will result in simpler models that are easier to understand.

1.1.2 Application Example of Variable Order Derivatives: Bone Remodeling

Bone is constantly being renewed, being destroyed, and formed, due to cells termed osteoclasts and osteoblasts, respectively. In the adult skeleton, both processes are in balance and tightly coupled through autocrine and paracrine factors between bone cells that allow for a constant bone density to be maintained. As this specific microenvironment provides the necessary conditions for the growth and proliferation of tumor cells, bone is a common site for the development of metastases, mainly from primary breast and prostate cancer.

Mathematical and computational models, with differential equations that represent the control mechanisms involved, can replicate this remodeling process (Komarova et al., 2003). These models have been extended to include the effects of tumor disruptive pathologies in the bone dynamics, as metastases contribute to the decoupling between bone resorption and formation and to the self-perpetuating tumor growth cycle (Ayati et al., 2010). Counteraction effects of currently used therapies were also contemplated (Ayati et al., 2010), through the pharmocokinetic (PK) and pharmacodynamic (PD) combination of anticancer (chemotherapy) with antiresorptive treatments (bisphosphonates or monoclonal antibodies) (Christ et al., 2018; Coelho et al., 2016, 2015).

Fractional and variable order derivatives can be successfully used in modeling the dynamics of bone remodeling. Christ et al. (2018) implemented fractional derivatives in the differential equations of bone remodeling presented in Ayati et al. (2010), and analyzed its dynamic behavior in the absence and presence of tumor and PK/PD applied treatment, for a discretized single point and a one-dimensional bone. Valério et al. (2016) applied the same methodology to the more accurate biochemical model of Coelho et al. (2016). Variable order derivatives have also been introduced in Neto et al. (2017), as a simplification technique, in the models of Ayati et al. (2010) and Coelho et al. (2015), in an effort to replicate the same bone microenvironment response but recurring to less parameters to impose the known bone behavior. It is the latter that is here revisited and further analyzed.

1.1.3 Chapter Organization

The remaining sections of this chapter are organized as follows. Variable order concepts and definitions are addressed in Section 1.2. Bone remodeling physiology, PK/PD concepts, and published integer mathematical models are presented in Section 1.3. Variable order bone models are address in Section 1.4. Finally, conclusions are presented in Section 1.5.

1.2 VARIABLE ORDER DERIVATIVES

This section introduces a variable order differential operator $D^{\alpha(t)}f(t)$, where $f(t)$ is the differentiated function, and $\alpha(t)$ is a time-variable differentiation order. The reader is presumed to already know what $D^{\alpha}f(t),\ \alpha\in R$ is. The variable order operator is introduced informally in Section 1.2.1 for integer orders only, and then more formally for arbitrary real orders in Section 1.2.2.

1.2.1 Intuitive Results for Integer Orders

If $\alpha(t)\in Z, \forall\ t$ then its variations with time must be steps, since it can only assume values in a discrete set. Still, this simple case illustrates an important issue when dealing with variable order derivatives: the existence or not of a memory of past values of the order. In fact, considering for instance the case

$$f(t)=t^2 \tag{1.1}$$

$$\alpha(t)=\begin{cases} 2 \text{ if } t\in[0,1[\ \cup\ [2,3[\ \cup\ [4,5[\ \cup\ \ldots \\ 1 \text{ if } t\in[1,2[\ \cup\ [3,4[\ \cup\ [5,6[\ \cup\ \ldots \end{cases} \tag{1.2}$$

it can intuitively be seen that the reasonable result of operator $D^{\alpha(t)}f(t)$, in this case, when orders are positive, should be

$$D^{\alpha(t)}f(t)=\begin{cases} 2 \text{ if } t\in[0,1[\ \cup\ [2,3[\ \cup\ [4,5[\ \cup\ \ldots \\ 2t \text{ if } t\in[1,2[\ \cup\ [3,4[\ \cup\ [5,6[\ \cup\ \ldots \end{cases} \tag{1.3}$$

In other words, the result simply jumps from one derivative to another. But things are different if there are negative orders. Consider now the case

$$f(t)=t \tag{1.4}$$

$$\alpha(t)=\begin{cases} -1 \text{ if } t\in[0,1[\ \cup\ [2,3[\ \cup\ [4,5[\ \cup\ \ldots \\ 0 \text{ if } t\in[1,2[\ \cup\ [3,4[\ \cup\ [5,6[\ \cup\ \ldots \end{cases} \tag{1.5}$$

If the result should again merely jump between the derivative of order -1 (the integral of the function) and the derivative of order 0 (the function itself), then

$$D^{\alpha(t)}f(t)=\begin{cases} \frac{1}{2}t^2 \text{ if } t\in[0,1[\ \cup\ [2,3[\ \cup\ [4,5[\ \cup\ \ldots \\ t \text{ if } t\in[1,2[\ \cup\ [3,4[\ \cup\ [5,6[\ \cup\ \ldots \end{cases} \tag{1.6}$$

Notice, consequently, that the integration goes on even when it is not being used. The question naturally arises of whether it should, or not. Without such a memory, the result would be

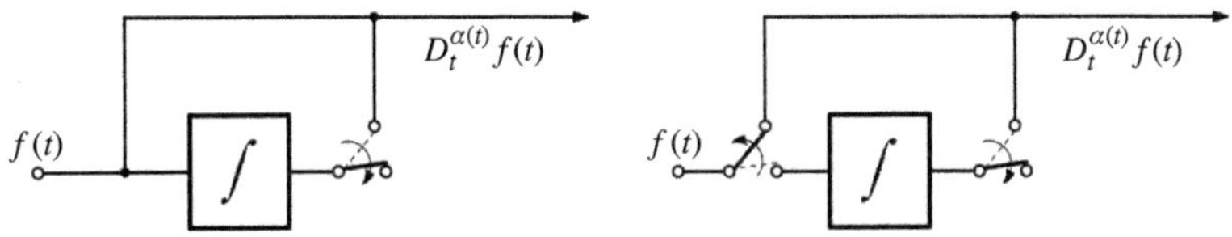

FIGURE 1.1 Left: block diagram for Eq. (1.6); right: block diagram for Eq. (1.7).

$$D^{\alpha(t)} f(t) = \begin{cases} \frac{1}{2}t^2 \text{ if } t \in [0,1[\\ t \text{ if } t \in [1,2[\\ \frac{1}{2}t^2 - \frac{3}{2} \text{ if } t \in [2,3[\\ t \text{ if } t \in [3,4[\\ \frac{1}{2}t^2 - 5 \text{ if } t \in [4,5[\\ t \text{ if } t \in [5,6[\\ \frac{1}{2}t^2 - \frac{11}{2} \text{ if } t \in [6,7[\\ t \text{ if } t \in [7,8[\\ \ldots \end{cases} \tag{1.7}$$

This result corresponds to an integral that, whenever it is not used, does not grow. When the -1 order branch is used again, the integral restarts at the same value it had when used the last time. In other words, the operator remembers that, for some past time intervals, the value of the order was not the current one. Notice that this memory of past values of the order is independent of the memory of past values of the differentiated function, which is always present for order -1 (and not for order 0).

Section 1.2.2 will introduce mathematical definitions of variable-order derivatives with and without memory of past values of the order. Before that, it is important to assert that both cases (in Eqs. 1.6 and 1.7) make sense: it is not that one definition is correct and the other not. And these results can be found in practice. The most obvious way is the use of electronic components, such as operational amplifiers and switches. The corresponding block diagrams are given in Fig. 1.1.

1.2.2 Defining Variable Order Derivatives

To generalize differentiation and integration notions of order $n \in \mathbb{N}$ to orders of $\alpha \in \mathbb{R}$, there are several alternative definitions: those of Grünwald–Letnikoff, Riemann–Liouville, Caputo, Atangana–Baleanu, etc. Here the first two are addressed. Notice that, save only for orders $\alpha = 1, 2, 3, \ldots$, operator D^α is non-local, as it always depends on the integration limits c and t. That is why it has memory of past values of $f(t)$ (Valério and Sá da Costa, 2013). Recall that, for constant differentiation orders, the Grünwald–Letnikoff (GL) definition is given by Eq. (1.8a), and the Riemann–Liouville (RL) definition is given by Eq. (1.8c).

$$
{}_cD_t^{\alpha}f(t) = \lim_{h\to 0^+} \frac{\sum_{k=0}^{\lfloor \frac{t-c}{h} \rfloor} (-1)^k \binom{\alpha}{k} f(t-kh)}{h^{\alpha}} \tag{1.8a}
$$

$$
\binom{\alpha}{k} = \begin{cases} \dfrac{\Gamma(\alpha+1)}{\Gamma(k+1)\Gamma(\alpha-k+1)}, & \text{if } \alpha, k, (\alpha-k) \in \mathbb{R}\ \mathbb{Z}^- \\ \dfrac{(-1)^k \Gamma(k-\alpha)}{\Gamma(k+1)\Gamma(-\alpha)}, & \text{if } \alpha \in \mathbb{Z}^- \wedge k \in \mathbb{Z}_0^+ \\ 0, & \text{if } (k \in \mathbb{Z}^- \vee (k-\alpha) \in \mathbb{N}) \wedge \alpha \notin \mathbb{Z}^- \end{cases} \tag{1.8b}
$$

$$
{}_cD_t^{\alpha}f(t) = \begin{cases} \int_c^t \dfrac{(t-\tau)^{-\alpha-1}}{\Gamma(-\alpha)} f(\tau)\, d\tau, & \text{if } \alpha \in \mathbb{R}^- \\ f(t), & \text{if } \alpha = 0 \\ \dfrac{d^{\lceil \alpha \rceil}}{dt^{\lceil \alpha \rceil}}\ {}_cD_t^{\alpha-\lceil \alpha \rceil} f(t), & \text{if } \alpha \in \mathbb{R}^+ \end{cases} \tag{1.8c}
$$

The fractional order of integrals and derivatives can be a function of time or some other variable (Lorenzo and Hartley, 2002b). Here we will express it a function of time; the other case is a straightforward generalization. If no memory of past values of the order is intended, the definition must only include its current value:

$$
{}_cD_t^{\alpha(t)}f(t) = \lim_{h\to 0^+} \frac{\sum_{k=0}^{\lfloor \frac{t-c}{h} \rfloor} (-1)^k \binom{\alpha(t)}{k} f(t-kh)}{h^{\alpha(t)}} \tag{1.9a}
$$

$$
{}_cD_t^{\alpha(t)}f(t) = \begin{cases} \int_c^t \dfrac{(t-\tau)^{-\alpha(t)-1}}{\Gamma(-\alpha(t))} f(\tau)\, d\tau, & \text{if } \alpha(t) \in \mathbb{R}^- \\ f(t), & \text{if } \alpha(t) = 0 \\ \dfrac{d^{\lceil \alpha(t) \rceil}}{dt^{\lceil \alpha(t) \rceil}}\ {}_cD_t^{\alpha(t)-\lceil \alpha(t) \rceil} f(t), & \text{if } \alpha(t) \in \mathbb{R}^+ \end{cases} \tag{1.9b}
$$

These GL and RL formulations without memory of α are equivalent (if the function is well-behaved enough so that both formulations can be applied). They are called type-1 variable order derivatives in Lorenzo and Hartley (2002a); Valério and Sá da Costa (2011b); Valério and Sá da Costa (2013), and type-$\mathcal{A}$ variable order derivatives in Sierociuk et al. (2015a,b).

It seems reasonable to think that, if a memory of past values of the order α is intended, then, in the summation of Eq. (1.8a), and in the integral of 8c, for each time instant, the value of the order that was available at that time instant should be used, as in Eq. (1.10).

$$ {}_cD_t^{\alpha(t)}f(t) = \lim_{h\to 0^+} \sum_{k=0}^{\lfloor \frac{t-c}{h} \rfloor} \frac{(-1)^k \binom{\alpha(t-kh)}{k} f(t-kh)}{h^{\alpha(t-kh)}} \tag{1.10a} $$

$$ {}_cD_t^{\alpha(t)}f(t) = \begin{cases} \int_c^t \frac{(t-\tau)^{-\alpha(\tau)-1}}{\Gamma(-\alpha(\tau))} f(\tau)\, d\tau, & \text{if } \alpha(t) \in \mathbb{R}^- \\ f(t), & \text{if } \alpha(t) = 0 \\ \frac{d^{\lceil \alpha(t) \rceil}}{dt^{\lceil \alpha(t) \rceil}}\, {}_cD_t^{\alpha(t)-\lceil \alpha(t) \rceil} f(t), & \text{if } \alpha(t) \in \mathbb{R}^+ \end{cases} \tag{1.10b} $$

This time, however, the GL definition of Eq. (1.10a) and the RL definition of Eq. (1.10b) do *not* provide the same result. The RL definition of Eq. (1.10b) will for instance correspond to the result of Eq. (1.7); the GL definition will not. In any case, these definitions are called type-2 variable order derivatives in Lorenzo and Hartley (2002a); Valério and Sá da Costa (2011b); Valério and Sá da Costa (2013), and type-$\mathcal{B}$ variable order derivatives in Sierociuk et al. (2015a,b).

Another way of obtaining a memory of past values of the order is using, in the RL definition, the order at time $t-\tau$, that appears in the numerator of the kernel. In the GL definition what corresponds to the difference between t (the current time) and τ (the time at which the function appears in the definition) turns out to be $t-(t-kh)=kh$, thereby resulting in a GL definition including a discrete time convolution. The expressions are those of Eq. (1.11), which are again not equivalent. They are called type-3 variable order derivatives in Lorenzo and Hartley (2002a); Valério and Sá da Costa (2011b); Valério and Sá da Costa (2013), and type-$\mathcal{C}$ variable order derivatives in Sierociuk et al. (2015a,b).

$$ {}_cD_t^{\alpha(t)}f(t) = \lim_{h\to 0^+} \sum_{k=0}^{\lfloor \frac{t-c}{h} \rfloor} \frac{(-1)^k \binom{\alpha(kh)}{k} f(t-kh)}{h^{\alpha(kh)}} \tag{1.11a} $$

$$ {}_cD_t^{\alpha(t)}f(t) = \begin{cases} \int_c^t \frac{(t-\tau)^{-\alpha(t-\tau)-1}}{\Gamma(-\alpha(t-\tau))} f(\tau)\, d\tau, & \text{if } \alpha(t) \in \mathbb{R}^- \\ f(t), & \text{if } \alpha(t) = 0 \\ \frac{d^{\lceil \alpha(t) \rceil}}{dt^{\lceil \alpha(t) \rceil}}\, {}_cD_t^{\alpha(t)-\lceil \alpha(t) \rceil} f(t), & \text{if } \alpha(t) \in \mathbb{R}^+ \end{cases} \tag{1.11b} $$

In all the models below, a GL variable order derivative of type-$\mathcal{D}$ was used (Sierociuk et al., 2015a,b). This is another possible formulation with a memory of past values of the order, and is derived from the approximations of the GL definition in Eqs. (1.12a) and (1.12b), which are respectively an iterative approximation and a recursive approximation. The step time is $h \in \mathbb{R}^+$, the lower integration limit is $c = 0$, and $n = \lfloor t/h \rfloor$. If implemented for the entire length of the simulation, without any truncation of the series, both formulations are equivalent (Sierociuk et al., 2015c).

$$ {}_0D_t^{\alpha}f(t) \approx \frac{1}{h^{\alpha}}\sum_{r=0}^{n}(-1)^r\binom{\alpha}{r}f(t-rh) \tag{1.12a} $$

$$ {}_0^{rec}D_t^{\alpha}f(t) \approx \left(\frac{f(t)}{h^{\alpha}} - \sum_{r=1}^{n}(-1)^r\binom{-\alpha}{r}\,{}_0^{r}D_{t-rh}^{\alpha}f(t-rh)\right) \tag{1.12b} $$

The GL type-$\mathcal{D}$ variable order derivative can be approximated according to Eq. (1.13a). This derivative was already successfully used for modeling the heat transfer process in a media with a time-varying structure (Sierociuk et al., 2013).

$$ {}_0^{\mathcal{D}}D_t^{\alpha(t)}f(t) \approx \left(\frac{f(t)}{h^{\alpha(t)}} - \sum_{r=1}^{n}(-1)^r\binom{-\alpha(t)}{r}\,{}_0^{\mathcal{D}}D_{t-rh}^{\alpha(t)}f(t)\right) \tag{1.13a} $$

$$ {}_{-\infty}^{\mathcal{D}}D_t^{\alpha(t)}f(t) \approx \left[\frac{f(t)}{h^{\alpha(t)}} - \sum_{j=1}^{n}(-1)^j\binom{-\alpha(t)}{j}\left({}_{-\infty}^{\mathcal{D}}D_{t-jh}^{\alpha(t-jh)}f(t)-c\right)+c\right] \tag{1.13b} $$

The type-$\mathcal{D}$ construction corresponds to an input-reductive strategy that assumes the rejection of input differentiators, which translates in an immediate effect of order switching (Fig. 1.2). Additionally, this construction also allows for the effect of initial conditions to be without memory of accumulated values, as formulated in Eq. (1.13b) where ${}_{-\infty}^{\mathcal{D}}D_l^{\alpha}f(t) = c = \text{constant}$, for $l = (-\infty, 0)$, and represented in the Fig. 1.3. These characteristics are

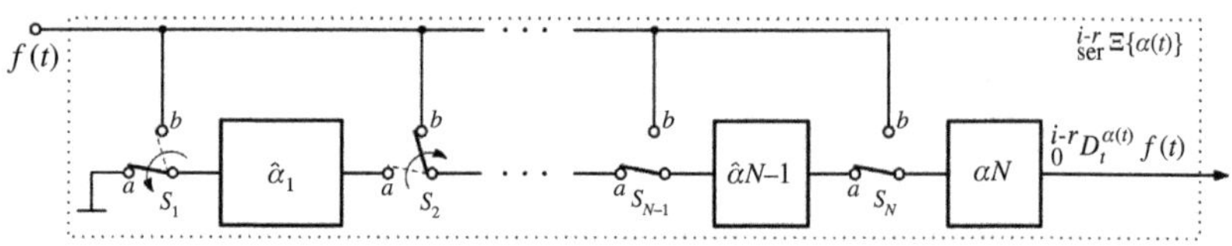

FIGURE 1.2 Schematic representation of the input-reductive switching order scheme, in serial form, from orders α_1 to α_2. *From Sierociuk, D., Malesza, W., Macias, M., 2015b. Fractional Variable Order Derivative Simulink User Guide, https://www.mathworks.com/matlabcentral/fileexchange/38801-fractional-variable-order-der.*

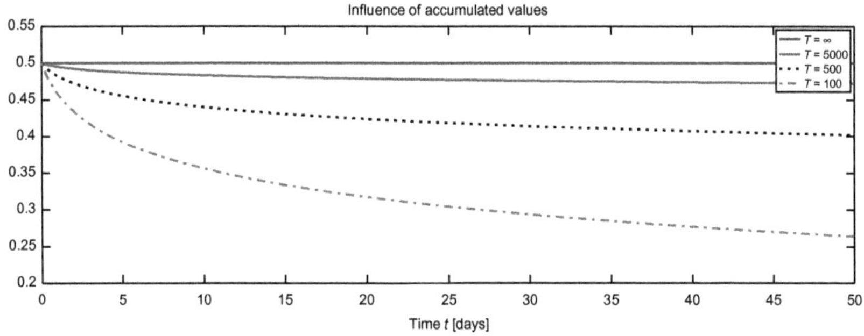

FIGURE 1.3 Testing the effects of accumulated values of order α. Obtained results for $\alpha = 0.5$ integrator output with initial conditions $c = 0.5$ and different values of T (the length of initial conditions function).

essential for the application of type-$\mathcal{D}$ variable order definition to the bone remodeling models presented next.

1.3 BONE REMODELING

This section begins by introducing bone physiology concepts, in Section 1.3.1. PK and PD models follow, in Section 1.3.2. Finally, a road-map through published integer models, that reflect bone dynamic, is presented in Section 1.3.3.

All models presented here use dimensionless variables and parameters, including the cell populations, except when explicitly said otherwise in Table 1.1. D^1 refers to the first order derivative in time, $\frac{d}{dt}$, and $D^{\alpha(t)}$ or $D^{\alpha(t,x)}$ refers to the Grünwald–Letnikoff type-$\mathcal{D}$ variable order derivative, ${}^{\mathcal{D}}_{-\infty}\mathrm{D}_t^{\alpha(t)}$ or ${}^{\mathcal{D}}_{-\infty}\mathrm{D}_t^{\alpha(t,x)}$, respectively.

1.3.1 Introducing Bone Physiology

The skeleton is an active metabolic tissue, besides providing support and protection to the vital organs (Chen et al., 2010). It is not, however, static, as it constantly undergoes remodeling. This process is spatially heterogeneous, with regular but asynchronous cycles that can take place in 5%–25% of the total bone surface available (Crockett et al., 2011). It is estimated that about 10% of the bone is renewed each year (Lerner, 2006). Cortical bone provides strength and protection while trabecular is the most metabolically active. Consequently, it is within the trabecular bone that most of bone turnover occurs, under normal conditions and in diseases of bone loss or formation.

This reconstruction occurs within a Basic Multicellular Unit (BMU), a temporary anatomical structure where bone is resorbed by cells termed *osteoclasts* and sequentially formed due to cells called *osteoblasts* (Parfitt, 1994). This process is remarkably well balanced, as a tightly controlled

TABLE 1.1 Summary and Description of the Variables and Parameters of the Models of Eqs. (1.19)–(1.22) (in the case of non-integer models, units of Day^{-1} are replaced by pseudo-units Day$^{-\alpha}$)

Variables	Description	Units
t	Time	Day
x	Distance	$x \in [0, 1]$
$C(t, x)$	Osteoclast population	—
$B(t, x)$	Osteoblast population	—
$z(t, x)$	Bone mass density	%
$T(t, x)$	Bone metastases density	%
$\alpha(t)/\alpha(t, x)$	Variable order expression	—
$d_1(t)$	Effect of denosumab	—
$d_2(t)$	Effect of zoledronic acid	—
$d_3(t)$	Effect of paclitaxel	—
Parameters	**Description**	**Units**
α_C	OC activation rate	Day^{-1}
α_B	OB activation rate	Day^{-1}
β_C	OC apoptosis rate	Day^{-1}
β_B	OB apoptosis rate	Day^{-1}
g_{CC}	OC autocrine regulator	—
g_{BC}	OC paracrine regulator	—
g_{CB}	OB paracrine regulator	—
g_{BB}	OB autocrine regulator	—
κ_C	Bone resorption rate	Day^{-1}
κ_B	Bone formation rate	Day^{-1}
σ_C	Diffusion coefficient for OC	Day^{-1}
σ_B	Diffusion coefficient for OB	Day^{-1}
σ_z	Diffusion coefficient for bone mass	Day^{-1}
σ_T	Diffusion coefficient for metastases	Day^{-1}
r_{CC}	OC tumorous autocrine regulation	—
r_{BC}	OC tumorous paracrine regulation	—
r_{CB}	OB tumorous paracrine regulation	—

(Continued)

TABLE 1.1 (Continued)

Parameters	Description	Units
r_{BB}	OB tumorous autocrine regulation	—
L_T	Maximum size of bone metastases	%
γ_T	Metastases growth rate	% Day^{-1}
$C(0)/C(0,x)$	Initial distribution of osteoclasts	—
$B(0)/B(0,x)$	Initial distribution of osteoblasts	—
$z(0)/z(0,x)$	Initial bone mass percentage	%
$T(0)/T(0,x)$	Initial tumorous mass percentage	%
C_{ss}	Steady-state OC number	—
B_{ss}	Steady-state OB number	—
D_0	Drug dosage	mg
τ	Drug administration time interval	Day
F	Bioavailability	—
V_d	Volume distribution	L
κ_g	Drug absorption rate	Day^{-1}
κ_p	Drug elimination rate	Day^{-1}
$C_p^{50/base}$	Initial drug concentration for 50% of its maximum effect	mg/L
K_r	Drug resistance capacity	—
K_{s_1}	Maximum effect of denosumab	—
K_{s_2}	Maximum effect of zoledronic acid	—
K_{i_3}	Maximum effect of paclitaxel	—

OC stands for Osteoclasts, and OB for Osteoblasts.

mechanism between the different clusters of bone cells is involved. It acts through the combined influence of autocrine and paracrine factors (Raggatt and Partridge, 2010). An active BMU can travel across the tissue at a constant speed of 20–40 μm/day for up to 6 months. It comprises 10–20 osteoclasts that remove old and damaged tissue, and around 1000–2000 osteoblasts that secrete and deposit unmineralized bone matrix, directing its formation and mineralization into mature lamellar bone (Ryser, 2011). The BMU can be seen as a mediator mechanism, as it bridges individual cellular activity to whole bone morphology (Raggatt and Partridge, 2010).

Osteoclasts are multinucleated cells, that result from mononucleated hematopoietic stem fused with progenitors cells. As they both express RANK (*Receptor Activator of Nuclear Factor* k*B*) and c-fms (*Macrophage Colony-stimulating Factor Receptor*), by biding to RANKL (*RANK-ligand*) and CSF-1 (*Colony-stimulating Factor 1*), respectively, these cells differentiate into active osteoclasts capable of bone resorption. OPG (*Osteoprotegerin*), a soluble decoy receptor for RANKL and a physiological negative regulator of osteoclastogenesis, also plays an important role in osteoclasts performance (Boyce, 2012; Raggatt and Partridge, 2010). It is their generation rate that dictates the BMU extension, whereas their life span determines the depth of the resorption (Bellido et al., 2014).

Osteoblasts are mononucleated cells that differentiate from MSC (*mesenchymal stem cells*). They are controlled by BMP (*bone morphogenetic protein*), Wnt-signaling, vitamin D, among other factors. PTH (*parathyroid hormone*) receptors in the osteoblasts upregulate the expression of RANKL, in the presence of the hormone, which binds to RANK expressed in osteoclasts precursors. Thus, PTH promotes their activation and bone resorption. These cells also produce OPG, which inhibits osteoclastogenesis by binding to RANKL. The secretion of OPG is reduced in response to PTH, which contributes further to osteoclastogenesis. Osteoblasts can undergo apoptosis, differentiate into osteocytes or into bone lining cells (Crockett et al., 2011; Roodman, 2004).

Bone resorption and formation is mainly regulated by the RANK/RANKL/OPG pathway and PTH. It is activated by either mechanical stimuli on the bone, or due to systemic changes in homeostasis which result in the production of estrogen or PTH (Raggatt and Partridge, 2010). The latter is triggered in response to a reduced calcium concentration, which leads to calcium release from the bone matrix, and is inhibited when an elevated calcium concentration is sensed (Silva and Bilezikian, 2015). The remodeling process begins when PTH triggers two mechanisms in the osteoblasts: (1) the existing PTH reduces the osteoblasts secretion of OPG that, by being a soluble decoy receptor for RANKL and allowing it to consequently bind to RANK, promotes osteoclastogenesis; (2) PTH receptors also upregulate the expression of RANKL, which again binds to RANK, further promoting osteoclast activation and bone resorption (Raggatt and Partridge, 2010). The initiation of the formation phase is coupled to the resorption phase in a process not yet fully understood, as factors released from the bone matrix during resorption (Insulin Growth Factors I and II (*IGF-I*, *IGF-II*), and Transforming Growth Factor-β (*TGF*-β)) may be involved in this coupling. Bone formation takes place even in the presence of malfunctioning osteoclasts, which has led to the hypothesis that osteoclasts produce the coupling factors responsible for attracting and regulating osteoblasts to the sites of bone resorption (Boyce, 2012). At the resorpted site, osteoblasts commence bone formation and replace the resorpted bone by the same amount, ending the bone remodeling cycle.

The presence of metastatic cancer cells (breast, prostate, lung, renal, and myeloma among others) accelerates the remodeling process and disturbs the balance between bone cells by disrupting its biochemical regulation (Lerner, 2006). Bone integrity is consequently lost. These sites of cancer metastasis are usually those where bone remodeling rates are high, such as the pelvis, the axial skeleton, or bones with abundant bone marrow (Boyce, 2012; Schneider et al., 2005).

Bone metastases can be osteolytic (increased bone resorption), or osteoblastic (bone formation is stimulated in an unstructured way). However, both are still present in any case, although out of balance, resulting in loss of bone resistance and integrity. Breast cancer metastases are prone to develop osteolytic metastasis and prostate cancer ones are usually osteoblastic (Suva et al., 2011).

For *osteolytic* metastases, tumor cells stimulate osteoclast activity and receive, in return, positive feedback from factors released by the bone microenvironment during bone destruction (Casimiro et al., 2016; Chen et al., 2010). As TGF-β is released from the bone matrix during resorption, it stimulates tumor growth and parathyroid hormone-related protein (*PTHrP*) production in metastatic cells. By binding to PTH receptors on cells of osteoblastic lineage, RANKL levels are then enhanced. Subsequently, osteoclasts are activated, leading to increased bone resorption (Casimiro et al., 2016). Osteoclasts activity, in turn, will result in the release of TGF-β from the degraded bone, which further stimulates tumor growth and PTHrP secretion, giving rise to the vicious cycle.

In *osteoblastic* metastases, tumorous cells grow as bone expresses endothelin-1 (*ET-1*). ET-1 stimulates osteoblasts through the endothelin A receptor (*ETR*), activating Wnt-signaling. Tumor-derived proteases contribute to the release of osteoblastic factors from the extracellular matrix, including TGF-β and IGF-I. RANKL is increased due to tumor-induced osteoblast activity, leading to the release of PTH and promoting osteoclast activity (Casimiro et al., 2016). Thus, tumor microenvironment leads to the accumulation of new formed bone.

Several approaches, that treat primary and metastatic bone tumors, have the potential to affect both tumor affected and healthy cells. However, strategies can be oriented to effectively inhibit tumor growth by targeting the bone and its microenvironment rather than the tumor alone. Antiresorptive therapy targets osteoclasts, when an osteolytic metastatic bone disease is present. *Bisphosphonates* such as alendronate or zoledronic acid (*Zometa® - Zoledronic Acid for Injection*, 2017; Chen et al., 2002), and *monoclonal antibodies* like denosumab (Sohn et al., 2014; Gibiansky et al., 2012), are effective treatments currently being administrated. While bisphosphonates lodge in bone and poison osteoclasts as they degrade bone, monoclonal antibodies in turn bind exclusively to RANKL, increasing the OPG/RANKL ratio and inhibiting osteoclast formation. For other diseases, such as multiple myeloma (*MM*), therapies include daily doses of PTH, endothelin, and proteasome inhibitors, that act by targeting osteoblasts to recover bone mass (Oyajobi et al., 2007). Of course, anticancer agents that target metastatic and primary tumor cells directly (chemotherapy, as paclitaxel (Perez et al., 2001), and hormone therapy) should be used in combination with the aforementioned therapies in either case (Casimiro et al., 2009).

1.3.2 Pharmacokinetics and Pharmacodynamics (PK/PD)

PK models characterize drug absorption and disposition within the body (Dhillon and Gill, 2006). The remaining drug concentration to be absorbed (C_g, in mg/L) and the effective drug concentration in the plasma (C_p, in mg/L) are described by the system of differential equations, given by Eq. (1.14). κ_g and κ_p are the absorption and elimination rate, respectively (Mager et al., 2003).

$$D^1 C_g(t) = -\kappa_g C_g(t) \tag{1.14a}$$

$$D^1 C_p(t) = \kappa_g C_g(t) - \kappa_p C_p(t) \tag{1.14b}$$

For a subcutaneous drug administration, the initial concentration is applied in the remaining drug to be absorbed ($C_g(0) = C_0$); for the intravenous case, the initial concentration goes directly to the plasma ($C_p(0) = C_0$). For a single dosage drug of initial concentration C_0, the plasma concentration can be determined by Eq. (1.15a). For multiple doses, the plasma concentration of the nth dose is given by Eq. (1.15b), for initial conditions $C_0 = \frac{D_0 F}{V_d}$ administrated at equally spaced intervals $t' = t - (n-1)\tau$, where D_0 is the dosage, F the bioavailability, and V_d the volume distribution. Multiple dosage is governed by the steady-state $C_{pss} = \frac{C_0}{\tau \kappa_p}$.

$$C_p = C_0 \frac{\kappa_g}{\kappa_g - \kappa_p} (e^{-\kappa_p t} - e^{\kappa_g t}) \tag{1.15a}$$

$$C_p(n, t') = C_0 \frac{\kappa_g}{\kappa_g - \kappa_p} \times \left(\frac{1 - e^{-n\kappa_p \tau}}{1 - e^{-\kappa_p \tau}} e^{-\kappa_p t'} - \frac{1 - e^{-n\kappa_g \tau}}{1 - e^{-\kappa_g \tau}} e^{-\kappa_g t'} \right) \tag{1.15b}$$

The PD consists in a drug effect. It can be described by a Hill function that depends on the drug's concentration, given by $d(t) = \frac{C_p(t)}{C_p^{50}(t) + C_p(t)}$ (Pinheiro et al., 2011). It varies between 0 and 1, where $C_{p^{50}}(t) = f(t) C_p^{50/base}$ represents the concentration at 50% of its maximum effect, $C_p^{50/base}$ is the initial value of $C_p^{50}(t)$, and resistance to a drug can be described by $f(t) = 1 + K_r \int_0^t \max[0, L_r - C_p(\lambda)] d\lambda$ (Pinheiro et al., 2011).

Different drugs can act in the same pathway and their combined action may differ from the isolated administration of each drug. To quantify the synergistic or antagonistic effect of the combination of drugs, the so-called Combination Index (CI) can be described by Eq. (1.16a), where C^*_{pi} is the necessary concentration of drug i, when combined with other drugs, to produce the same effect d as concentration C_{pi} of drug i when taken alone. The resulting combined effect of two drugs is given by d_{C12}, in Eq. (1.26).

$$CI = \frac{(C^*_p)_1}{(C_{50})_1 d/(1-d)} + \frac{(C^*_p)_2}{(C_{50})_2 d/(1-d)} \tag{1.16a}$$

$$d_{C_{12}} = \frac{\left[\frac{(C_p)_1}{(C_{50})_1} + \frac{(C_p)_2}{(C_{50})_2}\right]}{CI + \left[\frac{(C_p)_1}{(C_{50})_1} + \frac{(C_p)_2}{(C_{50})_2}\right]} \tag{1.16b}$$

A PK/PD model for a drug can be achieved through the combination of pharmacokinetic and pharmacodynamic models. A simulation of a PK/PD model can be found in Fig. 1.4, for multiple doses of a single drug administrated oral or subcutaneously.

Since a drug pathway can have either an inhibitory (i: $-$) or a stimulatory (s: $+$) effect on a given metabolism, a control action (CA) to the tumorous presence in the mathematical models for bone remodeling is given by $CA(t) = 1 \pm K_{i,s} d(t)$. Constants $K_i, K_s > 0$ represent the maximum effect of a drug in a specific mechanism, with $d(t)$ being the PD response of a single drug or a combination of drugs.

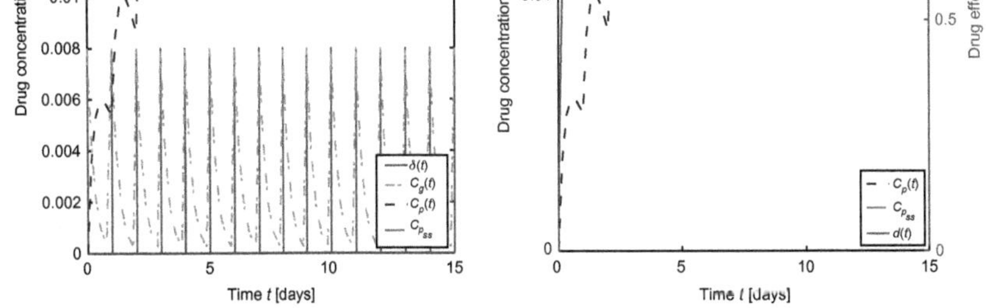

FIGURE 1.4 Multiple dosage administration. Left: evolution of the PK model; Right: evolution of the corresponding PD case. Used parameters were $C_g(0) = 0.008$ mg/L, $\kappa_g = 4$ days^{-1}, $\kappa_p = 0.5$ days^{-1}, $\tau = 1$ days, $C_{pss} = 0.016$ mg/L, and $C_p^{50}(t) = 0.003$ mg/L.

1.3.3 Modeling Bone Remodeling Cycles—Integer Models

Models based on differential equations have been applied to analyze and simulate biochemical and biomechanical interactions between the tumor cells with the bone microenvironment. They can be divided into local or nonlocal constructions (mainly one-dimensional models) and, within each category, three stages of bone behavior are encompassed: healthy bone microenvironment, tumor disrupted bone dynamics, and therapy counteraction.

The simplest model for bone remodeling was proposed in Komarova et al. (2003), taking an S-system form described by Eq. (1.17) (Savageau, 1988). Coupling of osteoclasts, $C(t)$, and osteoblasts, $B(t)$, behavior is done through biochemical autocrine (g_{CC}, g_{BB}) and paracrine (g_{BC}, g_{CB}) factors expressed implicitly in the system's exponents. Bone mass density, $z(t)$, is determined through the extent which values of $C(t)$ and $B(t)$ populations exceed their nontrivial steady state, C_{SS} and B_{SS}, respectively. Consequently, bone mass is but the reflection of the bone cells activities. Production and death rate of the bone cells are encompassed in $\alpha_{C,B}$ and $\beta_{C,B}$, respectively, and constants κ_C and κ_B represent the bone resorption and formation activity, respectively.

$$D^1 C(t) = \alpha_C C(t)^{g_{CC}} B(t)^{g_{BC}} - \beta_C C(t) \tag{1.17a}$$

$$D^1 B(t) = \alpha_B C(t)^{g_{CB}} B(t)^{g_{BB}} - \beta_B B(t) \tag{1.17b}$$

$$D^1 z(t) = -\kappa_C \max[0, C(t) - C_{ss}] + \kappa_B \max[0, B(t) - B_{ss}] \tag{1.17c}$$

This model is capable of representing single or periodical remodeling cycles by setting the autocrine and paracrine parameters to the appropriate values, specially the osteoblast-derived osteoclast paracrine regulator g_{BC}. The RANK/RANKL/OPG pathway is also implicitly encoded in this parameter. Response amplitude and frequency depend on the initial conditions, triggered by a deviation from the steady state, as seen in Fig. 1.5 for periodical cycles only. Parameter values are given in Table 1.2.

Disruptive pathologies to the bone microenvironment were also added. In Ayati et al. (2010) the previous model was extended to incorporate the effect of MM in the bone dynamics, as presented in Eq. (1.18). $T(t)$ represents the tumor cells density at time t, with a Gompertz form of constant growth $\gamma_T > 0$, and acts through the autocrine and paracrine regulations pathways in the form of r_{ij} parameters. The tumor action is considered independent of the bone mass, with a possible maximum tumor size of L_T. Periodic remodeling cycles are deregulated and bone mass density decreases. The bone mass equation is the same as that of Eq. (1.17c), remaining as a consequence of

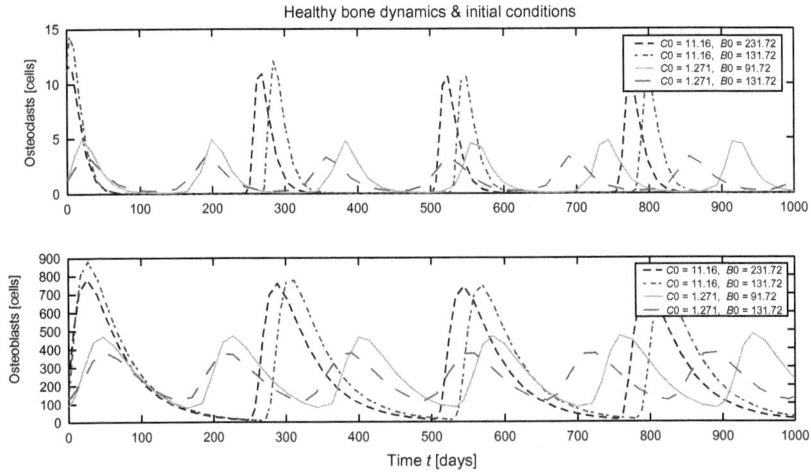

FIGURE 1.5 Osteoclasts and Osteoblasts evolutions, respectively, for a different set of initial conditions $C(0) = C_0$ and $B(0) = B_0$, according to the model of Eq. (1.17). Remaining parameters are the same as in Ayati et al. (2010) and presented in Table 1.2.

the activity of the bone cells. In Fig. 1.6, represented in **full lines**, a simulation of the bone microenvironment for Eq. (1.18) can be compared to a healthy case, represented in **dashed lines**.

$$D^1C(t) = \alpha_C C(t)^{g_{CC}\left(1+r_{CC}\frac{T(t)}{L_T}\right)} B(t)^{g_{BC}\left(1+r_{BC}\frac{T(t)}{L_T}\right)} - \beta_C C(t) \tag{1.18a}$$

$$D^1B(t) = \alpha_B C(t)^{\left(\frac{g_{CB}}{1+r_{CB}\frac{T(t)}{L_T}}\right)} B(t)^{\left(g_{BB}-r_{BB}\frac{T(t)}{L_T}\right)} - \beta_B B(t) \tag{1.18b}$$

$$D^1T(t) = \gamma_T T(t)\log\left(\frac{L_T}{T(t)}\right) \tag{1.18c}$$

In Coelho et al. (2015), a PK/PD based treatment of osteolytic bone metastases through anticancer and antiresorptive therapy is proposed, adapting the model of Ayati et al. (2010) as presented in Eq. (1.19). It corresponds to the administration of anticancer treatment (chemotherapy—paclitaxel, $d_3(t)$) and antiresorptive therapy (monoclonal anibodies—denosumab, $d_1(t)$; or bisphosphonates—zoledronic acid, $d_2(t)$). Bisphosphonates (e.g., zoledronic acid or alendronate) promote osteoclast apoptosis, and monoclonal antibodies (e.g., denosumab) indirectly inhibit osteoclast formation by acting as a decoy receptor for RANKL. Together with chemotherapy, the drug effect of this treatment was included in the model through their PK/PD action. The bone mass equation, $z(t)$, remains equal to Eq. (1.17c). Simulations can be found in Fig. 1.7.

TABLE 1.2 All Variables and Parameters Used for Simulation of the Complete Models With an Acting Tumor

Par.	*	**	Par.	*	**
α_C	3	3	γ_T	0.005	0.004
α_B	4	4	L_T	100	100
β_C	0.2	0.2	$C(0,x)$	15	(Ayati et al., 2010)
β_B	0.02	0.02	$B(0,x)$	316	316
g_{CC}	1.1	1.1	$T(0,x)$	1	(Ayati et al., 2010)
g_{BC}	−0.5	−0.5	$z(0,x)$	100	100
g_{CB}	1.0	1.0	C_{ss}	5	5
g_{BB}	0	0	B_{ss}	316	316
r_{CC}	0.005	0.005	β_{PTH}	—	—
r_{BC}	0	0	K_{PTH}	—	—
r_{CB}	0	0	r_{PTHrP}	—	—
r_{BB}	0.2	0.2	λ_W	—	—
σ_i	—	10^{-6}	k_W	—	—
κ_C	0.0748	0.45	κ_T	—	—
κ_B	6.39×10^{-4}	0.0048	λ_T	—	—
PK/PD Par.			d_1	d_2	d_3
D_0			120	4	176
τ			28	28	7
F			0.62	1	1
V_d			3.1508	536.3940	160.2570
κ_g			0.2568	—	—
κ_p			0.0248	0.1139	1.2797
$C_p^{50/base}$			1	0.0001	0.0002
K_r			—	—	—
$K_{s,i}$			0.004	0.058	1.70

* refers to Eq. (1.19), ** to Eq. (1.22) (both can be found in Ayati et al. (2010), mainly for the spatial distribution of $C(0,x)$ and $T(0,x)$. Parameter σ_i encompasses $i = C, B, T, z$. PK/PD models, for all models, included monoclonal antibodies, d_1 (denosumab), bisphosphonates, d_2 (zoledronic acid), and anticancer therapy, d_3 (paclitaxel). PK parameters (D_0, τ, F, V_d, κ_g, and κ_p), for denosumab can in found in Gibiansky et al. (2012); Sohn et al. (2014), zoledronic acid in *Zometa ® - Zoledronic Acid for Injection* (2017); Chen et al. (2002) and paclitaxel in Perez et al. (2001). PD parameters, $C_p^{50/base}$ and $K_{s,i}$, were chosen through simulation.

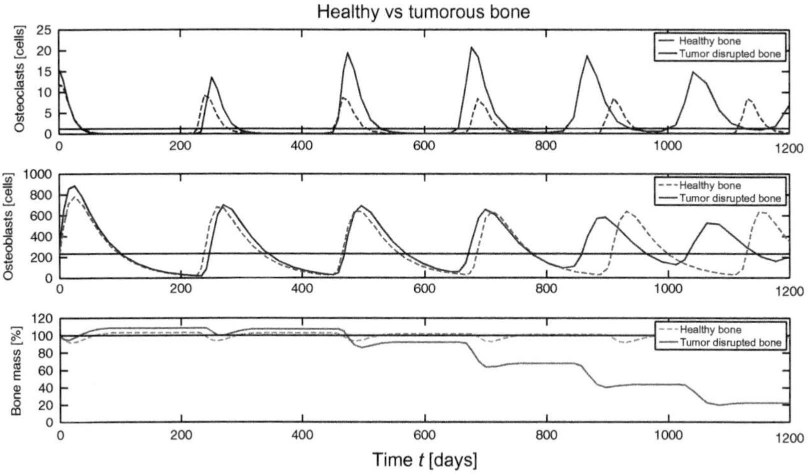

FIGURE 1.6 Osteoclasts, Osteoblasts, and Bone Mass evolutions, respectively. In **dashed lines**, simulation of Eq. (1.17) for the existing model replicating a healthy bone microenvironment. In **full lines**, simulation of Eq. (1.18), for a bone microenvironment disrupted by a developing tumor. Black lines represent stationary states. Tumor evolution is represented, in a full-blue line, on the left-side graphic of Fig. 1.10. Parameters were set according to Ayati et al. (2010), and can be found in Table 1.2.

$$D^1 C(t) = \alpha_C C(t)^{g_{CC}\left(1+r_{CC}\frac{T(t)}{L_T}\right)} B(t)^{g_{BC}\left(1+r_{BC}\frac{T(t)}{L_T}\right)\left(1+K_{s_1}d_1(t)\right)} - (1+K_{s_2}d_2(t))\beta_C C(t) \tag{1.19a}$$

$$D^1 B(t) = \alpha_B C(t)^{\left(\frac{g_{CB}}{1+r_{CB}\frac{T(t)}{L_T}}\right)} B(t)^{\left(g_{BB}-r_{BB}\frac{T(t)}{L_T}\right)} - \beta_B B(t) \tag{1.19b}$$

$$D^1 T(t) = (1-K_{i_{34}}d_{c_{34}}(t))\gamma_T T(t)\log\left(\frac{L_T}{T(t)}\right) \tag{1.19c}$$

The previous models can be extended to dimensional geometries, modeling diffusion processes in the bone through partial differential equations.

Ayati et al. (2010) also extended its model to Eq. (1.20) by allowing the diffusion over one-dimension, $\sigma_i \frac{\partial^2}{\partial x^2}$, of osteoclasts, osteoblasts, and bone mass. They now depend on both t and $x \in [0, 1]$. The diffusion of z accounts for the stochastic nature of bone dynamics and not necessarily migration of cells.

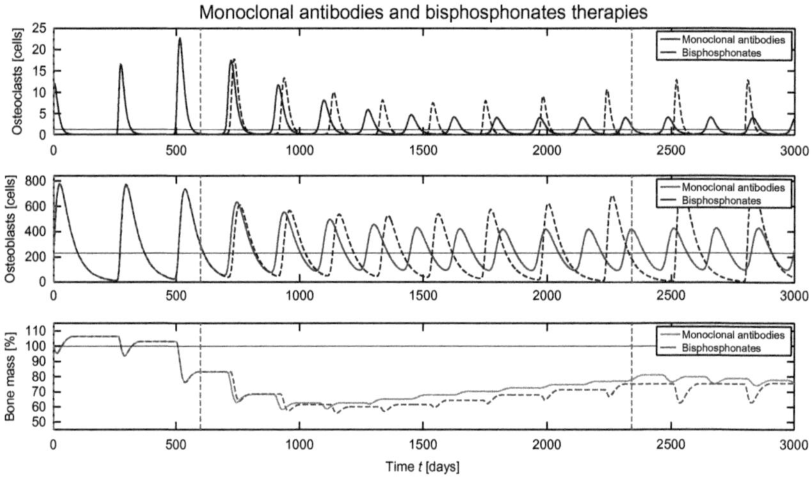

FIGURE 1.7 Osteoclasts, Osteoblasts, and Bone Mass evolutions, respectively, for the model of Eq. (1.19). **Full lines** represent the PK/PD combination of chemotherapy (paclitaxel—$d_3(t)$) with monoclonal antibodies (denosumab—$d_1(t)$). **Dashed lines** represent the combination of chemotherapy with bisphosphonates (zoledronic acid—$d_2(t)$). Again, black lines represent stationary state. For both cases, treatment begins at $t_{start} = 600$ days and it's interrupted at $t_{stop} = 2340$ days. Used parameters follow the work of Ayati et al. (2010), and PK/PD treatment parameters follow published literature (both can be found in Table 1.2). Tumor evolution counteracted with chemotherapy ($d_3(t)$) is represented, in a dashed-green line, on the left-side graphic of Fig. 1.10.

$$D^1C(t,x) = \sigma_c \frac{\partial^2}{\partial x^2} C(t,x) + \alpha_C C(t,x)^{g_{CC}} B(t,x)^{g_{BC}} - \beta_C C(t,x) \tag{1.20a}$$

$$D^1B(t,x) = \sigma_B \frac{\partial^2}{\partial x^2} B(t,x) + \alpha_B C(t,x)^{g_{CB}} B(t,x)^{g_{BB}} - \beta_B B(t,x) \tag{1.20b}$$

$$D^1z(t,x) = \sigma_z \frac{\partial^2}{\partial x^2} z(t,x) - \kappa_C \max\left[0, C(t,x) - \overline{C}_{SS}(x)\right] + \kappa_B \max[0, B(t,x) - \overline{B}_{SS}(x)] \tag{1.20c}$$

MM influence was also added, for Eq. (1.21), with tumor cells diffusing in $x \in [0, 1]$. The diffusion coefficient for the tumor is given by γ_T, which allows for its spatial growth. Regarding the bone mass equation, $z(t, x)$, the expression is the same as in Eq. (1.20c) and all variables are subjected to null Newmann boundary conditions. Initial conditions, now depending both on t and x, can be found in Ayati et al. (2010). Simulations for a healthy and tumor bone microenvironment, for these nonlocal models, is presented in the rows of Fig. 1.8, respectively. The tumor evolution, with an initial development on the right side on the normalized bone, can be found in the second graphic of Fig. 1.10.

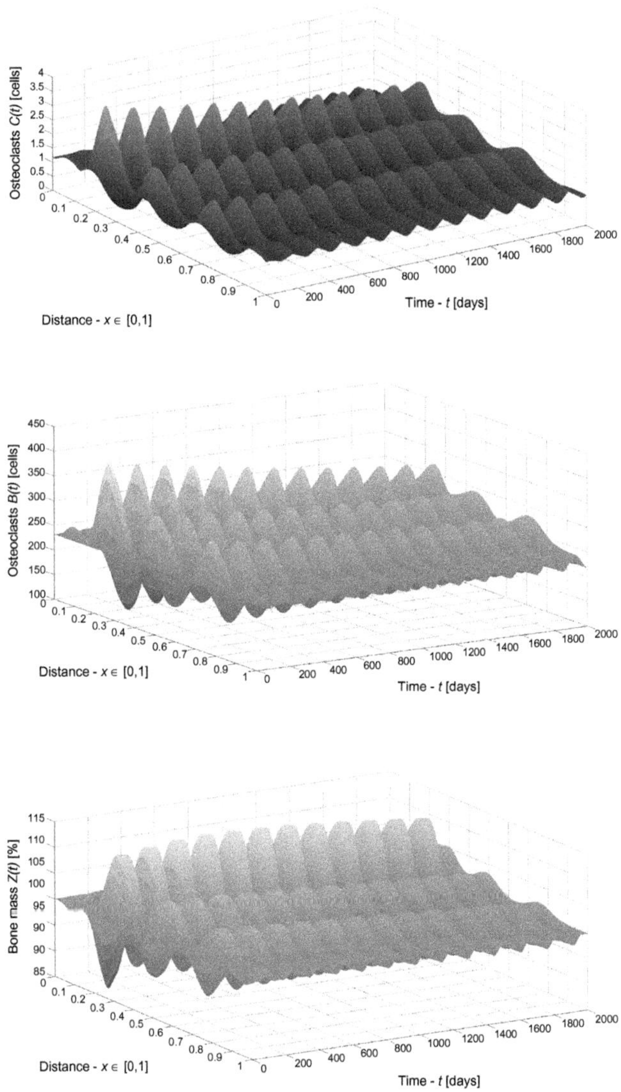

FIGURE 1.8 Nonlocal simulation of Osteoclasts, Osteoblasts and Bone Mass. **Fisrt row**, for healthy remodeling cycles (Eq. 1.20). **Second row**, for a tumor disrupted bone microenvironment (Eq. 1.21). Parameters, initial, and boundary conditions follow exactly what was presented in Ayati et al. (2010), and can be found in Table 1.2. Untreated tumor evolution, for all metastases disrupted models, is presented in the second graphic of Fig. 1.10.

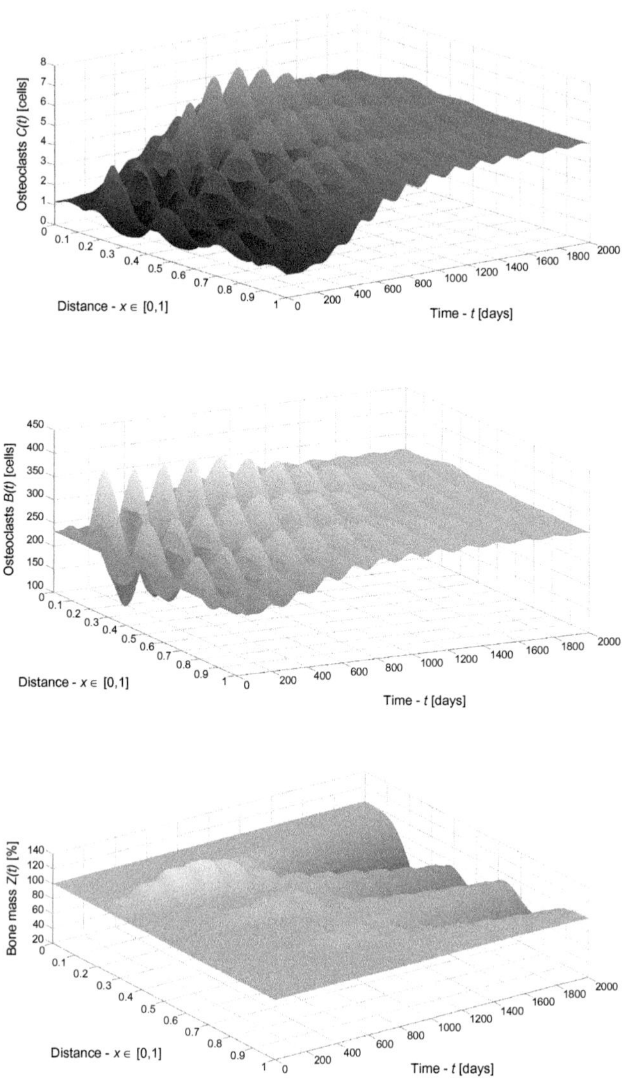

FIGURE 1.8 *(Continued)*.

$$D^1C(t,x) = \sigma_C \frac{\partial^2}{\partial x^2} C(t,x) - \beta_C C(t,x) + \\ + \alpha_C C(t,x)^{g_{CC}\left(1+r_{CC}\frac{T(t)}{L_T}\right)} B(t,x)^{g_{BC}\left(1+r_{BC}\frac{T(t)}{L_T}\right)} \tag{1.21a}$$

$$D^1B(t,x) = \sigma_B \frac{\partial^2}{\partial x^2} B(t,x) - \beta_B B(t,x) + \\ + \alpha_B C(t,x)^{\left(\frac{g_{CB}}{1+r_{CB}\frac{T(t)}{L_T}}\right)} B(t,x)^{\left(g_{BB}-r_{BB}\frac{T(t)}{L_T}\right)} \tag{1.21b}$$

$$D^1T(t,x) = \sigma_T \frac{\partial^2}{\partial x^2} T(t,x) + \gamma_T T(t,x) \log\left(\frac{L_T}{T(t,x)}\right) \tag{1.21c}$$

In Christ et al. (2018), the model of Ayati et al. (2010) adapted in Coelho et al. (2015), is further extended to include the PK/PD action of anticancer and antiresorptive therapy in the one-dimensional model of Eq. (1.22). Simulation results are presented in Fig. 1.9 for a two-case scenario.

$$D^1C(t,x) = \sigma_C \frac{\partial^2}{\partial x^2} C(t,x) - (1 + K_{s_2} d_2(t))\beta_C C(t,x) + \\ + \alpha_C C(t,x)^{g_{CC}\left(1+r_{CC}\frac{T(t)}{L_T}\right)} B(t,x)^{g_{BC}\left(1+r_{BC}\frac{T(t)}{L_T}\right)\left(1+K_{s_1} d_1(t)\right)} \tag{1.22a}$$

$$D^1B(t,x) = \sigma_B \frac{\partial^2}{\partial x^2} B(t,x) - \beta_B B(t,x) + \\ + \alpha_B C(t,x)^{\left(\frac{g_{CB}}{1+r_{CB}\frac{T(t)}{L_T}}\right)} B(t,x)^{\left(g_{BB}-r_{BB}\frac{T(t)}{L_T}\right)} \tag{1.22b}$$

$$D^1T(t,x) = \sigma_T \frac{\partial^2}{\partial x^2} T(t,x) + \left(1 - K_{i_{34}} d_{c_{34}}(t)\right)\gamma_T T(t,x) \log\left(\frac{L_T}{T(t,x)}\right) \tag{1.22c}$$

Involved variables and parameters are summarized in Tables 1.1 and 1.2.

1.4 VARIABLE ORDER MODELS—CREATING COMPACT BIOCHEMICAL BONE REMODELING MODELS

Fractional and variable order derivatives have been already successfully used in modeling the dynamics of bone remodeling. Christ et al. (2018) implemented fractional derivatives in the differential equations of bone remodeling

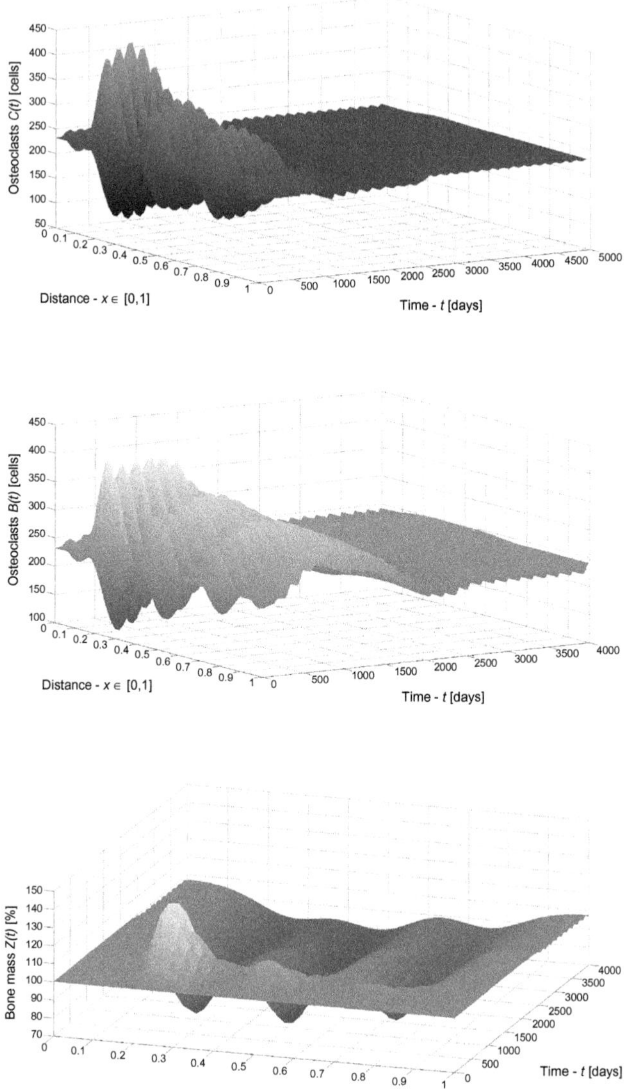

FIGURE 1.9 Nonlocal simulation of Osteoclasts, Osteoblasts, and Bone Mass. **First row**, for healthy remodeling cycles (Eq. 1.20). **Second row**, for a tumor disrupted bone microenvironment (Eq. 1.21). Parameters, initial and boundary conditions follow exactly what was presented in Ayati et al. (2010), and can be found in Table 1.2. Untreated tumor evolution, for all metastases disrupted models, is presented in the second graphic of Fig. 1.10.

of Ayati et al. (2010). These analyze the dynamic bone remodeling behavior in the absence and presence of tumor and treatment, for a discretized single point and a one-dimensional bone. More recently, in Neto et al. (2017), variable order derivatives have been introduced as a simplification technique in

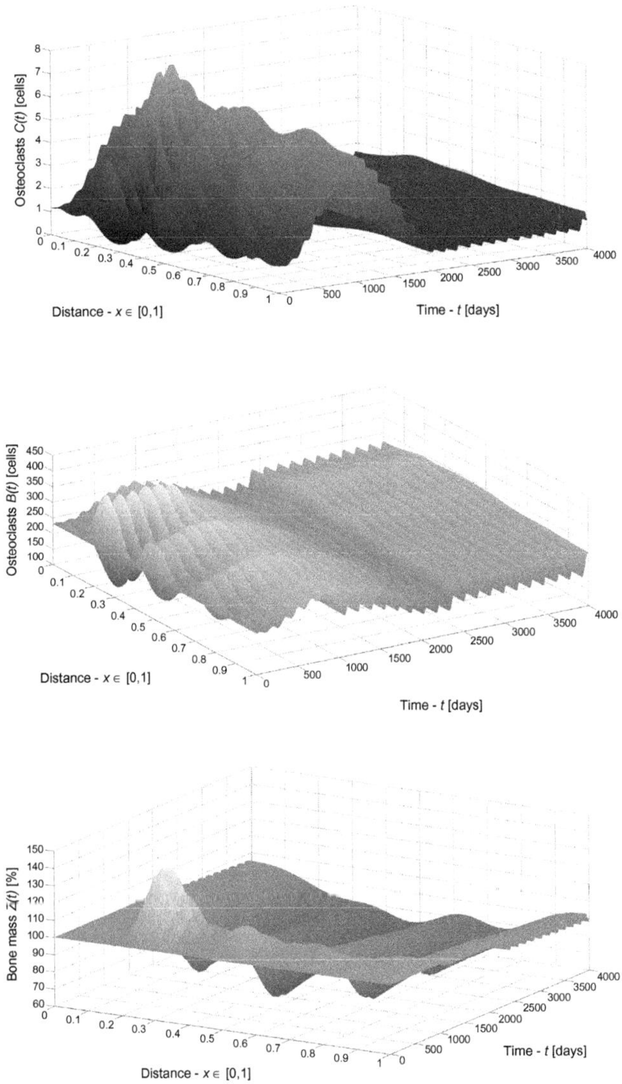

FIGURE 1.9 *(Continued).*

the same models of Ayati et al. (2010) and Coelho et al. (2015), in an effort to replicate the same bone microenvironment response but recurring to less parameters to impose the known bone behavior. Revisiting the latter, some previous considerations had to be taken into account.

The numerical implementation of the recursive type-$\mathcal{D}$ construction followed its matrix approach presented in Sierociuk and Malesza (2015), and available through the variable order derivatives *Simulink*-toolbox presented in Sierociuk et al. (2015b).

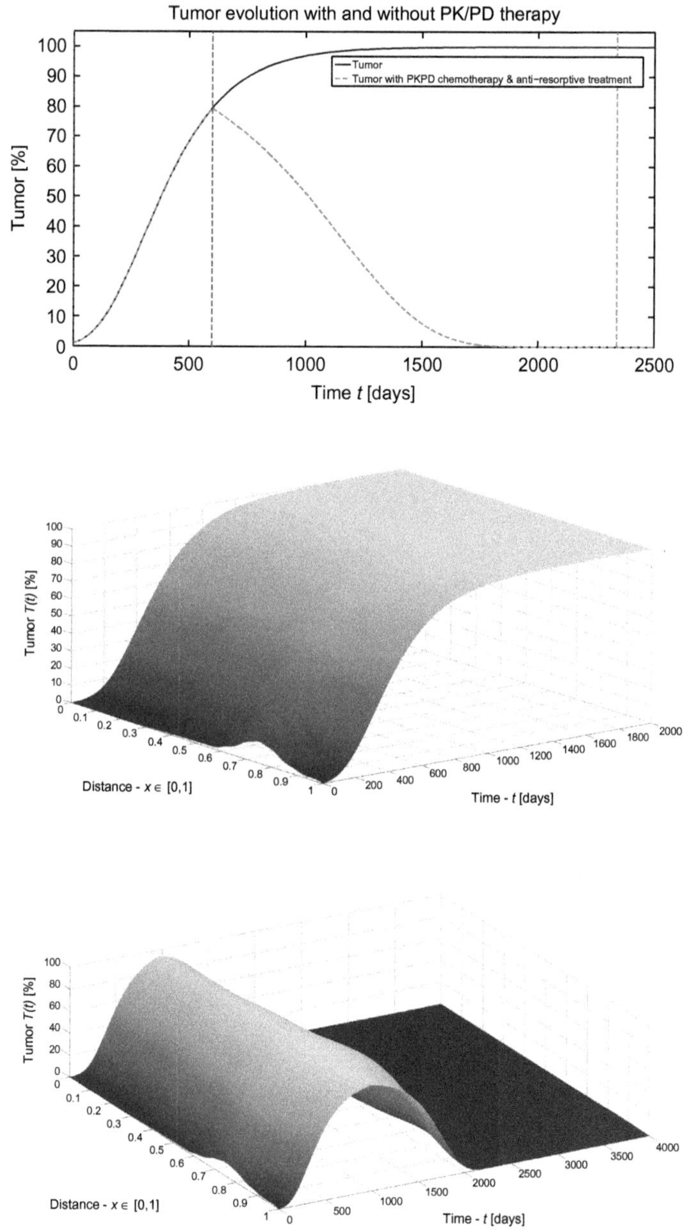

FIGURE 1.10 Local (Eq. 1.19) and nonlocal (Eq. 1.22) tumor evolutions. Left-side graphic: local evolution of the tumor's density, untreated (dashed green line) and treated with PK/PD chemotherapy of paclitaxel ($d_3(t)$) (full blue line, $t_{start} = 600$, $t_{stop} = 2340$). Middle graphic: untreated spatial tumor evolution for an initial location on the right side of the normalized bone. Right-side graphic: tumor counteracted with the same chemotherapy action of paclitaxel ($t_{start} = 980$, $t_{stop} = 2340$). Antitumor therapy parameters, either local or nonlocal, follow Table 1.2.

The order expression, $\alpha(t)$ or $\alpha(t,x)$, acts only on the osteoclasts and osteoblasts equations, as the tumor equations are said to be independent of the bone microenvironment (Ayati et al., 2010), and that bone mass variations are but a reflection of osteoclasts and osteoblasts activity (Komarova et al., 2003). The order is influenced by the tumor dynamics, according to Eq. (1.23), where θ is a constant term experimentally determined, and time t is related to the beginning of the tumor growth. It is now responsible for inducing in the original healthy model (Eq. 1.20) the same response as the tumor disrupted bone one (Eqs. 1.18 and 1.21).

$$\alpha(t) = 1 - \theta \times t \times T(t) \tag{1.23a}$$

$$\alpha(t,x) = 1 - \theta \times t \times T(t,x) \tag{1.23b}$$

It is noted that, being the osteoclasts and osteoblasts formulated in differential equations, the order will be subjected to an integration process represented by the term $1 - ...$ in the following formulation and translated in the integrator D^{-1}, as explained in Eq. (1.24) for a generic function $f(t)$ when a derivative's order $\alpha(t)$ is applied.

$$\begin{aligned} &D^{1-\alpha(t)}f(t) = \gamma[f(t)] \Leftrightarrow D^1 D^{-\alpha(t)}f(t) = \gamma[f(t)] \Leftrightarrow \\ &\Leftrightarrow D^1 f(t) = D^{\alpha}(t)[\gamma[f(t)]] \Leftrightarrow f(t) = D^{-1}[D^{\alpha}(t)[\gamma[f(t)]]] \end{aligned} \tag{1.24}$$

The last step in Eq. (1.24) is given by $f(t) = D^{-1}[D^{\alpha}(t)[\ldots]]$. According to the law of exponents of operator D, presented in Valério and Sá da Costa (2013) even if the order is time-varying the referred law is verified as long as the conditions of Eq. (1.25) are achieved. For such a case, coefficient correspondence is $\beta = -1$ and $\alpha = \alpha(t)$.

$${}_cD^{\alpha}\ {}_{tc}D_t^{\beta}f(t) = {}_cD_t^{\alpha+\beta}f(t) \quad \text{for} \quad \beta \leq 0 \wedge \alpha + \beta \leq 0 \tag{1.25}$$

The numerical Type-$\mathcal{D}$ formulation behavior, when the tumor is being extinguished, is said to provide inaccurate results by the authors (Sierociuk et al., 2015b). To solve such numerical behavior, when the tumor's mass is approximately zero, a switching scheme is applied bypassing the variable order *Simulink* block from the toolbox described. An acceptable value for the switching is for an achieved tumor density of 10^{-3}%. Not only that, due to intrinsic characteristics of its differential equation, the tumor is constantly growing until it reaches the maximum size of 100%. When treatment is applied, the tumor's density decreases to such small values it can be considered inexistent. However, when the referred therapy is halted, tumor density tends to retake its Gompertz form and grow again until maximum capacity is achieved. Physiologically speaking, such situation is in accordance to a tumor relapse, which is not meant to be mathematically replicated here. So, to solve such innate behavior when the tumor's density is approximatively null, after treatment, it should be maintained that way. Here, this was achieved through a simple *Simulink* step function.

In an effort to reduce parameters in the original models of Ayati et al. (2010), the action of tumor through coefficients r_{ij} is replaced by the variable order derivative $\alpha(t)$ applied in the operator D of $C(t)$ and $B(t)$ equations. However, since the resulting model's steady state, from a mathematical standpoint, is the same as the local healthy one, for all contemplated cases here, and most parameters remain with the same value, the associated activity of osteoclasts and osteoblasts in bone mass must differ from the original case. As such, the bone resorption–formation ratio, given by Eq. (1.26), is determined between 0 and $\bar{t}$, that corresponds to the completion time of a single cycle of $C(t)$ in the new model. When treatments are applied, the duration of an individual remodeling cycle is changed whenever an introduction is done. Consequently, three resorption rates must be determined: when the tumor begins, when treatment is applied, and when the tumor is extinguished and the treatment is consequently stopped. Each ratio is determined for the first complete cycle after the induced change. Such is the same method as used originally in Ayati et al. (2010). Bone resorption and formation activities are then given by $\kappa_C = rR$ and $\kappa_B = r$, respectively.

$$R = \frac{\int_0^{\bar{t}} \max[0, C(t) - C_{ss}]}{\int_0^{\bar{t}} \max[0, B(t) - B_{ss}]} \tag{1.26}$$

For the local and nonlocal tumor bone remodeling models, their simplification is presented in Eqs. (1.27) and (1.28), respectively.

$$D^{\alpha(t)}C(t) = \alpha_C C(t)^{g_{CC}} B(t)^{g_{BC}} - \beta_C C(t) \tag{1.27a}$$

$$D^{\alpha(t)}B(t) = \alpha_B C(t)^{g_{CB}} B(t)^{g_{BB}} - \beta_B B(t) \tag{1.27b}$$

$$D^1 z(t) = -\kappa_C \max[0, C(t) - C_{SS}] + \kappa_B \max[0, B(t) - B_{SS}] \tag{1.27c}$$

$$D^1 T(t) = \gamma_T T(t) \log\left(\frac{L_T}{T(t)}\right) \tag{1.27d}$$

$$\alpha(t) = 1 - \theta \times t \times T(t) \tag{1.27e}$$

Fig. 1.11 compares the results for a bone with tumor obtained with this simplified model with the results obtained with the model in Eq. (1.18). It is clear that the results are qualitatively similar, even though there are differences. Where the evolution of osteoclasts and osteoblasts is concerned, the simplified model is even more similar to the model with no tumor. However, it is not yet known if the tumorous-induced changes in the action of osteoblasts are the result of a reduction of their number or of a lessened capability

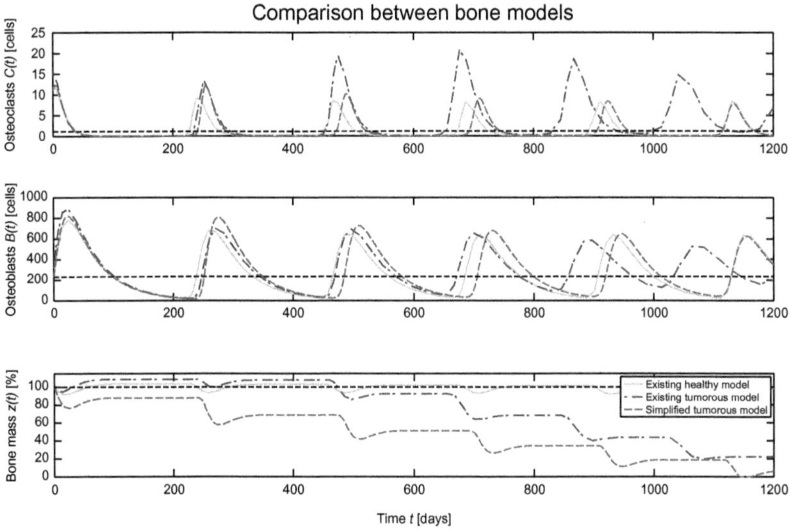

FIGURE 1.11 Osteoclasts, Osteoblasts, and Bone Mass evolutions, respectively. In **full lines**, simulation of Eq. (1.17) for the existing model replicating a healthy bone microenvironment. In dashed lines, simulation of Eq. (1.18) and (1.27), for a bone microenvironment disrupted by a developing tumor. **Black lines** represent stationary states. Tumor evolution is represented, in a full-blue line, on the left-side graphic of Fig. 1.10. Parameters were set according to Ayati et al. (2010), and can be found in Table 1.2.

of the existing cells to form bone (Holen, 2012). Consequently, these differences are immaterial, at least in the current state of knowledge about the phenomenon under study, even though future research may prove one of the alternatives the most correct.

The system of Eq. (1.27) is very sensitive to changes in order, as can be seen in Fig. 1.12. The value of this parameter must be set carefully to replicate the qualitative behavior desired, as was done in all other simulations in this chapter. On the other hand, this proves that a mere change in the differentiation order is in fact able to replicate different situations, and different evolutions of the bone mass, corresponding to different evolutions of a tumor, and different effects of the treatments that may be applied.

The extension of the simplified model to include diffusion over one dimension (as was done to the integer order model in its extension of Eq. 1.20) is trivial, and is given by the system of partial differential equations in Eq. (1.28). Simulating this system is not so trivial, but can be achieved simulating simultaneously several BMUs, and calculating the second order derivatives in order to space using centered finite differences.

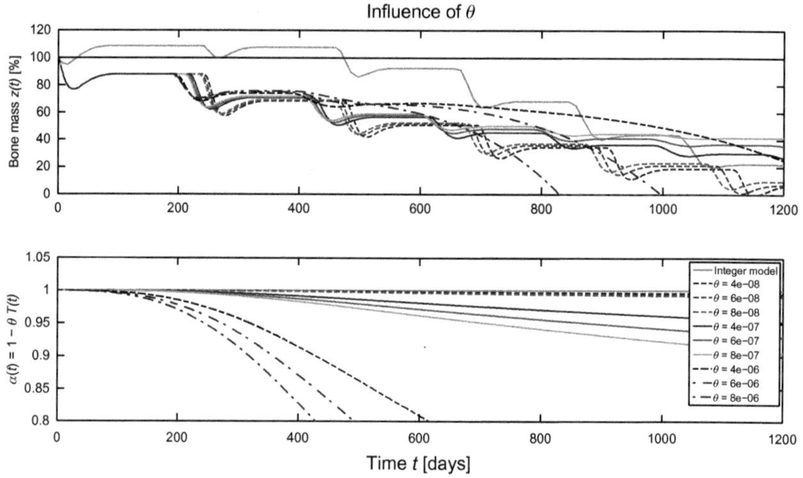

FIGURE 1.12 Influence of θ in the evolution of the bone mass according ot the model of Eq. (1.27). Parameters were set according to Ayati et al. (2010), and can be found in Table 1.2.

Results are shown in Fig. 1.13. The extension of this model to incorporate cancer treatment is obvious, and so is a further extension to a three-dimensional bone (which is, after all, the most realistic situation), which consists solely in replacing the second order derivative $\frac{\partial^2}{\partial x^2}$ with the Laplacian $\nabla^2 = \frac{\partial^2}{\partial x_1^2} + \frac{\partial^2}{\partial x_2^2} + \frac{\partial^2}{\partial x_3^2}$.

$$D^{\alpha(t,x)}C(t,x) = \sigma_C \frac{\partial^2}{\partial x^2} C(t,x) + \alpha_C C(t,x)^{g_{CC}} B(t,x)^{g_{BC}} - \beta_C C(t,x) \quad (1.28a)$$

$$D^{\alpha(t,x)}B(t,x) = \sigma_B \frac{\partial^2}{\partial x^2} B(t,x) + \alpha_B C(t,x)^{g_{CB}} B(t,x)^{g_{BB}} - \beta_B B(t,x) \quad (1.28b)$$

$$D^1 z(t,x) = \sigma_z \frac{\partial^2}{\partial x^2} z(t,x) - \kappa_C \max[0, C(t,x) - C_{SS}] + \kappa_B \max[0, B(t,x) - B_{SS}] \quad (1.28c)$$

$$D^1 T(t,x) = \sigma_T \frac{\partial^2}{\partial x^2} T(t,x) + \gamma_T T(t,x) \log\left(\frac{L_T}{T(t,x)}\right) \quad (1.28d)$$

$$\alpha(t,x) = 1 - \theta \times t \times T(t,x) \quad (1.28e)$$

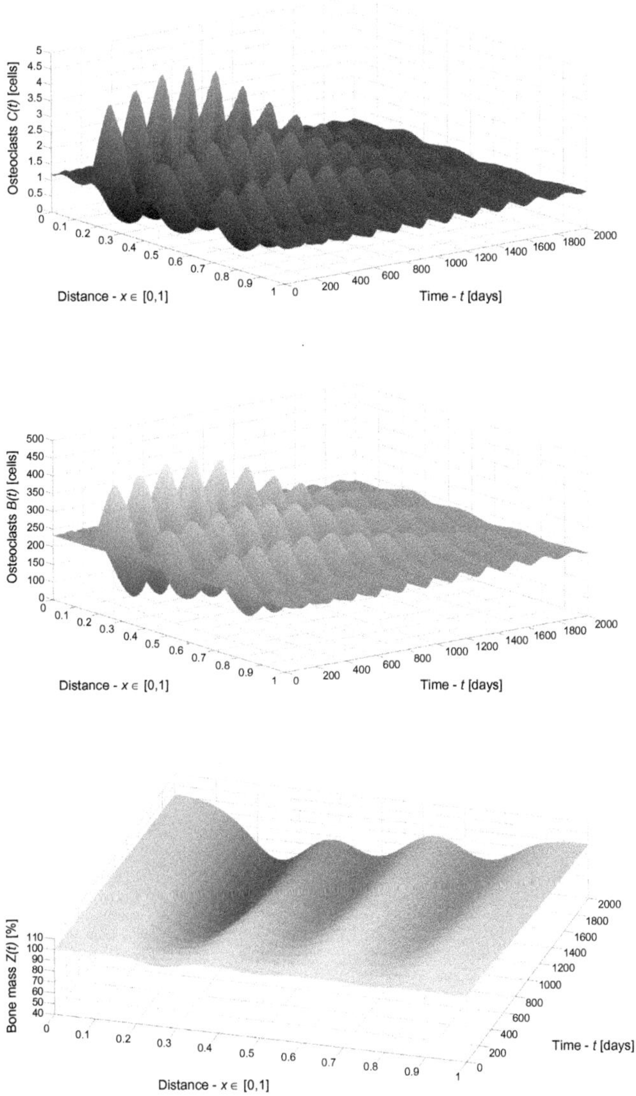

FIGURE 1.13 Nonlocal simulation of Osteoclasts, Osteoblasts, and Bone Mass, for the simplified model for bone remodeling with tumor (Eq. 1.28). Compare with Fig. 1.8. Parameters, initial, and boundary conditions follow exactly what was presented in Ayati et al. (2010), and can be found in Table 1.3. Variable order model parameters follow the integer healthy model values, except the actualized resorption rates of $\kappa_C = 0.1548$, $\kappa_B = 6.4924 \times 10^{-4}$ for $R = 238.43$ (equal for the analogous variable order local model), and $\theta = 2.5 \times 10^{-7}$. Untreated tumor evolution, for all metastases disrupted models, is presented in the second graphic of Fig. 1.10.

TABLE 1.3 Variables and Parameters

Unaltered Parameters			
α_C	3	σ_C	10^{-6}
α_B	4	σ_B	10^{-6}
β_C	0.2	σ_Z	10^{-6}
β_B	0.02	σ_T	10^{-6}
g_{CC}	1.1	r_{CC}	0.005
g_{BC}	−0.5	r_{BC}	0
g_{CB}	1.0	r_{CB}	0
g_{BB}	0	r_{BB}	0.2
$z(0)/z(0,x)$	100	L_T	100
Local versus Nonlocal		**Local**	**Nonlocal**
κ_C		0.0748	0.45
κ_B		6.39×10^{-4}	0.0048
γ_T		0.005	0.004
$T(0)/T(0,x)$		1	*
Healthy versus Metastatic Bone		**Healthy**	**Metastatic**
$C(0)/C(0,x)$		11.76*	15*
$B(0)/B(0,x)$		231.72	316
C_{ss}		1.16	5
B_{ss}		231.72	316

Source: From Ayati et al. (2010), used for the simulations and equations of local and non local models presented in this paper, except when explicitly said otherwise. Constant parameters, regardless of the simulation type, are presented in *Unaltered Parameters* section. Parameters that differ with local or nonlocal environments are presented in *Local vs Local*. Healthy or metastatic bone parameters are presented in *Healthy vs Metastatic Bone*.

1.5 CONCLUSIONS

This chapter introduced the concept of a variable order derivative, presenting the several different possible definitions, and stressing that these correspond to a different memory of past values of the order. This memory, which is independent of the memory of past values of the differentiated function, may not exist; or, if it exists, it may be weaker or stronger. These different definitions correspond to different manners of taking into account the changes in the differentiation order, and take place irrespective of the use of

Riemann–Liouville, Caputo, Grünwald–Letnikoff, etc., formulations of fractional derivatives.

It is important to retain that the different definitions, corresponding to different operators, all make sense. They can be implemented using electronic circuits, and the different memories of past values of the order corresponding to different ways of switching between constant orders. In other words, the way the order changes modifies the memory that remains of what happened before with the order. Different memories correspond to different phenomena found in applications.

An application to the modeling of the biochemical phenomena of bone remodeling has been presented. This application requires a memory of past values of the order; in particular, a type-$\mathcal{D}$ formulation from Sierociuk et al. (2015a,b) was used. The models simulate what happens to healthy bone tissue, to bone tissue with a tumor, and to bone tissue with a tumor subject to cancer treatments. It has been shown that variable order derivatives lead to models which are simpler and easier to understand. It is not difficult to see why, in this application, the variable order derivatives must have a memory of past values of the order. The memory is here reasonable, as the order models the effects of cancer, and these take time both to become manifest and to disappear even if the disease can be successfully treated. But there are other applications in which a different behavior of the order may be needed. That is why different definitions exist.

As to the models found, they can be used to develop personalized clinical decision systems for bone pathologies, aiming at the administration of the most efficient targeted therapies, adapted and tailored to each particular patient. This is a step towards achieving the impact that computational analysis of bone physiological models is expected to have on the development of clinical decision support systems in the future.

As to future developments of variable order derivatives, and beyond a further study of the operators, it is interesting to try to determine the conditions in which different memories of past values may be needed in applications, so as to more easily use the appropriate definitions in practice.

As to future developments of the bone remodeling equations, diffusion terms, and consequent boundary conditions, should be considered for the PK/PD treatments (because drugs are applied in a specific site, and will diffuse), and the biomechanical effects in the bone should be incorporated. This means that the changes in dynamics caused by mechanical solicitations should be added to the variable order differential equations (Belinha et al., 2015; Capacete, 2016). But the greatest challenge will likely be to measure coefficients from experimental data, possibly extrapolated from experiments with animals (Bonucci and Ballanti, 2014). While parameters found in the literature are not more than educated guesses by clinicians and oncobiologists, this approach may provide additional proof of the importance of variable order derivatives in this field.

ACKNOWLEDGMENTS

This work was supported by FCT, through IDMEC, under LAETA, projects UID/EMS/50022/2013, BoneSys, and IF/00653/2012.

REFERENCES

Ayati, B.P., Edwards, C.M., Webb, G.F., Wikswo, J.P., 2010. A mathematical model of bone remodeling dynamics for normal bone cell populations and myeloma bone disease. Biol. Direct 5 (1), 28.

Azar, A.T., Vaidyanathan, S., Ouannas, A. (Eds.), 2017. Fractional Order Control and Synchronization of Chaotic Systems. Springer, New York.

Belinha, J., Dinis, L.M.J.S., Natal Jorge, R.M., 2015. The meshless methods in the bone tissue remodelling analysis. Procedia Eng. 110, 51–58.

Bellido, T., Plotkin, L.I., Bruzzaniti, A., 2014. Chapter 2 - Bone Cells. Basic and Applied Bone Biology. Academic Press, San Diego, pp. 27–45, chapter 2.

Bonucci, E., Ballanti, P., 2014. Osteoporosis — bone remodeling and animal models. Toxicol. Pathol. 42, 957–969.

Boyce, B.F., 2012. Bone biology ad pathology. In: Coleman, R., Abrahamsson, P.-A., Hadji, P. (Eds.), Handbook of Cancer-Related Bone Disease, second ed. BioScientifica, pp. 5–6.

Capacete, J.D.C., 2016. Biochemical and Biomechanic Integrated Modeling ofBone, Master thesis, Instituto Superior Técnico.

Casimiro, S., Guise, T.A., Chirgwin, J., 2009. The critical role of the bone microenvironment in cancer metastases. Mol. Cell. Endocrinol. 310 (1-2), 71–81.

Casimiro, S., Ferreira, A.R., Mansinho, A., Alho, I., Costa, L., 2016. Molecular mechanisms of bone metastasis: which targets came from the bench to the bedside? Int. J. Mol. Sci. 17 (9), 1415.

Chen, T., Berenson, J., Vescio, R., Swift, R., Gilchick, A., Goodin, S., et al., 2002. 'Pharmacokinetics and pharmacodynamics of zoledronic acid in cancer patients with bone metastases'. J. Clin. Pharmacol. 42 (11), 1228–1236.

Chen, Y.-C., Sosnoski, D.M., Mastro, A.M., 2010. Breast cancer metastasis to the bone: mechanisms of bone loss. Breast Cancer Res. BCR 12 (6), 215.

Christ, L.F., Pedro, D., Oliveira, M.D., Almeida, S.D., Vinga, M., 2018. Modelling bone metastases using fractional derivatives. J. Appl. Nonlinear Dynam. 7 (1), 81–94.

Coelho, R.M., Vinga, S., Valério, D., 2015. CancerSys - Multiscale Modeling for Personalized Therapy of Bone Metastasis, Technical Report. Instituto Superior Técnico.

Coelho, R.M., Lemos, J.M., Alho, I., Valério, D., Ferreira, A.R., Costa, L., et al., 2016. Dynamic modeling of bone metastasis, microenvironment and therapy. Integrating parathyroid hormone (PTH) effect, anti-resorptive and anti-cancer therapy. J. Theor. Biol. 391, 1–12.

Crockett, J.C., Rogers, M.J., Coxon, F.P., Hocking, L.J., Helfrich, M.H., 2011. Bone remodelling at a glance. J. Cell Sci. 124, 991–998.

Dhillon, S., Gill, K., 2006. Basic pharmacokinetics, Clinical Pharmacokinetics, first ed. Pharmaceutical Press, Chicago, chapter Chapter 1.

Gibiansky, L., Sutjandra, L., Doshi, S., Zheng, J., Sohn, W., Peterson, M., et al., 2012. Population pharmacokinetic analysis of denosumab in patients with bone metastases from solid tumors. Clin. Pharmacokinet. 51 (4), 247–260.

Holen, I., 2012. Pathophysiology of bone metastases. In: Coleman, R., Abrahamsson, P.-A., Hadji, P. (Eds.), Handbook of Cancer-Related Bone Disease, second ed. BioScientifica, Sheffield, p. 49.

Komarova, S.V., Smith, R.J., Dixon, S.J., Sims, S.M., Wahl, L.M., 2003. Mathematical model predicts a critical role for osteoclast autocrine regulation in the control of bone remodeling. Bone 33 (2), 206–215.

Lerner, U., 2006. Bone Remodeling in Post-menopausal Osteoporosis. J. Dental Res. 85 (7), 584–595.

Lorenzo, C.F., Hartley, T.O.M.T., 2002a. Variable Order and distributed order fractional operators, Nonlinear Dynamics, Vol. 29. Kluwer Academic Publishers, pp. 57–98.

Lorenzo, C.F., Hartley, T.T., 2002b. Variable Fractional Order and Distributed Order Operators, Technical Report February. National Aeronautics and Space Administration (NASA).

Mager, D.E., Wyska, E., Jusko, W.J., 2003. Minireview on diversity of mechanism based pharmacodynamics models. Drug Metabol. Disposition 31 (5), 510–519.

Magin, R.L., 2006. Fractional Calculus in Bioengineering. Begell House Publishers Inc, Redding.

Neto, J.P., Valério, D., Vinga, S., 2017. Dynamic modeling of bone remodeling, osteolytic metastasis and PK / PD therapy: introducing variable order derivatives as a simplification technique.

Oyajobi, B.O., Garrett, I.R., Gupta, A., Flores, A., Esparza, J., Muñoz, S., et al., 2007. Stimulation of new bone formation by the proteasome inhibitor, bortezomib: implications for myeloma bone disease. Br. J. Haematol. 139 (3), 434–438.

Parfitt, A.M., 1994. Osteonal and hemi-osteonal remodeling: the spatial and temporal framework for signal traffic in adult human bone. J. Cell. Biochem. 55 (3), 273–286.

Perez, E.A., Vogel, C.L., Irwin, D.H., Kirshner, J.J., Patel, R., 2001. Multicenter Phase II Trial of Weekly Paclitaxel in Women With Metastatic Breast Cancer. J. Clin. Oncol. 19 (22), 4216–4223.

Petrás, I., 2009. Stability of fractional-order systems with fractional orders: a survey. Fract. Calcul. Appl. Anal. 12 (3), 269–298.

Pinheiro, J.V., Lemos, J.M., Vinga, S., 2011. Nonlinear MPC of HIV-1 infection with periodic inputs.pdf, in '50th IEEE Conference on Decision and control and European Control Conference (CDC-ECC)', pp. 65–70.

Raggatt, L.J., Partridge, N.C., 2010. Cellular and molecular mechanisms of bone remodeling. J. Biol. Chem. 285 (33), 25103–25108.

Rahimy, M., 2010. Applications of fractional differential equations. Appl. Math. Sci. 4 (50), 2453–2461.

Roodman, G.D., 2004. Mechanisms of bone metastasis. New England J. Med. 360 (16), 1655–1664.

Ryser, M.D. (2011), Of bones and noises, PhD thesis, McGill University.

Samko, S.G., 1995. Fractional integration and differentiation of variable order. Anal. Math. 21, 213–236.

Savageau, M.A., 1988. Introduction to S-systems and the underlying power law formalism. Math. Computer Model. II (3), 546–551.

Schneider, A., Kalikin, L.M., Mattos, A.C., Keller, E.T., Allen, M.J., Pienta, K.J., et al., 2005. Bone turnover mediates preferential localization of prostate cancer in the skeleton. Endocrinology 146 (4), 1727–1736.

Sierociuk, D., Malesza, W., 2015. On the recursive fractional variable-order derivative: equivalent switching strategy, duality, and analog modeling. Circuits Systems Signal Processing 34 (4), 1077–1113.

Sierociuk, D., Malesza, W., 2017. Fractional variable order discrete-time systems, their solutions and properties. Int. J. Systems Sci. 48 (14), 3098–3105.

Sierociuk, D., Dzieliński, A., Sarwas, G., Petras, I., Podlubny, I., Skovranek, T., 2013. Modelling heat transfer in heterogeneous media using fractional calculus. Philosop. Trans. Royal Soc. 371 (1990), 1–10.

Sierociuk, D., Malesza, W., Macias, M., 2015a. Derivation, interpretation, and analog modelling of fractional variable order derivative definition. Appl. Math. Model. 39 (13), 3876–3888.

Sierociuk, D., Malesza, W., Macias, M., 2015b. Fractional Variable Order Derivative Simulink User Guide, https://www.mathworks.com/matlabcentral/fileexchange/38801-fractional-variable-order-der

Sierociuk, D., Malesza, W., Macias, M., 2015c. On 685 the recursive fractional variable-order derivative: Equivalent switching strategy, duality, and analog modeling. Circuits Systems Signal Processing 34 (4), 1077–1113.

Silva, B., Bilezikian, J., 2015. Parathyroid hormone: anabolic and catabolic actions on the skeleton. Curr. Opin. Pharmacol. 22, 41–50.

Sohn, W., Simiens, M.A., Jaeger, K., Hutton, S., 2014. The pharmacokinetics and pharmacodynamics of denosumab in patients with advanced solid tumours and bone metastases: a systematic review. Bri. J. Clin. Pharmacol. 78 (3), 477–487.

Suva, L.J., Washam, C., Nicholas, R.W., Griffin, R.J., 2011. Bone metastasis: mechanisms and therapeutic opportunities. Nat. Rev. Endocrinol. 7 (4), 208–218.

Valério, D., Sá da Costa, J., 2006. Tuning of fractional PID controllers with Ziegler-Nichols type rules. Signal Processing 86 (10), 2771–2784.

Valério, D., Sá da Costa, J., 2011a. Introduction to single-input, single-output Fractional Control. IET Control Theory Applicat. 5 (8), 1033–1057.

Valério, D., Sá da Costa, J., 2011b. Variable-order fractional derivatives and their numerical approximations. Signal Processing 91 (3), 470–483.

Valério, D., Sá da Costa, J., 2012. Fractional reset control. Signal Image Video Processing 6 (3), 495–501.

Valério, D., Sá da Costa, J., 2013. An Introduction to Fractional Control. Institution of Engineering and Technology.

Valério, D., Coelho, R.M., Vinga, S., 2016. Fractional dynamic modelling of bone metastasis, microenvironment and therapy, in 'International Conference on Fractional Differentiation and its Applications', Novi Sad, Serbia.

Wheeler, N., 1997. Construction and physical application of the fractional calculus. Zometa® - Zoledronic Acid for Injection (2017). Novartis Pharmaceuticals Corporation, East Hanover, NJ.

Chapter 2

Nonlinear Fractional Order Boundary-Value Problems With Multiple Solutions

Mourad S. Semary[1,*], Hany N. Hassan[1,2] and Ahmed G. Radwan[3,4]
[1]*Benha Faculty of Engineering, Benha University, Benha, Egypt,* [2]*Imam Abdulrahman Bin Faisal University, Dammam, Saudi Arabia,* [3]*Faculty of Engineering, Cairo University, Egypt,* [4]*Nanoelectronics Integrated Systems Center (NISC), Nile University, Cairo, Egypt*

2.1 INTRODUCTION

The ordinary differential equations with boundary value problems play an essential role in both theory and applications. They are used to describe a large amount of biological, physic, heat transfer, circuits analysis, and chemical phenomena. It is very important not to lose any solutions of nonlinear differential equations with boundary conditions in engineering and physical sciences. Many authors constructed approaches that are based on the semianalytic methods for multiple solutions of nonlinear boundary value problems. These approaches are based on the framework of some semianalytical methods such as homotopy analysis method (HAM) and its modifications (Abbasbandy and Shivanian, 2010, 2011; Hassan and El-Tawil, 2011a; Liao, 2005, 2012; Soltani et al., 2017; Xu et al., 2010), variational iteration method (VIM) (Semary and Hassan, 2015; Wazwaz, 2017), and Adomian decomposition method (Wazwaz et al., 2016).

The idea of fractional calculus was proposed by L'Hopital and Leibniz in 1695. The first reasonable definitions for the fractional calculus were introduced by Liouville, Riemann, and Grnwald in 1834, 1847, and 1867, respectively (Podlubny, 1999). Recently, many theorems and generalized fundamentals have been introduced using fractional-order domain such as stability theorems, the generalized definition of fractional order resonance conditions, Chaotic Systems, new mathematical formulations, fractional order oscillators theorems, modeling of vegetables and fruits using

* Fully documented templates are available in the elsarticle package on CTAN.

Mathematical Techniques of Fractional Order Systems. DOI: https://doi.org/10.1016/B978-0-12-813592-1.00002-7

Cole–Cole model, generalized fractional order filters, and others in different engineering applications (Radwan et al., 2009a,b, 2008a,b; Radwan, 2013; Valerio et al., 2013; Odibat, 2010; Freeborn et al., 2013; Azar et al., 2017). Also, the fractional theory was extended to belong to the memristive elements (Fouda and Radwan, 2015).

Moreover, the fractional differential equations are a main and important subject in the area of physical sciences and engineering, as they have become popular in modeling processes in physical science and engineering, such as control theory, signal processing, bioengineering, circuit theory, and viscoplasticity (Das and Pan, 2012; Monje et al., 2010; Petras, 2011; Semary et al., 2016a,b). Therefore, the numerical methods for these equations have undergone fast growth in recent years. But only a few methods are available for solving the fractional order differential equations with multiple solutions (Arqub et al., 2014; Alomari et al., 2013; Semary et al., 2017a,b). This chapter introduces and discus the techniques based on the Picard method (Semary et al., 2017a,b) and HAM (Alomari et al., 2013) to predict the multiplicity of the solutions of fractional order differential equations with boundary conditions. These techniques are not only capable of predicting but also calculating all branches of the solutions simultaneously. The applied approaches detect multiple solutions to some of the practical models in fractional order domain, the first problem arising in mixed convection flows in a vertical channel (Abbasbandy and Shivanian, 2011; Alomari et al., 2013). The second model appears in a number of applications such as the model of thermal reaction process, the fuel ignition model of the thermal combustion theory, the Chandrasekhar model of the expansion of the Universe, questions in geometry and relativity about the Chandrasekhar model, nanotechnology, radiative heat transfer and chemical reaction theory, and this model is called the Bratu problem (Jacobsen and Schmitt, 2002; Jalilian, 2010; Wazwaz, 2005). The third problem describes the fractional order diffusion and reaction model in porous catalysts and also steady state heat conduction of the temperature distribution of a straight rectangular fin with a prescribed power-law temperature dependent surface heat flux (Abbasbandy and Shivanian, 2010; Semary et al., 2017a; Magyari, 2008). The fourth problem is a three-point nonlinear boundary-value problem (Graef et al., 2003, 2004). The main goal of this chapter is to illustrate the two methods procedures to predict and calculate all branches of the problem solutions in the fractional order domain. The advantages of these techniques are that they are very powerful for solving fractional order boundary value problems that admit multiple solutions. And the obtained series solutions are converge to multiple solutions with convergent region. The chapter symbols are listed in Table 2.1 and this chapter is organized as follows: Section 2.2 introduces the mathematical procedures for the used methods. Then, the numerical simulations for these methods are discussed for different classes of fractional order differential equations in Section 2.3. Finally, Section 2.4 concludes the presented work.

TABLE 2.1 Symbols

Symbol	Name
D_t^α	Caputo fractional derivative operator
$\mathcal{B},\mathcal{B}'$	Boundary operators
J^α	Riemann–Liouville fractional integral operator
α	Order
$\varphi(t, p)$, $\phi(t, p)$	Homotopy series
h	Control parameter
L	Auxiliary linear operator
p	Homotopy parameter
$u_0(t)$	Initial guess
$u_m(t)$	Successive approximation
$H(t)$	Auxiliary function
$U_M(t)$	mth-order approximate solution
d_k	Initial conditions values
$\lambda(\tau)$	A general Lagrange multiplier
Γ	Gamma function
ϵ	Unknown parameter
E	Dimensionless parameter
$u(t)$	Unknown function
ψ	Convective-conductive parameter
λ	Frank–Kamenetskii parameter
β	Dimensionless parameter

2.2 THE METHODS PROCEDURES

To illustrate the methods procedures consider the following nonlinear fractional differential equation:

$$D_t^\alpha u(t) + f(t, u(t)) = 0, n - 1 < \alpha \leq n, \tag{2.1}$$

with boundary conditions

$$\mathcal{B}\left(u, \frac{\partial u}{\partial n}\right) = 0, \tag{2.2}$$

where $\mathcal{B}$ is a boundary operator and D_t^α denotes Caputo fractional derivative which is defined as (Podlubny, 1999)

$$D_t^{\alpha}u(t) = \frac{1}{\Gamma(n-a)}\int_0^t (t-\tau)^{n-\alpha-1}u^{(n)}(\tau)d\tau, n-1<\alpha<n. \tag{2.3}$$

The important property of the Caputo fractional derivative operator D_t^{α} is (Podlubny, 1999)

$$J^{\alpha}D_t^{\alpha}u(t) = u(t) - \sum_{i=0}^{n-1} u^{(i)}(0)\frac{t^i}{i!}, \tag{2.4}$$

where J^{α} is the Riemann–Liouville fractional integral operator and defined as

$$J^{\alpha}u(t) = \frac{1}{\Gamma(\alpha)}\int_0^t (t-\tau)^{\alpha-1}u(\tau)d\tau. \tag{2.5}$$

2.2.1 Predictor Homotopy Analysis Method

Liao (1992) proposed a powerful and easy semianalytic tool for nonlinear problems, namely the HAM. The homotopy is a continuous transformation from one function to another (Liao, 2012, 2003). The homotopy between two continuous functions $f(t)$ and $g(t)$ from a topological space X to topological space Y is defined to be continuous function $\mathcal{H}(t, p)$

$$\mathcal{H}(t,p):X\times[0,1]\rightarrow Y, \tag{2.6}$$

where p is called the homotopy parameter and $p \in [0, 1]$ such that if $t \in X$ then $\mathcal{H}(t, 0) = f(t)$ and $\mathcal{H}(t, 1) = g(t)$. The mth-order homotopy derivative of $\mathcal{H}(t, p)$ is defined by

$$D^m\mathcal{H}(t,p) = \frac{1}{m!}\frac{\partial^m\mathcal{H}(t,p)}{\partial p^m}\Big|_{p=0} \tag{2.7}$$

where $m \geq 0$ is an integer. And has the following properties, for homotopy series

$$\varphi(t,p) = \sum_{i=0}^{\infty} u_i(t)p^i, \phi(t,p) = \sum_{i=0}^{\infty} v_i(t)p^i, \tag{2.8}$$

the following are held:

1. $D^m(\varphi) = u_m$ and $D^m(p^j\varphi) = D^{m-j}\varphi = u_{m-j}$.
2. Let L to be a linear operator independent of the homotopy parameter p then $D^m(L\varphi) = L\,[D^m(\varphi)]$
3. $D^m(\varphi\phi) = \sum_{k=0}^{m} u_{m-k}v_k$.
4. If f and g are functions independent of homotopy parameter p then $D^m(f\phi + g\varphi) = fv_m + gu_m$.

5. $D^m(\varphi^\sigma) = \sum_{r_1=0}^{m} u_{m-r_1} \sum_{r_2=0}^{r_1} u_{r_2-r_1} \sum_{r_3=0}^{r_2} u_{r_2-r_3} \ldots \sum_{r_{\sigma-1}=0}^{r_{\sigma-2}} u_{r_{\sigma-2}-r_{\sigma-1}} u_{r_{\sigma-1}}$,

where $m \geq 2$ and $\sigma \geq 2$ are positive integer numbers. The above properties are proved by Liao (2009). To illustrate the procedures of the HAM, let $N[u(t)] = 0$ denote a nonlinear differential equation. The HAM is based on a kind of continuous mapping $u(t) \rightarrow \varphi(t, p)$, constructed from the so-called the zero-order deformation equation as follows

$$(1-p)L[\varphi(t,p) - u_0(t)] = H(t)h^* pN[\varphi(t,p)], \tag{2.9}$$

where L is an auxiliary linear operator, $u_0(t)$ is an initial guess, $H(t)$ is the auxiliary function, and h is the nonzero control parameter. When $p = 0$, the zero-order deformation Eq. (2.9) becomes:

$$L[\varphi(t,0) - u_0(t)] = 0, \tag{2.10}$$

which gives $\varphi(t, 0) = u_0(t)$. And when $p = 1$, since $H(t) \neq 0$ and $h \neq 0$ Eq. (2.9) is equivalent to:

$$N[\varphi(t,1)] = 0, \tag{2.11}$$

which is exactly the same as the original equation $N[u(t)] = 0$, provided $\varphi(t, 1) = u(t)$. So, as the homotopy parameter p increases from 0 to 1, the solution $\varphi(t, p)$ of the zero-order deformation Eq. (2.9) varies from the initial guess $u_0(t)$ to the exact solution $u(t)$ of the nonlinear differential equation $N[u(t)] = 0$. By using Taylor theorem, $\varphi(t, p)$ expands in a power series of p as follows

$$\varphi(t,p) = \varphi(t,0) + \sum_{m=1}^{\infty} u_m(t) p^m, \tag{2.12}$$

where $u_m(t) = D^m \varphi(t, p)$. From $\varphi(t, 0) = u_0(t)$ then Eq. (2.12) becomes:

$$\varphi(t,p) = u_0(t) + \sum_{m=1}^{\infty} u_m(t) p^m. \tag{2.13}$$

When choosing the linear operator, the initial guess, the auxiliary function, the control parameter, and from $\varphi(t, 1) = u(t)$ then the power series $\varphi(t, p)$ (2.13) converges at $p = 1$, and the so-called homotopy series solution is given by:

$$u(t) = u_0(t) + \sum_{m=1}^{\infty} u_m(t). \tag{2.14}$$

which satisfies the original equation $N[u(t)] = 0$, as proved by Liao (2009). The terms $u_m(t)$ are unknown functions and determined by the so-called high-order deformation equations. Differentiating the zero-order deformation Eq. (2.9) m times with respective to the homotopy parameter p and then dividing it by $m!$ and finally setting $p = 0$, then the term $u_m(t)$ is given by

$$u_m(t) = \chi_m u_{m-1} + hL^{-1}\{H(t)D^m(pN[\varphi(t,p)])\}, \tag{2.15}$$

where χ_m is given by

$$\chi_m = \begin{cases} 0, m \leq 1 \\ 1, m > 1. \end{cases} \tag{2.16}$$

The HAM has been successfully employed to solve many types of nonlinear equations in science and engineering such as time-dependent Emden–Fowler type equations (Bataineh et al., 2007), the modified KdV-type equations (Liu and LiZ, 2009), the Boussinesq problem (Hassan and El-Tawil, 2011b), the Fitzhugh–Nagumo equation (Gorder, 2012), and the Sturm–Liouville problems (Abbasbandy and Shirzadi, 2011). Liao (2003) proved that the HAM logically contains Adomian decomposition method (ADM) in general. The homotopy perturbation method (HPM) is a special case from the HAM when setting $H(t) = 1$ and $h = -1$ (Hayat and Sajid, 2007; Sajid et al., 2007). Motsa et al. (2010) suggested modifying the HAM, the so-called spectral HAM (SHAM). This modified method is based on solving the higher order deformation equations by using the Chebyshev pseudospectral method and the auxiliary linear operator L is defined in terms of the Chebyshev spectral collocation differentiation matrix described by Don and Solomonoff (Don and Solomonoff, 1995). Semary and Hassan (2016) introduced the HAM to solve linear and nonlinear differential equations in q calculus. Abbasbandy and Shivanian (2011) suggested the predictor HAM (PHAM) to predict the multiplicity of solutions of nonlinear differential equations. And used it for finding the multiple solutions of nonlinear fractional differential equation with boundary conditions (Alomari et al., 2013). The core point in the PHAM for solving fractional differential Eqs. (2.1, 2.2) is to divide the boundary condition (2.2) to equivalent boundary conditions with unknown parameter ϵ as follows

$$\mathcal{B}'\left(u, \epsilon, \frac{\partial u}{\partial n}\right) = 0, u(a) = b, \tag{2.17}$$

where $u(a) = b$ is the forcing condition that arises from original boundary conditions (2.2). By applying the HAM on the problem (2.1) and with the condition $\mathcal{B}'\left(u, \epsilon, \frac{\partial u}{\partial n}\right) = 0$. Then the general zero-order deformation equation and the corresponding boundary conditions as follows:

$$(1-p)L[\varphi(t,\epsilon,p) - u_0(t,\epsilon)] = H(t)hp\{D_t^\alpha \varphi(t,\epsilon,p) + f(t, \varphi(t,\epsilon,p))\}, \tag{2.18}$$

$$\mathcal{B}'\left(\varphi(t,\epsilon,p), \epsilon, \frac{\partial \varphi(t,\epsilon,p)}{\partial n}\right) = 0, \tag{2.19}$$

where $u_0(t, \epsilon)$ is initial approximation guess of the exact solution $u(t, \epsilon)$ which satisfies the boundary conditions $\mathcal{B}'\left(u, \epsilon, \frac{\partial u}{\partial n}\right) = 0$. The general zero-order deformation Eq. (2.18) satisfies

$$\varphi(t,\epsilon,0) = u_0(t,\epsilon), \varphi(t,\epsilon,1) = u(t,\epsilon), \tag{2.20}$$

at $p = 0$ and $p = 1$ respectively. Taylor expansion of $\varphi(t, \epsilon, p)$ can be written in the form

$$\varphi(t,\epsilon,p) = \varphi(t,\epsilon,0) + \sum_{m=1}^{\infty} u_m(t,\epsilon)p^m, \tag{2.21}$$

where $u_m(t, \epsilon) = D^m\varphi(t, \epsilon, p)$. From the Eq. (2.20), the power series (2.21) becomes

$$\varphi(t,\epsilon) = u_0(t,\epsilon) + \sum_{m=1}^{\infty} u_m(t,\epsilon)p^m, \tag{2.22}$$

and they are solutions of the nonlinear problems (2.1) with boundary conditions $\mathcal{B}'\left(u,\epsilon,\frac{\partial u}{\partial n}\right) = 0$, but the functions $u_m(t,\epsilon)$ are still unknowns. These unknowns functions $u_m(t, \epsilon)$ are governed by the mth-order deformation equation. Take mth-order homotopy derivative (2.7) for both sides of the general zero-order deformation Eq. (2.18), we obtain

$$D^m\{(1-p)L[\varphi(t,\epsilon,p) - u_0(t,\epsilon)]\} = H(t)hD^m[p\{D_t^\alpha\varphi(t,\epsilon,p) + f(t,\varphi(t,\epsilon,p))\}], \tag{2.23}$$

since L is a linear operator independent of p and by using the properties of the mth-order homotopy derivative (2.7), then the left side of Eq. (2.23) becomes

$$\begin{aligned} D^m\{(1-p)L[\varphi(t,\epsilon,p) - u_0(t,\epsilon)]\} &= L[D^m\varphi(t,\epsilon,p) \\ - D^m(p\varphi(t,\epsilon,p)) - u_0(t,\epsilon)D^m(p)] &= L[u_m(t,\epsilon) - u_{m-1}(t,\epsilon) - u_0(t,\epsilon)D^m(p)] = \\ \begin{cases} L[u_m(t,\epsilon)], m \le 1 \\ L[u_m(t,\epsilon) - u_{m-1}(t,\epsilon)], m < 1 \end{cases} &= L[u_m(t,\epsilon) - \chi_m u_{m-1}(t,\epsilon)], \end{aligned} \tag{2.24}$$

and the mth-order deformation Eq. (2.23) becomes in the following form:

$$L[u_m(t,\epsilon) - \chi_m u_{m-1}(t,\epsilon)] = H(t)h\{D_t^\alpha u_{m-1}(t,\epsilon) + D^m[pf(t,\varphi(t,\epsilon,p))]\} \tag{2.25}$$

subject to the boundary conditions

$$\frac{\partial^m}{\partial p^m}\mathcal{B}'\left(\varphi(t,\epsilon,p),\epsilon,\frac{\partial\varphi(t,\epsilon,p)}{\partial n}\right)\Big|_{p=0} = 0. \tag{2.26}$$

It should be emphasized, the mth-order deformation Eq. (2.25) is just the linear ordinary differential equation with boundary condition (2.26) and can be easily solved by using some software programs such as Matlap, Mathematica, etc. Starting by $u_0(t,\epsilon)$ which satisfies the boundary conditions $\mathcal{B}'\left(u,\epsilon,\frac{\partial u}{\partial n}\right) = 0$ and from Eqs. (2.25) and (2.26), the unknowns functions $u_m(t, \epsilon)$ for $m : 1(1)M$ are obtained. The mth-order approximate solution of the problems (2.1) and with boundary conditions $\mathcal{B}'\left(u,\epsilon,\frac{\partial u}{\partial n}\right) = 0$ is given by

$$u(t,\epsilon) \approx U_M(t,h,\epsilon) = u_0(t,\epsilon) + \sum_{m=1}^{M} u_m(t,\epsilon). \tag{2.27}$$

2.2.2 Controlled Picard's Method

In 1890, Emile Picard introduced a basic tool for proving the existence of solutions of initial value problems regarding ordinary first order differential equations namely, the Picard's method. Recently, this method was used to solve and analyze the differential and integral equations with different definitions of the derivative (Azarnavid et al., 2015; El-Sayed et al., 2014; Salahshour et al., 2015; Vazquez-Leal et al., 2015). Semary et al. (2017a,b) suggested Picard iterative formula to solve the nonlinear fractional order differential equations. The advantage for this formula is that it contains an auxiliary parameter h which proves very effective in controlling the convergence region of an approximate solution. To drive controlled Picard's iterative formula for Eq. (2.1), multiply h and $H(t)$ to its both sides to become in the following form:

$$F[t,u(t),\alpha)] = H(t)h\{D_t^{\alpha}u(t) + f(t,u(t))\} = 0. \tag{2.28}$$

Adding and subtracting $D_t^{\alpha}u(t)$ from the Eq. (2.28) to become in the form:

$$D_t^{a}u(t) + F[t,u(t),\alpha] - D_t^{\alpha}u(t) = 0. \tag{2.29}$$

Applying the Riemann–Liouville integral of order α (J^{α}) on the problem (2.29) and after using of the property (2.4), the integrated form of Eq. (2.29) can be written as follows:

$$u(t) = \sum_{k=0}^{n-1} u^{(k)}(0)\frac{t^k}{k!} - J^{\alpha}\{F[t,u(t),\alpha] - D_t^{\alpha}u(t)\}. \tag{2.30}$$

Applying Picard method to the integral Eq. (2.30), the solution can be reconstructed as follows:

$$u_{m+1}(t) = \sum_{k=0}^{n-1} d_k\frac{t^k}{k!} - \frac{1}{\Gamma(\alpha)}\int_0^t (t-\tau)^{\alpha-1}\{F[\tau,u_m(\tau),\alpha] - D_{\tau}^{\alpha}u_m(\tau)\}d\tau, \tag{2.31}$$

where $d_k = u^{(k)}(0)$, $k = 0, 1, \ldots, n-1$ and $m \geq 0$. From Eq. (2.28), the iterative formula (2.31) becomes:

$$u_{m+1}(t) = \sum_{k=0}^{n-1} d_k\frac{t^k}{k!} - \frac{1}{\Gamma(\alpha)}\int_0^t (t-\tau)^{\alpha-1}H(\tau)h\{D_{\tau}^{\alpha}u_m(\tau) + f[\tau,u_m(\tau)]\}d\tau + J^{\alpha}D_t^{\alpha}u_m(t), \tag{2.32}$$

Using the property (2.4) of Caputo fractional order derivative, then Eq. (2.32) becomes:

$$\begin{aligned} u_{m+1}(t) &= \sum_{k=0}^{n-1} d_k \frac{t^k}{k!} - \frac{1}{\Gamma(\alpha)} \int_0^t (t-\tau)^{\alpha-1} H(\tau) h \{ D_\tau^\alpha u_m(\tau) + f[\tau, u_m(\tau)] \} d\tau \\ &+ u_m(t) - \sum_{k=0}^{n-1} u_m^{(k)} \frac{t^k}{k!}. \end{aligned} \tag{2.33}$$

The successive approximation $u_m(t)$ must satisfy the initial conditions, for that $u_m^{(k)} = d_k$ and the iterative formula (2.33) becomes:

$$u_{m+1}(t) = u_m(t) - \frac{h}{\Gamma(\alpha)} \int_0^t (t-\tau)^{\alpha-1} H(\tau) \{ D_\tau^\alpha u_m(\tau) + f[\tau, u_m(\tau)] \} d\tau. \tag{2.34}$$

The important property (2.4) for Caputo fractional order derivative is right for the integer order case ($\alpha = n$) and it is proved easily by using integration by parts. So, it should be emphasized that the iteration formula (2.34) is suitable to solve the problem (2.1) for integer and fractional orders. Starting by an initial approximation $u_0(t) = \sum_{k=0}^{n-1} d_k \frac{t^k}{k!}$ or which satisfies at least the initial conditions for the problem. Therefore, the approximate solution $u_{m+1}(t)$ for the nonlinear differential Eq. (2.1) can be obtained with the conditions $\mathcal{B}'\left(u, \epsilon, \frac{\partial u}{\partial n}\right) = 0$ after being converted to initial conditions as $u^{(k)}(0) = d_k$, $k = 0, 1, \ldots, n-1$. The VIM is one of the famous semianalytics techniques used to solve nonlinear differential equations (Semary and Hassan, 2015; He, 1999; Wazwaz, 2009). To solve the integer order differential Eq. (2.1) by the VIM (He, 1999; Wazwaz, 2009), one can construct an iteration formula for the problem (2.1) when $\alpha = n$ as follows:

$$u_{m+1}(t) = u_m(t) + \int_0^t \lambda(\tau) \{ D_\tau^n u_m(\tau) + f[\tau, u_m(\tau)] \} d\tau, \tag{2.35}$$

where $D_\tau^n u_m(\tau) = u_m^{(n)}(\tau)$ and $\lambda(\tau)$ is a general Lagrange multiplier and it is equal to $\lambda(\tau) = -(t-\tau)^{(n-1)} / n-1!$ (Wazwaz, 2009).

Remark: The Controlled Picard iterative formula (2.34) leads to a variational iterative formula generated by the variational iteration method (2.35) when $h = H(t) = 1$, $\alpha = n$,and the general Lagrange multiplier $\lambda(\tau) = -(t-\tau)^{(n-1)} / \Gamma(n)$.

2.2.3 The Prediction of Solutions Multiplicity

The approximate solutions (2.27) and (2.34) outputs of PHAM and controlled Picard's method still have two unknown parameters, namely h and ϵ, that should be determined. The existence of unique or multiple solutions for the original boundary value problem (2.1) depends on the fact of whether the condition $u(a) = b$ admits unique or multiple values for the formally introduced parameter ϵ in the boundary conditions (2.17). The condition $u(a) = b$ satisfies the series solutions (2.27) and (2.34) for the problem then

$$U_M(a, h, \epsilon) = b, u_{m+1}(a, h, \epsilon) = b. \tag{2.36}$$

The above equations have two unknown parameters h and ϵ which control the convergence of the obtained series solutions. If the value of ϵ is constant value, then the series solutions (2.27) and (2.34) converge at $\epsilon = r$ only in that range of h, where the parameter ϵ does not change with the variation of h. According to Eq. (2.36), ϵ is a function of h ($\epsilon(h)$), by drawing the Eq. (2.36) gives the famous way of the HAM, the so-called h-curve for each method. The number of such horizontal plateaus in h-curve where $\epsilon(h)$ constant will predict the multiplicity of the solution. The horizontal plateaus indicate the convergence because if the formally introduced $\epsilon(h)$ is a constant value then a horizontal line segment in h-curve which corresponds to the valid region of h for the series solutions approaches the exact solution for the problem (2.1).

2.3 APPLICATIONS

2.3.1 Mixed Convection Flows in a Vertical Channel

The aim of this section is to apply the PHAM and controlled Picard's method (CPM) to detect the multiple solutions of a kind of model in mixed convection flows in the asfractional order domain. The model in integer order domain is combined forced and free flow in the fully developed region of a vertical channel with isothermal walls having the same temperature (Barletta, 1999; Barletta et al., 2005). In this model, the viscous dissipation effect is taken into account and the fluid properties are assumed to be constant. The set of governing have led to a fourth order ordinary differential equation for the dimensionless velocity field and reads

$$\frac{d^4u}{dt} = \frac{E}{16}\left(\frac{du}{dt}\right)^2, \tag{2.37}$$

with the boundary conditions

$$u'(0) = u'''(0) = u(1) = 0, \qquad \int_0^1 u(t)dt = 1, \tag{2.38}$$

where E is dimensionless parameter coincides with the product of the Prandtl number, the Gebhart number, and the Reynolds number. In special case $E \to 0$, it can be easily verified that, the models (2.37) and (2.38) admit a unique solution, namely

$$u(t) = \frac{3}{2}(1-t)^2 \tag{2.39}$$

It has been shown (Abbasbandy and Shivanian, 2011; Semary and Hassan, 2015; Barletta, 1999; Barletta et al., 2005) by semianalytic and numerical methods that Eqs. (2.37), (2.38) admit dual solutions for any given E in the interval $(-\infty, 0) \cup (0, E_{max})$ in which $E_{max} \cong 228.128$. This study generalizes the boundary value problem (2.37), (2.38) in the fractional order domain as follows:

$$D_t^{\alpha} u(t) = \frac{E}{16}\left(\frac{du}{dt}\right)^2 \tag{2.40}$$

where $3 < \alpha \leq 4$ and subject to boundary conditions (2.38). The model (2.40) with the boundary conditions (2.38) has been solved using PHAM and power series method (Arqub et al., 2014; Alomari et al., 2013). Alomari et al. (2013) applied the PHAM explained in Section 2.2.1 by using the auxiliary linear operator $L[.] = \frac{\partial^4}{\partial t^4}[.]$. This section applies the same method for the model (2.40) but using fractional order linear operator L as follows

$$L[\varphi(t, \epsilon, p)] = D_t^{\alpha}[\varphi(t, \epsilon, p)], \tag{2.41}$$

where D_t^{α} defined in (2.3) and the operator L has the property

$$D_t^{\alpha}[c_0 + c_1 t + c_2 t^2 + c_3 t^3] = 0. \tag{2.42}$$

To construct the dual solution of the model (2.40) with conditions (2.38) can be replaced formally by the following problem

$$D_t^{\alpha} u(t) = \frac{E}{16}\left(\frac{du}{dt}\right)^2, u'(0) = u'''(0) = u(1) = 0, u''(0) = \epsilon, \tag{2.43}$$

and the forcing condition

$$\int_0^1 u(t)dt = 1. \tag{2.44}$$

From the conditions in Eq. (2.43), the initial approximation guess can be

$$u_0(t, \epsilon) = \frac{\epsilon}{2}(t^2 - 1). \tag{2.45}$$

According the linear operator (2.41) and choosing $H(t) = 1$, the mth-order deformation Eq. (2.25) of the problem (2.43) reads

$$D_t^{\alpha}\left[u_m(t, \epsilon) - \chi_m u_{m-1}(t, \epsilon)\right] = h\left\{ D_t^{\alpha} u_{m-1}(t, \epsilon) - \frac{E}{16} D^m\left[p\left(\frac{d\varphi(t, \epsilon, p)}{dt}\right)^2\right]\right\}, \tag{2.46}$$

applying the Riemann–Liouville integral of order $\alpha(J^{\alpha})$ on Eq. (2.46) and using the properties of the homotopy derivative (2.7), the mth-order deformation equation becomes

$$u_m(t,\epsilon) = \chi_m u_{m-1}(t,\epsilon) + J^\alpha h\left\{ D_t^\alpha u_{m-1}(t,\epsilon) - \frac{E}{16}\sum_{i=0}^{m-1} u_i'(t,\epsilon)u_{m-1-i}'(t,\epsilon)\right\} + c_0 + c_1 t + c_2 t^2 + c_3 t^3. \tag{2.47}$$

and the constants c_i, i: 0, 1, 2, 3 can be determined by the boundary conditions:

$$u_m'(0,\epsilon) = u''_m(0,\epsilon) = u'''_m(0,\epsilon) = u_m(1,\epsilon) = 0. \tag{2.48}$$

Using the software of Wolfram Mathematica, starting with $u_0(t, \epsilon)$ thus the first two terms of Eqs. (2.47, 2.48) are

$$u_1(t,\epsilon) = -\frac{Eh(-1+t^2+\alpha)\epsilon^2}{8\alpha(2+3\alpha+\alpha^2)\Gamma(\alpha)}, \tag{2.49}$$

$$u_2(t,\epsilon) = \frac{Eh\epsilon^2}{64\alpha(1+\alpha)(2+\alpha)\Gamma(\alpha)\Gamma(3+2\alpha)}(Eh(-1+t^{2+2\alpha})(2+\alpha)\,\epsilon\,\Gamma(3+\alpha) - 8(1+h)(-1+t^{2+\alpha})\Gamma(3+2\alpha)), \tag{2.50}$$

in this way can be obtained the mth-order approximate solution for the problem (2.43)

$$u(t) \approx U_M(t,h,\epsilon) = \sum_{m=0}^{M} u_m(t,\epsilon), \tag{2.51}$$

The homotopy solution (2.51) still contains two unknowns parameters, namely h and ϵ. So Eq. (2.36) and with the help of the forcing condition $\int_0^1 u(t)dt = 1$, becomes

$$\int_0^1 u(t)dt \approx \int_0^1 U_M(t,h,\epsilon)dt = 1, \tag{2.52}$$

by selecting some values of E and according to the above Eq. (2.52) in Figs. 2.1 and 2.2, the $h-\epsilon$ curves has been plotted in the range $[-2.5, 0]$ and when $E = 20$ and $E = -20$. The curve gives two lines that are parallel to the h-axes for each value of fractional order parameter α. This means the model (2.40) has two solutions in these cases and so is in full agreement with the solutions result reported by Arqub et al. (2014) and Alomari et al. (2013). On other hand, to apply the controlled Picard's method for the model (2.40), the model boundary conditions (2.38) can be replaced as follows

$$u'(0) = u'''(0) = 0, u''(0) = \epsilon, u(0) = \gamma, \tag{2.53}$$

where γ and ϵ are the unknown parameters and the two forcing conditions

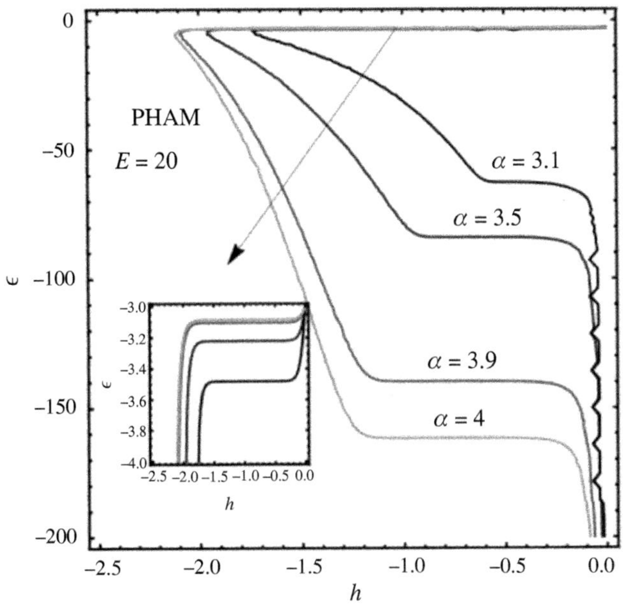

FIGURE 2.1 The $h-\epsilon$ curve for Eq. (2.52) with different values of α and $E=20$.

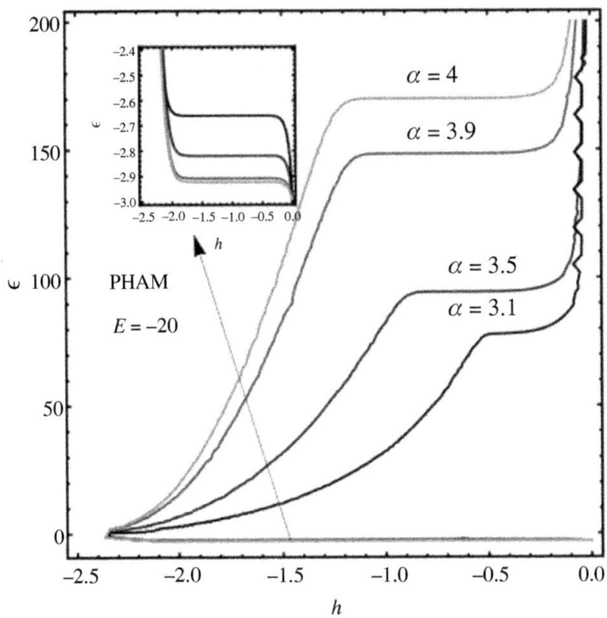

FIGURE 2.2 The $h-\epsilon$ curve for Eq. (2.52) with different values of α and $E=-20$.

$$u(1) = 0, \int_0^1 u(t)dt = 1. \tag{2.54}$$

By applying the iterative formula (2.34) in Eq. (2.40) with initial conditions (2.53) and by choosing $H(t) = 1$ then

$$u_{m+1}(t,h,\epsilon) = u_m(t,h,\epsilon) - \frac{h}{\Gamma(\alpha)}\int_0^t (t-\tau)^{\alpha-1}\left\{D_\tau^\alpha u_m(\tau,h,\epsilon) - \frac{E}{16}\left(\frac{du_m(\tau,h,\epsilon)}{d\tau}\right)^2\right\}d\tau. \tag{2.55}$$

According to the conditions (2.53), the initial approximation solution $u_0(t, \epsilon)$ can be written as

$$u_0(t,\epsilon)\frac{1}{2}(\epsilon t^2 + 2\gamma). \tag{2.56}$$

Starting with the initial approximation solution (2.56), the solution of the successive approximations (2.55), $u_{m+1}(t, h,\epsilon)$, $m \geq 0$ is as follows

$$u_1(t,h,\epsilon) = \gamma + \frac{t^2\epsilon}{2} + \frac{Eht^{2+\alpha}\epsilon^2}{8\alpha(2+3\alpha+\alpha^2)\Gamma(\alpha)} \tag{2.57}$$

$$u_2(t,h,\epsilon) = \gamma + \frac{t^2\epsilon}{2} + \frac{Eht^{2+\alpha}\epsilon^2}{8\alpha(2+3\alpha+\alpha^2)\Gamma(\alpha)} - \frac{Eht^{2+\alpha}\epsilon^2\left(\frac{64(-1+h)}{\alpha(2+3\alpha+\alpha^2)} - \frac{8Eht^\alpha(2+\alpha)\epsilon\Gamma(\alpha)}{\Gamma(3+2\alpha)} - \frac{3E^2h^2t^{2\alpha}\epsilon^2\Gamma(2+2\alpha)}{a^2\Gamma(\alpha)\Gamma(4+3\alpha)}\right)}{512\Gamma(\alpha)}, \tag{2.58}$$

and so on. With the help of the remaining two boundary conditions (2.54), then

$$\int_0^1 u(t)dt \approx \int_0^1 u_{m+1}(t,h,\epsilon)dt = 1, \tag{2.59}$$

$$u(1) \approx u_{m+1}(1,h,\epsilon) = 0, \tag{2.60}$$

By using Eq. (2.60) to delete the unknown parameter γ of the Eq. (2.59) so that it contains only two unknown parameters ϵ and h. When $\alpha = 3.9$, and according to the Eq. (2.59), the unknown parameter ϵ is a function of the auxiliary parameter h, which has been plotted in the h-range [0, 2.5] in Fig. 2.3 for different values of E and $m = 6$. Two ϵ-plateaus (two line segments give constant values of ϵ) can be identified in this Figure for each case; this means that there are two solutions. From this Figure it is clear that the valid value of h for the two solutions is one.

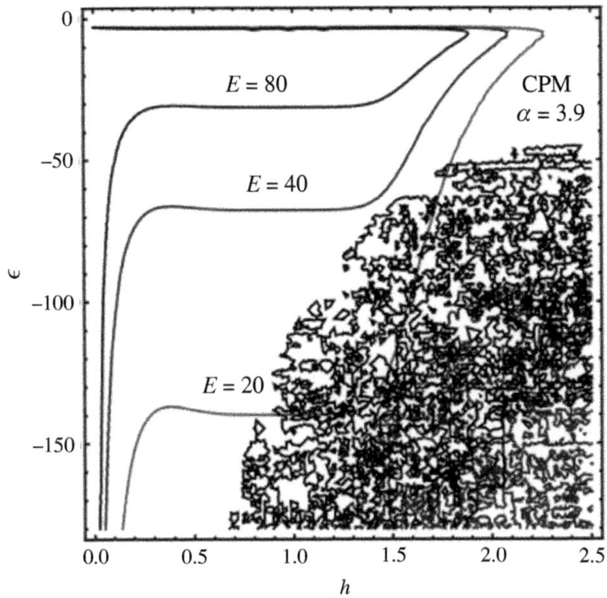

FIGURE 2.3 The $h-\epsilon$ curve for Eq. (2.59) with different values of E and $\alpha=3.9$.

Then when $h=1$, the multiplicity curves for different values of α are shown in Fig. 2.4. It is clear from this Figure that the model (2.40) admits dual solutions for $\alpha=4$, 3.9, 3.5 and any given E in the interval $(-\infty, 0) \cup (0, E_{max})$ in which $E_{max} \cong 228.128$ ($\alpha=4$), $E_{max} \cong 194.5$ ($\alpha=3.9$), and $E_{max} \cong 1106.1$ ($\alpha=3.5$). The value of E_{max} decreases with the decreasing value of α. Also, one can see, when $\alpha=3.9$ in Fig. 2.3 the dual solutions approach whenever they have the value of E close to the value of E_{max} (194.5). The problem solutions are graphically presented when $\alpha=3.5$, $E=20$ and $E=-20$ in Figs. 2.5 and 2.6. The present results are in a full agreement with the solutions reported by Arqub et al. (2014) and Alomari et al. (2013). Also, the two different positive solutions of the model (2.40) with different values of α and E are listed in Table 2.2.

2.3.2 Heat Transfer Model and Diffusion-Reaction Model

Finned surfaces are widely used in engineering, such as for the cylinders of aircraft engines, cooling electric transformers, and other heat transfer equipment. A one-dimensional steady state heat conduction equation for the temperature distribution of a straight rectangular fin with a power-law temperature dependent surface heat flux, in dimensionless form is given by

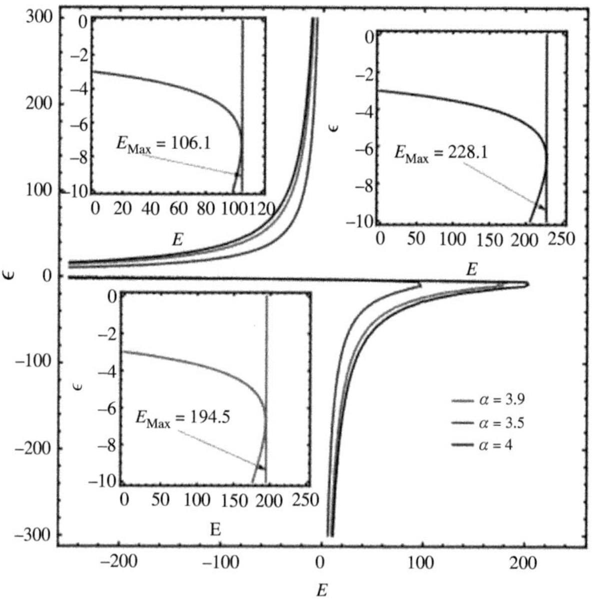

FIGURE 2.4 The multiplicity curve of the problem (2.40) for different value of α.

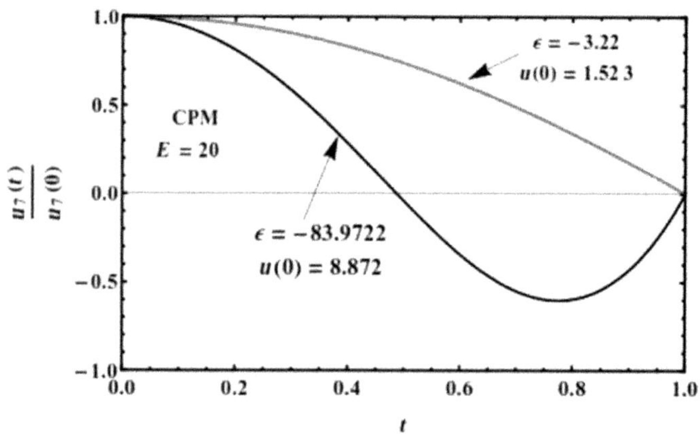

FIGURE 2.5 The dual solution of Eq. (2.40) with $\alpha = 3.5$ and $E = 20$.

(Abbasbandy and Shivanian, 2010; Chang, 2005; Chowdhury and Hashim, 2008)

$$u''(t) - \psi^2 u^{n+1}(t) = 0, \tag{2.61}$$

subject to the conditions

$$u'(0) = 0, u(1) = 1, \tag{2.62}$$

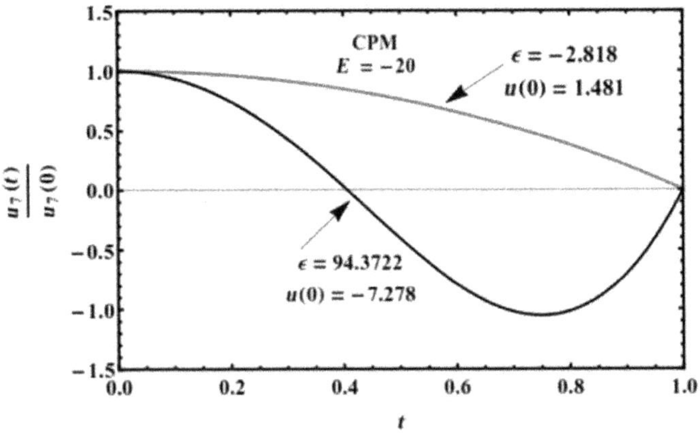

FIGURE 2.6 The dual solution of Eq. (2.40) with $\alpha = 3.5$ and $E = -20$.

where the axial distance t is measured from the fin tip, $u(t)$ is the temperature, and ψ is the convective-conductive parameter of the fin. The approximate analytical solution to the Eqs. (2.61) and (2.62) was presented using semianalytic methods such as Adomian decomposition method (Chang, 2005), HPM (Chowdhury and Hashim, 2008; Ganji, 2006), HAM (Abbasbandy and Shivanian, 2010; Hassan and Semary, 2014), and VIM (Hafez et al., 2007). It has been shown that this problem (2.61) with the conditions (2.62), when $-4 \leq n \leq -2$ either admits dual solutions or does not admit any solution based on the value of convective-conductive parameter ψ (Abbasbandy and Shivanian, 2011; Semary and Hassan, 2015; Ellery and Simpson, 2011). On other hand, for $n = -2$, Eq. (2.61) describes the steady diffusion-reaction regime in a porous slab with parallel plane boundaries. Where in this case $u(t)$ is the dimensionless concentration of the reactant and ψ is the Thiele modulus (Magyari, 2008; Sun et al., 2004; Abbasbandy, 2008). This two point boundary value problem (2.61) in fractional order domain can be considered as follows:

$$D_t^{1.9}u(t) - \psi^2 u^{n+1}(t) = 0, \tag{2.63}$$

subject to the boundary condition (2.62) and dual solutions of this model for the two cases can be examined when $n = -3$ and $n = -2$. To apply the procedures of the methods the boundary conditions (2.62) must be changed as follows:

$$u'(0) = 0, u(0) = \epsilon, \tag{2.64}$$

with the forcing condition

$$u(1) = 1. \tag{2.65}$$

According to the frame of the PHAM the linear operator can be

TABLE 2.2 The Value of $u''(0) = \epsilon$ of the Two Positive Solutions of (2.40) With Different Values of α and E

	$\alpha = 4$				$\alpha = 3.9$			
	Lower solution		Upper solution		Lower solution		Upper solution	
E	PHAM	CPM	PHAM	CPM	PHAM	CPM	PHAM	CPM
−100	−2.66818	−2.668181	36.994	36.99402	−2.6128	−2.6128	32.7057	32.7115
−60	−2.78639	−2.786393	59.2494	59.24944	−2.7482	−2.7482	52.0769	52.0817
−20	−2.923	−2.9230	170.039	170.03931	−2.9081	−2.9081	148.3602	148.3668
20	−161.726	−161.7259	−3.08411	−3.084108	−139.7624	−139.765	−3.1020	−3.1020
60	−50.8771	−50.87706	−3.2794	−3.279404	−43.3948	−43.3951	−3.3463	−3.3463
100	−28.493	−28.49296	−3.52581	−3.525808	−23.8322	−23.8322	−3.6734	−3.6734
160	−15.4805	−15.48053	−4.07735	−4.077351	−12.1291	−12.1291	−4.5436	−4.5436
200	−10.6349	−10.63485	−4.7648	−4.764798	–	–	–	–

$$L[\varphi(t,\epsilon,p)] = \frac{\partial^2\psi(t,\epsilon,p)}{\partial t^2}, \tag{2.66}$$

with the property

$$L[c_0 + c_1 t] = 0. \tag{2.67}$$

By choosing $H(t) = 1$ and using Eq. (2.25) the solution terms u_m is given as

$$u_m(t,\epsilon) = \chi_m u_{m-1}(t,\epsilon) + h\iint D^m p\{D_t^{1.9}[\varphi(t,\epsilon,p)](\varphi(t,\epsilon,p))^{-(n+1)} - \psi^2\}dtdt + c_0 + c_1 t, \tag{2.68}$$

where the constants c_0 and c_1 can be determined by the two boundary conditions:

$$u'_m(0,\epsilon) = u_m(0,\epsilon) = 0. \tag{2.69}$$

Using the properties of the homotopy derivative (2.7) the mth-order deformation Eq. (2.68) in simple form becomes

$$u_m(t,\epsilon) = \chi_m u_{m-1}(t,\epsilon) + h\iint R_{m,n}(u_{m-1},t,\epsilon)dtdt + c_0 + c_1 t, \tag{2.70}$$

where

$$R_{m,-3} = \sum_{i=0}^{m-1}\sum_{k=0}^{i} D_t^{1.9}[u_{m-1-i}(t,\epsilon)]u_{i-k}(t,\epsilon)u_i(t,\epsilon) - \psi^2(1-\chi_m), \tag{2.71}$$

and

$$R_{m,-2} = \sum_{i=0}^{m-1} D_t^{1.9}[u_{m-1-i}(t,\epsilon)]u_i(t,\epsilon) - \psi^2(1-\chi_m). \tag{2.72}$$

term of solutions for Eq. (2.70) when $n = -3$ are

$$u_1(t,\epsilon) = -\frac{1}{2}ht^2\psi^2, \tag{2.73}$$

$$u_2(t,\epsilon) = -0.455h(1.099t^2 + ht^{2.1}\epsilon^2)\psi^2, \tag{2.74}$$

$$u_3(t,\epsilon) = 0.0827h\psi^2 - (-6.0458t^2 - 11.0043ht^{2.1}\epsilon^2 - 4.9883h^2t^{2.2}\epsilon^4 + h^2t^{4.1}\epsilon\psi^2), \tag{2.75}$$

and so on. Also, when $n = -2$, the solution of Eq. (2.70) are given by

$$u_1(t,\epsilon) = -\frac{1}{2}ht^2\psi^2, \tag{2.76}$$

$$u_2(t,\epsilon) = -0.455h(1.099t^2 + ht^{2.1}\epsilon^2)\psi^2, \tag{2.77}$$

$$u_3(t,\epsilon) = 0.04134h\psi^2(-12.0917t^2 - 22.0087ht^{2.1}\epsilon - 9.9768h^2t^{2.2}\epsilon^2 + h^2t^{4.1}\psi^2), \tag{2.78}$$

and so on. In this way can be obtained mth-order approximate solution for the problem (2.63) with initial condition (2.64) as follows

$$u(t) \approx U_M(t,h,\epsilon,n) = \sum_{m=0}^{M} u_m(t,\epsilon). \tag{2.79}$$

According to the frame of the controlled Picard's method the iterative formula (2.34), for Eqs. (2.63), (2.64) by choosing $H(t) = 1$ is given by

$$u_{m+1}(t,h,\epsilon), u_m(t,h,\epsilon) - \frac{h}{\Gamma(\alpha)}\int_0^t (t-\tau)^{\alpha-1}\{D_t^{\alpha}[u_m(\tau,h,\epsilon)](u_m(t,\epsilon))^{-(n+1)} - \psi^2\}d\tau \tag{2.80}$$

Starting with the initial approximation solution $u_0(t, \epsilon) = \epsilon$ which satisfies the initial condition (2.64), the successive approximations $u_{m+1}(t, h, \epsilon)$ for $n = -3$ is

$$u_1(t,h,\epsilon) = \epsilon + 0.5472ht^{1.9}\psi^2, \tag{2.81}$$

$$\begin{aligned} u_2(t,h,\epsilon) = {} & 1.0945ht^{1.9}\psi^2 - 0.5472h^2t^{1.9}\epsilon^2\psi^2 - 0.01292h^4t^{5.7}\psi^6 \\ & + \epsilon(1 - 0.11212h^3t^{3.8}\psi^4), \end{aligned} \tag{2.82}$$

and for $n = -2$, the first two solutions of Eq. (2.80) are given by

$$u_1(t,h,\epsilon) = \epsilon + 0.5472\ ht^{1.9}\psi^2, \tag{2.83}$$

$$u_2(t,h,\epsilon) = 1.0944ht^{1.9}\psi^2 - 0.05606h^3t^{3.8}\psi^4 + \epsilon(1 - 0.54724h^2t^{1.9}\psi^2), \tag{2.84}$$

and similarly for other iterations. With the help of the remaining boundary conditions (2.65), then the relation between the parameters h and ϵ for the two methods is given by

$$u(1) \approx U_M(1,h,\epsilon,n) = 1, u(1) \approx u_{m+1}(1,h,\epsilon,n) = 1. \tag{2.85}$$

It has been shown in Magyari (2008) and Abbasbandy et al. (2009) that when $n = -2$, the model (2.61) admits dual solutions for $0 < \psi \leq 0.765152$. To show that the number of solutions is continuous for the model in the fractional order domain the value of ψ variation about 0.7 is selected. Fig. 2.7A and B shows the $h - \epsilon$ curve of Eq. (2.85), where two values of ϵ are obtained which reflect the dual solutions in this case. It is clear from Fig. 2.7B the two line segment where ϵ constant ended when $\psi = 0.74$, this

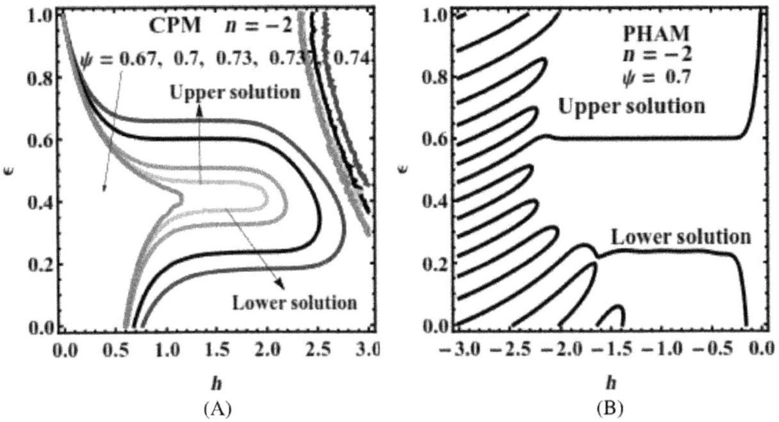

FIGURE 2.7 The $h-\epsilon$ curve for Eq. (2.85) when $n=-2$ and with different values of ψ (A) CPM and (B) PHAM.

means the maximum value of ψ is less than 0.74 to obtain dual solutions from the model (2.63). When $\psi=0.7$, the lower and upper values of ϵ are $\{0.2375, 0.59937\}$ by the PHAM when $h=-1$ and by the controlled Picard's method corresponds are $\{0.2392, 0.59933\}$ to $h=\{2.2, 1.5\}$. The dual solutions of the behavior of $u(t)$ are shown in Fig. 2.8A and B where both techniques give similar responses when $\psi=0.7$. The absolute difference $\Delta(t)$ between the methods solutions is defined by:

$$\Delta(t)=|U_M(t,h,\epsilon)-u_{m+1}(t,h,\epsilon)|. \tag{2.86}$$

In the value of $\psi=0.7$, the absolute difference $\Delta(t)$ between the PHAM solutions and CPM solutions are very small as shown in Fig. 2.9 which reflects high accuracy of matching. Note that although the two solutions give similar response, the required number of iterations is different. For the simulations shown in Fig. 2.9, the PHAM and the controlled Picard's method are carried out using 30 and 7 iterations respectively.

For $n=-3$, the model (2.61) with integer order 2 admits dual solutions for $0<\psi\leq 0.591611$. When the order for model (2.61) is changed from integer into fractional order 1.9 as in Eq. (2.63), this equation also has dual solutions when $\psi=\sqrt{3}$ as shown in Figs. 2.10A and B.

These figures show the $h-\epsilon$ curve according to Eq. (2.85) for the two methods where two ϵ-plateaus (two line segments give constant values of ϵ) can be identified which reflects the dual solutions in this case. When $\psi=\sqrt{3}$, the values of ϵ are $\{0.4588, 0.7276\}$ and $\{0.4594, 0.7278\}$ by the PHAM, and the controlled Picard's method corresponds to $h=-1$ and $h=\{2.2, 1.3\}$ respectively. The behaviors of dual solutions are shown in Fig. 2.11 where both techniques give similar responses.

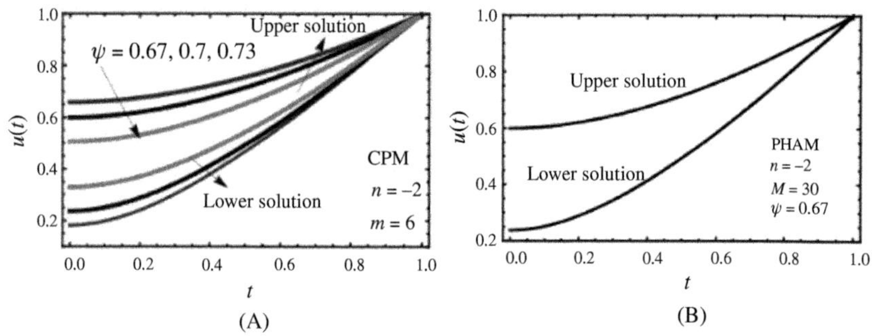

FIGURE 2.8 The dual solution of Eq. (2.63) when $n = -2$ and by (A) CPM and (B) PHAM Starting by the initial approximation solution $u_0(t, \epsilon) = \epsilon$, thus the three.

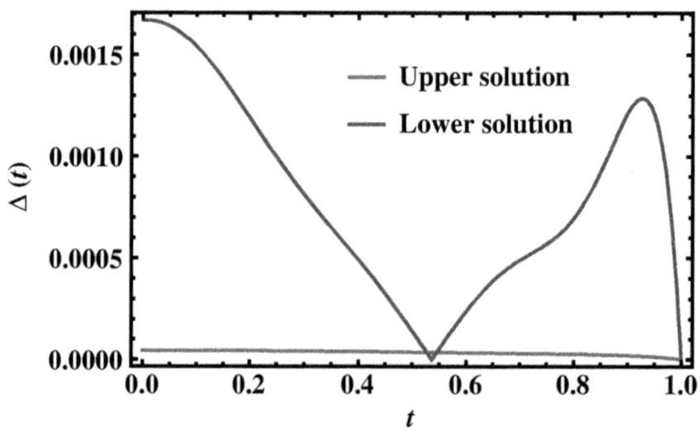

FIGURE 2.9 The difference error (2.86) between the two methods used.

2.3.3 The Bratu's Problem

The one-dimensional Bratu's equation is given by

$$\frac{d^2u(t)}{dt^2} + \lambda e^{u(t)} = 0, \tag{2.87}$$

with the boundary conditions

$$u(0) = u(1) = 0, \tag{2.88}$$

where λ is the Frank–Kamenetskii parameter and a positive number (Jacobsen and Schmitt, 2002). The Bratu's problem nonlinear two boundary value problem with parameter λ and strong nonlinear term e^u appears in a number of applications in physics and engineering sciences (Jacobsen and Schmitt, 2002; Jalilian, 2010; Wazwaz, 2005). The analytical solution of the problem (2.87) with boundary conditions (2.88) is given by (Abbasbandy and Shivanian, 2010; Wazwaz, 2005)

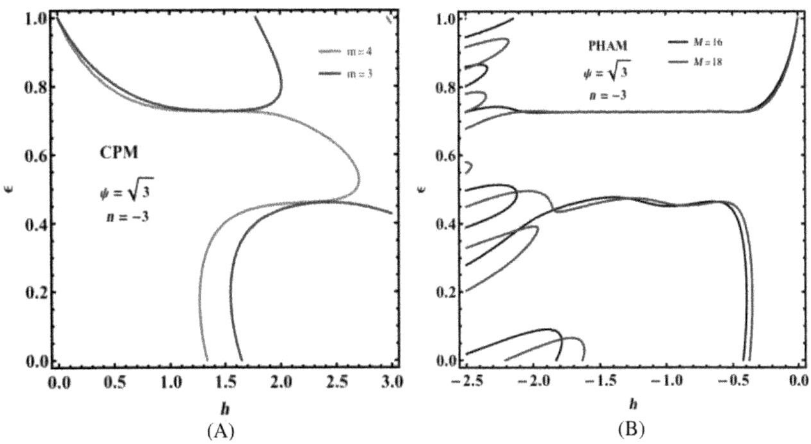

FIGURE 2.10 The $h-\epsilon$ curve for Eq. (2.85) when $n=-3$ and $\psi=\sqrt{3}$.

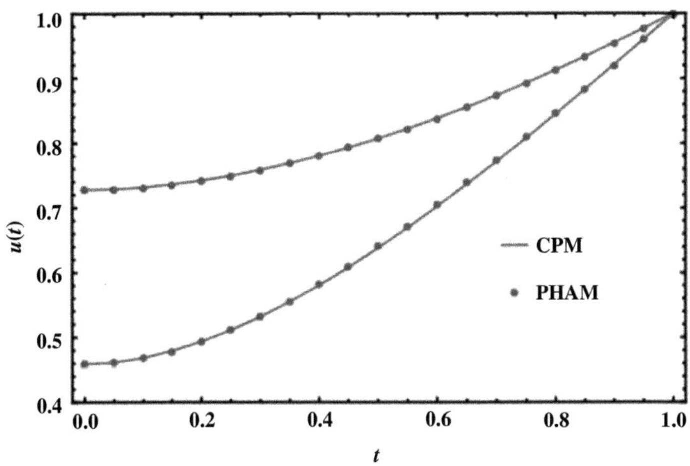

FIGURE 2.11 The dual solution of Eq. (2.63) when $n=-3$ and by CPM and PHAM.

$$u(t)=-2\ln\frac{\cosh\left((t-0.5)\frac{\theta}{2}\right)}{\cosh\left(\frac{\theta}{4}\right)}, \tag{2.89}$$

where θ is a solution $\theta=\sqrt{2\lambda}\cosh\left(\frac{\theta}{4}\right)$. The problem has no, one, or two solutions when $\lambda>\lambda_c$, $\lambda=\lambda_c$ and $\lambda<\lambda_c$. respectively, where the critical value λ_c given by $\lambda_c=3.513830719$ and $u'(0)=\theta\tanh\left(\frac{\theta}{4}\right)$ as shown in Fig. 2.13. The problem was solved using the numerical and semianalytic techniques using HAM (Abbasbandy and Shivanian, 2010; Hassan and Semary, 2013; Yang and Liao, 2017), spline method (Jalilian, 2010), VIM (Semary and Hassan, 2015), differential quadrature method (Ragb et al., 2017), and other methods. The purpose of this section is to solve and to show how one can find out existence of dual solutions for the problem (2.87) in the fractional order domain. The one-dimensional fractional order Bratu equation is given by

$$D_t^\alpha u(t) + \lambda e^{u(t)} = 0, \tag{2.90}$$

where $1 < \alpha \leq 2$. To apply the Controlled Picard's method, suppose that $u'(0) = \epsilon$, so the boundary conditions (2.88) become:

$$u(0) = 0, u'(0) = \epsilon, \tag{2.91}$$

with additional forcing condition

$$u(1) = 0. \tag{2.92}$$

By applying the iterative formula for Eq. (2.90) with initial conditions (2.91) and by taking $H(t) = 1$, then

$$u_{m+1}(t,h,\epsilon) = u_m(t,h,\epsilon) - \frac{h}{\Gamma(\alpha)}\int_0^t (t-\tau)^{\alpha-1}\{D_\tau^\alpha[u_m(\tau,h,\epsilon)] + \lambda e^{u_m(\tau,h,\epsilon)}\}d\tau \tag{2.93}$$

and using the property (2.4) then the Eq. (2.93) becomes

$$u_{m+1}(t,h,\epsilon) = (1-h)u_m(t,h,\epsilon) + h\epsilon t - \frac{h}{\Gamma(\alpha)}\int_0^t (t-\tau)^{\alpha-1}\lambda e^{u_m(\tau,h,\epsilon)}d\tau \tag{2.94}$$

The term $\int_0^t (t-\tau)^{\alpha-1}\lambda e^{u_m(\tau,h,\epsilon)}d\tau$ in Eq. (2.94) is difficult to obtain the exact value. Therefore, the value of this term is approximated by using composite Simpson method (Atkinson, 1989). This method is as follows:

$$I(t) = \int_0^t z(\tau)d\tau \approx \frac{t}{6l}\left\{ y_0 + y_{2l} + \sum_{i-1}^{l} 4y_{2i-1} + \sum_{i=1}^{l-1} 2y_{2i} \right\} \tag{2.95}$$

where $2l$ is the number of subintervals, $z_0 = z(0)$, $z_{2l} = z(t)$, and $z_i = z\left(\frac{it}{2l}\right)$ $i{:}1(1)2l-1$. Then the iterative formula (2.94) to solve the problem (2.90) and (2.91) becomes:

$$u_{m+1}(t,h) = (1-h)u_m(t,h,\epsilon) + h\epsilon t - \frac{th}{6l}\left\{ z_0 + z_{2l} + \sum_{i-1}^{l} 4z_{2i+1} + \sum_{i=1}^{l-1} 2z_{2i} \right\}, \tag{2.96}$$

where $\frac{(t-\tau)^{\alpha-1}}{\Gamma(\alpha)}\left\{\lambda e^{u_m(t,h,\epsilon)}\right\}\Big|_{\tau=\frac{it}{2l}}, \forall i{:}0(1)2l.$

By selecting $l = 5$ and using the initial solution $u_0(t,\epsilon) = \epsilon t - \frac{\lambda t^\alpha}{\Gamma(\alpha+1)} - \frac{\lambda \epsilon t^\alpha + 1}{\Gamma(\alpha+2)}$ which at least satisfies the initial conditions

(2.91) which can be used to obtain the mth-successive approximations (2.96). The first approximation solution $u_1(t, h, \epsilon)$ is given by:

$$
\begin{aligned}
u_1(t,h,\epsilon) \quad &= t\epsilon - \frac{0.033ht^{\alpha}\lambda}{\Gamma[\alpha]} - \frac{ht^{\alpha}\lambda}{\Gamma[\alpha]}\{f_1+f_2+f_3+f_4+f_5+f_6+f_7+f_8+f_9\} \\
&\quad - \frac{t^{\alpha}\lambda}{\Gamma[1+\alpha]} + \frac{ht^{\alpha}\lambda}{\Gamma[1+\alpha]} - \frac{t^{1+\alpha}\epsilon\lambda}{\Gamma[2+\alpha]} + \frac{ht^{1.+\alpha}\epsilon\lambda}{\Gamma[2+\alpha]}
\end{aligned}
\tag{2.97}
$$

where,

$f_1 = 0.1481e^{-0.105\alpha+0.1t\epsilon-\frac{e^{-2.30\alpha}t^{\alpha}\lambda}{\Gamma[1+\alpha]}-\frac{0.1e^{-2.30\alpha}t^{1.+\alpha}\epsilon\lambda}{\Gamma[1+\alpha]}}$,-

$f_2 = 0.083e^{-0.223\alpha+0.2t\epsilon-\frac{e^{-1.60\alpha}t^{\alpha}\lambda}{\Gamma[1+\alpha]}-\frac{0.2e^{-1.609\alpha}t^{1.+\alpha}\epsilon\lambda}{\Gamma[2+\alpha]}}$,

$f_3 = 0.190e^{-0.356\alpha+0.3t\epsilon-\frac{e^{-1.20\alpha}t^{\alpha}\lambda}{\Gamma[1+\alpha]}-\frac{0.3e^{-1.203\alpha}t^{1.+\alpha}\epsilon\lambda}{\Gamma[1+\alpha]}}$,-

$f_4 = 0.111e^{-0.510\alpha+0.4t\epsilon-\frac{e^{-0.916\alpha}t^{\alpha}\lambda}{\Gamma[1.+\alpha]}-\frac{0.4e^{-0.916\alpha}t^{1.+\alpha}\epsilon\lambda}{\Gamma[2.+\alpha]}}$,

$f_5 = 0.266e^{-0.693\alpha+0.5t\epsilon-\frac{e^{-0.693\alpha}t^{\alpha}\lambda}{\Gamma[1+\alpha]}-\frac{0.5e^{-0.69\alpha}t^{1.+\alpha}\epsilon\lambda}{\Gamma[2+\alpha]}}$,

$f_6 = 0.166e^{-0.916\alpha+0.6t\epsilon-\frac{e^{-0.510\alpha}t^{\alpha}\lambda}{\Gamma[1.+\alpha]}-\frac{0.6e^{-0.510\alpha}t^{1.+\alpha}\epsilon\lambda}{\Gamma[2+\alpha]}}$,

$f_7 = 0.444e^{-1.203\alpha+0.7t\epsilon-\frac{e^{-0.35\alpha}t^{\alpha}\lambda}{\Gamma[1+\alpha]}-\frac{0.7e^{-0.356\alpha}t^{1.+\alpha}\epsilon\lambda}{\Gamma[2+\alpha]}}$,

$f_8 = 0.33e^{-1.609\alpha+0.8t\epsilon-\frac{e^{-0.223\alpha}t^{\alpha}\lambda}{\Gamma[1.+\alpha]}-\frac{0.8e^{-0.223\alpha}t^{1.+\alpha}\epsilon\lambda}{\Gamma[2.+\alpha]}}$,

$f_9 = 1.333e^{-2.302\alpha+0.9t\epsilon-\frac{e^{-0.1053\alpha}t^{\alpha}\lambda}{\Gamma[1.+\alpha]}-\frac{0.899e^{-0.1053\alpha}t^{1.+\alpha}\epsilon\lambda}{\Gamma[2.+\alpha]}}$, and so on. With the help of forcing condition $u(1) = 0$, then

$$u(1) = u_{m+1}(1, \epsilon, h) = 0. \tag{2.98}$$

According to the above equation and after six iterations calculated ($m = 5$), in Fig. 2.12 the ϵ as a function of auxiliary parameter h, has been plotted in the h range [0, 2] implicitly, for different values of α and λ. Two

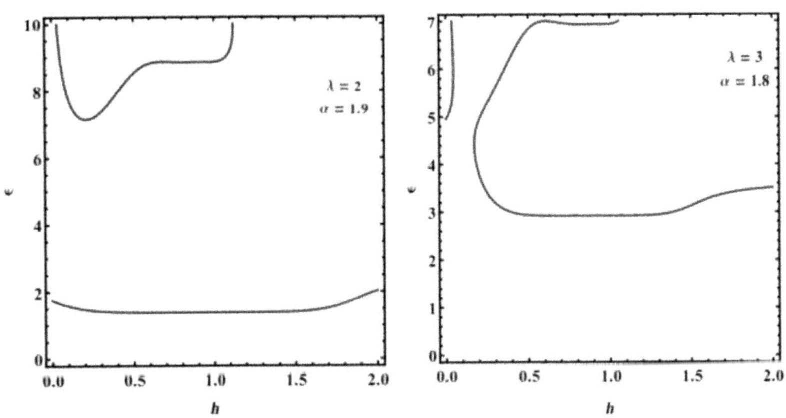

FIGURE 2.12 The $h - \epsilon$ curves of Eq. (2.98) for different value of λ and α.

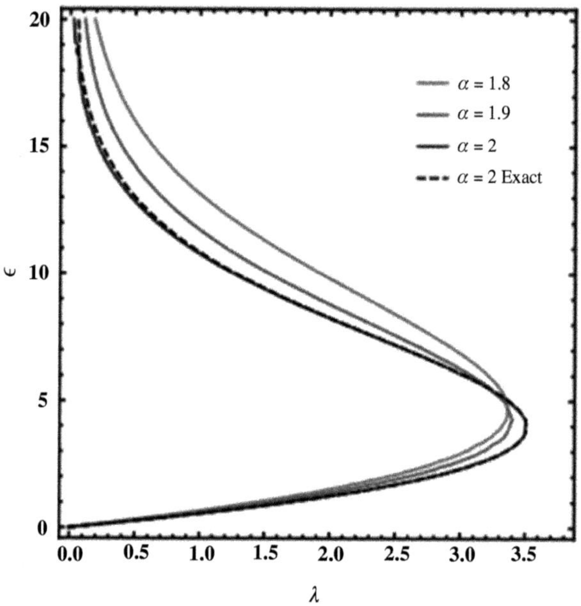

FIGURE 2.13 The multiplicity curve of the Bratu's problem (2.90) of different values of α.

TABLE 2.3 The Values of λ_c for Different Values of α

α	λ_c	λ_c exact
1.8	3.3693	–
1.85	3.3739	–
1.9	3.4015	–
1.95	3.4488	–
2	3.5140	3.5138

line segments giving constant values of ϵ can be identified in this Figure for each case, this means that there are two solutions for these cases and the used method is convergent when $h = 1$. In general, the multiplicity curves for different values of the fractional order α are shown in Fig. 2.13. It is clear from this Figure the approximate solution when $\alpha = 2$ is fully agreement with the exact solution. Also, The problem in the fractional order domain has no, one, or two solutions when $\lambda > \lambda_c$, $\lambda = \lambda_c$ and $\lambda < \lambda_c$. respectively, where the critical value λ_c is listed in Table 2.3 for different values of α. For $\alpha = 2$, the exact values of ϵ from the analytic solution (2.89) are 2.319602 and 6.103 for $\lambda = 3$. By the present method, the

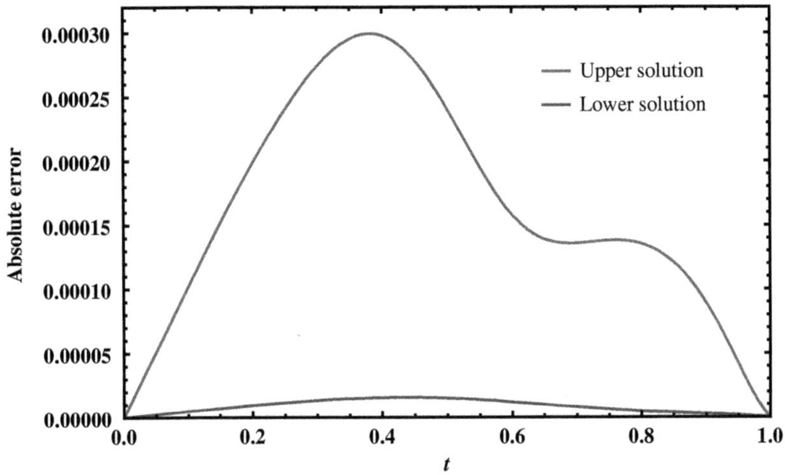

FIGURE 2.14 Absolute error of Eq. (2.90) with $\alpha = 2$ and for the lower and the upper solution.

approximate values of ϵ are 2.31965 and 6.104. In this case they show the accuracy of these dual approximate solutions when $\alpha = 2$, the absolute errors for first and second solutions are shown in Fig. 2.14. For $\lambda = 3$, the approximate values of ϵ when $\alpha = 1.9$ and $\alpha = 1.8$ are 2.6326, 6.295 and 2.92033, 6.939, respectively. The problem solutions when $\lambda = 3$ for different values of α are given in Table 2.4. Also some solutions graphically are shown in Figs. 2.15 and 2.16 where the lower and upper values of ϵ when $\alpha = 1.9$ and $\lambda = 2$ are 1.38134 and 8.8603, respectively.

2.3.4 Three-Point Boundary Value Problem

Consider a three-point nonlinear boundary value problem

$$D_t^{\alpha} u(t) = \beta t(1 + u^2), \tag{2.99}$$

with the following conditions

$$u(0) = u'(1) = u''(1) = 0, u''(0) - u''(\gamma) = 0, \tag{2.100}$$

where $3 < \alpha \leq 4$. For $\alpha = 4$, the above equation has at least two positive solutions when $\gamma = 10$ and $\beta = 1/5$ (Graef et al., 2003, 2004). This section discuss finding multiple solutions of this problem when $\alpha \neq 4$ using the PHAM. To apply this method, suppose $u(1) - c$ and change the boundary conditions (2.100) into

TABLE 2.4 The Numerical Solutions of Eq. (2.90) for Different Values of α

	Lower solution			Upper solution		
t	$\alpha=2$	$\alpha=1.9$	$\alpha=1.8$	$\alpha=2$	$\alpha=1.9$	$\alpha=1.8$
0.1	0.2157	0.2406	0.2607	0.5919	0.6037	0.6573
0.2	0.3943	0.43498	0.4645	1.1284	1.1375	1.2224
0.3	0.5284	0.57644	0.60665	1.571	1.558	1.6401
0.4	0.6118	0.65929	0.68304	1.8693	1.8131	1.8512
0.5	0.6401	0.6805	0.69319	1.9755	1.8646	1.8318
0.6	0.6118	0.64097	.64133	1.8691	1.7133	1.6154
0.7	0.5284	0.54504	0.53564	1.5708	1.4	1.2702
0.8	0.3943	0.4004	0.38652	1.1283	0.9808	0.8601
0.9	0.2157	0.21568	0.2047	0.5919	0.50342	0.42861

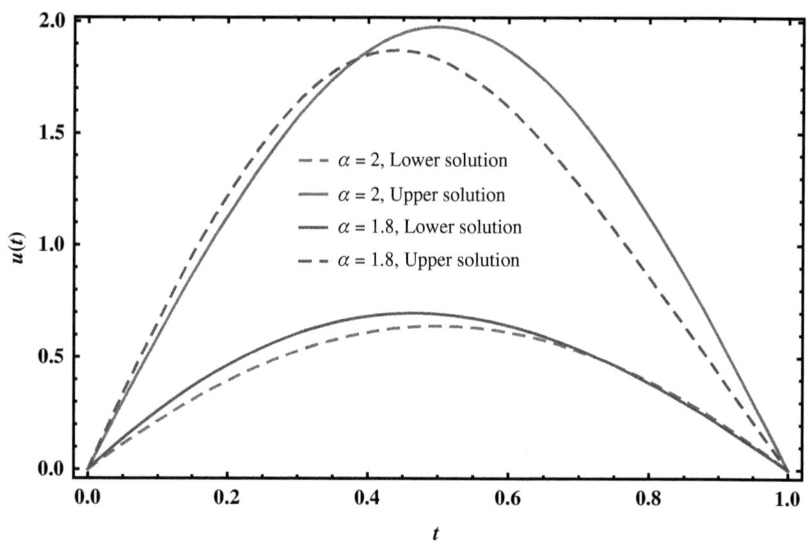

FIGURE 2.15 The approximate solutions of the Bratu's problem (2.90) when $\lambda = 3$ and different values of α.

$$u(0) = u'(1) = u''(1) = 0, u(1) = \epsilon, \tag{2.101}$$

and the condition

$$u''(0) - u''(\gamma) = 0. \tag{2.102}$$

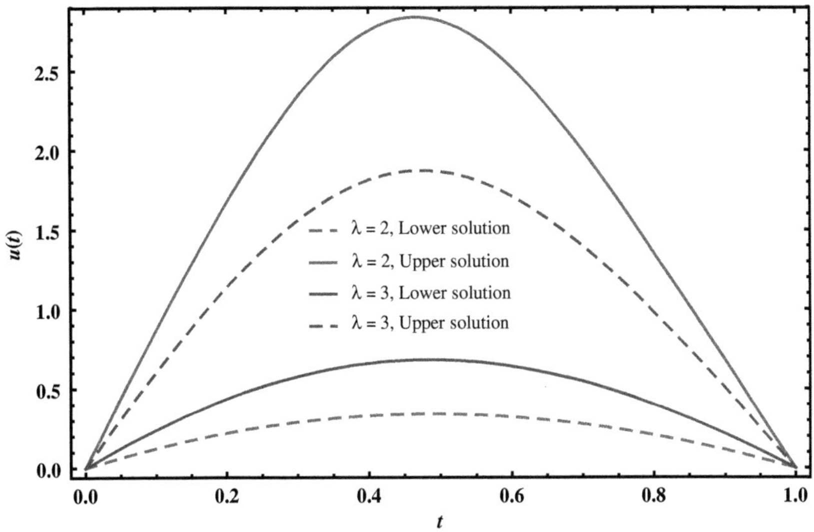

FIGURE 2.16 The approximate solutions of the Bratu's problem (2.90) when $\alpha = 1.9$ and different values of λ.

The mth-order deformation Eq. (2.25) of the problem (2.99) reads as

$$L[u_m(t,\epsilon) - \chi_m u_{m-1}(t,\epsilon)] = h\{D_t^\alpha u_{m-1}(t,\epsilon) - \beta D^m[pt(1+(\varphi(t,\epsilon,p))^2)]\} \tag{2.103}$$

By selecting $L = \frac{\partial^4}{\partial t^4}$ and using the properties of homotopy derivative D^m (2.7) thus the explicit form of the mth-order deformation Eq. (2.103) is given by

$$u_m(t,\epsilon) = \chi_m u_{m-1}(t,\epsilon) + h\int\int\int\int R_m(u_{m-1},t,\epsilon)dtdtdtdt + c_0 + c_1 t + c_2 t^2 + c_3 t^3, \tag{2.104}$$

where

$$R_m = D_t^\alpha u_{m-1}(t,\epsilon) - \beta t\sum_{i=0}^{m-1} u_i(t,\epsilon)u_{m-1-i}(t,\epsilon) - \beta t(1-\chi_m), \tag{2.105}$$

and the integration constants c_i, i : 0, 1, 2, 3 can be determined by the boundary conditions:

$$u_m(1,\epsilon) = u_m'(0,\epsilon) = u_m'(1,\epsilon) = u''_m(1,\epsilon) = 0. \tag{2.106}$$

Starting by the following initial solution

$$u_0(t,\epsilon) = \epsilon + \epsilon(t-1)^3, \tag{2.107}$$

which satisfies the conditions (2.101), thus the first solution of Eq. (2.104) is given by

$$u_1((t,\epsilon)) = -\frac{1}{55440}\left\{h(-1+t)^3 t\beta(1386 + 462t + 1209\epsilon^2 + 559t\epsilon^2 + 125t^2\epsilon^2 - 93t^3\epsilon^2 - 95t^4\epsilon^2 + 119t^5\epsilon^2 - 45t^6\epsilon^2 + 7t^7\epsilon^2)\right\} \tag{2.108}$$

and so on. The mth-order approximate solution for the problem (2.99) with conditions (2.101) is

$$u(t) \approx U_M(t,h,\epsilon) = \sum_{m=0}^{M} u_m(t,\epsilon), \tag{2.109}$$

and with the help of the remaining condition (2.102), then

$$U_M''(0,h,\epsilon) - U_M''(\gamma,h,\epsilon) = 0. \tag{2.110}$$

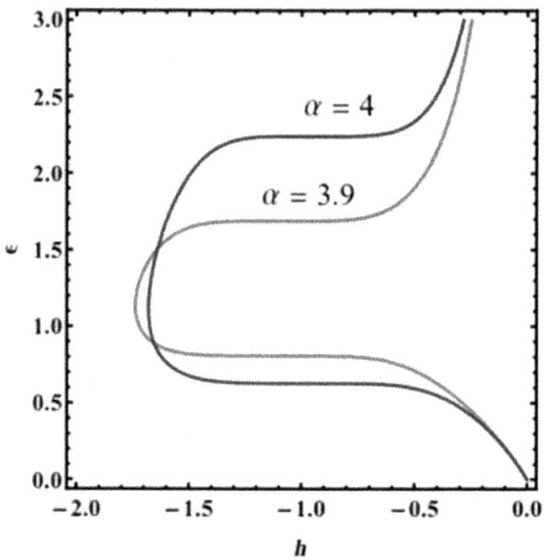

FIGURE 2.17 The $h-\epsilon$ curves of Eq. (2.110) for different value of and when $\gamma = 10$, $\beta = 1/5$.

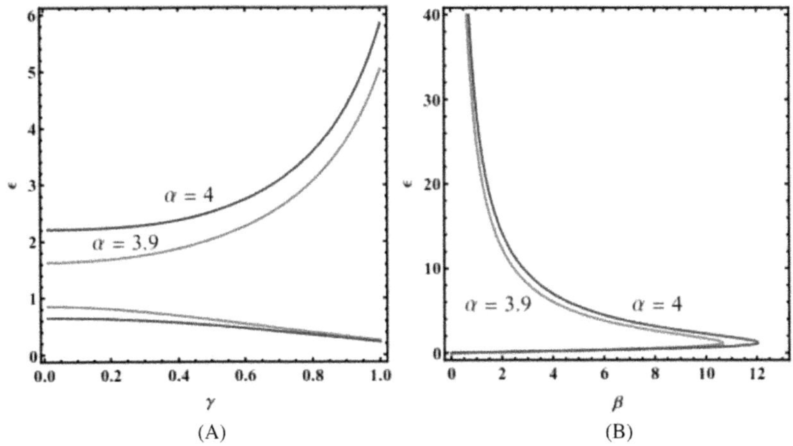

FIGURE 2.18 The dual solutions of Eq. (2.99) for different value of α and when (A) $\beta = 10$ (B) $\gamma = 1/5$.

By selecting $\gamma = 1/5$, $\beta = 10$ and according to the above equation in Fig. 2.17, ϵ as a function of control parameter h, has been plotted in the h range $[-2, 0]$, for $M = 5$ and different value of fractional order parameter α. Graef et al. (2003, 2004) proved that the Eq. (2.99) with integer order ($\alpha = 4$) has at least two positive solutions when $\beta = 10$ and $\gamma = 1/5$. From Fig. 2.17, a dual solution of ϵ can be identified in this Figure for each case. It means the duality solution is continuous in the fractional order domain. Also, the two approximate solutions obtained by this approach are convergent around $h = -1$. Therefore, two different positive solutions are obtained of the three-point nonlinear boundary value Eqs. (2.99), (2.100) when $\gamma = 1/5$, $\beta = 10$, as shown in Fig. 2.18 and listed in Tables 2.5 and 2.6. For $\gamma = 1/5$, it clear from this Figure and Table 2.5, the critical value of β is 10.68 when $\alpha = 3.9$. For $\beta = 10.68$, the two solutions are very close together and not when $\beta > 10.68$ as shown in Fig. 2.18B and Table 2.6. It is very important to notice the huge dependency between the two fractional orders, e.g., when $\alpha = 3.8$, the value of the critical value of β is 9.51 as shown in Fig. 2.19. For integer order case ($\alpha = 4$), the present method solutions are agreement with Liao solutions in Liao (2012) as shown in Tables 2.5 and 2.6. The problem solutions are graphically presented when $\gamma = 1/5$, $\beta = 10$ and different value of α in Fig. 2.20.

TABLE 2.5 The Value of $u(1) = \epsilon$ of the Two Positive Solutions of (2.99) When $\gamma = 1/5$ With Different Values of α and β

	Lower solution			Upper solution		
	(Liao, 2012)	Present method		(Liao, 2012)	Present method	
β	$\alpha = 4$	$\alpha = 4$	$\alpha = 3.9$	$\alpha = 4$	$\alpha = 4$	$\alpha = 3.9$
0.25	0.012224	0.012223	0.013593	115.0270	115.0268	100.0405
0.5	0.024455	0.024454	0.027197	57.49488	57.49481	49.9996
1	0.048974	0.0489735	0.054486	28.71013	28.71009	24.9583
4	0.201271	0.2012711	0.225716	6.98569	6.98586	6.0246
8	0.44755	0.4475553	0.523608	3.14143	3.141428	2.5969
10	0.627315	0.6273146	0.805389	2.24117	2.241175	1.6883
10.68	–	0.71297	1.121579	–	1.97189	1.21232
11	0.762509	0.76250904	–	1.843780	1.843778	–
12.05	1.10240	1.10241	–	1.275268	1.275266	–

TABLE 2.6 The Value of $u(1) = \epsilon$ of the Two Positive Solutions of (2.99) When $\beta = 10$ With Different Values of α and γ

	Lower solution			Upper solution		
	(Liao, 2012)	Present method		(Liao, 2012)	Present method	
γ	$\alpha = 4$	$\alpha = 4$	$\alpha = 3.9$	$\alpha = 4$	$\alpha = 4$	$\alpha = 3.9$
0.01	0.648176	0.6481758	0.85253	2.213774	2.213772	1.6342
0.1	0.643033	0.64303303	0.83984	2.219512	2.219510	1.6476
0.2	0.627315	0.6273146	0.80539	2.241177	2.241175	1.6883
0.4	0.564806	0.5648057	0.68975	2.377541	2.377539	1.8760
0.8	0.354329	0.3543293	0.39874	3.583159	3.583148	3.0667
1	70.231157	0.2311565	0.25648	5.848855	5.84865	5.0513

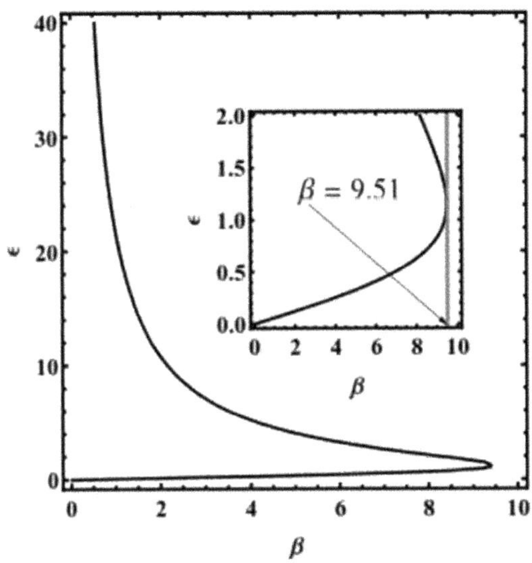

FIGURE 2.19 The dual solutions of Eq. (2.99) for $\alpha = 3.8$ and $\gamma = 1/5$.

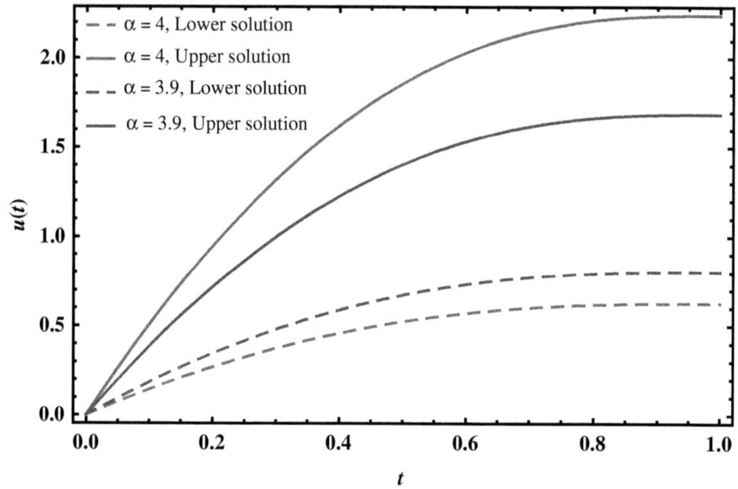

FIGURE 2.20 The approximate solutions of the problem (2.99) when $\beta = 10$, $\gamma = 1/5$, and different values of α.

2.4 CONCLUSION

In this chapter, general procedures for two semianalytic techniques to predict and obtain multiple solutions of nonlinear fractional order boundary value problems are introduced. Unknown parameter ϵ is introduced into the boundary conditions of the problems as a tool to search for multiple solutions. The advantage of the PHAM is the ability to deal with many different boundary conditions as the conditions problems 2.39 and 2.98. The controlled Picard method deals with initial conditions or the boundary conditions after converting into initial conditions as the conditions problems 2.39, 2.62, and 2.89. The controlled Picard method is simple to construct and there is no need to calculate Adomian polynomial and a general Lagrange multiplier, like in the Adomian decomposition method and the VIM, respectively. The advantages of the methods used that contain a simple way to control convergence region and the rate of the approximate series solution are unlike other semianalytic methods such as Adomian decomposition method and HPM. The methods are capable of finding the dual solutions for nonlinear fractional order model with boundary conditions at the same time. The controlled Picard method was successful in finding dual solutions of the Bratu's problem in the fractional order domain. In future work as an application these methods can be used to solve unbounded domain problems with fractional order which have applications in fluids problems.

REFERENCES

Abbasbandy, S., 2008. Approximate solution for the nonlinear model of diffusion and reaction in porous catalysts by means of the homotopy analysis method. Chem. Eng. J. 136, 144–150.

Abbasbandy, S., Shirzadi, A., 2011. A new application of the homotopy analysis method: solving the sturm-liouville problems. Commun. Nonlinear Sci. Num. Simul. 16 (1), 112–126.

Abbasbandy, S., Shivanian, E., 2010. Prediction of multiplicity of solutions of nonlinear boundary value problems: novel application of homotopy analysis method. Commun. Nonlinear Sci. Num. Simul. 15 (12), 3830–3846.

Abbasbandy, S., Shivanian, E., 2011. Predictor homotopy analysis method and its application to some nonlinear problems. Commun. Nonlinear Sci. Num. Simul. 16 (6), 2456–2468.

Abbasbandy, S., Magyari, E., Shivanian, E., 2009. The homotopy analysis method for multiple solutions of nonlinear boundary value problems. Commun. Nonlinear Sci. Num. Simul. 14 (9-10), 3530–3536.

Alomari, A., Awawdeh, F., Tahat, N., Ahmad, F.B., Shatanawi, W., 2013. Multiple solutions for fractional differential equations: analytic approach. Appl. Math. Comput. 219, 8893–8903.

Arqub, O.A., El-Ajou, A., Zhour, Z.A., Momani, S., 2014. Multiple solutions of nonlinear boundary value problems of fractional order: a new analytic iterative technique. Entropy 16, 471–493.

Atkinson, K.E., 1989. An Introduction to Numerical Analysis. John Wiley & Sons, Hoboken.

Azar, A.T., Vaidyanathan, S., Ouannas, A., 2017. Fractional Order Control and Synchronization of Chaotic Systems. Studies in Computational Intelligence, Vol. 688. Springer-Verlag, Germany.

Azarnavid, B., Parvaneh, F., Abbasbandy, S., 2015. Picard-reproducing kernel hilbert space method for solving generalized singular nonlinear lane-emden type equations. Math. Model. Anal. 20 (6), 754–767.

Barletta, A., 1999. Laminar convection in a vertical channel with viscous dissipation and buoyancy effects. Int. Commun. Heat Mass Transfer 26, 153–164.

Barletta, A., Magyari, E., Keller, B., 2005. Dual mixed convection flows in a vertical channel. Int. J. Heat Mass Transfer 48, 4835–4845.

Bataineh, A., Noorani, M., Hashim, I., 2007. Solutions of time-dependent emden fowler type equations by homotopy analysis method. Phys. Lett. A 371 (1 2), 72–82.

Chang, M.H., 2005. A decomposition solution for fins with temperature dependent surface heat flux. Int. J. Heat Mass Transfer 48 (9), 1819–1824.

Chowdhury, M., Hashim, I., 2008. Analytical solutions to heat transfer equations by homotopy-perturbation method revisited. Phys. Lett. A 372, 1240–1243.

Das, S., Pan, I., 2012. Fractional Order Signal Processing: Introductory Concepts and Applications. Springer, New York.

Don, W.S., Solomonoff, A., 1995. Accuracy and speed in computing the cheby-shev collocation derivative. SIAM J. Sci. Comput. 16, 1253–1268.

El-Sayed, A.M.A., Hashem, H.H.G., Ziada, E.A.A., 2014. Picard and adomian decomposition methods for a quadratic integral equation of fractional order. Comput. Appl. Math. 33 (1), 95–109.

Ellery, A., Simpson, M., 2011. An analytical method to solve a general class of nonlinear reactive transport models. Chem. Eng. J. 169, 313–318.

Fouda, M.E., Radwan, A.G., 2015. Fractional-order memristor response under dc and periodic signals. Circ. Syst. Signal Proc. 34 (3), 961–970.

Freeborn, T.J., Maundy, B., Elwakil, A.S., 2013. Cole impedance extractions from the step-response of a current excited fruit sample. Comput. Electron. Agric. 98, 100–108.

Ganji, D.D., 2006. The application of he's homotopy perturbation method to nonlinear equations arising in heat transfer equations by optimal homotopy analysis method. Phys. Lett. A 355, 337–341.

Gorder, R., 2012. Gaussian waves in the fitzhugh-nagumo equation demonstrate one role of the auxiliary function h(x,t) in the homotopy analysis method. Commun. Nonlinear Sci. Num. Simul. 17 (3), 1233–1240.

Graef, J.R., Qian, C., Yang, B., 2003. A three point boundary value problem for nonlinear forth order differential equations. J. Math. Anal. Applicat. 287, 217–233.

Graef, J.R., Qian, C., Yang, B., 2004. Multiple positive solutions of a boundary value prolem for ordinary differential equations. Electr. J. Qualit. Theory Different. Eq. 11, 1–13.

Hafez, T., Ganji, D.D., Babazadeh, H., 2007. The application of he's variational iteration method to nonlinear equations arising in heat transfer. Phys. Lett. A 363, 213–217.

Hassan, H.N., El-Tawil, M.A., 2011a. An efficient analytic approach for solving two-point nonlinear boundary value problems by homotopy analysis method. Math. Methods Appl. Sci. 34 (8), 977–989.

Hassan, H.N., El-Tawil, M.A., 2011b. A new technique of using homotopy analysis method for solving high-order nonlinear differential equations. Math Methods Appl. Sci. 34 (6), 728–742.

Hassan, H.N., Semary, M.S., 2013. Analytic approximate solution for the bratus problem by optimal homotopy analysis method. Commun. Numer. Anal. 2013, 1–14.

Hassan, H.N., Semary, M.S., 2014. The application of he's homotopy perturbation method to nonlinear equations arising in heat transfer equations by optimal homotopy analysis method. Walailak J. Sci. Technol. 11 (8), 659–677.

Hayat, T., Sajid, M., 2007. On analytic solution for thin film flow of a fourth grade fluid down a vertical cylinder. Phys. Lett. A 361 (4-5), 316–322.

He, J.H., 1999. Variational iteration method - a kind of non-linear analytical technique: some examples. Int. J. Non-Linear Mech. 34, 699–708.

Jacobsen, J., Schmitt, K., 2002. The liouville-bratu-gelfand problem for radial operators. J. Diff. Eq. 184, 283–298.

Jalilian, R., 2010. Non-polynomial spline method for solving bratu's problem. Comput. Phys. Commun. 181, 1868–1872.

Liao, S., 1992. The proposed homotopy analysis technique for the solution of nonlinear problems. Ph.D. thesis, Shanghai Jiao Tong University.

Liao, S., 2003. Beyond Perturbation: Introduction to the Homotopy Analysis Method. Chapman & Hall/CRC Press, London/Boca Raton, FL.

Liao, S., 2005. A new branch of solutions of boundary-layer flows over an impermeable stretched plate. Int. J. Heat Mass Tran. 48, 977–989.

Liao, S., 2009. Notes on the homotopy analysis method: some definitions and theorems. Commun. Nonlinear Sci. Num. Simul. 14 (4), 983–997.

Liao, S., 2012. Homotopy Analysis Method in Nonlinear Differential Equations. Springer, Heidelberg Dordrecht London New York.

Liu, Y., LiZ, B., 2009. The homotopy analysis method for approximating the solution of the modified korteweg-de vries equation. Chaos Solitons Fractals 39 (1), 1–8.

Magyari, E., 2008. Exact analytical solution of a nonlinear reaction-diffusion model in porous catalysts. Chem. Eng. J. 143 (1-2), 167–171.

Monje, C.A., Chen, Y., Vinagre, B., Xue, D., Fileu, V., 2010. Fractional Order Controls: Fundamentals and Application. Springer Verlag, Berlin.

Motsa, S.S., Sibanda, P., Shateyi, S., 2010. A new spectral homotopy analysis method for solving a nonlinear second order bvp. Commun. Nonlinear Sci. Num. Simul. 15, 2293–2302.

Odibat, Z.M., 2010. Analytic study on linear systems of fractional differential equations. Comput. Math. Applicat. 59 (3), 1171–1183.

Petras, I., 2011. Fractional-Order Nonlinear Systems: Modelling, Analysis and Simulation. Springer Verlag, Berlin.

Podlubny, I., 1999. Fractional Differential Equations. Academic Press, San Diego.

Radwan, A.G., 2013. Resonance and quality factor of the *rlc* fractional circuit. IEEE J. Emerg. Selected Topics Circuits Systems 3 (3), 377–385.

Radwan, A.G., Soliman, A.M., Elwakil, A.S., 2008a. Design equations for fractional-order oscillators: four practical design examples. Int. J. Circuit Theory Applicat. 36 (4), 473–492.

Radwan, A.G., Soliman, A.M., Elwakil, A.S., 2008b. International journal of circuit theory and applications. Circ. Syst. Signal Proc. 36 (4), 473–492.

Radwan, A.G., Elwakil, A.S., Soliman, A.M., 2009a. On the generalization of second-order filters to the fractional-order domain. J. Circuits Systems Comput. 18 (2), 361–386.

Radwan, A.G., Soliman, A.M., Elwakil, A.S., Sedeek, A., 2009b. On the stability of linear systems with fractional order elements. Chaos Solitons Fractals 40, 2317–2328.

Ragb, O., Seddek, L.F., Matbuly, M., 2017. Iterative differential quadrature solutions for bratu problem. Comput. Math. Applicat. 74, 249–257.

Sajid, M., Hayat, T., Asghar, S., 2007. Comparison between the ham and hpm solutions of thin film flows of non-newtonian fluids on a moving belt. Nonlinear Dynamics 50 (1-2), 27–35.

Salahshour, S., Ahmadian, A., Chan, C.S., 2015. Successive approximation method for caputo q-fractional ivps. Commun. Nonlinear Sci. Num. Simul. 24 (1-3), 153–158.

Semary, M.S., Hassan, H.N., 2015. A new approach for a class of nonlinear boundary value problems with multiple solutions. J. Associat. Arab Univ. Basic Appl. Sci. 17, 27–35.

Semary, M.S., Hassan, H.N., 2016. The homotopy analysis meyhod for q- difference equation. Ain Shams Eng. J. In press.

Semary, M.S., AbdelMalek, H.L., Hassan, H.N., Radawn, A.G., 2016a. An optimal linear system approximation of nonlinear fractional-order memristor-capacitor charging circuit. Microelectron. J. 51, 58–66.

Semary, M.S., Radawn, A.G., Hassan, H.N., 2016b. Fundamentals of fractional order lti circuits and systems: number of poles, stability, time and frequency responses. Int. J. Circuit Theory Applicat. 44 (12), 2114–2133.

Semary, M.S., Hassan, H.N., Radwan, A.G., 2017a. Controlled picard method for solving nonlinear fractional reaction-diffusion models in porous catalysts. Chem. Eng. Commun. 204 (6), 635–647.

Semary, M.S., Hassan, H.N., Radwan, A.G., 2017b. Single and dual solutions of fractional order differential equations based on controlled picard method and simpson rule. J. Associat. Arab Univ. Basic Appl. Sci. In press.

Soltani, L.A., Shivanian, E., Ezzati, R., 2017. Shooting homotopy analysis method: a fast method to find multiple solutions of nonlinear boundary value problems arising in fluid mechanics. Eng. Comput. 34 (2), 471–498.

Sun, Y.P., Liu, S.B., Keith, S., 2004. Approximate solution for the nonlinear model of diffusion and reaction in porous catalysts by the decomposition method. Chem. Eng. J. 102, 1–10.

Valerio, D., Trujillo, J.J., Rivero, M., Machado, J.A.T., Baleanu, D., 2013. Fractional calculus: a survey of useful formulas. Eur. Phys. J. Spec. Topics 222 (8), 1827–1846.

Vazquez-Leal, H., Rashidinia, J., Hernandez-Martinez, L., Daei-Kasmaei, H., 2015. A comparison of hpm, ndhpm, picard and picard-pade methods for solving michaelis-menten equation. J. King Saud Univ. Sci. 27, 7–14.

Wazwaz, A.M., 2005. Adomian decomposition method for a reliable treatment of the bratu type equations. Appl. Math. Comput. 166 (3), 652–663.

Wazwaz, A.M., 2009. The variational iteration method for analytic treatment for linear and nonlinear odes. Appl. Math. Comput. 212, 120–134.

Wazwaz, A.M., 2017. Dual solutions for nonlinear boundary value problems by the variational iteration method. Int. J. Num. Methods Heat Fluid Flow 27 (1), 210–220.

Wazwaz, A.M., Rach, R., Bougoff, L., 2016. Dual solutions for nonlinear boundary value problems by the adomian decomposition method. Int. J. Num. Methods Heat Fluid Flow 26 (8), 2393–2409.

Xu, H., Lin, Z.L., Liao, S., Wu, J.Z., Majdalani, J., 2010. Homotopy-based solutions of the navier-stokes equations for a porous channel with orthogonally moving walls. Phys. Fluids 22, 053601. Available from: https://doi.org/10.1063/1.3392770.

Yang, Z., Liao, S., 2017. A ham-based wavelet approach for nonlinear ordinary differential equations. Commun. Nonlinear Sci. Num. Simul. 48, 439–453.

Chapter 3

Fractional Order System Forced-response Decomposition and Its Application

Daniele Casagrande[1], Wiesław Krajewski[2] and Umberto Viaro[1]
[1]*University of Udine, Udine, Italy,* [2]*Systems Research Institute, Polish Academy of Sciences, Warsaw, Poland*

3.1 INTRODUCTION

The theory of fractional order systems has already attained sufficient maturity to allow its systematic presentation in several books (Azar et al., 2017; Tepljakov, 2017; Kaczorek, 2011; Caponetto et al., 2010; Monje et al., 2010) and to be the subject of many special journal issues (Caponetto et al., 2016; Ionescu et al., 2016; Psychalinos et al., 2016). Despite its increasing popularity, however, some important aspects need further investigation, among them the detailed analysis of the system forced response to inputs with rational order transform, of which the harmonic and singularity inputs (integrals of the impulse) are distinctive cases. In particular, relatively little attention has been paid to the separate study of the transient and asymptotic responses with some notable exceptions limited to canonical inputs (Monje et al., 2010; Trigeassou et al., 2012; Jakubowska and Walczak, 2016; Semary et al., 2016; Kesarkar and Selvaganesan Narayanasamy, 2016). A more thorough characterization of the system dynamic behavior requires the consideration of both the short- and long-term behavior of the responses to more general inputs. Such an analysis is particularly meaningful in the derivation of simplified models that retain essential properties of the original system, such as stability and performance indices, and in the synthesis of controllers that satisfy both transient and asymptotic specifications, e.g., on overshoot and steady-state error.

The present contribution aims at a more systematic study of the constituent parts of the forced-response that characterize different aspects of the system

Mathematical Techniques of Fractional Order Systems. DOI: https://doi.org/10.1016/B978-0-12-813592-1.00003-9

behavior and can conveniently be considered separately. To this purpose, reference is made to the fairly numerous class of inputs with rational order transform. Following a path similar to that taken in Casagrande et al. (2017) for integer order systems and based on Dorato et al. (1994), the forced response of a stable fractional order system to a persistent input of this kind is decomposed into the sum of two components: (1) a component with the same pseudo-polynomial denominator as the system transfer function; and (2) a component with the same pseudo-polynomial denominator as the input transform. The first is called the *system component* of the forced response and the second is called the *input component* because the first is characterized by the same evolution modes of the impulse response, which depends on the system only, and the second by elementary functions that exhibit similar structure (Mittag–Leffler functions) but depend only on the input and can therefore be called "input modes." If there are common modes between the input and the system, a third *resonant component* is also present. However, for the sake of simplicity, this possibility is ruled out (which is necessarily true when the system is stable, so that all of its modes tend to zero as time tends to infinity, and the input is antistable, so that all of its modes are persistent).

To ascertain the stability of the rational order system, resort can be made either to efficient numerical algorithms or to Routh–Hurwitz-like criteria for polynomials with real and complex coefficients. A section of this chapter is dedicated to this problem. As is known, it entails determining the distribution of the roots of a characteristic polynomial with respect to two radii delimiting a sector of the right half-plane (instead of the entire right half-plane, as is the case for integer order systems).

Robust stability issues are outside the scope of the present contribution and, therefore, are not treated in the sequel. Let us only observe, in this regard, that many results concerning the so-called $\mathcal{D}$-stability (Levkovich et al., 1996; Tempo, 1989) can be extended to fractional order systems.

The rest of this chapter is organized as follows. Section 3.2 introduces some essential notation and specifies the families of fractional order systems and inputs to which the aforementioned decomposition of the forced response applies. Section 3.3 shows how such a decomposition can uniquely be obtained from the Laplace transform of the forced response and, for stable systems, defines its transient and steady-state components. Section 3.4 presents some simple stability conditions. Section 3.5 shows how the decomposition can be used to find simplified models that reproduce the asymptotic response of original complex systems while still approximating well the transient behavior. Some illustrative examples are worked out in Section 3.6. The results are discussed in Section 3.7 where the relationship between the suggested response decomposition and the model-matching problem, strictly related to controller synthesis, is also pointed out. Possible directions of future research are indicated in Section 3.8.

3.2 NOTATION AND PRELIMINARIES

The transfer function of a continuous-time LTI strictly-proper rational order system can be written as

$$\widehat{G}(s) = \frac{b_m s^{\frac{m}{q}} + b_{m-1} s^{\frac{m-1}{q}} + \ldots + b_1 s^{\frac{1}{q}} + b_0}{a_n s^{\frac{n}{q}} + a_{n-1} s^{\frac{n-1}{q}} + \ldots + a_1 s^{\frac{1}{q}} + a_0}, \tag{3.1}$$

where q, m, n are positive integers, $m < n$, $q \geq 1$ is the least common denominator (lcd) of the (commensurate) fractional exponents of the Laplace variable s. The numerator and denominator coefficients of (3.1) are assumed to be real.

Consider now the class of inputs whose rational order Laplace transform can be written as

$$\widehat{U}(s) = \frac{d_k s^{\frac{k}{q}} + d_{k-1} s^{\frac{k-1}{q}} + \ldots + d_1 s^{\frac{1}{q}} + d_0}{c_\ell s^{\frac{\ell}{q}} + c_{\ell-1} s^{\frac{\ell-1}{q}} + \ldots + c_1 s^{\frac{1}{q}} + c_0}, \tag{3.2}$$

where k and ℓ are positive integers, $k < \ell$, and the numerator and denominator coefficients are real. It follows that the lcd of the fractional exponents of both (3.1) and (3.2) is q. This assumption is not much restrictive because it is always possible to express the fractional powers of s in (3.1) and (3.2) in terms of the same lcd, even if this lcd may be larger than the lcd of either function. The class of inputs (3.2) is fairly numerous and includes all inputs whose Laplace transform has only integer powers of s, such as the singularity and harmonic inputs.

By the change of variable

$$w = s^{\frac{1}{q}}, \tag{3.3}$$

functions (3.1) and (3.2) are transformed, respectively, into the following strictly-proper *rational* functions of w:

$$G(w) = \frac{B(w)}{A(w)}, \quad U(w) = \frac{D(w)}{C(w)}, \tag{3.4}$$

where

$$B(w) = b_m w^m + b_{m-1} w^{m-1} + \ldots + b_1 w + b_0, \tag{3.5}$$

$$A(w) = a_n w^n + a_{n-1} w^{n-1} + \ldots + a_1 w + a_0, \tag{3.6}$$

$$D(w) = d_k w^k + d_{k-1} w^{k-1} + \ldots + d_1 w + d_0, \tag{3.7}$$

$$C(w) = c_\ell w^\ell + c_{\ell-1} w^{\ell-1} + \ldots + c_1 w + c_0. \tag{3.8}$$

Correspondingly, the Laplace transform $\widehat{Y}(s)$ of the forced response of the system with transfer function (3.1) to the input with transform (3.2) is converted into the following rational function of w:

$$Y(w) = G(w)U(w) = \frac{B(w)D(w)}{A(w)C(w)}. \tag{3.9}$$

With some abuse of terminology, $G(w)$, $U(w)$, and $Y(w)$ will simply be referred to as system, input, and output functions, respectively, because they are directly related via (3.3) to $\widehat{G}(s)$, $\widehat{U}(s)$, and $\widehat{Y}(s) = \widehat{G}(s)\widehat{U}(s)$.

It has been long recognized that the denominator of the rational order function (3.1) is a multivalued function of s which becomes a single-valued function on a Riemann surface consisting of q sheets with branch cuts along the negative real semi-axis. The first, or principal, sheet contains the *physical* poles of (3.1) (Radwan et al., 2009) corresponding to the so-called structural, or relevant, roots of its denominator (Petráš, 2009). The stability of the rational order system depends on their location with respect to the imaginary axis. The right half of the first sheet, corresponding to the unstable region, maps into the (minor) sector of the w plane defined by

$$\mathcal{S} \triangleq \left\{ w = \rho e^{j\phi} : \rho \in \mathbb{R}_+, \phi \in \left[-\frac{\pi}{2q}, \frac{\pi}{2q} \right] \right\}. \tag{3.10}$$

As is known, the time-domain expressions of the fractional order system responses are easily obtained from the partial fraction expansions of the w-domain expressions (Semary et al., 2016; Valerio et al., 2013). It has rightfully been observed in this regard that the Mittag–Leffler functions play for fractional order systems a role analogous to that played by the exponential modes characterizing the time-domain response of integer order systems (Rivero et al., 2013; Trzaska, 2008).

The next section shows how the expression (3.9) of the forced-response of a fractional order system can be separated into a component consisting of the same modes as $G(w)$ and a component consisting of the same modes as $U(w)$.

3.3 DECOMPOSITION OF THE FORCED RESPONSE

The rational output (3.9) can be expanded into elementary partial fractions corresponding to its poles which are poles of either $G(w)$ or $U(w)$. If no pole-zero cancellation occurs, *all* (and only) the poles of $G(w)$ and $U(w)$ are poles of $Y(w)$. However, if p is a pole of *both* $G(w)$ and $U(w)$, the multiplicity of pole p in $Y(w)$ is the sum of the multiplicities of the same pole in $G(w)$ and $U(w)$, so that at least one elementary fraction appears in the expansion of $Y(w)$ that is not present in $G(w)$ or $U(w)$. This typically happens in the presence of resonance phenomena where an input frequency coincides with a natural frequency of the system. For the sake of simplicity, in this chapter the following assumption is made. (Indications on the extension of the following procedure to the general case are given in Remark 1 at the end of this section.)

Assumption 1: *Polynomials* $A(w)$, $B(w)$, $C(w)$, *and* $D(w)$ *have no common factors.*

Therefore the (strictly-proper) representations (3.4) and (3.9) are irreducible and no cancellation occurs in (3.9); in particular, resonance phenomena are not possible. Under Assumption 1, (3.9) can *uniquely* be decomposed as

$$Y(w) = \frac{X_A(w)}{A(w)} + \frac{X_C(w)}{C(w)} \tag{3.11}$$

where $X_A(w)$ and $X_C(w)$ are the solutions of the polynomial Diophantine equation (Ferrante et al., 2000; Kučera, 1993)

$$X_A(w)C(w) + X_C(w)A(w) = B(w)D(w) \tag{3.12}$$

with $\deg[X_A(w)] < \deg[A(w)] = n$, $\deg[X_C(w)] < \deg[C(w)] = k$ and, by the strict properness of $G(w)$ and $U(w)$, $\deg[B(w)D(w)] < \deg[A(w)] + \deg[C(w)]$. In this case, in fact, equation (3.12) is equivalent to a set of $n+k$ linear equations in the $n+k$ unknown coefficients x_i and y_i of polynomials:

$$X_A(w) = x_{A,n-1}w^{n-1} + x_{A,n-2}w^{n-2} + \ldots + x_{A,1}w + x_{A,0}, \tag{3.13}$$

$$X_C(w) = x_{C,n-1}w^{k-1} + x_{C,n-2}w^{k-2} + \ldots + x_{C,1}w + x_{C,0}, \tag{3.14}$$

obtained by equating the coefficients of the equal powers of w on both sides of (3.12). By properly ordering the unknowns, this set can be written in a matrix form where the coefficient matrix is nonsingular (Antsaklis and Michel, 2006; Henrion, 1998) (being the Sylvester matrix associated with the polynomials $A(w)$ and $C(w)$ that are co-prime by Assumption 1). It follows from Cramer's rule (Brunetti, 2014) that the aforementioned set of equations admits one, and only one, solution. For clarity of exposition, this result is restated next in the form of a proposition.

Proposition 1: *If* $A(w)$ *and* $C(w)$ *are co-prime and if* $\deg[B(w)D(w)] < \deg[A(w)] + \deg[C(w)]$, *there is a unique pair of polynomials* $X_A(w)$ *and* $X_C(w)$ *with* $\deg[X_A(w)] < \deg[A(w)]$ *and* $\deg[X_C(w)] < \deg[C(w)]$ *that solves the polynomial Diophantine equation (3.12).*

The two addenda in (3.11) will be denoted by

$$Y_\Sigma(w) = \frac{X_A(w)}{A(w)}, \quad Y_U(w) = \frac{X_C(w)}{C(w)} \tag{3.15}$$

and, borrowing the terminology adopted for integer order systems (Dorato et al., 1994), will be called the *system component* and *input component* of the output, respectively, because $Y_\Sigma(w)$ is characterized by exactly the same modes as the system (3.1) and $Y_U(w)$ by exactly the same modes as the input (3.2).

If the fractional order system is asymptotically stable (Petráš, 2009), so is also the system component, and its time-domain counterpart, obtainable by inverse Laplace transformation of the s-domain expression corresponding to $Y_\Sigma(w)$ via (3.3), tends asymptotically to zero. In this case, $Y_\Sigma(w)$ can rightfully be referred to as the *transient response* to input $U(w)$. Also, if the input is persistent, then the time-domain counterpart of the input component $Y_U(w)$ also persists and can rightfully be referred to as the *steady-state response* or, more in general, the *asymptotic response*.

Since very efficient and fast algorithms exist today to find the roots of a polynomial (Akritas et al., 2008; Jenkins and Traub, 1970) and the related computer programs are readily available, the easiest way to check the stability of a fractional order system is probably to determine numerically the precise location of the roots of $A(w)$ and see whether some of them lie in the instability sector (3.10). Nevertheless, the problem of finding the root distribution with respect to suitable contours (in particular, the perimeter of circular sectors with *bounded* radius, because upper bounds on the "size" of polynomial roots can be determined easily (Hirst and Macey, 1997)) is certainly of interest for other purposes, such as root clustering or $\mathcal{D}$-stability analysis (see Yedavalli, 2014; Gutman and Jury, 1981 and bibliographies therein), transient characterization, and stability margin evaluation. This problem is discussed in the following section.

Remark 1: *If, contrary to Assumption 1, $A(w)$ and $C(w)$ are not co-prime, they may be factored as:*

$$A(w) = \overline{A}(w)I_A(w), \quad C(w) = \overline{C}(w)I_C(w), \tag{3.16}$$

where $I_A(w)$ is the factor of $A(w)$ containing all and only the roots of $A(w)$ that are roots of $C(w)$ too (with their multiplicities), and $I_C(w)$ is the factor of $C(w)$ containing all and only the roots of $C(w)$ that are roots of $A(w)$ too (with their multiplicities). Clearly, if all of the common roots are simple $I_A(w) = I_C(w)$. Let $I(w) \triangleq I_A(w)I_C(w)$. Since the three pairs $[\overline{A}(w), \overline{C}(w)]$, $[\overline{A}(w), I(w)]$ and $[I(w), \overline{C}(w)]$ are co-prime, $Y(w)$ can uniquely be expressed (Ferrante et al., 2000) as

$$Y(w) = G(w)U(w) = \frac{X_{\overline{A}}(w)}{\overline{A}(w)} + \frac{X_{\overline{C}}(w)}{\overline{C}(w)} + \frac{X_I(w)}{I(w)}, \tag{3.17}$$

where the first addendum is a combination of modes proper to $G(w)$, the second is a combination of modes proper to $U(w)$, and the third is a combination of modes proper to the "interaction" or "resonant" component $Y_R(w) \triangleq X_I(w)/I(w)$. Some of the modes of $Y_R(w)$ are not contained in both $G(w)$ and $U(w)$ because the multiplicities of the roots of $I(w)$ are greater than the multiplicities of the same roots in $A(w)$ and $C(w)$.

As already said, the possibility of decomposing the forced response into a transient and a steady-state component depends on the system stability. The next section deals with the problem of checking this fundamental property.

3.4 STABILITY CONDITIONS

Although Routh–Hurwitz-like conditions have been derived to determine how the roots of the characteristic pseudo-polynomial of a fractional order system are distributed among the LHP and RHP half-planes of its principal Riemann sheet or its sectors (Liang et al., 2017), no simple rules are as yet available to establish *directly from polynomial* $A(w)$ whether some of its roots belong to given sectors of the w-plane (for arc angles different from π), except for those given in (Ahmed et al., 2006) that deal with very special cases. Indeed, conditions for *all* of the roots of a polynomial, or the eigenvalues of a matrix, to lie inside a minor LHP sector symmetric with respect to the real axis were obtained in the 1970s (Gutman, 1979; Anderson et al., 1974) from properties of Kronecker products of matrices (Graham, 1981) or rational maps (Gutman and Jury, 1981). However, the same result, i.e., the confinement of all the roots in the aforementioned minor LHP sector (no roots in the corresponding major sector), had already been obtained well before by means of Routh–Hurwitz arguments (Usher, 1957; Lüthi, 1942-43) or could easily have been achieved based on generalizations of the Routh–Hurwitz criteria (Hurwitz, 1895; Routh, 1877) to polynomials with complex coefficients (Frank, 1946; Billarz,1944). New formulations, extensions, and improvements of similar algebraic conditions, *including the analysis of the critical cases* and different tabular-form presentations, can be found in Sivanandam and Sreekala (2012), Chen and Tsai (1993), Benidir and Picinbono (1991), Agashe (1985), Hwang and Tripathi (1970) and, more recently, in Bistritz (2013) where numerically very efficient variants are presented. A different approach has been followed in Kaminski et al. (2015) where, for $q > 1$, a test based on regular chains for semi-algebraic sets (Chen et al., 2013) has been suggested. Here, some simple conditions based on the direct application of the Routh test to $A(w)$ are suggested to check whether some (not necessarily all) roots of $A(w)$ lie in an RHP sector symmetric with respect to the real axis.

To this purpose, consider the "forbidden" RHP sector defined by (3.10) (similar considerations hold, of course, for the opposite sector). If no roots of $A(w)$ are in the RHP, which may be checked by means of the standard Routh test, then this sector, as well as the two sectors containing the points of the RHP which do not belong to $\mathcal{S}$, do not contain any root, either. Therefore, the method in (Usher, 1957) can be adopted to determine the number of roots inside any LHP sector symmetric with respect to the negative real semi-axis.

Also, since the number of the real roots in any interval of the real axis can be found easily on the basis of the classic Sturm algorithm (see, e.g., the lecture notes in Jia, 2016), whose computational complexity is not greater than that of the Routh algorithm (i.e., $\mathcal{O}(n^2)$), only the root distribution of the roots with a nonzero imaginary part need actually be determined. Therefore, for notational simplicity and without loss of generality, the following assumption is made.

Assumption 2: *The real polynomial $A(w)$ has no real roots.*

It follows that the degree n of $A(w)$ is even because its complex roots with nonzero imaginary part are in conjugate pairs (In other words, $A(w)$ can be though of as the even-degree factor containing all of the complex conjugate roots of an original polynomial $\tilde{A}(w) = A(w)A_r(w)$, where $A_r(w)$ is the factor containing all of the real roots of $\tilde{A}(w)$. For simplicity, the following analysis refers to $A(w)$ only. The results, however, can easily be extended to include the real roots).

By combining the information on the root distribution with respect to the imaginary axis with the information on the root distributions with respect to each of the two slanted straight lines with slope $\pm\pi/2q$ (see again (3.10)), some interesting results can be established straightaway. To state them in a compact form, the following notation, illustrated in Fig. 3.1, is introduced:

1. n_{1u}, n_{1l} denote the number of roots above and, respectively, below the slanted line through the origin with *positive* slope $\pi/2q$;
2. n_{2u}, n_{2l} denote the number of roots above and, respectively, below the slanted line through the origin with *negative* slope $-\pi/2q$; and
3. n_+, n_- denote the number of roots in the closed RHP and in the open LHP, respectively;
4. the *difference* between the number of roots in $\mathcal{S}$ (unstable roots) and the LHP sector (symmetric of $\mathcal{S}$) is denoted by δ.

Clearly,

$$\delta = n_{2u} - n_{1u} = n_{1l} - n_{2l} = n_{2u} - n_{2l} = n_{1l} - n_{1u}. \tag{3.18}$$

The following result is obvious.

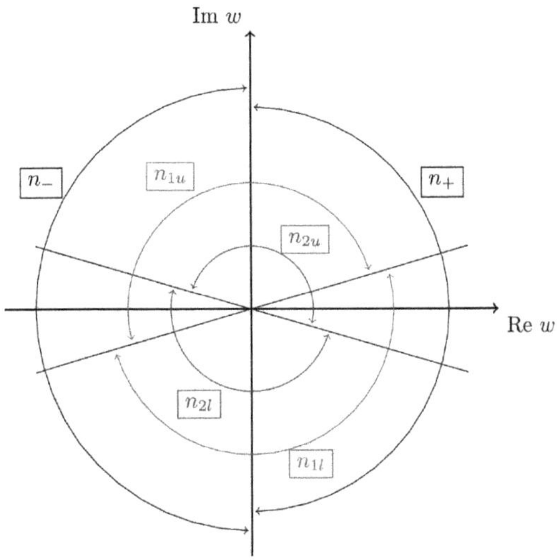

FIGURE 3.1 Notation for: (i) the number of roots in each of the two half-planes separated by each of the two slanted straight lines through the origin with opposite slope (n_{1u}, n_{1l} and n_{2u}, n_{2l}), and (ii) the number of roots in each of the two half-planes separated by the vertical axis (n_+ and n_-).

Proposition 2: *If $\delta > 0$ the fractional order system is unstable.*

When $\delta \leq 0$, the system may be stable, but the stability conditions also depend on n, n_+, and $n_- = n - n_+$. By considering that, in the absence of real roots, *all of the above numbers are even*, the following two stability conditions can easily be proved. The first only requires the knowledge of n.

Proposition 3: *If $\delta = -i$, i nonnegative, and $n < i + 4$, the fractional order system is stable.*

Proof. Assume that the fractional order system is unstable. Since all roots appear in conjugate pairs, the number of roots in the instability sector is 2 or more. Therefore, in order for $\delta = -i$, the sector opposite to the instability sector must contain at least $i + 2$ roots, and the polynomial degree n, which is greater than, or equal to, the sum of the roots in both sectors, must be equal, at least, to $i + 4$, contrary to the assumption that $n < i + 4$.

Proposition 4: *If $\delta = -i$, i nonnegative, $n_- = i + j$, j nonnegative, and $n < i + j + 2$, the fractional order system is stable.*

Proof. It is enough to consider that, under the adopted assumptions, $n - n_- < 2$ so that no root may lie in the instability sector.

More general stability conditions require the acquisition of additional information, which in some cases may be worthwhile. For instance, to determine whether some roots lie inside the instability sector $\mathcal{S}$, the slope of the two slanted lines (i.e., the angle $\frac{\pi}{2q}$) can gradually be taken to zero. If the difference between the numbers of roots in the LHP and RHP sectors decreases *monotonically* to zero as the sectors angle tends to zero, then the system is stable. A similar procedure can be applied to detect roots with damping factor in a given range.

By simple adaptation to fractional order systems of the classic Mikhailov stability criterion for integer order systems (Busłowicz, 2008; Mikhailov, 1938), the following graphically-based criterion also holds.

Proposition 5: *The fractional order system is stable if and only if the phase variation of the nth degree polynomial*

$$\tilde{A}(\rho) \triangleq A\left(\rho\ e^{j\frac{\pi}{2q}}\right) \tag{3.19}$$

as ρ varies from 0 to $+\infty$ is equal to $n\frac{\pi}{2q}$:

$$\Delta \arg[\tilde{A}(\rho)]_{[0,\infty)} = n\frac{\pi}{2q}. \tag{3.20}$$

Proof. Express $A\left(\rho\ e^{j\frac{\pi}{2q}}\right)$ in factored form as

$$A\left(\rho\ e^{j\frac{\pi}{2q}}\right) = a_n \prod_{i=1}^{n} \left(\rho\ e^{j\frac{\pi}{2q}} - p_i\right), \tag{3.21}$$

where the p_i, $i = 1, 2, \ldots, n$, are the (possibly repeated) roots of $A(w)$. The phase variation of (3.21) as point $\rho e^{j\frac{\pi}{2q}}$ moves along the slanted half-line leaving the origin and making an angle $\pi/(2q)$ with the positive real axis is the sum of the phase variations of its factors which can be regarded as vectors applied at the p_is and pointing to $\rho e^{j\frac{\pi}{2q}}$. Now, if a *real* root, say p_k, is outside the instability sector and thus negative, the initial phase of the corresponding factor (when ρ is equal to zero and the point $\rho e^{j\frac{\pi}{2q}}$ coincides with the origin) is zero and its final phase (when $\rho e^{j\frac{\pi}{2q}}$ tends to infinity along the aforementioned slanted half-line) is $\pi/(2q)$. If a *complex* root, say p_h, is outside the instability sector, consider the two factors associated with the pair of conjugate poles p_h and $p_h^\star$. The sum of the initial phases of the two vectors $\rho e^{j\frac{\pi}{2q}} - p_h$ and $\rho e^{j\frac{\pi}{2q}} - p_h^\star$ is also zero, while the final sum of their phases is

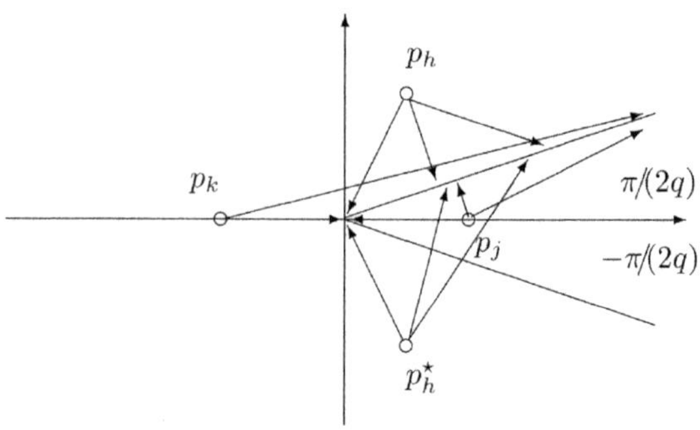

FIGURE 3.2 Vector representation of the phase variation of the factors in (3.20) associated with either real or complex roots outside or inside the RHP minor sector with central angle π/q straddling the positive real axis.

$2 \cdot \pi/(2q)$, as shown in Fig. 3.2. Therefore, if all the n roots are outside the instability sector, the overall phase variation is (3.20), which proves necessity. The sufficiency of (3.20) can be proved by contradiction. To this purpose, assume that (3.20) holds true but that a real root, say p_j, lies inside the instability sector. The phase variation of the factor associated with p_j, i.e., $\rho e^{j\frac{\pi}{2q}} - p_j$, is $-[\pi - \pi/(2q)]$ (see Fig. 3.2) so that the overall phase variation is less than (3.20), which contradicts the assumption that (3.20) holds true. A similar reasoning can be used for a pair of conjugate roots inside the instability sector.

Once system stability has been ascertained, efficient approximation methods can be applied to simplify an original complex model. The next section is devoted to such a problem.

3.5 MODEL REDUCTION

The separate consideration of the two components (3.15) of (3.9) can be used for analysis, synthesis, and approximation purposes. In this section, it is shown how to obtain reduced order models that retain the original asymptotic response along the lines of Casagrande et al. (2017). Essentially, the suggested procedure operates as follows.

3.5.1 Approximation Procedure

1. Find the fractional order system transfer function (3.1) and determine the Laplace transform (3.2) of the input whose response is of interest.
2. Via the change of variable (3.3), convert the Laplace transform of the related response into the rational function $Y(w)$ (see (3.9)).
3. Decompose $Y(w)$ according to (3.11) into a system component $Y_\Sigma(w)$ and an input component $Y_U(w)$ (see (3.15)).
4. Find a rational function $Y_\Sigma^\nu(w)$ of order $\nu < n$ (usually, $\nu \ll n$) approximating the original system component $Y_\Sigma(w)$ according to any criterion for rational approximation.
5. Form the reduced order w-domain transfer function $G_r(w)$ of the reduced order model in such a way that the reduced order model response $Y_r(w)$ to $U(w)$ admits $Y_U(w)$ as its input component and $Y_\Sigma^\nu(w)$ as its system component up to an auxiliary additive term of negligible importance $Y_\Sigma^a(w)$, namely:

$$Y_r(w) = G_r(w)\,U(w) = Y_U(w) + Y_\Sigma^\nu(w) + Y_\Sigma^a(w). \tag{3.22}$$

6. Construct the simplified transfer function $\widehat{G}_r(s)$ from $G_r(w)$ using again (3.3).

To clarify step (v), some remarks are opportune.

Remark 2: *As explained in detail in* Casagrande et al. (2017), *the introduction of the auxiliary term is necessary to make the number of unknowns in (3.22) equal to the number of equations. If the poles of $Y_{\Sigma}^{a}(w)$ are fixed, this result is obtained when the orders of $Y_{\Sigma}^{a}(w)$ and $U(s)$ are equal, i.e., the degree of the denominator of $Y_{\Sigma}^{a}(w)$ coincides with the degree n_u of the denominator of $U(w)$. The solution is then obtained by equating the coefficients of the equal powers of w at the numerators of the product $W^r(w)U(w)$ and of $Y_U(w) + Y_{\Sigma}^{\nu}(w) + Y_{\Sigma}^{a}(w)$, respectively. The problem turns out to be linear.*

Remark 3: *If the poles of $Y_{\Sigma}^{a}(w)$ are located far to the left of the imaginary axis, this additional term does not alter appreciably the transient dynamics of the system while it leaves unchanged the input component.*

Remark 4: *Due to the introduction of the auxiliary term, the order r of $W^r(w)$ is greater than the order ν of the function $Y_{\Sigma}^{\nu}(w)$ approximating the original system component $Y_{\Sigma}(w)$. However, since usually $\nu \ll n$, the order $r = \nu + n_u$ is still much smaller than n for the canonical inputs ($n_u \leq 2$ for steps, ramps and sinusoids).*

Procedure 3.5.1 is schematically represented in Fig. 3.3. It has been applied to several benchmark examples with considerable success; three of them are illustrated in the next section.

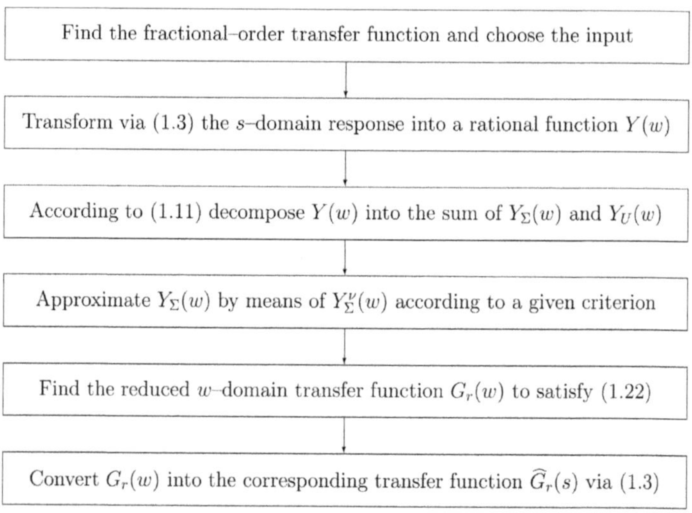

FIGURE 3.3 Basic flow chart of Procedure 3.5.1.

3.6 EXAMPLES

The following examples show that the response of the simplified model to the desired input matches closely the original response even during the transient.

3.6.1 Example 1

Consider the fractional order system described by the transfer function

$$\widehat{G}(s) = \frac{s^4 + 9s^{3.2} + 31s^{2.4} + 58.01s^{1.6} + 60.01s^{0.8} + 16.03}{s^{4.8} + 6s^4 + 48s^{3.2} + 286s^{2.4} + 935s^{1.6} + 1580s^{0.8} + 888}, \tag{3.23}$$

which has also been adopted in Tavakoli–Kakhki and Haeri (2009) and Jiang and Xiao (2015), and assume that the system is driven by the input

$$\widehat{U}(s) = \frac{1}{s^{0.8}}. \tag{3.24}$$

Remark 5: *Input (3.24) can also be written as*

$$\widehat{U}(s) = s^{0.2} \cdot \frac{1}{s} \tag{3.25}$$

whose time-domain counterpart is

$$\hat{u}(t) = D^{0.2}\mathcal{H}(t) \tag{3.26}$$

which is the fractional derivative of order 0.2 *of the usual step function* $\mathcal{H}(t)$. *It seems reasonable to consider the inputs with Laplace transform*:

$$\widehat{U}(s) = \frac{1}{s^{\alpha}} \tag{3.27}$$

and time-domain expression (Caponetto et al., 2010)

$$u(t) = \frac{t^{\alpha-1}}{\Gamma(\alpha)}, \tag{3.28}$$

as the fractional order equivalents of the canonical inputs for integer order systems (also called singularity inputs (Tewari, 2011)). Fig. 3.4 *shows the time course of (3.28) for three values of* α.

By the change of variable

$$w = s^{\frac{1}{5}} \tag{3.29}$$

the response of system (3.23) to input (3.24) is given in the w-domain by the 24th order rational function

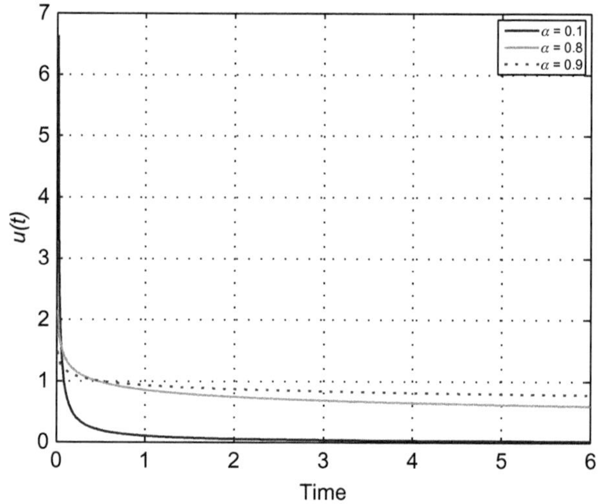

FIGURE 3.4 Input signals of form (3.28) with $\alpha = 0.1$ (lower solid line), $\alpha = 0.8$ (upper solid line), and $\alpha = 0.9$ (dotted line).

$$Y(w) = G(w)\ U(w) = \\ = \frac{w^{20} + 9w^{16} + 31w^{12} + 58.01w^8 + 60.01w^4 + 16.03}{w^{24} + 6w^{20} + 48w^{16} + 286w^{12} + 935w^8 + 1580w^4 + 888} \cdot \frac{1}{w^4} \tag{3.30}$$

whose poles are outside the minor sector delimited by the radii $\rho e^{\pm j\frac{\pi}{10}}$ (see equation (3.10)) so that the fractional order system is stable.

Since, in this special case, all powers of w are multiples of the same integer 4, the decomposition and reduction procedures can more conveniently be applied to the 6th order rational function

$$\tilde{Y}(z) = \frac{z^5 + 9z^4 + 31z^3 + 58.01z^2 + 60.01z + 16.03}{z^6 + 6z^5 + 48z^4 + 286z^3 + 935z^2 + 1580z + 888} \cdot \frac{1}{z}, \tag{3.31}$$

obtained from (3.30) by setting $z = w^4$. Clearly, the poles of (3.31) are the fourth powers of the poles of (3.30), which means, in particular, that the instability sector in the z-plane is delimited by the radii $\rho e^{\pm j\frac{4\pi}{10}}$ instead of the radii $\rho e^{\pm j\frac{\pi}{10}}$ that enclose the instability sector in the w-plane. Of course, (3.31) could directly be obtained from (3.23) by setting $z = s^{4/5}$.

Function (3.31) can be decomposed into a system-dependent and an input-dependent component as

$$\widetilde{Y}(z) = \\ = \frac{-0.0181z^5 + 0.8917z^4 + 8.1335z^3 + 25.8372z^2 + 41.1316z + 31.4882}{z^6 + 6z^5 + 48z^4 + 286z^3 + 935z^2 + 1580z + 888} \\ + \frac{0.0181}{z}. \tag{3.32}$$

Applying the shifted Padé approximation method suggested in Tavakoli–Kakhki and Haeri (2009) with $s_0 = 133$, we find the following 2nd order approximation of the system-dependent component (first addendum at the right-hand side of (3.32)):

$$\widetilde{Y}^2_{\Sigma}(z) = \frac{-0.0181z + 1.0407}{z^2 - 2.2538z + 40.6760}. \tag{3.33}$$

Adding the input-dependent component (second addendum at the right-hand side of (3.32)) to (3.33) as well as an auxiliary 1st order term with the far-off pole at $z = -100$ (step (v) of Procedure 3.5.1), we find the following 3rd order approximation of the z-domain system transfer function:

$$\widetilde{G}_r(z) = \frac{z^2 + 100.7402z + 73.4276}{z^3 + 97.7z^2 - 184.7z + 4067.6} \tag{3.34}$$

which, via the change of variable $z = s^{4/5}$, corresponds to the stable fractional order transfer function:

$$\widehat{G}_r(s) = \frac{s^{1.6} + 100.7402s^{0.8} + 73.4276}{s^{2.4} + 97.7s^{1.6} - 184.7s^{0.8} + 4067.6}. \tag{3.35}$$

Instead, by applying the shifted Padé method with $s_0 = 133$ directly to (3.23), as in Tavakoli–Kakhki and Haeri (2009), without consideration of the input component of the forced response, the following simplified stable fractional order model is obtained:

$$\widehat{G}_{r,TH}(s) = \frac{s^{1.6} + 5.0349s^{0.8} + 0.3743}{s^{2.4} + 2.0349s^{1.6} + 29.2696s^{0.8} + 145.3930}. \tag{3.36}$$

The responses to input (3.24) of the reduced system (3.35) obtained according to the method suggested in this section and of the reduced system (3.36) are compared in Fig. 3.5 with the response to the same input of the original system (3.23). It is apparent that the retention of the steady-state component leads to a better approximation in the medium to long run.

Since fractional order systems are infinite-dimensional, or long-memory, systems (Sabatier et al., 2014), it might be argued that the transfer function of a fractional order system can be considered "simple" if it contains a small number of parameters. From this point of view, functions (3.35) and (3.36) in the previous subsection are indeed simpler than the original transfer

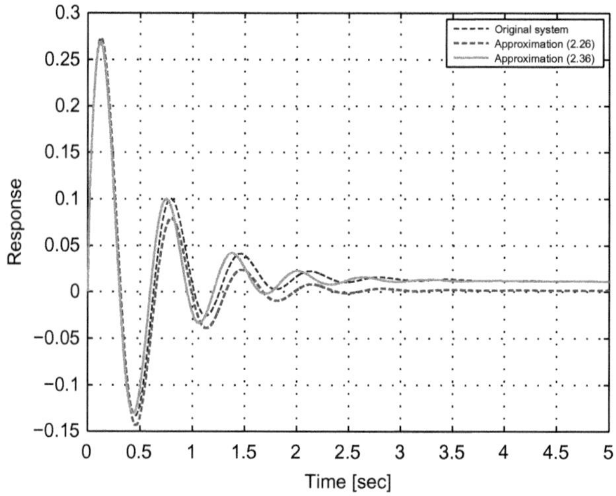

FIGURE 3.5 Responses to $\widehat{U}(s) = 1/s^{0.8}$ of: (i) the original system (3.23) (blue upper dashed line), (ii) the approximation (3.36) (red lower dashed line), and (iii) the approximation (3.35) retaining the steady-state component (green solid line).

function (3.23). Instead, from the point of view of system dimensionality, any integer order input–output or state–space model simulating the behavior of a given fractional order system can be regarded as a simplified model of the fractional order system (see, e.g., Krajewski and Viaro, 2014). In a sense, the w-domain rational function $G(w)$ itself, having a finite integer order, represents the original fractional system in a more compact form.

In this section, simplification is considered to be achieved if the maximum fractional degree of the s-domain approximation is smaller than the maximum fractional degree of the original transfer function, which corresponds to the fact that the (integer) degree of the denominator of $G_r(w)$ is smaller than that of the denominator of $G(w)$, even if the number of parameters in $G_r(w)$ might sometimes be greater than that in $G(w)$. The last situation typically occurs when a number of intermediate coefficients (between the leading term and the term of lowest degree, usually a constant) in the numerator and denominator of $G(w)$ are missing. Of course, also the choice of the input whose asymptotic term should be preserved influences the model complexity because the fractional powers of s in $\widehat{U}(s)$ contribute to the determination of the minimum common denominator of all fractional exponents of $\widehat{Y}(s) = \widehat{G}(s)\widehat{U}(s)$.

3.6.2 Example 2

As a second example, consider the system analyzed in Tavazoei (2016, 2011) whose transfer function turns out to be

$$\widehat{G}(s) = \frac{s^{2.7} + 2s^{1.8} + s^{0.9} + 2}{s^{6.3} + 4.9s^{5.4} + 11.05s^{4.5} + 14.07s^{3.6} + 10.53s^{2.7} + 4.55s^{1.8} + 1.05s^{0.9} + 0.1}. \tag{3.37}$$

By the change of variable $w = s^{1/10}$, we obtain

$$G(w) = \frac{w^{27} + 2w^{18} + w^9 + 2}{w^{63} + 4.9w^{54} + 11.05w^{45} + 14.07w^{36} + 10.53w^{27} + 4.55w^{18} + 1.05w^9 + 0.1} \tag{3.38}$$

whose poles are outside the instability sector. Since all powers are multiples of 9, in this case too, the reduction procedure can profitably be applied to an integer order rational function of smaller degree in the variable $z = w^9$, namely:

$$\widetilde{G}(z) = \frac{z^3 + 2z^2 + z + 2}{z^7 + 4.9z^6 + 11.05z^5 + 14.07z^4 + 10.53z^3 + 4.55z^2 + 1.05z + 0.1} \tag{3.39}$$

whose 3rd order optimal Hankel-norm approximation (Glover, 1984) turns out to be

$$\widetilde{G}_{r,HN}(z) = \frac{0.5592z^2 - 0.4066z + 0.4178}{z^3 + 0.5841z^2 + 0.1885z + 0.0204} \tag{3.40}$$

and in the s-domain with $z = s^{9/10}$

$$\widehat{G}_{r,HN}(s) = \frac{0.5592s^{1.8} - 0.4066s^{0.9} + 0.4178}{s^{2.7} + 0.5841s^{1.8} + 0.1885s^{0.9} + 0.0204}. \tag{3.41}$$

Let us apply now the procedure based on the retention to the asymptotic response to the input

$$\widehat{U}(s) = \frac{1}{s^{0.9}}. \tag{3.42}$$

To this purpose, the original forced response to input (3.42) is decomposed into the sum of a system-dependent component and an input-dependent component, which in the domain of $z = w^9 = (s^{1/10})^9$ turn out to be, respectively,

$$\widetilde{Y}_{\Sigma}(z) = \frac{-20z^6 - 98z^5 - 221z^4 - 281.4z^3 - 209.6z^2 - 89z - 20}{z^7 + 4.9z^6 + 11.05z^5 + 14.07z^4 + 10.53z^3 + 4.55z^2 + 1.05z + 0.1}, \tag{3.43}$$

$$\widetilde{Y}_U(z) = \frac{20}{z}. \tag{3.44}$$

The 2nd order optimal Hankel-norm approximation of (3.43) is

$$\tilde{Y}^{2}_{\Sigma}(z) = \frac{-15.9384z - 9.9977}{z^2 + 0.3796z + 0.0511}. \tag{3.45}$$

By adding to (3.45) an auxiliary term with a far-off pole at -100 (step (v) of Procedure 3.5.1), and combining the resulting sum with the original input-dependent component (3.44), the reduced system transfer function in the z-domain turns out to be

$$\tilde{G}_r(z) = \frac{402.2097z^2 - 239.7748z + 102.1383}{z^3 + 100.3796z^2 + 38.0102z + 5.1069} \tag{3.46}$$

and in the s-domain with $z = s^{9/10}$

$$\hat{G}_r(s) = \frac{402.2097s^{1.8} - 239.7748s^{0.9} + 102.1383}{s^{2.7} + 100.3796s^{1.8} + 38.0102s^{0.9} + 5.1069}. \tag{3.47}$$

The responses to (3.42) of the original system (3.37) and of the approximating models (3.41) and (3.47) are shown in Fig. 3.6.

3.6.3 Example 3

Consider finally the transfer function

$$\hat{G}(s) = \frac{5s^{0.6} + 2}{s^{3.3} + 3.1s^{2.6} + 2.89s^{1.9} + 2.5s^{1.4} + 1.2} \tag{3.48}$$

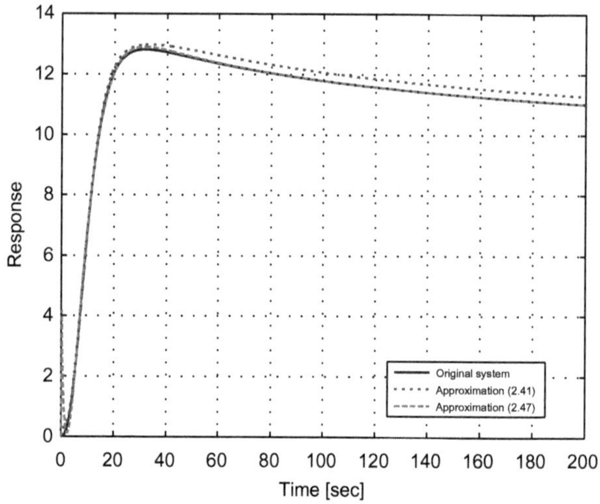

FIGURE 3.6 Responses to $\hat{U}(s) = 1/s^{0.9}$ of: (i) the original system (3.37) (blue solid line), (ii) the approximating model (3.41) (red dotted line), and (iii) the approximating model (3.47) retaining the asymptotic component of the response (green dashed line).

taken from (Xue and Chen, 2007), and assume that the input whose asymptotic response must be retained is

$$\widehat{U}(s) = \frac{10}{s^{0.2} - 0.7s^{0.1}} \tag{3.49}$$

which has been chosen, rather arbitrarily, to test the system long-term response to nondecaying inputs. The corresponding time-domain signal $\widehat{u}(t)$ is shown in Fig. 3.7. It can be obtained as the step response of a filter with transfer function

$$\widehat{F}(s) = \frac{10s^{0.9}}{s^{0.1} - 0.7}. \tag{3.50}$$

For $w = s^{0.1}$ function (3.48) becomes

$$G(w) = \frac{5w^6 + 2}{w^{33} + 3.1w^{26} + 2.89w^{19} + 2.5w^{14} + 1.2}, \tag{3.51}$$

A 4th order approximation of (3.51) has been obtained by interpolating (3.51) at $s = 1$ and $s = 2$ with intersection number 2 (retention of 2 time moments) according to the multipoint Padé technique via Lanczos' iteration method (Gallivan et al., 1996). The corresponding simplified fractional order transfer function turns out to be

$$\widehat{G}_{r,PL}(s) = \frac{9.6597s^{0.3} + 50.106s^{0.2} + 56.107s^{0.1} + 4.753}{s^{0.4} + 35.5s^{0.3} + 161.08s^{0.2} + 173.02s^{0.1} + 14.037}. \tag{3.52}$$

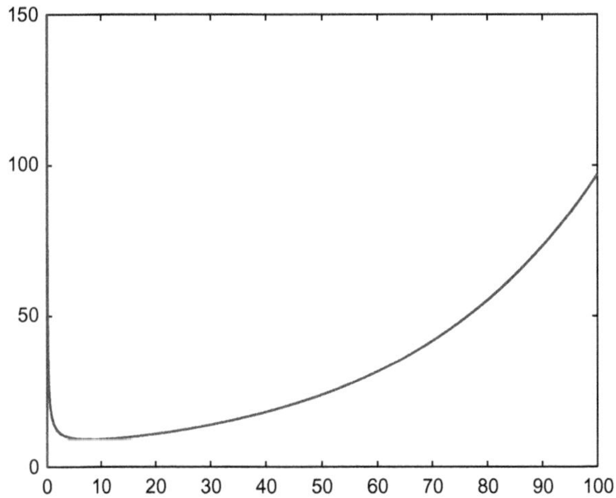

FIGURE 3.7 Time-domain input signal $\widehat{u}(t)$ corresponding to (3.49).

By applying instead the suggested reduction method based on:

1. the decomposition of $Y(w) = G(w)U(w)$ into a system-dependent component

$$Y_{\Sigma}(w) = \frac{X_A(w)}{w^{33} + 3.1w^{26} + 2.89w^{19} + 2.5w^{14} + 1.2} \tag{3.53}$$

and an input-dependent component

$$Y_U(w) = \frac{X_C(w)}{w^2 + w + 100}, \tag{3.54}$$

2. the approximation of (3.53) by means of the same method used to find (3.52) from (3.48), and
3. the retention of (3.54),

the following approximating fractional order transfer function is obtained (after substituting $s^{0.1}$ for w)

$$\widehat{G}_r(s) = \frac{68.447s^{0.3} + 357.21s^{0.2} + 528.5s^{0.1} + 157.17}{s^{0.4} + 3.7838s^{0.3} + 103.73s^{0.2} + 279.32s^{0.1} + 94.3}. \tag{3.55}$$

Fig. 3.8 shows the responses to the input with Laplace transform (3.49) of: (i) the original system (3.48), (ii) the approximating model (3.52), and (iii) the approximating model (3.55) retaining the asymptotic response.

The previous examples show that the suggested response decomposition can be applied satisfactorily in many approximation problems. On the other

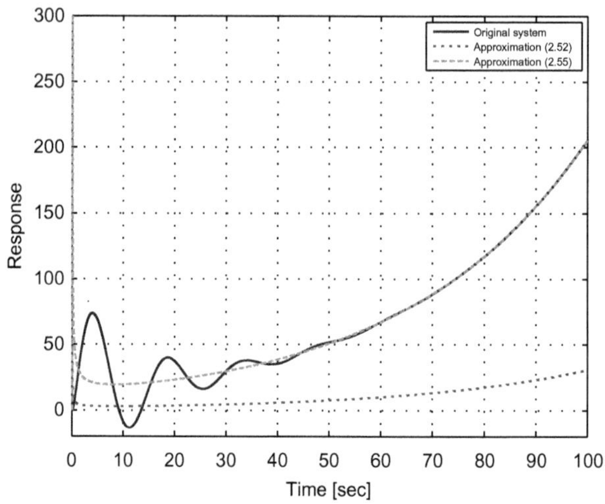

FIGURE 3.8 Responses of: (i) the original model (3.48) (blue solid line), (ii) the approximating model (3.52) (red dotted line), and (iii) the approximating model (3.55) retaining the asymptotic component of the response (green dashed line)

hand, as outlined in the next section, some difficulties arise in achieving a simplification in terms of number of model parameters. A similar problem arises in the use of the suggested decomposition for solving the model-matching problem, strictly related to controller synthesis, as the example in the next section will show.

3.7 DISCUSSION AND EXTENSIONS

In the previous Section 3.5 the decomposition of the forced response has profitably been applied to the derivation of "simplified" models of fractional order systems. It has been observed, in this regard, that the definition of model complexity is to some extent arbitrary. It may be related to the (finite) dimension of the integer order models associated with the fractional order systems via the variable transformation (3.3), or to the "compactness" of the fractional order transfer functions, in particular, the number of nonzero parameters that appear in them, or to the maximum degree of the denominator of the transfer functions. The models obtained in the previous section can be considered simpler from all of these points of view. It should be observed, however, that the results strongly depend on the input whose asymptotic component of the forced response must be retained. If the minimum common denominator (mcd) of the fractional exponents of the input transform does not coincide with the mcd of fractional exponents of the original transfer function, forcing $U(w)$ and $G(w)$ to have a common q might entail a considerable increase of the order of the integer order model obtained via (3.3) from the fractional order transforms.

Also the reduction criterion adopted to approximate the system component of the forced response is rather arbitrary. Even if its choice is outside the scope of the present contribution, it should be noted that not all methods cannot be applied. In fact, most reduction methods suggested in the literature for integer order systems are directly applicable only to (stable) systems with poles in the open LHP. To overcome this problem, it is sometimes suggested to preliminarily separate the stable and unstable parts of the systems with RHP poles and then apply the reduction procedure only to the first. Further difficulties arise in the case of fractional order systems, because their stability is compatible with the presence of RHP poles in the integer order function derived from the fractional one via (3.3), provided these poles are outside the instability sector.

In this book chapter attention has been focused on the model reduction problem, but the relevance of the response decomposition goes beyond model simplification. Suffice it to recall, in this regard, the interpolation problem, strictly related to the model matching problem (Doyle et al., 1992) or, more generally, the moment matching problem (Astolfi, 2010). Indeed, forcing the coincidence of the input components of two different systems in the response to a given input entails interpolating the values taken by the transfer function of either system at the roots of the denominator of the input transform. To clarify this, consider two systems whose respective transfer functions are $G_1(w) = B_1(w)/A_1(w)$ and $G_2(w) = B_2(w)/A_2(w)$. Under suitable co-primeness assumptions, their responses to $U(w) = D(w)/C(w)$ can be decomposed as

$$Y_1(w) = \frac{B_1(w)}{A_1(w)} \frac{D(w)}{C(w)} = \frac{X_{A_1}(w)}{A_1(w)} + \frac{X_{C_1}(w)}{C(w)} \tag{3.56}$$

and

$$Y_2(w) = \frac{B_2(w)}{A_2(w)} \frac{D(w)}{C(w)} = \frac{X_{A_2}(w)}{A_2(w)} + \frac{X_{C_2}(w)}{C(w)}. \tag{3.57}$$

For $X_{C_1}(w) = X_{C_2}(w) = X_C(w)$ (equality of the input-dependent components), from (3.56) and (3.57) we get

$$B_1(w)D(w) = X_{A_1}(w)C(w) + X_C(w)A_1(w), \tag{3.58}$$

$$B_2(w)D(w) = X_{A_2}(w)C(w) + X_C(w)A_2(w), \tag{3.59}$$

so that, at the roots of $C(w)$, i.e., for $C(w) = 0$, we have

$$G_2(w) = \frac{B_2(w)}{A_2(w)} = \frac{B_1(w)}{A_1(w)} = G_1(w), \tag{3.60}$$

which means that $G_2(w)$ interpolates $G_1(w)$ at the poles of $U(w)$. As is well known, if $G_2(w)$ and $G_1(w)$ are realized in a unity-feedback fashion, this means, in turn, that their forward paths include an internal model of the (common) input (Francis and Wonham, 1976).

The previous considerations have obvious implications on the so-called *direct* or *analytic* synthesis of control systems (Ferrante et al., 2000) whose first step consists in choosing an overall, or total, or complementary sensitivity, system transfer function $T(w)$ that satisfies the specifications, the next step being its realization, possibly by means of a feedback structure with controller $G_c(w)$ and plant $G_p(w)$ located in the forward path so that

$$T(w) = \frac{G_c(w)G_p(w)}{1 + G_c(w)G_p(w)} \tag{3.61}$$

and

$$G_c(w) = \frac{1}{G_p(w)} \frac{T(w)}{1 - T(w)}. \tag{3.62}$$

In the case of fractional order systems, to profit by the efficient techniques developed for integer order systems, the rational function $T(w)$ will be obtained, via (3.3), from an original fractional order transfer function. To facilitate the synthesis procedure, it is convenient to choose the lcd of the fractional powers in this function equal to the lcd of the powers in the fractional order process transfer function. Correspondingly, also the lcd of the powers in the resulting fractional order controller transfer function obtained from (3.62) via (3.3) will be the same.

3.7.1 Example 4

Let the fractional order transfer function of a given process be

$$\widehat{G}_p(s) = \frac{1}{1 + 10s^{0.8}} \tag{3.63}$$

whose response to the input $\hat{u}(t)$ with Laplace transform

$$\widehat{U}(s) = 1/s^{0.8} \tag{3.64}$$

(see Fig. 3.4) is shown in Fig. 3.9 (solid line). Assume that it is desired to speed up the response by resorting to a unity-feedback control system. To this purpose, the complementary sensitivity function of such a feedback system is chosen to be

$$\widehat{T}(s) = \frac{1}{1 + s^{0.8}} \tag{3.65}$$

whose response to $\hat{u}(t)$ is compared to that of $\widehat{G}_p(s)$ in Fig. 3.9.

This choice also ensures that the response to the aforementioned input contains an input-dependent component equal to the input itself because $\widehat{Y}(s) = \widehat{T}(s)\widehat{U}(s)$ can be decomposed as

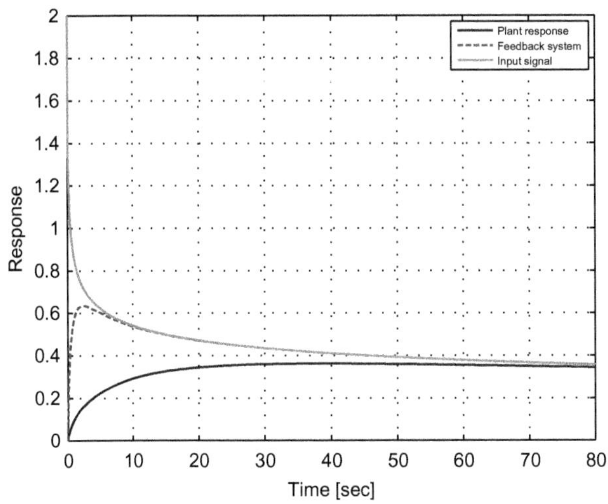

FIGURE 3.9 Responses to input (3.64) of: (i) the plant (3.63) (blue lower solid line), and (ii) the feedback control system (3.65) (red dashed line). The input (3.64) is represented by the green upper solid line.

$$\widehat{Y}(s) = -\frac{1}{1+s^{0.8}} + \frac{1}{s^{0.8}},$$

so that it tends quickly to the chosen reference input.

By the change of variable $w = s^{0.1}$, $\widehat{G}_p(s)$ is transformed into

$$G_p(w) = \frac{1}{1+10w^8} \tag{3.66}$$

and $\widehat{T}(s)$ into

$$T(w) = \frac{1}{1+w^8} \tag{3.67}$$

so that, according to (3.62), the controller transfer function in w-domain turns out to be

$$G_c(w) = \frac{1+10w^8}{w^8}, \tag{3.68}$$

whence

$$\widehat{G}_c(s) = \frac{1+10s^{0.8}}{s^{0.8}}. \tag{3.69}$$

As is expected, this controller contains an internal model of the input transform (Francis and Wonham, 1976).

3.8 CONCLUSIONS

It has been shown that the forced response of a fractional order system to an input belonging to a very numerous class can uniquely be decomposed into a system component and an input component, as is the case also for integer order systems. The first is characterized by the same modes as the system and the second by the same modes as the input. Therefore, if the system is asymptotically stable and the input is persistent, the input component corresponds to the steady-state or asymptotic response to the selected input whereas the system component corresponds to the transient response.

To ascertain whether the fractional order system is BIBO stable without computing numerically the system poles, resort can be made to the Routh–Hurwitz criteria for complex polynomials which allow us to determine the root distribution with respect to any straight line of the complex plane. On the basis of these criteria, simple stability and instability conditions have been provided that partly extend previous results of the same kind presented in the literature.

The response decomposition can be used in various contexts, ranging from system approximation to system analysis and synthesis. In particular, it has been applied to find a simplified model that retains the asymptotic behavior of the original system in the response to characteristic inputs, a result that most model-reduction techniques do not ensure even for integer order systems. As shown by some examples, in most cases the approximation of the corresponding system component is not appreciably affected by the requirement of steady-state retention, which only entails a usually small increase of the reduced model order depending on the input complexity.

Future research directions along the same lines include: (1) the characterization of the transient output component in the response to suitable inputs; (2) the synthesis of feedback controllers that ensure the desired asymptotic behavior; and (3) the extension of the decomposition procedure to systems of nonrational order.

REFERENCES

Agashe, S., 1985. A new general Routh–like algorithm to determine the number of RHP roots of a real or complex polynomial. IEEE Trans. Automat. Contr 30 (4), 406–409.

Ahmed, E., El–Sayed, A.M.A., El–Saka, H.A. A., 2006. On some Routh–Hurwitz conditions for fractional order differential equations and their applications in Lorenz, Rössler, Chua and Chen Systems. Phys. Lett. A 358 (1), 1–4.

Akritas, A.G., Strzeboński, A.W., Vigklas, P.S., 2008. Improving the performance of the continued fractions method using new bounds of positive roots. Nonlinear Anal. Model. Control 13 (3), 265–279.

Anderson, B.D.O., Bose, N.K., Jury, E.I., 1974. A simple test for zeros of a complex polynomial in a sector. IEEE Trans. Automat. Contr. 19 (4), 437–438.

Antsaklis, P.J., Michel, A.N., 2006. Linear Systems. Birkhäuser, Boston, MA.

Astolfi, A., 2010. Model reduction by moment matching for linear and nonlinear systems. IEEE Trans. Automat. Contr. 55 (10), 2321–2336.

Azar, A.T., Vaidyanathan, S., Ouannas, A., 2017. Fractional Order Control and Synchronization of Chaotic Systems. Springer, Cham, Switzerland.

Benidir, M., Picinbono, B., 1991. The extended Routh's table in the complex case. IEEE Trans. Automat. Contr. 36 (2), 253–256.

Billarz, H., 1944. Bemerkung zu einem Staze von Hurwitz. Zeitschrif. für Mathematik and Mechanik 24, 77–82.

Bistritz, Y., 2013. Optimal Fraction–free routh tests for complex and real integer polynomials. IEEE Trans. Circuits Systems I: Regular Papers 60 (9), 2453–2464.

Brunetti, M., 2014. Old and new proofs of cramer's rule. Appl. Math. Sci. 8 (133), 6689–6697.

Busłowicz, M., 2008. Stability of linear continuous-time fractional order systems with delays of the retarded type. Bull. Polish Acad. Sci. 56 (4), 319–324.

Caponetto, R., Dongola, G., Fortuna, L., Petráš, I., 2010. Fractional Order Systems – Modelling and Control Applications. World Scientific, New Jersey, USA.

Special issue: theory and applications of fractional order systems 2016. In: Caponetto, R., Trujillo, J.J., Tenreiro Machado, J.A. (Eds.), Math. Probl. Eng.

Casagrande, D., Krajewski, W., Viaro, U., 2017. On the asymptotic accuracy of reduced–order models. Int. J. Control, Automation Systems. Available from: https://doi.org/10.1007/s12555-015-0443-y.

Chen, C., Davenport, J.H., May, J.P., Maza, M.M., Xia, B., Xiao, R., 2013. Triangular Decomposition of Semi-algebraic systems. J. Symbolic Comput. 49, 3–26.

Chen, S.S., Tsai, J.S.H., 1993. A new tabular form for determining root distribution of a complex polynomial with respect to the imaginary axis. IEEE Trans. Automat. Contr. 38 (10), 1536–1541.

Dorato, P., Lepschy, A.M., Viaro, U., 1994. Some comments on steady–state and asymptotic responses. IEEE Trans. Education 37 (3), 264–268.

Doyle, J., Francis, B., Tannenbaum, A., 1992. Feedback Control Theory. Macmillan, New York, USA.

Ferrante, A., Lepschy, A.M., Viaro, U., 2000. *Introduzione ai controlli automatici* [*Introduction to Automatic Control*]. UTET, Torino, Italy.

Francis, B.A., Wonham, W.M., 1976. The internal model principle of control theory. Automatica 12 (5), 457–465.

Frank, E., 1946. On the zeros of polynomials with complex coefficients. Bull. Amer. Math. Soc. 52 (2), 144–157.

Gallivan, K., Grimme, E., Van Dooren, P., 1996. A rational lanczos algorithm for model reduction. Numer. Algor. 12 (1), 33–63.

Glover, K., 1984. All optimal hankel–norm approximations of linear multivariable systems and their L^{∞} error bounds. Int. J. Control 39 (6), 1145–1193.

Graham, A., 1981. Kronecker Products and Matrix Calculus With Applications. Ellis Horwood, Chichester, UK.

Gutman, S., 1979. Root clustering of a complex matrix in an algebraic region. IEEE Trans. Automat. Contr. 24 (4), 647–650.

Gutman, S., Jury, E.I., 1981. A general theory for matrix root clustering in subregions of complex plane. IEEE Trans. Automat. Contr. 26 (4), 853–863.

Henrion, D., 1998. Reliable algorithms for polynomial matrices. Ph.D. thesis, Institute of Information Theory and Automation, Czech Academy of Sciences, Prague, Czech Republic.

Hirst, H.P., Macey, W.T., 1997. Bounding the roots of polynomials. College Math. J. 28 (4), 292–295.

Hurwitz, A., 1895. Über die Bedingungen, unter welchen eine Gleichung nur Wurzeln mit negativen reellen Teilen besitzt. Math. Annalen Leipzig 46, 273–284.

Hwang, H.H., Tripathi, P.C., 1970. Generalisation of the Routh-Hurwitz criterion and its applications. Electr. Lett. 6 (13), 410–411.

Ionescu, C., Zhou, Y., Tenreiro Machado, J.A., eds., 2016. Special issue: advances in fractional dynamics and control. J. Vibrat. Control 22(8).

Jakubowska, A., Walczak, J., 2016. Analysis of the transient state in a series circuit of the class $RL_{\beta}C_{\alpha}$. Circuits Systems Signal Processing 35 (6), 1831–1853.

Jenkins, M.A., Traub, J.F., 1970. A three–stage variables–shift iteration for polynomial zeros and its relation to generalized rayleigh iteration. Numer. Math. 14 (3), 252–263.

Jia, Y.B., 2016. Roots of polynomials. Com S 477/577 Notes, Department of Computer Science, Iowa State Univ., Ames, IA, USA (http://web.cs.iastate.edu/cs577/handouts/polyroots.pdf).

Jiang, Y.L., Xiao, Z.H., 2015. Arnoldi–based model reduction for fractional order linear systems. Int. J. Systems Sci 46 (8), 1411–1420.

Kaczorek, T., 2011. Selected Problems of Fractional Systems Theory. Springer, Berlin, Germany.

Kaminski, J.Y., Shorten, R., Zeheb, E., 2015. Exact stability test and stabilization for fractional systems. Systems Control Lett. 85, 95–99.

Kesarkar, A.A., Selvaganesan Narayanasamy, N., 2016. Asymptotic magnitude bode plots of fractional–order transfer functions. IEEE/CAA J. Automatica Sinica (99), 1–8. Available from: https://doi.org/10.1109/JAS.2016.7510196. PP.

Krajewski, W., Viaro, U., 2014. A method for the integer-order approximation of fractional-order systems. J. Franklin Inst. 351 (1), 555–564.
Kučera, V., 1993. Diophantine equations in control - a survey. Automatica 29 (6), 1361–1375.
Levkovich, A., Cohen, N., Zeheb, E., 1996. A root–distribution criterion for an interval polynomial in a sector. IMA J. Math. Control Inf. 13 (4), 321–333.
Liang, S., Wang, S.G., Wang, Y., 2017. Routh–type table test for zero distribution of polynomials with commensurate fractional and integer degrees. J. Franklin Inst. 354 (1), 83–104.
Lüthi, A., 1942. Damping conditions for regulator equations of any order. Escher Wyss News 15-16, 90–95.
Mikhailov, A.V., 1938. The methods of harmonic analysis in the theory of control. Avtomat. i Telemekh 3, 27–81.
Monje, C.A., Chen, Y.Q., Vinagre, B.M., Xue, D., Feliu–Batlle, V., 2010. Fractional–Order Systems and Controls – Fundamentals and Applications. Springer, London, UK.
Petráš, I., 2009. Stability of fractional–order systems with rational orders: a survey. Fract. Calculus Appl. Anal. 12 (3), 269–298.
Psychalinos, C., Elwakil, A.S., Radwan, A.G., Biswas, K., eds., 2016. Special issue: fractional-order circuits and systems: theory, design, and applications. *Circuits Systems Signal Processing* 35(6):1807–2281.
Radwan, A.G., Soliman, A.M., Elwakil, A.S., Sedeek, A., 2009. On the stability of linear systems with fractional order elements. Chaos Solitons Fractals 40 (5), 2317–2328.
Rivero, M., Rogosin, S.V., Tenreiro Machado, J.A., Trujillo, J.J., 2013. Stability of fractional order systems. Math. Probl. Eng. . Available from: https://doi.org/10.1155/2013/356215Article ID 356215, 14 pages.
Routh, E.J., 1877. A Treatise on the Stability of a Given State of Motion, Particularly Steady Motion. Macmillan, London, UK.
Sabatier, J., Farges, C., Trigeassou, J.C., 2014. Fractional systems state space description: some wrong ideas and proposed solutions. J. Vibrat.Control 20 (7), 1076–1084.
Semary, M.S., Radwan, A.G., Hassan, H.N., 2016. Fundamentals of fractional–order LTI circuits and systems: number of poles, stability, time and frequency responses. Int. J. Circuit Theory Applicat. 44 (12), 2114–2133.
Sivanandam, S.N., Sreekala, K., 2012. An algebraic approach for stability analysis of linear systems with complex coefficients. Int. J. Computer Applicat. 44 (3), 13–16.
Tavakoli–Kakhki, M., Haeri, M., 2009. Model reduction in commensurate fractional–order linear systems. Proc. Inst. Mech. Eng. Part I: J. Syst. Control Eng. 223 (4), 493–505.
Tavazoei, M.S., 2011. On monotonic and nonmonotonic step responses in fractional order systems. IEEE Trans. Circuits Systems II: Express Briefs 58 (7), 447-ĂŞ451.
Tavazoei, M.S., 2016. Criteria for response monotonicity preserving in approximation of fractional order systems. IEEE/CAA J. Automatica Sinica 3 (4), 422–429.
Tempo, R., 1989. A Simple Test for Schur Stability of a Diamond of Complex Polynomials. In: Proc. 28th Conf. Decision Control. Tampa, FL, U.S.A., pp. 1892-1895.
Tepljakov, A., 2017. Fractional–Order Modeling and Control of Dynamic Systems. Springer, Cham, Switzerland.
Tewari, A., 2011. Automatic Control of Atmospheric and Space Flight Vehicles. Birkhäuser, Basel, Switzerland.
Trigeassou, J.C., Maamri, N., Sabatier, J., Oustaloup, A., 2012. Transients of fractional–order integrator and derivatives. Signal Image Video Proces. 6 (3), 359–372.
Trzaska, Z., 2008. Fractional-order systems: their properties and applications. Elektronika 49 (10), 137–144.

Usher Jr., T., 1957. A new application of the Hurwitz–Routh stability criteria. Trans. Amer. Inst. Electr. Eng. Part I: Commun. Electr. 76 (5), 530–533.

Valerio, D., Trujillo, J.J., Rivero, M., Tenreiro Machado, J.A., Baleanu, D., 2013. Fractional calculus: a survey of useful formulas. Europ. Phys. J. Special Topics 222 (8), 1827–1846.

Xue, D., Chen, Y.Q., 2007. Suboptimum H_2 pseudo-rational approximations to fractional–order linear time invariant systems. In: Sabatier, J., Agrawal, O.P., Machado, J.A.T. (Eds.), Advances in Fractional Calculus: Theoretical Developments and Applications in Physics and Engineering. Springer, Dordrecht, The Netherlands, pp. 61–75.

Yedavalli, R.K., 2014. Robust Control of Uncertain Dynamic Systems. Springer, New York, USA.

Chapter 4

Exact Solution of Linear Fractional Distributed Order Systems With Exponential Order Weight Functions

Hamed Taghavian and Mohammad Saleh Tavazoei
Sharif University of Technology, Tehran, Iran

4.1 INTRODUCTION

Fractional calculus is now omnipresent in different realms of science and technology, such as in control systems theory (Azar et al., 2017; Dumlu and Erenturk, 2014), signal processing (Tavazoei, 2015; Aslam and Raja, 2015), and differential equations (Cao et al., 2016; Bekir et al., 2017; Jin et al., 2016). In fact fractional differential operators help in modeling physical processes with long-lasting memory and anamolous behavior. As famous instances of fractional differential equations, one could refer to anomolous diffusion equations (Lv and Xu, 2016; Vabishchevich, 2016) describing the diffusion behavior occurring mainly in biological systems and fractional relaxation equations (Garrappa et al., 2014; Garra et al., 2014) describing the delayed reaction of a system to matter condensation.

Nowadays fractional operators have undergone an even further generalization by the advent of distributed order operators. It has been shown that using these operators is essential for modeling some physical phenomena in an accurate way, among which one could refer to ultraslow diffusion equations whose square displacement growth occurs at a logarithmic rate and distributed order space diffusion equations describing accelerating superdiffusion phenomena (Chechkin et al., 2002). In fact alongside diffusion equations (Garra et al., 2014; Chechkin et al., 2002; Sandev et al., 2015; Mainardi et al., 2008; Chechkin et al., 2003), differential equations involving derivatives of distributed order are encountered in different physical problems, including wave equations (Gorenflo et al., 2013) and relaxation phenomena (Meerschaert and Toaldo, 2015). In addition, distributed order

Mathematical Techniques of Fractional Order Systems. DOI: https://doi.org/10.1016/B978-0-12-813592-1.00004-0

systems of differential equations are getting more attention in engineering as well, due to the fact that their behavior is sometimes irreplaceable by common fractional or integer order dynamics. For instance, one could refer to distributed order filters which are studied in Chapter 4 of Jiao et al. (2012). Distributed order operators have also been successfully incorporated in PID controllers (Zhou et al., 2013; Jakovljevic et al., 2014) leading to an improvement of modeling of uncertainties tolerance and lead-lag compensators (Li et al., 2010). This, along with the fact that nowadays more physical phenomena are being modeled by using distributed order operators (Lazović et al., 2014; Saxena et al., 2014; Petrovic et al., 2015; Caputo and Carcione, 2013; Su, 2012), more attention is paid to differential equations of distributed order in the literature in the recent years. The major part of the literature however attends numerical methods for solving distributed order differential equations (Hu et al., 2016; Ye et al., 2015; Gao et al., 2015; Li and Wu, 2016; Gao and Sun, 2016), whereas there are less analytic approaches towards the solution of these equations (Kochubei, 2009; Naber, 2004; Mainardi and Pagnini, 2007). Thus, the analytical solution of distributed order equations is the main concern of the present chapter, where we focus on the case of exponential weight functions and present the exact solution of a system of linear time invariant (LTI) distributed order differential equations associated with such weight functions. It shall be remarked that the exact solution of a system of multiple distributed order differential equations in the time domain has been rarely discussed before. Of course the conventional Laplace transform approach used for the scalar case (with a single differential equation and a single unknown variable) in the literature may be still extendable to the problem here which would eventually lead to an integral representation of the solution through the Mellin's inverse formula. However in this chapter, we consider the problem as two separate Volterra integral equations of the second type and use convolution calculus to tackle the problem. This leads to a series type representation of the solution which is applicable to both scalar and matrix cases of the problem. Besides, some new results in calculation of convolution powers show up by the unprecedented use of convolution calculus for this problem. Moreover, as a special case, a relatively simple expression of the solution is presented for the equations with a negative scalar as the dynamic matrix. The expression presented here involves Gamma functions and is different from the regular Laplace-type integral expression derived by using Laplace transform in Naber (2004). Investigation of the role of fractional calculus special functions in distributed order differential equations has previously intrigued scientists seeking to construct a clearer relationship between fractional calculus and its extension, i.e., distributed order calculus (Mainardi and Pagnini, 2007). This new simple representation contributes to this problem to some extent by providing the exact solution of a single-term distributed order differential equation with a constant coefficient which describes anomalous nonexponential relaxation processes (Kochubei, 2009) for exponential weight functions.

By the advent of distributed order elements, distributed order calculus is hoped to soon appear in different problem contexts by opening paths in various engineering fields, such as electrical circuits, signal processing, and control systems. Therefore, the results of this chapter provide a small step towards initial contributions in studying the behaviors of various systems containing such elements by presenting the exact solution of a class of systems of distributed order differential equations. In order to emphasize this notion, a simple electrical circuit containing a distributed order capacitor is considered in this chapter whose exact response in the time domain is immediately obtained by using the results of the chapter. In addition, stability analysis comes first in studying an emerging system or designing a circuit performing a particular task which could be controlling a wheeled robot for instance. In this respect, a stability discussion is provided at the end of this chapter giving some insights about the poles of distributed order systems, stability and the relationship that exists between them. Some important results on this matter are brought from the literature which introduces a general graphical approach towards the stability problem of distributed order systems. Then it is shown how stability analysis of distributed order systems considered in this chapter shares some common traits with that of retarded systems with time delays. In particular, it is shown that the characteristic equations of both systems are made up of Lambert W functions and both generate infinitely many poles on the complex plane. It is also noticed that both sets of poles are located on the left side of a vertical line. This is actually an important property to observe, since it allows one to determine the stability of the systems by merely checking the location of characteristic equations roots associated with them.

The rest of this chapter is organized into six sections. At first, we provide a review on distributed order calculus and some other essential preliminaries in Section 4.2. The main results are then presented in Section 4.3, which are followed by a numerical simulation in Section 4.4 to test the obtained results. In addition, an electrical circuit with a distributed order element is presented in Section 4.5 to give some insights about the physical applications of the results. Stability analysis of LTI distributed order systems is briefly discussed in Section 4.6 and the chapter is finally concluded in Section 4.7. We end this section by providing a table containing definitions of all the symbols used in the chapter for the reader's convenience (Table 4.1).

4.2 PRELIMINARIES

This section starts with some essential definitions and continues with a brief review on fractional order and distributed order operators, which will come in useful in deriving the main results in the next section. We commence by

TABLE 4.1 Symbols Used in the Chapter Alongside Their Definitions

Symbol	Definition	Symbol	Definition
$E_1(.)$	Exponential integral function	${}_0I_t^{\alpha}$	Fractional integral
$H(.)$	Heaviside function	${}_0^CD_t^{w(\alpha)}$	Fractional derivative of Caputo type
$lW_k(.)$	The kth branch of Lambert W function	${}_0^CD_t^{w(\alpha)}$	Distributed order derivative of Caputo type
$\Gamma(.)$	Gamma function	${}_0J_t^{w(\alpha)}$	Distributed order integral
$E_{\alpha,\beta}(.)$	Mittag–Leffler function	$L_{t\to s}\{.\}$	Laplace transform
$f^{*k}(.)$	The kth convolution power of function $f(.)$	$L_{s\to t}^{-1}\{.\}$	Laplace inverse transform

recalling definitions of some special functions used in this chapter. Exponential integral function is defined as (Bell, 2004, p. 218)

$$E_1(t) = \int_t^{+\infty} \frac{e^{-\tau}}{\tau} d\tau, \quad t > 0 \tag{4.1}$$

The lower branch of Lambert W function (Banwell and Jayakumar, 2000) is the function $y = lW_{-1}(x)$ satisfying

$$ye^y = x \tag{4.2}$$

for $y < -1$. Heaviside step function is defined as

$$H(t) = \begin{cases} 1, & t > 0 \\ 0, & t < 0 \end{cases} \tag{4.3}$$

Gamma function is presented by (Podlubny, 1998, p. 1)

$$\Gamma(z) = \int_0^{+\infty} e^{-\tau} \tau^{z-1} d\tau \tag{4.4}$$

for $z \in \mathrm{C}, Re\{z\} > 0$. Gamma function can also be defined on the whole complex plane except nonpositive integers by mathematical continuation (Podlubny, 1998, Chapter 1). Let $s = re^{i\theta}$ $(r > 0)$ be an arbitrary complex number with $-\pi < \theta \leq \pi$. We denote the principal value of complex logarithm function by

$$\ln s = \ln|r| + i\theta \tag{4.5}$$

Let $f(t)$:$\mathrm{R}^{>0} \rightarrow \mathrm{R}$. We denote iterated self-convolution of function $f(t)$ by

$$f^{*k}(t) = \int_0^t f(\tau) f^{*(k-1)}(t-\tau) d\tau, \qquad k \in \mathrm{N} \tag{4.6}$$

and for the initial function we define $f^{*0}(t) = \delta(t)$. Definition (4.6) is sometimes referred to as the convolution power of function $f(t)$. Fractional integral operator is defined by the convolution integral (Podlubny, 1998, p. 65)

$$_0I_t^{\alpha} f(t) = \int_0^t \frac{(t-\tau)^{\alpha-1}}{\Gamma(\alpha)} f(\tau) d\tau \tag{4.7}$$

The fractional derivatives considered in the present chapter is in the Caputo sense, which is defined as (Podlubny, 1998, p. 78)

$$_0^cD_t^{\alpha} f(t) = \begin{cases} f(t), & \alpha = 0 \\ _0I_t^{1-\alpha} \dot{f}(t), & 0 < \alpha < 1 \\ \dot{f}(t), & \alpha = 1 \end{cases} \tag{4.8}$$

Caputo also introduced the distributed order differential operator, which is in fact defined by a weighted integration of fractional order differential operator over the order of differentiation. This definition generalizes common fractional order derivatives and is represented by (Jiao et al., 2012, p. 6)

$$_0^cD_t^{w(\alpha)} x(t) = \int_0^1 w(\alpha) {}_0^cD_t^{\alpha} x(t) d\alpha \tag{4.9}$$

In the above definition, $w(\alpha)$ $(\alpha \in [0, 1])$ denotes the weight function. In this chapter, an exponential weight function is considered, though the methodology and some of the results and procedures could be applied to more generally defined weight functions equally well. As a matter of fact, a generalized calculus has emerged by introduction of definition (4.9), which is called distributed order calculus. Prior to any attempt at investigation of the problems that this calculus potentially has to offer, it is needed to review distributed order integrals and derivatives properties more closely. In order to start, let us write (4.9) in the Laplace domain as

$$L_{t\rightarrow s}\left\{ {}_0^cD_t^{w(\alpha)} x(t) \right\} = \int_0^{+\infty} \hat{w}(\alpha) s^{\alpha} X(s) d\alpha - \int_0^{+\infty} \hat{w}(\alpha) s^{\alpha-1} x(0) d\alpha \tag{4.10}$$

where $L_{t\rightarrow s}\{.\}$ is the Laplace transform operator, $X(s)$ is the Laplace transform of $x(t)$, and

$$\hat{w}(\alpha) = \begin{cases} w(\alpha) & \alpha \in [0, 1] \\ 0, & \alpha \notin [0, 1] \end{cases} \tag{4.11}$$

Rewriting (4.10) in the form

$$L_{t\rightarrow s}\left\{ {}_0^cD_t^{w(\alpha)} x(t) \right\} = \int_0^{+\infty} \hat{w}(\alpha) e^{\alpha \ln(s)} X(s) d\alpha - \frac{x(0)}{s} \int_0^{+\infty} \hat{w}(\alpha) e^{\alpha \ln(s)} d\alpha \tag{4.12}$$

gives

$$L_{t\to s}\left\{{}_{0}^{c}D_{t}^{w(\alpha)}x(t)\right\} = W(-\ln s)(X(s) - x(0)/s) \tag{4.13}$$

where $W(s)$ is the Laplace transform of $\hat{w}(t)$. Thus, the impulse response of distributed order derivative operator (4.9) is given by

$$d(t) = L_{s\to t}^{-1}\left\{W(-\ln s)\right\}, \qquad t>0, \tag{4.14}$$

in which $L_{s\to t}^{-1}\{.\}$ denotes the inverse Laplace transform operator. Now, let us define the distributed order integrator as

$${}_{0}J_{t}^{w(\alpha)}x(t) = x(t)^{*}i(t) \tag{4.15}$$

in which

$$i(t) = L_{s\to t}^{-1}\left\{1/W(-\ln s)\right\} \tag{4.16}$$

and * denotes the convolution operator. As it can be seen, the impulse response of distributed order integrator given by (4.16) is the convolution inverse of that of the corresponding distributed order differentiator (4.14). By definition, a linear time-invariant system of differential equations of distributed order is represented by (Jiao et al., 2012, p. 11)

$${}_{0}^{c}D_{t}^{w(\alpha)}x(t) = Ax(t) + Bu(t), \tag{4.17}$$

in which $x(t)\in R^{n}$ is called the pseudo state and $u(t)\in R$ denotes the input function of the system. Also, $A\in R^{n\times n}$ is a constant matrix called the dynamic matrix and $B\in R^{n\times 1}$ is the input vector. The aim of this chapter is to present the analytical solution of this system in the case that the associated weight function is in exponential form, i.e., $w(\alpha) = ca^{\alpha}, \alpha\in[0, 1]$ in which $a\in R^{>0}$ and $c\in R - \{0\}$.

Lemma 1: Let c be a real constant. Solution of (4.17) with the weight function $w(\alpha) = cw_{1}(\alpha)$, dynamic matrix $A = A_{1}$ and input function $u(t) = u_{1}(t)$ is equal to the solution of (4.17) with the weight function $w(\alpha) = w_{1}(\alpha)$, dynamic matrix $A = A_{1}/c$, and input function $u(t) = u_{1}(t)/c$.

Proof: This Lemma is immediately proved by inserting the definition of operator ${}_{0}^{c}D_{t}^{w(\alpha)}$ from (4.9) in (4.17).

We conclude this section by recalling a Laplace pair which will be used later to derive the main results in this chapter.

Lemma 2: (Erdelyi et al., 1954, p. 144): The following Laplace pair holds true

$$L_{t\to s}\left\{\frac{e^{t}t^{k-1}}{(k-1)!}\right\} = \frac{1}{(s-1)^{k}} \tag{4.18}$$

4.3 MAIN RESULTS

The cornerstone of the analytical solution of (4.17) is in fact the impulse response of the integrator with the same weight function as in the differentiator involved in this equation. This is similar to the case of fractional order and integer order systems. Recall that the analytical solution of the fractional order system of linear differential equations ${}_0^cD_t^{\alpha}x(t) = Ax(t) + Bu(t)$ is written by using Mittag–Leffler functions $E_{\alpha,\beta}(t^{\alpha}) = \sum_{k=0}^{+\infty} \frac{t^{\alpha k}}{\Gamma(\alpha k + \beta)}$. Mittag–Leffler functions are constructed on the basis of iterated self-convolutions of fractional power function, i.e., $t^{\alpha-1}/\Gamma(\alpha)$ which is the impulse response of fractional integration operator (4.7). A general form of the analytical solution is obtained in a similar way for distributed order equations. This idea is in fact analogous to the analytic approach to integral equations using resolvent kernels. Using this concept here makes sense due to the fact that it is always possible to turn Eq. (4.17) into the following two Volterra integral equations of the second kind

$$\begin{cases} x_1 - Ax_1{*}i = x(0) \\ x_2 - Ax_2{*}i = Bu{*}i \end{cases}, \quad x(t) = x_1(t) + x_2(t) \tag{4.19}$$

in which $i(t)$ is defined in (4.16). In order to show this, let us take the Laplace transform of (4.19) which yields

$$\begin{cases} X_1(s) - AX_1(s)I(s) = x(0)/s \\ X_2(s) - AX_2(s)I(s) = BU(s)I(s) \end{cases} \tag{4.20}$$

Dividing each relation by $I(s)$ and replacing $W(-\ln s) = 1/I(s)$ afterwards results

$$\begin{cases} X_1(s)W(-\ln s) - AX_1(s) = x(0)W(-\ln s)/s \\ X_2(s)W(-\ln s) - AX_2(s) = BU(s) \end{cases} \tag{4.21}$$

Solving the resultant equations for $X_1(s)$ and $X_2(s)$ gives

$$\begin{cases} X_1(s) = (W(-\ln s)I - A)^{-1}x(0)W(-\ln s)/s \\ X_2(s) = (W(-\ln s)I - A)^{-1}BU(s) \end{cases} \tag{4.22}$$

In which I is the identity matrix with appropriate dimensions. The solution of (4.17) is then obtained by adding the homogeneous and particular solutions $X(s) = X_1(s) + X_2(s)$ as

$$X(s) = (W(-\ln s)I - A)^{-1}\left(x(0)W(-\ln s)/s + BU(s)\right) \tag{4.23}$$

Multiplying each side of (4.23) by $W(-\ln s)I - A$ yields

$$(W(-\ln s)I - A)X(s) = x(0)W(-\ln s)/s + BU(s) \tag{4.24}$$

By rewriting (4.24) in the form

$$W(-\ln s)\big(X(s)-x(0)/s\big)=AX(s)+BU(s) \tag{4.25}$$

and taking the inverse Laplace transform afterwards reveals the fact that $x(t)=x_1(t)+x_2(t)$ satisfies (4.17). Eq. (4.23) indicates that the solution of distributed order systems in the Laplace domain may be obtained by simply substituting the Laplace variable s in the solution of their corresponding integer order systems with $W(-\ln s)=\int_0^1 w(\alpha)s^\alpha d\alpha$. In the following Lemma the impulse response of distributed order integrator with a unitary weight function is presented. This result will be used to derive the analytical solution of (4.17) later.

Lemma 3: Impulse response of distributed order integrator with the unitary weight function $w(\alpha)=1$ ($\alpha\in[0,1]$) is given by

$$i(t)=e^tE_1(t),\qquad t>0 \tag{4.26}$$

in which $E_1(t)$ is the exponential integral function as defined in (4.1).

Proof: Noting the Laplace pair $L_{t\to s}\{E_1(t)\}=\frac{\ln(s+1)}{s}$ which is presented in Erdelyi et al. (1954, p. 178), and the fact that the Laplace transform of $i(t)$ is given by $I(s)=1/\int_0^1 s^\alpha d\alpha=\frac{\ln s}{s-1}$ in the case of a unitary weight function, the proof is easily obtained via the frequency shift property of the Laplace transform.

Lemma 4: Let $lW_{-1}(.)$ denote the lower branch of Lambert W function defined by (4.2) and define

$$r_0=-\lambda lW_{-1}\left(-e^{-\pi-1/\lambda}/\lambda\right) \tag{4.27}$$

for some $\lambda>\frac{-1}{lW_{-1}(-e^{-\pi-1})}$. If $|s|>r_0$ then $\frac{|\ln s|}{|s-1|}<\frac{1}{\lambda}$.

Proof: Define $g(\lambda)=-\frac{e^{-\pi-\frac{1}{\lambda}}}{\lambda}$. It can be easily verified that this function has the minimum of $-e^{-\pi-1}$ occurring at $\lambda=1$. Therefore, the inequality

$$-e^{-\pi-1}\le-\frac{e^{-\pi-\frac{1}{\lambda}}}{\lambda}<0 \tag{4.28}$$

holds for $\lambda>0$. This, would result in

$$-\infty<lW_{-1}\left(-e^{-\pi-1/\lambda}/\lambda\right)\le lW_{-1}\left(-e^{-\pi-1}\right) \tag{4.29}$$

as $lW_{-1}(.)$ is strictly decreasing. Therefore, considering the condition $\lambda>\frac{-1}{lW_{-1}(-e^{-\pi-1})}$ from (4.27) and (4.28) one could write

$$1<r_0<+\infty \tag{4.30}$$

It is immediately followed by definition (4.2) that

$$r_0 e^{\frac{r_0}{-\lambda}} = e^{-\pi-\frac{1}{\lambda}} \tag{4.31}$$

which is equivalent to

$$e^{\frac{r_0}{\lambda}} = r_0 e^{\pi+\frac{1}{\lambda}} \tag{4.32}$$

Writing (4.32) in a logarithmic form and a little calculation gives

$$\underbrace{\frac{\ln r_0}{r_0 - 1}}_{f_1(r_0)} = \underbrace{\frac{1}{\lambda} - \frac{\pi}{r_0 - 1}}_{f_2(r_0)} \tag{4.33}$$

It can be verified that $f_1(r_0)$ ($f_2(r_0)$) is strictly decreasing (increasing) in the range $r_0 \in (1, +\infty)$. Therefore, for $r > r_0$ one could write

$$\frac{\ln r}{r-1} < \frac{1}{\lambda} - \frac{\pi}{r-1} \tag{4.34}$$

or equivalently

$$\frac{\ln r + \pi}{\sqrt{r^2 - 2r + 1}} < \frac{1}{\lambda} \tag{4.35}$$

Note that for any $\theta \in (-\pi, \pi]$ there holds $\frac{\ln r + \pi}{\sqrt{r^2 - 2\cos(\theta) r + 1}} < \frac{\ln r + \pi}{\sqrt{r^2 - 2r + 1}}$. Thus by replacing the left side of (4.35) with its lower bound we get

$$\frac{\ln r + \pi}{\sqrt{r^2 - 2\cos(\theta) r + 1}} < \frac{1}{\lambda} \tag{4.36}$$

It is possible to write the left side of (4.36) in the form

$$\frac{\sqrt{(\ln r + \pi)^2}}{\sqrt{(r\cos\theta - 1)^2 + (r\sin\theta)^2}} < \frac{1}{\lambda} \tag{4.37}$$

As $\ln r > 0$, it is obvious that $\sqrt{(\ln r)^2 + \pi^2} < \sqrt{(\ln r + \pi)^2}$. On the other hand, θ generally lies in the interval $(-\pi, \pi]$. Therefore, as a more conservative inequality, we obtain the following inequality from (4.37)

$$\frac{\sqrt{(\ln r)^2 + \theta^2}}{\sqrt{(r\cos\theta - 1)^2 + (r\sin\theta)^2}} < \frac{1}{\lambda} \tag{4.38}$$

Finally, writing (4.38) in terms of $s = re^{i\theta}$ gives

$$\frac{|\ln s|}{|s-1|} < \frac{1}{\lambda} \tag{4.39}$$

in which using the relation $\frac{s-1}{\ln s} = \int_0^1 s^\alpha d\alpha$ results in

$$\left|\int_0^1 s^\alpha d\alpha\right| > \lambda \tag{4.40}$$

Lemma 4: helps us to start constructing the new representation of the exact solution of (4.17) with the aid of the Laplace transform as shown in the next Lemma.

Lemma 5: Consider distributed order system of Eq. (4.17) with a unitary weight function, i.e., $w(\alpha)=1$ $(\alpha \in [0,1])$ and assume $\|A\| > \frac{-1}{lW_{-1}(-e^{-\pi-1})}$, where $\|.\|$ is any submultiplicative matrix norm satisfying the inequality $\|A_1A_2\| \le \|A_1\|\,\|A_2\|$ for any square matrices A_1 and A_2. The exact solution of (4.17) is given by

$$x(t) = \phi_1(t)x(0) + \int_0^t \phi_2(t-\tau)Bu'(\tau)d\tau \tag{4.41}$$

in which

$$\phi_1(t) = \int_0^t \sum_{k=0}^{+\infty} A^k e^\tau E_1^{*k}(\tau)d\tau \tag{4.42}$$

$$\phi_2(t) = \int_0^t \sum_{k=0}^{+\infty} A^k e^\tau E_1^{*(k+1)}(\tau)d\tau \tag{4.43}$$

$$u'(t) = L_{s\to t}^{-1}\{sU(s)\} \tag{4.44}$$

and $E_1(t)$ is the exponential integral function.

Proof: The Laplace transforms of (4.42) and (4.43) are given by

$$\Phi_1(s) = \frac{1}{s}\sum_{k=0}^{+\infty}\left(A\frac{\ln s}{s-1}\right)^k, \quad \Phi_2(s) = \frac{\ln s}{s(s-1)}\sum_{k=0}^{+\infty}\left(A\frac{\ln s}{s-1}\right)^k \tag{4.45}$$

Let $r_0 = -\|A\|\,lW_{-1}\left(-e^{-\pi-1/\|A\|}/\|A\|\right)$ in which $lW_{-1}(.)$ denote the lower branch of the Lambert W function. According to Lemma 4, provided that the condition $|s| > r_0$ is satisfied, the inequality $\frac{|\ln s|}{|s-1|} < \frac{1}{\|A\|}$ holds true. Therefore, the series involved in the both functions in (4.45) is convergent to

$$\sum_{k=0}^{+\infty}\left(A\frac{\ln s}{s-1}\right)^k = \left(I - A\frac{\ln s}{s-1}\right)^{-1} \tag{4.46}$$

in which I is the identity matrix of appropriate dimensions. Thus, functions (4.45) could also be written in the following form

$$\Phi_1(s) = \frac{1}{s}\left(I - A\frac{\ln s}{s-1}\right)^{-1}, \quad \Phi_2(s) = \frac{\ln s}{s(s-1)}\left(I - A\frac{\ln s}{s-1}\right)^{-1} \tag{4.47}$$

Taking the Laplace transform of Eq. (4.41) and substituting (4.47) in it gives

$$X(s) = \frac{1}{s}\left(I - A\frac{\ln s}{s-1}\right)^{-1} x(0) + \frac{\ln s}{s(s-1)}\left(I - A\frac{\ln s}{s-1}\right)^{-1} BsU(s) \tag{4.48}$$

By a little manipulation of (4.48), we obtain

$$X(s) = \frac{1}{s}\frac{s-1}{\ln s}\left(\frac{s-1}{\ln s}I - A\right)^{-1} x(0) + \left(\frac{s-1}{\ln s}I - A\right)^{-1} BU(s) \tag{4.49}$$

Left multiplying both sides of (4.49) by $\left(\frac{s-1}{\ln s}I - A\right)$ gives

$$\left(\frac{s-1}{\ln s}I - A\right)X(s) = \frac{1}{s}\frac{s-1}{\ln s}x(0) + BU(s) \tag{4.50}$$

Thereby

$$\left(\frac{s-1}{\ln s}\right)\left(X(s) - x(0)/s\right) = AX(s) + BU(s) \tag{4.51}$$

Taking the inverse Laplace transform of (4.51) results in Eq. (4.17) which proves this Lemma.

It is worth mentioning that the initial impulse resulted from the calculation of the derivative $u'(t)$ when $u(0) \neq 0$ in (4.45) is taken into account in the solution.

In order to derive $\phi_1(t)$ in a simpler form, define

$$p_k(t) = \int_0^t e^{\tau} E_1^{*k}(\tau) d\tau \tag{4.52}$$

By means of (4.52), it is possible to write $\phi_1(t)$ as

$$\phi_1(t) = \sum_{k=0}^{+\infty} A^k p_k(t) \tag{4.53}$$

Also, from (4.43) it is revealed that

$$\phi_2(t) = \sum_{k-1}^{+\infty} A^{k-1} p_k(t) \tag{4.54}$$

It is observed from (4.53) and (4.54) that obtaining closed form expressions of $\phi_1(t)$ and $\phi_2(t)$ is contingent on the calculation of function $p_k(t)$. In the following Lemma, a new Laplace pair is introduced which will be utilized to provide a method for calculation of $p_k(t)$ in Lemma 7.

Lemma 6: Define the lower triangular matrix

$$E = \begin{pmatrix} e_{1,1} & 0 & 0 & \cdots & 0 \\ e_{2,1} & e_{2,2} & 0 & \cdots & 0 \\ e_{3,1} & e_{3,2} & e_{3,3} & \cdots & 0 \\ \vdots & \vdots & \vdots & \ddots & \vdots \\ e_{N,1} & e_{N,2} & e_{N,3} & \cdots & e_{N,N} \end{pmatrix}, \quad e_{k,i} = \binom{k}{i}(-1)^i \Gamma^{(k-i)}(1) \tag{4.55}$$

in which $\Gamma^{(k-i)}(1)$ denotes the $(k-i)$-th derivative of Gamma function at point 1 and $N \geq k$ is any natural number. We have

$$\sum_{i=1}^{k} b_{k,i}\left(L_{t\to s}\{\ln^i t\} - \Gamma^{(i)}(1)/s\right) = \ln^k(s)/s \tag{4.56}$$

where $b_{k,i}$ are the elements of the matrix $E^{-1} = [b_{k,i}]$.

Proof: Consider the Laplace transform of the power function as

$$\int_0^{+\infty} t^a e^{-st} dt = \frac{\Gamma(a+1)}{s^{a+1}}, \qquad a > -1 \tag{4.57}$$

Inspired by Pahikkala (2013), we differentiate (4.57) with respect to a, which gives

$$\int_0^{+\infty} t^a \ln t e^{-st} dt = \frac{\Gamma^{(1)}(a+1) - \ln s \Gamma(a+1)}{s^{a+1}} \tag{4.58}$$

A mathematical induction indicates that for the k-th derivative one would have

$$\int_0^{+\infty} t^a (\ln t)^k e^{-st} dt = \frac{1}{s^{a+1}} \sum_{i=0}^{k} \binom{k}{i} \Gamma^{(i)}(a+1)(-\ln s)^{k-i} \tag{4.59}$$

In order to prove this, assume (4.59) holds true for some $k > 0$. Differentiating both sides of (4.59) with respect to a yields

$$\begin{aligned} & s^{a+1} \int_0^{+\infty} t^a (\ln t)^{k+1} e^{-st} dt = \\ & \sum_{i=0}^{k} \binom{k}{i} \Gamma^{(i+1)}(a+1)(-\ln s)^{k-i} + \sum_{i=0}^{k} \binom{k}{i} \Gamma^{(i)}(a+1)(-\ln s)^{k-i+1} \\ & = \sum_{i=1}^{k+1} \binom{k}{i-1} \Gamma^{(i)}(a+1)(-\ln s)^{k-i+1} + \\ & \sum_{i=0}^{k} \binom{k}{i} \Gamma^{(i)}(a+1)(-\ln s)^{k-i+1} \\ & = \Gamma^{(k+1)}(a+1) + \\ & \sum_{i=1}^{k} \left(\binom{k}{i-1} + \binom{k}{i}\right) \Gamma^{(i)}(a+1)(-\ln s)^{k-i+1} + \Gamma(a+1)(-\ln s)^{k+1} \end{aligned} \tag{4.60}$$

It can be shown that the relation $\binom{k}{i-1}+\binom{k}{i}=\binom{k+1}{i}$ holds true. By using this identity we obtain

$$\begin{aligned}&s^{a+1}\int_0^{+\infty} t^a(\ln t)^{k+1}e^{-st}dt=\\&\Gamma^{(k+1)}(a+1)+\sum_{i=1}^{k}\binom{k+1}{i}\Gamma^{(i)}(a+1)(-\ln s)^{k-i+1}+\\&\Gamma(a+1)(-\ln s)^{k+1}=\\&\sum_{i=0}^{k+1}\binom{k+1}{i}\Gamma^{(i)}(a+1)(-\ln s)^{k-i+1}\end{aligned}\tag{4.61}$$

This indicates that (4.59) holds true for $k+1$ too, which in turn proves the induction. Setting $a=0$ in expression (4.59) and a little calculation results in the following formula

$$L_{t\to s}\{(\ln t)^k\}=\sum_{i=0}^{k}\binom{k}{i}(-1)^i\Gamma^{(k-i)}(1)(\ln s)^i/s \tag{4.62}$$

or equivalently

$$L_{t\to s}\{(\ln t)^k\}-\Gamma^{(k)}(1)/s=\sum_{i=1}^{k}\binom{k}{i}(-1)^i\Gamma^{(k-i)}(1)(\ln s)^i/s \tag{4.63}$$

for calculation of the Laplace transform of the logarithmic functions with integer powers. By defining $y_k=\frac{(\ln s)^k}{s}$ and $c_k=L_{t\to s}\{(\ln t)^k\}-\Gamma^{(k)}(1)/s$ for $k=1,2,\ldots,N$ it is possible to write (4.63) in the form of a system of linear algebraic equations as

$$c=Ey \tag{4.64}$$

in which $y=\begin{pmatrix}y_1 & y_2 & \cdots & y_N\end{pmatrix}^T$, $c=\begin{pmatrix}c_1 & c_2 & \cdots & c_N\end{pmatrix}^T$ and E is defined by (4.55). Note that since $e_{kk}=(-1)^k$, matrix E is always nonsingular which makes it possible to uniquely solve (4.64) for y to give

$$y=E^{-1}c \tag{4.65}$$

From which the relation (4.56) is concluded.

Now Lemma 6 will be used in the next Lemma to obtain $p_k(t)$ in explicit form. On the basis of this result, we would be prepared to rewrite the solution of (4.17) accordingly.

Lemma 7: Let $p_k(t)$ and E be defined as in (4.52) and (4.55), respectively. We have

$$p_k(t)=\sum_{i=1}^{k}\int_0^t\frac{b_{k,i}}{(k-1)!}e^{\tau}\tau^{k-1}\left(\ln^i(t-\tau)-\Gamma^{(i)}(1)\right)d\tau \tag{4.66}$$

where $b_{k,i}$ denote the elements of matrix $E^{-1}=\left[b_{k,i}\right]$.

Proof: First of all, it is possible to write (4.66) in the form

$$p_k(t) = \sum_{i=1}^{k} \int_0^{+\infty} \frac{b_{k,i}}{(k-1)!} e^{\tau} \tau^{k-1} \left(\ln^i(t-\tau) - \Gamma^{(i)}(1)\right) H(t-\tau) d\tau \tag{4.67}$$

in which $H(.)$ denotes the Heaviside unit step function. Taking the Laplace transform of (4.67) and using the time shift property of the Laplace transform yields

$$\begin{aligned} P_k(s) &= \sum_{i=1}^{k} \int_0^{+\infty} \frac{b_{k,i}}{(k-1)!} e^{\tau} \tau^{k-1} e^{-s\tau} L_{t\to s}\left\{ \left(\ln^i(t) - \Gamma^{(i)}(1)\right) H(t) \right\} d\tau \\ &= \int_0^{+\infty} \frac{1}{(k-1)!} e^{\tau} \tau^{k-1} e^{-s\tau} \sum_{i=1}^{k} b_{k,i} \left(L_{t\to s}\{\ln^i(t)\} - \Gamma^{(i)}(1)/s \right) d\tau \end{aligned} \tag{4.68}$$

Using Lemma 6 gives

$$P_k(s) = \int_0^{+\infty} \frac{1}{(k-1)!} e^{\tau} \tau^{k-1} e^{-s\tau} \frac{(\ln s)^k}{s} d\tau = \frac{(\ln s)^k}{s} \int_0^{+\infty} \frac{e^{\tau} \tau^{k-1}}{(k-1)!} e^{-s\tau} d\tau \tag{4.69}$$

Note that the integral in the right side of (4.69) is actually the Laplace transform $L_{\tau\to s}\left\{\frac{e^{\tau}\tau^{k-1}}{(k-1)!}\right\}$. Therefore, using Lemma 2 gives

$$P_k(s) = \frac{(\ln s)^k}{s(s-1)^k} \tag{4.70}$$

On the other hand, note that taking the Laplace transform of $p_k(t)$ by directly using its definition (i.e., (4.52)) would result

$$\begin{aligned} &L_{t\to s}\left\{ \int_0^t e^{\tau} E_1^{*k}(\tau) d\tau \right\} = \\ &\frac{1}{s} L_{t\to s}\left\{ E_1^{*k}(t) \right\}_{|s\to s-1} = \frac{1}{s} \left(L_{t\to s}\{E_1(t)\} \right)^k_{|s\to s-1} = \frac{\ln^k(s)}{s(s-1)^k} \end{aligned} \tag{4.71}$$

By comparing (4.70) and (4.71), the proof of this Lemma is completed.

Prior to proceeding, it is worth mentioning that calculation of convolutions involving exponential integral functions are of interest in some papers (Fisher and Al-Sirehy, 2015; Geller and Ng, 1969). This problem is treated in the following Remark as a side result of Lemma 7, which concerns the k-fold convolution of exponential integral function with itself. This generalizes the work done in Geller and Ng (1969).

Remark 1: Let E be defined as in (4.55). From (4.52), it is revealed that $E_1^{*k}(t) = e^{-t}\frac{d}{dt}p_k(t)$. Thus, according to Lemma 7, it is deduced that

$$E_1^{*k}(t) = \sum_{i=1}^{k} \frac{b_{k,i}}{(k-1)!}\left(\int_0^t e^{\tau-t}\tau^{k-2}(\tau+k-1)\ln^i(t-\tau)d\tau - t^{k-1}\Gamma^{(i)}(1)\right) \tag{4.72}$$

where $k \geq 2$ and $b_{k,i}$ are the elements of matrix $E^{-1} = \left[b_{k,i}\right]$.

The results of the previous Lemmas are now gathered to obtain the exact solution of (4.17) in the time domain as stated in the following theorem.

Theorem 1: Assume $\|A\| > -1/lW_{-1}\left(-e^{-\pi-1}\right)$ holds where $lW_{-1}(.)$ denotes the lower branch of the Lambert W function. Then the exact solution of (4.17) with a unitary weight function $w(\alpha)=1, \alpha \in [0,1]$ is given by

$$\begin{aligned} x(t) &= \sum_{k=0}^{+\infty}\sum_{i=1}^{k}\int_0^t c_{k,i}A^k e^{\tau}\tau^{k-1}\left(\ln^i(t-\tau) - \Gamma^{(i)}(1)\right)d\tau x(0) \\ &+ \left(\sum_{k=1}^{+\infty}\sum_{i=1}^{k}\int_0^t c_{k,i}A^{k-1}e^{\tau}\tau^{k-1}\left(\ln^i(t-\tau) - \Gamma^{(i)}(1)\right)d\tau\right) * Bu'(t) \end{aligned} \tag{4.73}$$

in which

$$c_{k,i} = \frac{b_{k,i}}{(k-1)!} \tag{4.74}$$

Coefficients $b_{k,i}$ are the elements of the matrix $E^{-1} = \left[b_{k,i}\right]$ in which E is given by (4.55) and $u'(t)$ is the derivative of $u(t)$ defined by (4.44).

Proof: Substituting $p_k(t)$ obtained from Lemma 7 in (4.53) and (4.54) gives $\psi_1(t)$ and $\phi_2(t)$, respectively. Solution (4.74) is deduced by replacing these values in (4.41).

Theorem 1 can be immediately extended to the case in which the weight function $w(\alpha)$ is of exponential type. This is stated in the following corollary.

Corollary 1: Assume $\|A\| > -1/lW_{-1}\left(-e^{-\pi-1}\right)$ holds where $lW_{-1}(.)$ denotes the lower branch of Lambert W function. Then the exact solution of system (4.17) with exponential weight function $w(\alpha) = ca^{\alpha}$, $\alpha \in [0,1]$ in which $a \in \mathrm{R}^{>0}$ and $c \in \mathrm{R} - \{0\}$ is given by

$$\begin{aligned} x(t) &= \sum_{k=0}^{+\infty}\sum_{i=1}^{k}\int_0^{t/a} c_{k,i}\left(A\tau/c\right)^k \frac{e^{\tau}}{\tau}\left(\ln^i\left(t/a-\tau\right) - \Gamma^{(i)}(1)\right)d\tau x(0) \\ &+ \left(\sum_{k=1}^{+\infty}\sum_{i=1}^{k}\int_0^{t/a} \frac{c_{k,i}}{c}\left(A\tau/c\right)^{k-1}e^{\tau}\left(\ln^i\left(t/a-\tau\right) - \Gamma^{(i)}(1)\right)d\tau\right) * Bu'(t) \end{aligned} \tag{4.75}$$

in which $c_{k,i}$ is defined as (4.74) and $u'(t)$ is defined by (4.44).

Proof: Let $w_1(\alpha) = 1$, $\alpha \in [0, 1]$ be a unitary weight function and $w_2(\alpha) = a^\alpha$, $\alpha \in [0, 1]$ be an exponential weight function. At first, we derive the solution of (4.17) associated with the weight function $w(\alpha) = w_2(\alpha)$ and then prove this corollary by using the mutual relationship which holds between the two weight functions in Laplace domain. Considering $w(\alpha) = w_2(\alpha)$, taking the Laplace transform of (4.17) gives

$$W_2(-\ln s)X(s) - W_2(-\ln s)\frac{x(0)}{s} = AX(s) + BU(s) \tag{4.76}$$

Solving (4.76) for $X(s)$ results

$$X(s) = (W_2(-\ln s)I - A)^{-1}\frac{W_2(-\ln s)}{s}x(0) + (W_2(-\ln s)I - A)^{-1}BU(s) \tag{4.77}$$

Note that the relation $w_2(\alpha) = e^{\alpha \ln a} w_1(\alpha)$ holds between the two weight functions. From the frequency shift property of the Laplace transform we obtain

$$\begin{aligned} W_2(s) &= W_1(s - \ln a) \\ W_2(-\ln s) &= W_1(-\ln(as)) \end{aligned} \tag{4.78}$$

Thereby, rewriting (4.77) in terms of W_1 gives

$$\begin{aligned} X(s) = (W_1(-\ln as)I - A)^{-1}\frac{W_1(-\ln as)}{s}x(0) + \\ (W_1(-\ln as)I - A)^{-1}BU(s) \end{aligned} \tag{4.79}$$

Replacing $W_1(-\ln as) = \frac{as-1}{\ln as}$ in (4.79) and writing the resultant expression in terms of $\Phi_1(as)$ and $\Phi_2(as)$ which are derived from (4.47) yield

$$X(s) = a\Phi_1(as)x(0) + a\Phi_2(as)BsU(s) \tag{4.80}$$

By using the time scale property of the Laplace transform we obtain

$$x(t) = \phi_1(t/a)x(0) + \phi_2(t/a) * Bu'(t) \tag{4.81}$$

Note that the original weight function involves a scaling term, i.e., $w(\alpha) = cw_2(\alpha)$. Substituting $p_k(t)$ obtained from Lemma 7 in (4.53) and (4.54) gives $\phi_1(t)$ and $\phi_2(t)$, respectively. Therefore, replacing $\phi_1(t/a)$ and $\phi_2(t/a)$ with their expressions in (4.81) and using Lemma 1 afterwards, verifies the solution (4.75).

Consider the scalar differential equation

$${}_0^C D_t^{w(\alpha)} x(t) = -\lambda x(t), \quad \lambda > 0 \tag{4.82}$$

which is in fact a special case of the more general system (4.17) and therefore its solution is readily available from Corollary 1 in case of an exponential weight function. However, it is possible to obtain a simpler expression for the solution of (4.82) which is done in the next Theorem. This system describes anomalous nonexponential distributed order relaxation processes (Kochubei, 2009). The typical method used to treat (4.82) in the literature is through the Laplace transform. In this method, the solution is first derived in the Laplace domain. Then the Mellin's inverse formula is utilized to return the obtained solution back to time domain, which involves solving a complex integral (Naber, 2004). This approach leads to a solution expressed by a Laplace-type integral eventually (Naber, 2004; Mainardi and Pagnini, 2007). For instance, in case of a unitary weight function, substituting $w(\alpha)=1,\ \alpha\in[0,1]$ in the solution obtained by (Naber, 2004), gives

$$x(t)=x(0)\int_0^{+\infty}\frac{e^{-rt}\lambda(r+1)}{r\lambda^2\pi^2+r(r+1-\lambda\ln r)^2}dr \tag{4.83}$$

In the following Theorem we present an alternative representation of this solution expressed in terms of Gamma functions, which may be more coherent with the concept of distributed order differential equations as the generalization of the fractional order counterparts.

Theorem 2: Exact solution of (4.82) under the assumption $w(\alpha)=1,\ \alpha\in[0,1]$ is given by

$$x(t)=x(0)\int_0^t\left(\frac{(t-\tau)^{\lambda\tau-1}}{\Gamma(\lambda\tau)}-\frac{(t-\tau)^{\lambda\tau}}{\Gamma(\lambda\tau+1)}\right)e^{\tau}d\tau,\qquad t>0 \tag{4.84}$$

Proof: Let us define the auxiliary function

$$v(t)=\int_0^t\frac{(t-\tau)^{\lambda\tau}}{\Gamma(\lambda\tau+1)}e^{\tau}d\tau \tag{4.85}$$

In terms of (4.85) we will derive the actual solution of (4.82) as follows. At first, note that it is possible to write (4.85) in the form

$$v(t)=\int_0^{+\infty}\frac{(t-\tau)^{\lambda\tau}}{\Gamma(\lambda\tau+1)}e^{\tau}H(t-\tau)d\tau \tag{4.86}$$

in which $H(.)$ is the Heaviside unit step function. Using the variable change $\tau=-\ln\alpha$ in the integral (4.86) gives

$$v(t)=\int_0^1\frac{(t+\ln\alpha)^{-\lambda\ln\alpha}}{\Gamma(-\lambda\ln\alpha+1)}e^{-2\ln\alpha}H(t+\ln\alpha)d\alpha \tag{4.87}$$

Taking the Laplace transform of (4.87) and using the time shift property of the Laplace transform, we obtain

$$L_{t\to s}\{v(t)\} = \int_0^1 e^{-2\ln\alpha} L_{t\to s}\left\{\frac{(t+\ln\alpha)^{-\lambda\ln\alpha}}{\Gamma(-\lambda\ln\alpha+1)} H(t+\ln\alpha)\right\} d\alpha$$

$$= \int_0^1 e^{-2\ln\alpha} \frac{1}{s^{-\lambda\ln\alpha+1}} e^{s\ln\alpha} d\alpha \tag{4.88}$$

$$= \int_0^1 e^{-2\ln\alpha} e^{-(-\lambda\ln\alpha+1)\ln s} e^{s\ln\alpha} d\alpha$$

$$= \int_0^1 e^{(-2+\lambda\ln s+s)\ln\alpha} d\alpha/s$$

$$= \int_0^1 \alpha^{(-2+\lambda\ln s+s)} d\alpha/s$$

The resultant expression under the condition $Re\{s-2+\lambda\ln s\}>0$ equals

$$L_{t\to s}\{v(t)\} = \frac{1}{s(s-1+\lambda\ln s)} \tag{4.89}$$

On the other hand, it can be shown that solving (4.82) for $X(s)$ in the Laplace domain results

$$X(s) = \frac{s-1}{s(s-1+\lambda\ln s)} x(0) \tag{4.90}$$

From (4.89) and (4.90), it is deduced that

$$x(t) = (v'(t) - v(t))x(0) \tag{4.91}$$

Calculating the derivative $\frac{d}{dt}v(t)$ from (4.85) results

$$v'(t) = \int_0^t \frac{(t-\tau)^{\lambda\tau-1}}{\Gamma(\lambda\tau)} e^{\tau} d\tau \tag{4.92}$$

in which $t>0$. This Theorem is proved by replacing $v(t)$ and $v'(t)$ in (4.91) with their expressions derived in (4.85) and (4.92), respectively.

A similar result may be stated in case of exponential weight functions as in the following corollary.

Corollary 2: Exact solution of (4.82) under the assumption that $w(\alpha) = ca^{\alpha}$, $\alpha \in [0, 1]$, $a \in R^{>0}$, and $c \in R - \{0\}$ holds, is given by

$$x(t) = x(0)\int_0^{t/a} \left(\frac{(t/a-\tau)^{\frac{\lambda\tau}{c}-1}}{\Gamma(\lambda\tau/c)} - \frac{(t/a-\tau)^{\frac{\lambda\tau}{c}}}{\Gamma(\lambda\tau/c+1)} \right) e^{\tau} d\tau \tag{4.93}$$

Proof: Solution (4.93) is verified by following a procedure similar to the one presented in Corollary 1.

We conclude this section by presenting a new integral identity that directly follows from the results of this section as a side result. In fact

Theorem 2 reveals an identity between two expressions (4.83) and (4.84) as stated in the following Remark.

Remark 2: : The following identity holds true for all $\lambda, t > 0$

$$\int_0^t \frac{(t-\tau)^{\lambda\tau-1}((\lambda+1)\tau - t)}{\Gamma(\lambda\tau+1)} e^{\tau} d\tau = \int_0^{+\infty} \frac{\lambda(r+1)}{r\lambda^2\pi^2 + r(r+1-\lambda\ln r)^2} e^{-rt} dr \quad (4.94)$$

4.4 NUMERICAL EXAMPLES

In this section, four examples are presented to evaluate the obtained solutions of distributed order differential Eqs. (4.17) and (4.82) for different coefficients and weight functions.

Example 1: Assume $x(0) = 1$. Solutions of (4.82) with

$$\lambda \in \{12, 10, 8, 6, 4\} \quad (4.95)$$

and the unitary weight function $w(\alpha) = 1 \quad (\alpha \in [0, 1])$ are plotted in Fig. 4.1. As it is expected, decay in solutions is more intense for greater values of λ.

Example 2: The first elements of the solutions of (4.17) with $u(t) = H(t)$, a zero initial condition and different matrix pairs (A, B) given in Table 4.2 are depicted in Fig. 4.2.

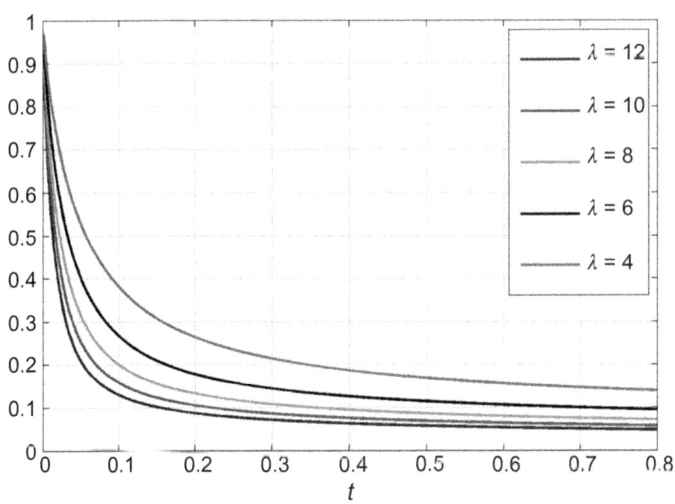

FIGURE 4.1 Solution of (4.82) with $x(0) = 1$ and the weight function $w(\alpha) = 1, \alpha \in [0, 1]$ for different values of λ given by (4.95).

TABLE 4.2 Matrices A and B, and the Eigenvalues of Matrix A in Example 2

A	B	$eig(A)$
$\begin{pmatrix} 41.4286 & 172.3810 \\ -42.8571 & -21.4286 \end{pmatrix}$	$\begin{pmatrix} 303.3333 \\ 0 \end{pmatrix}$	$10 \pm 80i$
$\begin{pmatrix} 21.4286 & 172.3810 \\ -42.8571 & -41.4286 \end{pmatrix}$	$\begin{pmatrix} 156.8966 \\ 0 \end{pmatrix}$	$-10 \pm 80i$
$\begin{pmatrix} 33.5714 & 129.2857 \\ -32.1429 & -13.5714 \end{pmatrix}$	$\begin{pmatrix} 272.6316 \\ 0 \end{pmatrix}$	$10 \pm 60i$
$\begin{pmatrix} 23.5714 & 129.2857 \\ -32.1429 & -23.5714 \end{pmatrix}$	$\begin{pmatrix} 152.7273 \\ 0 \end{pmatrix}$	$\pm 60i$

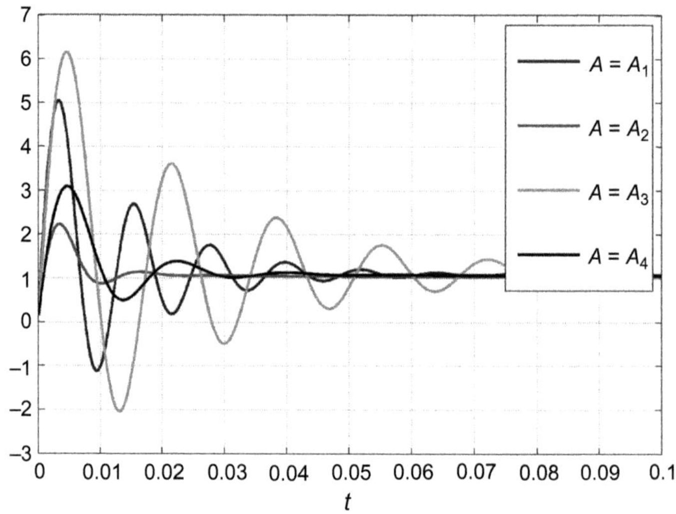

FIGURE 4.2 The first elements of the solutions of (4.17) with constant matrices given in Table 4.2, zero initial conditions and the unitary weight function (The Heaviside step function is considered as the input function $u(t)$).

Example 3: Solutions of (4.82) with $x(0) = 1$, $\lambda = 8$ and exponential weight functions $w(\alpha) = a^{\alpha}, \alpha \in [0, 1]$ in which

$$a \in \{0.4, 0.8, 1.2, 1.4\} \tag{4.96}$$

are plotted in Fig. 4.3.

Example 4: The first elements of the solutions of (4.17) with the same weight functions as in Example 3 and $u(t) = H(t)$ are plotted in Fig. 4.4. In this example a zero initial condition and the matrix pair $(A, B) = (A_1, B_1)$ given in Table 4.2, are considered.

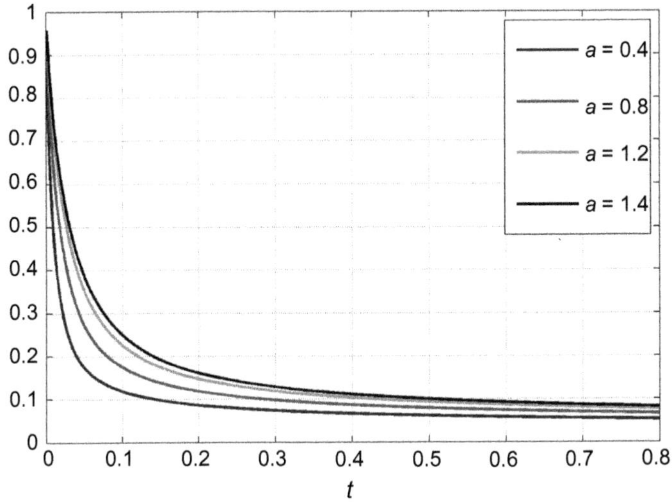

FIGURE 4.3 Solution of (4.82) with $x(0) = 1$, $\lambda = 8$ and the weight functions $w(\alpha) = a^{\alpha}, \alpha \in [0, 1]$ for $a \in \{0.4, 0.8, 1.2, 1.4\}$.

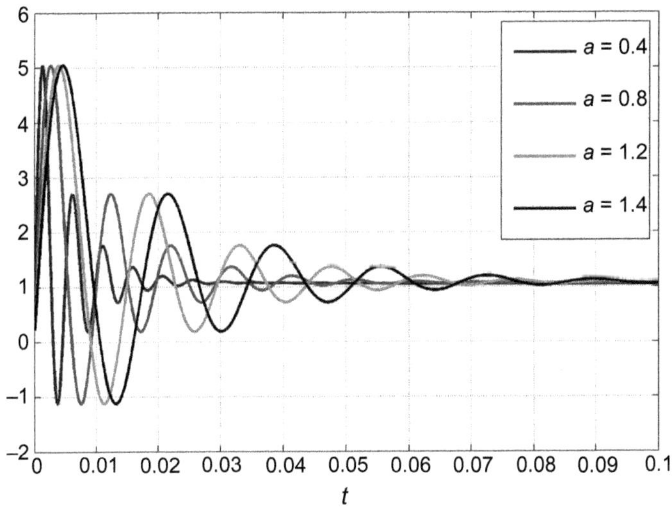

FIGURE 4.4 The first elements of the solutions of (4.17) with $x(0) = 0$, $u(t) = H(t)$, $(A, B) = (A_1, B_1)$ and the weight functions $w(\alpha) = a^{\alpha}, \alpha \in [0, 1]$ in which constants a are given in (4.96).

4.5 EXACT RESPONSE OF A DISTRIBUTED ORDER RC ELECTRICAL CIRCUIT

In this section, we consider an RC electrical circuit with a distributed order capacitor, and use Corollary 2 to describe its current step response in the time domain. To this end, consider the voltage-driven RC circuit shown in Fig. 4.5.

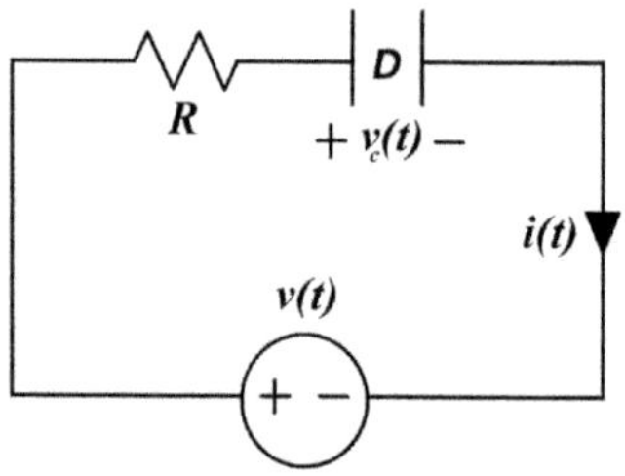

FIGURE 4.5 An RC circuit with a distributed order capacitor.

The voltage–current relationship of a distributed order capacitor in the Laplace domain is described by (Li and Chen, 2011)

$$I(s) = h(s)V_c(s) \tag{4.97}$$

where $h(s) = \int_0^1 w(\alpha)s^\alpha d\alpha$ denotes the admittance, $V_c(s) = L_{t\to s}\{v_c(t)\}$ denotes the voltage and $I(s) = L_{t\to s}\{i(t)\}$ denotes the current of the capacitor in the Laplace domain. It is immediately followed that the current response of the circuit shown in Fig. 4.5 in the Laplace domain is given by

$$I(s) = V(s)/\left(R + 1/h(s)\right) \tag{4.98}$$

where $V(s) = L_{t\to s}\{v(t)\}$ is the Laplace transform of the source voltage function. Assume the weight function of the capacitor to be of exponential type, i.e., $w(\alpha) = ca^\alpha$, $\alpha \in [0, 1]$, $a \in \mathrm{R}^{>0}$, $c \in \mathrm{R} - \{0\}$ and that the source voltage is a step Heaviside function (i.e., $v(t) = H(t)$). In this case, the admittance of the capacitor becomes $h(s) = c(as - 1)/\ln(as)$, and the current response (4.98) is then obtained as

$$I(s) = \frac{c(as-1)/Rs}{c(as-1) + \left(1/R\right)\ln as} \tag{4.99}$$

On the other hand, it can be shown that the solution of (4.82) with the same weight function in the Laplace domain is given by

$$X(s) = \frac{c(as-1)x(0)/s}{c(as-1) + \lambda \ln as} \tag{4.100}$$

Comparing (4.99) with (4.100) reveals that the current response $i(t)$ in the time domain is obtained by replacing $x(0) = 1/R$ and $\lambda = 1/R$ in Corollary 2. Hence,

$$i(t) = \int_0^{t/a} \left(\frac{\left(t/a-\tau\right)^{\frac{\tau}{Rc}-1}}{\Gamma\left(\tau/Rc\right)} - \frac{\left(t/a-\tau\right)^{\frac{\tau}{Rc}}}{\Gamma\left(\tau/Rc+1\right)} \right) e^\tau d\tau/R, \qquad t > 0 \tag{4.101}$$

is the exact step response of the electrical circuit shown in Fig. 4.5 in the time domain.

4.6 STABILITY

Stability analysis of the dynamic system represented by the set of linear differential equations of distributed order (4.17) is of interest in various applications, including control systems. Among several definitions of stability used in the literature, the following definitions are within the scope of this section.

Considering the pseudo state as the output, system (4.17) is BIBO stable, if a bounded output $x(t)$ is generated for every bounded input $u(t)$ in case of a zero initial condition $x(0) = 0$.

The origin in system (4.17) is said to be asymptotically stable, if for every $\varepsilon > 0$ there exists some $\delta > 0$ such that if $\|x(0)\| < \delta$ holds, then $\|x(t)\| < \varepsilon$ holds for $t \geq 0$, and there exists a neighborhood of the origin such that if $x(0)$ is located inside it, then the limit $\lim_{t \to +\infty} x(t) = 0$ is guaranteed.

It can be shown that the same conditions on matrix A and the weight function $w(\alpha)$ would guarantee stability of system (4.17) with respect to both of the above definitions. Therefore, the term "stability" is used to refer to either of the two stability definitions in the rest of the discussion for convenience.

In order to analyze stability, it is possible to consider system (4.17) as the two Volterra integral Eq. (4.19) and use the Lyapunov based approaches available for stability analysis of these equations (Messina et al., 2015). However, in order to obtain necessary and sufficient conditions of stability, considering the characteristic equation of (4.17) is essential. The exact stability criterion of system (4.17) stems from the characteristic equation of (4.17) which is obtained from (4.23) as

$$\det(W(-\ln s)I - A) = 0 \tag{4.102}$$

System (4.17) is stable if and only if all the roots of (4.102) are located on the open left half plane. Since the characteristic Eq. (4.102) is irrational in general, the stability analysis of distributed order systems is significantly more complicated than the classical systems. All the same, distributed order systems are not the only systems with this trait. Delay differential equations have also been known to possess irrational characteristic equations. Today, the stability analysis approaches available for systems with time delays are generally divided into two main categories: the methods based on Lyapunov–Krasovskii functionals and characteristic equations related approaches. Since the former is notorious for excessive conservatism, handling irrational characteristic equations of time-delay systems is inevitable. This is often realized by using graphical approaches discriminating the stable region from the unstable one in a parameter space. The same idea can be utilized for distributed order systems. This would result in a standard graphical method for stability analysis of system (4.17) as presented by Jiao et al. (2012). In order to explain this approach, denote the eigenvalues of matrix A by λ_i where $i\in\{1,2,...,n\}$. This allows one to write Eq. (4.102) in the following form

$$W(-\ln s) = \lambda_i, \qquad i = 1,2,...,n \tag{4.103}$$

Assuming $w(\alpha) = ca^{\alpha}$, $\alpha\in[0,1]$, $a\in\mathrm{R}^{>0}$, and $c\in\mathrm{R}-\{0\}$ would result in the following characteristic equation

$$c\int_0^1 (as)^{\alpha}d\alpha = \lambda_i, \quad i = 1,2,\ldots,n \tag{4.104}$$

from (4.103). In the following discussion, we restrict ourselves to the principal branch of the complex logarithm function (4.5) since a physical significance is only associated with its first Riemann sheet (Jiao et al., 2012, p. 12). If A is singular, then system (4.17) is clearly unstable from (4.104) as the root $s = 0$ is derived for the zero eigenvalue. Also, $s = 1/a$ satisfies (4.104) if and only if $c = \lambda_i$ holds for some $i\in\{1,2,...,n\}$ which also indicates another unstable case of the system. Therefore, assume matrix A is nonsingular and $c\neq\lambda_i$ holds for all $i\in\{1,2,...,n\}$. Based on this assumption, Eq. (4.104) could be written in the following form

$$c\frac{as-1}{\ln as} = \lambda_i, \quad i = 1,2,...,n \tag{4.105}$$

Setting $s = j\omega$, $\omega\in(0,+\infty)$ in (4.105) results in a boundary curve in the λ-plane, separating the stable and unstable regions. Therefore, the stability boundary curve of system (4.17) with respect to eigenvalues of matrix A in this case is derived as

$$\left\{s = x\pm jy|x = c\frac{-4\ln a\omega + 2\pi a\omega}{4(\ln a\omega)^2+\pi^2},\ \ y = c\frac{2\pi + 4a\omega\ln a\omega}{4(\ln a\omega)^2+\pi^2},\ \ \omega\in(0,+\infty)\right\} \tag{4.106}$$

The region on the left (right) side of the boundary curve (4.106) is called the stable (unstable) region. In fact, system (4.17) is stable, if and only if all the eigenvalues of matrix A are located within the stable region. It is worthwhile mentioning that, only the location of eigenvalues with nonnegative imaginary parts needs to be checked with respect to the curve (4.106) to determine stability. This is due to the fact that the curve (4.106) is symmetrical with respect to the real axis. The parametric curve (4.106) is plotted in Fig. 4.6 for $c = 1$ and $a = 1$. As it can be seen, the curve is tangent to the real axis at the origin and it can be shown that its slope increases infinitely as $\omega\rightarrow+\infty$. This somehow implies the fact that the weight function here comprises the fractional orders from zero to one.

In the following, the roots of the characteristic Eq. (4.105) are found by using the complex Lambert W function. Lambert W function is also used to express the characteristic roots of time-delay systems in some cases

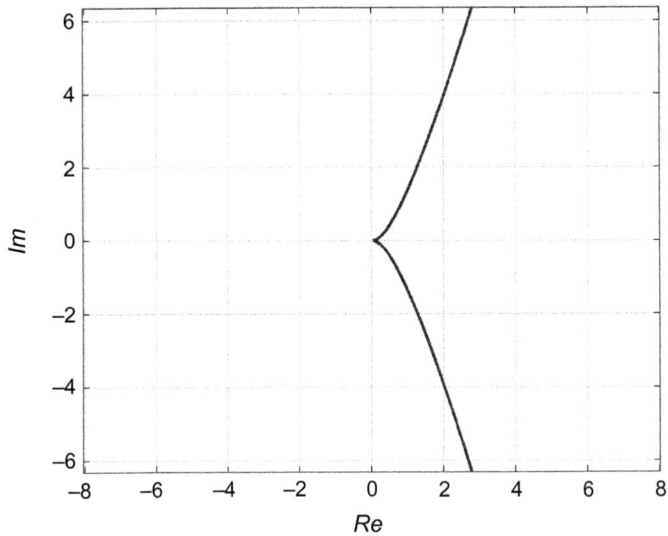

FIGURE 4.6 stability boundary curve (4.106) for $c = 1$ and $a = 1$.

(Corless et al., 1996). Again, assume matrix A is nonsingular and $c \neq \lambda_i$ holds for all $i \in \{1, 2, \ldots, n\}$. The characteristic Eq. (4.105) may be written in the form

$$c\frac{as-1}{\lambda_i} = \ln as$$
$$e^{c\frac{(as-1)}{\lambda_i}} = as \qquad (4.107)$$
$$e^{-\frac{c}{\lambda_i}} = ase^{-c\frac{as}{\lambda_i}}$$

Multiplying both sides of (4.107) by $-\frac{c}{\lambda_i}$ gives

$$-ce^{-\frac{c}{\lambda_i}}/\lambda_i = -case^{-c\frac{as}{\lambda_i}}/\lambda_i \qquad (4.108)$$

From (4.108) it is deduced that

$$\frac{-cas}{\lambda_i} = lW_k\left(-ce^{-\frac{c}{\lambda_i}}/\lambda_i\right), \qquad k \in \mathrm{Z} \qquad (4.109)$$

where lW_k denotes the kth branch of the complex valued Lambert W function. The ranges of different branches of lW_k are separated by the following set of curves on the complex plane (Corless et al., 1996)

$$\{s = x + jy | x = -y\cot y,\ 2k\pi < |y| < (2k+1)\pi,\ k \in \mathrm{N} \cup \{0\}\} \qquad (4.110)$$

and the ray $\{s = x + jy | x \in (-\infty, -1], y = 0\}$ on the real axis, which are plotted in Fig. 4.7.

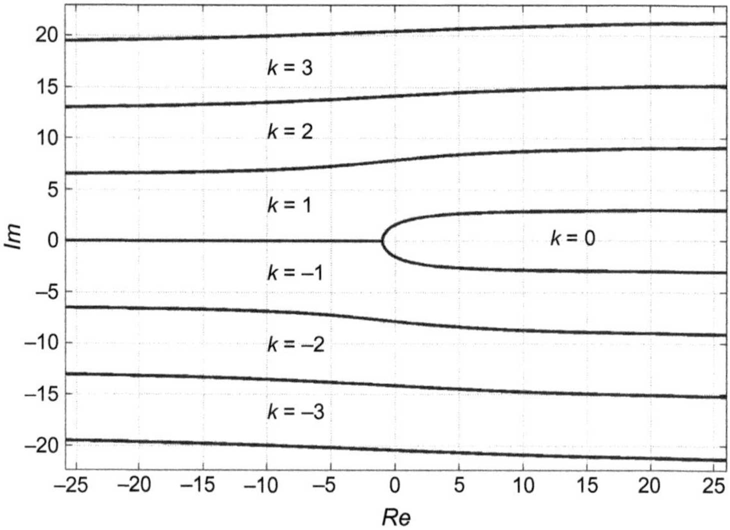

FIGURE 4.7 Ranges of different branches of W_k.

Denote the branch index of the Lambert W function whose range includes the complex number $-\frac{c}{\lambda_i}$ by $\hat{k}_i$. This branch produces the trivial value $lW_{\hat{k}_i}\left(-ce^{-\frac{c}{\lambda_i}}/\lambda_i\right) = -\frac{c}{\lambda_i}$ in (4.109) resulting in the root $s = 1/a$ which is invalid due to our initial assumption. Removing this branch from our scope, for each eigenvalue λ_i of matrix A, an infinite number of roots are obtained via (4.109) as follows

$$s = \frac{-\lambda_i}{ca} lW_k\left(-ce^{-\frac{c}{\lambda_i}}/\lambda_i\right), \qquad k \in \mathrm{Z} - \left\{\hat{k}_i\right\} \tag{4.111}$$

Thus it can be said that system (4.17) with the exponential weight function $w(\alpha) = ca^{\alpha}$, $\alpha \in [0, 1]$, $a \in \mathrm{R}^{>0}$, and $c \in \mathrm{R} - \{0\}$ is stable, if and only if

$$Re\left\{\frac{-\lambda_i}{ca} lW_k\left(-ce^{-\frac{c}{\lambda_i}}/\lambda_i\right)\right\} < 0 \tag{4.112}$$

holds for all $i \in \{1, 2, \cdots, n\}$ and $k \in \mathrm{Z} - \left\{\hat{k}_i\right\}$. Note that due to the property $lW_k(\bar{z}) = \overline{(lW_{-k}(z))}$ which is valid for $z \in \mathrm{C} - \mathrm{R} \times \{0\}$ (Corless et al., 1996), the root obtained from (4.111) for $\lambda_i = l$ and $k = k_0$ where $l \in \mathrm{C} - \mathrm{R} \times \{0\}$ is the complex conjugate of the root obtained from (4.111) with $\lambda_i = \bar{l}$ and $k = -k_0$. Thus among the complex eigenvalues, assessment of criterion (4.112) for eigenvalues that are above the real axis suffices to decide on stability of the system in practice. For the case $c = 1$, $a = 1$ and the scalar matrix $A = -4$, the characteristic equation roots (4.111) are depicted in Fig. 4.8.

As it can be seen in Fig. 4.8, all of the roots are located on the open left half plane, indicating that the system is stable in this case. This was expected

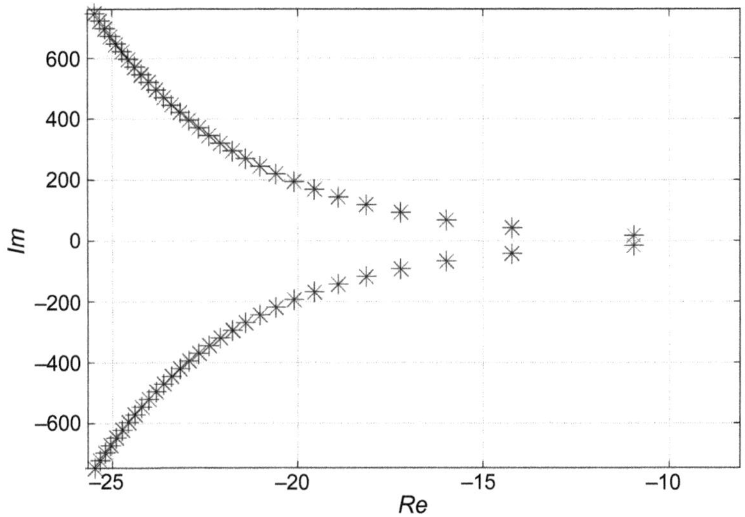

FIGURE 4.8 Characteristic roots (4.111) with $c = 1$, $a = 1$ and the eigenvalue $\lambda = -4$.

since the system in this case is immediately proven to be stable considering the relative position of its only eigenvalue and the parametric curve (4.106) shown in Fig. 4.6.

4.7 CONCLUSION

The exact solution of a system of LTI distributed order differential equations in the Caputo sense was found in the case of exponential weight functions in this chapter. The method employed in this regard is fundamentally similar to the resolvent formalism technique in integral equations. The solution was derived by using the convolution powers of the impulse response of the associated distributed order integration operator eventually resulting in a series involving powers of the dynamic matrix. In this trait, the new expression is analogous to the known solutions of linear fractional order differential equations. The results of this chapter provide an alternative expression of the solution to the already known Laplace-type integral representations derived from the Laplace transform technique in the literature. Besides, as a special case, the exact solution was derived in a simpler expression for anomalous nonexponential distributed order relaxation equation. After some numerical examples, the results were shown to be useful in describing the responses of analog electrical circuits incorporating distributed order elements. A brief discussion about stability of the responses of LTI distributed order systems with exponential weight functions was also addressed in this chapter. It was shown that characteristic equations of delay differential equations and LTI distributed order differential equations share the common trait of suffering from irrationality and the roots of both can be expressed by means of Lambert W functions in certain cases. It was also pointed out that LTI distributed order systems with exponential weight functions have infinitely many poles just like the infinite-dimensional time-delay systems.

The methodology employed in this chapter may be extended to obtain exact representations of the solutions of distributed order systems with more general weight functions. Moreover, the graphical technique offered by Jiao et al. (2012) and used in stability analysis of distributed order systems in this chapter can also be applied to systems with more general weight functions. Nevertheless, the characteristic functions become more complex and are no longer able to be expressed by means of Lambert W functions. Studying distribution of system poles on the complex plane however is not possible without analyzing the characteristic function of the system. Thus a rigorous study of the characteristic functions of distributed order systems which can be expressed in terms of Lambert W functions in the special case of exponential weight functions is an interesting path for future research.

REFERENCES

Aslam, M.S., Raja, M.A.Z., 2015. A new adaptive strategy to improve online secondary path modeling in active noise control systems using fractional signal processing approach. Signal Proc. 107, 433–443.

Azar, A.T., Vaidyanathan, S., Ouannas, A., 2017. Fractional order control and synchronization of chaotic systems, Stud. Comput. Intelligence, Vol. 688. Springer-Verlag, Germany.

Banwell, T.C., Jayakumar, A., 2000. Exact analytical solution for current flow through diode with series resistance. Electron. Lett. 36 (4), 291–292.

Bekir, A., Guner, O., Cevikel, A., 2017. The exp-function method for some time-fractional differential equations. IEEE/CAA J Automat Sinica 4 (2), 315–321.

Bell, W.W., 2004. Special Functions for Scientists and Engineers. Courier Corporation, London.

Cao, W., Zeng, F., Zhang, Z., Karniadakis, G.E., 2016. Implicit-explicit difference schemes for nonlinear fractional differential equations with nonsmooth solutions. SIAM J. Sci. Comput. 38 (5), A3070–A3093.

Caputo, M., Carcione, J.M., 2013. A memory model of sedimentation in water reservoirs. J. Hydrol. 476, 426–432.

Chechkin, A., Gorenflo, R., Sokolov, I., 2002. Retarding subdiffusion and accelerating superdiffusion governed by distributed-order fractional diffusion equations. Phys. Rev. E 66 (4), 046129.

Chechkin, A.V., Gorenflo, R., Sokolov, I.M., Gonchar, V.Y., 2003. Distributed order time fractional diffusion equation. Fract. Calculus Appl. Anal. 6 (3), 259–280.

Corless, R.M., Gonnet, G.H., Hare, D.E., Jeffrey, D.J., Knuth, D.E., 1996. On the lambertw function. Adv. Computat. Math. 5 (1), 329–359.

Dumlu, A., Erenturk, K., 2014. Trajectory tracking control for a 3-dof parallel manipulator using fractional-order $pi^{\lambda}d^{\mu}$ control. IEEE Trans. Ind. Electron. 61 (7), 3417–3426.

Erdelyi, A., Magnus, W., Oberhettinger, F., Tricomi, F.G., 1954. Tables of Integral Transforms, Vol. 2. McGraw-Hill Book Company, Incorporated, London.

Fisher, B., Al-Sirehy, F., 2015. Convolutions involving the exponential function and the exponential integral. Math. Moravica 19 (2), 65–73.

Gao, G.H., Sun, Z.Z., 2016. Two unconditionally stable and convergent difference schemes with the extrapolation method for the one-dimensional distributed-order differential equations. Numer. Methods Partial Different. Eq. 32 (2), 591–615.

Gao, G.H., Sun, H.W., Sun, Z.Z., 2015. Some high-order difference schemes for the distributed-order differential equations. J. Comput. Phys. 298, 337–359.

Garra, R., Giusti, A., Mainardi, F., Pagnini, G., 2014. Fractional relaxation with time-varying coefficient. Fract. Calculus Appl. Anal. 17 (2), 424–439.

Garrappa, R., Maione, G., Popolizio, M., 2014. Time-domain simulation for fractional relaxation of havriliak-negami type, in: Fractional Differentiation and Its Applications (ICFDA), 2014 International Conference on, IEEE, pp. 1–6.

Geller, M., Ng, E.W., 1969. A table of integrals of the exponential integral. J. Res. Natl Bureau Standards 71, 1–20.

Gorenflo, R., Luchko, Y., Stojanovic, M., 2013. Fundamental solution of a distributed order time-fractional diffusion-wave equation as probability density. Fract. Calculus Appl. Anal. 16 (2), 297–316.

Hu, X., Liu, F., Turner, I., Anh, V., 2016. An implicit numerical method of a new time distributed-order and two-sided space-fractional advection-dispersion equation. Num. Alg. 72 (2), 393–407.

Jakovljevic, B.B., Rapaic, M.R., Jelicic, Z.D., Sekara, T.B., 2014. Optimization of distributed order fractional pid controller under constraints on robustness and sensitivity to measurement noise, in: Fractional Differentiation and Its Applications (ICFDA), International Conference on, IEEE, pp. 1–6.

Jiao, Z., Chen, Y., Podlubny, I., 2012. Distributed-order dynamic systems: stability, simulation, applications and perspectives. springerbriefs in electrical and computer engineering/springerbriefs in control, automation and robotics.

Jin, B., Lazarov, R., Zhou, Z., 2016. A petrov–galerkin finite element method for fractional convection-diffusion equations. SIAM J. Numer. Anal. 54 (1), 481–503.

Kochubei, A.N., 2009. Distributed order derivatives and relaxation patterns. J. Phys. A: Math. Theor. 42 (31), 315203.

Lazović, G., Vosika, Z., Lazarević, M., Simić-Krstić, J., Koruga, Đ., 2014. Modeling of bioimpedance for human skin based on fractional distributed-order modified Cole model. FME Trans. 42 (1), 74–81.

Li, Y., Chen, Y., 2011. Theory and implementation of distributed-order element networks. ASME 2011 International Design Engineering Technical Conferences and Computers and Information in Engineering Conference. American Society of Mechanical Engineers, pp. 361–367.

Li, X.Y., Wu, B.Y., 2016. A numerical method for solving distributed order diffusion equations. Appl. Math. Lett. 53, 92–99.

Li, Y., Sheng, H., Chen,Y.Q., 2010. On distributed order lead-lag compensator, in: Proceedings of the 4th IFAC Workshop, Badajoz, Spain.

Lv, C., Xu, C., 2016. Error analysis of a high order method for time-fractional diffusion equations. SIAM J. Sci. Comput. 38 (5), A2699–A2724.

Mainardi, F., Pagnini, G., 2007. The role of the Fox–Wright functions in fractional subdiffusion of distributed order. J. Comput. Appl. Math. 207 (2), 245–257.

Mainardi, F., Mura, A., Pagnini, G., Gorenflo, R., 2008. Time-fractional diffusion of distributed order. J. Vibration Control 14 (9-10), 1267–1290.

Meerschaert, M.M., Toaldo, B., 2015. Relaxation patterns and semi-markov dynamics. arXiv preprint arXiv 1506, 02951.

Messina, E., Russo, E., Vecchio, A., 2015. Volterra integral equations on time scales: stability under constant perturbations via liapunov direct method. Ricerche di Matematica 64 (2), 345–355.

Naber, M., 2004. Distributed order fractional sub-diffusion. Fractals 12 (01), 23–32.

Pahikkala, J., 2013. Laplace transform of logarithm. [online] planetmath.org. Available at: http://planetmath.org/laplacetransformoflogarithm [Accessed: 20 April. 2017].

Petrovic, L.M., Zorica, D.M., Stojanac, I.L., Krstonosic, V.S., Hadnadjev, M.S., Janev, M.B., et al., 2015. Viscoelastic properties of uncured resin composites: dynamic oscillatory shear test and fractional derivative model. Dent. Mater. 31 (8), 1003–1009.

Podlubny, I., 1998. Fractional Differential Equations: An Introduction to Fractional Derivatives, Fractional Differential Equations, to Methods of Their Solution and Some of Their Applications, Vol. 198. Academic Press, San Diego.

Sandev, T., Chechkin, A.V., Korabel, N., Kantz, H., Sokolov, I.M., Metzler, R., 2015. Distributed-order diffusion equations and multifractality: models and solutions. Phys. Rev. E 92 (4), 042117.

Saxena, R.K., Mathai, A.M., Haubold, H.J., 2014. Distributed order reaction-diffusion systems associated with Caputo derivatives. J. Math. Phys. 55 (8), 083519.

Su, N., 2012. Distributed-order infiltration, absorption and water exchange in mobile and immobile zones of swelling soils. J. Hydrol. 468, 1–10.

Tavazoei, M.S., 2015. Reduction of oscillations via fractional order pre-filtering. Signal Proc. 107, 407–414.

Vabishchevich, P.N., 2016. Numerical solution of nonstationary problems for a space-fractional diffusion equation. Fract. Calculus Appl. Anal. 19 (1), 116–139.

Ye, H., Liu, F., Anh, V., 2015. Compact difference scheme for distributed-order time-fractional diffusion-wave equation on bounded domains. J. Comput. Phys. 298, 652–660.

Zhou, F., Zhao, Y., Li, Y., Chen, Y., 2013. Design, implementation and application of distributed order pi control. ISA Trans. 52 (3), 429–437.

Chapter 5

Fractional Order Time-Varying-Delay Systems: A Delay-Dependent Stability Criterion by Using Diffusive Representation

Y. Boukal[1,2,3], M. Zasadzinski[1], M. Darouach[1] and N.E. Radhy[2]
[1]*Université de Lorraine, Cosnes et Romain, France,* [2]*Université Hassan II, Casablanca, Maroc,* [3]*Université de Valenciennes et du Hainaut-Cambrésis, Famars, France*

5.1 INTRODUCTION

Since the physical interpretation of the fractional order derivatives given in Podlubny (2002) has became clear to the researchers and engineers, the modeling of physical (Battaglia et al., 2000; Vinagre, 2001; Ortigueira and Machado, 2003; Sabatier et al., 2007; Sheng et al., 2012; Sierociuk et al., 2013; Azar et al., 2017), biological (Magin, 2006; Freed and Diethelm, 2006), and chemical (Darling and Newman, 1997; Audounet et al., 1998; Lederman et al., 2002) phenomena employing fractional order differentiation–integration and controllers have been studied by many researchers and scientists (Bagley and Calico, 1991; Chen, 2006). Specially, the fractional order time-delay systems can characterize a class of chaotic behaviors (Deng et al., 2007; Lin and Lee, 2011; Yuan et al. 2013).

Generally, the time-delay phenomena included in the dynamics of fractional order systems (FOS) are due to transportation of material, energy, or information. Furthermore, the presence of time-delays, also called dead-time or after-effect, can cause the plant instability. In fact, the problems of stability analysis, control, and observer designs for this kind of system have attracted the attention of many researchers and scientists. Recently, FOS with time-varying delays or constant delays have been subject to some research and studies (see e.g., Tarbouriech, 1997; Bonnet and Partington, 2001, 2002; Chyi and Yi-Cheng, 2006; Busłowicz, 2008; Farshad and

Mathematical Techniques of Fractional Order Systems. DOI: https://doi.org/10.1016/B978-0-12-813592-1.00005-2

Masoud, 2009; Mihailo and Aleksandar, 2009; Si-Ammour et al., 2009; Liqiong and Shouming, 2010; Lianglin et al., 2011; Boukal et al., 2014a,b, 2015a, 2017a,c).

The main goal of the feedback controller design is to stabilize unstable systems (Boyd et al., 1994; Ahmad et al., 2004; Amato et al., 2007; Ahn et al., 2007; Amato et al., 2009, 2010; Boukal et al., 2015b, 2016b, 2017b) or to improve the stability in the presence of transient phenomena which do not die out quickly. From a practical point of view, it is very important to be able to analyze the stability of dynamic FOS. The fundamentals of the rigorous mathematical theory of stability were laid down in the works of the prominent Russian mathematician A.M. Lyapunov 100 years ago.

Recently Trigeassou et al. (2011) proposed the application of Lyapunov's method to linear and nonlinear fractional derivative equations (FDEs) by choosing a specific Lyapunov function candidate. Their approach is based on the concept of diffusive representation of the fractional integration operator.

In this work, the fractional order time-varying delay systems stability analysis is studied with the help of the diffusive representation of the fractional order derivative and Lyapunov theory. Moreover, new sufficient delay-independent and delay-dependent stability criteria are derived in a linear matrix inequality (LMI) formulation, which can be easily solved. In addition, based on this result, sufficient conditions of the existence of stabilizing pseudo-state feedback controllers for an unstable plant can be checked. The controller gain matrix can be computed by solving one of the obtained matrices inequalities. An unstable FOS-TVD (time-varying delay) and stable FOS-TVD for a bounded delay are studied as numerical examples to illustrate the effectiveness of both the proposed stability analysis method and the controller synthesis.

The chapter will be organized as follows. In Section 5.2, some preliminaries on the fractional order derivative and some useful inequalities which will be used later are provided. In Section 5.3, the problem that motivated the present work is described. Section 5.4 is dedicated to the stability analysis of FOS in presence of time-varying delays by using a Lyapunov function candidate based on the concept of diffusive representation of the fractional integration operator. Using the stability conditions formulated in terms of LMIs in Section 5.4, pseudo-state-feedback controllers are derived in Section 5.5. Numerical simulation results demonstrate the effectiveness of the proposed method are shown in Section 5.6. Finally, concluding remarks on the presented results and open problems are given in Section 5.7.

Notation: In the sequel of the chapter, the following notations are used. $\mathbb{R}^n$ and $\mathbb{R}^{n\times m}$ to denote the n dimensional Euclidean space and the set of all $n\times m$ real matrices, respectively; A^T denotes the transpose of a matrix A; matrix A is symmetric positive definite if and only if $A^T = A$ and $A > 0$.

I and 0 denote the identity matrix and zero matrix of appropriate dimensions, respectively. I_n is the identity matrix of dimensions $n \times n$; $Sym\{X\}$ is used to denote $X^T + X$. The symbol "*" corresponds to the convolution operator; and "*" denotes the corresponding part of a symmetric matrix.

5.2 PRELIMINARY RESULTS

This section will be started by providing a brief recall of some useful definitions of the fractional order derivatives and Newton–Leibniz formula.

The most known definitions of noninteger order derivatives, which are used in the literature, are divided into two main classes. On one hand, the Riemann–Liouville derivative defined as (Podlubny, 1999; Das, 2008).

$$ {}_{a}^{RL}D_t^{\alpha}f(t) = \frac{1}{\Gamma(n-\alpha)}\frac{d^n}{dt^n}\int_a^t \frac{f(\tau)}{(t-\tau)^{\alpha-n+1}}d\tau, \quad (n-1) \leq \alpha < n \tag{5.1} $$

Or the Caputo derivative on the other (Das, 2008; Podlubny, 1999), which arise as a result of a simple permutation of integral and derivative in the Eq. (5.1). The noninteger order derivative of a causal time function introduced by Caputo is then defined as

$$ {}_{a}^{C}D_t^{\alpha}f(t) = \frac{1}{\Gamma(n-\alpha)}\int_a^t \frac{\frac{d^n f(\tau)}{dt^n}}{(t-\tau)^{\alpha-n+1}}d\tau, \quad (n-1) < \alpha < n \tag{5.2} $$

with $n \in \mathrm{N}$ and $\alpha \in \mathbb{R}^+$, where the Gamma function $\Gamma\colon (0,\infty) \to \mathbb{R}$ is defined as

$$ \Gamma(\alpha) = \int_0^{\infty} \tau^{\alpha-1}\exp(-\tau)d\tau, \tag{5.3} $$

The physical interpretation of the fractional derivatives and the solution of fractional differential equations are given in Podlubny (2002) and Das (2008). In Podlubny (1999), the author explain the differences between these definitions, and when they are equivalent. In this work, the definition introduced by Riemann–Liouville, given in (5.1), is considered.

The diffusive representation of the fractional integral operator I^n given in Trigeassou and Maamri (2009) which plays an important role to derive our main result, is stated in the following definition.

Definition 1: *Consider a linear system such as:*

$$ y(t) = h(t)^{*}u(t) \tag{5.4} $$

*where * is the convolution operator.*

Its impulse response $h(t)$ verify the pseudo-Laplace transform definition is given by (Matignon, 1994):

$$h(t) = \int_0^\infty \mu(\omega) e^{-\omega t} d\omega \tag{5.5}$$

where $\mu(\omega)$ is called the diffusive representation (Montseny, 1998) (or frequency weighting function) of the impulse response $h(t)$, which has the following form

$$\mu(\omega) = \frac{\sin(\alpha\pi)}{\pi}\omega^{-\alpha} \tag{5.6}$$

□

Lemma 1: *(Trigeassou et al., 2011) Consider a nonlinear fractional derivative system:*

$${}^{RL}_{t_0}D_t^{\alpha} x(t) = f(x(t)) \tag{5.7}$$

due to the continuous frequency distributed model of the fractional integrator, the nonlinear fractional system (5.7) *can be expressed as*:

$$\frac{\partial z(\omega, t)}{\partial t} = -\omega z(\omega, t) + f(x(t)) \tag{5.8a}$$

$$x(t) = \int_0^\infty \mu(\omega) z(\omega, t) d\omega \tag{5.8b}$$

where $\mu(\omega)$ is given by (5.6)

Remark 1: *In the sequel, the initial time t_0 is equal to zero. Then, the notation ${}^{RL}_{a}D_t^{\alpha}$ can be replaced by D^{α} without loss of generality.*

In Komornik (2016), it is shown that the following Newton–Leibniz formula

$$\int_{t-\tau(t)}^{t} \dot{x}(s) ds = x(t) - x(t - \tau(t)) \tag{5.9}$$

is one of the most important formula used in the theorems proofs of classical analysis, and its validity it was extended to the Lebesgue integrable functions. This formula will be used in this work later to derive a delay-dependent condition stability.

In addition, the following relationship between the terms of the equality (5.9)

$$2\left(x^T(t)\mathcal{Y} + x^T(t-\tau(t))\mathcal{T}\right) \times \underbrace{\left(x(t) - x(t-\tau(t)) - \int_{t-\tau(t)}^{t} \dot{x}(s) ds\right)}_{0} = 0 \tag{5.10}$$

is satisfied, for any matrices $\mathcal{Y}$ and $\mathcal{T}$ with appropriate dimension.

Here, a useful inequality is given, which will be used later in the proof of the main results. For any semipositive definite matrix $\Gamma = \begin{bmatrix} \Gamma_{11} & \Gamma_{12} \\ * & \Gamma_{22} \end{bmatrix} \geq 0$, the inequality

$$\tau_{\max}\eta^T(t)\Gamma\eta(t) - \int_{t-\tau(t)}^{t} \eta^T(t)\Gamma\eta(t)ds \geq 0 \tag{5.11}$$

holds, where $\eta(t) = \begin{bmatrix} x^T(t) & x^T(t-\tau(t)) \end{bmatrix}^T$.

5.3 PROBLEM FORMULATION

Firstly, the following fractional order linear time-invariant systems with a constant time-delay of retarded type is considered

$$D^{\alpha}x(t) = A_0 x(t) + A_{\tau}x(t-\tau) \tag{5.12a}$$

$$x(t) = \psi(t), t \in [-\tau, 0], \quad 0 < \alpha < 1 \tag{5.12b}$$

where $x(t) \in \mathbb{R}^n$ is the pseudo-state vector (for an introduction about the pseudo-state space description see Sabatier et al. (2014)). $\tau \in \mathbb{R}^+$ is a constant time-delay. The matrices $A_0, A_{\tau} \in \mathbb{R}^{n \times n}$ are known and constant. The associated function $\psi(t)$ represents a continuous vector-valued initial pseudo-state.

The characteristic equation of the system (5.12), which is quasi-polynomial in s^{α} and $e^{-s\tau}$, is given by

$$\det(s^{\alpha}I - A - A_{\tau}e^{-s\tau}) = 0 \tag{5.13}$$

Generally, the characteristic Eq. (5.13) can be rewritten as follows

$$M(s,\tau) = p(s^{\alpha}) + \sum_{i=1}^{K} q_i(s^{\alpha})e^{-is\tau} \tag{5.14}$$

where $\tau \in \mathbb{R}^+$ is the delay parameter, $p(s^{\alpha})$ and $q_i(s^{\alpha})$ for $i = 1, 2..., K$ are polynomials functions of s^{α} and $\deg p \geq \deg q_i$. The considered polynomials degrees are the degrees in the variable s^{α}, then they are integers.

Lemma 2: *(Fioravanti et al., 2012) Let G be a strictly proper system with characteristic equation given by (5.14) satisfying* $\deg p \geq \deg q_i$ *for* $i = 1, 2, \ldots, K$ *being thus of retarded type. Then G is BIBO-stable if and only if G has no poles in* $\Re(s) \geq 0$ *(in particular, no poles of fractional order at* $s = 0$*).*

It is shown in Fioravanti et al. (2012), that by applying a variable substitution as $\varsigma = s^{\alpha}$, a practical test for stability can achieved. Using this substitution, the characteristic Eq. (5.14) becomes

$$M(\varsigma,\tau) = p(\varsigma) + \sum_{i=1}^{K} q_i(\varsigma)e^{-i\varsigma^{\frac{1}{\alpha}}\tau} \tag{5.15}$$

which can be rewritten as

$$M(\varsigma,\tau) = p(\varsigma) + \sum_{i=1}^{K} q_i(\varsigma) \tag{5.16}$$

in the case when $\tau = 0$.

The advantage of this substitution is to transform the system domain from the multisheeted Riemann surface into the complex plane. In fact, the stability region of the original system is not given by the left half-plane, but by the stability condition given in Matignon (1998) as

$$Arg(\varsigma) \geq \alpha\frac{\pi}{2} \tag{5.17}$$

where $\varsigma \in \mathbb{C}$.

By using this transformation, it is easier to remark that the imaginary axis in the s-domain is mapped into the following lines

$$Arg(\varsigma) = \pm\alpha\frac{\pi}{2} \tag{5.18}$$

in the new ς-domain.

The frequency-domain methods provide necessary and sufficient conditions to analyze the stability of a system with or without delays. Despite the advantages of the frequency-domain analysis, the Eq. (5.13) causes some difficulties in order to resolve it, this is due to the presence of the exponential type transcendental terms and the noninteger orders. In addition, the presence of time-varying delays or uncertainties make the solution more complicated, which proves that the frequency-domain analysis has some limitations.

In this work, only the time-domain methods are considered, which allows us to obtain a sufficient condition for the stability of the system (5.12) and its most general cases. The main objective of the next section, is to improve the stability analysis of this kind of system by using the Lyapunov theory, where the Lyapunov–Krasovskii functional and the diffusive representation of the fractional integral operator I^n will be used.

5.4 STABILITY ANALYSIS IN PRESENCE OF TIME-VARYING DELAYS

In the time domain, the stability analysis of time-varying delay systems are based mainly on two known theorems, the Lyapunov–Krasovskii stability theorem and the Razumikhin theorem. The key idea of these theorems consists of choosing the appropriate candidate Lyapunov function or Lyapunov–Krasovskii functional to get sufficient conditions for analyzing the system (5.21) stability. These conditions are divided into two basic categories. One of them is independent of the time-delay length, and it is called the delay-independent condition. On the other hand, there is the condition that takes into account the time-delay length, and it is known as the delay-dependent condition.

In this section, stability analysis of fractional order time-varying delay systems will be dealt with. The obtained conditions are very useful in stability analysis and controller synthesis. In addition, they will be used in the next section to design pseudo-state feedback controller law, where the designed controller law must stabilize the considered unstable fractional order time-varying delay systems.

The mathematical model for a linear forced fractional order time-varying delay system can be written as

$$D^{\alpha}x(t) = A_0 x(t) + A_{\tau} x(t - \tau(t)) + B_u u(t) \tag{5.19a}$$

$$x(t) = \psi(t), t \in [-\tau_m, 0], \quad 0 < \alpha < 1 \tag{5.19b}$$

where $x(t) \in \mathbb{R}^n$ is the pseudo-state vector (For an introduction about the pseudo-state space description see Sabatier et al. (2014)), $u(t) \in \mathbb{R}^m$ is the input vector.

Assumption 1: *The time-delay function* $\tau(t) \geq 0$ *is assumed to be continuous, bounded, and satisfies*

$$0 < \tau(t) < \tau_m < \infty \tag{5.20a}$$

$$0 \leq \frac{\partial \tau(t)}{\partial t} \leq \partial \tau_m < 1 \tag{5.20b}$$

where τ_{max} and $\partial \tau_{max}$ *are two constant scalars.*

The matrices A_0, A_{τ}, and B_u are known and constant with appropriate dimensions. The associated function $\psi(t)$ represents a continuous vector-valued initial pseudo-states.

5.4.1 Time-Delay-Independent Stability

The purpose of this part is to investigate how the stability of fractional order time-varying delay systems can be improved by the indirect Lyapunov method. The authors focus on the case where the time-delay length is not needed to obtain the delay-independent stability condition.

For analyzing the stability of a linear unforced (without input) fractional order time-varying delay systems, a special case of the system (Eq. 5.19) is considered, which can be represented by the following pseudo-state space equations

$$D^{\alpha}x(t) = A_0 x(t) + A_{\tau} x(t - \tau(t)) \tag{5.21a}$$

$$x(t) = \psi(t), t \in [-\tau_m, 0], \quad 0 < \alpha < 1 \tag{5.21b}$$

In the following theorem, the stability analysis of the system (5.21), by using the indirect Lyapunov approach, leads to a sufficient time-delay-independent stability condition given in a LMI formulation, which can be solved easily.

Theorem 1: *The unforced fractional order time-delay system* (5.21) *is asymptotically stable if there exists a positive definite matrix P which satisfies the following LMI*

$$\Omega = \begin{bmatrix} PA_0 + A_0^T P + 2\Lambda_1 & PA_\tau \\ * & -\Lambda_2 \end{bmatrix} < 0 \tag{5.22}$$

where

$$\Lambda_1 = \frac{1}{2(1 - \partial\tau_{\max})} I, \quad \Lambda_2 = \frac{1}{(1 - \partial\tau_{\max})} I.$$

and $*$ *denotes the corresponding part of a symmetric matrix.*

□

Proof 1: *Firstly, The Lyapunov–Krasovskii functional candidate as*

$$V(t) = V_1(t) + V_2(t) \tag{5.23}$$

is chosen, where the first term $V_1(t)$ *of the Lyapunov function candidate* $V(t)$ *is defined as*

$$V_1(t) = \int_0^\infty \mu(\omega) v_1(\omega, t) d\omega \tag{5.24a}$$

by summing all the monochromatic $v_1(\omega, t)$ *with the weighting function* $\mu(\omega)$, *and the second term* $V_2(t)$ *is defined as*

$$V_2(t) = \frac{1}{(1 - \partial\tau_{\max})} \int_{t-\tau(t)}^{t} x^T(s) x(s) ds \tag{5.24b}$$

Now, to analyze the stability of system (5.21), *the following monochromatic Lyapunov function candidate for the elementary frequency* ω

$$v_1(\omega, t) = z^T(\omega, t) P z(\omega, t) \tag{5.25}$$

is considered, where $P \in \mathbb{R}^{n \times n}$ *is a symmetric positive definite matrix.*

Following the idea of Trigeassou et al. (2011), by introducing the diffusive representation of the fractional integral operator I^n *given in Trigeassou and Maamri (2009), the system* (5.21) *is rewritten in a diffusive form as given by* Lemma 1

$$\frac{\partial z(\omega,t)}{\partial t} = -\omega z(\omega,t) + A_0 x(t) + A_\tau x(t-\tau(t)) \tag{5.26a}$$

$$x(t) = \int_0^\infty \mu(\omega) z(\omega,t) d\omega \tag{5.26b}$$

Then, the derivatives of the Lyapunov function $v_1(t)$ *with respect to* $z(\omega,t)$ *and t are given by*

$$\frac{\partial v_1(\omega,t)}{\partial z(\omega,t)} = 2z^T(\omega,t)P \tag{5.27}$$

and

$$\begin{aligned}\frac{\partial v_1(\omega,t)}{\partial t} &= \frac{\partial v_1(\omega,t)}{\partial z(\omega,t)}\frac{\partial z(\omega,t)}{\partial t}\\ &= 2z^T(\omega,t)P(-\omega z(\omega,t) + A_0 x(t) + A_\tau x(t-\tau(t)))\end{aligned} \tag{5.28}$$

The dynamics of the Lyapunov function (5.24a) *along the solution trajectories of* (5.26) *is given as*

$$\begin{aligned}\frac{dV_1(t)}{dt} &= \int_0^\infty \mu(\omega)\frac{\partial v_1(\omega,t)}{\partial t} d\omega\\ &= \int_0^\infty \mu(\omega) 2z^T(\omega,t)P(-\omega z(\omega,t) + A_0 x(t) + A_\tau x(t-\tau(t)))\ d\omega\\ &= -2\int_0^\infty \omega\mu(\omega) z^T(\omega,t)Pz(\omega,t)\ d\omega + 2\int_0^\infty \mu(\omega) z^T(\omega,t)P\ d\omega\quad (A_0 x(t)\\ &\quad + A_\tau x(t-\tau(t)))\end{aligned} \tag{5.29}$$

By substituting the pseudo-state giving by diffusive representation (5.26) in the Eq. (5.29) *leads to*

$$\frac{dV_1(t)}{dt} = -2\int_0^\infty \omega\mu(\omega) z^T(\omega,t)Pz(\omega,t)d\omega + 2x^T(t)P(A_0 x(t) + A_\tau x(t-\tau(t))) \tag{5.30}$$

As a result, sufficient conditions are given such that the following inequalities $V_1(t) > 0$ *and* $\frac{dV_1(t)}{dt} < 0$ *are fulfilled. These two inequalities are equivalent to*

$$P > 0 \tag{5.31a}$$

$$2x^T(t)P(A_0 x(t) + A_\tau x(t-\tau(t))) < 0 \tag{5.31b}$$

Now, the time derivative of $V_2(t)$ *along the solution of* (5.21a) *is given as follows*

$$\frac{dV_2(t)}{dt} = \frac{1}{(1-\partial\tau_{\max})}x^T(t)x(t) - \frac{1}{(1-\partial\tau_{\max})}\left(1-\frac{\partial\tau(t)}{\partial t}\right)x^T(t-\tau(t))x(t-\tau(t))) \tag{5.32}$$

Under the Assumption (5.20)*, there exists an upper bound on the delay first derivative. This allows us to obtain an upper bound of the* Eq. (5.32)

$$\frac{dV_2(t)}{dt} \le \frac{1}{(1-\partial\tau_{\max})}\{x^T(t)x(t) - x^T(t-\tau(t))x(t-\tau(t)) \tag{5.33}$$

Summing the two time derivatives of the Lyapunov functions $V_1(t)$ *and* $V_2(t)$ *yields to*

$$\begin{aligned}\frac{dV(t)}{dt} \le & -2\int_0^{\infty}\omega\mu(\omega)z^T(\omega,t)Pz(\omega,t)d\omega + 2x^T(t)(PA_0+\Lambda_1)x(t) \\ & + 2x^T(t)PA_\tau x(t-\tau(t)) - x^T(t-\tau(t))\Lambda_2 x(t-\tau(t))\end{aligned} \tag{5.34}$$

where $\Lambda_1 = \frac{1}{2(1-\partial\tau_{\max})}I$ *and* $\Lambda_2 = \frac{1}{(1-\partial\tau_{\max})}I.$

The inequality (5.34) *is equivalent to*

$$X^T(t)\Omega_1 X(t) \le 0 \tag{5.35}$$

where

$$\Omega_1 = \begin{bmatrix} 2(PA_0+\Lambda_1) & 2PA_\tau \\ 0 & -\Lambda_2 \end{bmatrix} \tag{5.36}$$

and

$$X(t) = \begin{bmatrix} x(t) \\ x(t-\tau(t)) \end{bmatrix} \tag{5.37}$$

Notice that the matrix Ω_1 *is not a symmetric matrix. That means inequality* (5.35) *can not be solved by using LMI solvers of Matlab, because it is not a semidefinite programming (SDP) problem.*

To overcome this situation, the matrix Ω_1 *can be replaced by an equivalent matrix* Ω *given by*

$$\Omega = \left(\frac{\Omega_1+\Omega_1^T}{2}\right) \tag{5.38}$$

where

$$\Omega = \begin{bmatrix} PA_0 + A_0^T P + 2\Lambda_1 & PA_\tau \\ A_\tau^T P & -\Lambda_2 \end{bmatrix} \tag{5.39}$$

A simple proof of the equality of $X^T(t)\Omega X(t) = X^T(t)\Omega_1 X(t)$ *can be found in Boroujeni and Momeni (2012a,b).*

Then, the sufficient stability condition of the system (5.21) *is given in a LMI formulation* (5.22). *That means the system* (5.21) *is asymptotically stable if the matrix* Ω *is negative definite. It completes the proof.*

5.4.2 Time-Delay-Dependent Stability

The time-delay-dependent stability analysis for time-varying delay systems has been attracting increasing researcher attention. In this case, the stability analysis typically requires the use of the Lyapunov–Krasovskii functional method in the time domain. In addition, some useful inequalities and the Newton–Leibnitz formula (5.9) are used to overcome the existence of delayed pseudo-state in the Lyapunov–Krasovskii functional dynamics, and replace it by pseudo-states without delay and its integral. This approach allows us to obtain time-delay-dependent stability criterion which will be used hereafter to synthesize a controller law.

In the following theorem, a sufficient time-delay-dependent stability condition for the system (5.21) is derived. The obtained condition is formulated in LMI expression. New decision variables are added to the Lyapunov matrices of the functional candidate functional, which add more degrees of freedom in the optimization problem.

Theorem 2: *The unforced fractional order time-delay system* (5.21) *is asymptotically stable if there exist four symmetric positive definite matrices* $\mathcal{P}$, $\mathcal{Q}$, $\mathcal{Z}$ *and* $\boldsymbol{\Gamma}$, *and two matrices* $\mathcal{T}$ and $\mathcal{Y}$ *with appropriate dimensions such that the following LMIs*

$$\Upsilon = \begin{bmatrix} \Upsilon_{11} & \Upsilon_{12} & \Upsilon_{13} \\ * & \Upsilon_{22} & \Upsilon_{23} \\ * & * & \Upsilon_{33} \end{bmatrix} < 0 \tag{5.40a}$$

$$\Omega = \begin{bmatrix} \Gamma_{11} & \Gamma_{12} & \mathcal{Y} \\ * & \Gamma_{22} & \mathcal{T} \\ * & * & \mathcal{Z} \end{bmatrix} \geq 0 \tag{5.40b}$$

where

$$\begin{aligned}
\Upsilon_{11} &= \mathcal{P}A_0 + A_0^T\mathcal{P}^T + \mathcal{Y} + \mathcal{Y}^T + \tau_m\boldsymbol{\Gamma}_{11} + \mathcal{Q} \\
\Upsilon_{12} &= \mathcal{P}A_\tau - \mathcal{Y} + \mathcal{T}^T + \tau_m\boldsymbol{\Gamma}_{12} \\
\Upsilon_{12} &= \tau_m A_0^T \mathcal{Z} \\
\Upsilon_{22} &= -\Lambda_1\mathcal{Q} - \mathcal{T} - = \mathcal{T}^T + \tau_m\boldsymbol{\Gamma}_{22} \\
\Upsilon_{23} &= \tau_m A_\tau^T \mathcal{Z} \\
\Upsilon_{33} &= -\tau_m\mathcal{Z}
\end{aligned}$$

hold for a given scalars τ_m and $\partial\tau_m$ satisfy the Assumption (5.20), where $*$ denotes the corresponding part of a symmetric matrix.

□

Proof 2: *From Lemma 1, Eq. (5.21) can be expressed as*

$$\frac{\partial z(\omega,t)}{\partial t} = -\omega z(\omega,t) + A_0 x(t) + A_\tau x(t-\tau(t)) \tag{5.41a}$$

$$x(t) = \int_0^\infty \mu(\omega) z(\omega,t) d\omega \tag{5.41b}$$

Now, to analyze the stability of system (5.21), the following monochromatic Lyapunov function candidate for the elementary frequency ω

$$v_1(\omega,t) = z^T(\omega,t)\mathcal{P}z(\omega,t) \tag{5.42}$$

is considered, where $\mathcal{P} = \mathcal{P}^T \in \mathbb{R}^{n\times n}$ is a symmetric positive definite matrix.

Then, the derivative of the Lyapunov function $v_1(t)$ with respect to $z(\omega,t)$ and t are given by

$$\frac{\partial v_1(\omega,t)}{\partial z(\omega,t)} = 2z^T(\omega,t)\mathcal{P} \tag{5.43}$$

and

$$\begin{aligned}\frac{\partial v_1(\omega,t)}{\partial t} &= \frac{\partial v_1(\omega,t)}{\partial z(\omega,t)}\frac{\partial z(\omega,t)}{\partial t} \\ &= 2z^T(\omega,t)\mathcal{P}(-\omega z(\omega,t) + A_0 x(t) + A_\tau x(t-\tau(t)))\end{aligned} \tag{5.44}$$

Furthermore, a Lyapunov function candidate $V_1(t)$ is defined by summing all the monochromatic $v_1(\omega,t)$ with the weighting function $\mu(\omega)$, i.e., $V_1(t)$ is given by

$$V_1(t) = \int_0^\infty \mu(\omega) v_1(\omega,t) d\omega \tag{5.45}$$

The Lyapunov function (5.45) dynamics along the solution trajectories of (5.41) is given as

$$\begin{aligned}\frac{\mathrm{d}V_1(t)}{\mathrm{d}t} &= \int_0^\infty \mu(\omega)\frac{\partial v_1(\omega,t)}{\partial t}\mathrm{d}\omega \\ &= \int_0^\infty \mu(\omega)2z^T(\omega,t)\mathcal{P}(-\omega z(\omega,t)+A_0x(t)+A_\tau x(t-\tau(t)))\,\mathrm{d}\omega \\ &= -2\int_0^\infty \omega\mu(\omega)z^T(\omega,t)\mathcal{P}z(\omega,t)\,\mathrm{d}\omega \\ &\quad +2\int_0^\infty \mu(\omega)z^T(\omega,t)\mathcal{P}\,\mathrm{d}\omega\ (A_0x(t)+A_\tau x(t-\tau(t)))\end{aligned} \tag{5.46}$$

By substituting in (5.46) *the expression of the pseudo-state giving by* (5.41), *the following expression*

$$\frac{dV_1(t)}{dt} = -2\int_0^{\infty} \omega\mu(\omega)z^T(\omega,t)\mathcal{P}z(\omega,t)d\omega + 2x^T(t)\mathcal{P}(A_0x(t) + A_\tau x(t-\tau(t))) \tag{5.47}$$

is obtained.

It is easy to remark that the following inequality

$$-2\int_0^{\infty} \omega\mu(\omega)z^T(\omega,t)\mathcal{P}z(\omega,t)d \leq 0 \tag{5.48}$$

holds, for any symmetric positive definite matrix $\mathcal{P}$.

As a result, sufficient conditions are given such that the following inequalities $V_1(t) > 0$ *and* $\frac{\mathrm{d}V_1(t)}{\mathrm{d}t} \leq 0$ *are fulfilled. These two inequalities are equivalent to*

$$\mathcal{P} > 0 \tag{5.49a}$$

$$2x^T(t)\mathcal{P}(A_0x(t) + A_\tau x(t-\tau(t))) \leq 0 \tag{5.49b}$$

In addition, a second Lyapunov–Krasovskii function $V_2(t)$ *is defined as*

$$V_2(t) = \int_{t-\tau(t)}^{t} x^T(s)\mathcal{Q}x(s)ds \tag{5.50}$$

Thus, its time derivative along the solution of (5.21a) *is given as follows*

$$\frac{\mathrm{d}V_2(t)}{\mathrm{d}t} = x^T(t)\mathcal{Q}x(t) - \left(1 - \frac{\partial\tau(t)}{\partial t}\right)x^T(t-\tau(t))\mathcal{Q}x(t-\tau(t))) \tag{5.51}$$

and is bounded by

$$\frac{\mathrm{d}V_2(t)}{\mathrm{d}t} \leq x^T(t)\mathcal{Q}x(t) - \underbrace{(1-\partial\tau_m)}_{\Lambda_1} x^T(t-\tau(t))\mathcal{Q}x(t-\tau(t)) \tag{5.52}$$

Now, a third Lyapunov functional candidates $V_3(t)$

$$V_3(t) = \int_{-\tau_m}^{0}\int_{t+\theta}^{t} \dot{x}(s)\mathcal{Z}\dot{x}(s)\mathrm{d}s\ \mathrm{d}\theta \tag{5.53}$$

is defined, which has a time derivative given as

$$\frac{\mathrm{d}V_3(t)}{\mathrm{d}t} = \int_{-\tau_m}^{0}\int_{t+\theta}^{t} \dot{x}(s)\mathcal{Z}\dot{x}(s)\mathrm{d}s\ \mathrm{d}\theta \tag{5.54}$$

$$= \tau_m\dot{x}^T(t)\mathcal{Z}\dot{x}(t) - \int_{t-\tau_m}^{t} \dot{x}(s)\mathcal{Z}\dot{x}(s)ds \tag{5.55}$$

$$= \tau_m \left(x^T(t)A_0^T + x^T(t-\tau(t))A_\tau^T \right) \mathcal{Z} \left(A_0 x(t) + A_\tau x(t-\tau(t)) \right) - \int_{t-\tau_m}^{t} \dot{x}(s) \mathcal{Z} \dot{x}(s) \mathrm{d}s \tag{5.56}$$

On the other hand, by adding the left sides of (5.10) *and* (5.11) *to the time derivative of the Lyapunov functional candidates* $V(t)$, *where* $V(t) = V_1(t) + V_2(t) + V_3(t)$, *yields*

$$\begin{aligned}\frac{dV(t)}{dt} \leq & -2\int_0^\infty \omega\mu(\omega) z^T(\omega,t)\mathcal{P}z(\omega,t)d\omega + 2x^T(t)\mathcal{P}(A_0x(t) + A_\tau x(t-\tau(t))) \\ & + x^T(t)\mathcal{Q}x(t) - (1-\partial\tau_m)x^T(t-\tau(t))\mathcal{Q}x(t-\tau(t)) \\ & + \tau_m\left(x^T(t)A_0^T + x^T(t-\tau(t))A_\tau^T\right)\mathcal{Z}(A_0x(t) + A_\tau x(t-\tau(t))) \\ & - \int_{t-\tau_m}^{t} \dot{x}(s)\mathcal{Z}\dot{x}(s) + \tau_m\eta^T(t)\Gamma\eta(t) - \int_{t-\tau(t)}^{t} \eta^T(t)\Gamma\eta(t)dsds \\ & + 2(x^T(t)\mathcal{Y} + x^T(t-\tau(t))\mathcal{T}) \times \left(x(t) - x(t-\tau(t)) - \int_{t-\tau(t)}^{t} \dot{x}(s)ds \right)\end{aligned} \tag{5.57a}$$

$$\leq -2\int_0^\infty \omega\mu(\omega)z^T(\omega,t)Pz(\omega,t)\mathrm{d}\omega + \eta(t)^T\Upsilon_1\eta(t) - \int_{t-\tau(t)}^{t} \eta(t,s)^T\mathbf{W}\eta(t,s)\mathrm{d}s \tag{5.57b}$$

where

$$\Upsilon_1 = \begin{bmatrix} \mathbf{\Phi}_{11} & \mathbf{\Phi}_{12} \\ \mathbf{\Phi}_{21} & \mathbf{\Phi}_{22} \end{bmatrix} \tag{5.58a}$$

$$\Omega_1 = \begin{bmatrix} \mathbf{\Gamma}_{11} & \mathbf{\Gamma}_{12} & 2\mathcal{Y} \\ \mathbf{\Gamma}_{12}^T & \mathbf{\Gamma}_{22} & 2\mathcal{T} \\ 0 & 0 & \mathcal{Z} \end{bmatrix} \tag{5.58b}$$

$$\eta(t,s) = \left[x^T(t) \quad x^T(t-\tau(t)) \quad \dot{x}^T(t) \right]^T \tag{5.58c}$$

and

$$\begin{aligned}\mathbf{\Phi}_{11} &= 2\mathcal{P}A_0 + \tau_m A_0^T\mathcal{Z}A_0 + 2\mathcal{Y} + \tau_m\mathbf{\Gamma}_{11} + \mathcal{Q} \\ \mathbf{\Phi}_{12} &= 2\mathcal{P}A_\tau + \tau_m A_0^T\mathcal{Z}A_\tau - 2\mathcal{Y} + \tau_m\mathbf{\Gamma}_{12} \\ \mathbf{\Phi}_{21} &= \tau_m A_\tau^T\mathcal{Z}A_0 + \tau_m\mathbf{\Gamma}_{12}^T + 2\mathcal{T} \\ \mathbf{\Phi}_{22} &= -\Lambda_1\mathcal{Q} - 2\mathcal{T} + \tau_m A_\tau^T\mathcal{Z}A_\tau + \tau_m\Gamma_{22}\end{aligned}$$

As a result, sufficient conditions are given such that the inequalities $V(t) > 0$ *and* $\frac{dV(t)}{dt} < 0$ *are fulfilled for any* $\eta(t) \neq 0$ *and* $\eta(t,s) \neq 0$. *These two inequalities are equivalent to* $\Upsilon_1 < 0$ *and* $\Omega_1 \geq 0$.

The inequality $\Upsilon_1 < 0$ is equivalent by using the Schur complement formula to

$$\tilde{\Upsilon}_1 = \begin{bmatrix} \tilde{\Phi}_{11} & \tilde{\Phi}_{12} & \tilde{\Phi}_{13} \\ \tilde{\Phi}_{21} & \tilde{\Phi}_{22} & \tilde{\Phi}_{23} \\ \tilde{\Phi}_{31} & \tilde{\Phi}_{32} & \tilde{\Phi}_{33} \end{bmatrix} < 0 \tag{5.59}$$

where

$$\begin{aligned}
\tilde{\mathbf{\Phi}}_{11} &= 2\mathcal{P}A_0 + 2\mathcal{Y} + \tau_m \mathbf{\Gamma}_{11} + \mathcal{Q} \\
\tilde{\mathbf{\Phi}}_{12} &= 2\mathcal{P}A_\tau - 2\mathcal{Y} + \tau_m \mathbf{\Gamma}_{12} \\
\tilde{\mathbf{\Phi}}_{13} &= \tau_m A_0^T \mathcal{Z} \\
\tilde{\mathbf{\Phi}}_{21} &= + \tau_m \mathbf{\Gamma}_{12}^T + 2\mathcal{T} \\
\tilde{\mathbf{\Phi}}_{22} &= - \Lambda_1 \mathcal{Q} - 2\mathcal{T} + \tau_m \mathbf{\Gamma}_{22} \\
\tilde{\mathbf{\Phi}}_{23} &= \tau_m A_\tau^T \mathcal{Z}
\end{aligned}$$

$$\begin{aligned}
\tilde{\Phi}_{31} &= \tau_m \mathcal{Z} A_0 \\
\tilde{\Phi}_{32} &= \tau_m \mathcal{Z} A_\tau \\
\tilde{\Phi}_{33} &= - \tau_m \mathcal{Z}
\end{aligned}$$

Notice that at this stage the matrices $\tilde{\mathbf{\Upsilon}}_1$ and $\mathbf{\Omega}_1$ are not symmetric ones. That means, inequalities $\tilde{\mathbf{\Upsilon}}_1 < 0$ and $\mathbf{\Omega}_1 \geq 0$ can not be solved by using LMI solvers of Matlab, because it is not an SDP problem.

To overcome this situation, matrices $\tilde{\mathbf{\Upsilon}}_1$ and $\mathbf{\Omega}_1$ can be replaced by equivalent ones as

$$\begin{aligned}
\Upsilon &= \frac{(\tilde{\Upsilon}_1 + \tilde{\Upsilon}_1^T)}{2} \\
&= \begin{bmatrix} \mathcal{P}A_0 + A_0^T\mathcal{P}^T + \mathcal{Y} + \mathcal{Y}^T + \tau_m\Gamma_{11} + \mathcal{Q} & \mathcal{P}A_\tau - \mathcal{Y} + \mathcal{T}^T + \tau_m\Gamma_{12} & \tau_m A_0^T \mathcal{Z} \\ * & -\Lambda_1\mathcal{Q} - \mathcal{T} - \mathcal{T}^T + \tau_m\Gamma_{22} & \tau_m A_\tau^T \mathcal{Z} \\ * & * & -\tau_m \mathcal{Z} \end{bmatrix} < 0
\end{aligned} \tag{5.60a}$$

$$\begin{aligned}
\mathbf{\Omega} &= \frac{(\mathbf{\Omega}_1 + \mathbf{\Omega}_1^T)}{2} \\
&= \begin{bmatrix} \mathbf{\Gamma}_{11} & \mathbf{\Gamma}_{12} & \mathcal{Y} \\ * & \mathbf{\Gamma}_{22} & \mathcal{T} \\ * & * & \mathcal{Z} \end{bmatrix} \geq 0
\end{aligned} \tag{5.60b}$$

Then, under the assumption (5.20), *the system* (5.21) *is asymptotically stable if the LMI conditions given by Eq. (5.60) are satisfied. This completes the proof.*

Remark 2: *By requiring that the matrices $\mathcal{T}$, $\mathcal{Y}$ and Γ are equal to zero, and the spectral radius of the matrix Z is chosen small enough, the inequalities* (5.60) *given in* Theorem 2 *are reduced to the delay-independent stability condition as given in* Theorem 1 in Boukal et al. (2016b).

5.5 FEEDBACK STABILIZATION

The main objective of this section is to investigate how the system (5.19) can be asymptotically stabilized by using a pseudo-state feedback controller. In Boukal et al. (2016b), a delay-independent stability condition has been proposed. On the basis of the results given in Boukal et al. (2016b), a simple pseudo-state feedback has been used to stabilize the considered class of linear systems with time-varying delay. Therefore, the delay-dependent stabilization may be directly extended form the stabilization methodology and the delay-dependent stability given in Theorem 2.

Assumption 2: *In the following, all the pseudo-states are assumed to be available in order to establish the control law.*

Over the past decade, the problem of controller synthesis to ensure the stabilization of time-delay systems has been considered by many researchers in a large number of papers concerning time-delay systems with integer order. In this work, a simple linear pseudo-state feedback control law of the form

$$u(t) = K_0 x(t) \tag{5.61}$$

is considered to prove the stabilization of FOS in the form (5.19), where $K_0 \in \mathbb{R}^{m \times n}$ is the feedback gain matrix.

In fact, the closed-loop given by the system (5.19) and (5.61) is described by

$$D^{\alpha}x(t) = (A_0 + B_u K_0)x(t) + A_{\tau}x(t - \tau(t)) \tag{5.62a}$$

$$x(t) = \psi(t), t \in [-\tau_m, 0], \quad 0 < \alpha < 1 \tag{5.62b}$$

Definition 2: *The control law* (5.61) *is an asymptotic pseudo-state feedback control of the system* (5.19) *if the closed loop system* (5.62) *is asymptotically stable.*

5.5.1 Feedback Stabilization Based on Time-Delay-Independent Stability Condition

The Theorem 3 will lead us to an optimization problem, which provides as a solution the parameters of the considered control law (5.61). The following Theorem 3 gives a sufficient condition on the stabilization of the system (5.19) by the pseudo-state feedback controller gain matrix, expressed in terms of LMIs, which can be solved easily by using any free solvers based, e.g., on Interior Point techniques. So, the computed pseudo-state feedback controller ensures the stabilization of an unstable system given by (5.62).

Theorem 3: *There exists a pseudo-state feedback controller* (5.61) *that stabilizes an unstable fractional order time-delay system in the form* (5.19) *if there exists a positive definite matrix P and a matrix Y_0 which satisfy the following LMI*

$$\Omega = \begin{bmatrix} A_0P + PA_0^T + B_0Y_0 + Y_0^TB_0^T + 2\Lambda_1 & A_\tau P \\ * & -\Lambda_2 \end{bmatrix} < 0 \tag{5.63}$$

where

$$\Lambda_1 = \frac{1}{2(1-\partial\tau_{\max})}I, \quad \Lambda_2 = \frac{1}{(1-\partial\tau_{\max})}I.$$

The pseudo-state feedback controller gain matrix is given as $K_0 = Y_0P$.
□

Proof 3: *The proof is similar to the one of* Theorem 1 *and is omitted. Indeed, its suffice to replace A_0 in the proof of* Theorem 1 *by $A_0 + B_uK_0$. Then, the inequality* (5.63) *is obtained by pre- and postmultiplication of the above inequality with the following matrices* $\begin{bmatrix} P^{-1} & 0 \\ 0 & I \end{bmatrix}$ and $\begin{bmatrix} P^{-1T} & 0 \\ 0 & I \end{bmatrix}$, *respectively, with $P^{-1} = P^{-1T}$. Thus, using this change of variable $Y_0 = K_0P^{-1}$, the obtained expression becomes* (5.63). *It completes the proof.*

5.5.2 Feedback Stabilization Based on Time-Delay-Dependent Stability Condition

In this part, the procedure given in the subsection 5.1 is extended to the case where the time-delay-dependent stability condition is used.

Theorem 4: *There exists a pseudo-state feedback controller* (5.61) *that stabilizes an unstable fractional order time-delay system in the form* (5.19) *if there exist four symmetric positive definite matrices $\tilde{\mathcal{P}}$, $\mathcal{Q}$, $\mathcal{Z}$ and Γ, and three matrices $\mathcal{T}$, $\mathcal{Y}_0$ and $\mathcal{Y}$ with appropriate dimensions such that the following matrices inequalities*

$$\Upsilon = \begin{bmatrix} \Upsilon_{11} & \Upsilon_{12} & \Upsilon_{13} \\ * & \Upsilon_{22} & \Upsilon_{23} \\ * & * & \Upsilon_{33} \end{bmatrix} < 0 \tag{5.64a}$$

$$\Omega = \begin{bmatrix} \mathbf{\Gamma}_{11} & \mathbf{\Gamma}_{12} & \mathcal{Y} \\ * & \mathbf{\Gamma}_{22} & \mathcal{T} \\ * & * & \mathcal{Z} \end{bmatrix} \geq 0 \tag{5.64b}$$

where

$$\begin{aligned}
\Upsilon_{11} &= A_0\tilde{\mathcal{P}} + \tilde{\mathcal{P}}A_0^T + B_0\mathcal{Y}_0 + \mathcal{Y}_0^T B_0^T + \tilde{\mathcal{Y}} + \tilde{\mathcal{Y}}^T + \tau_m\tilde{\Gamma}_{11} + \tilde{\mathcal{Q}} \\
\Upsilon_{12} &= A_\tau - \tilde{\mathcal{P}}\mathcal{Y} + \tilde{\mathcal{T}}^T + \tau_m\tilde{\Gamma}_{12} \\
\Upsilon_{13} &= \tau_m\tilde{\mathcal{P}}A_0^T\mathcal{Z} \\
\Upsilon_{22} &= -\Lambda_1\mathcal{Q} - \mathcal{T} - \mathcal{T}^T + \tau_m\Gamma_{22} \\
\Upsilon_{23} &= \tau_m A_\tau^T\mathcal{Z} \\
\Upsilon_{33} &= -\tau_m\mathcal{Z}
\end{aligned}$$

and

$$\tilde{\mathcal{Y}} = \tilde{\mathcal{P}}\mathcal{Y}, \quad \tilde{\mathcal{Q}} = \tilde{\mathcal{P}}\mathcal{Q}, \quad \tilde{\mathcal{T}} = \mathcal{T}\tilde{\mathcal{P}}, \quad \tilde{\mathbf{\Gamma}}_{11} = \tilde{\mathcal{P}}\mathbf{\Gamma}_{11}, \quad \tilde{\mathbf{\Gamma}}_{12} = \tilde{\mathcal{P}}\mathbf{\Gamma}_{12}.$$

hold for a given scalars τ_m *and* $\partial\tau_m$ *satisfying the Assumption* (5.20), *where* $*$ *denotes the corresponding part of a symmetric matrix.*

The pseudo-state feedback controller gain matrix is given as $K_0 = \mathcal{Y}_0\tilde{\mathcal{P}}$.

Proof 4: *The proof of this Theorem is similar to the one given for the* Theorem 2, *the explanation here is omitted for brevity. After replacing the matrix* A_0 *by* $A_0 + B_uK_0$ *and following the same procedure as the proof of* Theorem 3, *i.e., pre- and postmultiply the obtained inequality by the following* matrices $\begin{bmatrix} \mathcal{P}^{-1} & 0 & 0 \\ 0 & I & 0 \\ 0 & 0 & I \end{bmatrix}$ and $\begin{bmatrix} \mathcal{P}^{-1} & 0 & 0 \\ 0 & I & 0 \\ 0 & 0 & I \end{bmatrix}^T$, *respectively, and making some change of variables, where* $\tilde{\mathcal{P}} = \mathcal{P}^{-1}$ *yields the inequality* (5.64). *It completes the proof.*

We can remark that the feedback stabilization problem based on the time-delay-dependent stability condition given by the inequality (5.64) is a nonconvex problem. This is due to the presence of decision variable $\mathcal{Z}$alone and in the product $\tilde{\mathcal{P}}A_0^T\mathcal{Z}$, such that the matrix A_0 is not necessarily nonsingular, which leads to a bilinear matrix inequality (BMI) structure. Then, the inequality (5.64) can not be solved for the given decision variables in the same time. Thanks to the Schur lemma, all diagonal components of a matrix must satisfy the inequality. Therefore, a two-step procedure is proposed to resolve this problem, as in Boukal et al. (2016a). Firstly, solving the first component of the inequality (5.64) allows us to compute $\tilde{\mathcal{P}}$ and $\tilde{\mathcal{Y}}_0$. Then, after obtaining $\tilde{\mathcal{P}}$ and $\tilde{\mathcal{Y}}_0$, replacing them into the inequality (5.4) by their value leads to a feasible LMI.

5.6 NUMERICAL EXAMPLE

This section describes simulation results to show the performances of the proposed stabilization method. The simulation and implementation of fractional derivative and integrator present many difficulties. There are some methods in the literature that are used to approximate the fractional order integrals in the frequency domain (see Oustaloup, 1991; Matsuda and Fuji, 1993; Krishna and Reddy, 2008). In this work, the simulations are based on the CRONE Toolbox, developed by the CRONE research group (for more details see Oustaloup, 1991; Malti et al., 2011 and references therein).

5.6.1 Approximation of the Fractional Order Derivatives and Consistent Initialization

One of the difficulties that have been seen in applications and simulations of fractional order derivative and integrator terms are the complications in practicable implementation of this kind of operators. This explains the use of approximations of fractional order derivative and integrator terms by using integer order derivatives and integrals only (Oustaloup, 1991; Vinagre et al., 2000).

Among the several existing approximations in the literature, the most used is the Crone approximation which is the one considered in this paper (see Oustaloup, 1995; Malti et al., 2012; Lanusse et al., 2013 and the references therein). This approximation makes use of a recursive distribution of poles and zeros within a frequency range $[w_l, w_h]$. A more precise approximation of the fractional differentiator in the frequency band $[w_l, w_h]$ is obtained by increasing N. Based on this approximation, the original mathematical concepts developed were integrated in the CRONE toolbox, developed since the 1990s by the CRONE team—a Matlab toolbox dedicated to fractional calculus and its applications in signal processing and automatic control.

For the fractional order differential equations, the initialization of this kind of system is different from the integer order. These systems are known for the fact that they have effectively infinite memory. In fact, the initial conditions must be replaced by initial condition functions, or history functions. For an extensive overview, refer to Sabatier et al. (2010) and the references therein.

The initialization method used in this work to create a consistent system initial behavior is given as:

- for $t < -t_0$ the system is at rest;
- for $-t_0 < t < t_1$, an input is applied to the system
- for $-t_1 < t < 0$, an input equal to 0 is applied to the system, that reach time $t = 0$ with a nonnull initialization.

where $-t_0 < -t_1 < 0$.

5.6.2 Example 1

Consider an unstable fractional order time-varying delay system given by

$$D^{0.5}x(t) = \begin{bmatrix} 1 & 2 \\ 1 & -2 \end{bmatrix} x(t) + \begin{bmatrix} -0.2 & 0.8 \\ -0.75 & 0.5 \end{bmatrix} x(t-\tau(t)) + \begin{bmatrix} 2 \\ 1 \end{bmatrix} u(t) \quad (5.65)$$

The pseudo-state time-varying delay function is chosen as

$$\tau(t) = \tau_1 \sin(\varpi t - \varphi) + \gamma,$$

where $\tau_1 = 0.3$, $\varpi = 50\text{rad/s}$, $\gamma = 0.5$ and $\varphi = 0$.

The instability of the considered system (5.65) is shown in Fig. 5.1, where the evolution of the pseudo-state vector $x(t)$ in open loop with $u(t) = 0$ is drawn.

The solvability of the LMI problem (5.22) given in Theorem 1 can be checked by using any (SDP solver. Therefore, the pseudo-state feedback controller gain matrix K_0 can be deduced from the LMI (5.63) constraint solution.

A feasible solution of the LMI (5.63) is given by

$$P = \begin{bmatrix} 1.2275 & 0.45528 \\ 0.45528 & 1.0637 \end{bmatrix}, \quad Y_0 = \begin{bmatrix} -2.1154 & -0.40352 \end{bmatrix}$$

Then, from the matrices P and Y_0, the controller gain matrix K_0

$$K_0 = Y_0 P = \begin{bmatrix} -2.7805 & -1.3923 \end{bmatrix}$$

is deduced.

By adding the pseudo-state feedback control law, in Fig. 5.2, the evolution of the pseudo-state vector $x(t)$ converges asymptotically to zero, which is in agreement with what is proved in Theorem 3.

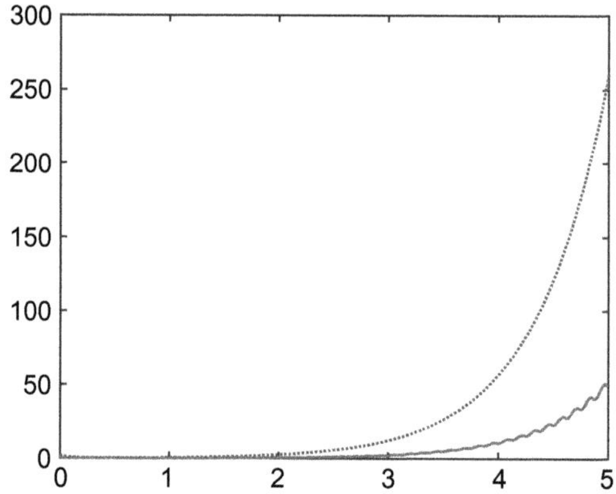

FIGURE 5.1 Evolution of the unforced system pseudo-state vector $x(t)$.

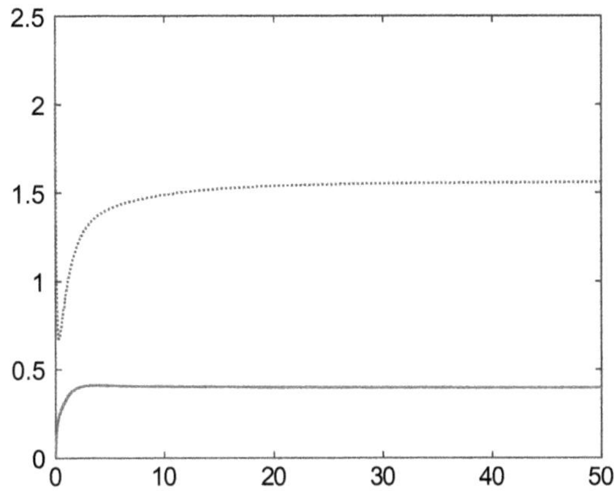

FIGURE 5.2 Evolution of the forced system pseudo-state vector $x(t)$ under a stabilizing control low.

5.6.3 Example 2

For a comparison between the two approaches summarized in the Theorems 1 and 2. In this example, the following fractional order time-varying delay system given by

$$D^{0.75}x(t) = \begin{bmatrix} -1 & 2 \\ 0 & -1 \end{bmatrix} x(t) + \begin{bmatrix} 0.5 & 0 \\ 0 & 1 \end{bmatrix} x(t-\tau(t)) + \begin{bmatrix} 2 \\ 1 \end{bmatrix} u(t) \tag{5.66}$$

is considered.

The pseudo-state time-varying delay function is chosen as

$$\tau(t) = \tau_1 \sin(\varpi t - \varphi) + \gamma,$$

where $\tau_1 = 0.5$, $\varpi = 50$ rad/s, $\gamma = 1$ and $\varphi = 0$.

The LMI constraint (5.22) given in the Theorem 1 was found infeasible. For the same parameters of the system and the same solver, the LMI constraint (5.40) given in the theorem 2 was found feasible, and the solution is given by the following matrices

$$\mathcal{P} = \begin{bmatrix} 4.1969 & 2.0055 \\ 2.0055 & 14.953 \end{bmatrix}, \qquad \Upsilon_{11} = \begin{bmatrix} 0.65502 & -0.53683 \\ -0.53683 & 2.8218 \end{bmatrix},$$

$$\Upsilon_{12} = \begin{bmatrix} -0.31046 & -0.15357 \\ -0.15357 & -2.6286 \end{bmatrix}, \qquad \Upsilon_{22} = \begin{bmatrix} 5.0806 & 0.52675 \\ 0.52675 & 7.8508 \end{bmatrix},$$

$$\mathcal{T} = \begin{bmatrix} -0.0039364 & -0.067655 \\ -0.067655 & -0.042041 \end{bmatrix}, \qquad \mathcal{Y} = \begin{bmatrix} -0.059691 & 0.029161 \\ 0.029161 & -0.078605 \end{bmatrix},$$

$$\mathcal{Z} = \begin{bmatrix} 0.21828 & 0.2245 \\ 0.2245 & 0.66927 \end{bmatrix}.$$

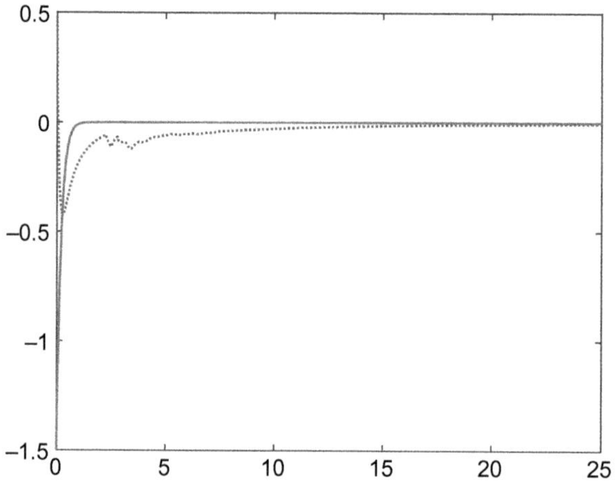

FIGURE 5.3 Evolution of the unforced system pseudo-state vector $x(t)$.

The condition given by the LMI (5.63) is not satisfied. This is not sufficient to confirm the nonstability of the system (5.66), which is proved by the existence of a solution to the LMI (5.64) (see the Fig. 5.3).

Now, to achieve more stability robustness, the synthesis algorithm presented in Section 5.2 is applied to solve the delay-independent stabilization problems given by the LMI (5.64).

A feasible solution of the LMI (5.63) is given by

$$P = \begin{bmatrix} 2.119 & -0.088516 \\ -0.088516 & 1.2144 \end{bmatrix}, \quad Y_0 = \begin{bmatrix} -0.12171 & -1.2753 \end{bmatrix}$$

Then, from the matrices P and Y_0, the controller gain matrix K_0

$$K_0 = Y_0 P = \begin{bmatrix} -0.14502 & -1.5379 \end{bmatrix}$$

is deduced.

In Fig. 5.4, the evolution of the controlled system by using pseudo-state vector $x(t)$ under a stabilizing controller low is shown. Due to scale effect, Fig. 5.4 is zoomed in Fig. 5.5, with time interval $\forall t \in [0, 5]$ in order to show the behavior near to the origin.

5.7 CONCLUSION

In this chapter, the stability and stabilization of fractional order time-varying delay systems are investigated. The indirect Lyapunov and Lyapunov–Krasovskii approaches are used to derive dependent and independent delay time sufficient conditions for the asymptotic stability of fractional order time-varying delay systems. In addition, based on the obtained

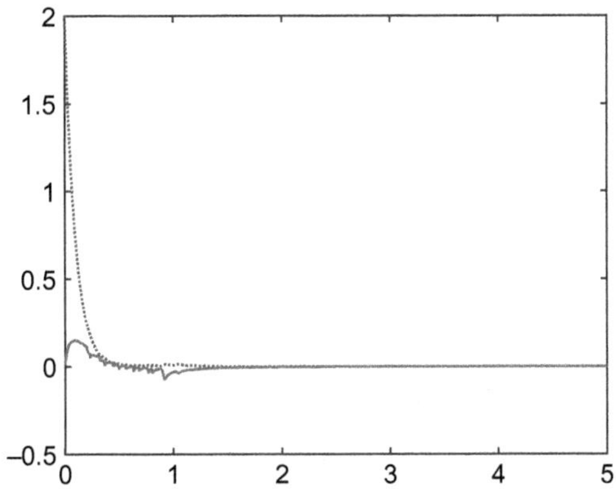

FIGURE 5.4 Evolution of the forced system pseudo-state vector $x(t)$ under a stabilizing control low.

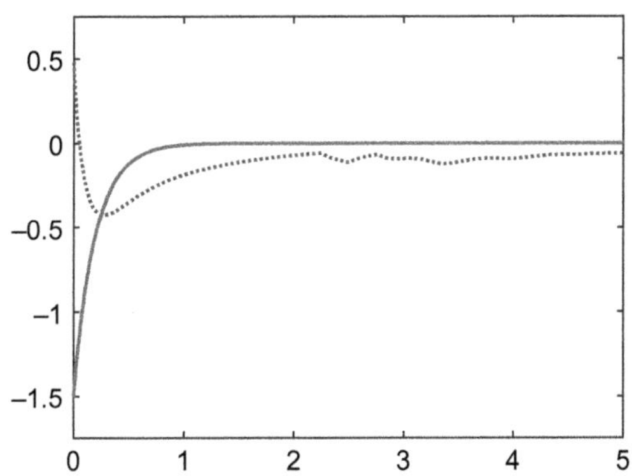

FIGURE 5.5 Zoom on the part near the origin of the evolution of the unforced system pseudo-state vector $x(t)$.

conditions, new conditions guaranteeing the existence and the synthesis of pseudo-state feedback stabilizing controller are deduced. The computed controller law ensures the stabilization of unstable fractional order time-varying delay systems. Examples with an unstable and stable fractional order time-varying delay systems are studied to illustrate the effectiveness of the proposed approaches.

REFERENCES

Ahmad, W., El-Khazali, R., Al-Assaf, Y., 2004. Stabilization of generalized fractional order chaotic systems using state feedback control. Chaos Solitons Fractals 22, 141–150.

Ahn, H., Chen, Y., Podlubny, I., 2007. Robust stability test of a class of linear time-invariant interval fractional-order system using Lyapunov inequality. Appl. Math. Comput. 187, 27–34.

Amato, F., Ambrosino, R., Ariola, M., Cosentino, C., Merola, A., 2007. State feedback control of nonlinear quadratic systems, in: Proc. IEEE Conf. Decision & Control, New Orleans, USA. pp. 1699–1703.

Amato, F., Cosentino, C., Fiorillo, A., Merola, A., 2009. Stabilization of bilinear systems via linear state-feedback control. IEEE Trans. Circ. Syst. II : Express Briefs 56, 76–80.

Amato, F., Ariola, M., Cosentino, C., Merola, A., 2010. Output feedback control of nonlinear quadratic systems, in: Proc. IEEE Conf. Decision & Control, Atlanta, USA. pp. 3349–3354.

Audounet, J., Giovangigli, V., Roquejoffre, J.M., 1998. A threshold phenomenon in the propagation of a point source initiated flame. Phys. D: Nonlinear Phenomena 121, 295–316.

Azar, A., Vaidyanathan, S., Ouannas, A., 2017. Fractional Order Control and Synchronization of Chaotic Systems. Springer-Verlag, Germany.

Bagley, R., Calico, R., 1991. Fractional order state equations for the control of viscoelastically damped structures. J. Guidance, Contr. Dynamics 14, 304–311.

Battaglia, J., Le Lay, L., Batsale, J., Oustaloup, A., Cois, O., 2000. Heat flux stimation through inverted non integer identification models. Int. J. Ther. Sci. 39, 374–389.

Bonnet, C., Partington, J., 2001. Stabilization of fractional exponential systems including delays. Kybernetika 37, 345–353.

Bonnet, C., Partington, J., 2002. Analysis of fractional delay systems of retarded and neutral type. Automatica 38, 1133–1138.

Boroujeni, E., Momeni, H., 2012a. Observer based control of a class of nonlinear fractional-order systems using LMI. Int. J. Sci. Eng. Investigat. 1, 48–52.

Boroujeni, E.A., Momeni, H.R., 2012b. Non-fragile nonlinear fractional order observer design for a class of nonlinear fractional order systems. Signal Process. 92, 2365–2370.

Boukal, Y., Darouach, M., Zasadzinski, M., Radhy, N., 2014a. Design of functional fractional-order observers for linear time-delay fractional-order systems in the time domain. Fractional Differentiation and Its Applications (ICFDA). International Conference on, Catania, Italy, pp. 1–6.

Boukal, Y., Darouach, M., Zasadzinski, M., Radhy, N., 2014b. H_∞ observer design for linear fractional-order systems in time and frequency domain, in: Proc. European Contr. Conf., Strasbourg, France. pp. 2975–2980.

Boukal, Y., Darouach, M., Zasadzinski, M., Radhy, N., 2015a. Unknown input observer design for linear fractional-order time-delay systems. J. Appl. Nonlinear Dynam. 4, 117–130.

Boukal, Y., Zasadzinski, M., Darouach, M., Radhy, N., 2015b. H_∞ filters design for fractional-order time-varying delay systems, in: Proc. European Contr. Conf., Linz, Austria. pp. 1243–1248.

Boukal, Y., Zasadzinski, M., Darouach, M., Radhy, N., 2016a. Robust H_∞ observer-based stabilization of disturbed uncertain fractional-order systems using a two-step procedure. Theoretical Developments and Applications of Non-Integer Order Systems. Springer, pp. 167–180, Lecture Notes in Electrical Engineering.

Boukal, Y., Zasadzinski, M., Darouach, M., Radhy, N.E., 2016b. Robust functional observer design for uncertain fractional-order time-varying delay systems, in: 2016 American Control Conference (ACC), pp. 2741–2746.

Boukal, Y., Darouach, M., Zasadzinski, M., Radhy, N.E., 2017a. H_∞ observer-based-controller for fractional-order time-varying-delay systems, in: 2017 25th Mediterranean Conference on Control and Automation (MED), pp. 1287–1292.

Boukal, Y., Darouach, M., Zasadzinski, M., Radhy, N.E., 2017b. Robust H_∞ observer-based control of fractional-order systems with gain parametrization. IEEE Trans. Autom. Control to appear . Available from: https://doi.org/10.1109/TAC.2017.2690140.

Boukal, Y., Zasadzinski, M., Darouach, M., Radhy, N., 2017c. H_∞ dynamic output feedback controller design for disturbed fractional-order system, in: 20th IFAC World Congress, Elsevier. pp. 10134–10139.

Boyd, S., El Ghaoui, L., Féron, E., Balakrishnan, V., 1994. Linear Matrix Inequalities in Systems and Control Theory. SIAM, Philadelphia.

Busłowicz, M., 2008. Stability of linear continuous-time fractional order systems with delays of the retarded type. Bull. Polish Acad. Sci. Tech. Sci. 56, 237–240.

Chen, Y., 2006. Ubiquitous fractional order controls? in: Proc. IFAC Workshop on Fractional Differentiation and its Application, Porto, Portugal, pp. 481–492.

Chyi, H., Yi-Cheng, C., 2006. A numerical algorithm for stability testing of fractional delay systems. Automatica 42, 825–831.

Darling, R., Newman, J., 1997. On the short behaviour of porous interaction electrodes. J. Electrochem. Soc. 144, 3057–3063.

Das, S., 2008. Functional Fractional Calculus for System Identification and Controls. Springer-Verlag, Heidelberg.

Deng, W., Li, C., Lü, J., 2007. Stability analysis of linear fractional differential system with multiple time delays. Nonlinear Dynam. 48, 409–416.

Farshad, M.B., Masoud, K.G., 2009. An efficient numerical algorithm for stability testing of fractional-delay systems. {ISA} Transactions 48, 32–37.

Fioravanti, A.R., Bonnet, C., Özbay, H., Niculescu, S.I., 2012. A numerical method for stability windows and unstable root-locus calculation for linear fractional time-delay systems. Automatica 48, 2824–2830.

Freed, A., Diethelm, K., 2006. Fractional calculus in biomechanics: a 3D viscoelastic model using regularized fractional derivative kernels with application to the human calcaneal fat pad. Biomech. Model. Mechanobiol. 5, 203–215.

Komornik, V., 2016. Generalized newton–leibniz formula. Lectures on Functional Analysis and the Lebesgue Integral. Springer, pp. 197–209.

Krishna, B., Reddy, K., 2008. Active and passive realization of fractance device of order 1/2. Active and passive electronic components 2008.

Lanusse, P., Malti, R., Melchior, P., 2013. Crone control system design toolbox for the control engineering community: tutorial and case study. Phil. Trans. R. Soc. A 371.

Lederman, C., Roquejoffre, J.M., Wolanski, N., 2002. Mathematical justification of a nonlinear integro-differential equation for the propagation of spherical flames. Compt. Rendus Math. 334, 569–574.

Lianglin, X., Yun, Z., Tao, J., 2011. Stability analysis of linear fractional order neutral system with multiple delays by algebraic approach. World Acad. Sci. Eng. Technol. 5, 758–761.

Lin, T., Lee, T., 2011. Chaos synchronization of uncertain fractional-order chaotic systems with time delay based on adaptive fuzzy sliding mode control. Fuzzy Syst. IEEE Trans. 19, 623–635.

Liqiong, L., Shouming, Z., 2010. Finite-time stability analysis of fractional-order with multi-state time delay. Int. J. Informat. Math. Sci. 6, 237–240.

Magin, R.L., 2006. Fractional calculus in bioengineering. Begell House Redding.

Malti, R., Melchior, P., Lanusse, P., Oustaloup, A., 2011. Towards an object oriented crone toolbox for fractional differential systems, in: Proceedings of 18th IFAC World Congress, pp. 10830–10835.

Malti, R., Melchior, P., Lanusse, P., Oustaloup, A., 2012. Object-oriented crone toolbox for fractional differential signal processing. Signal Image Video Process. 6, 393–400.

Matignon, D., 1994. Représentation en Variables d'État de Modèles de Guides d'Ondes avec Dérivation Fractionnaire. Ph.D. thesis. Université de Paris XI, Orsay, France.

Matignon, D., 1998. Generalized fractional differential and difference equations: stability properties and modelling issues, in: Proc. Mathematical Theory of Networks and Systems Symposium, Padova, Italy. pp. 503–506.

Matsuda, K., Fuji, H., 1993. H_∞ optimized wave-absorbing control: analytical and experimental results. J. Guidance, Contr. Dynamics 16, 1146–1153.

Mihailo, P.L., Aleksandar, M.S., 2009. Finite-time stability analysis of fractional order time-delay systems: Gronwall's approach. Math. Computer Modelling 49, 475–481.

Montseny, G., 1998. Diffusive representation of pseudo-differential time-operators, in: ESAIM: Proceedings, Fractional Differential Systems: Models, Methods and Applications, EDP Sciences, pp. 159–175.

Ortigueira, M.D., Machado, J.A.T., 2003. Editorial: fractional signal processing and applications. Signal Process. 83, 2285–2286.

Oustaloup, A., 1991. La Commande CRONE. Hermes, Paris.

Oustaloup, A., 1995. La Dérivation Non Entière: Synthése et Applications. Hermes, Paris.

Podlubny, I., 1999. Fractional Differential Equations. Academic, New York.

Podlubny, I., 2002. Geometric and physical interpretation of fractional integration and fractional differentiation. Fract. Calculus Appl. Anal. 5, 367–386.

Sabatier, J., Agrawal, O.P., Machado, J.A.T., 2007. Advances in Fractional Calculus: Theoretical Developments and Applications in Physics and Engineering. Springer, New York.

Sabatier, J., Merveillaut, M., Malti, R., Oustaloup, A., 2010. How to impose physically coherent initial conditions to a fractional system? Commun Nonlinear Sci. Numer. Simulat. 10, 1318–1326.

Sabatier, J., Farges, C., Trigeassou, J., 2014. Fractional systems state space description: some wrong ideas and proposed solutions. J. Vibration Control 20, 1076–1084.

Sheng, H., Chen, Y., Qiu, T., 2012. Fractional Processes and Fractional-Order Signal Processing - Techniques and Applications. Springer, Berlin.

Si-Ammour, A., Djennoune, S., Bettayeb, M., 2009. A sliding mode control for linear fractional systems with input and state delays. Commun. Nonlinear Sci. Numer. Simul. 14, 2310–2318.

Sierociuk, D., Dzieliński, A., Sarwas, G., Petras, I., Podlubny, I., Skovranek, T., 2013. Modelling heat transfer in heterogeneous media using fractional calculus. Philos. Trans. R. Soc., A 371.

Tarbouriech, S., 1997. Local stabilization of continuous-time delay systems with bounded inputs, in: Proc. European Contr. Conf., Brussels, Belgium. pp. 3684–3689.

Trigeassou, J., Maamri, N., Sabatier, J., Oustaloup, A., 2011. A Lyapunov approach to the stability of fractional differential equations. Signal Process. 91, 437–445.

Trigeassou, J.C., Maamri, N., 2009. State space modeling of fractional differential equations and the initial condition problem, in: Systems, Signals and Devices, 2009. SSD '09. 6th International Multi-Conference on, pp. 1–7.

Vinagre, B., 2001. Modeling and Control of Dynamic Systems characterized by Integro-differential Equations of Fractional Order. Ph.D. thesis. University of Distance Education. Spain.

Vinagre, B.M., Podlubny, I., Hernández, A., Feliu, V., 2000. Some Approximations of fractional order operators used in control theory and applications. Fract. Calculus Appl. Anal. 3, 231–248. chapter 3.

Yuan, L., Yang, Q., Zeng, C., 2013. Chaos detection and parameter identification in fractional-order chaotic systems with delay. Nonlinear Dynam. 73, 439–448.

Chapter 6

Fractional Order Error Models With Parameter Constraints

Norelys Aguila-Camacho, Manuel A. Duarte-Mermoud and Matías G. Mayol-Suárez
University of Chile, Santiago, Chile

6.1 INTRODUCTION

Error models (EM) were first introduced in adaptive control literature by Narendra and Kudva (1974) and then in a unified form in Narendra (1976), for the analysis of integer order (IO) adaptive systems. They have become an important approach to face numerous adaptive control and identification problems. Four EM have been recognized in the IO case and completely studied, denoted as IOEM1 to IOEM4, where the adaptive laws (AL) are of IO type. For details on these IOEM, the reader is referred to Chapter 6 of the book by Narendra and Annaswamy (2005) where a complete study of them is performed. This viewpoint has been successfully used for continuous-time systems (Narendra and Kudva, 1974; Narendra, 1976; Narendra and Annaswamy, 2005) (which will be the focus of this study), but it has also been applied for discrete-time systems (Narendra and Lin, 1980) and even for hybrid systems (systems with two different time-scales) (Narendra et al., 1983, 1985). Its importance lies in the fact that the solution of any adaptive control or identification problem can be readily obtained if the problem is stated within this framework. The conditions under which stability and convergence of IOEM are attained, were established in Narendra and Annaswamy (2005) by defining suitable AL for each case.

A further step in the study of IOEM was done in 1988 when the analysis of two IOEM, whose constant but unknown parameters are related through a linear algebraic relationship, was completely solved in Duarte and Narendra (1988a, 1996a), where conditions for stability and convergence of the control/identification error and parameter error were derived. It is well known that in adaptive systems the ideal controller parameters and the true plant parameters are related by linear algebraic equations through the certainty equivalence principle or the Bezout identity. The importance of this subject

Mathematical Techniques of Fractional Order Systems. DOI: https://doi.org/10.1016/B978-0-12-813592-1.00006-4

is that any control scheme that is performing a parameter estimation, which is simultaneously being used in the controller parameter updating, leads to the situation of two IOEM with a constraint on their true parameters. Thus, AL taking into account this parameter linkage are called for. This is precisely the problem solved in Duarte and Narendra (1988a, 1996a) in a general manner that leads to better solutions than those where the parameters of the controller and the identifier are independently updated. It is worth mentioning that the so-called combined direct and indirect adaptive control introduced by Duarte & Narendra (e.g., combined model reference adaptive control) (Duarte and Narendra, 1987a,b, 1989a,b) and the dynamical indirect adaptive control (e.g., dynamical indirect MRAC) (Duarte and Narendra, 1988b, 1996b) belong to this class of problems.

On the other hand, a great deal of attention has been paid on the study of the FO systems (FOS) where FO operators (integrals and/or derivatives) rather than IO operators are used either in the plant description (modeling) and/or in the controller design. This fact has allowed to expand the controllers/observers designs to include FO operators providing additional degrees of freedom in the design. For example, the IOPID controller contains three design parameters (proportional, integral, and derivative constants) whereas the FOPID contains five parameters (proportional, integral, and derivative constants; plus the integral and derivative orders) augmenting the search space for tuning the controller allowing to obtain better solutions according to prespecified performance indexes. As an example, in Valério and Da Costa (2006) is presented a Ziegler–Nichols type of method to tune FOPID controllers, whereas in Aguila-Camacho and Duarte-Mermoud (2013) an application of FOPID to an automatic voltage regulator (AVR) is discussed. The AVR is a controller whose main purpose is to maintain the voltage level in an electric generator by adjusting the generator exciter voltage. In that work a genetic optimization algorithm was used to tune a FOPID. Several other applications of FO controllers and observers have been lately reported. It is worth mentioning the work by (Kavuran et al., 2016) where an experimental test platform of a low-cost coaxial rotor was developed and a model reference adaptive controller with FO AL was explored and implemented in MATLAB/Simulink. In Huan et al. (2016) a method is proposed to solve all possible FOPD controllers to stabilize a given system. Robust stability with phase margin and gain margin is considered and the parameters regions are described on the delay-parameter plane. The rather complex dynamics of lithium-ion battery is described in Takamatsu and Ohmori (2015) as a fractional order (FO) system and then the Kreisselmeier-type of adaptive observer is used to estimate the state and parameters battery. The results obtained by numerical simulations indicate that the proposed model represents adequately the system behavior.

On the theoretical side, several new results have been recently published on the FO controllers and observers in the book (Azar et al., 2017), where several methods for FO control are treated in Part I: Fractional Order Control Systems. Also, in Part II: Applications of Fractional Order Chaotic Systems of Azar et al. (2017), numerous applications on control and synchronization of chaotic systems are discussed. Other recent analytical developments worth to mention are Merrikh-Bayat (2013), where the FO unstable pole-zero cancellation problem in linear feedback systems is presented; Ladaci and Bensafia (2016), in which the indirect adaptive control using FO pole assignment is discussed; Rapaic and Pisano (2014), where the case of variable derivation order is analyzed through an estimation process; Liu et al. (2017), in which an interesting adaptive fuzzy backstepping control technique of FO nonlinear systems is presented; and Wei et al. (2017), where a discussion on FO adaptive observers is presented, just to mention a few of them.

The next step in this sequence of studying FOEM is to study two EM whose true and unknown parameters are coupled by a linear relationship, but the AL for updating the unknown parameters are done using FOAL. This situation occurs when a FO plant is being controlled by an adaptive control strategy of FO type (i.e., FOAL) and simultaneously, the AL for estimating plant parameters are of FO type. This problem was already solved for the case of the fractional order EM (FOEM) of type 1 (FOEM1) in Aguila-Camacho and Duarte-Mermoud (2015, 2017). In this Chapter, the cases of the so-called FOEM2 and FOEM3 when the true parameters of both FOEM are coupled are treated in detail.

FOEM2 defines a special case of FOEM, where the whole state error vector is accessible to the designer. The analytical results presented in this chapter show that it is possible to find coupled fractional AL, such that the overall adaptive system is globally stable, when the FO is in the interval $\alpha \in (0, 1]$. Also, it is analytically proved that the mean value of the squared norm of the state error vector converges asymptotically to zero. The same analysis is performed for the FOEM3, where the main difference with FOEM2 lies in the fact that only one component of the state error vector is available (output error), proving the same properties as in the FOEM2.

The presentation is organized as follows. After the introductory Section 6.1, some preliminary results used in this study are presented in Section 6.2. They form the basis for understanding the mathematical developments in analyzing the coupled FOEM2 (CFOEM2) and the CFOEM3 performed in Sections 6.3 and 6.4, respectively. In Section 6.5 some simulation examples on a CFOEM2 of dimension 2 are presented to illustrate the benefit of using CFOEM theory under ideal conditions as well as in the presence of noise. Finally, some general conclusions on this study are drawn in Section 6.6.

6.2 PRELIMINARIES

In this section, some Definitions, Lemmas, and Theorems are presented, which will be used in the analyses performed throughout the chapter.

6.2.1 Fractional Calculus

Since the first known mention of derivatives of noninteger order, made by Leibinz in a note to L'Hopital in 1695, the theory of fractional calculus was developed mainly as a pure theoretical field of mathematics, useful only for mathematicians. However, during the last decades of the 20th century, some applications of fractional operators in science appeared, and the interest of using these operators in more applied fields started to grow significantly. Nowadays, it has been recognized that fractional calculus plays an important role in science, with an increased use of fractional operators in engineering.

In the time domain, the fractional order derivative (FOD) and the fractional order integral (FOI) operators are defined by a convolution operation. The Riemann–Liouville FOI is one of the main concepts of fractional calculus, and is presented in Definition 1.

Definition 1: Riemann–Liouville FOI (Kilbas et al., 2006). *The Riemann–Liouville FOI of order* $\alpha > 0$ *of a function* $f(t) \in \mathbb{R}$ *is defined as*

$$I_{t_0}^{\alpha} f(t) = \frac{1}{\Gamma(\alpha)} \int_{t_0}^{t} \frac{f(\tau)}{(t-\tau)^{1-\alpha}} d\tau, \quad t > t_0, \tag{6.1}$$

where $\Gamma(\alpha)$ *corresponds to the Gamma Function* (Kilbas et al., 2006).

Regarding the FOD of order $\alpha > 0$ of a function $f(t) \in \mathbb{R}$, there exist several definitions. The results presented in this work use the Caputo definition given in Definition 2, which has been extensively used in the literature for systems modeling and control.

Definition 2: Caputo FOD (Diethelm, 2010). *Let* $\alpha \geq 0$ *and* $n = [\alpha]$, *i.e., the integer part of* α. *According to* (Diethelm, 2010, Def. 3.1), *the Caputo fractional derivative of order* $\alpha > 0$ *of a function* $f(t) \in \mathbb{R}$ *is defined as*

$${}^{C}D_{t_0}^{a} f(t) = \frac{1}{\Gamma(n-\alpha)} \int_{t_0}^{t} \frac{f^{(n)}(\tau)}{(t-\tau)^{\alpha-n+1}} d\tau \tag{6.2}$$

whenever $f^{(n)} \in L_1[t_0, t]$.

Some active researchers in the field have stated that the Caputo definition does not allow to take into account initial conditions properly, and for that reason they state that it should not be used to model fractional systems. However, still there is no consensus about this topic, since there are other works where real systems have been modeled using fractional differential equations with the Caputo's definition (see for instance Freed and Diethelm, 2007; Tejado et al., 2014). Thus, it can be stated that there are systems described using this kind of fractional differential equations, and for that reason as control engineers it is worth to study this kind of systems, as well as proposing control strategies for them. In the sequel, Caputo's FOD will be used in this work.

6.2.2 Stability of Fractional Order Systems

The known stability methods for integer order systems (IOS) are not directly applicable to the stability analysis of fractional order systems (FOS). However, many of them have been generalized to the FOS case. The conditions under which linear time-invariant FOS are stable were studied by Matignon (1994). However, in the case of fractional order adaptive systems (FOAS) this analysis is not valid, since they are nonlinear and time-varying.

Lyapunov's direct method provides an effective way of analyzing the stability of nonlinear and time-varying IOS. The FO extension of Lyapunov direct method was proposed by Li et al. (2010). This method allows concluding asymptotic stability and Mittag–Leffler stability for FOS. However, it doesn't address the frequent case of adaptive schemes where the FOD of the Lyapunov function is only negative semidefinite, and conclusions about stability or uniform stability can be drawn. In that case, Theorem 1 is useful, and it will be used in this work.

Theorem 1: (Duarte-Mermoud et al., 2015). *Let $x = 0$ be an equilibrium point for the nonautonomous FOS* (6.3).

$$ {}^{C}D_{t_0}^{a}\, x(t) = f(x(t), t), \quad \alpha \in (0, 1) \tag{6.3} $$

It is assumed that there exists a continuous function $V(x(t), t)$ such that (Duarte-Mermoud et al., 2015, *Thm. 3 & Rmk. 1),*

- *$V(x(t), t)$ is positive definite, and*
- *${}^{C}D_{t_0}^{\beta} V(x(t), t)$, with $\beta \in (0, 1]$, is negative semidefinite.*
  Then the origin of system (6.3) *is Lyapunov stable.*
- *Furthermore, if $V(x(t), t)$ is decrescent, then the origin of system* (6.3) *is Lyapunov uniformly stable.*

6.2.3 Additional Tools for the Analysis of FOS

The following lemma will be used in this chapter to prove boundedness of the signals in the fractional order error models (FOEM) analyzed here.

Lemma 1: (Duarte-Mermoud et al., 2015). *Let $x(t) \in \mathbb{R}^n$ be a vector of differentiable functions. Then, for all $t > t_0$, the following relationship holds* (Duarte-Mermoud et al, 2015, Lemma 4)

$$\frac{1}{2}\,{}^{C}D_{t_0}^{a}\{x^T P\, x\}(t) \le x^T(t) P\ {}^{C}D_{t0}^{a} x(t), \tag{6.4}$$

where $\alpha \in (0, 1]$ and $P \in \mathbb{R}^{n \times n}$ is a constant, square, symmetric and positive definite matrix.

Particular cases of relation (6.4) (Duarte-Mermoud et al., 2015, *Lemma 4) were proposed in* Aguila-Camacho et al. (2014, *Lemma 1) and* Alikhanov (2010, *Lemma 1).*

In order to analyze the evolution of the output error in the FOEM treated in this chapter, the following lemma will be used.

Lemma 2: (Aguila-Camacho and Duarte-Mermoud, 2016) *L. et $x(\cdot): \mathbb{R}^+ \to \mathbb{R}$ be a bounded nonnegative function. According to* Aguila-Camacho and Duarte-Mermoud (2016, Lemma 4), *if there exists some $\alpha \in (0, 1]$ such that*

$$\frac{1}{\Gamma(\alpha)}\int_{t_0}^{t} \frac{x(\tau)}{(t-\tau)^{1-\alpha}} d\tau < M, \quad \forall t \ge t_0, \quad \text{with } M \in (0, \infty) \tag{6.5}$$

then

$$\lim_{t \to \infty} \left[t^{\alpha - \varepsilon} \ \frac{\int_{t_0}^{t} x(\tau) d\tau}{t} \right] = 0, \qquad \forall \varepsilon > 0 \tag{6.6}$$

6.3 ANALYSIS OF FRACTIONAL ORDER ERROR MODEL 2 WITH PARAMETER CONSTRAINTS

From stability and performance viewpoints, many adaptive systems (identification and/or adaptive control schemes) can be described by a couple of differential equations; one describing the evolution of output error and the other that of the parameter error. This pair of equations are called Error Model (EM). Thus, all the conclusions obtained from the study of this couple of equations regarding stability and performance, are applicable to any adaptive scheme that can be put in this form. This is precisely the main attraction of EM, since it gives a general framework to analyze adaptive systems.

In the IO case, four EM have been completely studied (Narendra and Annaswamy, 2005). With the introduction of fractional operators in adaptive schemes, the generalization of these four EM to the FO case arises. Important advances in the analysis of FOEM have been made in the last years (Aguila-Camacho and Duarte-Mermoud, 2016), although there are some topics currently under investigation. The so called FOEM1 with parameter constraints has already been thoroughly studied in Aguila-Camacho and Duarte-Mermoud (2015,2017) and its main properties were analyzed. In this chapter the focus is on the FOEM2 and FOEM3 with parameters constraints, which will be stated in what follows.

6.3.1 Fractional Order Error Model 2

One of the FOEM studied in Aguila-Camacho and Duarte-Mermoud (2016) is the FOEM2, where the output error $e(t) \in \mathbb{R}^n$ is given by the following fractional order differential equation

$$ {}^{C}D^{\alpha} e(t) = Ae(t) + k \;\; b \;\; \phi^{T}(t)\omega(t), \quad e(t_0) = e_0, \tag{6.7} $$

where $A \in \mathbb{R}^{n \times n}$ is a matrix whose eigenvalues have negative real parts, i.e., positive definite symmetric matrices $P, Q \in \mathbb{R}^{n \times n}$ exist such that $A^T P + PA = -Q$. $e: \mathbb{R}^+ \rightarrow \mathbb{R}^n$ corresponds to the output error, which is assumed to be accessible. k_p is an unknown constant whose sign is assumed to be known (usually the high frequency gain of the plant to be controlled/identified). The sign of the high frequency gain can be also unknown, and in this case an alternative solution using the Nussbaum gain has been proposed for integer order error models (IOEM) (Narendra and Annaswamy, 2005) and it is being currently investigated for the FOEM2 case. $b \in \mathbb{R}^n$, $\phi: \mathbb{R}^+ \rightarrow \mathbb{R}^m$ is the parameter error vector, defined as $\phi(t) = \theta(t) - \theta^*(t)$, with $\theta: \mathbb{R}^+ \rightarrow \mathbb{R}^m$ the estimated parameters and $\theta^* \in \mathbb{R}^m$ the true but unknown parameters. $\omega: \mathbb{R}^+ \rightarrow \mathbb{R}^m$ is a function vector of known signals usually called information vector. The analyses in this paper will be performed for $\alpha \in (0, 1]$.

It was proved in Aguila-Camacho and Duarte-Mermoud (2016) that adaptive law (6.8) can be used to estimate the unknown parameters $\theta(t)$ through

$$ {}^{C}D^{\alpha} \phi(t) = {}^{C}D^{\alpha} \theta(t) = -\gamma \;\; \mathrm{sgn}(k) e^{T}(t) P \;\; b \;\; \omega(t), \quad \phi(t_0) = \phi_0, \tag{6.8} $$

where $\gamma \in \mathbb{R}^+$ is the adaptive gain that allows to handle the parameters convergence. A straightforward extension of this adaptive law is the case when the adaptive gain is given by a matrix $\Gamma = \Gamma^T > 0$, with $\Gamma \in \mathbb{R}^{m \times m}$. Time-varying adaptive gains $\Gamma(t)$ satisfying certain conditions are also an option. This approach successfully used in the IOEM case is a topic currently under research for the FOEM case. Using adaptive law (6.8), it was proved in Aguila-Camacho and Duarte-Mermoud (2016) that all the signals in the adaptive scheme (6.7), (6.8) remain bounded and that the mean value of the squared norm of the output error ($\|e(t)\|^2$) converges asymptotically to zero.

An interesting problem arises when the true unknown parameters of two adaptive systems represented by FOEM2 like (6.7), (6.8), are related through a known linear relationship. In that case, both adaptive schemes can be treated separately, using AL as (6.8) for every system. However, one can wonder if the inclusion of the additional knowledge contained in the linear relationship in the AL could improve their behaviors. This is precisely the idea behind the FOEM2 with parameter constraints, which will be introduced and thoroughly analyzed in the next subsection.

6.3.2 FOEM2 With Parameter Constraints

Two FOEM2 will be consider in what follows where their true unknown parameters satisfy a linear matrix relationship. Let consider two FOEM2, whose output error equations are defined as follows

$$^C D^{\alpha} e_1(t) = A_1 e_1(t) + k_1 \ b_1 \ \phi_1^T(t)\omega_1(t), \quad e_1(t_0) = e_{1_0}, \tag{6.9}$$

$$^C D^{\alpha} e_2(t) = A_2 e_2(t) + k_2 \ b_2 \ \phi_2^T(t)\omega_2(t), \quad e_2(t_0) = e_{2_0}, \tag{6.10}$$

where $A_1, A_2 \in \mathbb{R}^{n \times n}$ are matrices whose eigenvalues have negative real parts, i.e., positive definite symmetric matrices $P_1, P_2, Q_1, Q_2 \in \mathbb{R}^{n \times n}$ exist such that

$$\begin{aligned} A_1^T P_1 + P_1 A_1 &= -Q_1 \\ A_2^T P_2 + P_2 A_2 &= -Q_2. \end{aligned} \tag{6.11}$$

Vectors $e_1, e_2 : \mathbb{R}^+ \rightarrow \mathbb{R}^n$ correspond to the output errors, which are assumed to be accessible. k_1, k_2 are unknown constants whose signs are assumed to be known (usually the high frequency gains of the plants to be controlled/identified), $b_1, b_2 \in \mathbb{R}^n$, $\phi_1, \phi_2 : \mathbb{R}^+ \rightarrow \mathbb{R}^m$ are the parameter errors, defined as

$$\begin{aligned} \phi_1(t) &= \theta_1(t) - \theta_1^*(t) \\ \phi_2(t) &= \theta_2(t) - \theta_2^*(t) \end{aligned} \tag{6.12}$$

with $\theta_1, \theta_2 : \mathbb{R}^+ \rightarrow \mathbb{R}^m$ the estimated parameters and $\theta_1^*, \theta_2^* \in \mathbb{R}^m$ the real but unknown parameters. On the other hand, $\omega_1, \omega_2 : \mathbb{R}^+ \rightarrow \mathbb{R}^m$ are information vectors of available signals and $\alpha \in (0, 1]$.

Let's assume that the unknown parameters $\theta_1^*, \theta_2^*, k_1$ and k_2 are not independent but related through the following linear matrix relationship

$$k_1 R_1 \theta_1^* + k_2 R_2 \theta_2^* = W \tag{6.13}$$

where $R_1, R_2 \in \mathbb{R}^{n \times n}$ are known constant matrices and $W \in \mathbb{R}^n$ is a known constant vector.

The question becomes now how to adjust the parameter estimates $\theta_1(t), \theta_2(t)$ using the information contained in (6.13). One solution that does not use the relationship (6.13) is to adjust independently each error model, i.e., to say, using AL with the same structure as in (6.8). This solution, however, does not consider the additional information contained in (6.13), which could be valuable improving the behavior of the resulting adaptive system. In order to consider this information in the problem solution, let's define an auxiliary error $\xi:\mathbb{R}^+ \to \mathbb{R}^n$, using the same idea as in Duarte and Narendra (1996a). Thus it is defined $\xi(t)$ as

$$\xi(t) = \hat{k}_1(t)R_1\theta_1(t) + \hat{k}_2(t)R_2\theta_2(t) - W \tag{6.14}$$

where $\hat{k}_1, \hat{k}_2:\mathbb{R}^+ \to \mathbb{R}$ are additional estimated parameters, which estimate k_1 and k_2 respectively. This auxiliary equation somehow tells how far are the estimates from the true, unknown, and coupled values at any time instant.

Defining additional parameter errors for k_1, k_2 it is obtained

$$\begin{aligned} \phi_{k_1}(t) &= \hat{k}_1(t) - k_1 \\ \phi_{k_2}(t) &= \hat{k}_1(t) - k_2 \end{aligned} \tag{6.15}$$

where $\phi_{k_1}, \phi_{k_2}:\mathbb{R}^+ \to \mathbb{R}$. Subtracting (6.13) from (6.14), and using the definition of the parameter errors (6.12), (6.15) it can be rewritten the auxiliary error (6.14) as

$$\xi(t) = k_1R_1\phi_1(t) + \phi_{k_1}(t)R_1\theta_1(t) + k_2R_2\phi_2(t) + \phi_{k_2}(t)R_2\theta_2(t) \tag{6.16}$$

Note that Eq. (6.16) can be used only for analytical purposes, because the parameter errors $\phi_1(t)$, $\phi_2(t)$, $\phi_{k_1}(t)$, and $\phi_{k_2}(t)$ cannot be implemented due to the fact that the true parameters θ_1^*, θ_2^*, k_1, and k_2 are unknown. For implementation purposes, the expression (6.14) should be used.

Now, coupled fractional AL are proposed of the form

$$\begin{aligned} {}^CD^\alpha\phi_1(t) &= {}^CD^\alpha\theta_1(t) = -\gamma\text{sgn}(k_1)[e_1^T(t)P_1\ b_1\ \omega_1(t) + R_1^T\xi(t)], & \phi_1(t_0) &= \phi_{1_0} \\ {}^CD^\alpha\phi_{k_1}(t) &= {}^CD^\alpha\hat{k}_1(t) = -\gamma\theta_1(t)R_1^T\xi(t), & \phi_{k_1}(t_0) &= \phi_{k1_0} \\ {}^CD^\alpha\phi_2(t) &= {}^CD^\alpha\theta_2(t) = -\gamma\text{sgn}(k_2)[e_2^T(t)P_2\ b_2\ \omega_2(t) + R_2^T\xi(t)], & \phi_2(t_0) &= \phi_{2_0} \\ {}^CD^\alpha\phi_{k_2}(t) &= {}^CD^\alpha\hat{k}_2(t) = -\gamma\theta_2(t)R_2^T\xi(t), & \phi_{k_2}(t_0) &= \phi_{k2_0} \end{aligned} \tag{6.17}$$

where $\gamma \in \mathbb{R}^+$ corresponds to a constant adaptive gain. Nevertheless, adaptive gains γ can be chosen all different for each parameter adjustment in (6.17). For simplicity in this study all adaptive gains will be chosen constant and equal. Furthermore, as mentioned in Section 6.3.1, it is also possible to use time-varying adaptive gains for parameter's adjustment in (6.17) and this extension is being currently being investigated.

As can be seen, FOAL (6.17) include the additional information given by the linear constraint (6.13), through the inclusion of the auxiliary error $\xi(t)$.

It is interesting to note that by making $\xi(t) \equiv 0$ in (6.17) the independent adjustment of estimates is obtained (noncoupled fractional order adaptive laws, NCFOAL). Stability analysis of the adaptive systems (6.9), (6.10) using coupled fractional order adaptive laws (CFOAL) (6.17) will be addressed in the next subsection.

6.3.3 Stability Analysis of FOEM2 With Parameter Constraints

Since the two adaptive systems whose output error described by (6.9), (6.10) are now linked through the relationship (6.17), combined FOAL as stated in (6.17) will be used. Thus it can be defined an enlarged adaptive system given by the set of Eqs. (6.9), (6.10), (6.16), (6.17), for analysis purposes.

In order to prove the stability of the adaptive scheme (6.9), (6.10), (6.16), and (6.17) let's use the fractional extension of Lyapunov direct method, specifically Theorem 1, proposing the following Lyapunov function candidate, which is positive definite and decrescent,

$$\begin{aligned} V = {} & e_1^T(t)P_1\ e_1(t) + \frac{|k_1|}{\gamma}\ \phi_1^T(t)\phi_1(t) + \frac{1}{\gamma}\phi_{k_1}^2(t) + e_2^T(t)P_2\ e_2(t) \\ & + \frac{|k_2|}{\gamma}\ \frac{1}{\gamma}\phi_2^T(t)\phi_2(t) + \phi_{k_2}^2(t) \end{aligned} \tag{6.18}$$

Using the results by Miller and Feldstein (1971) it can be proved that $e_1, \phi_1, \phi_{k_1}, e_2, \phi_2$, and ϕ_{k_2} are differentiable, thus Lemma 1 can be used allowing to write

$$\begin{aligned} {}^C D^\alpha V \ & \le 2e_1^T(t)P_1\ {}^C D^\alpha e_1(t) + 2|k_1|\ \phi_1^T(t)\ {}^C D^\alpha \phi_1(t) + 2\phi_{k_1}(t)\ {}^C D^\alpha \phi_{k_1}(t) + \\ & + 2e_2^T(t)P_2\ {}^C D^\alpha e_2(t) + 2|k_2|\ \phi_2^T(t)\ {}^C D^\alpha \phi_2(t) + 2\phi_{k_2}(t)\ {}^C D^\alpha \phi_{k_2}(t) \end{aligned} \tag{6.19}$$

Evaluating the derivative (6.19) along (6.9), (6.10), and (6.17) and grouping similar terms, the following expression is obtained

$$\begin{aligned} {}^C D^\alpha V \ & \le e_1^T(t)[A_1^T P_1 + P_1 A_1]e_1(t) + e_2^T(t)[A_2^T P_2 + P_2 A_2]e_2(t) - \\ & - 2\ k_1\ \phi_1^T(t)R_1^T\ \xi(t) - 2\ k_2\ \phi_2^T(t)R_2^T\ \xi(t) - \\ & - 2\ \phi_{k_1}(t)\ \theta_1^T(t)R_1^T\ \xi(t) - 2\ \phi_{k_2}(t)\ \theta_2^T(t)R_2^T\ \xi(t) \end{aligned} \tag{6.20}$$

Replacing expressions (6.11) and (6.16) in (6.20) it can be rewritten that

$${}^C D^\alpha V \le -e_1^T(t)Q_1\ e_1(t) - e_2^T(t)Q_2\ e_2(t) - 2\xi^T(t)\xi(t) \tag{6.21}$$

As can be seen from Eq. (6.21), the fractional derivative of the Lyapunov function is negative semidefinite, so from Theorem 1 it can be concluded that the origin of the system (6.9), (6.10), (6.16), (6.17) is uniformly stable. That is, $e_1, e_2, \phi_1, \theta_1, \phi_2, \theta_2, \phi_{k_1}, \hat{k}_1, \phi_{k_2}, \hat{k}_2 \in \mathcal{L}^\infty$. Using this result in (6.16) it can be concluded also that $\xi \in \mathcal{L}^\infty$.

Additional conclusions regarding the evolution on time of output errors $e_1(t), e_2(t)$ and the auxiliary error $\xi(t)$ can be drawn from the following analysis.

Let's apply the FOI of order α to expression (6.21), from which it is obtained

$$I_{t_0}^{\alpha}\{e_1^T Q_1 \ e_1\}(t) + I_{t_0}^{\alpha}\{e_2^T Q_2 \ e_2\}(t) + 2I_{t_0}^{\alpha}\{\xi^T \xi\}(t) \le V(t_0) - V(t) \qquad (6.22)$$

Since it was proved that all the signals of the FOAS remain bounded, from expression (6.18) it can be concluded that V remains bounded $\forall t \ge t_0$. Thus, from (6.22) it can be concluded that the three FOI in the left-hand side of (6.22) remain bounded $\forall t \ge t_0$.

This result, together with the fact that $e_1(t), e_2(t)$ and $\xi(t)$ are bounded and the arguments of the FOI are nonnegative, allow using Lemma 2 to conclude that, $\forall \varepsilon > 0$, it holds that

$$\begin{aligned} \lim_{t\to\infty}\left[t^{\alpha-\varepsilon} \ \frac{\int_{t_0}^{t} \|e_1(\tau)\|^2 d\tau}{t}\right] &= \lim_{t\to\infty}\left[t^{\alpha-\varepsilon} \ \frac{\int_{t_0}^{t} \|e_2(\tau)\|^2 d\tau}{t}\right] \\ &= \lim_{t\to\infty}\left[t^{\alpha-\varepsilon} \ \frac{\int_{t_0}^{t} \|\xi(\tau)\|^2 d\tau}{t}\right] = 0 \end{aligned} \qquad (6.23)$$

Expression (6.23) implies that the mean values of the squared norm of the output errors $e_1(t), e_2(t)$ and the auxiliary error $\xi(t)$ are $o(t^{\varepsilon-\alpha})$, $\forall \varepsilon > 0$, which means that they converge asymptotically to zero, with a convergence speed higher than $t^{-\alpha}$.

6.4 ANALYSIS OF FRACTIONAL ORDER ERROR MODEL 3 WITH PARAMETER CONSTRAINTS

This section presents the same analysis performed in Section 6.3 but for two FOEM3 whose true and unknown parameters satisfy a matrix linear constraint.

6.4.1 Fractional Order Error Model 3

FOEM3 has the same structure as FOEM2, but the main difference is that only one of the components of the error $e(t) \in \mathbb{R}^n$ is accessible. This situation is very common when access to the whole pseudo state of the plant to be controlled/identified is not available but only one component is measurable (plant output). This makes FOEM3 applicable to a much wider class of problems than FOEM2.

In FOEM3, the evolution of the output error $e_1(t) \in \mathbb{R}$ has the form

$$\begin{aligned} {}^{C}D^{\alpha}e(t) &= Ae(t) + b\ \phi^{T}(t)\omega(t), & e(t_0) &= e_0 \\ e_m(t) &= k\ h^{T}e(t), & e_m(t_0) &= e_{m0}, \end{aligned} \tag{6.24}$$

where $A \in \mathbb{R}^{n \times n}$ is a matrix whose eigenvalues have negative real parts. $e: \mathbb{R}^{+} \to \mathbb{R}^{n}$ corresponds to the nonaccessible pseudo state error vector and $e_m: \mathbb{R}^{+} \to \mathbb{R}$ is the measurable output error. k is an unknown constant with known sign (usually the high frequency gain of the plant to be controlled/identified), $b, h \in \mathbb{R}^{n}$, $\phi: \mathbb{R}^{+} \to \mathbb{R}^{m}$ is the parameter error, defined as $\phi(t) = \theta(t) - \theta^{*}(t)$, with $\theta: \mathbb{R}^{+} \to \mathbb{R}^{m}$ the estimated parameters and $\theta^{*} \in \mathbb{R}^{m}$ the real and unknown parameter vector. $\omega: \mathbb{R}^{+} \to \mathbb{R}^{m}$ is a vector of available signals and $\alpha \in (0, 1]$.

As might be expected, more stringent conditions have to be imposed on the FOEM3, in order to compensate for the fact that only $e_m(t)$ is accessible. In this case, positive definite symmetric matrices $P, Q \in \mathbb{R}^{n \times n}$ must exist such that $A^{T}P + PA = -Q$ and $Pb = h$, which implies that the triplet $\{A, b, h\}$ satisfies the Kalman–Yakubovich–Popov lemma (Narendra and Annaswamy, 2005).

As in the case of FOEM2, when $\alpha = 1$ in (6.24), the case of classic (integer) error model 3 (IOEM3) arises, which has been completely studied (Narendra and Annaswamy, 2005). In the case when $\alpha \in (0, 1)$, it was proved in Aguila-Camacho and Duarte-Mermoud (2016) that the following AL

$${}^{C}D^{\alpha}\phi(t) = {}^{C}D^{\alpha}\theta(t) = -\gamma\ \mathrm{sgn}(k)e_m(t)\omega(t), \quad \phi(t_0) = \phi_0, \tag{6.25}$$

where $\gamma \in \mathbb{R}^{+}$ is the adaptive gain, guarantee that all the signals in the adaptive scheme (6.24), (6.25) remain bounded and that the mean value of the squared norm of error $e(t)$ converges asymptotically to zero. As pointed out in Section 6.3.1 for the FOEM2, the sign of the high frequency gain need not to be known and in this case the solution makes use of the concept of Nussbaum gain proposed for IOEM (Narendra and Annaswamy, 2005). This extension is currently under investigation for the FOEM3 case.

After the analysis of FOEM2 with parameter constraints in Section 6.3.2, it is natural to wonder if coupled fractional order AL can be derived also for two FOEM3 whose real unknown parameters are related through a linear relationship. This question is addressed in the next subsection.

6.4.2 FOEM3 With Parameter Constraints

The problem of two FOEM3, as described in the previous subsection 6.4.1, whose true and unknown parameters are constraints through a linear matrix relationship is stated in this subsection.

Let consider two FOEM3, whose output error equations are defined as follows

$$\begin{aligned} {}^{C}D^{\alpha}e_1(t) &= A_1e_1(t) + b_1\ \phi_1^T(t)\omega_1(t) \\ e_{m_1}(t) &= k_1h_1^Te_1(t), \qquad e_{m_1}(t_0) = e_{m_{1_0}} \end{aligned} \tag{6.26}$$

$$\begin{aligned} {}^{C}D^{\alpha}e_2(t) &= A_2e_2(t) + b_2\ \phi_2^T(t)\omega_2(t) \\ e_{m_2}(t) &= k_2h_2^Te_2(t), \qquad e_{m_2}(t_0) = e_{m_{2_0}} \end{aligned} \tag{6.27}$$

where $A_1, A_2 \in \mathbb{R}^{n\times n}$ are matrices whose eigenvalues have negative real parts, $b_1, b_2, h_1, h_2 \in \mathbb{R}^n$ and the triplets (A_1, b_1, h_1) and (A_2, b_2, h_2) satisfy the Kalman–Yakubovich–Popov lemma, i.e., positive definite symmetric matrices $P_1, P_2, Q_1, Q_2 \in \mathbb{R}^{n\times n}$ exist such that

$$\begin{aligned} &A_1^TP_1 + P_1A_1 = -Q_1 \\ &P_1b_1 = h_1 \\ &A_2^TP_2 + P_2A_2 = -Q_2 \\ &P_2b_2 = h_2. \end{aligned} \tag{6.28}$$

The error vectors $e_1, e_2{:}\mathbb{R}^+ \to \mathbb{R}^n$ are not accessible in this fractional EM, but only the output errors $e_{m_1}, e_{m_2}{:}\mathbb{R}^+ \to \mathbb{R}$ are assumed to be accessible. k_1, k_2 are unknown constants whose signs are assumed to be known (usually the high frequency gains of the plants to be controlled/identified), $\phi_1, \phi_2{:}\mathbb{R}^+ \to \mathbb{R}^m$ are the parameter errors, defined as

$$\begin{aligned} \phi_1(t) = \theta_1(t) - \theta_1^*(t) \\ \phi_2(t) = \theta_2(t) - \theta_2^*(t) \end{aligned} \tag{6.29}$$

with $\theta_1, \theta_2{:}\mathbb{R}^+ \to \mathbb{R}^m$ the estimated parameters and $\theta_1^*, \theta_2^* \in \mathbb{R}^m$ the real unknown parameters. On the other hand, $\omega_1, \omega_2{:}\mathbb{R}^+ \to \mathbb{R}^m$ are vectors of available signals and $\alpha \in (0, 1]$.

Let's assume that the unknown parameters $\theta_1^*, \theta_2^*, k_1$, and k_2 are not independent but related through the following linear matrix relationship

$$\frac{1}{k_1}R_1\theta_1^* + \frac{1}{k_2}R_2\theta_2^* = W \tag{6.30}$$

where $R_1, R_2 \in \mathbb{R}^{n\times n}$ are known matrices and $W \in \mathbb{R}^n$ is a known vector.

Using the same definition for the auxiliary error than in FOEM2 with parameter constraints (6.14), where $\hat{k}_1, \hat{k}_2{:}\mathbb{R}^+ \to \mathbb{R}$ are additional estimated parameters, which in this case estimate $\frac{1}{k_1}$ and $\frac{1}{k_2}$, respectively, then additional parameter errors can be defined as

$$\begin{aligned} \phi_{k_1}(t) = \hat{k}_1(t) - \frac{1}{k_1} \\ \phi_{k_2}(t) = \hat{k}_2(t) - \frac{1}{k_2}, \end{aligned} \tag{6.31}$$

which allows writing the auxiliary error equation like in (6.16).

Thus, the following CFOAL for FOEM3 with parameter constraints are proposed

$$\begin{array}{ll} {}^{C}D^{\alpha}\phi_1(t) = {}^{C}D^{\alpha}\theta_1(t) = -\gamma \mathrm{sgn}(k_1)[e_{m_1}(t)\omega_1(t) + R_1^T\xi(t)], & \phi_1(t_0) = \phi_{10} \\ {}^{C}D^{\alpha}\phi_{k_1}(t) = {}^{C}D^{\alpha}\hat{k}_1(t) = -\gamma\theta_1^T(t)R_1^T\xi(t), & \phi_{k_1}(t_0) = \phi_{k_10} \\ {}^{C}D^{\alpha}\phi_2(t) = {}^{C}D^{\alpha}\theta_2(t) = -\gamma \mathrm{sgn}(k_2)[e_{m_2}(t)\omega_2(t) + R_2^T\xi(t)], & \phi_2(t_0) = \phi_{20} \\ {}^{C}D^{\alpha}\phi_{k_2}(t) = {}^{C}D^{\alpha}\hat{k}_2(t) = -\gamma\theta_2^T(t)R_2^T\xi(t), & \phi_{k_2}(t_0) = \phi_{k_20} \end{array} \tag{6.32}$$

where $\gamma \in \mathbb{R}^+$ corresponds to the adaptive gain.

6.4.3 Stability Analysis of FOEM3 With Parameter Constraints

The stability analysis of two FOEM3 whose true and unknown parameters are constraints through a linear matrix relationship, as described in the previous subsection 6.4.2, is presented and discussed in this subsection.

Let consider the FOEM3 with parameter constraints, given by (6.26), (6.27), (6.16), and (6.32). The stability analysis can be done along the same lines used for FOEM2 with parameter constraints in subsection. Thus, only the main steps of the proofs are detailed in the following.

Let's propose the following Lyapunov function candidate, which is positive definite and decrescent

$$\begin{aligned} V = {}& e_1{}^T(t)P_1\ e_1(t) + \frac{|k_1|}{\gamma}\ \phi_1^T(t)\phi_1(t) + \frac{1}{\gamma}\phi_{k_1}^2(t) + e_2{}^T(t)P_2\ e_2(t) \\ & + \frac{|k_2|}{\gamma}\ \phi_2^T(t)\phi_2(t) + \frac{1}{\gamma}\phi_{k_2}^2(t) \end{aligned} \tag{6.33}$$

Using the results by Miller and Feldstein (1971) it can be proved that $e_1, \phi_1, \phi_{k_1}, e_2, \phi_2$, and ϕ_{k_2} are differentiable, thus Lemma 1 can be used in a similar way to the case of FOEM2, which together with (6.26), (6.27), (6.32), (6.28), and (6.16) allows writing

$${}^{C}D^{\alpha}V \le -e_1{}^T(t)Q_1\ e_1(t) - e_2{}^T(t)Q_2\ e_2(t) - 2\xi^T(t)\xi(t) \tag{6.34}$$

As can be seen from Eq. (6.34), the fractional derivative of the Lyapunov function is negative semidefinite, so it can be concluded from Theorem 1 that the origin of the system (6.26), (6.27), (6.32) is uniformly stable. That is, $e_1, e_2, e_{m_1}, e_{m_2}, \phi_1, \theta_1, \phi_2, \theta_2, \phi_{k_1}, \hat{k}_1, \phi_{k_2}, \hat{k}_2 \in \mathcal{L}^{\infty}$. Using this result in (6.16) it can be concluded also that $\xi \in \mathcal{L}^{\infty}$.

Also, using the same arguments than in the stability analysis of FOEM2, it can be concluded here that, $\forall \varepsilon > 0$, it holds that

$$\lim_{t\to\infty}\left[t^{\alpha-\varepsilon}\frac{\int_{t_0}^{t}\|e_1(\tau)\|^2 d\tau}{t}\right]=\lim_{t\to\infty}\left[t^{\alpha-\varepsilon}\frac{\int_{t_0}^{t}\|e_2(\tau)\|^2 d\tau}{t}\right]$$
$$=\lim_{t\to\infty}\left[t^{\alpha-\varepsilon}\frac{\int_{t_0}^{t}\|\xi(\tau)\|^2 d\tau}{t}\right]=0 \qquad (6.35)$$

which also allows concluding that

$$\lim_{t\to\infty}\left[t^{\alpha-\varepsilon}\frac{\int_{t_0}^{t}e_{m_1}^2(\tau)d\tau}{t}\right]=\lim_{t\to\infty}\left[t^{\alpha-\varepsilon}\frac{\int_{t_0}^{t}e_{m_2}^2(\tau)d\tau}{t}\right]=0 \qquad (6.36)$$

Expressions (6.35) and (6.36) imply that the mean values of the squared norm of the errors $e_1(t), e_2(t)$, of the auxiliary error $\xi(t)$ and the mean value of the squared output errors $e_{m_1}^2(t), e_{m_2}^2(t)$ are $o(t^{\varepsilon-\alpha})$, $\forall\varepsilon>0$, which means that they converge asymptotically to zero, with a convergence speed higher than $t^{-\alpha}$.

6.5 SIMULATION EXAMPLES

This section presents simulation examples, which show that under different scenarios, the coupled FOAL can improve the behavior of two FOEM2 with parameter constraints, as compared with the independent adjustment of parameters.

No simulation example is presented in this chapter for FOEM3 with parameter constraints, since the behavior and conclusions remain the same to that obtained from the following example for FOEM2 with parameter constraints.

6.5.1 Simulation Specifications

Two second order FO systems will be analyzed in the simulations. Let's consider two FOEM2 as described by (6.9), (6.10), and (6.12), with the following parameters.

$$A_1=\begin{bmatrix}-1 & 0\\ 2 & -5\end{bmatrix}\quad b_1=\begin{bmatrix}1\\0\end{bmatrix}\quad k_1=1\quad \theta_1^*=\begin{bmatrix}5 & -1\end{bmatrix}^T$$
$$A_2=\begin{bmatrix}-3 & 1\\ 0 & -4\end{bmatrix}\quad b_2=\begin{bmatrix}0\\1\end{bmatrix}\quad k_2=1\quad \theta_2^*=\begin{bmatrix}-2 & 3\end{bmatrix}^T$$

It can be noted that relationship (6.11) holds for the following matrices

$$P_1=\begin{bmatrix}0.567 & 0.0333\\ 0.0333 & 0.1\end{bmatrix}\quad P_2=\begin{bmatrix}0.1667 & 0.0238\\ 0.0238 & 0.1310\end{bmatrix}\quad Q_1=Q_2=\begin{bmatrix}1 & 0\\ 0 & 1\end{bmatrix}$$

It will be assumed that the unknown parameter vectors θ_1^*, θ_2^*, k_1, and k_2, satisfy relationship (6.13) for known matrices and vector

$$R_1 = \begin{bmatrix} 1 & 1 \\ 2 & 0 \end{bmatrix} \quad R_2 = \begin{bmatrix} 1 & 2 \\ 0 & 3 \end{bmatrix} \quad W = \begin{bmatrix} 8 \\ 19 \end{bmatrix}$$

It will be simulated the two FOEM2 using noncoupled fractional order adaptive laws (NCFOAL) (6.8) and compare it with the case using coupled fractional adaptive laws (CFOAL) (6.17).

In these simulations, the following initial conditions are used for the different variables in the FOEM2.

$$e_1(0) = \begin{bmatrix} 5 & -1 \end{bmatrix}^T \quad \theta_1(0) = \begin{bmatrix} 0 & 2 \end{bmatrix}^T \quad e_2(0) = \begin{bmatrix} 1 & 0 \end{bmatrix}^T \quad \theta_2(0) = \begin{bmatrix} 1 & 0 \end{bmatrix}^T$$

For those cases where CFOAL are implemented, the following additional initial conditions are used for the estimated parameters

$$\hat{k}_1(0) = 0 \quad \hat{k}_2(0) = 0$$

In all the simulations, unity adaptive gains are used for all the AL.

The fractional operators were implemented using the NID block included in the Ninteger toolbox for Matlab (Valério and Da Costa, 2004). The Oustaloup numerical approximation (Oustaloup, 1991) was used for the fractional operator in the NID block, where it is included as the Crone approximation.

6.5.2 Simulation Results: Ideal Conditions

Simulations under ideal conditions will be presented first together with two input (information) vectors; one constant and one time-varying. Let consider the case when the input vectors $\omega_1(t)$ and $\omega_2(t)$ are given by

$$\omega_1(t) = \begin{bmatrix} 1 & 2 \end{bmatrix}^T \quad \omega_2(t) = \begin{bmatrix} 1 & \sin(t) \end{bmatrix}^T$$

A simulation time of $T = 1000\ s$ will be used, although for $\|e_1(t)\|$ only the first 50 s are shown in the figures.

Using the parameters given previously, the FOEM2 are simulated in both cases, using NCFOAL (6.8) and using CFOAL (6.17).

Fig. 6.1 shows the evolution of $\|e_1(t)\|$ and $\|e_2(t)\|$ for different values of α. For every α used, the corresponding results using NCFOAL and CFOAL have been plotted in the same graph, for comparison purposes.

In Fig. 6.1A, the time axis has been zoomed for the first 50 seconds in the case of $\|e_1(t)\|$, for the sake of clarity. It can be seen from Fig. 6.1 that $\|e_1(t)\|$ converges to zero for all α, and the convergence speed is not dramatically different in the four cases presented. However, it can be seen that the

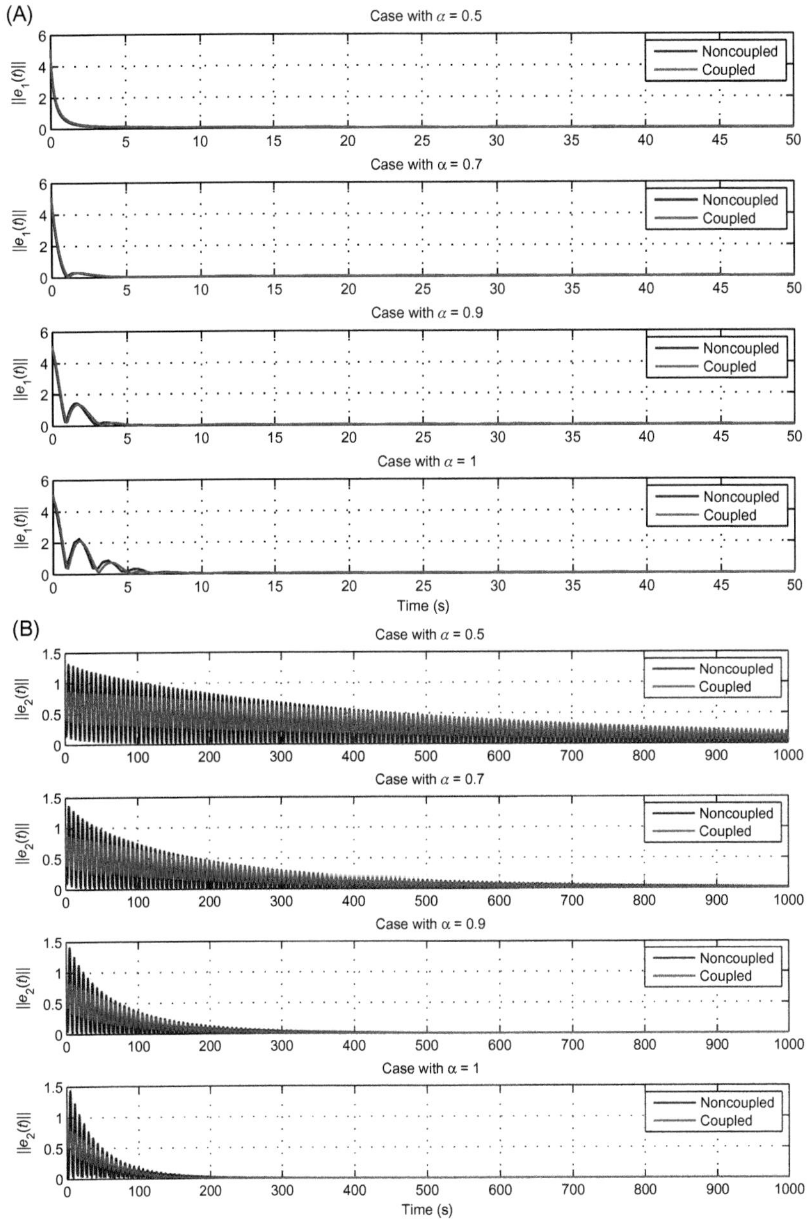

FIGURE 6.1 Evolution for different values of α, using NCFOAL and CFOAL of the (A) $\|e_1(t)\|$, and (B) $\|e_2(t)\|$. Ideal conditions.

number of initial oscillations in the transient response increases as α increases, being the case that the lower α, the smoother the response.

In the case of $\|e_2(t)\|$, it can be seen from Fig. 6.1B that it converges to zero for every α used but in a more oscillatory way, although the convergence speed is clearly different according to the FO used.

Regarding the differences between the noncoupled and the coupled case, it can be seen from Fig. 6.1A that for the first FOEM2, the differences in the response of both strategies are almost imperceptible in the plot, thus there is no an evident advantage in using CFOAL.

However, for the case of the second FOEM2, it can be seen from Fig. 6.1B that those cases using CFOAL present lower amplitudes of $\|e_2(t)\|$ than the cases with NCFOAL. This, together with the fact that for a fixed α the convergence speed is pretty much the same for both strategies, represents an advantage for the CFOAL.

On the other hand, Fig. 6.2 shows the evolution of $\|\phi_1(t)\|$ and $\|\phi_2(t)\|$, for this simulation scenario. It can be seen from Fig. 6.2A that $\|\phi_1(t)\|$ does not converge to zero for the first FOEM2. This behavior was expected, since it has been observed from several simulation studies that when the input vector $\omega_1(t)$ does not contain enough spectral lines, the convergence of the parameter error is not achieved. Although this is an empirical observation from simulation results, some advances in the analytical support can be found in Gallegos and Duarte-Mermoud (2016, 2017a,b).

It can be seen from Fig. 6.2A that $\|\phi_1(t)\|$ converges to a final value that is lower for the case of CFOAL. This result can be useful in adaptive schemes, since the final values of the parameters is usually related to the magnitude of, for instance, the control signal generated in adaptive control schemes, which means that this behavior could decrease the energy used by the control.

In the case of $\|\phi_2(t)\|$ in the second FOEM2, it can be seen from Fig. 6.2B that it converges to zero for both coupled and noncoupled FOAL. It was also expected in this case, since the input vector $\omega_2(t)$ includes enough spectral frequencies for achieving convergence (see Gallegos and Duarte-Mermoud, 2016, 2017a,b for preliminary results). Nevertheless, it must be mentioned that the initial magnitude of $\|\phi_2(t)\|$ is lower for the case using CFOAL, obtaining the same conclusion stated for $\|\phi_1(t)\|$.

6.5.3 Simulation Results: Noisy Environment

Since under real conditions processes are usually subjected to noise, it is interesting to check how the adaptive systems behave under nonideal conditions. To that extent, let's now consider the case when the input vectors $\omega_1(t)$ and $\omega_2(t)$ are given by

$$\omega_1(t) = \begin{bmatrix} 1+n(t) & 2 \end{bmatrix}^T \quad \omega_2(t) = \begin{bmatrix} 1+n(t) & \sin(t) \end{bmatrix}^T$$

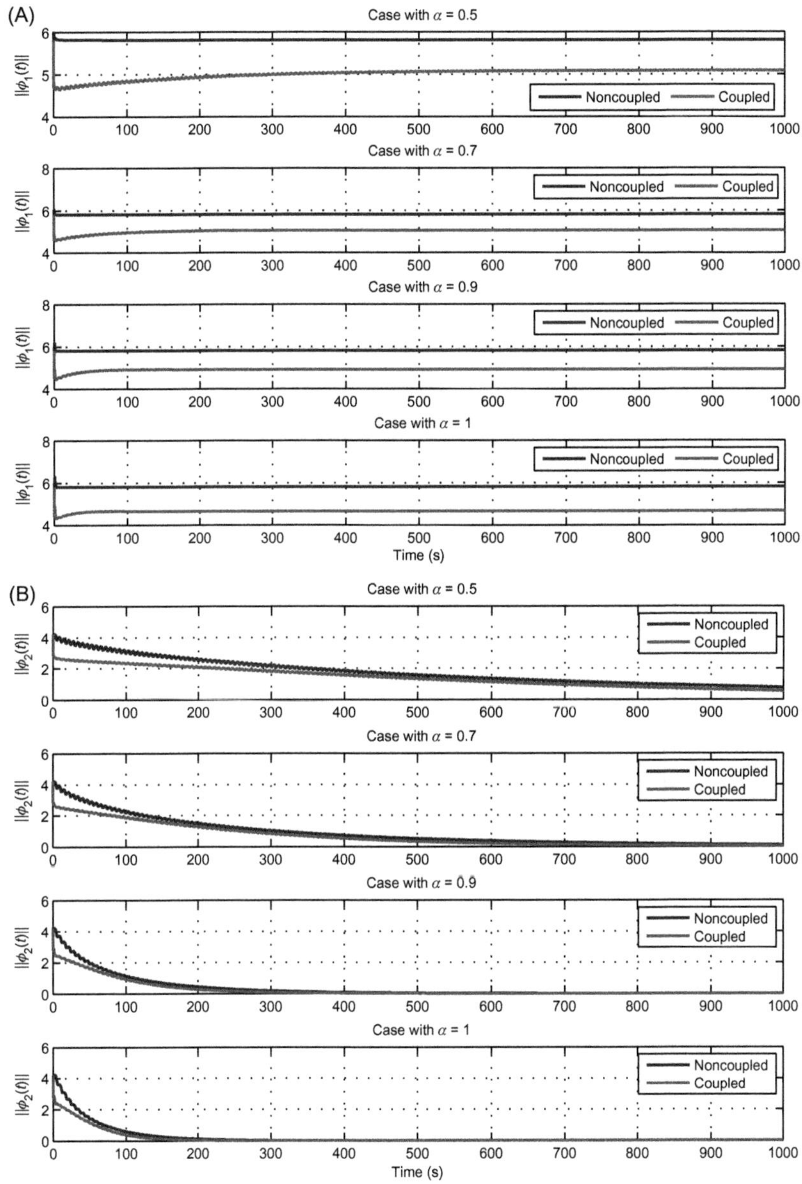

FIGURE 6.2 Evolution for different values of α, using NCFOAL and CFOAL of the (A) $\|\phi_1(t)\|$, and (B) $\|\phi_2(t)\|$. Ideal conditions.

where $n(t)$ corresponds to a random noise uniformly distributed, with maximum amplitude of 0.5, which represents 50% of the magnitude of the ideal reference. A simulation time of $T = 1500\ s$ will be used, although for $\|e_1(t)\|$ only the first 50 s are shown in the figures. The the evolution of $\|e_1(t)\|$, $\|e_2(t)\|$ in this noisy scenario is presented in Fig. 6.3. The evolution of the norm of the parameter errors $\|\phi_1(t)\|$, $\|\phi_2(t)\|$, on the other hand, is presented in Fig. 6.4.

It can be seen from Figs. 6.3 and 6.4 that the noise indeed affect the evolution of the signals in the scheme. In order to characterize and have a better insight on the behavior of the adaptive schemes when noise is present, the following two indexes are calculated

$$ISN_1 = \int_{t=0}^{T} \|e_1(\tau)\|^2 d\tau \tag{6.37}$$

$$ISN_2 = \int_{t=0}^{T} \|e_2(\tau)\|^2 d\tau \tag{6.38}$$

Table 6.1 shows the resulting values of indexes (6.37) and (6.38) for the simulations, which helps appreciating the results numerically.

As can be seen from Table 6.1, for the first FOEM2 the ISN_1 is lower for the case of using CFOAL compared with the case of using NCFOAL, for every α used, which represents an important improvement in the system behavior. In the case of the second FOEM2, it can be seen from Table 6.1 that for $\alpha = 0.5$, the NCFOAL behaves clearly better than the CFOAL, e.g., the ISN_2 is lower for the NCFOAL. However, this behavior starts changing as α increases, and for $\alpha = 0.7$ both approaches have similar ISN_2. For the cases $\alpha = 0.9$ and $\alpha = 1$ the CFOAL behaves better than the NCFOAL.

Summarizing, it can be concluded that the use of CFOAL can lead to an improvement in the adaptive systems behavior. Thus, they should be considered as a possible alternative in the design stage, wherever it is possible to implement them.

6.6 CONCLUSION

In this chapter, it has been proved that given two fractional order EM 2 with linear constraints on their true unknown parameters, it is possible to derive coupled fractional AL including the information of the constraint, guaranteeing stability of the resulting adaptive system for $\alpha \in (0, 1]$. The same conclusions have been established for two fractional order EM 3, also with linear parameter constraints. Also, it was analytically proved that the mean value of the squared norm of the state error vector converges asymptotically to zero in both cases.

Besides, simulation studies showed that the inclusion of the additional information in the coupled fractional AL results in a better dynamic behavior

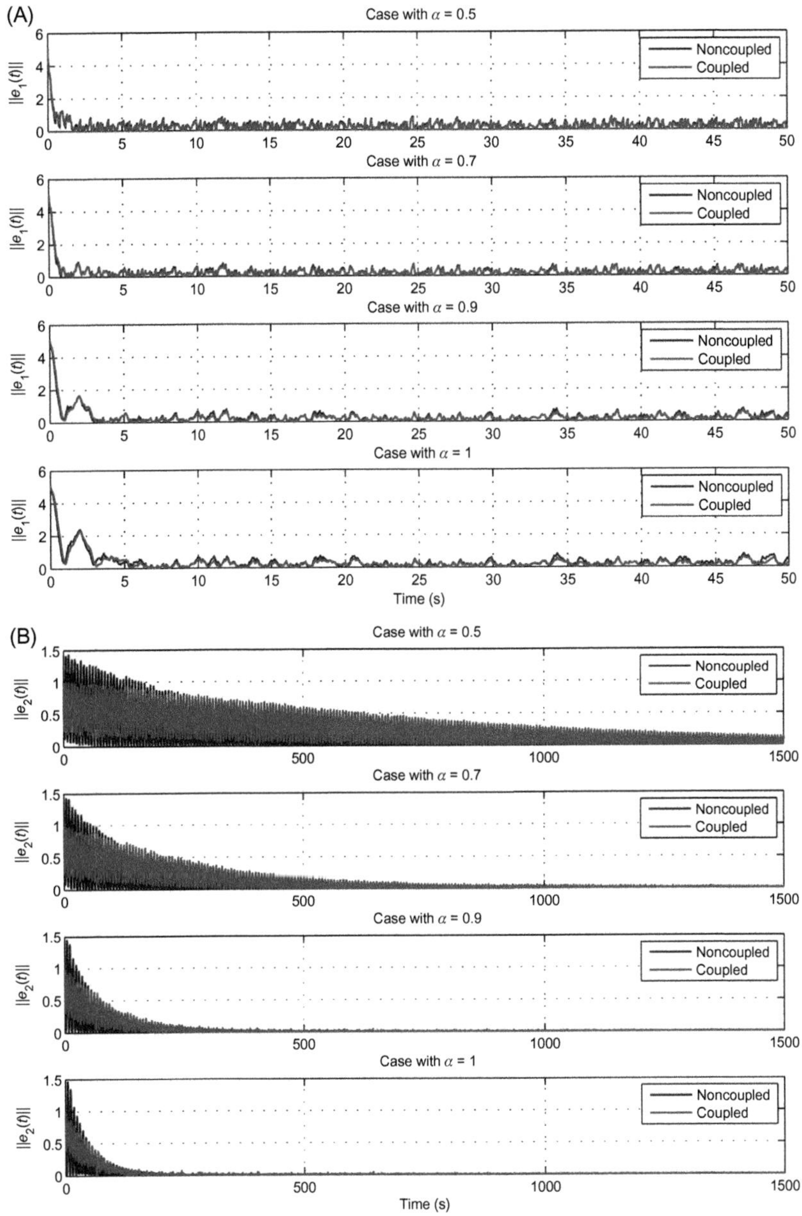

FIGURE 6.3 Evolution for different values of α, using NCFOAL and CFOAL of the (A) $\|e_1(t)\|$, and (B) $\|e_2(t)\|$. Noisy environment.

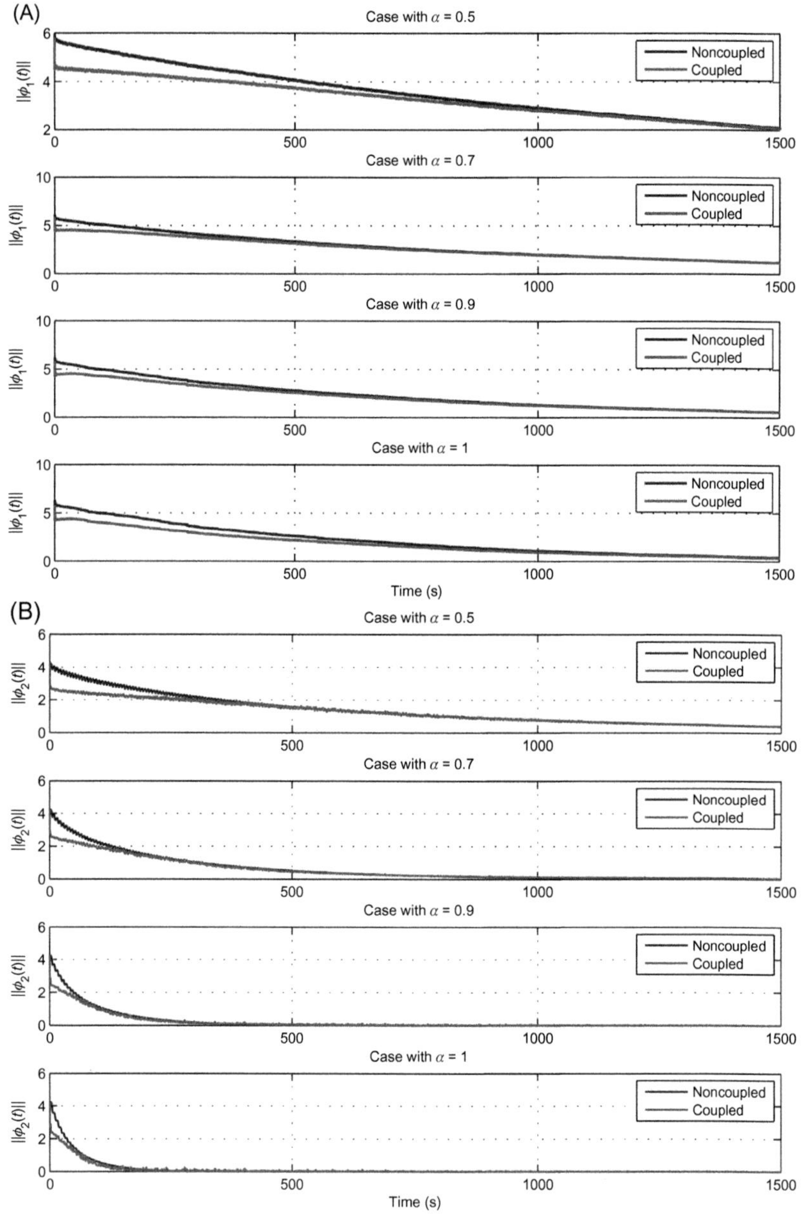

FIGURE 6.4 Evolution for different values of α, using NCFOAL & CFOAL of the (A) $\|\phi_1(t)\|$, and (B) $\|\phi_2(t)\|$. Noisy environment.

TABLE 6.1 Integral of the Squared Norm of Output Errors for the Noisy Simulation Scenario

		CFOAL	NCFOAL
$^*\alpha = 0.5$	ISN1	62.2957	68.0988
	ISN2	160.2580	142.9454
$^*\alpha = 0.7$	ISN1	37.6153	41.1030
	ISN2	59.7030	57.7515
$^*\alpha = 0.9$	ISN1	32.2745	37.8421
	ISN2	22.5958	25.4242
$^*\alpha = 1$	ISN1	35.1002	47.7553
	ISN2	14.7588	18.1608

of the overall adaptive system, compared to the case when classic decoupled fractional AL are used, under ideal conditions as well as in the presence of noise in the inputs.

ACKNOWLEDGMENTS

The results reported in this chapter have been financed by CONICYT- Chile, under the Basal Financing Program FB0809 "Advanced Mining Technology Center," FONDECYT Project 1150488 "Fractional Error Models in Adaptive Control and Applications," and FONDECYT 3150007 "Postdoctoral Program 2015."

REFERENCES

Aguila-Camacho, N., Duarte-Mermoud, M.A., 2013. Fractional adaptive control for an automatic voltage regulator. ISA Trans. 52, 807–815.

Aguila-Camacho, N., Duarte-Mermoud, M.A., 2015. Fractional Error Models 1 with Parameter Constraints. In: Proceedings of the 2015 International Symposium on Fractional Signals and Systems (FSS2015), Cluj-Napoca, Romania, 1-3 October, pp. 18–23.

Aguila-Camacho, N., Duarte-Mermoud, M.A., 2016. Boundedness of the solutions for certain classes of fractional differential equations with application to adaptive systems. ISA Trans. 60, 82–88.

Aguila-Camacho, N., Duarte-Mermoud, M.A., 2017. Improved adaptive laws for Fractional Error Models 1 with parameter constraints. Int. J. Dynam. Control 5, 198–207.

Aguila-Camacho, N., Duarte-Mermoud, M.A., Gallegos, J., 2014. Lyapunov functions for fractional order systems. Commun. Nonlinear Sci. Numer. Simul. 19, 2951–2957.

Alikhanov, A., 2010. A priori estimates for solutions of boundary value problems for fractional-order equations. Differ. Eq. 46, 660–666.

Azar, A.T., Vaidyanathan, S., Ouannas, A., 2017. Fractional Order Control and Synchronization of Chaotic Systems. Studies in Computational Intelligence, Vol. 688. Springer-Verlag, Germany.

Diethelm, K., 2010. The Analysis of Fractional Differential Equations, first ed. Springer-Verlag, Berlin Heidelberg.

Duarte, M.A., Narendra, K.S., 1987a. Combined Direct and Indirect Approach to Adaptive Control, Technical Report 8711. Center for Systems Science, Yale University, New Haven, Connecticut, USA.

Duarte, M.A., Narendra, K.S., 1987b. Combined Direct and Indirect Adaptive Control of Plants with Relative Degree Greater than one, Technical Report 8715. Center for Systems Science, Yale University, New Haven, Connecticut, USA.

Duarte, M.A., Narendra, K.S., 1988a. Error Models for adaptive Control With Parameter Constraints, Technical Report 8814. Center for Systems Science, Yale University, New Haven, Connecticut, USA.

Duarte, M.A., Narendra, K.S., 1988b. Indirect Adaptive Control using Dynamical Adjustment of Control Parameters, Technical Report 8813. Center for Systems Science, Yale University, New Haven, Connecticut, USA.

Duarte, M.A., Narendra, K.S., 1989a. Combined direct and indirect approach to adaptive control. IEEE Trans. Autom.c Control AC-34, 1071–1075.

Duarte, M.A., Narendra, K.S., 1989b. A new approach to model reference adaptive control. Int. J. Adaptive Control Signal Processing 3, 53–73.

Duarte, M.A., Narendra, K.S., 1996a. Error models with parameter constraints. Int. J. Control 64, 1089–1111.

Duarte, M.A., Narendra, K.S., 1996b. Indirect model reference adaptive control with dynamic adjustment of parameters. Int. J. Adaptive Control Signal Processing 10, 603–621.

Duarte-Mermoud, M.A., Aguila-Camacho, N., Gallegos, J., Castro-Linares, R., 2015. Using general quadratic lyapunov functions to prove lyapunov uniform stability for fractional order systems. Commun. Nonlinear Sci. Numer. Simul. 22, 650–659.

Freed, A., Diethelm, K., 2007. Caputo derivatives in viscoelasticity: a non-linear finite deformation theory for tissue. Frac. Calculus Appl. Analysis 10, 219–248.

Gallegos, J., Duarte-Mermoud, M.A., Robustness and convergence of fractional systems and their applications to adaptive systems. https://arxiv.org/abs/1609.05544, accessed 14 March 2017a.

Gallegos, J., Duarte-Mermoud, M.A., 2017b. Convergence of fractional adaptive systems using gradient approach. ISA Trans. 69, 31–42.

Huan, C., Li, Y., Liu, D., Guan, X., 2016. Stability analysis for fractional-order PD controlled delayed systems. J. Franklin Institute 353, 3118–31328.

Kavuran, G., Alagoz, B.B., Ates, A., Yeroglu, C., 2016. Implementation of model reference adaptive controller with fractional order adjustment rules for coaxial rotor control test system. Balkan J. Electr. Computer Eng. 4, 84–88.

Kilbas, A., Srivastava, H., Trujillo, J., 2006. Theory and Applications of Fractional Differential Equations, first ed. Elsevier, Amsterdam.

Ladaci, S., Bensafia, Y., 2016. Indirect fractional order pole assignment based adaptive control. Eng. Sci. Technol. 19, 518–530.

Li, Y., Chen, Y.Q., Podlubny, I., 2010. Stability of fractional-order nonlinear dynamic systems: Lyapunov direct method and generalized Mittag Leffler stability. Comput. Math. Applications 59, 1810–1821.

Liu, H., Pan, Y., Li, S., Chen, Y., 2017. Variable-order fractional operators for adaptive order and parameter estimation. IEEE Trans. Syst. Man Cyber. 47, 2209–2217.

Matignon, D., 1994. Représentations en variables d état de modéles de guides d ondes avec dérivation fractionnaire. PhD thesis, Université Paris.

Merrikh-Bayat, F., 2013. Fractional-order unstable pole-zero cancellation in linear feedback systems. J. Process Contr. 23, 817–825.

Miller, R.K., Feldstein, A., 1971. Smoothness of solutions of volterra integral equations with weakly singular kernels. SIAM J. Math. Anal. 2, 242–258.

Narendra, K.S., 1976. Stable Identification Schemes. Academic Press, New York.

Narendra, K.S., Annaswamy, A.M., 2005. Stable Adaptive Systems, second ed. Dover Publications Inc, Mineola.

Narendra, K.S., Kudva, P., 1974. Stable adaptive schemes for system identification and control - Parts I and II. IEEE Trans. Syst. Man Cyber. SMC-4, 542–560.

Narendra, K.S., Lin, Y.H., 1980. Stable discrete adaptive control. IEEE Trans. Automatic Control AC-25, 456–461.

Narendra, K.S., Khalifa, I.H., Annaswamy, A.M., 1983. Error Models for Stable Hybrid Adaptive Systems. In: Proceedings of the 1983 American Control Conference, pp. 559–564.

Narendra, K.S., Khalifa, I.H., Annaswamy, A.M., 1985. Error models for stable hybrid adaptive systems. IEEE Trans. Automatic Control AC-30, 339–347.

Oustaloup, A., 1991. La commande CRONE: commande robuste d'ordre non entier. Hermes, Paris.

Rapaic, M.R., Pisano, A., 2014. Variable-order fractional operators for adaptive order and parameter estimation. IEEE Trans. Automatic Control 59, 798–803.

Takamatsu, T., Ohmori, H., July 2015. State and parameter estimation of lithium ion battery by Kreisselmeier-type adaptive observer for fractional calculus system. In: Proceedings of the 54th Annual Conference of the Society of Instrument and Control Engineers of Japan, SICE 2015, Hangzhou, China, pp. 86–90.

Tejado, I., Valerio, D., Valerio, N., June 2014. Fractional calculus in economic growth modeling. The Portuguese case. In: International Conference in Fractional Differentiation and its Applications ICFDA14, Catania, Italy.

Valério, D., Da Costa, J.S., July 2004. Ninteger: a non-integer control toolbox for Matlab. In: Fractional Derivatives and Applications, Bordeaux, France.

Valério, D., Da Costa, J.S., 2006. Tuning of fractional PID controllers with Ziegler-Nichols-type rules. Signal Processing (Special Issue: Fractional calculus applications in signals and systems) 86, 2771–2784.

Wei, Y., Sun, Z., Hu, Y., Wang, Y., 2017. On fractional order adaptive observer. Int. J. Automation Computing 12, 664–670.

Chapter 7

Controllability of Single-valued and Multivalued Fractional Stochastic Differential Equations

Peachimuthu Tamilalagan[1] and Pagavathigounder Balasubramaniam[2]
[1]*Amrita Vishwa Vidyapeetham (Deemed to be University), Coimbatore, Tamil Nadu, India,*
[2]*Gandhigram Rural Institute (Deemed to be University), Dindigul, Tamil Nadu, India*

7.1 INTRODUCTION

The fractional calculus instills new dimensions to understand or describe basic nature of the real life phenomena arising in science and engineering in a better way. Hence, it serves as an eminent tool for providing more accurate and successful results than integer order differential equations in the modeling of several real-life phenomena. Accordingly, the fractional abstract Cauchy problems attain its necessity of studying the existence, controllability, and other qualitative and quantitative properties (see Azar et al., 2017; Bajlekova, 2001; Srivastava et al., 2006; Podlubny, 1998).

It is well known that, random differential and integral equations play an important role in characterizing many social, physical, biological, and engineering problems. The modeling of natural phenomena by stochastic differential equations (SDEs) plays an important role, wherever we encounter the fluctuations in nature (see Da Prato and Zabczyk, 2014; Mao, 2007). Hence, it is of great significant to study fractional differential equations (FDEs) with stochastic effects.

Further, the notion of multimaps arises in many branches of mathematics, namely mathematical economics, theory of games, convex analysis, etc. The multimaps play a significant role in the description of process in control theory, since the presence of control provides an intrinsic multivalence in the evolution of the system (Kamenskii et al., 2001). Differential inclusions have wide applications in economics, engineering, and so on. The theory of differential inclusions has been developed accordingly in the past three decades (see Aubin and Cellina, 1984; Balasubramaniam and Ntouyas, 2006; Balasubramaniam et al., 2005; Deimling, 1992; Hu, 1997). The extension of

Mathematical Techniques of Fractional Order Systems. DOI: https://doi.org/10.1016/B978-0-12-813592-1.00007-6

the study of differential inclusions to the world of fractional calculus is essential based on the widespread applications of multivalued analysis in science and engineering. The study of fractional differential inclusions was initiated by El-Sayed and Ibrahim (1995) and much interest has been given along this line (see, Henderson and Ouahab, 2010; Vijayakumar et al., 2014; Zhou, 2016; Balasubramaniam and Tamilalagan, 2015).

In the theory of abstract differential equations of integer order, C_0-semigroup, also known as a strongly continuous one-parameter family of semigroup, is a generalization of the exponential function, which provides solutions of integer order differential equations in Banach spaces, such as delay differential equations and partial differential equations etc. It is worthwhile to mention that the Riemann–Liouville and Caputo fractional operators do not possess semigroup and commutative properties, which are inherent to the derivatives of integer order (Balachandran and Kiruthika, 2011). In order to study the existence of solutions, controllability, and other qualitative properties of FDEs, various approaches are proposed, namely by employing the method of solution operators, (a,k)-regularized families of bounded linear operators and α-resolvent operators, etc. (see Kexue and Jigen, 2012; Li et al., 2012; Lizama and N'Guérékata, 2013; Peng and Li, 2012).

In particular, Hernández et al. (2013, 2010) noted that the concept of a solution is not realistic, while using variation of constant formula, also they pointed out that similar problem arises, where the operator A is taken as a Hille–Yosida type operator. Further, they have studied the existence result for a general class of abstract FDEs by using the well-developed theory of resolvent operators for integral equations.

In the case of FDEs of order $1<\alpha<2$, Dos Santos et al. (2013) established the existence of mild solutions for the nonlocal Cauchy problem of abstract fractional neutral integro-differential equations with unbounded delay by using the theory of resolvent operators. The functional equation associated with general (a,k) regularized families of bounded linear operator, which can replace the property of semigroups has been studied by Lizama and Poblete (2012). The variation of constants formula obtained for FDEs through (a,k)-regularized families of bounded linear operators is considered to be an unified functional analytic approach, which covers the theories of C_0-semigroups and cosine families as particular cases. Lizama and N'Guérékata (2013) proved the existence of solutions for some classes of FDEs with nonlocal conditions of order $1<\alpha<2$ by using (a,k)-regularized families.

Further, during the last decades fractional optimal control problems gained the attention of the researchers (see Agrawal, 2004; Agrawal et al., 2010; Balasubramaniam and Tamilalagan, 2016). The solvability and optimal controls of a class of fractional integro-differential evolution systems with infinite delay in Banach spaces has been investigated by Wang et al. (2012). Wang and Zhou (2011a) studied the existence of mild solutions and the existence of optimal pairs for semilinear fractional evolution equations in α-norm by means of singular version Gronwall inequality and Leray–Schauder fixed point theorem. Fan and Mophou (2014) established the existence of optimal control for semilinear composite fractional relaxation equations under suitable conditions. Liu et al. (2013) studied the solvability and optimal controls for some fractional impulsive differential equations by using fractional calculus, Gronwall inequality, and Leray-Schauder fixed point theorem.

Highly inspired by the above research findings, this chapter deals with the approximate controllability, solvability, and existence of optimal control for some classes of single-valued and multivalued fractional stochastic differential equation (FSDEs). It is mandatory to mention that, except for standard notation all other notations are uniquely defined for each subsection.

This chapter is organized as follows. In section 7.2, the approximate controllability of multivalued fractional stochastic integro-differential equation, solvability and optimal controls for FSDEs of order $1<\alpha<2$ are studied in Hilbert space by using (a,k)-regularized families of bounded linear operators. Section 7.3 investigates the solvability and optimal controls for fractional stochastic integro-differential equations with infinite delay in Hilbert space by using analytic resolvent operators. Finally, section 7.4 deals with some conclusions and future directions of these theoretical results.

7.2 CONTROLLABILITY RESULTS OF SINGLE-VALUED AND MULTIVALUED FSDEs BY USING (a, k)-REGULARIZED FAMILIES OF BOUNDED LINEAR OPERATORS

This section deals with controllability results of single-valued and multivalued FDEs by using (a,k)-regularized families of bounded linear operators. A new set of sufficient conditions is formulated for the approximate controllability of a class of multivalued fractional stochastic integro-differential equation of order $1<\alpha<2$ in Hilbert space by using Bohnenblust–Karlin's fixed point theorem . Further, the solvability and optimal control results are investigated for FSDEs of order $1<\alpha<2$ in Hilbert space by using the classical Banach contraction mapping principle.

7.2.1 Approximate Controllability of Multivalued Fractional Stochastic Integro-differential Equation

This subsection concerns the approximate controllability of a class of multivalued fractional stochastic integro-differential equations of the form

$$
{}^{c}D_t^{\alpha}x(t) \in Ax(t) + {}^{c}D_t^{\alpha-1}\left(Bu(t) + F\left(t, x_t, \int_0^t g(t,s,x_s)ds\right) + \int_0^t \sigma(t,s,x_s)dW(s)\right), t \in J \tag{7.1}
$$

$$x_0 = \phi \in \mathcal{B}, x'(0) = x_1 \tag{7.2}$$

where $J := [0, b]$, ${}^cD^\alpha$ is the Caputo fractional derivative of order $1 < \alpha < 2$, A generates an $\left(\frac{t^{\alpha-1}}{\Gamma(\alpha)}, 1\right)$-regularized family $\mathcal{S}_\alpha(t)$. Here $x(\cdot)$ takes values in the separable Hilbert space H with inner product $\langle \cdot, \cdot \rangle_H$ and norm $\| \cdot \|_H$. Let K be another separable Hilbert space with inner product $\langle \cdot, \cdot \rangle_K$ and norm $\| \cdot \|_K$. Suppose that W is a given K-valued Wiener process with a finite trace nuclear covariance operator $Q \geq 0$ defined on the filtered complete probability space $(\Omega, \mathcal{F}, \{\mathcal{F}_t\}_{t \geq 0}; \mathbb{P})$. The control function $u(\cdot)$ is given in $L_2(J, U)$ of admissible control functions, where $L_2(J, U)$ is the Hilbert space of all $\mathcal{F}_t$-adapted, square integrable process, U is a Hilbert space. B is a bounded linear operator from U into H. The history $x_t : (-\infty, 0] \to H$, $x_t(\theta) = x(t + \theta)$ for $t \geq 0$ belongs to some abstract space $\mathcal{B}$ defined axiomatically. Here $F : J \times \mathcal{B} \times H \to \mathcal{P}(H)$, $g : J \times J \times \mathcal{B} \to H$, and $\sigma : J \times J \times \mathcal{B} \to L(K, H)$ are the appropriate functions, where $L(K, H)$ denotes the space of all bounded linear operators from K into H.

7.2.1.1 Preliminaries

Let $(\Omega, \mathcal{F}, \{\mathcal{F}\}_{t \geq 0}; \mathbb{P})$ be a filtered complete probability space satisfying the usual conditions, which means that the filtration $\{\mathcal{F}_t\}_{t \geq 0}$ is a right continuous increasing family and $\mathcal{F}_0$ contains all P-null sets. Suppose $\{W(t) : t \geq 0$ be Q- Wiener process defined on $(\Omega, \mathcal{F}, \{\mathcal{F}_t\}_{t \geq 0}; \mathbb{P})$ with a finite trace nuclear covariance operator $Q \geq 0$, such that $Tr(Q) < \infty$. We assume that there exists a complete orthonormal basis $\{e_k\}_{k \geq 1}$ in K, a bounded sequence of nonnegative real numbers λ_k such that $Qe_k = \lambda_k e_k$, $k = 1, 2, \cdots$ and a sequence of independent Brownian motions $\{\beta_k\}_{k \geq 1}$ such that

$$\langle w(t), e \rangle_K = \sum_{k=1}^{\infty} \sqrt{\lambda_k} \langle e_k, e \rangle_K \beta_k(t), \quad e \in K, \quad t \geq 0.$$

We employ the axiomatic definition of the phase space $\mathcal{B}$ introduced by Hale and Kato (1978). The axioms of the space $\mathcal{B}$ are established for $\mathcal{F}_0$ measurable functions from $(-\infty, 0]$ into H, endowed with a seminorm $\| \cdot \|_{\mathcal{B}}$. We assume that $\mathcal{B}$ satisfies the following axioms

- **(ai)** If $x : (-\infty, a) \to H, a > 0$ is continuous on $[0,a)$ and x_0 in $\mathcal{B}$, then for every $t \in [0,a)$ the following conditions hold:
  - **(i)** x_t is in $\mathcal{B}$;
  - **(ii)** $\| x(t) \| \leq \hat{L} \| x_t \|_{\mathcal{B}}$;
  - **(iii)** $\| x_t \|_{\mathcal{B}} \leq \hat{K}(t) \sup\{ \| x(s) \| : \ 0 \leq s \leq t\} + \hat{N}(t) \| x_0 \|_{\mathcal{B}}$, where $\hat{L} > 0$ is a constant; $\hat{K}$, $\hat{N} : [0, \infty) \to [0, \infty)$, $\hat{K}$ is continuous, $\hat{N}$ is locally bounded and $\hat{L}$, $\hat{K}$, $\hat{N}$ are independent of $x(\cdot)$.
- **(aii)** For the function $x(\cdot)$ in (ai), x_t is a $\mathcal{B}$-valued function.
- **(aiii)** The space $\mathcal{B}$ is complete.

Let $\mathcal{P}(H)$ denote the class of all nonempty subsets of H. Let $\mathcal{P}_{bd,cl}(H)$, $\mathcal{P}_{bd,cl,cv}(H)$ denote, respectively, the family of all nonempty bounded-closed, bounded-closed-compact subsets of H (see Fitzpatrick and Petryshyn, 1974). For $x \in H$, and $Y,\ Z \in \mathcal{P}_{bd,cl}(H)$, we denote by $d(x,Y) = \inf\{\|x-y\|:\ y \in Y\}$ and $\rho(Y,Z) = \sup_{a \in Y} d(a,Z)$ and the Hausdorff metric $\hat{H}:\mathcal{P}_{bd,cl}(H) \times \mathcal{P}_{bd,cl}(H) \to \mathbb{R}^+$ by $\hat{H}(A,B) = \max\{\kappa(A,B), \kappa(B,A)\}$.

Definition 7.1: (Balasubramaniam and Ntouyas, 2006) *A multivalued map $F:H \to \mathcal{P}(H)$ is convex (closed) valued if $F(x)$ is convex (closed) for all $x \in H$. F is bounded on bounded sets, if $F(V) = \cup_{x \in V} F(x)$ is bounded in H, for any bounded set V of H, i.e.,* $\sup_{x \in V}\{\sup\{\|y\|:\ y \in F(x)\}\} < \infty$.

Definition 7.2: (Balasubramaniam and Ntouyas, 2006) *F is upper semicontinuous (u.s.c) on H, if for each $x_* \in H$ the set $F(x_*)$ is a nonempty, closed subset of H, and if for each open set V of H containing $F(x_*)$, there exists an open neighborhood N of x_* such that $F(N) \subseteq V$.*

Definition 7.3: (Balasubramaniam and Ntouyas, 2006) *The multivalued operator F is said to be completely continuous if $F(V)$ is relatively compact for every bounded subset $V \subseteq H$.*

Definition 7.4: (Balasubramaniam and Ntouyas, 2006) *A multivalued map $F:J \to \mathcal{P}_{bd,cl}(H)$ is said to be measurable if for each $x \in H$ the function $t \mapsto d(x, F(t))$ is measurable on J.*

If the multivalued map F is completely continuous with nonempty compact values, then F is u.s.c if and only if F has a closed graph. i.e $x_n \to x_*$, $y_n \to y_*$, $y_n \in Fx_n$ imply $y_* \in Fx_*$.

F has fixed point if there is $x \in H$ such that $x \in Fx$.

For each $(\phi, y) \in \mathcal{B} \times H$, define the set of selections of F by

$$f \in N_{F,\phi} = \left\{ f \in L^1(J,H):\ f(t) \in F\left(t, \phi, \int_0^t g(t,s,\phi)ds\right) \text{ for a.e } t \in J \right\}.$$

Lemma 7.1: (Losta and Opial, 1965) *Let J be a compact interval and H be a Hilbert space. Let F be an L_2-Caratheodory multivalued map with $N_{F,x} \neq \emptyset$ and let Θ be a linear continuous mapping from $L_2(J,H)$ to $C(J,H)$, then the operator*

$$\Theta \circ N_F : C(J,H) \to \mathcal{P}_{bd,cl,cv}(H),\quad x \mapsto (\Theta \circ N_F)(x) := \Theta(N_{F,x})$$

is a closed graph operator in $C(J,H) \times C(J,H)$.

Lemma 7.2: (Wang and Zhou, 2011b) *Let H be a Hilbert space, $\tilde{S}$ be a nonempty subset of H, which is bounded, closed and convex. Suppose $F:\tilde{S}\to\mathcal{P}_{bd,cl,cv}(H)$ is u.s.c with closed, convex values and such that $F(\tilde{S})\subset\tilde{S}$ and $F(\tilde{S})$ is compact, then F has a fixed point.*

Definition 7.5: (Podlubny, 1998) *The Riemann–Liouville fractional integral of order $\alpha>0$ for the function f: $(0,b]\to H$ is defined by*

$$J_t^{\alpha}f(t)=\frac{1}{\Gamma(\alpha)}\int_0^t(t-s)^{\alpha-1}f(s)ds$$

where Γ is the Euler's Gamma function.

Definition 7.6: (Podlubny, 1998) *The Caputo fractional derivative of order $\alpha>0$ for the function $f:J\to H$ is defined by*

$${}_a^cD_t^{\alpha}f(t)=\frac{1}{\Gamma(n-\alpha)_a}\int\frac{f^n(s)}{(t-s)^{\alpha-n+1}}ds.$$

Definition 7.7: (Lizama and N'Guérékata, 2013) *Let H be a Hilbert space, $k\in C(\mathbb{R}^+)$, $k\neq 0$ and let $a\in L^1_{loc}(\mathbb{R}^+)$, $a\neq 0$. Assume that A is a linear operator with domain D(A). A strongly continuous family $\{\mathcal{S}_\alpha(t)\}_{t\geq 0}$ of bounded linear operators from H into H is called an (a,k)-regularized resolvent family on H having A as a generator if the following hold*

(i) $\mathcal{S}_\alpha(0)=k(0)I$
(ii) $\mathcal{S}_\alpha(t)x\in D(A)$ *and* $\mathcal{S}_\alpha(t)Ax=A\mathcal{S}_\alpha(t)x$ *for all* $x\in D(A)$ *and* $t\geq 0$
(iii) $\mathcal{S}_\alpha(t)x=k(t)x+\int_0^t a(t-s)A\mathcal{S}_\alpha(s)xds,\ \ t\geq 0,\ \ x\in D(A).$

Assume that a and k are both positive and one of them is nondecreasing. Let $\{\mathcal{S}_\alpha(t)\}_{t\geq 0}$ be an (a,k)-regularized family with generator A such that $\|\mathcal{S}_\alpha(t)\|\leq Mk(t),\ t\geq 0$ for some constant $M>0$. Then, we have

$$Ax=\lim_{t\to 0^+}\frac{\mathcal{S}_\alpha(t)x-k(t)x}{(a^*k)(t)},\quad x\in\overline{D(A)}.$$

We say that $\{\mathcal{S}_\alpha(t)\}_{t\geq 0}$ is of type (M,ω), if there exist constants $M\geq 0$ and $\omega\in\mathbb{R}$ such that

$$\|\mathcal{S}_\alpha(t)\|\leq Me^{\omega t},\quad \forall t\geq 0.$$

Theorem 7.1: (Lizama, 2000) *Let A be a closed linear densely defined operator in a Hilbert space H. Then $\{\mathcal{S}_\alpha(t)\}_{t\geq 0}$ is an (a,k)-regularized family of type (M,ω) if and only if the following conditions hold*

(i) *$\hat{a}(\lambda) \neq 0$ and $\frac{1}{\hat{a}(\lambda)} \in \rho(A)$ for all $\lambda > \omega$*
(ii) *$\overline{H}(\lambda) := \hat{k}(\lambda)(I - \hat{a}(\lambda)A)^{-1}$ satisfies the estimates*

$$\|\overline{H}^{(n)}(\lambda)\| \leq \frac{Mn!}{(\lambda-\omega)^{n+1}}, \quad \lambda > \omega, \quad n \in \mathbb{N}.$$

Let A be a closed linear operator and $\{\mathcal{S}_\alpha(t)\}_{t \geq 0}$ be an exponentially bounded and strongly continuous operator family in $L(H)$ such that the Laplace transform $\hat{\mathcal{S}}_\alpha(\lambda)$ exists for $\lambda > \omega$. It is proved in Lizama (2000) that $\mathcal{S}_\alpha(t)$ is an (a,k)-regularized family with generator A if and only if for every $\lambda > \omega$, $(I - \hat{a}(\lambda)A)^{-1}$ exists in $L(H)$ and

$$\overline{H}(\lambda)x = \frac{\hat{k}(\lambda)}{\hat{a}(\lambda)}\left(\frac{1}{\hat{a}(\lambda)}I - A\right)^{-1} x = \int_0^\infty e^{-\lambda s}\mathcal{S}_\alpha(s)x ds, \quad x \in H.$$

Consider the following form of FSDEs

$${}^cD_t^\alpha x(t) = Ax(t) + {}^cD_t^{\alpha-1}\left(Bu(t) + F\left(t, x_t, \int_0^t g(t,s,x_s)ds\right) + \int_0^t \sigma(t,s,x_s)dW(s)\right)$$

by taking the Laplace transform on both sides of the above equation, we have

$$\lambda^\alpha \hat{x}(\lambda) - \lambda^{\alpha-1}x(0) - \lambda^{\alpha-2}x'(0) = A\hat{x}(\lambda) + \frac{1}{\lambda^{1-\alpha}}\left[\hat{u}(\lambda) + \hat{F}(\lambda) + \hat{\sigma}(\lambda)\right] \quad (7.3)$$

where

$$\hat{u}(\lambda) = \int_0^\infty e^{-\lambda t}Bu(t)dt, \quad \hat{\sigma}(\lambda) = \int_0^\infty e^{-\lambda t}\left(\int_0^t \sigma(t,s,x_s)dW(s)\right)dt$$
$$\hat{x}(\lambda) = \int_0^\infty e^{-\lambda t}x(t)dt, \quad \hat{F}(\lambda) = \int_0^\infty e^{-\lambda t}F\left(t, x_t, \int_0^t g(t,s,x_s)ds\right)dt$$

it follows from (7.3) that

$$\begin{aligned}(\lambda^\alpha I - A)\hat{x}(\lambda) &= \lambda^{\alpha-1}\phi(0) + \lambda^{\alpha-2}x_1 + \lambda^{\alpha-1}\hat{u}(\lambda) + \lambda^{\alpha-1}\hat{F}(\lambda) + \lambda^{\alpha-1}\hat{\sigma}(\lambda)\\ \hat{x}(\lambda) &= \lambda^{\alpha-1}(\lambda^\alpha I - A)^{-1}\phi(0) + \lambda^{\alpha-2}(\lambda^\alpha I - A)^{-1}x_1\\ &\quad + \lambda^{\alpha-1}(\lambda^\alpha I - A)^{-1}[\hat{u}(\lambda) + \hat{F}(\lambda) + \hat{\sigma}(\lambda)]\end{aligned} \quad (7.4)$$

now, one needs to find the Laplace transformable and strongly continuous family of bounded linear operators, say $\mathcal{S}_\alpha(t)$ such that $\hat{\mathcal{S}}_\alpha(\lambda) = \lambda^{\alpha-1}(\lambda^\alpha I - A)^{-1}$. In other words, one searches for the scalar functions $a(t)$ and $k(t)$ such that

$$\frac{\hat{k}(\lambda)}{\hat{a}(\lambda)}\left(\frac{1}{\hat{a}(\lambda)}I - A\right)^{-1} = \lambda^{\alpha-1}(\lambda^\alpha I - A)^{-1}. \quad (7.5)$$

In order to have the identity (7.5), one necessarily has $\hat{a}(\lambda) = \frac{1}{\lambda^\alpha}$ and $\hat{k}(\lambda) = \frac{1}{\lambda}$. By using inverse Laplace transformation, one gets $a(t) = \frac{t^{\alpha-1}}{\Gamma(\alpha)}$ and

$k(t) = 1$. It can be concluded that the appropriate family $\mathcal{S}_\alpha(t)$ corresponds to an (a,k)-regularized family with $a(t)$ & $k(t)$ as precisely defined above.

By setting $\hat{\mathcal{S}}_\alpha(\lambda) = \lambda^{\alpha-1}(\lambda^\alpha I - A)^{-1}$ in (7.4) and using the inverse Laplace transformation, one can get

$$\begin{aligned} x(t) = \mathcal{S}_\alpha(t)\phi(0) &+ \textstyle\int_0^t \mathcal{S}_\alpha(s)x_1 ds + \int_0^t \mathcal{S}_\alpha(t-s)Bu(s)ds \\ &+ \textstyle\int_0^t \mathcal{S}_\alpha(t-s)F\left(s, x_s, \int_0^s g(s,\tau,x_\tau)d\tau\right)ds \\ &+ \textstyle\int_0^t \mathcal{S}_\alpha(t-s)\left(\int_0^s \sigma(s,\tau,x_\tau)dW(\tau)\right)ds. \end{aligned}$$

From the above observations, one can define the mild solution for fractional stochastic differential inclusions (7.1)–(7.2) as follows

Definition 7.8: *An H-valued stochastic process* $\{x(t),\ t\in(-\infty,b]\}$ *is said to be a mild solution of the fractional stochastic differential inclusion* (7.1)–(7.2) *if*

(i) *$x(t)$ is $\mathcal{F}_t$-adapted and measurable for all $t \geq 0$*
(ii) *$x(t)$ is continuous on J almost surely and the following stochastic integral is verified*

$$\begin{aligned} x(t) = \mathcal{S}_\alpha(t)\phi(0) + \textstyle\int_0^t \mathcal{S}_\alpha(s)x_1 ds + \int_0^t \mathcal{S}_\alpha(t-s)Bu(s)ds + \int_0^t \mathcal{S}_\alpha(t-s)f(s)ds \\ + \textstyle\int_0^t \mathcal{S}_\alpha(t-s)\left(\int_0^s \sigma(s,\tau,x_\tau)dW(\tau)\right)ds,\ f \in N_{F,x}. \end{aligned}$$

Let $x_b(x_0; u)$ be the state value of the fractional order system (7.1)–(7.2) at terminal time b corresponding to the control u and the initial value $x_0 = \phi(t) \in \mathcal{B}$. Set

$$\mathcal{R}(b, x_0) = \{x_b(x_0, u)(0):\ u(\cdot) \in L_2(J, U)\}$$

which is called the reachable set of the system (7.1)–(7.2) at terminal time b. Its closure in H is denoted by $\overline{\mathcal{R}(b, x_0)}$.

Definition 7.9: *The fractional order system* (7.1)–(7.2) *is said to be approximately controllable on J, if $\overline{\mathcal{R}(b, x_0)} = H$.*

It is convenient at this point to introduce the operators

$$\begin{aligned} \Pi_0^b\{\cdot\} \quad &= \textstyle\int_0^b \mathcal{S}_\alpha(b-s)BB^*\mathcal{S}_\alpha^*(b-s)\mathbb{E}\{\cdot|\mathcal{F}_s\}ds \\ \text{and} \quad &R(\varepsilon, \Pi_0^b) = (\varepsilon I + \Pi_0^b)^{-1}\ \textit{for}\ \epsilon > 0 \end{aligned}$$

where B^* denotes the adjoint of B and $\mathcal{S}_\alpha^*(t)$ is the adjoint of $\mathcal{S}_\alpha(t)$. It is straightforward that the operator Π_0^b is a linear bounded operator.

Lemma 7.3: (Mahmudov and Zorlu, 2003) *The linear system corresponding to the fractional order system* (7.1)–(7.2) *is approximately controllable on J if and only if $\varepsilon(\varepsilon I + \Pi_0^b)^{-1} \to 0$ as $\varepsilon \to 0^+$ in the strong operator topology.*

The approximate controllability of the considered system (7.1)–(7.2) is studied under the following hypotheses

(H_1) A generates an $\left(\frac{t^{\alpha-1}}{\Gamma(\alpha)}, 1\right)$-regularized family $\mathcal{S}_\alpha(t)$ such that there exist $\omega > 0$ and $M > 0$ such that

$$\|\mathcal{S}_\alpha(t)\| \leq Me^{\omega t}, \quad \forall t \in J.$$

(H_2) For each $(t,s) \in J \times J$, the function $g(t,s,\cdot):\mathcal{B} \to H$ is continuous and for each $\phi \in \mathcal{B}$ the function $g(\cdot,\cdot,\phi):J \times J \to H$ is strongly measurable.

(H_3) The multivalued map $F:J \times \mathcal{B} \times H \to \mathcal{P}_{bd,cl,cv}(H)$ satisfies the following conditions

(i) for each $t \in J$, the map $F(t,\cdot,\cdot):\mathcal{B} \times H \to \mathcal{P}_{bd,cl,cv}(H)$ is u.s.c.

(ii) for each $(\phi, y) \in \mathcal{B} \times H$, the map $F(\cdot,\phi,y):J \to \mathcal{P}_{bd,cl,cv}(H)$ is measurable and the set

$$N_{F,\phi} = \left\{ f \in L^1(J,H): \ f(t) \in F\left(t, \phi, \int_0^t g(t,s,\phi)ds\right) \ for \ a.e \ t \in J \right\}$$

is nonempty.

(H_4) For each positive number $r > 0$, there exists a positive function $\mu(r)$ depending on r such that

$$\sup_{\|\phi\|_{\mathcal{B}}^2 \leq r} \|F(t,\phi,\hat{L}\phi)\|^2 \leq \mu(r), \qquad \lim_{r\to\infty} \frac{\mu(r)}{r} = \delta < \infty$$

and there exists constant $\hat{d} > 0$ such that

$$0 \leq \lim_{\|\phi\|_{\mathcal{B}}^2 \to \infty} \left(\sup_{t\in J} \frac{\mathbb{E}\|F(t,\phi,\hat{L}\phi)\|^2}{\|\phi\|_{\mathcal{B}}^2} \right) \leq \hat{d}.$$

(H_5) The function $\sigma:J \times J \times \mathcal{B} \to L(K,H)$ satisfies the following

(i) for each $(t,s) \in J \times J$, the function $\sigma(t,s,\cdot):\mathcal{B} \to L(K,H)$ is continuous and for each $\phi \in \mathcal{B}$, the function $\sigma(\cdot,\cdot,\phi):J \times J \to L(K,H)$ is strongly measurable.

(ii) there is a positive integrable function $m \in L^1([0,b])$ and a continuous nondecreasing function $\Lambda_\sigma: [0,\infty) \to (0,\infty)$ such that for every $(t,s,x) \in J \times J \times \mathcal{B}$, we have

$$\int_0^t \mathbb{E}\|\sigma(t,s,x)\|_Q^2 ds \leq m(t)\Lambda_\sigma(\|x\|_{\mathcal{B}}^2), \qquad \lim_{r\to\infty} \frac{\Lambda_\sigma(4r)}{r} = \Delta < \infty.$$

(H_6) The linear system corresponding to the fractional order system (7.1)–(7.2) is approximately controllable.

(H_7) The functions $F(t,\psi,y):J \times \mathcal{B} \times H \to \mathcal{P}_{bd,cl,cv}(H)$, $g(t,s,\psi):J \times J \times \mathcal{B} \to H$, and $\sigma(t,s,\psi):J \times J \times \mathcal{B} \to L(K,H)$ are uniformly bounded for all $t \in J$, $\psi \in \mathcal{B}$ and $y \in H$.

7.2.1.2 Approximate Controllability Result

Theorem 7.2: *Suppose that the hypotheses* $(H_1)-(H_5)$ *are satisfied then the system* (7.1)–(7.2) *has a mild solution on* J, *provided that*

$$4M^2b\left(\frac{e^{2\omega b}-1}{2\omega}\right)\left(\delta+4(\hat{d}+\beta)+\Delta Tr(Q)\sup_{t\in J}m(t)\right)\left(\frac{5\|B\|^4M^4b(e^{2\omega b}-1)}{2\omega\varepsilon^2}+1\right)<1 \tag{7.6}$$

Proof: Let $\mathcal{B}_b$ be the space of all functions $x:(-\infty,b]\to H$ such that $x_0\in\mathcal{B}$ and the restriction x: $[0,b]\to H$ is continuous. Let $\|\cdot\|_b$ be a seminorm in $\mathcal{B}_b$ defined by

$$\|x\|_b=\|x_0\|_{\mathcal{B}}+\sup_{s\in[0,b]}\left(\mathbb{E}\|x(s)\|^2\right)^{\frac{1}{2}}.$$

For any $\epsilon>0$, consider the operator $\Phi^{\varepsilon}:\mathcal{B}_b\to\mathcal{P}(\mathcal{B}_b)$ defined by $\Phi^{\varepsilon}x$, the set of $\rho\in\mathcal{B}_b$ such that

$$\rho(t)=\begin{cases}\phi(t),\ t\in(-\infty,0];\\ \mathcal{S}_\alpha(t)\phi(0)+\int_0^t\mathcal{S}_\alpha(s)x_1ds+\int_0^t\mathcal{S}_\alpha(t-s)Bu^{\varepsilon}(s,x)ds+\int_0^t\mathcal{S}_\alpha(t-s)f(s)ds\\ \quad+\int_0^t\mathcal{S}_\alpha(t-s)\left(\int_0^s\sigma(s,\tau,x_\tau)dW(\tau)\right)ds,t\in J,\ f\in N_{F,x}\end{cases}$$

where

$$\begin{aligned}u^{\epsilon}(s,x)&=B^*\mathcal{S}_\alpha^*(b-s)(\epsilon I+\Pi_0^b)^{-1}\Big\{\mathbb{E}\tilde{x}_b+\int_0^b\tilde{\phi}(s)dW(s)-\mathcal{S}_\alpha(b)\phi(0)\\&\quad-\int_0^b\mathcal{S}_\alpha(s)x_1ds-\int_0^b\mathcal{S}_\alpha(b-s)f(s)ds\\&\quad-\int_0^b\mathcal{S}_\alpha(b-s)\left(\int_0^s\sigma(s,\tau,x_\tau)dW(\tau)\right)ds\Big\},\quad f\in N_{F,x}.\end{aligned}$$

For $\phi\in\mathcal{B}$, define

$$\hat{\phi}(t)=\begin{cases}\phi(t),\ t\in(-\infty,0]\\ \mathcal{S}_\alpha(t)\phi(0),\ t\in J\end{cases}$$

then $\hat{\phi}\in\mathcal{B}_b$. Let $x(t)=\hat{\phi}(t)+y(t)$, $-\infty<t\le b$. It is easy to check that x satisfies (7.1)–(7.2) if and only if $y_0=0$ and

$$\begin{aligned} y(t) &= \int_0^t \mathcal{S}_\alpha(s)x_1 ds + \int_0^t \mathcal{S}_\alpha(t-s)Bu^\varepsilon(s,\hat{\phi}+y)ds + \int_0^t \mathcal{S}_\alpha(t-s)f(s)ds \\ &+ \int_0^t \mathcal{S}_\alpha(t-s)\Big(\int_0^s \sigma(s,\tau,\hat{\phi}_\tau + y_\tau)dW(\tau)\Big)ds, t\in J, \ f\in N_{F,x} \end{aligned}$$

where

$$\begin{aligned} u^\varepsilon(s,\hat{\phi}+y) = B^*\mathcal{S}^*_\alpha(b-s)(\epsilon I+\Pi_0^b)^{-1}\Big\{&\mathbb{E}\tilde{x}_b + \int_0^b \tilde{\phi}(s)dW(s) - S_\alpha(b)\phi(0) \\ &- \int_0^b \mathcal{S}_\alpha(s)x_1 ds - \int_0^b \mathcal{S}_\alpha(b-s)f(s)ds \\ &- \int_0^b \mathcal{S}_\alpha(b-s)\Big(\int_0^s \sigma(s,\tau,\hat{\phi}_\tau + y_\tau)dW(\tau)\Big)ds\Big\}, \ f\in N_{F,x}. \end{aligned}$$

Set $\mathcal{B}_b^0 = \{y\in\mathcal{B}_b, \ y_0 = 0\in\mathcal{B}\}$. For any $y\in\mathcal{B}_b^0$, we have

$$\|y\|_b = \|y_0\|_\mathcal{B} + \sup_{s\in[0,b]}\left(\mathbb{E}\|y(s)\|^2\right)^{\frac{1}{2}} = \sup_{s\in[0,b]}\left(\mathbb{E}\|y(s)\|^2\right)^{\frac{1}{2}}$$

thus $(\mathcal{B}_b^0, \|\cdot\|_b)$ is a Banach space.

For any positive number r, set $B_r = \{y\in\mathcal{B}_b^0\colon \ \|y\|_b^2\le r\}$, then for each r, B_r is clearly a bounded closed convex set in $\mathcal{B}_b^0$, and for $y\in B_r$, we have

$$\begin{aligned} \|y_t + \hat{\phi}_t\|_\mathcal{B}^2 &\le 2(\|y_t\|_\mathcal{B}^2 + \|\hat{\phi}_t\|_\mathcal{B}^2) \\ &\le 4\Big(\sup_{s\in[0,t]}\mathbb{E}\|y(s)\|^2 + \|y_0\|_\mathcal{B}^2 + \sup_{s\in[0,t]}\mathbb{E}\|\hat{\phi}(s)\|^2 + \|\hat{\phi}_0\|_\mathcal{B}^2\Big) \\ &\le 4\big(r + M^2e^{2\omega b}\mathbb{E}\|\phi(0)\|_H^2\big) + 4\|\phi\|_\mathcal{B}^2 = 4r + r^*, \end{aligned}$$

where $r^* = 4M^2e^{2\omega b}\mathbb{E}\|\phi(0)\|_H^2 + 4\|\phi\|_\mathcal{B}^2$.

Define the multivalued map $\Psi\colon\mathcal{B}_b^0\to\mathcal{P}(\mathcal{B}_h^0)$ by Ψy the set of $\overline{\rho}\in\mathcal{B}_b^0$ such that

$$\overline{\rho}(t) = \begin{cases} 0, \ t\in(-\infty,0] \\ \int_0^t \mathcal{S}_\alpha(s)x_1 ds + \int_0^t \mathcal{S}_\alpha(t-s)Bu^\varepsilon(s,\hat{\phi}+y)ds + \int_0^t \mathcal{S}_\alpha(t-s)f(s)ds \\ \quad + \int_0^t \mathcal{S}_\alpha(t-s)\Big(\int_0^s \sigma(s,\tau,\hat{\phi}_\tau + y_\tau)dW(\tau)\Big)ds, t\in J \end{cases}$$

where $f\in N_{F,x}$. Obviously the operator Φ^ε has a fixed point if and only if the operator Ψ has a fixed point.

Step 1: For each $y\in B_r$, Ψ is convex.

Let $\overline{\rho}_1$, $\overline{\rho}_2\in\Psi y$ then there exist $f_1,f_2\in N_{F,x}$ such that for each $t\in J$, one has

$$\begin{aligned} \overline{\rho}_i(t) &= \int_0^t \mathcal{S}_\alpha(s)x_1 ds + \int_0^t \mathcal{S}_\alpha(t-s)Bu^\varepsilon(s,\hat{\phi}+y)ds + \int_0^t \mathcal{S}_\alpha(t-s)f_i(s)ds \\ &+ \int_0^t \mathcal{S}_\alpha(t-s)\Big(\int_0^s \sigma(s,\tau,\hat{\phi}_\tau + y_\tau)dW(\tau)\Big)ds, \ i = 1,2. \end{aligned}$$

Let $0 \leq \lambda \leq 1$, then for each $t \in J$, one can get

$$\begin{aligned}(\lambda\overline{\rho}_1 + (1-\lambda)\overline{\rho}_2)(t) = & \int_0^t \mathcal{S}_\alpha(s)x_1 ds + \int_0^t \mathcal{S}_\alpha(t-s)BB^*\mathcal{S}^*_\alpha(b-s)(\epsilon I + \Pi_0^b)^{-1} \\ & \Big\{\mathbb{E}\tilde{x}_b + \int_0^b \tilde{\phi}(s)dW(s) - \mathcal{S}_\alpha(b)\phi(0) - \int_0^b \mathcal{S}_\alpha(s)x_1 ds \\ & - \int_0^b \mathcal{S}_\alpha(b-s)[\lambda f_1(s) + (1-\lambda)f_2(s)]ds \\ & - \int_0^b \mathcal{S}_\alpha(b-s)\Big(\int_0^s \sigma(s,\tau,\hat{\phi}_\tau + y_\tau)dW(\tau)\Big)ds\Big\}ds \\ & + \int_0^t \mathcal{S}_\alpha(t-s)[\lambda f_1(s) + (1-\lambda)f_2(s)]ds \\ & + \int_0^t \mathcal{S}_\alpha(t-s)\Big(\int_0^s \sigma(s,\tau,\hat{\phi}_\tau + y_\tau)dW(\tau)\Big)ds\end{aligned}$$

It is easy to see that $N_{F,x}$ is convex, since F has convex values, so $\lambda f_1 + (1-\lambda)f_2 \in N_{F,x}$. Thus $\lambda\overline{\rho}_1 + (1-\lambda)\overline{\rho}_2 \in \Psi y$.

Step 2: Next, one can show that there exists a positive number r such that $\Psi(B_r) \subseteq B_r$. If it is not true, then there exists $\epsilon > 0$ such that for every positive number r and $t \in J$, there exists a function $y^r(\cdot) \in B_r$ but $\Psi(y_r) \notin B_r$, i.e., $\mathbb{E}\|(\Psi y^r)(t)\|^2 \equiv \{\|\overline{\rho}^r\|_b^2 : \overline{\rho}^r \in (\Psi y^r)\} > r$. For such $\epsilon > 0$,

$$\begin{aligned} r & \leq \mathbb{E}\|(\Psi y^r)(t)\|^2 \\ & \leq 4\mathbb{E}\Big\|\int_0^t \mathcal{S}_\alpha(s)x_1 ds\Big\|^2 + 4\mathbb{E}\Big\|\int_0^t \mathcal{S}_\alpha(t-s)Bu^\epsilon(s,\hat{\phi}+y^r)ds\Big\|^2 \\ & + 4\mathbb{E}\Big\|\int_0^t \mathcal{S}_\alpha(t-s)f^r(s)ds\Big\|^2 + 4\mathbb{E}\Big\|\int_0^t \mathcal{S}_\alpha(t-s)\Big(\int_0^s \sigma(s,\tau,\hat{\phi}_\tau + y^r_\tau)dW(\tau)\Big)ds\Big\|^2 \\ & \sum_{i=1}^4 I_i. \end{aligned} \tag{7.7}$$

By using $(H_1) - (H_5)$, one can obtain

$$\begin{aligned} I_1 & = 4\mathbb{E}\Big\|\int_0^t \mathcal{S}_\alpha(s)x_1 ds\Big\|^2 \leq 4M^2 b x_1 \left(\frac{e^{2\omega b} - 1}{2\omega}\right) \\ I_2 & = 4\mathbb{E}\Big\|\int_0^t \mathcal{S}_\alpha(t-s)Bu^\epsilon(s,\hat{\phi}+y^r)ds\Big\|^2 \\ & \leq 4M^2 b\|B\|^2 \int_0^t e^{2\omega(t-s)}\mathbb{E}\|u^\epsilon(s,\hat{\phi}+y^r)\|^2 ds. \end{aligned} \tag{7.8}$$

Now, in view of (H_4), there exist positive constants, β, γ, such that, for all $\|\phi\|_{\mathcal{B}}^2 > \gamma$, $\mathbb{E}\|F(t,\phi,\hat{L}\phi)\|^2 \leq (\hat{d} + \beta)\|\phi\|_{\mathcal{B}}^2$. Let $F_1 = \{\phi : \|\phi\|_{\mathcal{B}}^2 \leq \gamma\}$, $F_2 = \{\phi : \|\phi\|_{\mathcal{B}}^2 > \gamma\}$ thus

$$\mathbb{E}\|F(t,\phi,\hat{L}\phi)\|^2 \leq \mu(\gamma) + (\hat{d} + \beta)\|\phi\|_{\mathcal{B}}^2.$$

and

$$
\begin{aligned}
\mathbb{E}\|u^{\varepsilon}(s,\hat{\phi}+y^{r})\|^2 \leq & \frac{5\|B\|^2M^2}{\varepsilon^2}\Big\{\mathbb{E}\Big\|\mathbb{E}\tilde{x}_b+\int_0^b\tilde{\phi}(s)dW(s)\Big\|^2+\mathbb{E}\|S_\alpha(b)\phi(0)\|^2 \\
&+\mathbb{E}\Big\|\int_0^b\mathcal{S}_\alpha(s)x_1ds\Big\|^2+\mathbb{E}\Big\|\int_0^b\mathcal{S}_\alpha(b-s)f(s)ds\Big\|^2 \\
&+\mathbb{E}\Big\|\int_0^b\mathcal{S}_\alpha(b-s)\Big(\int_0^s\sigma(s,\tau,\hat{\phi}_\tau+y_\tau)dW(\tau)\Big)ds\Big\|^2\Big\} \\
\leq & \frac{5\|B\|^2M^2}{\epsilon^2}\Big\{2\mathbb{E}\|\tilde{x}_b\|^2+2\int_0^b\mathbb{E}\|\tilde{\phi}(s)\|^2ds+M^2e^{2\omega b}\mathbb{E}\|\phi(0)\|^2 \\
&+M^2bx_1\left(\frac{e^{2\omega b}-1}{2\omega}\right)+M^2b\int_0^b e^{2\omega(b-s)}\mathbb{E}\|f(s)\|^2ds \\
&+M^2b\int_0^b e^{2\omega(b-s)}\mathbb{E}\Big\|\int_0^s\sigma(s,\tau,\hat{\phi}_\tau+y_\tau)dW(\tau)\Big\|^2ds \\
\leq & \frac{5\|B\|^2M^2}{\varepsilon^2}\Big\{2\mathbb{E}\|\tilde{x}_b\|^2+2\int_0^b\mathbb{E}\|\tilde{\phi}(s)\|^2ds+M^2e^{2\omega b}\mathbb{E}\|\phi(0)\|^2 \\
&+M^2bx_1\left(\frac{e^{2\omega b}-1}{2\omega}\right)+M^2b\int_0^b e^{2\omega(b-s)}(\mu(r)+(\hat{d}+\beta)\mathbb{E}\|\hat{\phi}_s+y_s\|_{\mathcal{B}}^2)ds \\
&+M^2b\int_0^b e^{2\omega(b-s)}Tr(Q)\int_0^s\mathbb{E}\Big\|\sigma(s,\tau,\hat{\phi}_\tau+y_\tau)\Big\|_Q^2d\tau ds\Big\} \\
\leq & \frac{5\|B\|^2M^2}{\varepsilon^2}\Big\{2\mathbb{E}\|\tilde{x}_b\|^2+2\int_0^b\mathbb{E}\|\tilde{\phi}(s)\|^2ds+M^2e^{2\omega b}\mathbb{E}\|\phi(0)\|^2 \\
&+M^2bx_1\left(\frac{e^{2\omega b}-1}{2\omega}\right)+M^2b(\mu(r)+(\hat{d}+\beta)(4r+r^*))\left(\frac{e^{2\omega b}-1}{2\omega}\right) \\
&+M^2b\int_0^b e^{2\omega(b-s)}Tr(Q)m(s)\Lambda_\sigma(\|\hat{\phi}_s+y_s\|_{\mathcal{B}}^2)ds\Big\} \\
\leq & \frac{5\|B\|^2M^2}{\varepsilon^2}\Big\{2\mathbb{E}\|\tilde{x}_b\|^2+2\int_0^b\mathbb{E}\|\tilde{\phi}(s)\|^2ds+M^2e^{2\omega b}\mathbb{E}\|\phi(0)\|^2 \\
&+M^2bx_1\left(\frac{e^{2\omega b}-1}{2\omega}\right)+M^2b(\mu(r)+(\hat{d}+\beta)(4r+r^*))\left(\frac{e^{2\omega b}-1}{2\omega}\right) \\
&+M^2bTr(Q)\sup_{t\in J}m(s)\int_0^b e^{2\omega(b-s)}\Lambda_\sigma(4r+r^*)ds\Big\}:=\Xi
\end{aligned}
$$

thus from (7.8), one has

$$
\begin{aligned}
I_2 \leq & \frac{20\|B\|^4M^4b}{\varepsilon^2}\left(\frac{e^{2\omega b}-1}{2\omega}\right)\Big\{2\mathbb{E}\|\tilde{x}_b\|^2+2\int_0^b\mathbb{E}\|\tilde{\phi}(s)\|^2ds+M^2e^{2\omega b}\mathbb{E}\|\phi(0)\|^2 \\
&+M^2bx_1\left(\frac{e^{2\omega b}-1}{2\omega}\right)+M^2b(\mu(r)+(\hat{d}+\beta)(4r+r^*))\left(\frac{e^{2\omega b}-1}{2\omega}\right) \\
&+M^2bTr(Q)\sup_{t\in J}m(t)\int_0^b e^{2\omega(b-s)}\Lambda_\sigma(4r+r^*)ds\Big\}
\end{aligned}
$$

$$
\begin{aligned}
I_3 &= 4\mathbb{E}\left\| \int_0^t \mathcal{S}_\alpha(t-s)f^r(s)ds \right\|^2 \\
&\leq 4M^2 b \int_0^t e^{2\omega(t-s)}\mathbb{E}\| f^r(s)\|^2 ds \leq 4M^2 b \int_0^t e^{2\omega(t-s)}(\mu(r) + (\hat{d} + \beta)(4r + r^*))ds \\
&\leq 4M^2 b(\mu(r) + (\hat{d} + \beta)(4r + r^*))\left(\frac{e^{2\omega b} - 1}{2\omega}\right),
\end{aligned}
$$

$$
\begin{aligned}
I_4 &= 4\mathbb{E}\left\| \int_0^t \mathcal{S}_\alpha(t-s)\left(\int_0^s \sigma(s,\tau,\hat{\phi}_\tau + y_\tau^r)dW(\tau)\right)ds \right\|^2 \\
&\leq 4M^2 b \int_0^t e^{2\omega(t-s)}\mathbb{E}\left\| \int_0^s \sigma(s,\tau,\hat{\phi}_\tau + y_\tau^r)dW(\tau)\right\|^2 ds \\
&\leq 4M^2 b \int_0^t e^{2\omega(t-s)} Tr(Q) \int_0^s \mathbb{E}\left\| \sigma(s,\tau,\hat{\phi}_\tau + y_\tau^r)\right\|_Q^2 d\tau ds \\
&\leq 4M^2 b Tr(Q) \int_0^t e^{2\omega(t-s)} m(s)\Lambda_\sigma(\|\hat{\phi}_\tau + y_\tau^r\|_{\mathcal{B}}^2)ds \\
&\leq 4M^2 b Tr(Q) \int_0^t e^{2\omega(t-s)} m(s)\Lambda_\sigma(4r + r^*)ds \\
&\leq 4M^2 b Tr(Q) \sup_{t\in J} m(t)\Lambda_\sigma(4r + r^*)\left(\frac{e^{2\omega b} - 1}{2\omega}\right)
\end{aligned}
$$

by combining the estimated I_1-I_4 together with (7.7), one arrives at

$$
\begin{aligned}
r \leq 4M^2 b\left(\frac{e^{2\omega b} - 1}{2\omega}\right)\Bigg[x_1 &+ \frac{5M^2\|B\|^4}{\varepsilon^2}\Bigg\{2\mathbb{E}\|\tilde{x}_b\|^2 + 2\int_0^b \mathbb{E}\|\tilde{\phi}(s)\|^2 ds \\
&+ M^2 e^{2\omega b}\mathbb{E}\|\phi(0)\|^2 + M^2 b x_1\left(\frac{e^{2\omega b} - 1}{2\omega}\right) + M^2 b(\mu(r) + (\hat{d} + \beta)(4r + r^*)) \\
&\times\left(\frac{e^{2\omega b} - 1}{2\omega}\right) + M^2 b Tr(Q)\sup_{t\in J} m(t)\int_0^b e^{2\omega(b-s)}\Lambda_\sigma(4r + r^*)ds\Bigg\} \\
&+ (\mu(r) + (\hat{d} + \beta)(4r + r^*)) + Tr(Q)\sup_{t\in J} m(t)\Lambda_\sigma(4r + r^*)\Bigg]
\end{aligned}
$$

dividing both sides of the above inequality by r and taking $r \to \infty$, one gets

$$
\begin{aligned}
&4M^2 b\left(\frac{e^{2\omega b} - 1}{2\omega}\right)\left(\delta + 4(\hat{d} + \beta) + \Delta Tr(Q)\sup_{t\in J} m(t)\right) \\
&\qquad\left(\frac{5\|B\|^4 M^4 b(e^{2\omega b} - 1)}{2\omega\varepsilon^2} + 1\right) \geq 1
\end{aligned}
$$

which is a contradiction to our assumption (7.6). Thus, for some $r > 0$, $\Psi(B_r) \subseteq B_r$.

Step 3: $\Psi(B_r)$ is equicontinuous. Indeed, let $\epsilon > 0$ be small, $0 < t_1 < t_2 \leq b$. For each $y \in B_r$ and $\overline{\rho}$ belonging to Ψy, there exists $f \in N_{F,x}$ such that for each $t \in J$,

$$
\begin{aligned}
\mathbb{E}\|\overline{p}(t_2)-\overline{p}(t_1)\|^2 \leq 12\Big\{&\mathbb{E}\Big\|\int_0^{t_1-\epsilon}[\mathcal{S}_\alpha(t_2)-\mathcal{S}_\alpha(t_1)]x_1ds\Big\|^2\\
&+\mathbb{E}\Big\|\int_{t_1-\epsilon}^{t_1}[\mathcal{S}_\alpha(t_2)-\mathcal{S}_\alpha(t_1)]x_1ds\Big\|^2+\mathbb{E}\Big\|\int_{t_1}^{t_2}\mathcal{S}_\alpha(t_2)x_1ds\Big\|^2\\
&+\mathbb{E}\Big\|\int_0^{t_1-\epsilon}[\mathcal{S}_\alpha(t_2-s)-\mathcal{S}_\alpha(t_1-s)]Bu^\epsilon(s,\hat{\phi}+y)ds\Big\|^2\\
&+\mathbb{E}\Big\|\int_{t_1-\epsilon}^{t_1}[\mathcal{S}_\alpha(t_2-s)-\mathcal{S}_\alpha(t_1-s)]Bu^\epsilon(s,\hat{\phi}+y)ds\Big\|^2\\
&+\mathbb{E}\Big\|\int_{t_1}^{t_2}\mathcal{S}_\alpha(t_2-s)Bu^\epsilon(s,\hat{\phi}+y)ds\Big\|^2+\mathbb{E}\Big\|\int_0^{t_1-\epsilon}[\mathcal{S}_\alpha(t_2-s)-\mathcal{S}_\alpha(t_1-s)]f(s)ds\Big\|^2\\
&+\mathbb{E}\Big\|\int_{t_1-\epsilon}^{t_1}[\mathcal{S}_\alpha(t_2-s)-\mathcal{S}_\alpha(t_1-s)]f(s)ds\Big\|^2+\mathbb{E}\Big\|\int_{t_1}^{t_2}\mathcal{S}_\alpha(t_2-s)f(s)ds\Big\|^2\\
&+\mathbb{E}\Big\|\int_0^{t_1-\epsilon}[\mathcal{S}_\alpha(t_2-s)-\mathcal{S}_\alpha(t_1-s)]\Big(\int_0^s\sigma(s,\tau,\hat{\phi}_\tau+y_\tau)dW(\tau)\Big)ds\Big\|^2\\
&+\mathbb{E}\Big\|\int_{t_1-\epsilon}^{t_1}[\mathcal{S}_\alpha(t_2-s)-\mathcal{S}_\alpha(t_1-s)]\Big(\int_0^s\sigma(s,\tau,\hat{\phi}_\tau+y_\tau)dW(\tau)\Big)ds\Big\|^2\\
&+\mathbb{E}\Big\|\int_{t_1}^{t_2}\mathcal{S}_\alpha(t_2-s)\Big(\int_0^s\sigma(s,\tau,\hat{\phi}_\tau+y_\tau)dW(\tau)\Big)ds\Big\|^2\Big\}\\
\leq 12\Big\{&(t_1-\epsilon)\int_0^{t_1-\epsilon}\|\mathcal{S}_\alpha(t_2)-\mathcal{S}_\alpha(t_1)\|^2\mathbb{E}\|x_1\|^2ds\\
&+\epsilon\int_{t_1-\epsilon}^{t_1}\|\mathcal{S}_\alpha(t_2)-\mathcal{S}_\alpha(t_1)\|^2\mathbb{E}\|x_1\|^2ds+(t_2-t_1)M^2\int_{t_1}^{t_2}e^{2\omega t}\mathbb{E}\|x_1\|^2ds\\
&+(t_1-\epsilon)\|B\|^2\int_0^{t_1-\epsilon}\|\mathcal{S}_\alpha(t_2-s)-\mathcal{S}_\alpha(t_1-s)\|^2\Xi ds\\
&+\epsilon\|B\|^2\int_{t_1-\epsilon}^{t_1}\|\mathcal{S}_\alpha(t_2-s)-\mathcal{S}_\alpha(t_1-s)\|^2\Xi ds+(t_2-t_1)M^2\|B\|^2\int_{t_1}^{t_2}e^{2\omega t}\Xi ds\\
&+(t_1-\epsilon)\int_0^{t_1-\epsilon}\|\mathcal{S}_\alpha(t_2-s)-\mathcal{S}_\alpha(t_1-s)\|^2(\mu(r)+(\hat{d}+\beta)(4r+r^*))ds\\
&+\epsilon\int_{t_1-\epsilon}^{t_1}\|\mathcal{S}_\alpha(t_2-s)-\mathcal{S}_\alpha(t_1-s)\|^2(\mu(r)+(\hat{d}+\beta)(4r+r^*))ds\\
&+(t_2-t_1)M^2\int_{t_1}^{t_2}e^{2\omega t}(\mu(r)+(\hat{d}+\beta)(4r+r^*))ds\\
&+(t_1-\epsilon)Tr(Q)\sup_{t\in J}m(t)\int_0^{t_1-\epsilon}\|\mathcal{S}_\alpha(t_2-s)-\mathcal{S}_\alpha(t_1-s)\|^2\Lambda_\sigma(4r+r^*)ds\\
&+\epsilon Tr(Q)\sup_{t\in J}m(t)\int_{t_1-\epsilon}^{t_1}\|\mathcal{S}_\alpha(t_2-s)-\mathcal{S}_\alpha(t_1-s)\|^2\Lambda_\sigma(4r+r^*)ds\\
&+(t_2-t_1)M^2Tr(Q)\sup_{t\in J}m(t)\int_{t_1}^{t_2}e^{2\omega t}\Lambda_\sigma(4r+r^*)ds
\end{aligned}
$$

for ϵ sufficiently small, we can verify that the right-hand side of the above inequality tends to zero as $t_2\to t_1$. Furthermore, the compactness of $\mathcal{S}_\alpha(t)$ implies the continuity in the uniform operator topology. Thus, Ψ maps B_r into an equicontinuous family of functions.

Step 4: Next, one needs to prove that $V(t)=\{(\Psi y)(t);\ \ y\in B_r\}$ is relatively compact in H. Obviously $V(t)$ is relatively compact for $t=0$. Let $0<t\leq b$ be fixed and ϵ be a real number satisfying $0<\varepsilon<t$. For $y\in B_r$, define

$$
\begin{aligned}
\overline{p}_\varepsilon = &\int_0^{t-\epsilon}\mathcal{S}_\alpha(s)x_1ds+\int_0^{t-\epsilon}\mathcal{S}_\alpha(t-s)Bu^\epsilon(s,\hat{\phi}+y)ds+\int_0^{t-\epsilon}\mathcal{S}_\alpha(t-s)f(s)ds\\
&+\int_0^{t-\epsilon}\mathcal{S}_\alpha(t-s)\Big(\int_0^s\sigma(s,\tau,\hat{\phi}_\tau+y_\tau)dW(\tau)\Big)ds
\end{aligned}
$$

where $f\in N_{F,x}$. Since $\mathcal{S}_\alpha(t)$ is a compact operator, the set $V_\epsilon(t)=\{\overline{p}_\epsilon(t){:}y\in B_r\}$ is relatively compact in H for each ϵ, $0<\epsilon<t$. Moreover, for each $0<\epsilon<t$,

$$\begin{aligned}\mathbb{E}\|\overline{\rho}(t)-\overline{\rho}_\epsilon(t)\|^2 \le 4\Big\{&\mathbb{E}\Big\|\int_{t-\epsilon}^t \mathcal{S}_\alpha(s)x_1 ds\Big\|^2 + \mathbb{E}\Big\|\int_{t-\epsilon}^t \mathcal{S}_\alpha(t-s)Bu^\epsilon(s,\hat{\phi}+y)ds\Big\|^2\\ &+\mathbb{E}\Big\|\int_{t-\epsilon}^t \mathcal{S}_\alpha(t-s)f(s)ds\Big\|^2\\ &+\mathbb{E}\Big\|\int_{t-\epsilon}^t \mathcal{S}_\alpha(t-s)\Big(\int_0^s \sigma(s,\tau,\hat{\phi}_\tau+y_\tau)dW(\tau)\Big)ds\Big\|^2\Big\}.\end{aligned}$$

Hence, there exist relatively compact sets arbitrarily close to the set $V(t)=\{(\Psi y)(t):y\in B_r\}$, and the set $V(t)$ is relatively compact in H for all $t\in[0,b]$. Since, it is compact at $t=0$, hence $V(t)$ is relatively compact in H for all $t\in[0,b]$.

Step 5: Ψ has a closed graph. Let $y^{(n)}\to y^*$ as $n\to\infty$, $\overline{\rho}^n\in\Psi y^{(n)}$, for each $y^{(n)}\in B_q$ and $\overline{\rho}^n\to\overline{\rho}^*$ as $n\to\infty$. We prove that $\overline{\rho}^*\in\Psi y^*$. Since $\overline{\rho}^n\in\Psi y^{(n)}$, there exists $f^{(n)}\in N_{F,y^{(n)}}$ such that

$$\overline{\rho}^n(t)=\begin{cases}0, \ t\in(-\infty,0]\\ \int_0^t \mathcal{S}_\alpha(s)x_1 ds+\int_0^t \mathcal{S}_\alpha(t-s)Bu^\epsilon(s,\hat{\phi}+y^{(n)})ds+\int_0^t \mathcal{S}_\alpha(t-s)f^{(n)}(s)ds\\ \quad+\int_0^t \mathcal{S}_\alpha(t-s)\Big(\int_0^s \sigma(s,\tau,\hat{\phi}_\tau+y^{(n)}_\tau)dW(\tau)\Big)ds, t\in J.\end{cases}$$

one must show that there exists $f^*\in N_{F,y^*}$ such that for each $t\in J$

$$\overline{\rho}^*(t)=\begin{cases}0, \ t\in(-\infty,0]\\ \int_0^t \mathcal{S}_\alpha(s)x_1 ds+\int_0^t \mathcal{S}_\alpha(t-s)Bu^\epsilon(s,\hat{\phi}+y^*)ds+\int_0^t \mathcal{S}_\alpha(t-s)f^*(s)ds\\ \quad+\int_0^t \mathcal{S}_\alpha(t-s)\Big(\int_0^s \sigma(s,\tau,\hat{\phi}_\tau+y^*_\tau)dW(\tau)\Big)ds, t\in J.\end{cases}$$

Now, for every $t\in J$, one has

$$\begin{aligned}\Big\|\overline{\rho}^n(t)&-\int_0^t \mathcal{S}_\alpha(s)x_1 ds-\int_0^t \mathcal{S}_\alpha(t-s)Bu^\epsilon(s,\hat{\phi}+y^{(n)})ds\\ &-\int_0^t \mathcal{S}_\alpha(t-s)f^{(n)}(s)ds-\int_0^t \mathcal{S}_\alpha(t-s)\Big(\int_0^s \sigma(s,\tau,\hat{\phi}_\tau+y^{(n)}_\tau)dW(\tau)\Big)ds\\ &-\Big(\overline{\rho}^*(t)-\int_0^t \mathcal{S}_\alpha(s)x_1 ds-\int_0^t \mathcal{S}_\alpha(t-s)Bu^\epsilon(s,\hat{\phi}+y^*)ds\\ &-\int_0^t \mathcal{S}_\alpha(t-s)f^*(s)ds-\int_0^t \mathcal{S}_\alpha(t-s)\Big(\int_0^s \sigma(s,\tau,\hat{\phi}_\tau+y^*_\tau)dW(\tau)\Big)ds\Big)\Big\|_b^2\to 0\end{aligned}$$

as $n\to\infty$. Consider the linear continuous operator $\Theta:L_2(J,H)\to C(J,H)$

$$\begin{aligned}(\Theta\overline{\rho})(t)=&\int_0^t \mathcal{S}_\alpha(t-s)f(s)ds+\int_0^t \mathcal{S}_\alpha(t-s)BB^*\mathcal{S}^*_\alpha(b-t)\\ &\times\Big(\int_0^b(\epsilon I+\Pi_\tau^b)^{-1}\mathcal{S}_\alpha(b-\tau)f(\tau)d\tau\Big)ds\end{aligned}$$

clearly it follows from Lemma 7.1 that $\Theta\circ N_{F,x}$ is a closed graph. Also from the definition of Θ, one has that for every $t\in J$

$$\begin{aligned}\overline{\rho}^n(t)&-\int_0^t \mathcal{S}_\alpha(s)x_1 ds-\int_0^t \mathcal{S}_\alpha(t-s)Bu^\epsilon(s,\hat{\phi}+y^{(n)})ds\\ &-\int_0^t \mathcal{S}_\alpha(t-s)\Big(\int_0^s \sigma(s,\tau,\hat{\phi}_\tau+y^{(n)}_\tau)dW(\tau)\Big)ds\in\Theta(N_{F,y^{(n)}})\end{aligned}$$

since $y^{(n)} \to y^*$ for some $f^* \in N_{F,y^*}$, it follows from Lemma 7.1 that

$$\begin{aligned}\overline{\rho}^*(t) &- \int_0^t S_\alpha(s)x_1 ds - \int_0^t S_\alpha(t-s)Bu^\epsilon(s,\hat{\phi}+y^*)ds \\ &- \int_0^t S_\alpha(t-s)\Big(\int_0^s \sigma(s,\tau,\hat{\phi}_\tau + y_\tau^*)dW(\tau)\Big)ds \in \Theta(N_{F,y^*}).\end{aligned}$$

Therefore Ψ has a closed graph.

It can be concluded from Step 1–Step 5 together with the Arzela–Ascoli theorem that Ψ is a compact multivalued map, u.s.c with convex closed values. As a consequence of Lemma 7.2, one can deduce that Ψ has a fixed point, which is a mild solution of the fractional order system (7.1)–(7.2).

Theorem 7.3: *Suppose that the hypotheses* $(H_1)-(H_7)$ *hold, then the nonlinear multivalued fractional stochastic differential equation* (7.1)–(7.2) *is approximately controllable on* J.

Proof: Let $x^\epsilon(\cdot)$ be a fixed point of Φ^ε in B_r. By Theorem 7.2 any fixed point of Φ^ϵ is a mild solution of (7.1)–(7.2) under the control function $u(s,x^\epsilon)$ and by stochastic Fubini theorem it satisfies, that for some $f^\varepsilon \in N_{F,x^\epsilon}$

$$\begin{aligned}x^\epsilon(b) &= \tilde{x}_b - \epsilon(\epsilon I+\Pi_0^b)^{-1}\Big[\mathbb{E}\tilde{x}_b + \int_0^b \tilde{\phi}(s)dW(s) - \int_0^t S_\alpha(s)x_1 ds\Big] \\ &+ \epsilon\int_0^t (\epsilon I+\Pi_s^b)^{-1}S_\alpha(t-s)f^\epsilon(s)ds \\ &+ \epsilon\int_0^t (\epsilon I+\Pi_s^b)^{-1}S_\alpha(t-s)\big(\int_0^s \sigma(s,\tau,x_\tau^\epsilon)dW(\tau)\big)ds\end{aligned}$$

Moreover, by hypothesis (H_7) on F & σ and the Dounford–Pettis theorem, one has that the sequences $\{f^\epsilon(s)\}$ and $\{\sigma(s,\tau,x_\tau^\epsilon)\}$ are weakly compact, respectively, in $L_2(J,H)$ and $L_2(L(K,H))$, so they are subsequences still denoted by $\{f^\epsilon(s)\}$ and $\{\sigma(s,\tau,x_\tau^\epsilon)\}$ that are weakly converges to say f and σ, respectively, in $L_2(J,H)$ and $L_2(L(K,H))$. Thus, one has

$$\begin{aligned}\mathbb{E}\|x^\epsilon(b)-\tilde{x}_b\|^2 \le 7\Big\{ &\mathbb{E}\|\epsilon(\epsilon I+\Pi_0^b)^{-1}\tilde{x}_b\|^2 + \mathbb{E}\Big\|\epsilon(\epsilon I+\Pi_0^b)^{-1}\Big(\int_0^b S_\alpha(s)x_1 ds\Big)\Big\|^2 \\ &+ \mathbb{E}\Big(\int_0^b \|\epsilon(\epsilon I+\Pi_0^b)^{-1}\tilde{\phi}(s)\|_{L_2(K,H)}^2 ds\Big)^2 \\ &+ \mathbb{E}\Big\|\int_0^b \epsilon(\epsilon I+\Pi_s^b)^{-1}S_\alpha(b-s)[f^\epsilon(s)-f(s)]ds\Big\|^2 \\ &+ \mathbb{E}\Big\|\int_0^b \epsilon(\epsilon I+\Pi_s^b)^{-1}S_\alpha(b-s)f(s)ds\Big\|^2 \\ &+ \mathbb{E}\Big\|\int_0^b \epsilon(\epsilon I+\Pi_s^b)^{-1}S_\alpha(b-s) \\ &\times\big(\int_0^s[\sigma(s,\tau,x_\tau^\epsilon)-\sigma(s,\tau,x_\tau)]dW(\tau)\big)ds\Big\|^2 \\ &+ \mathbb{E}\Big\|\int_0^b \epsilon(\epsilon I+\Pi_s^b)^{-1}S_\alpha(b-s)\big(\int_0^s \sigma(s,\tau,x_\tau)dW(\tau)\big)ds\Big\|^2.\end{aligned}$$

On the other hand by hypothesis (H_6), for all $0 \le s \le b$ the operator $\epsilon(\epsilon I+\Pi_0^b)^{-1} \to 0$ strongly as $\epsilon \to 0^+$ and moreover $\|\epsilon(\epsilon I+\Pi_0^b)^{-1}\| \le 1$. It follows from Lebesgue dominated convergence theorem and the compactness of $S_\alpha(t)$ that $\mathbb{E}\|x^\epsilon(b)-\tilde{x}_b\|^2 \to 0$ as $\epsilon \to 0^+$. This proves the approximate controllability of the multivalued fractional differential equation (7.1)–(7.2).

7.2.2 Solvability and Optimal Control Results for FSDEs

This subsection investigates the solvability and optimal controls for FSDEs of order $1<\alpha<2$ in Hilbert space by using (a,k)-regularized families of bounded linear operators. Sufficient conditions are formulated to prove that the system has a unique mild solution by using the classical Banach contraction mapping principle. Then, the existence of optimal control for the corresponding Lagrange optimal control problem is investigated. Consider the following form of FSDE

$$^{c}D_t^{\alpha}x(t)=Ax(t)+{}^{c}D_t^{\alpha-1}\left(B(t)u(t)+f(t,x(t))+\int_0^t\sigma(s,x(s))dW(s)\right),\ t\in J \quad (7.9)$$

$$x(0)=x_0,\ x'(0)=0 \quad (7.10)$$

where $1<\alpha<2$, A generates an $\left(\frac{t^{\alpha-1}}{\Gamma(\alpha)},1\right)$-regularized family $\mathcal{S}_{\alpha}(t)$, u is a given control function, it takes values from another separable reflexive Hilbert space U, B is a linear operator from U into H. Here $f:J\times H\to H$ and $\sigma:J\times H\to L(K,H)$ are the appropriate functions.

7.2.2.1 Preliminaries

The collection of all strongly-measurable, p-integrable H-valued random variables, denoted by $L_p(\Omega,\mathcal{F},\mathbb{P};H)\equiv L_p(\Omega;H)$ is a Banach space equipped with norm $\|x(\cdot)\|_{L_p}=(\mathbb{E}\|x(\cdot;w)\|_H^p)^{\frac{1}{p}}$. Let $C(J,L_p(\Omega;H))$ be the Banach space of all continuous maps from J to $L_p(\Omega;H)$ satisfying the condition

$$\sup_{t\in J}\mathbb{E}\|x(t)\|^p<\infty.$$

Let $L_2^0=L_2(Q^{\frac{1}{2}}K,H)$ be the space of all Hilbert-Schmidt operators from $Q^{\frac{1}{2}}K$ to H with inner product $\langle\psi,\pi\rangle_{L_2^0}=tr(\psi Q\pi^*)$.

Let $\mathcal{C}$ be the closed subspace of all continuous process x that belong to the space $C(J,L_p(\Omega;H))$ consisting of $\mathcal{F}_t$-adapted measurable processes such that the $\mathcal{F}_0$-adapted processes $x(0)$ with a seminorm $\|\cdot\|_{\mathcal{C}}$ is defined by $\|x\|_{\mathcal{C}}=\left(\sup_{t\in J}\|x(t)\|_{L_p}^p\right)^{\frac{1}{p}}$. It is easy to verify that $\mathcal{C}$ furnished with the norm topology as defined above is a Banach space.

Consider the following form of FSDEs

$$^{c}D_t^{\alpha}x(t)=Ax(t)+{}^{c}D_t^{\alpha-1}\left(B(t)u(t)+f(t,x(t))+\int_0^t\sigma(s,x(s))dW(s)\right)$$

by taking the Laplace transform on both sides of the above equation, one can get

$$\lambda^{\alpha}\hat{x}(\lambda)-\lambda^{\alpha-1}x(0)-\lambda^{\alpha-2}x'(0)=A\hat{x}(\lambda)+\frac{1}{\lambda^{1-\alpha}}\left[\hat{u}(\lambda)+\hat{f}(\lambda)+\hat{\sigma}(\lambda)\right] \quad (7.11)$$

where

$$\hat{u}(\lambda) = \int_0^\infty e^{-\lambda t}B(t)u(t)dt, \ \hat{f}(\lambda) = \int_0^\infty e^{-\lambda t}f(t,x(t))dt,$$
$$\hat{g}(\lambda) = \int_0^\infty e^{-\lambda t}\big(\int_0^t \sigma(s,x(s))dW(s)\big)dt$$

It follows from (7.11) that

$$\begin{aligned}(\lambda^\alpha I - A)\hat{x}(\lambda) &= \lambda^{\alpha-1}x_0 + \lambda^{\alpha-1}\hat{u}(\lambda) + \lambda^{\alpha-1}\hat{f}(\lambda) + \lambda^{\alpha-1}\hat{\sigma}(\lambda)\\ \hat{x}(\lambda) &= \lambda^{\alpha-1}(\lambda^\alpha I - A)^{-1}x_0 + \lambda^{\alpha-1}(\lambda^\alpha I - A)^{-1}\hat{u}(\lambda)\\ &\quad + \lambda^{\alpha-1}(\lambda^\alpha I - A)^{-1}\hat{f}(\lambda) + \lambda^{\alpha-1}(\lambda^\alpha I - A)^{-1}\hat{\sigma}(\lambda)\end{aligned} \tag{7.12}$$

one needs to find the Laplace transformable and strongly continuous family of bounded linear operators, say $\mathcal{S}_\alpha(t)$ such that $\hat{\mathcal{S}}_\alpha(\lambda) = \lambda^{\alpha-1}(\lambda^\alpha I - A)^{-1}$. In other words, one searches for the scalar functions $a(t)$ and $k(t)$ such that

$$\frac{\hat{k}(\lambda)}{\hat{a}(\lambda)}\left(\frac{1}{\hat{a}(\lambda)}I - A\right)^{-1} = \lambda^{\alpha-1}(\lambda^\alpha I - A)^{-1}. \tag{7.13}$$

In order to have the identity (7.13), we necessarily have $\hat{a}(\lambda) = \frac{1}{\lambda^\alpha}$ and $\hat{k}(\lambda) = \frac{1}{\lambda}$. By using the inverse Laplace transformation, one obtains $a(t) = \frac{t^{\alpha-1}}{\Gamma(\alpha)}$ and $k(t) = 1$. It can be concluded that the appropriate family $\mathcal{S}_\alpha(t)$ corresponds to an (a,k)-regularized family with $a(t)$ & $k(t)$ as precisely defined above.

From (7.12), one has

$$\hat{x}(\lambda) = \hat{\mathcal{S}}_\alpha(\lambda)x_0 + \hat{\mathcal{S}}_\alpha(\lambda)\hat{u}(\lambda) + \hat{\mathcal{S}}_\alpha(\lambda)\hat{f}(\lambda) + \hat{\mathcal{S}}_\alpha(\lambda)\hat{\sigma}(\lambda)$$

by employing the inverse Laplace transformation on both sides of the above equation, one can get

$$\begin{aligned}x(t) &= \mathcal{S}_\alpha(t)x_0 + \int_0^t \mathcal{S}_\alpha(t-s)B(s)u(s)ds + \int_0^t \mathcal{S}_\alpha(t-s)f(s,x(s))ds\\ &\quad + \int_0^t \mathcal{S}_\alpha(t-s)\big(\int_0^s \sigma(\tau,x(\tau))dW(\tau)\big)ds.\end{aligned}$$

Definition 7.10: *An H-valued stochastic process* $\{x(t),\ t\in J\}$ *is said to be a mild solution of the fractional Cauchy problem* (7.9)–(7.10) *if*

(i) *$x(t)$ is $\mathcal{F}_t$-adapted and measurable for all $t \geq 0$.*
(ii) *$x(t)$ is continuous on J almost surely and the following stochastic integral is verified*

$$\begin{aligned}x(t) &= \mathcal{S}_\alpha(t)x_0 + \int_0^t \mathcal{S}_\alpha(t-s)B(s)u(s)ds + \int_0^t \mathcal{S}_\alpha(t-s)f(s,x(s))ds\\ &\quad + \int_0^t \mathcal{S}_\alpha(t-s)\big(\int_0^s \sigma(\tau,x(\tau))dW(\tau)\big)ds.\end{aligned}$$

The following hypotheses are considered to prove the main results

(H_8) A generates an $\left(\frac{t^{\alpha-1}}{\Gamma(\alpha)}, 1\right)$-regularized family $\mathcal{S}_\alpha(t)$ such that there exist $\omega > 0$ and $M > 0$ such that $\|\mathcal{S}_\alpha(t)\| \leq Me^{\omega t}$, $\forall t \in J$.

(H_9) The function $f: J \times H \to H$ satisfies the following

(i) for each $t \in J$, $f(t, \cdot): H \to H$ is continuous and for each $x \in H$, $f(\cdot, x): J \to H$ is strongly measurable;

(ii) there exist constants $M_f > 0$, $\hat{M}_f > 0$ such that

$$\mathbb{E}\|f(t,x) - f(t,y)\|_H^p \leq M_f \mathbb{E}\|x - y\|_H^p, \quad \forall x, y \in H$$
$$\text{and } \mathbb{E}\|f(t,x)\|_H^p \leq \hat{M}_f(1 + \mathbb{E}\|x\|_H^p), \quad \forall x \in H.$$

(H_{10}) The function $\sigma: J \times H \to L(K,H)$ satisfies the following

(i) for each $t \in J$, $\sigma(t, \cdot): H \to L(K,H)$ is continuous and for each $x \in H$, $\sigma(\cdot, x): J \to L(K,H)$ is strongly measurable;

(ii) there exist constants $M_\sigma > 0$, $\hat{M}_\sigma > 0$ such that

$$\mathbb{E}\|\sigma(t,x) - \sigma(t,y)\|_H^p \leq M_\sigma \mathbb{E}\|x - y\|_H^p, \quad \forall x, y \in H$$
$$\text{and } \mathbb{E}\|\sigma(t,x)\|_H^p \leq \hat{M}_\sigma(1 + \mathbb{E}\|x\|_H^p), \quad \forall x \in H.$$

(H_{11}) Let $u \in U$ be the control function and the operator $B \in L_\infty(J, L(U,H))$, $\|B\|$ stands for the norm of the operator B.

(H_{12}) The multivalued map $\mathcal{A}: J \to 2^U$ $\{\emptyset\}$ has closed, convex, and bounded values. $\mathcal{A}(\cdot)$ is graph measurable and $\mathcal{A}(\cdot) \subseteq \Sigma$, where Σ is a bounded subset of U.

7.2.2.2 Existence of Mild Solution

Theorem 7.4: *Suppose that the hypotheses* $(H_8) - (H_{12})$ *hold, then the fractional control problem* (7.9)–(7.10) *has a unique mild solution on J provided that*

$$2^{p-1} M^p b^{p-1}\left(\frac{e^{p\omega b} - 1}{p\omega}\right)(M_f + b^{\frac{p}{2}} c_p M_\sigma) < 1. \tag{7.14}$$

Proof: Consider the map Φ on $\mathcal{C}$ defined by

$$(\Phi x)(t) = \mathcal{S}_\alpha(t)x_0 + \int_0^t \mathcal{S}_\alpha(t-s)B(s)u(s)ds + \int_0^t \mathcal{S}_\alpha(t-s)f(s,x(s))ds$$
$$+ \int_0^t \mathcal{S}_\alpha(t-s)\left(\int_0^s \sigma(\tau, x(\tau))dW(\tau)\right)ds.$$

Now,

$$\begin{aligned}\mathbb{E}\|(\Phi x)(t)\|_H^p &\leq 4^{p-1}\mathbb{E}\|\mathcal{S}_\alpha(t)x_0\|^p + 4^{p-1}\mathbb{E}\Big\|\int_0^t \mathcal{S}_\alpha(t-s)\big(\int_0^s \sigma(\tau,x(\tau))dW(\tau)\big)ds\Big\|^p\\
&+ 4^{p-1}\mathbb{E}\Big\|\int_0^t \mathcal{S}_\alpha(t-s)B(s)u(s)ds\Big\|^p + 4^{p-1}\mathbb{E}\Big\|\int_0^t \mathcal{S}_\alpha(t-s)f(s,x(s))ds\Big\|^p\\
&\leq 4^{p-1}M^p e^{p\omega b}\mathbb{E}\|x_0\|^p + 4^{p-1}M^p b^{p-1}\int_0^t e^{p\omega(t-s)}\mathbb{E}\Big\|\int_0^s \sigma(\tau,x(\tau))dW(\tau)\Big\|^p ds\\
&+ 4^{p-1}M^p\|B\|^p\left[\left(\int_0^t e^{\frac{p\omega(t-s)}{p-1}}ds\right)^{1-\frac{1}{p}}\left(\int_0^t \mathbb{E}\|u(s)\|^p ds\right)^{\frac{1}{p}}\right]^p\\
&+ 4^{p-1}M^p b^{p-1}\int_0^t e^{p\omega(t-s)}\mathbb{E}\|f(s,x(s))\|^p ds\\
&\leq 4^{p-1}M^p e^{p\omega b}\mathbb{E}\|x_0\|^p + 4^{p-1}M^p b^{p-1}\int_0^t e^{p\omega(t-s)}c_p\left[\int_0^s \left(\mathbb{E}\|\sigma(\tau,x(\tau))\|_{L_2^0}^p\right)^{\frac{2}{p}}d\tau\right]^{\frac{p}{2}}ds\\
&+ 4^{p-1}M^p\|B\|^p\left[\left(\frac{p-1}{p\omega}\right)^{\frac{p-1}{p}}(e^{p\omega b}-1)^{\frac{p-1}{p}}\|u\|_{L_p(J,U)}\right]^p\\
&+ 4^{p-1}M^p b^{p-1}\int_0^t e^{p\omega(t-s)}\mathbb{E}\|f(s,x(s))\|^p ds\\
&\leq 4^{p-1}M^p e^{p\omega b}\mathbb{E}\|x_0\|^p + 4^{p-1}M^p b^{p-1}\int_0^t e^{p\omega(t-s)}c_p\left[\int_0^s \left(\hat{M}_\sigma(1+\mathbb{E}\|x(\tau)\|^p)\right)^{\frac{2}{p}}d\tau\right]^{\frac{p}{2}}ds\\
&+ 4^{p-1}M^p\|B\|^p\left(\frac{p-1}{p\omega}\right)^{p-1}(e^{p\omega b}-1)^{p-1}\|u\|_{L_p(J,U)}^p\\
&+ 4^{p-1}M^p b^{p-1}\int_0^t e^{p\omega(t-s)}\hat{M}_f(1+\mathbb{E}\|x(s)\|^p)ds\end{aligned}$$

Thus,

$$\begin{aligned}\sup_{t\in J}\mathbb{E}\big\|(\Phi x)(t)\big\|_H^p &\leq 4^{p-1}M^p e^{p\omega b}\mathbb{E}\|x_0\|^p + 4^{p-1}M^p\|B\|^p\left(\frac{p-1}{p\omega}\right)^{p-1}(e^{p\omega b}-1)^{p-1}\|u\|_{L_p(J,U)}^p\\
&+ 4^{p-1}M^p b^{p-1}\left(\frac{e^{p\omega b}-1}{p\omega}\right)(\hat{M}_f + b^{\frac{p}{2}}c_p\hat{M}_\sigma)\sup_{t\in J}(1+\mathbb{E}\|x(t)\|^p)\end{aligned}$$

it follows from the fact that $u(\cdot), f(\cdot,\cdot)$ are continuous, Φ is well-defined on $\mathcal{C}$. Moreover, for $x,\ y\in\mathcal{C}$ and $t\in J$, one has

$$\begin{aligned}\mathbb{E}\|(\Phi x)(t)-(\Phi y)(t)\|_H^p &\leq 2^{p-1}\mathbb{E}\Big\|\int_0^t \mathcal{S}_\alpha(t-s)[f(s,x(s))-f(s,y(s))]ds\Big\|^p\\
&+ 2^{p-1}\mathbb{E}\Big\|\int_0^t \mathcal{S}_\alpha(t-s)\big(\int_0^s[\sigma(\tau,x(\tau))-\sigma(\tau,y(\tau))]dW(\tau)\big)ds\Big\|^p\\
&\leq 2^{p-1}M^p b^{p-1}\int_0^t e^{p\omega(t-s)}\mathbb{E}\|f(s,x(s))-f(s,y(s))\|^p ds\\
&+ 2^{p-1}M^p b^{p-1}\int_0^t e^{p\omega(t-s)}\mathbb{E}\Big\|\int_0^s[\sigma(\tau,x(\tau))-\sigma(\tau,y(\tau))]dW(\tau)\Big\|^p ds\\
&\leq 2^{p-1}M^p b^{p-1}\int_0^t e^{p\omega(t-s)}M_f\mathbb{E}\|x(s)-y(s)\|^p ds\\
&+ 2^{p-1}M^p b^{p-1}\int_0^t e^{p\omega(t-s)}c_p\left[\int_0^s \left(M_\sigma\mathbb{E}\|x(\tau)-y(\tau)\|^p\right)^{\frac{2}{p}}d\tau\right]^{\frac{p}{2}}ds\end{aligned}$$

Thus,

$$\sup_{t\in J}\mathbb{E}\|(\Phi x)(t)-(\Phi y)(t)\|_H^p \le 2^{p-1}M^p b^{p-1}\left(\frac{e^{p\omega b}-1}{p\omega}\right)(M_f + b^{\frac{p}{2}}c_p M_\sigma)\sup_{t\in J}\mathbb{E}\|x(t)-y(t)\|^p$$

which implies by inequality (7.14) that Φ is a contraction and hence by contraction mapping principle there exists a unique fixed point x of Φ, which is a mild solution of the fractional control problem (7.9)–(7.10).

7.2.2.3 Existence of Optimal Control Result

Consider the following Lagrange problem $(\mathcal{P})$:

Find a control $u^0\in\mathcal{A}_{ad}$ such that

$$\mathcal{J}(u^0)\le\mathcal{J}(u),\quad \forall\ u\in\mathcal{A}_{ad}$$

where

$$\mathcal{J}(u)=\mathbb{E}\left\{\int_0^b \mathcal{L}(t,x^u(t),u(t))dt\right\}$$

and x^u denotes the mild solution of (7.9)–(7.10) corresponding to the control $u\in\mathcal{A}_{ad}$. For the existence of solution for problem $(\mathcal{P})$, one shall introduce the following hypothesis

(H_{13})

- **(i)** The functional $\mathcal{L}:J\times H\times U\to\mathbb{R}\cup\{\infty\}$ is $\mathcal{F}_t$- measurable;
- **(ii)** $\mathcal{L}(t,\cdot,\cdot)$ is sequentially lower semicontinuous on $H\times U$ for almost all $t\in J$;
- **(iii)** L(t,x,ċ) is convex on U for each $x\in H$ and almost all $t\in J$;
- **(iv)** there exist constants $d\ge 0,\ e>0,\ \mu$ is nonnegative and $\mu\in L^1(J,\mathbb{R})$ such that

$$\mathcal{L}(t,x,u)\ge\mu(t)+d\mathbb{E}\|x\|_H^p+\mathbb{E}\|u\|_U^p.$$

Theorem 7.5: *Let the hypotheses* $(H_8)-(H_{13})$ *hold. Suppose that* B *is a strongly continuous operator. Then the Lagrange problem* $(\mathcal{P})$ *admits at least one optimal pair, i.e., there exists an admissible control* $u^0\in\mathcal{A}_{ad}$ *such that*

$$\mathcal{J}(u^0)=\mathbb{E}\left(\int_0^b \mathcal{L}(t,x^0(t),u^0(t))dt\right)\le\mathcal{J}(u),\quad\forall\ u\in\mathcal{A}_{ad}.$$

Proof: If $\inf\{\mathcal{J}(u)|u\in\mathcal{A}_{ad}\}=+\infty$, there is nothing to prove. Without loss of generality, one can assume that $\inf\{\mathcal{J}(u)|u\in\mathcal{A}_{ad}\}=\epsilon<+\infty$. Using (H_{13}), one has $\epsilon>-\infty$. By the definition of infimum there exists a minimizing sequence feasible pair $\{(x^m,u^m)\}\subset\mathcal{P}_{ad}$, where $\mathcal{P}_{ad}=\{(x,u)$: x is a mild solution of the system (7.9)–(7.10) corresponding to $u \in \mathcal{A}_{ad}\}$ such that

$$\mathcal{J}(x^m, u^m) \to \epsilon \text{ as } m \to +\infty.$$

Since $\{u^m\} \subseteq \mathcal{A}_{ad}$, $m = 1, 2, \cdots$, $\{u^m\}$ is a bounded subset of the separable reflexive Banach space $L_p(J, U)$, there exists a subsequence, relabeled as $\{u^m\}$ and $u^0 \in L_p(J, U)$ such that $u^m \longrightarrow^w u^0$ in $L_p(J, U)$.

Since $\mathcal{A}_{ad}$ is closed and convex, then by Marzur lemma $u^0 \in \mathcal{A}_{ad}$. Let $\{x^m\} \subset C(J, L_p^{\mathcal{F}}(\Omega; H))$ denote the corresponding sequence of solutions of the integral equation

$$\begin{aligned} x^m(t) &= \mathcal{S}_\alpha(t)x_0 + \int_0^t \mathcal{S}_\alpha(t-s)B(s)u^m(s)ds + \int_0^t \mathcal{S}_\alpha(t-s)f(s, x^m(s))ds \\ &\quad + \int_0^t \mathcal{S}_\alpha(t-s)\left(\int_0^s \sigma(\tau, x^m(\tau))dW(\tau)\right)ds. \end{aligned}$$

It follows from the boundedness of $\{u^m\}$ and Theorem 7.4, one can easily check that there exists a $\rho > 0$ such that

$$\mathbb{E}\|x^m\|^p \leq \rho, \quad m = 0, 1, 2, \cdots.$$

For $t \in J$, one has

$$\begin{aligned} \mathbb{E}\|x^m(t) - x^0(t)\|_H^p &\leq 3^{p-1}\mathbb{E}\left\| \int_0^t \mathcal{S}_\alpha(t-s)[f(s, x^m(s)) - f(s, x^0(s))]ds \right\|^p \\ &\quad + 3^{p-1}\mathbb{E}\left\| \int_0^t \mathcal{S}_\alpha(t-s)[B(s)u^m(s) - B(s)u^0(s)]ds \right\|^p \\ &\quad + 3^{p-1}\mathbb{E}\left\| \int_0^t \mathcal{S}_\alpha(t-s)\left(\int_0^s [\sigma(\tau, x^m(\tau)) - \sigma(\tau, x^0(\tau))]dW(\tau)\right)ds \right\|^p \\ &\leq 3^{p-1}M^p b^{p-1} \int_0^t e^{p\omega(t-s)}\mathbb{E}\|f(s, x^m(s)) - f(s, x^0(s))\|^p ds \\ &\quad + 3^{p-1}M^p \left[\left(\int_0^t e^{\frac{p\omega(t-s)}{p-1}} ds \right)^{1-\frac{1}{p}} \left(\int_0^t \mathbb{E}\|B(s)u^m(s) - B(s)u^0(s)\|^p ds \right)^{\frac{1}{p}} \right]^p \\ &\quad + 3^{p-1}M^p b^{p-1} \int_0^t e^{p\omega(t-s)}\mathbb{E}\left\| \int_0^s [\sigma(\tau, x^m(\tau)) - \sigma(\tau, x^0(\tau))]dW(\tau) \right\|^p ds \\ &\leq 3^{p-1}M^p b^{p-1} \int_0^t e^{p\omega(t-s)} M_f \mathbb{E}\|x^m(s) - x^0(s)\|^p ds \\ &\quad + 3^{p-1}M^p \left(\frac{p-1}{p\omega} \right)^{p-1} (e^{p\omega b} - 1)^{p-1} \|Bu^m - Bu^0\|_{L_p(J,U)}^p \\ &\quad + 3^{p-1}M^p b^{p-1} \int_0^t e^{p\omega(t-s)} c_p \left[\int_0^s (M_\sigma \mathbb{E}\|x^m(\tau) - x^0(\tau)\|^p)^{\frac{2}{p}} d\tau \right]^{\frac{p}{2}} ds. \end{aligned}$$

Thus,

$$\begin{aligned} \sup_{t \in J} \mathbb{E}\|(\Phi x)(t) - (\Phi y)(t)\|_H^p &\leq 3^{p-1}M^p b^{p-1} \left(\frac{e^{p\omega b} - 1}{p\omega} \right) (M_f + b^{\frac{p}{2}} c_p M_\sigma) \sup_{t \in J} \mathbb{E}\|x^m(t) - x^0(t)\|^p \\ &\quad + 3^{p-1}M^p \left(\frac{p-1}{p\omega} \right)^{p-1} (e^{p\omega b} - 1)^{p-1} \|Bu^m - Bu^0\|_{L_p(J,U)}^p \end{aligned}$$

which implies that there exists a constant $N^* > 0$ such that

$$\sup_{t\in J}\mathbb{E}\left\|x^m(t)-x^0(t)\right\|^p \le N^*\left\|Bu^m-Bu^0\right\|^p_{L_p(J,U)} \quad \text{for } t\in J$$

where

$$N^* = \frac{3^{p-1}M^p\left(\frac{p-1}{p\omega}\right)^{p-1}(e^{p\omega b}-1)^{p-1}}{1-3^{p-1}M^pb^{p-1}\left(\frac{e^{p\omega b}-1}{p\omega}\right)(M_f+b^{\frac{p}{2}}c_pM_\sigma)}$$

and $3^{p-1}M^pb^{p-1}\left(\frac{e^{p\omega b}-1}{p\omega}\right)(M_f+b^{\frac{p}{2}}c_pM_\sigma)<1$.

Since B is strongly continuous, we have $\left\|Bu^m-Bu^0\right\|^p_{L_p(J,U)}\xrightarrow{s}0$ as $m\to\infty$ thus, we have $\mathbb{E}\|x^m-x^0\|^p\xrightarrow{s}0$ as $m\to\infty$, this yields that $x^m\xrightarrow{s}x^0$ in $C(J,L_p(\Omega;H))$ as $m\to\infty$. Note that (H_{13}) implies the assumptions of Balder (1987) are satisfied. Hence by Balder's theorem, one can conclude that $(x,u)\to\mathbb{E}\left(\int_0^b\mathcal{L}(t,x(t),u(t))dt\right)$ is sequentially lower semicontinuous in the strong topology of $L_1(J,H)$. Since $L_p(J,U)\subset L_1(J,U)$, $\mathcal{J}$ is weakly lower semicontinuous on $L_p(J,U)$, and since by $(H_{13})(iv)$, $\mathcal{J}>-\infty$, $\mathcal{J}$ attains its infimum at $u^0\in\mathcal{A}_{ad}$ that is

$$\epsilon=\lim_{m\to\infty}\mathbb{E}\left(\int_0^b\mathcal{L}(t,x^m(t),u^m(t))dt\right)\ge\mathbb{E}\left(\int_0^b\mathcal{L}(t,x^0(t),u^0(t))dt\right)=\mathcal{J}(x^0,u^0)\ge\epsilon.$$

7.3 CONTROLLABILITY RESULT OF SINGLE-VALUED FRACTIONAL STOCHASTIC DIFFERENTIAL EQUATION BY USING ANALYTIC RESOLVENT OPERATORS

This section investigates the solvability and optimal controls for fractional stochastic integro-differential equations with infinite delay in Hilbert space by using analytic resolvent operators. Some suitable conditions are established to guarantee the existence of mild solutions with the help of Leray-Schauder nonlinear alternative fixed point theorem. Then the existence of optimal control is investigated for the corresponding Lagrange problem.

Consider the following form of fractional stochastic integro-differential equation with infinite delay

$${}^cD_t^\alpha x(t)=Ax(t)+B(t)u(t)+J_t^{1-\alpha}f(t,x_t)+\int_{-\infty}^t\sigma(s,x_s)dW(s),\quad t\in J \tag{7.15}$$

$$x(t)=\phi(t)\in\mathcal{C}_h \tag{7.16}$$

where $0<\alpha<1$, $J_t^{1-\alpha}$ is the $(1-\alpha)$ order Riemann–Liouville fractional integral operator, u is a given control function, it takes values from separable reflexive Hilbert space U. B is a linear operator from U into H. The history $x_t:\Omega\to\mathcal{C}_h$ is defined by $x_t(\theta)=\{x(t+\theta),\ \theta\in(-\infty,0]\}$, which belongs to an abstract space $\mathcal{C}_h$. The functions $f{:}J\times\mathcal{C}_h\to H$ and $\sigma{:}J\times\mathcal{C}_h\to L_2^0(K,H)$ are the appropriate functions, where $L_2^0(K,H)$ denotes the space of all Q-Hilbert-Schmidt operators from K into H.

7.3.1 Preliminaries

The abstract phase space $\mathcal{C}_h$ is defined by

$$\mathcal{C}_h=\left\{\psi{:}(-\infty,0]\to H{:}\text{for any } a>0,\left(\mathbb{E}|\psi(\theta)|^2\right)^{\frac{1}{2}} \text{ is bounded and measurable function on}[-a,0] \text{ with } \int_{-\infty}^{0}\hat{h}(s)\sup_{s\le\theta\le 0}\left(\mathbb{E}|\psi(\theta)|^2\right)^{\frac{1}{2}}ds<\infty\right\}.$$

If $\mathcal{C}_h$ is endowed with the norm

$$\|\psi\|_{\mathcal{C}_h}=\int_{-\infty}^{0}\hat{h}(s)\sup_{s\le\theta\le 0}\left(\mathbb{E}|\psi(\theta)|^2\right)^{\frac{1}{2}}ds,\quad \text{for all } \psi\in\mathcal{C}_h$$

then $(\mathcal{C}_h,\|\cdot\|_{\mathcal{C}_h})$ is a Banach space [Li and Liu (2007)].

Let $C((-\infty,b],H)$ be the space of all continuous H-valued stochastic process $\{\xi(t),\ t\in(-\infty,b]\}$. Let $\mathcal{C}_b=\{x{:}x\in C((-\infty,b],H),\ x_0=\phi\in\mathcal{C}_h\}$. Set $\|\cdot\|_b$ to be a seminorm defined by

$$\|x\|_b=\|x_0\|_{\mathcal{C}_h}+\sup_{s\in[0,b]}\left(\mathbb{E}|x(s)|^2\right)^{\frac{1}{2}},\quad x\in\mathcal{C}_b.$$

Lemma 7.4: (Li and Liu, 2007) *Assume that $x\in\mathcal{C}_b$, then for all $t\in J$, $x_t\in\mathcal{C}_h$. Moreover,*

$$l\left(\mathbb{E}|x(t)|^2\right)^{\frac{1}{2}}\le\|x_t\|_{\mathcal{C}_h}\le l\sup_{s\in[0,t]}\left(\mathbb{E}|x(s)|^2\right)^{\frac{1}{2}}+\|x_0\|_{\mathcal{C}_h}$$

where $l=\int_{-\infty}^{0}\hat{h}(s)ds<\infty$.

Consider the following fractional stochastic integro-differential equation

$${}^cD_t^\alpha x(t)=Ax(t)+B(t)u(t)+J_t^{1-\alpha}f(t,x_t)+\int_{-\infty}^{t}\sigma(s,x_s)dW(s)$$

the above equation is equivalent to the following integral equation

$$\begin{aligned}x(t)&=\phi(0)+\frac{1}{\Gamma(\alpha)}\int_0^t(t-s)^{\alpha-1}Ax(s)ds+\frac{1}{\Gamma(\alpha)}\int_0^t(t-s)^{\alpha-1}B(s)u(s)ds\\&\quad+\int_0^t f(s,x_s)ds+\frac{1}{\Gamma(\alpha)}\int_0^t(t-s)^{\alpha-1}\left(\int_{-\infty}^{s}\sigma(\tau,x_\tau)dW(\tau)\right)ds\end{aligned}$$

this integral equation can be written in the following form

$$x(t) = h(t) + \frac{1}{\Gamma(\alpha)}\int_0^t (t-s)^{\alpha-1}Ax(s)ds, \quad t \geq 0 \tag{7.17}$$

where

$$\begin{aligned} h(t) &= \phi(0) + \frac{1}{\Gamma(\alpha)}\int_0^t (t-s)^{\alpha-1}B(s)u(s)ds + \int_0^t f(s, x_s)ds \\ &+ \frac{1}{\Gamma(\alpha)}\int_0^t (t-s)^{\alpha-1}\left(\int_{-\infty}^{s} \sigma(\tau, x_\tau)dW(\tau)\right)ds. \end{aligned}$$

Let us assume that the integral Eq. (7.17) has an associated resolvent operator $\{S(t)\}_{t \geq 0}$ on H.

Definition 7.11: (Prüss, 2013) *A one parameter family of bounded linear operators* $\{S(t)\}_{t \geq 0}$ *on* H *is called a resolvent operator for* (7.17) *if the following conditions hold*

(i) $S(\cdot)x \in C([0, \infty), H)$ *and* $S(0)x = x$ *for all* $x \in H$,
(ii) $S(t)D(A) \subset D(A)$ *and* $AS(t)x = S(t)Ax$, $\forall x \in D(A)$ *and every* $t \geq 0$,
(iii) *for every* $x \in D(A)$ *and* $t \geq 0$, $S(t)x = x + \frac{1}{\Gamma(\alpha)}\int_0^t (t-s)^{\alpha-1}AS(s)xds$.

Definition 7.12: (Prüss, 2013) *A resolvent operator* $\{S(t)\}_{t \geq 0}$ *for* (7.17) *is called differentiable if* $S(\cdot)x \in W^{1,1}_{loc}(\mathbb{R}^+, H)$ *for all* $x \in D(A)$ *and there exists* $\varphi_A \in L^1_{loc}(\mathbb{R}^+)$ *such that* $\|S'(t)x\| \leq \varphi_A(t)\|x\|_{[D(A)]}$ *for all* $x \in D(A)$, *where the notation* $[D(A)]$ *stands the domain of the operator A provided with the graph norm* $\|x\|_{[D(A)]} = \|x\| + \|Ax\|$.

Definition 7.13: (Prüss, 2013) *A resolvent operator* $\{S(t)\}_{t \geq 0}$ *for* (7.17) *is called analytic if the operator function* $S(\cdot) : (0, \infty) \to L(H)$ *admits an analytic extension to a sector* $\sum_{0,\theta_0} = \{\lambda \in \mathbb{C} : |\arg(\lambda)| < \theta_0\}$ *for some* $0 < \theta_0 \leq \frac{\pi}{2}$.

Definition 7.14: (Prüss, 2013) *A function* $x \in C(J, H)$ *is called a mild solution of the integral* Eq. (7.17) *on* J, if $\int_0^t (t-s)^{\alpha-1}x(s)ds \in D(A)$ *for all* $t \in J$, $h(t) \in C(J, H)$ *and*

$$x(t) = \frac{1}{\Gamma(\alpha)}\int_0^t (t-s)^{\alpha-1}Ax(s)ds + h(t), \quad \forall t \in J$$

Lemma 7.5: (Hernández et al., 2013) *Under the above conditions the following properties are valid*

(i) *if* $x(\cdot)$ *is a mild solution of* (7.17) *on* J, *then the function* $t \to \int_0^t S(t-s)h(s)ds$ *is differentiable on* J *and*

$$x(t) = \frac{d}{dt}\int_0^t S(t-s)h(s)ds, \quad \forall t \in J;$$

(ii) *if* $\{S(t)\}_{t\geq 0}$ *is analytic and* $h \in C^{\gamma}(J,H)$ *for some* $\gamma \in (0,1)$*, then the function defined by*

$$x(t) = S(t)(h(t)-h(0)) + \int_0^t S'(t-s)[h(s)-h(t)]ds + S(t)h(0), \quad t \in J$$

is a mild solution of (7.17) *on J;*

(iii) *if* $\{S(t)\}_{t\geq 0}$ *is differentiable and* $h \in C(J,[D(A)])$ *then the function* $x{:}J \to H$ *defined by*

$$x(t) = \int_0^t S'(t-s)h(s)ds + h(t), \quad t \in J$$

is a mild solution of (7.17).

Lemma 7.6: *Let H be a Banach space, G is a closed, convex subset of H,* $\hat{U}$ *an open subset of G and* $0 \in \hat{U}$*. Suppose that* $F{:}\hat{U} \to G$ *is a continuous, compact (i.e.,* $F(\hat{U})$ *is a relatively compact subset of G) map, then either*

(i) *F has a fixed point in* $\hat{U}$*; or*
(ii) *there is a* $y \in \partial\hat{U}$ *and* $\lambda \in (0,1)$ *with* $y = \lambda F(y)$.

The following hypotheses are considered to prove the main results

(H_{14}) $S(t)$ is compact for all $t > 0$.
(H_{15}) $f{:}J \times \mathcal{C}_h \to H$ satisfies the following

(i) $f(t,\cdot){:}\mathcal{C}_h \to H$ is continuous for each $t \in J$ and for each $x \subset \mathcal{C}_h$, $f(\cdot,x){:}J \to H$ is strongly measurable.
(ii) There is a positive integrable function $n \in L^1([0,b])$ and a continuous nondecreasing function Λ_f: $[0,\infty) \to (0,\infty)$ such that for every $(t,x) \in J \times \mathcal{C}_h$,

$$\mathbb{E}|f(t,x)|_H^2 \leq n(t)\Lambda_f(\|x\|_{\mathcal{C}_h}^2), \quad \lim_{r\to\infty}\inf\frac{\Lambda_f(r)}{r} = \Upsilon_1 < \infty.$$

(iii) For arbitrary $\xi_1, \xi_2 \in \mathcal{C}_h$, satisfying $\|\xi_1\|_{\mathcal{C}_h}^2$, $\|\xi_2\|_{\mathcal{C}_h}^2 \leq \rho$, there exists a $M_f(\rho) > 0$ such that

$$\mathbb{E}\left|f(t,\xi_1) - f(t,\xi_2)\right|^2 \leq M_f(\rho)\left\|\xi_1 - \xi_2\right\|_{\mathcal{C}_h}^2.$$

(H_{16}) The function $\sigma{:}J \times \mathcal{C}_h \to L(K,H)$ satisfies the following

(i) For each $t \in J$, $\sigma(t,\cdot){:}\mathcal{C}_h \to L(K,H)$ is continuous and for each $x \in \mathcal{C}_h$, $\sigma(\cdot,x){:}J \to L(K,H)$ is strongly measurable.

(ii) There is a positive integrable function $m \in L^1([0,b])$ and a continuous nondecreasing function $\Lambda_\sigma : [0,\infty) \to (0,\infty)$ such that for every $(t,x) \in J \times \mathcal{C}_h$,

$$\int_0^t \mathbb{E}|\sigma(t,x)|^2_{L_2^0} ds \leq m(t)\Lambda_\sigma(\|x\|^2_{\mathcal{C}_h}), \quad \liminf_{r\to\infty} \frac{\Lambda_\sigma(r)}{r} = \Upsilon_2 < \infty.$$

(iii) For arbitrary $\xi_1, \xi_2 \in \mathcal{C}_h$, satisfying $\|\xi_1\|^2_{\mathcal{C}_h}$, $\|\xi_2\|^2_{\mathcal{C}_h} \leq \rho$, there exists a $M_\sigma(\rho) > 0$ such that

$$\int_0^t \mathbb{E}\left|\sigma(s,\xi_1) - \sigma(s,\xi_2)\right|^2 ds \leq M_\sigma(\rho)\left\|\xi_1 - \xi_2\right\|^2_{\mathcal{C}_h}.$$

(H_{17}) For each $\phi \in \mathcal{C}_h$, $R(t) = \lim_{a\to\infty} \int_{-a}^0 \sigma(t,\phi)dW(s)$ exists and is continuous. Further there exists a positive constant M_R such that $\mathbb{E}|R(t)|^2_H \leq M_R$.

(H_{18}) Let U be a separable reflexive Hilbert space from which the control u takes the values. Operator $B \in L_\infty(J, L(U,H))$, $\|B\|$ stands for the norm of operator B on Hilbert space $L_\infty(J, L(U,H))$.

(H_{19}) The multivalued map $\mathcal{A}: J \to 2^U\ \{\emptyset\}$ has closed, convex, and bounded values. $\mathcal{A}(\cdot)$ is graph measurable and $\mathcal{A}(\cdot) \subseteq \Sigma$, where Σ is a bounded set of U.

Let the admissible set be,

$$\mathcal{A}_{ad} = \left\{u : J \times \Omega \to H /\ u \text{ is } \mathcal{F}_t - \text{adapted stochastic process and } E\int_0^b \|u(t)\|^2 dt < \infty\right\}.$$

7.3.1.1 Existence of Mild Solution

The subsection deals with the existence result for (7.15)–(7.16), this problem is equivalent to the following integral equation

$$x(t) = \begin{cases} \phi(t), \ t \in (-\infty, 0] \\ \phi(0) + \dfrac{1}{\Gamma(\alpha)}\displaystyle\int_0^t (t-s)^{\alpha-1} Ax(s)ds + \dfrac{1}{\Gamma(\alpha)}\displaystyle\int_0^t (t-s)^{\alpha-1} B(s)u(s)ds \\ \quad + \int_0^t f(s,x_s)ds + \dfrac{1}{\Gamma(\alpha)}\displaystyle\int_0^t (t-s)^{\alpha-1}\left(\int_{-\infty}^s \sigma(\tau, x_\tau)dW(\tau)\right)ds, \ t \in J \end{cases}$$

By Lemma 7.5, and the above representation, the mild solution of (7.15)–(7.16) can be defined as follows

Definition 7.15: *An H-valued stochastic process* $\{x(t),\ t\in(-\infty,b]\}$ *is said to be a mild solution of the problem* (7.15)–(7.16) *if*

- $x(t)$ *is measurable and* $\mathcal{F}_t$ *adapted for* $t\geq 0$;
- $x(t)=\phi(t)$ *on* $(-\infty,0]$ *satisfying* $\|\phi\|^2_{\mathcal{C}_h}<\infty$;
- $\int_0^t (t-s)^{\alpha-1}x(s)ds\in D(A)$ *for all* $t\in J$; *and*

$$x(t) = \phi(0)+\frac{1}{\Gamma(\alpha)}\int_0^t (t-s)^{\alpha-1}Ax(s)ds+\frac{1}{\Gamma(\alpha)}\int_0^t (t-s)^{\alpha-1}B(s)u(s)ds + \int_0^t f(s,x_s)ds+\frac{1}{\Gamma(\alpha)}\int_0^t (t-s)^{\alpha-1}\left(\int_{-\infty}^{s}\sigma(\tau,x_\tau)dW(\tau)\right)ds.$$

Suppose that there exists a resolvent set $\{S(t)\}_{t\geq 0}$ which is differentiable and the functions $f,\ \sigma$ are continuous, then by Lemma 7.5 (*iii*), if x is a mild solution of (7.15–7.16), then

$$x(t)=\begin{cases}\phi(t),\ t\in(-\infty,0]\\ \phi(0)+\dfrac{1}{\Gamma(\alpha)}\displaystyle\int_0^t (t-s)^{\alpha-1}B(s)u(s)ds+\dfrac{1}{\Gamma(\alpha)}\int_0^t (t-s)^{\alpha-1}\left(\int_{-\infty}^{s}\sigma(\tau,x_\tau)dW(\tau)\right)ds\\ +\displaystyle\int_0^t f(s,x_s)ds+\int_0^t S'(t-s)\left[\phi(0)+\dfrac{1}{\Gamma(\alpha)}\int_0^s (s-\tau)^{\alpha-1}B(\tau)u(\tau)d\tau\right.\\ \left.+\displaystyle\int_0^s f(\tau,x_\tau)d\tau+\dfrac{1}{\Gamma(\alpha)}\int_0^s (s-\tau)^{\alpha-1}\left(\int_{-\infty}^{\tau}\sigma(\eta,x_\eta)dW(\eta)\right)d\tau\right]ds,\ t\in J.\end{cases}$$

Theorem 7.6: *If hypotheses* $(H_{14})-(H_{17})$ *hold, then the fractional order control problem* (7.15)–(7.16) *has a mild solution on* $(-\infty,b]$, *provided that, there exists a constant* $M^*>0$ *such that*

$$\frac{M^*}{4l^2\left[8\mathbb{E}|\phi(0)|^2+8\|B\|^2 b^{\frac{2\alpha p-2}{p}}\left(\frac{p-1}{\alpha p-1}\right)^{\frac{p-1}{p}}(b^2+1)\|u\|^2_{L_p(J,U)}+\Theta\right]+4\|\phi\|^2_{\mathcal{C}_h}}>1 \qquad (7.18)$$

where

$$\Theta=8b^2\|n\|_\infty(b^2\|\varphi_A\|^2+1)\Lambda_f(M^*)+\left[\frac{16b^{2\alpha}+16\|\varphi_A\|^2_{L^1}b^{2\alpha+2}}{\Gamma^2(\alpha+1)}\right]\times[M_R+Tr(Q)\|m\|_\infty\Lambda_\sigma(M^*)].$$

Proof: Consider the operator $\Phi:\mathcal{C}_b\to\mathcal{C}_b$ defined by

$$(\Phi x)(t)=\begin{cases}\phi(t), & t\in(-\infty,0]\\ \phi(0)+\frac{1}{\Gamma(\alpha)}\int_0^t(t-s)^{\alpha-1}B(s)u(s)ds \\ +\frac{1}{\Gamma(\alpha)}\int_0^t(t-s)^{\alpha-1}\left(\int_{-\infty}^s\sigma(\tau,x_\tau)dW(\tau)\right)ds+\int_0^t f(s,x_s)ds \\ +\int_0^t S'(t-s)\Big[\phi(0)+\frac{1}{\Gamma(\alpha)}\int_0^s(s-\tau)^{\alpha-1}B(\tau)u(\tau)d\tau \\ +\int_0^s f(\tau,x_\tau)d\tau+\frac{1}{\Gamma(\alpha)}\int_0^s(s-\tau)^{\alpha-1}\left(\int_{-\infty}^\tau\sigma(\eta,x_\eta)dW(\eta)\right)d\tau\Big]ds, & t\in J.\end{cases}$$

Let $y(\cdot):(-\infty,b]\to H$ be the function defined by $y(t)=\begin{cases}\phi(t), & \text{if } t\in(-\infty,0]\\ 0, & \text{if } t\in J\end{cases}$ then $y_0=\phi$, we define the function $\overline{z}$ by

$$\overline{z}(t)=\begin{cases}0, & \text{if } t\in(-\infty,0]\\ z(t), & if\ t\in J.\end{cases}$$

If $x(t)$ satisfies

$$\begin{aligned}x(t)&=\phi(0)+\frac{1}{\Gamma(\alpha)}\int_0^t(t-s)^{\alpha-1}B(s)u(s)ds\\&+\frac{1}{\Gamma(\alpha)}\int_0^t(t-s)^{\alpha-1}\left(\int_{-\infty}^s\sigma(\tau,x_\tau)dW(\tau)\right)ds+\int_0^t f(s,x_s)ds\\&+\int_0^t S'(t-s)\Big[\phi(0)+\frac{1}{\Gamma(\alpha)}\int_0^s(s-\tau)^{\alpha-1}B(\tau)u(\tau)d\tau+\int_0^s f(\tau,x_\tau)d\tau\\&+\frac{1}{\Gamma(\alpha)}\int_0^s(s-\tau)^{\alpha-1}\left(\int_{-\infty}^\tau\sigma(\eta,x_\eta)dW(\eta)\right)d\tau\Big]ds,\end{aligned}$$

one can decompose it as $x(t)=\overline{z}(t)+y(t),\ t\in J$ which implies $x_t=\overline{z}_t+y_t$, $t\in J$ and the function $z(\cdot)$ satisfies $z_0=0$ and

$$\begin{aligned}z(t)&=\phi(0)+\frac{1}{\Gamma(\alpha)}\int_0^t(t-s)^{\alpha-1}B(s)u(s)ds\\&+\frac{1}{\Gamma(\alpha)}\int_0^t(t-s)^{\alpha-1}\left(\int_{-\infty}^s\sigma(\tau,\overline{z}_\tau+y_\tau)dW(\tau)\right)ds+\int_0^t f(s,\overline{z}_s+y_s)ds\\&+\int_0^t S'(t-s)\Big[\phi(0)+\frac{1}{\Gamma(\alpha)}\int_0^s(s-\tau)^{\alpha-1}B(\tau)u(\tau)d\tau\\&+\int_0^s f(\tau,\overline{z}_\tau+y_\tau)d\tau+\frac{1}{\Gamma(\alpha)}\int_0^s(s-\tau)^{\alpha-1}\left(\int_{-\infty}^\tau\sigma(\eta,\overline{z}_\eta+y_\eta)dW(\eta)\right)d\tau\Big]ds.\end{aligned}$$

Set $\mathcal{C}_b^0=\{z\in\mathcal{C}_b,\ \ z_0=0\in\mathcal{C}_h\}$. For any $z\in\mathcal{C}_b^0$,

$$\|z\|_b=\|z_0\|_{\mathcal{C}_h}+\sup_{s\in[0,b]}\left(\mathbb{E}|z(s)|^2\right)^{\frac{1}{2}}=\sup_{s\in[0,b]}\left(\mathbb{E}|z(s)|^2\right)^{\frac{1}{2}}$$

thus $(\mathcal{C}_b^0,\|\cdot\|_b)$ is a Banach space. For each positive number q, set $B_q=\{y\in\mathcal{C}_b^0,\ \ \|y\|_b^2\le q\}$, then for each q, B_q is clearly a bounded closed convex set in $\mathcal{C}_b^0$. For $z\in B_q$, from Lemma 7.4, one can get

$$\begin{aligned}\|\bar{z}_t+y_t\|_{\mathcal{C}_h}^2&\le 2(\|\bar{z}_t\|_{\mathcal{C}_h}^2+\|y_t\|_{\mathcal{C}_h}^2)\\&\le 4\left(l^2\sup\nolimits_{s\in[0,t]}\mathbb{E}|z(s)|^2+\|z_0\|_{\mathcal{C}_h}^2+l^2\sup\nolimits_{s\in[0,t]}\mathbb{E}|y(s)|^2+\|y_0\|_{\mathcal{C}_h}^2\right)\\&\le 4l^2q+4\|\phi\|_{\mathcal{C}_h}^2=q^*.\end{aligned}$$

Let the operator Π on $\mathcal{C}_b^0$ be defined by

$$(\Pi z)(t)=\begin{cases}0,\ t\in(-\infty,0]\\ \phi(0)+\dfrac{1}{\Gamma(\alpha)}\displaystyle\int_0^t(t-s)^{\alpha-1}B(s)u(s)ds+\int_0^t f(s,\bar{z}_s+y_s)ds\\ +\dfrac{1}{\Gamma(\alpha)}\displaystyle\int_0^t(t-s)^{\alpha-1}\left(\int_{-\infty}^s\sigma(\tau,\bar{z}_\tau+y_\tau)dW(\tau)\right)ds+\int_0^t S'(t-s)\Bigg[\phi(0)\\ +\dfrac{1}{\Gamma(\alpha)}\displaystyle\int_0^s(s-\tau)^{\alpha-1}B(\tau)u(\tau)d\tau+\int_0^s f(\tau,\bar{z}_\tau+y_\tau)d\tau\\ +\dfrac{1}{\Gamma(\alpha)}\displaystyle\int_0^s(s-\tau)^{\alpha-1}\left(\int_{-\infty}^\tau\sigma(\eta,\bar{z}_\eta+y_\eta)dW(\eta)\right)d\tau\Bigg]ds,\ t\in J.\end{cases}$$

It is obvious that the operator Φ has a fixed point if and only if Π has a fixed point. In order to prove that Π is continuous, decompose Π as $\Pi=\Pi_1+\Pi_2$, where the operators Π_1 and Π_2 are defined respectively by

$$\begin{aligned}(\Pi_1z)(t)&=\phi(0)+\frac{1}{\Gamma(\alpha)}\int_0^t(t-s)^{\alpha-1}B(s)u(s)ds+\int_0^t f(s,\bar{z}_s+y_s)ds\\&\quad+\frac{1}{\Gamma(\alpha)}\int_0^t(t-s)^{\alpha-1}\left(\int_{-\infty}^s\sigma(\tau,\bar{z}_\tau+y_\tau)dW(\tau)\right)ds\\(\Pi_2z)(t)&=\int_0^t S'(t-s)(\Pi_1z)(s)ds.\end{aligned}$$

One needs to prove that Π_1 and Π_2 are completely continuous. It follows from the hypothesis (H_{14}) that $S'(t)$ is compact for all $t>0$ (see Hernández et al., 2010).

Step 1: Π_1 is completely continuous. First, to prove that Π_1 is continuous. Let $\{z_n\}$ be a xsequence such that $z_n \to z$ in C_b^0 as $n \to \infty$. Then for $t \in J$,

$$\begin{aligned}\mathbb{E}|(\Pi_1 z_n)(t) - (\Pi_1 z)(t)|^2 &\leq 2\mathbb{E}\left|\int_0^t [f(s, \overline{z}_{n_s} + y_s) - f(s, \overline{z}_s + y_s)]ds\right|^2 \\ &\quad + 2\mathbb{E}\left|\frac{1}{\Gamma(\alpha)}\int_0^t (t-s)^{\alpha-1}\left(\int_{-\infty}^s \sigma(\tau, \overline{z}_\tau + y_\tau)dW(\tau)\right)ds\right|^2 \\ &\leq 2b\int_0^t \mathbb{E}\left|f(s, \overline{z}_{n_s} + y_s) - f(s, \overline{z}_s + y_s)\right|^2 ds \\ &\quad + \frac{2b^\alpha}{\alpha\Gamma^2(\alpha)}\int_0^t (t-s)^{\alpha-1}\mathbb{E}\left|\int_{-\infty}^s [\sigma(\tau, \overline{z}_{n_\tau} + y_\tau) - \sigma(\tau, \overline{z}_\tau + y_\tau)]dW(\tau)\right|^2 ds\end{aligned}$$

since f and σ are continuous, one has $\|(\Pi_1 z_n)(t) - (\Pi_1 z)(t)\|_b^2 \to 0\ as\ n \to \infty$. Now, to prove Π_1 maps bounded sets into bounded sets in C_b^0. Indeed, it is enough to prove that for any $q > 0$, there exists a positive constant δ such that for each $z \in B_q$, one has $(\Pi_1 z) \in B_\delta$. Let $z \in B_q$, since $f\ and\ \sigma$ are continuous, for each $t \in J$

$$\begin{aligned}\mathbb{E}|(\Pi_1 z)(t)|^2 &\leq 4\mathbb{E}|\phi(0)|^2 + 4\mathbb{E}\left|\frac{1}{\Gamma(\alpha)}\int_0^t (t-s)^{\alpha-1}B(s)u(s)ds\right|^2 + 4\mathbb{E}\left|\int_0^t f(s, \overline{z}_s + y_s)ds\right|^2 \\ &\quad + \frac{4b^\alpha}{\alpha\Gamma^2(\alpha)}\int_0^t (t-s)^{\alpha-1}\mathbb{E}\left|\int_{-\infty}^s \sigma(\tau, \overline{z}_\tau + y_\tau)dW(\tau)\right|^2 ds \\ &\leq 4\mathbb{E}|\phi(0)|^2 + \frac{4\|B\|^2}{\Gamma^2(\alpha)}\left[\left(\int_0^t (t-s)^{\frac{(\alpha-1)p}{p-1}} ds\right)^{1-\frac{1}{p}}\left(\int_0^t \mathbb{E}|u(s)|^p ds\right)^{\frac{1}{p}}\right]^2 \\ &\quad + 4b\int_0^t \mathbb{E}\left|f(s, \overline{z}_s + y_s)\right|^2 ds \\ &\quad + \frac{4b^\alpha}{\alpha\Gamma^2(\alpha)}\int_0^t (t-s)^{\alpha-1}\left[2M_R + 2\mathbb{E}\left|\int_0^s \sigma(\tau, \overline{z}_\tau + y_\tau)dW(\tau)\right|^2\right]ds \\ &\leq 4\mathbb{E}|\phi(0)|^2 + 4\frac{\|B\|^2}{\Gamma^2(\alpha)}b^{\frac{2\alpha p-2}{p}}\left(\frac{p-1}{\alpha p-1}\right)^{\frac{2p-2}{p}}\|u\|_{L_p(J,U)}^2 \\ &\quad + 4b\int_0^t n(s)\Lambda_f(\|\overline{z}_s + y_s\|_{\mathcal{C}_h}^2)ds \\ &\quad + \frac{8b^\alpha}{\alpha\Gamma^2(\alpha)}\int_0^t (t-s)^{\alpha-1}\left[M_R + Tr(Q)m(s)\Lambda_\sigma(\|\overline{z}_s + y_s\|_{\mathcal{C}_h}^2)\right]ds \\ &\leq 4\mathbb{E}|\phi(0)|^2 + 4\frac{\|B\|^2}{\Gamma^2(\alpha)}b^{\frac{2\alpha p-2}{p}}\left(\frac{p-1}{\alpha p-1}\right)^{\frac{2p-2}{p}}\|u\|_{L_p(J,U)}^2 + 4b^2\|n\|_\infty\Lambda_f(q^*) \\ &\quad + \frac{8b^{2\alpha}}{\Gamma^2(\alpha+1)}[M_R + Tr(Q)\|m\|_\infty\Lambda_\sigma(q^*)] = \delta < \infty\end{aligned}$$

thus $\mathbb{E}|(\Pi_1 z)(t)|^2 \leq \delta$ and hence $\Pi_1 z \in B_\delta$. Now, to prove that Π_1 maps bounded sets into equicontinuous sets of C_b^0. Let $t_1, t_2 \in J,\ t_2 > t_1$ and let B_q be a bounded set. Let $z \in B_q$, then if $\epsilon > 0$ and $\epsilon \leq t_1 \leq t_2$, one has

$$
\begin{aligned}
\mathbb{E}\left|(\Pi_1 z)(t_2)-(\Pi_1 z)(t_1)\right|^2 \le\, & 7\mathbb{E}\left|\frac{1}{\Gamma(\alpha)}\int_0^{t_1-\epsilon}\left[(t_2-s)^{\alpha-1}-(t_1-s)^{\alpha-1}\right]B(s)u(s)ds\right|^2 \\
& +7\mathbb{E}\left|\frac{1}{\Gamma(\alpha)}\int_{t_1-\epsilon}^{t_1}\left[(t_2-s)^{\alpha-1}-(t_1-s)^{\alpha-1}\right]B(s)u(s)ds\right|^2 \\
& +7\mathbb{E}\left|\frac{1}{\Gamma(\alpha)}\int_{t_1}^{t_2}(t_2-s)^{\alpha-1}B(s)u(s)ds\right|^2 +7\mathbb{E}\left|\int_{t_1}^{t_2} f(s,\bar{z}_s+y_s)ds\right|^2 \\
& +7\mathbb{E}\left|\frac{1}{\Gamma(\alpha)}\int_0^{t_1-\epsilon}\left[(t_2-s)^{\alpha-1}-(t_1-s)^{\alpha-1}\right]\left(\int_{-\infty}^{s}\sigma(\tau,\bar{z}_\tau+y_\tau)dW(\tau)\right)ds\right|^2 \\
& +7\mathbb{E}\left|\frac{1}{\Gamma(\alpha)}\int_{t_1-\epsilon}^{t_1}\left[(t_2-s)^{\alpha-1}-(t_1-s)^{\alpha-1}\right]\left(\int_{-\infty}^{s}\sigma(\tau,\bar{z}_\tau+y_\tau)dW(\tau)\right)ds\right|^2 \\
& +7\mathbb{E}\left|\frac{1}{\Gamma(\alpha)}\int_{t_1}^{t_2}(t_2-s)^{\alpha-1}\left(\int_{-\infty}^{s}\sigma(\tau,\bar{z}_\tau+y_\tau)dW(\tau)\right)ds\right|^2 \\
\le\, & \frac{7(t_1-\epsilon)\|B\|^2}{\Gamma^2(\alpha)}\int_0^{t_1-\epsilon}\left[(t_2-s)^{\alpha-1}-(t_1-s)^{\alpha-1}\right]^2\mathbb{E}\left|u(s)\right|^2 ds \\
& +\frac{7\epsilon\|B\|^2}{\Gamma^2(\alpha)}\int_{t_1-\epsilon}^{t_1}\left[(t_2-s)^{\alpha-1}-(t_1-s)^{\alpha-1}\right]^2\mathbb{E}\left|u(s)\right|^2 ds \\
& +\frac{7(t_2-t_1)^{\alpha}\|B\|^2}{\alpha\Gamma^2(\alpha)}\int_{t_1}^{t_2}(t_2-s)^{\alpha-1}\mathbb{E}\left|u(s)\right|^2 ds+7(t_2-t_1)\int_{t_1}^{t_2}\mathbb{E}\left|f(s,\bar{z}_s+y_s)\right|^2 ds \\
& +\frac{7(t_1-\epsilon)}{\Gamma^2(\alpha)}\int_0^{t_1-\epsilon}\left[(t_2-s)^{\alpha-1}-(t_1-s)^{\alpha-1}\right]^2\mathbb{E}\left|\int_{-\infty}^{s}\sigma(\tau,\bar{z}_\tau+y_\tau)dW(\tau)\right|^2 ds \\
& +\frac{7\epsilon}{\Gamma^2(\alpha)}\int_{t_1-\epsilon}^{t_1}\left[(t_2-s)^{\alpha-1}-(t_1-s)^{\alpha-1}\right]^2\mathbb{E}\left|\int_{-\infty}^{s}\sigma(\tau,\bar{z}_\tau+y_\tau)dW(\tau)\right|^2 ds \\
& +\frac{7(t_2-t_1)^{\alpha}}{\alpha\Gamma^2(\alpha)}\int_{t_1}^{t_2}(t_2-s)^{\alpha-1}\mathbb{E}\left|\int_{-\infty}^{s}\sigma(\tau,\bar{z}_\tau+y_\tau)dW(\tau)\right|^2 ds
\end{aligned}
$$

as $t_2 \to t_1$ and ϵ sufficiently small, the right-hand side of the above inequality tends to zero. By the Arzela–Ascoli theorem it suffices to show that Π_1 maps B_q into a precompact sets in H. Let $0<t<b$ be fixed and let ϵ be a real number satisfying $0<\epsilon<t$. For $z\in B_q$, define

$$
\begin{aligned}
(\Pi_{1\epsilon} z)(t) &= \phi(0)+\frac{1}{\Gamma(\alpha)}\int_0^{t-\epsilon}(t-\epsilon-s)^{\alpha-1}B(s)u(s)ds+\int_0^{t-\epsilon} f(s,\bar{z}_s+y_s)ds \\
&+\frac{1}{\Gamma(\alpha)}\int_0^{t-\epsilon}(t-\epsilon-s)^{\alpha-1}\left(\int_{-\infty}^{s}\sigma(\tau,\bar{z}_\tau+y_\tau)dW(\tau)\right)ds.
\end{aligned}
$$

Note that, the set

$$
\begin{aligned}
\Bigg\{\phi(0) &+\frac{1}{\Gamma(\alpha)}\int_0^{t-\epsilon}(t-\epsilon-s)^{\alpha-1}B(s)u(s)ds+\int_0^{t-\epsilon} f(s,\bar{z}_s+y_s)ds \\
&+\frac{1}{\Gamma(\alpha)}\int_0^{t-\epsilon}(t-\epsilon-s)^{\alpha-1}\left(\int_{-\infty}^{s}\sigma(\tau,\bar{z}_\tau+y_\tau)dW(\tau)\right)ds\Bigg\}
\end{aligned}
$$

is bounded, since

$$\mathbb{E}\left|\phi(0) + \frac{1}{\Gamma(\alpha)}\int_0^{t-\epsilon}(t-\epsilon-s)^{\alpha-1}B(s)u(s)ds + \int_0^{t-\epsilon}f(s,\bar{z}_s+y_s)ds + \frac{1}{\Gamma(\alpha)}\int_0^{t-\epsilon}(t-\epsilon-s)^{\alpha-1}\left(\int_{-\infty}^{s}\sigma(\tau,\bar{z}_\tau+y_\tau)dW(\tau)\right)ds\right|$$

$$\leq 4\mathbb{E}|\phi(0)|^2 + 4\mathbb{E}\left|\tfrac{1}{\Gamma(\alpha)}\int_0^{t-\epsilon}(t-\epsilon-s)^{\alpha-1}B(s)u(s)ds\right|^2 + 4\mathbb{E}\left|\int_0^{t-\epsilon}f(s,\bar{z}_s+y_s)ds\right|^2 + 4\mathbb{E}\left|\tfrac{1}{\Gamma(\alpha)}\int_0^{t-\epsilon}(t-\epsilon-s)^{\alpha-1}\left(\int_{-\infty}^{s}\sigma(\tau,\bar{z}_\tau+y_\tau)dW(\tau)\right)ds\right|^2$$

$$\leq 4\mathbb{E}|\phi(0)|^2 + \frac{4\|B\|^2}{\Gamma^2(\alpha)}\left[\left(\int_0^{t-\epsilon}(t-\epsilon-s)^{\frac{(\alpha-1)p}{p-1}}ds\right)^{1-\frac{1}{p}}\left(\int_0^{t-\epsilon}\mathbb{E}|u(s)|^p ds\right)^{\frac{1}{p}}\right]^2 + 4(t-\epsilon)\int_0^{t-\epsilon}\mathbb{E}|f(s,\bar{z}_s+y_s)|^2 ds + \frac{4(t-\epsilon)^\alpha}{\alpha\Gamma^2(\alpha)}\int_0^{t-\epsilon}(t-\epsilon-s)^{\alpha-1}\left[2M_R + 2Tr(Q)\int_0^s \mathbb{E}\|\sigma(\tau,\bar{z}_\tau+y_\tau)\|_{L_2^0}^2 d\tau\right]ds$$

$$\leq 4\mathbb{E}|\phi(0)|^2 + 4\frac{\|B\|^2}{\Gamma^2(\alpha)}(t-\varepsilon)^{\frac{2\alpha p-2}{p}}\left(\frac{p-1}{\alpha p-1}\right)^{\frac{2p-2}{p}}\|u\|_{L_p(J,U)}^2 + 4(t-\epsilon)^2\|n\|_\infty \Lambda_f(q^*) + \frac{8(t-\epsilon)^{2\alpha}}{\Gamma^2(\alpha+1)}\left[M_R + Tr(Q)\|m\|_\infty \Lambda_\sigma(q^*)\right]$$

then for $t>0$, the set $Z_\epsilon(t) = \left\{(\Pi_{1\epsilon}z)(t),\ z\in B_q\right\}$ is precompact in H for every $0<\epsilon<t$. Furthermore

$$\mathbb{E}|(\Pi_1 z)(t)-(\Pi_{1\epsilon}z)(t)|^2 \leq \frac{5\|B\|^2}{\Gamma^2(\alpha)}\mathbb{E}\left|\int_0^{t-\epsilon}\left[(t-s)^{\alpha-1}-(t-\epsilon-s)^{\alpha-1}\right]u(s)ds\right|^2 + \frac{5\|B\|^2}{\Gamma^2(\alpha)}\mathbb{E}\left|\int_{t-\epsilon}^{t}(t-s)^{\alpha-1}u(s)ds\right|^2 + 5\epsilon\int_{t-\epsilon}^{t}\mathbb{E}|f(s,\bar{z}_s+y_s)|^2 ds + \frac{5}{\Gamma^2(\alpha)}\mathbb{E}\left|\int_0^{t-\epsilon}\left[(t-s)^{\alpha-1}-(t-\epsilon-s)^{\alpha-1}\right]\left(\int_{-\infty}^{s}\sigma(\tau,\bar{z}_\tau+y_\tau)dW(\tau)\right)\right|ds + \frac{5}{\Gamma^2(\alpha)}\mathbb{E}\left|\int_{t-\epsilon}^{t}(t-s)^{\alpha-1}\left(\int_{-\infty}^{s}\sigma(\tau,\bar{z}_\tau+y_\tau)dW(\tau)\right)\right|ds$$

therefore, the set $Z(t) = \{(\Pi_1 z)(t){:}\ z\in B_q\}$ is precompact in H. Hence the operator is completely continuous.

Step 2: Π_2 is completely continuous.

By Step 1, Π_1 is continuous. Now, to prove the operator Π_2 is continuous, let B_q be a bounded set as in Step 1. For $z\in B_q$, one has

$$\mathbb{E}|(\Pi_2 z)(t)|^2 = \mathbb{E}\left|\int_0^t S'(t-s)(\Pi_1 z)(s)ds\right|^2 \leq b\int_0^t(\varphi_A(t-s))^2\mathbb{E}|(\Pi_1 z)(s)|^2 ds \leq b\|\varphi_A\|_{L^1}^2\int_0^t \mathbb{E}|(\Pi_1 z)(s)|^2 ds \leq b^2\|\varphi_A\|_{L^1}^2\delta = \delta'$$

there exists a positive number δ' such that $\mathbb{E}|(\Pi_2 z)| \leq \delta'$, which means that $\Pi_2(z) \in B_{\delta'}$. Now, to prove that Π_2 maps bounded sets into equicontinuous sets in C_b^0. Let $t_1, t_2 \in J$, $t_2 > t_1$, and consider B_q as a bounded set as in Step 1. Let $z \in B_q$, then if $\epsilon > 0$ and $\epsilon \leq t_1 \leq t_2$, one has

$$\begin{aligned}
\mathbb{E}|(\Pi_2 z)(t_2) - (\Pi_2 z)(t_1)|^2 &\leq 3\mathbb{E}\left|\int_0^{t_1-\epsilon}[S'(t_2-s)-S'(t_1-s)](\Pi_1 z)(s)ds\right|^2 \\
&+ 3\mathbb{E}\left|\int_{t_1-\varepsilon}^{t_1}[S'(t_2-s)-S'(t_1-s)](\Pi_1 z)(s)ds\right|^2 \\
&+ 3\mathbb{E}\left|\int_{t_1}^{t_2} S'(t_2-s)(\Pi_1 z)(s)ds\right|^2 \\
&\leq 3(t_1-\epsilon)\delta \int_0^{t_1-\epsilon} |S'(t_2-s) - S'(t_1-s)|^2 ds \\
&+ 3\epsilon\delta \int_{t_1-\varepsilon}^{t_1} |S'(t_2-s) - S'(t_1-s)|^2 \\
&+ 3(t_2-t_1)\delta \int_{t_1}^{t_2} |S'(t_2-s)|^2 ds
\end{aligned}$$

as $t_2 \to t_1$ and ε sufficiently small, the right-hand side of the above inequality tends to zero. By the Arzela–Ascoli theorem it suffices to show that Π_2 maps B_q into a precompact set in H. Let $0 < t < b$ be fixed and let ε be a real number satisfying $0 < \varepsilon < t$. For $z \in B_q$,

$$(\Pi_{2\epsilon} z)(t) = S'(\epsilon) \int_0^{t-\epsilon} S'(t-s-\epsilon)(\Pi_1 z)(s)ds$$

since $S'(t)$ is a compact operator for $t > 0$, the set $Z_\epsilon(t) = \{(\Pi_{2\epsilon} z)(t): z \in B_q\}$ is precompact in H for every ϵ, $0 < \epsilon < t$. Furthermore

$$\begin{aligned}
\mathbb{E}|(\Pi_2 z)(t) - (\Pi_{2\epsilon} z)(t)|^2 &\leq 2\mathbb{E}\left|\int_0^{t-\epsilon} S'(t-s)[(\Pi_1 z)(s)-(\Pi_{1\epsilon} z)(s)]ds\right|^2 \\
&+ 2\mathbb{E}\left|\int_{t-\varepsilon}^{t} S'(t-s)(\Pi_1 z)(s)ds\right|^2 \\
&\leq 2\|\varphi_A\|_{L^1}^2 (t-\epsilon) \int_0^{t-\epsilon} \mathbb{E}|(\Pi_1 z)(s) - (\Pi_{1\epsilon} z)(s)|^2 ds \\
&+ 2\|\varphi_A\|_{L^1}^2 \epsilon \int_{t-\varepsilon}^{t} \mathbb{E}|(\Pi_1 z)(s)|^2 ds \\
&\leq 2\|\varphi_A\|_{L^1}^2 (t-\varepsilon)^2 \Xi + 2\|\varphi_A\|_{L^1}^2 \varepsilon^2 \delta
\end{aligned}$$

so $Z(t) = \{(\Pi_2 z)(t): z \in B_q\}$ is precompact in H. Hence Π_2 is completely continuous.

Step 3: To prove that there exists an open set $\hat{U} \subset C(J,H)$ with $z \notin \lambda\Pi(z)$ for $\lambda \in (0,1)$ and $y \in \partial\hat{U}$. Let $\lambda \in (0,1)$ and

$$\begin{aligned}
z(t) = \lambda(\Pi z)(t) &= \lambda\phi(0) + \frac{\lambda}{\Gamma(\alpha)}\int_0^t (t-s)^{\alpha-1}B(s)u(s)ds + \lambda\int_0^t f(s,\bar{z}_s + y_s)ds \\
&+ \frac{\lambda}{\Gamma(\alpha)}\int_0^t (t-s)^{\alpha-1}\left(\int_{-\infty}^s \sigma(\tau,\bar{z}_\tau + y_\tau)dW(\tau)\right)ds \\
&+ \lambda\int_0^t S'(t-s)\left[\phi(0) + \frac{\lambda}{\Gamma(\alpha)}\int_0^s (s-\tau)^{\alpha-1}B(\tau)u(\tau)d\tau\right. \\
&\left. + \int_0^s f(\tau,\bar{z}_\tau + y_\tau)d\tau + \frac{1}{\Gamma(\alpha)}\int_0^s (s-\tau)^{\alpha-1}\left(\int_{-\infty}^\tau \sigma(\eta,\bar{z}_\eta + y_\eta)dW(\eta)\right)d\tau\right] ds.
\end{aligned}$$

then

$$
\begin{aligned}
\mathbb{E}|z(t)|^2 \le\ & 8\mathbb{E}|\phi(0)|^2 + 8\mathbb{E}\left|\frac{1}{\Gamma(\alpha)}\int_0^t (t-s)^{\alpha-1}B(s)u(s)ds\right|^2 \\
& + 8\mathbb{E}\left|\int_0^t f(s,\bar{z}_s+y_s)ds\right|^2 + 8\mathbb{E}\left|\int_0^t S'(t-s)\phi(0)ds\right|^2 \\
& + 8\mathbb{E}\left|\frac{1}{\Gamma(\alpha)}\int_0^t (t-s)^{\alpha-1}\left(\int_{-\infty}^s \sigma(\tau,\bar{z}_\tau+y_\tau)dW(\tau)\right)ds\right|^2 \\
& + 8\mathbb{E}\left|\int_0^t S'(t-s)\left(\frac{1}{\Gamma(\alpha)}\int_0^s (s-\tau)^{\alpha-1}B(\tau)u(\tau)d\tau\right)ds\right|^2 \\
& + 8\mathbb{E}\left|\int_0^t S'(t-s)\left(\int_0^s f(\tau,\bar{z}_\tau+y_\tau)d\tau\right)ds\right|^2 \\
& + 8\mathbb{E}\left|\int_0^t S'(t-s)\left(\frac{1}{\Gamma(\alpha)}\int_0^s (s-\tau)^{\alpha-1}\left(\int_{-\infty}^\tau \sigma(\eta,\bar{z}_\eta+y_\eta)dW(\eta)\right)d\tau\right)ds\right|^2 \\
\le\ & 8\mathbb{E}|\phi(0)|^2 + \frac{8\|B\|^2}{\Gamma^2(\alpha)} b^{\frac{2\alpha p-2}{p}}\left(\frac{p-1}{\alpha p-1}\right)^{\frac{2p-2}{p}}\|u\|^2_{L_p(J,U)} \\
& + 8b\int_0^t n(s)\Lambda_f(\|\bar{z}_s+y_s\|^2_{\mathcal{C}_h})ds + \frac{8b^\alpha}{\alpha\Gamma^2(\alpha)}\int_0^t (t-s)^{\alpha-1} \\
& \times\left[2M_R + 2Tr(Q)\int_0^s \mathbb{E}|\sigma(\tau,\bar{z}_\tau+y_\tau)|^2 d\tau\right]ds + 8b^2\|\varphi_A\|^2_{L^1}\mathbb{E}|\phi(0)|^2 \\
& + \frac{8b\|B\|^2\|\varphi_A\|^2_{L^1}}{\Gamma^2(\alpha)}\int_0^t b^{\frac{2\alpha p-2}{p}}\left(\frac{p-1}{\alpha p-1}\right)^{\frac{2p-2}{p}}\|u\|^2_{L_p(J,U)}ds \qquad (7.19)\\
& + 8b^2\|\varphi_A\|^2_{L^1}\int_0^t\int_0^s \mathbb{E}|f(\tau,\bar{z}_\tau+y_\tau)|^2 d\tau ds \\
& + \frac{8\|\varphi_A\|^2_{L^1}b^{\alpha+1}}{\alpha\Gamma^2(\alpha)}\int_0^t\int_0^s (s-\tau)^{\alpha-1}\mathbb{E}\left|\int_{-\infty}^\tau \sigma(\eta,\bar{z}_\eta+y_\eta)dW(\eta)\right|^2 d\tau ds \\
\le\ & 8\mathbb{E}|\phi(0)|^2 + \frac{8\|B\|^2}{\Gamma^2(\alpha)} b^{\frac{2\alpha p-2}{p}}\left(\frac{p-1}{\alpha p-1}\right)^{\frac{2p-2}{p}}\|u\|^2_{L_p(J,U)} \\
& + 8b\int_0^t n(s)\Lambda_f(\|\bar{z}_s+y_s\|^2_{\mathcal{C}_h})ds + \frac{16b^{2\alpha}M_R}{\Gamma^2(\alpha+1)} \\
& + \frac{16b^\alpha Tr(Q)}{\alpha\Gamma^2(\alpha)}\int_0^t (t-s)^{\alpha-1}m(s)\Lambda_\sigma(\|\bar{z}_s+y_s\|^2_{\mathcal{C}_h})ds \\
& + 8b^2\|\varphi_A\|^2_{L^1}\mathbb{E}|\phi(0)|^2 + \frac{8b^2\|\varphi_A\|^2_{L^1}\|B\|^2}{\Gamma^2(\alpha)} b^{\frac{2\alpha p-2}{p}}\left(\frac{p-1}{\alpha p-1}\right)^{\frac{2p-2}{p}}\|u\|^2_{L_p(J,U)} \\
& + 8b^3\|\varphi_A\|^2_{L_1}\int_0^t n(s)\Lambda_f(\|\bar{z}_s+y_s\|^2_{\mathcal{C}_h})ds + \frac{16\|\varphi_A\|^2_{L^1}b^{2\alpha+2}M_R}{\Gamma^2(\alpha+1)} \\
& + \frac{16\|\varphi_A\|^2_{L^1}b^{2\alpha+1}Tr(Q)}{\Gamma^2(\alpha+1)}\int_0^t m(s)\Lambda_\sigma(\|\bar{z}_s+y_s\|^2_{\mathcal{C}_h})ds.
\end{aligned}
$$

But $\|\bar{z}_s + y_s\|^2_{\mathcal{C}_h} \leq 4l^2 \sup_{s\in[0,t]} \mathbb{E}|z(s)|^2 + 4\|\phi\|^2_{\mathcal{C}_h}$ as proved in Step 1. If one lets $q^*(t)$ be the right-hand side of the above inequality then, $\|\bar{z}_s + y_s\|^2_{\mathcal{C}_h} \leq q^*(t),\ \ t\in J$ and therefore (7.19) becomes

$$
\begin{aligned}
\mathbb{E}|z(t)|^2 \ &\leq 8\mathbb{E}|\phi(0)|^2 + \frac{8\|B\|^2}{\Gamma^2(\alpha)} b^{\frac{2\alpha p-2}{p}} \left(\frac{p-1}{\alpha p-1}\right)^{\frac{2p-2}{p}} \|u\|^2_{L_p(J,U)} \\
&+ 8b\int_0^t n(s)\Lambda_f(q^*(s))ds + \frac{16b^{2\alpha}M_R}{\Gamma^2(\alpha+1)} \\
&+ \frac{16b^{\alpha}Tr(Q)}{\alpha\Gamma^2(\alpha)} \int_0^t (t-s)^{\alpha-1} m(s)\Lambda_\sigma(q^*(s))ds + 8b^2\|\varphi_A\|^2_{L^1}\mathbb{E}|\phi(0)|^2 \\
&+ \frac{8b^2\|\varphi_A\|^2_{L^1}\|B\|^2}{\Gamma^2(\alpha)} b^{\frac{2\alpha p-2}{p}} \left(\frac{p-1}{\alpha p-1}\right)^{\frac{2p-2}{p}} \|u\|^2_{L_p(J,U)} \\
&+ 8b^3\|\varphi_A\|^2_{L_1}\int_0^t n(s)\Lambda_f(q^*(s))ds + \frac{16\|\varphi_A\|^2_{L^1} b^{2\alpha+2}M_R}{\Gamma^2(\alpha+1)} \\
&+ \frac{16\|\varphi_A\|^2_{L^1} b^{2\alpha+1}Tr(Q)}{\Gamma^2(\alpha+1)} \int_0^t m(s)\Lambda_\sigma(q^*(s))ds
\end{aligned}
\tag{7.20}
$$

using (7.20) in the definition of $q^*(s)$, one has

$$
\begin{aligned}
q^*(t) &= 4l^2 \sup_{s\in[0,t]} \mathbb{E}|z(s)|^2 + 4\|\phi\|^2_{\mathcal{C}_h} \\
&\leq 4l^2 \Bigg[8\mathbb{E}|\phi(0)|^2 + \frac{8\|B\|^2}{\Gamma^2(\alpha)} b^{\frac{2\alpha p-2}{p}} \left(\frac{p-1}{\alpha p-1}\right)^{\frac{2p-2}{p}} \|u\|^2_{L_p(J,U)} \\
&+ 8b\int_0^t n(s)\Lambda_f(q^*(s))ds + \frac{16b^{2\alpha}M_R}{\Gamma^2(\alpha+1)} \\
&+ \frac{16b^{\alpha}Tr(Q)}{\alpha\Gamma^2(\alpha)} \int_0^t (t-s)^{\alpha-1} m(s)\Lambda_\sigma(q^*(s))ds \\
&+ 8b^2\|\varphi_A\|^2_{L^1}\mathbb{E}|\phi(0)|^2 + \frac{8b^2\|\varphi_A\|^2_{L^1}\|B\|^2}{\Gamma^2(\alpha)} b^{\frac{2\alpha p-2}{p}} \left(\frac{p-1}{\alpha p-1}\right)^{\frac{2p-2}{p}} \|u\|^2_{L_p(J,U)} \\
&+ 8b^3\|\varphi_A\|^2_{L_1}\int_0^t n(s)\Lambda_f(q^*(s))ds + \frac{16\|\varphi_A\|^2_{L^1} b^{2\alpha+2}M_R}{\Gamma^2(\alpha+1)} \\
&+ \frac{16\|\varphi_A\|^2_{L^1} b^{2\alpha+1}Tr(Q)}{\Gamma^2(\alpha+1)} \int_0^t m(s)\Lambda_\sigma(q^*(s))ds \Bigg] + 4\|\phi\|^2_{\mathcal{C}_h}
\end{aligned}
$$

consequently

$$
\frac{\|q^*\|}{4l^2\left[8\mathbb{E}|\phi(0)|^2 + 8\frac{\|B\|^2}{\Gamma^2(\alpha)} b^{\frac{2\alpha p-2}{p}} \left(\frac{p-1}{\alpha p-1}\right)^{\frac{2p-2}{p}} (b^2+1)\|u\|^2_{L_p(J,U)} + \Theta\right] + 4\|\phi\|^2_{\mathcal{C}_h}} \leq 1
$$

where

$$\Theta = 8b^2\|n\|_\infty(b^2\|\varphi_A\|^2 + 1)\Lambda_f(q^*) + \left[\frac{16b^{2\alpha} + 16\|\varphi_A\|^2_{L^1}b^{2\alpha+2}}{\Gamma^2(\alpha+1)}\right][M_R + Tr(Q)\|m\|_\infty\Lambda_\sigma(q^*)]$$

by (7.18) there exists a M^* such that $\|q^*\| \neq M^*$. Let us set $\hat{U} = \{x \in C(J,H)\text{: } \|y\| < M^*\}$, from the choice of $\hat{U}$, there is no $y \in \partial\hat{U}$ such that $y = \lambda\Pi(y)$ for some $\lambda \in (0,1)$. Consequently, by the nonlinear alternative of Leray-Schauder type, one can deduce that Π has a fixed point $x \in \hat{U}$, which is a solution of (7.15) and (7.16) on $(-\infty, b]$.

7.3.1.2 Existence of Optimal Control

Consider the following Lagrange problem ($\mathcal{P}$):

Find a control $u^0 \in \mathcal{A}_{ad}$ such that $\mathcal{J}(u^0) \le \mathcal{J}(u)$, *for all* $u \in \mathcal{A}_{ad}$ where

$$\mathcal{J}(u) = \mathbb{E}\left\{\int_0^b L(t, x_t^u, x^u(t), u(t))dt\right\}$$

and x^u denotes the mild solution of (7.15) and (7.16) corresponding to the control $u \in \mathcal{A}_{ad}$. For the existence of solutions of Lagrange problem $\mathcal{P}$, one shall introduce the following hypothesis

(H_{20})

1. The functional $L\text{:}J \times \mathcal{C}_h \times H \times U \to \mathbb{R} \cup \{\infty\}$ is $\mathcal{F}_t$- measurable.
2. $L(t,\cdot,\cdot,\cdot)$ is sequentially lower semicontinuous on $\mathcal{C}_h \times H \times U$ for almost all $t \in J$.
3. $L(t,x,y,\cdot)$ is convex on U for each $x \in \mathcal{C}_h$, $y \in H$ and all $t \in J$.
4. There exists constants d, $e \ge 0$, $j > 0$, μ is nonnegative and $\mu \in L^1(J,\mathbb{R})$ such that $L(t,x,y,u) \ge \mu(t) + d\|x\|^2_{\mathcal{C}_h} + e\mathbb{E}|y|^2 + j\|u\|^p_U$.

Theorem 7.7: *Let the hypotheses* $(H_{14}) - (H_{20})$ *hold. Suppose that B is a strongly continuous operator. Then, the Lagrange problem* ($\mathcal{P}$) *admits at least one optimal pair, i.e., there exists an admissible control* $u^0 \in \mathcal{A}_{ad}$ *such that*

$$\mathcal{J}(u^0) = \mathbb{E}\left(\int_0^b L(t, x_t^0, x^0(t), u^0(t))dt\right) \le \mathcal{J}(u), \ \textit{for all} \ u \in \mathcal{A}_{ad}.$$

Proof: If $\inf\{\mathcal{J}(u)|u \in \mathcal{A}_{ad}\} = \infty$, there is nothing to prove. Without loss of generality, assume that $\inf\{\mathcal{J}(u)|u \in \mathcal{A}_{ad}\} = \hat{\epsilon} < +\infty$. Using ($H_{20}$), one has $\hat{\epsilon} > -\infty$. By the definition of infimum there exists a minimizing sequence feasible pair $\{(x^m, u^m)\} \subset \mathcal{P}_{ad}$, where

$$\mathcal{P}_{ad} = \Big\{(x,u)\text{: } x \text{ is a mild solution of the system (7.15)} - \text{(7.16)} \text{ corresponding to} u \in \mathcal{A}_{ad}\Big\}$$

such that $\mathcal{J}(x^m, u^m) \to \hat{\varepsilon}$ *as* $m \to \infty$. Since $\{u^m\} \subseteq \mathcal{A}_{ad}$, $m = 1, 2, \cdots, \{w^m\}$ is a bounded subset of the separable reflexive Banach space $L^p(J, U)$, there exists a subsequence, relabeled as $\{u^m\}$ and $u^0 \in L^p(J, U)$ such that $u^m \xrightarrow{w} u^0$ in $L^p(J, U)$.

Since $\mathcal{A}_{ad}$ is closed and convex, then by Marzur lemma $u^0 \in \mathcal{A}_{ad}$. Let $\{x^m\} \subset \mathcal{C}_b$ denote the corresponding sequence of solutions of the integral equation

$$x^m(t) = \begin{cases} \phi(t), \ t \in (-\infty, 0] \\ \phi(0) + \dfrac{1}{\Gamma(\alpha)} \displaystyle\int_0^t (t-s)^{\alpha-1} B(s) u^m(s) ds + \int_0^t f(s, x_s^m) ds \\ + \dfrac{1}{\Gamma(\alpha)} \displaystyle\int_0^t (t-s)^{\alpha-1} \left(\int_{-\infty}^s \sigma(\tau, x_\tau^m) dW(\tau) \right) ds \\ + \int_0^t S'(t-s) \left[\phi(0) + \dfrac{1}{\Gamma(\alpha)} \displaystyle\int_0^s (s-\tau)^{\alpha-1} B(\tau) u^m(\tau) d\tau + \int_0^s f(\tau, x_\tau^m) d\tau \right. \\ \left. + \dfrac{1}{\Gamma(\alpha)} \displaystyle\int_0^s (s-\tau)^{\alpha-1} \left(\int_{-\infty}^\tau \sigma(\eta, x_\eta^m) dW(\eta) \right) d\tau \right] ds, \ t \in J \end{cases}$$

by Theorem 7.6, there exists a $\rho > 0$ such that $\mathbb{E}|x^m|^2 \le \rho$, $m = 0, 1, 2, \cdots$. Let $x^m(t) = \overline{z}^m(t) + y(t)$. For $t \in J$,

$$\begin{aligned} \mathbb{E}\left|\overline{z}^m(t) - \overline{z}^0(t)\right|^2 \le\, & 6\mathbb{E}\left| \tfrac{1}{\Gamma(\alpha)} \textstyle\int_0^t (t-s)^{\alpha-1} \left[B(s)u^m(s) - B(s)u^0(s)\right] ds \right|^2 \\ & + 6\mathbb{E}\left| \textstyle\int_0^t [f(s, \overline{z}_s^m + y_s) - f(s, \overline{z}_s^0 + y_s)] ds \right|^2 \\ & + 6\mathbb{E}\left| \tfrac{1}{\Gamma(\alpha)} \textstyle\int_0^t (t-s)^{\alpha-1} \left(\int_{-\infty}^s [\sigma(\tau, \overline{z}_\tau^m + y_\tau) - \sigma(\tau, \overline{z}_\tau^0 + y_\tau)] dW(\tau) \right) ds \right|^2 \\ & + 6\mathbb{E}\left| \textstyle\int_0^t S'(t-s) \left[\tfrac{1}{\Gamma(\alpha)} \int_0^s (s-\tau)^{\alpha-1} \left[B(\tau)u^m(\tau) - B(\tau)u^0(\tau)\right] d\tau \right] ds \right|^2 \\ & + 6\mathbb{E}\left| \textstyle\int_0^t S'(t-s) \left[\int_0^s [f(\tau, \overline{z}_\tau^m + y_\tau) - f(\tau, \overline{z}_\tau^0 + y_\tau)] d\tau \right] ds \right|^2 \\ & + 6\mathbb{E}\left| \textstyle\int_0^t S'(t-s) \left[\dfrac{1}{\Gamma(\alpha)} \displaystyle\int_0^s (s-\tau)^{\alpha-1} \right.\right. \\ & \left.\left. \times \left(\textstyle\int_{-\infty}^\tau [\sigma(\eta, \overline{z}_\eta^m + y_\eta) - \sigma(\eta, \overline{z}_\eta^0 + y_\eta)] dW(\eta) \right) d\tau \right] ds \right|^2 \le 6 \sum_{i=1}^6 I_i. \end{aligned}$$

Now,

$$\begin{aligned}
I_1 &= \mathbb{E}\left|\frac{1}{\Gamma(\alpha)}\int_0^t (t-s)^{\alpha-1}\left[B(s)u^m(s)-B(s)u^0(s)\right]ds\right|^2 \\
&\le \frac{1}{\Gamma^2(\alpha)}\left[\left(\frac{p-1}{\alpha p-1}\right)^{\frac{p-1}{p}} b^{\frac{\alpha p-1}{p}}\left(\int_0^t \mathbb{E}\left|B(s)u^m(s)-B(s)u^0(s)\right|^p ds\right)^{\frac{1}{p}}\right]^2 \\
&\le \frac{1}{\Gamma^2(\alpha)}\left(\frac{p-1}{\alpha p-1}\right)^{\frac{2p-2}{p}} b^{\frac{2\alpha p-2}{p}} \left\| B(s)u^m - B(s)u^0 \right\|^2_{L^p(J,U)}
\end{aligned}$$

$$\begin{aligned}
I_2 &= \mathbb{E}\left|\int_0^t [f(s,\bar{z}_s^m+y_s)-f(s,\bar{z}_s^0+y_s)]ds\right|^2 \\
&\le b\int_0^t \mathbb{E}\left|f(s,\bar{z}_s^m+y_s)-f(s,\bar{z}_s^0+y_s)\right|^2 ds \\
&\le bM_f(\rho)\int_0^t \| \bar{z}_s^m - \bar{z}_s^0 \|^2_{\mathcal{C}_h} ds,
\end{aligned}$$

$$\begin{aligned}
I_3 &= \mathbb{E}\left|\frac{1}{\Gamma(\alpha)}\int_0^t (t-s)^{\alpha-1}\left(\int_{-\infty}^s [\sigma(\tau,\bar{z}_\tau^m+y_\tau)-\sigma(\tau,\bar{z}_\tau^0+y_\tau)]dW(\tau)\right)ds\right|^2 \\
&\le \frac{b^\alpha}{\alpha\Gamma^2(\alpha)}\int_0^t (t-s)^{\alpha-1}\mathbb{E}\left|\int_{-\infty}^s [\sigma(\tau,\bar{z}_\tau^m+y_\tau)-\sigma(\tau,\bar{z}_\tau^0+y_\tau)]dW(\tau)\right|^2 ds \\
&\le \frac{2b^{2\alpha}M_R}{\Gamma^2(\alpha+1)} + \frac{2b^\alpha Tr(Q)M_\sigma(\rho)}{\alpha\Gamma^2(\alpha)}\int_0^t (t-s)^{\alpha-1}\| \bar{z}_s^m - \bar{z}_s^0 \|^2_{\mathcal{C}_h} ds
\end{aligned}$$

$$\begin{aligned}
I_4 &= \mathbb{E}\left|\int_0^t S'(t-s)\left[\frac{1}{\Gamma(\alpha)}\int_0^s (s-\tau)^{\alpha-1}\left[B(\tau)u^m(\tau)-B(\tau)u^0(\tau)\right]d\tau\right]ds\right|^2 \\
&\le \frac{\| \varphi_A \|^2}{\Gamma^2(\alpha)}\int_0^t \left(\frac{p-1}{\alpha p-1}\right)^{\frac{2p-2}{p}} b^{\frac{2\alpha p-2}{p}} \left\| B(s)u^m - B(s)u^0 \right\|^2_{L^p(J,U)} ds
\end{aligned}$$

$$\begin{aligned}
I_5 &= \mathbb{E}\left|\int_0^t S'(t-s)\left[\int_0^s [f(\tau,\bar{z}_\tau^m+y_\tau)-f(\tau,\bar{z}_\tau^0+y_\tau)]d\tau\right]ds\right|^2 \\
&\le \| \varphi_A \|^2 bM_f(\rho)\int_0^t\int_0^s \| \bar{z}_\tau^m - \bar{z}_\tau^0 \|^2_{\mathcal{C}_h} d\tau ds
\end{aligned}$$

$$\begin{aligned}
I_6 = \mathbb{E}\Bigg| \int_0^t S'(t-s)\Bigg[\frac{1}{\Gamma(\alpha)}\int_0^s (s-\tau)^{\alpha-1} \\
\times\left(\int_{-\infty}^\tau [\sigma(\eta,\bar{z}_\eta^m+y_\eta)-\sigma(\eta,\bar{z}_\eta^0+y_\eta)]dW(\eta)\right)d\tau\Bigg]ds\Bigg|^2 \\
\le \frac{2\| \varphi_A \|^2 b^{2\alpha+2}}{\Gamma^2(\alpha+1)} + \frac{2\| \varphi_A \|^2 b^{\alpha+1} Tr(Q)M_\sigma(\rho)}{\alpha\Gamma^2(\alpha)}\int_0^t\int_0^s (s-\tau)^{\alpha-1}\| \bar{z}_\tau^m - \bar{z}_\tau^0 \|^2_{\mathcal{C}_h} d\tau ds.
\end{aligned}$$

Thus,

$$
\begin{aligned}
\mathbb{E}\left|z^m(t)-z^0(t)\right|^2 \leq 6\Bigg\{ & \frac{1}{\Gamma^2(\alpha)}\left(\frac{p-1}{\alpha p-1}\right)^{\frac{2p-2}{p}} b^{\frac{2\alpha p-2}{p}} \left\|Bu^m-Bu^0\right\|^2_{L^p(J,U)} \\
& + bM_f(\rho)\int_0^t\int_0^t \| z_s^m - z_s^0 \|^2_{\mathcal{C}_h} ds + \frac{2b^{2\alpha}M_R}{\Gamma^2(\alpha+1)} \\
& + \frac{2b^{\alpha}Tr(Q)M_\sigma(\rho)}{\alpha\Gamma^2(\alpha)}\int_0^t (t-s)^{\alpha-1}\| z_s^m - z_s^0 \|^2_{\mathcal{C}_h} ds \\
& + \frac{\|\varphi_A\|^2}{\Gamma^2(\alpha)}\left(\frac{p-1}{\alpha p-1}\right)^{\frac{2p-2}{p}} b^{\frac{2\alpha p+p-2}{p}} \left\|Bu^m-Bu^0\right\|^2_{L^p(J,U)} \\
& + \|\varphi_A\|^2 bM_f(\rho)\int_0^t\int_0^s \| z_\tau^m - z_\tau^0 \|^2 d\tau ds + \frac{2\|\varphi_A\|^2 b^{2\alpha+2}}{\Gamma^2(\alpha+1)} \\
& + \frac{2\|\varphi_A\|^2 b^{\alpha+1}Tr(Q)M_\sigma(\rho)}{\alpha\Gamma^2(\alpha)}\int_0^t\int_0^s (s-\tau)^{\alpha-1}\| z_\tau^m - z_\tau^0 \|^2_{\mathcal{C}_h} d\tau ds \Bigg\}
\end{aligned}
$$

which implies that there exists a constant $\hat{N}>0$ such that

$$\sup_{t\in J}\mathbb{E}\left|z^m(t)-z^0(t)\right|^2 \leq \hat{N}\left\|Bu^m-Bu^0\right\|^2_{L^p(J,U)} \quad \text{for } t\in J$$

since B is strongly continuous, one has $\left\|Bu^m-Bu^0\right\|^2_{L^p(J,U)} \xrightarrow{s} 0$ as $m\to\infty$ thus, $\mathbb{E}|z^m(t)-z^0(t)|^2 \xrightarrow{s} 0$ as $m\to\infty$, which is equivalent to $\|x^m-x^0\|_{\mathcal{C}_b}\to 0$ as $m\to\infty$, this yields that $x^m \xrightarrow{s} x^0$ in $\mathcal{C}_b$ as $m\to\infty$. Note that (H_{20}) implies the assumptions of Balder (1987). Hence by Balder's theorem, one can conclude that $(x,x_t,u)\to\mathbb{E}\left(\int_0^t L(t,x_t,x(t),u(t))dt\right)$ is sequentially lower semicontinuous in the strong topology of $L^1(J,H)$. Since, $L^p(J,U)\subset L^1(J,U)$, $\mathcal{J}$ is weakly lower semicontinuous on $L^p(J,U)$, and since by $(H_{20})(iv)$, $\mathcal{J}>-\infty$, $\mathcal{J}$ attains its infimum at $u^0\in\mathcal{A}_{ad}$, i.e.,

$$\hat{\epsilon}=\lim_{m\to\infty}\mathbb{E}\left(\int_0^b L(t,x_t^m,x^m(t),u^m(t))dt\right)\geq\mathbb{E}\left(\int_0^b L(t,x_t^0,x^0(t),u^0(t))dt\right)=\mathcal{J}(u^0)\geq\hat{\epsilon}.$$

This completes the proof.

7.4 CONCLUSION

This chapter studied the controllability results of different classes of single-valued and multivalued FSDEs. In the first part of this chapter, the approximate controllability of some class of fractional stochastic integro-differential inclusion, and solvability and existence of optimal control of FSDEs of order

$1\alpha < 2$ have been studied in Hilbert space by using (a, k)-regularized families of bounded linear operators. The peculiarity of the obtained theoretical results lies in the fact that (a, k)-regularized families of bounded linear operators cover the theories of C_0-semigroups and cosine families as particular cases. The second part of this chapter dealt with the solvability and existence of optimal control for a class of fractional stochastic integro-differential equations with infinite delay in Hilbert space by employing the well developed theory of resolvent operators for integral equations.

Our future investigations are mainly focused on stability analysis of new classes of single and multivalued FSDEs in the sense of Caputo–Fabrizio's or Atangana–Baleanu's fractional derivative by employing fixed point technique. Since, the fractional derivatives are nonlocal and have weakly singular kernels, hence the stability analysis of FDEs is more complex than that of classical integer order differential equations. However, the new definitions of fractional derivative, namely Caputo–Fabrizio's and Atangana–Baleanu's fractional derivative can be employed to study the stability analysis, since they are defined without singular kernel. Caputo–Fabrizio's fractional derivative is based on exponential kernel and Atangana–Baleanu's fractional derivative used Mittag–Leffler function as nonlocal kernels.

REFERENCES

Agrawal, O.P., 2004. A general formulation and solution scheme for fractional optimal control problems. Nonlinear Dynam. 38 (1-4), 323–337.

Agrawal, O.P., Defterli, O., Baleanu, D., 2010. Fractional optimal control problems with several state and control variables. J. Vibrat. Control 16 (13), 1967–1976.

Aubin, J., Cellina, A., 1984. Differential Inclusions. Heidelberg, Berlin.

Azar, A.T., Vaidyanathan, S., Ouannas, A., 2017. Fractional Order Control and Synchronization of Chaotic Systems, Vol. 688. Springer, New York.

Bajlekova, E.G., 2001. Fractional evolution equations in banach spaces. Ph.D. thesis, University Press Facilities, Eindhoven University of Technology Eindhoven, The Netherlands.

Balachandran, K., Kiruthika, S., 2011. Existence results for fractional integrodifferential equations with nonlocal condition via resolvent operators. Comput. Math. Applicat. 62 (3), 1350–1358.

Balasubramaniam, P., Ntouyas, S., 2006. Controllability for neutral stochastic functional differential inclusions with infinite delay in abstract space. J. Math. Anal. Applicat. 324 (1), 161–176.

Balasubramaniam, P., Tamilalagan, P., 2015. Approximate controllability of a class of fractional neutral stochastic integro-differential inclusions with infinite delay by using mainardis function. Appl. Math. Comput. 256, 232–246.

Balasubramaniam, P., Tamilalagan, P., 2016. The solvability and optimal controls for impulsive fractional stochastic integro-differential equations via resolvent operators. J. Optimizat. Theory Applicat. 1–17.

Balasubramaniam, P., Ntouyas, S., Vinayagam, D., 2005. Existence of solutions of semilinear stochastic delay evolution inclusions in a hilbert space. J. Math. Anal. Applicat. 305 (2), 438–451.

Balder, E.J., 1987. Necessary and sufficient conditions for l1-strong-weak lower semicontinuity of integral functionals. Nonlinear Anal. Theory Methods Applicat. 11 (12), 1399–1404.

Da Prato, G., Zabczyk, J., 2014. Stochastic Equations in Infinite Dimensions. Cambridge University Press, Cambridge.

Deimling, K., 1992. Multivalued Differential Equations, Vol. 1. Walter de Gruyter, Berlin.

Dos Santos, J., Vijayakumar, V., Murugesu, R., et al., 2013. Existence of mild solutions for non-local cauchy problem for fractional neutral integro-differential equation with unbounded delay. Commun. Math. Anal. 14 (1), 59–71.

El-Sayed, A.M., Ibrahim, A., 1995. Multivalued fractional differential equations. Appl. Math. Comput. 68 (1), 15–25.

Fan, Z., Mophou, G., 2014. Existence of optimal controls for a semilinear composite fractional relaxation equation. Rep. Math. Phys. 73 (3), 311–323.

Fitzpatrick, P., Petryshyn, W., 1974. Fixed point theorems for multivalued noncompact acyclic mappings. Pacif. J. Math. 54 (2), 17–23.

Hale, J.K., Kato, J., 1978. Phase space for retarded equations with infinite delay. Funkcial. Ekvac 21 (1), 11–41.

Henderson, J., Ouahab, A., 2010. Impulsive differential inclusions with fractional order. Comput. Math. Appl. 59 (3), 1191–1226.

Hernández, E., ORegan, D., Balachandran, K., 2010. On recent developments in the theory of abstract differential equations with fractional derivatives. Nonlinear Anal. Theory Methods Applicat. 73 (10), 3462–3471.

Hernández, E., ORegan, D., Balachandran, K., 2013. Existence results for abstract fractional differential equations with nonlocal conditions via resolvent operators. Ind. Math. 24 (1), 68–82.

Hu, S.P., 1997. Ns-handbook of multivalued analysis, vol. i.

Kamenskii, M.I., Obukhovskii, V.V., Zecca, P., 2001. Condensing Multivalued Maps and Semilinear Differential Inclusions in Banach Spaces., Vol. 7. Walter de Gruyter, Berlin.

Kexue, L., Jigen, P., 2012. Fractional resolvents and fractional evolution equations. Appl. Math. Lett. 25 (5), 808–812.

Li, K., Peng, J., Jia, J., 2012. Cauchy problems for fractional differential equations with riemann–liouville fractional derivatives. J. Funct. Anal. 263 (2), 476–510.

Li, Y., Liu, B., 2007. Existence of solution of nonlinear neutral stochastic differential inclusions with infinite delay. Stochastic Anal. Applicat. 25 (2), 397–415.

Liu, X., Liu, Z., Han, J., 2013. The solvability and optimal controls for some fractional impulsive equation, Abstract and Applied Analysis., Vol. 2013. Hindawi Publishing Corporation.

Lizama, C., 2000. Regularized solutions for abstract volterra equations. J. Math. Anal. Applicat. 243 (2), 278–292.

Lizama, C., N'Guérékata, G.M., 2013. Mild solutions for abstract fractional differential equations. Applicable Analysis 92 (8), 1731–1754.

Lizama, C., Poblete, F., 2012. On a functional equation associated with ( )-regularized resolvent families, Abstract and Applied Analysis, Vol. 2012. Hindawi Publishing Corporation.

Losta, A., Opial, Z., 1965. Application of the kakutani-ky-fan theorem in the theory of ordinary differential equations or noncompact acyclic-valued map. Bull. Acad. Pol. Sci., Sér. Sci. Math. Astron. Phys 13, 781–786.

Mahmudov, N., Zorlu, S., 2003. Controllability of non-linear stochastic systems. Int. J. Control 76 (2), 95–104.

Mao, X., 2007. Stochastic Differential Equations and Applications. Elsevier.

Peng, J., Li, K., 2012. A novel characteristic of solution operator for the fractional abstract cauchy problem. J. Math. Anal. Applicat. 385 (2), 786–796.

Podlubny, I., 1998. Fractional Differential Equations: An Introduction to Fractional Derivatives, Fractional Differential Equations, to Methods of Their Solution and Some of Their Applications, Vol. 198. Academic press.

Prüss, J., 2013. Evolutionary Integral Equations and Applications., Vol. 87. Birkhäuser.

Srivastava, H.M., Trujillo, J.J., et al., 2006. Theory and applications of fractional differential equations.

Vijayakumar, V., Ravichandran, C., Murugesu, R., Trujillo, J., 2014. Controllability results for a class of fractional semilinear integro-differential inclusions via resolvent operators. Appl. Math. Computation 247, 152–161.

Wang, J., Zhou, Y., 2011a. A class of fractional evolution equations and optimal controls. Nonlinear Anal. Real World Applicat. 12 (1), 262–272.

Wang, J., Zhou, Y., 2011b. Existence and controllability results for fractional semilinear differential inclusions. Nonlinear Anal. Real World Applicat. 12 (6), 3642–3653.

Wang, J., Zhou, Y., Medved', M., 2012. On the solvability and optimal controls of fractional integrodifferential evolution systems with infinite delay. J. Optimiz. Theory Applicat. 152 (1), 31–50.

Zhou, Y., 2016. Fractional Evolution Equations and Inclusions: Analysis and Control. Academic Press, London.

Chapter 8

Controllability of Fractional Higher Order Stochastic Integrodifferential Inclusions

T. Sathiyaraj and Pagavathigounder Balasubramaniam
Gandhigram Rural Institute (Deemed to be University), Dindigul, Tamil Nadu, India

8.1 INTRODUCTION

Fractional differential equations serve as an appropriate phenomenon such that it can even describe the real-world problems which are impossible to do using classical integer order differential equations. Over the past decades, the theory of fractional differential equation has gained more attention, and has obtained a prior position in the field of physics, signal processing, fluid mechanics, viscoelasticity, mathematical biology, electro chemistry, and many other science and engineering fields (for detail one may refer to the books and monographs: Azar et al., 2017; Miller and Ross, 1993; Oldham and Spanier, 1974; Podlubny, 1998; Samko et al., 1993; Sabatier et al., 2007).

Controllability of dynamical systems is one of the fundamental notions of modern control theory. Generally speaking, controllability enables one to steer the control system from an arbitrary initial state to an arbitrary final state using the set of admissible controls. This concept leads to some important conclusions regarding the behavior of linear and nonlinear dynamical systems. Controllability of fractional order deterministic and stochastic differential equations and inclusions in finite dimensional space have been studied by Balachandran and Kokila (2012) and Sathiyaraj and Balasubramaniam (2016). Balasubramaniam and Ntouyas (2006) studied controllability for neutral stochastic functional differential inclusions with infinite delay in abstract space. Controllability of linear stochastic systems has been investigated by Mahmudov and Denker (2000).

It is natural to define a solution as continuously differentiable functions satisfying the fractional differential equations $D^q x(t) = f(x(t))$ in all points of some interval with continuous function in the right-hand side. In many applied problems, there is a need to consider differential equations with an

Mathematical Techniques of Fractional Order Systems. DOI: https://doi.org/10.1016/B978-0-12-813592-1.00008-8

upper semicontinuous of the right-hand side, but the class of continuously differentiable functions is not large enough to guarantee the existence of solution. Thus, the fractional differential inclusions described by

$$D^q x(t) \in f(x(t))$$

are required.

On the other hand, the theory of differential inclusions has become an active area of investigation due to its applications in various fields such as mechanics, electrical engineering, medicine biology, ecology, and so on (Abbas and Mouffak, 2013; Balasubramaniam, 2002). There are many applied problems in mathematics that induce the reader to the study of dynamical systems having velocities nonuniquely determined by the state of the system but depending loosely upon it.

Filippov (1988) systemized the theory of differential inclusions and introduced the main properties of differential inclusions. As a matter of fact, there exist extensive literature concerning differential and integral inclusions in deterministic cases (see Aubin and Cellina, 1984; Hu and Papageorgiou, 1997; Deimling, 1992). Kree (1982) introduced stochastic differential inclusions as a separate theory. Consequently there have been substantial works, which have dealt with stochastic differential inclusions with different right-hand sides and provided sufficient conditions for the existence of solutions of corresponding stochastic differential inclusions.

However, there are only limited works considering the existence of solutions, and controllability results of integer order stochastic differential inclusions in finite and infinite dimensional space (Balasubramaniam and Ntouyas, 2006; Balasubramaniam, 2002). Existence and controllability results for fractional semilinear differential inclusions have been proposed by Wang and Zhou (2011). Sakthivel et al. (2013) formulated a new set of sufficient conditions for the approximate controllability of fractional nonlinear differential inclusions. Fractional order Riemann–Liouville integral inclusions with two independent variables and multiple delays have been illustrated by Abbas and Mouffak (2013). So far no work has been reported in the literature about the complete controllability of fractional higher order stochastic integrodifferential inclusions.

Inspired by the abovementioned works, the aim of this chapter is to fill this gap. The purpose of this chapter is to show the controllability of fractional order stochastic integrodifferential inclusions. The sufficient conditions are acquired via fixed point theorems, namely the Bohnenblust–Karlin fixed point theorem for the convex case and the Covitz–Nadler fixed point theorem for the nonconvex case where the controllability Grammian matrix is defined by using Mittag–Leffler matrix function. In particular, the complete controllability results are investigated for the following nonlinear fractional higher order stochastic integrodifferential inclusions.

$$
\begin{aligned}
{}^{C}D^{q}x(t) &\in A\,{}^{C}D^{p}x(t) + Bu(t) + F(t,x(t)) + \int_0^t G(s,x(s))dw(s), \quad t\in[0,T]: = J,\\
& x(0) = x_0, \ x'(0) = x_0'
\end{aligned}
\tag{8.1}
$$

where ${}^{C}D^{q}$ and ${}^{C}D^{p}$ denote Caputo fractional derivatives of order $0<p\leq 1<q\leq 2$, A and B are matrices of dimensions $n\times n$ and $n\times m$, respectively, $x\in\mathbb{R}^n$, $u\in\mathbb{R}^m$ are the state and control vectors. The nonlinear functions f, G are appropriate functions to be defined later.

The chapter is organized as follows: In Section 8.2, some essential results on the basic definitions of fractional integral and derivatives, Lemmas, propositions, and some hypotheses are given to obtain the controllability results successfully. In Section 8.3, the controllability results for the fractional system (8.1) are studied under the fixed point theorems. In Section 8.4, a discussion and future work is given. Finally, conclusions are drawn in Section 8.5.

Notations: Throughout this chapter, $(\Omega, \mathfrak{F}, P)$ denotes the complete probability space with a right continuous and complete filtration $\{\mathfrak{F}_t, t\in J\}$ ($\mathfrak{F}_t$ the $\sigma-$algebra generated by the random variables $\left\{W^H_{(s)}, s\in[0,t]\right\}$ and $P-$null set) and satisfying $\mathfrak{F}_t\subset\mathfrak{F}$. Let $L_2(\Omega, \mathfrak{F}_T, \mathbb{R}^n)$ be the Hilbert space of all $\mathfrak{F}_T-$ measurable square-integrable random variables with values in $\mathbb{R}^n$. Let $L_2^{\mathfrak{F}}(J,\mathbb{R}^n)$ be the Hilbert space of all square integrable and $\mathfrak{F}_t-$ measurable processes with values in $\mathbb{R}^n$. Let $\mathcal{B}=C(J,\mathbb{R}^n)$ be the Banach space. Denote the class of $\mathbb{R}^n-$ valued stochastic processes $\{\xi(t):t\in J\}$ which are $\mathfrak{F}_t-$ adapted and have a finite second moments, i.e.,

$$
\|\xi\| = \sup_t \left(\mathbb{E}|\xi(t)|^2\right)^{\frac{1}{2}} < \infty .
$$

For convenience, define the following notations

$$
n_1 = \sup\left\|E_{q-p}(At^{q-p})x_0\right\|^2, n_2 = \sup\left\|At^{q-p}E_{q-p,q-p+1}(At^{q-p})x_0\right\|^2,
$$

$$
n_3 = \sup\left\|tE_{q-p,2}(At^{q-p})x_0'\right\|^2, n_4 = \sup\left\|E_{q-p,q}(A(t-s)^{q-p})\right\|^2, l = \left\|\mathbb{W}^{-1}\right\|^2.
$$

8.2 PRELIMINARIES

Let $\mathbb{R}^n$ be the n-dimensional Euclidean space, $\mathbb{R}_+ = (0, \infty)$, $f(t)$ be a suitable function and the fractional order $q>0$, with $n-1<q\leq n$, $n\in\mathbb{N}$. Then the following results are well known (for more details, see Miller and Ross, 1993; Oldham and Spanier, 1974; Samko et al., 1993; Sabatier et al., 2007; Kilbas et al., 2006; Kexue and Jigen, 2011).

Caputo fractional derivative is defined by

$$
({}^{C}D^{q}f)(t) = (I_{0+}^{n-q}D^{n}f)(t),
$$

where

$$(I^q_{0+}f)(t) = \frac{1}{\Gamma(q)}\int_0^t (t-s)^{q-1}f(s)ds.$$

Its Laplace transform is given by

$$\mathcal{L}\{^C D^q f(t)\}(s) = s^q \mathcal{F}(s) - \sum_{k=0}^{n-1} f^{(k)}(0^+)s^{q-1-k}.$$

In particular, if $0 < q \leq 1$, then

$$\mathcal{L}\{^C D^q f(t)\}(s) = s^q \mathcal{F}(s) - f(0^+)s^{q-1},$$

and, if $1 < q \leq 2$, then

$$\mathcal{L}\{^C D^q f(t)\}(s) = s^q \mathcal{F}(s) - f(0^+)s^{q-1} - f'(0^+)s^{q-2}.$$

For an $n \times n$ matrix A the Mittag–Leffler function is defined by

$$E_{q,p}(A) = \sum_{k=0}^{\infty} \frac{A^k}{\Gamma(kq+p)}.$$

For positive q, p, its Laplace transform is given by

$$\mathcal{L}\left\{t^{p-1}E_{q,p}(\pm At^q)\right\}(s) = \frac{s^{q-p}}{(s^q I \mp A)}.$$

Consider the following linear fractional deterministic control system

$$\begin{aligned} &^C D^q x(t) = A\,^C D^p x(t) + Bu(t) + f(t), \quad t \in J, \\ &x(0) = x_0, \quad x'(0) = x_0', \end{aligned} \tag{8.2}$$

where p, q, x, u, A and B are defined as above, and $f: J \to \mathbb{R}^n$ is a continuous function. In order to find the solution of the above problem (8.2), take the Laplace transform on both sides, one can get

$$s^q \mathcal{X}(s) - s^{q-1}x(0) - s^{q-2}x'(0) - As^p \mathcal{X}(s) + As^{p-1}x(0) = B\mathcal{U}(s) + \mathcal{F}(s).$$

Substituting the Laplace transformation of the Mittag–Leffler function and the Laplace convolution operator, one can obtain the solution of the given problem as

$$x(t) = E_{q-p}(At^{q-p})x_0 - At^{q-p}E_{q-p,q-p+1}(At^{q-p})x_0 + tE_{q-p,2}(At^{q-p})x_0'$$
$$+ \int_0^t (t-s)^{q-1}E_{q-p,q}(A(t-s)^{q-p})[Bu(s) + f(s)]ds.$$

Similarly, it is easy to obtain the following solution of Eq. (8.1) as a function x, defined on J with $x(0) = x_0$, $x'(0) = x_0'$ such that there exists $v(t) \in F(t, x(t))$ almost everywhere on J

$$x(t) = E_{q-p}(At^{q-p})x_0 - At^{q-p}E_{q-p,q-p+1}(At^{q-p})x_0 + tE_{q-p,2}(At^{q-p})x_0'$$
$$+ \int_0^t (t-s)^{q-1}E_{q-p,q}(A(t-s)^{q-p})\left[Bu(s) + v(s) + \int_0^s G(\theta, x(\theta))dw(\theta)\right]ds.$$

The system (8.2) is controllable on J if and only if the controllability Grammian matrix

$$\mathbb{W} = \int_0^T (T-s)^{q-1}[E_{q-p,q}(A(T-s)^{q-p})B][E_{q-p,q}(A(T-s)^{q-p})B]^* ds$$

is positive definite, for some $T > 0$ (see, Sakthivel et al., 2013). Here B^* denotes the transpose of B.

Lemma 8.1: (Mahmudov, 2003) *Let $G:J \times \Omega \to L_2^0$ be a strongly measurable mapping such that*

$$\int_0^T \mathbb{E}\|G(t)\|_{L_2^0}^p dt < \infty.$$

Then

$$\mathbb{E}\left\|\int_0^t G(s)dw(s)\right\|^p \leq L_G \int_0^t \mathbb{E}\|G(s)\|_{L_2^0}^p ds,$$

for all $t \in J$ and $p \geq 2$, where L_G is the constant involving p and T.

Let $(X, \|\cdot\|)$ be a Banach space. Denote $\mathcal{P}_{cl}(X) = \{Y \in \mathcal{P}(X): Y \text{ is closed}\}$, $\mathcal{P}_{bd}(X) = \{Y \in \mathcal{P}(X): Y \text{ is bounded}\}$, $\mathcal{P}_{cp}(X) = \{Y \in \mathcal{P}(X): Y \text{ is compact}\}$, and $\mathcal{P}_{cp,cv}(X) = \{Y \in \mathcal{P}(X): Y \text{ is compact and convex}\}$.

For more details on multivalued maps readers can refer to the books (Aubin and Cellina, 1984; Gorniewicz, 1999; Hu and Papageorgiou, 2013; Kisielewicz, 1991).

Definition 8.1: *A multivalued map $\Phi: X \to \mathcal{P}(X)$ is convex (closed) valued if $\Phi(x)$ is convex (closed) for all $x \in X$. Φ is bounded on bounded sets if $\Phi(B) = \cup_{x\in B}\Phi(x)$ is bounded in X for all $B \in \mathcal{P}_{bd}(X)$(i.e., $\sup_{x\in B}\sup_{y\in\Phi(x)} \|y\| < \infty$).*

Definition 8.2: *Φ is called upper semicontinuous (u.s.c.) on X if for each $x_0 \in X$, the set $\Phi(x_0)$ is a nonempty closed subset of X, and if for each open set N of X containing $\Phi(x_0)$, there exists an open neighborhood N_0 of x_0 such that $\Phi(N_0) \subseteq N$.*

Definition 8.3: *Φ is said to be completely continuous if $\Phi(B)$ is relatively compact for every $B \in \mathcal{P}_{bd}(X)$. Φ has a fixed point if there is $x \in X$ such that $x \in \Phi(x)$. The fixed point set of the multivalued operator Φ will be denoted by FixΦ.*

Definition 8.4: *A multivalued map* $\Phi:X\to\mathcal{P}_{cl}(\mathbb{R}^n)$ *is said to be measurable if for every* $v\in\mathbb{R}^n$, *the function* $x\mapsto d(v,\Phi(x))=\inf\{\|v-z\|:z\in\Phi(x)\}$ *is measurable.*

For each $x\in L_2(J,\mathbb{R}^n)$, define the set of selections of F by

$$v\in N_{F,x}=\{v\in L_2(J,\mathbb{R}^n):v(t)\in F(t,x(t))\ \text{for a.e. } t\in J\}.$$

Lemma 8.2: (Hu and Papageorgiou, 2013) *Let* Φ *be a completely continuous multivalued map with nonempty compact values, then* Φ *is u.s.c. if and only if* Φ *has a closed graph (i.e.,* $x_n\to x, y_n\to y, y_n\in\Phi(x_n)$ *imply* $y\in\Phi(x)$*).*

Definition 8.5: *A multivalued map* $\Phi:J\times\mathbb{R}^n\to\mathcal{P}(\mathbb{R}^n)$ *is said to be* L_2- *Carathéodory if*

1. $t\mapsto\Phi(t,x)$ *is measurable for each* $x\in\mathbb{R}^n$,
2. $x\mapsto\Phi(t,x)$ *is u.s.c. for almost all* $t\in J$,
3. *for each* $\rho>0$, *there exists* $\varphi_\rho\in L^1(J,\mathbb{R}^+)$ *such that*

$$\|\Phi(t,x)\|^2:=\sup\{\mathbb{E}\|v\|^2:v\in\Phi(t,x)\}\le\varphi_\rho(t)\ \text{for all } \|x\|^2_{\mathbb{R}^n}\le\rho \text{ and for a.e } t\in J.$$

Lemma 8.3: (Lasota and Opial, 1965) *Let* X *be a Banach space. Let* $\Phi:J\times X\to\mathcal{P}_{cp,cv}(X)$ *be a* L_2- *Carathéodory multivalued map with* $N_{F,x}\neq\phi$ *and let* Λ *be a linear continuous mapping from* $L_2(J,X)$ *to* $\mathcal{C}(J,X)$, *then the operator*

$$\Lambda\circ N_F:\mathcal{C}(J,X)\to\mathcal{P}_{cp,cv}(\mathcal{C}(J,X)),\quad x\mapsto(\Lambda\circ N_F)(x):=\Lambda(N_{F,x})$$

is a closed graph operator in $\mathcal{C}(J,X)\times\mathcal{C}(J,X)$.

Proposition 8.1: (Castaing and Valadier, 2006) *Let* X *be a separable Banach space. Let* $F_1,F_2:J\to\mathcal{P}_{cp}(X)$ *be measurable multivalued maps, then the multivalued map* $t\mapsto F_1(t)\cap F_2(t)$ *is measurable.*

Theorem 8.1: (Castaing and Valadier, 2006) *Let* X *be a separable metric space,* $(\Phi,\mathcal{L})$ *be a measurable space,* F *is a multivalued map from* Φ *to complete nonempty subset of* X. *If for each open set* U *in* X, $\overline{F}(U)=\{t:F(t)\cap U\neq\phi\}\in\mathcal{L}$, *then* F *admits a measurable selection.*

Now introduce the set of all states attainable from x_0 in time $t>0$ denoted as $\mathcal{R}_t(x_0)=\{x(t;x_0,u):u(\cdot)\in L_2(\Omega,\mathfrak{F}_T,\mathbb{R}^m)\}$, where $x(t,x_0,u)$ is the solution of the system (8.1) corresponding to $x_0\in\mathbb{R}^n$, $u(\cdot)\in L_2(\Omega,\mathfrak{F}_T,\mathbb{R}^m)$.

Definition 8.6: *The stochastic system* (8.1) *is completely controllable on* J *if*

$$\mathcal{R}_T(x_0)=L_2(\Omega,\mathfrak{F}_T,\mathbb{R}^n),$$

that is if all points in $L_2(\Omega,\mathfrak{F}_T,\mathbb{R}^n)$ *can be reached from the point* x_0 *at time* T.

Let (X,d) be a metric space induced from $(X,\|\cdot\|)$ be a square normed space. Consider $H_d:\mathcal{P}(X)\times\mathcal{P}(X)\rightarrow\mathbb{R}_+\cup\{\infty\}$ given by

$$H_d(A,B)=\max\{\sup_{a\in A} d(a,B),\sup_{b\in B} d(A,b)\}$$

where

$$d(A,b)=\inf_{a\in A} d(a,b),\quad d(a,B)=\inf_{b\in B} d(a,b).$$

Then $(\mathcal{P}_{bd,cl}(X),H_d)$ is a metric space and $(\mathcal{P}_{cl}(X),H_d)$ is a generalized metric space (see Kisielewicz, 1991).

Definition 8.7: *A multivalued operator $\Phi:X\rightarrow\mathcal{P}_{cl}(X)$ is called*

1. $\gamma-$ *Lipschitz if and only if there exists $\gamma>0$ such that*

$$H_d(\Phi(x),\Phi(y))\leq\gamma d(x,y)$$

for each $x,y\in X$,

2. *a contraction if and only if it is $\gamma-$ Lipschitz with $\gamma<1$.*

Lemma 8.4: (Bohnenblust and Karlin, 1950). *Let X be a Banach space and $K\in\mathcal{P}_{cl,cv}(X)$ and suppose that the operator $\Phi:K\rightarrow\mathcal{P}_{cl,cv}(K)$ is upper semi-continuous and the set $\Phi(K)$ is relatively compact in X. Then Φ has a fixed point in K.*

Lemma 8.5: (Covitz and Nadler, 1970). *Let (X,d) be a complete metric space. If $\Phi:X\rightarrow\mathcal{P}_{cl}(X)$ is a contraction, then Φ has fixed points.*

8.3 MAIN RESULTS

In this section, the controllability criteria are discussed for the considered system (8.1). In order to prove the controllability results assume the following hypotheses hold

(H1) The multivalued map $F:J\times\mathbb{R}^n\rightarrow\mathcal{P}(\mathbb{R}^n)$ be an L_2- Carathéodory function and satisfies the following conditions: for each $t\in J$, the function $F(t,\cdot):\mathbb{R}^n\rightarrow\mathcal{P}_{bd,cl,cv}(\mathbb{R}^n)$ is u.s.c. and for each $x\in\mathbb{R}^n$ the function $F(\cdot,x)$ is measurable; for each fixed $x\in\mathbb{R}^n$ the set

$$N_{F,x}=\{v\in L_2(J,\mathbb{R}^n):v(t)\in F(t,x(t))\ \text{for a.e}\ t\in J\}$$

is nonempty. There exists a positive function $\varphi_\rho:J\rightarrow\mathbb{R}_+$ such that

$$\sup\left\{\mathbb{E}\|v(t)\|^2:v(t)\in F(t,x(t))\right\}\leq\varphi_\rho(t)$$

for a.e $t\in J$ and

$$\lim_{\rho\rightarrow\infty}\inf\frac{\varphi_\rho(t)}{\rho}=\Upsilon<\infty.$$

(**H2**) The function G is continuous and there exists constant M_1 such that,

$$\left\|G(t,x)\right\|^2 \leq M_1(1+\left\|x\right\|^2)$$

(**H4**) The multifunction $F{:}J\times\mathbb{R}^n\to\mathcal{P}_{cp}(\mathbb{R}^n)$ has the property that $F(\cdot,x){:}J\to\mathcal{P}_{cp}(\mathbb{R}^n)$ is measurable for each $x\in\mathbb{R}^n$.

(**H5**) There exists a nonnegative function $m\in L_2(J)$ such that

$H_d(F(t,x_1),F(t,x_2))\leq m(t)\left[\left\|x_1-x_2\right\|^2\right]$ for every $x_1,x_2\in\mathbb{R}^n$, and $d(0,F(t,0))\leq m(t)$ a.e. $t\in J$.

Theorem 8.2: *(Convex Case) Suppose that the hypotheses (H1)–(H2) are satisfied, then the system* (8.1) *has at least one solution, provided that the following holds:*

$$\begin{aligned}1>\;&6n_1\left(1+6\frac{T^{2q}}{q^2}n_4^2\left\|B\right\|^2\left\|B^*\right\|^2 l\right)+36\frac{T^{2q}}{q^2}n_4^2\left\|B\right\|^2\left\|B^*\right\|^2 l\mathbb{E}\left\|x_1\right\|^2\\&+6n_2\left(1+6\frac{T^{2q}}{q^2}n_4^2\left\|B\right\|^2\left\|B^*\right\|^2 l\right)+6n_3\left(1+6\frac{T^{2q}}{q^2}n_4^2\left\|B\right\|^2\left\|B^*\right\|^2 l\right)\\&+6\frac{T^{2q}}{q^2}n_4\Upsilon\left(1+6\frac{T^{2q}}{q^2}n_4^2\left\|B\right\|^2\left\|B^*\right\|^2 l\right)+6\left(1+6\frac{T^{2q}}{q^2}n_4^2\left\|B\right\|^2\left\|B^*\right\|^2 l\right)\\&\times\frac{T^{2q}}{q^2}Tn_4M_1\widehat{M}_G\left(1+\sup_{t\in J}\mathbb{E}\left\|x(t)\right\|^2\right).\end{aligned} \tag{8.3}$$

Proof: For any arbitrary function $x\in\mathbb{R}^n$, define the following control function

$$\begin{aligned}u_x(t)=&B^*E_{q-p,q}(A^*(T-t)^{q-p})\mathbb{W}^{-1}\Big\{\mathbb{E}x_1-E_{q-p}(AT^{q-p})x_0\\&+AT^{q-p}E_{q-p,q-p+1}(AT^{q-p})x_0-TE_{q-p,2}(AT^{q-p})x_0'\\&-\int_0^T(T-s)^{q-1}E_{q-p,q}(A(T-s)^{q-p})\left(v(s)+\int_0^s G(\theta,x(\theta))dw(\theta)\right)ds\Big\}\end{aligned}$$

where $t\in J$, $v\in N_{F,x}$. Using the above control, first show that the operator $\Phi{:}\mathcal{B}\to\mathcal{P}(\mathcal{B})$, defined as

$$\begin{aligned}\Phi(x)=&\Big\{\Psi\in\mathcal{B}{:}\Psi(t)=E_{q-p}(At^{q-p})x_0-At^{q-p}E_{q-p,q-p+1}(At^{q-p})x_0+tE_{q-p,2}(At^{q-p})x_0'\\&+\int_0^t(t-s)^{q-1}E_{q-p,q}(A(t-s)^{q-p})\left[Bu_x(s)+v(s)+\int_0^s G(\theta,x(\theta))dw(\theta)\right]ds\Big\}\end{aligned}$$

has a fixed point, which is a solution of the system (8.1).

One needs to show that Φ satisfies all the conditions of Lemma 8.4. For the sake of convenience, subdivide the proof into four steps.

Step 1. For each $x\in\mathcal{B}$, Φ is convex.

In fact, if $\Psi_1,\Psi_2\in\Phi(x)$, then for each $t\in J$, there exists $v_1,v_2\in N_{F,x}$ such that

$$\begin{aligned}\Psi_i(t)=&E_{q-p}(At^{q-p})x_0-At^{q-p}E_{q-p,q-p+1}(At^{q-p})x_0+tE_{q-p,2}(At^{q-p})x_0'\\&+\int_0^t(t-s)^{q-1}E_{q-p,q}(A(t-s)^{q-p})B\Bigg[B^*E_{q-p,q}(A^*(T-s)^{q-p})\mathbb{W}^{-1}\\&\Bigg\{(\mathbb{E}x_1-E_{q-p}(AT^{q-p})x_0+AT^{q-p}E_{q-p,q-p+1}(AT^{q-p})x_0-TE_{q-p,2}(AT^{q-p})x_0'\\&-\int_0^T(T-s)^{q-1}E_{q-p,q}(A(T-s)^{q-p})\\&\times\left(v_i(s)+\int_0^s G(\theta,x(\theta))dw(\theta)\right)ds\Bigg\}\Bigg](s)ds+\int_0^t(t-s)^{q-1}E_{q-p,q}(A(t-s)^{q-p})\\&\times\left[v_i(s)+\int_0^s G(\theta,x(\theta))dw(\theta)\right]ds,\ i=1,2.\end{aligned}$$

Let $0\le\lambda\le1$, then for each $t\in J$, one can have

$$\begin{aligned}[\lambda\Psi_1+(1-\lambda)\Psi_2](t)=&E_{q-p}(At^{q-p})x_0-At^{q-p}E_{q-p,q-p+1}(At^{q-p})x_0+tE_{q-p,2}(At^{q-p})x_0'\\&+\int_0^t(t-s)^{q-1}E_{q-p,q}(A(t-s)^{q-p})B\Bigg[B^*E_{q-p,q}(A^*(T-s)^{q-p})\mathbb{W}^{-1}\\&\Bigg\{\mathbb{E}x_1-E_{q-p}(AT^{q-p})x_0\\&+AT^{q-p}E_{q-p,q-p+1}(AT^{q-p})x_0-TE_{q-p,2}(AT^{q-p})x_0'\\&-\int_0^T(T-s)^{q-1}E_{q-p,q}(A(T-s)^{q-p})\\&\times\left[(\lambda v_1(s)+(1-\lambda)v_2(s))+\int_0^s G(\theta,x(\theta))dw(\theta)\right]ds\Bigg\}\Bigg](s)ds\\&+\int_0^t(t-s)^{q-1}E_{q-p,q}(A(t-s)^{q-p})\\&\times\left[(\lambda v_1(s)+(1-\lambda)v_2(s))+\int_0^s G(\theta,x(\theta))dw(\theta)\right]ds.\end{aligned}$$

It is easy to see that $N_{F,x}$ is convex since F has convex values. So, $\lambda v_1 + (1-\lambda)v_2 \in N_{F,x}$. Thus, $\lambda\Psi_1 + (1-\lambda)\Psi_2 \in \Phi(x)$.

Step 2. For each positive number $\rho > 0$, let $\mathcal{B}_\rho = \{x \in \mathcal{B} : \|x\|_{\mathcal{B}}^2 \leq \rho\}$. Obviously, $\mathcal{B}_\rho$ is a bounded, closed, and convex set of $\mathcal{B}$. It is to claim that there exists a positive number ρ such that $\Phi(\mathcal{B}_\rho) \subset \mathcal{B}_\rho$.

If this is not true, then for each positive number ρ, there exists a function $x_\rho \in \mathcal{B}_\rho$, but $\Phi(x_\rho) \notin \mathcal{B}_\rho$ that is $\|\Phi(x_\rho)\|_{\mathcal{B}}^2 \equiv \sup\{\|\Psi\|_{\mathcal{B}}^2 : \Psi_\rho \in (\Phi x_\rho)\} > \rho$ and

$$\Psi_\rho(t) = E_{q-p}(At^{q-p})x_0 - At^{q-p}E_{q-p,q-p+1}(At^{q-p})x_0 + tE_{q-p,2}(At^{q-p})x_0' \\ + \int_0^t (t-s)^{q-1}E_{q-p,q}(A(t-s)^{q-p})\left[Bu_\rho(s) + v_\rho(s) + \int_0^s G(\theta, x_\rho(\theta))dw(\theta)\right]ds$$

for some $v_\rho \in N_{F,x_\rho}$. Using Lemma 8.4 one can have

$$\mathbb{E}\|u_\rho(t)\|^2 \leq 6\|B^*\|^2\|E_{q-p,q}(A^*(T-t)^{q-p})\|^2\|\mathbb{W}^{-1}\|^2\Big\{\{\mathbb{E}\|x_1\|^2 + \|E_{q-p}(AT^{q-p})x_0\|^2 \\ + \|AT^{q-p}E_{q-p,q-p+1}(AT^{q-p})x_0\|^2 + \|TE_{q-p,2}(AT^{q-p})x_0'\|^2 \\ + \mathbb{E}\left\|\int_0^T (T-s)^{q-1}E_{q-p,q}(A(T-s)^{q-p})v(s)ds\right\|^2 \\ + \mathbb{E}\left\|\int_0^T (T-s)^{q-1}E_{q-p,q}(A(T-s)^{q-p})\left(\int_0^s G(\theta, x(\theta))dw(\theta)\right)ds\right\|^2\Big\} \\ \leq 6\|B^*\|^2\|E_{q-p,q}(A^*(T-t)^{q-p})\|^2\|\mathbb{W}^{-1}\|^2\{\mathbb{E}\|x_1\|^2 + \|E_{q-p}(AT^{q-p})x_0\|^2 \\ + \|AT^{q-p}E_{q-p,q-p+1}(AT^{q-p})x_0\|^2 + \|TE_{q-p,2}(AT^{q-p})x_0'\|^2 \\ + \frac{T^{2q}}{q^2}\|E_{q-p,q}(A(T-t)^{q-p})\|^2\|v(t)\|^2 \\ + \frac{T^{2q}}{q^2}\|E_{q-p,q}(A(T-t)^{q-p})\|^2\left(\int_0^s G(\theta, x(\theta))dw(\theta)\right)ds\Big\} \\ \leq 6\|B^*\|^2 n_4 l\Big\{\mathbb{E}\|x_1\|^2 + n_1 + n_2 + n_3 + \frac{T^{2q}}{q^2}n_4\varphi_\rho(t) \\ + \frac{T^{2q}}{q^2}Tn_4M_1\widehat{M}_G\left(1 + \sup_{t\in J}\mathbb{E}\|x(t)\|^2\right)\Big\}$$

also find that

$$
\begin{aligned}
\rho<&\mathbb{E}\left\|\Phi(x_\rho)(t)\right\|^2\\
&=\Big\|E_{q-p}(At^{q-p})x_0-At^{q-p}E_{q-p,q-p+1}(At^{q-p})x_0+tE_{q-p,2}(At^{q-p})x_0'\\
&+\int_0^t(t-s)^{q-1}E_{q-p,q}(A(t-s)^{q-p})Bu_\rho(s)ds+\int_0^t(t-s)^{q-1}E_{q-p,q}(A(t-s)^{q-p})v_\rho(s)ds\\
&+\int_0^t(t-s)^{q-1}E_{q-p,q}(A(t-s)^{q-p})\left(\int_0^s G(\theta,x_\rho(\theta))dw(\theta)\right)ds\Big\|^2\\
&\le 6\left\|E_{q-p}(At^{q-p})x_0\right\|^2+6\left\|At^{q-p}E_{q-p,q-p+1}(At^{q-p})x_0\right\|^2\\
&+6\left\|tE_{q-p,2}(At^{q-p})x_0'\right\|^2+6\mathbb{E}\left\|\int_0^t(t-s)^{q-1}E_{q-p,q}(A(t-s)^{q-p})Bu_\rho(s)ds\right\|^2\\
&+6\mathbb{E}\left\|\int_0^t(t-s)^{q-1}E_{q-p,q}(A(t-s)^{q-p})v_\rho(s)ds\right\|^2\\
&+6\mathbb{E}\left\|\int_0^t(t-s)^{q-1}E_{q-p,q}(A(t-s)^{q-p})\left(\int_0^s G(\theta,x_\rho(\theta))dw(\theta)\right)ds\right\|^2\\
&\le 6n_1+6n_2+6n_3+36\frac{T^{2q}}{q^2}n_4^2\|B\|^2\|B^*\|^2 l\Big\{\mathbb{E}\|x_1\|^2+n_1+n_2+n_3\\
&+\frac{T^{2q}}{q^2}n_4\varphi_\rho(t)+\frac{T^{2q}}{q^2}Tn_4M_1\widehat{M}_G\left(1+\sup_{t\in J}\mathbb{E}\|x_\rho(t)\|^2\right)\Big\}+6\frac{T^{2q}}{q^2}n_4\varphi_\rho(t)\\
&+6\frac{T^{2q}}{q^2}Tn_4M_1\widehat{M}_G\left(1+\sup_{t\in J}\mathbb{E}\|x_\rho(t)\|^2\right)\\
&\le 6n_1\left(1+6\frac{T^{2q}}{q^2}n_4^2\|B\|^2\|B^*\|^2 l\right)+36\frac{T^{2q}}{q^2}n_4^2\|B\|^2\|B^*\|^2 l\mathbb{E}\|x_1\|^2\\
&+6n_2\left(1+6\frac{T^{2q}}{q^2}n_4^2\|B\|^2\|B^*\|^2 l\right)+6n_3\left(1+6\frac{T^{2q}}{q^2}n_4^2\|B\|^2\|B^*\|^2 l\right)\\
&+6\frac{T^{2q}}{q^2}n_4\varphi_\rho(t)\left(1+6\frac{T^{2q}}{q^2}n_4^2\|B\|^2\|B^*\|^2 l\right)+6\left(1+6\frac{T^{2q}}{q^2}n_4^2\|B\|^2\|B^*\|^2 l\right)\\
&\times\frac{T^{2q}}{q^2}Tn_4M_1\widehat{M}_G\left(1+\sup_{t\in J}\mathbb{E}\|x_\rho(t)\|^2\right).
\end{aligned}
$$

Dividing both sides of the above inequality by ρ and taking the limit as $\rho\to\infty$, one can get

$$1 \leq 6n_1\left(1+6\frac{T^{2q}}{q^2}n_4^2\left\|B\right\|^2\left\|B^*\right\|^2 l\right)+36\frac{T^{2q}}{q^2}n_4^2\left\|B\right\|^2\left\|B^*\right\|^2 l\mathbb{E}\left\|x_1\right\|^2$$

$$+6n_2\left(1+6\frac{T^{2q}}{q^2}n_4^2\left\|B\right\|^2\left\|B^*\right\|^2 l\right)$$

$$+6n_3\left(1+6\frac{T^{2q}}{q^2}n_4^2\left\|B\right\|^2\left\|B^*\right\|^2 l\right)+6\frac{T^{2q}}{q^2}n_4\Upsilon\left(1+6\frac{T^{2q}}{q^2}n_4^2\left\|B\right\|^2\left\|B^*\right\|^2 l\right)$$

$$+6\left(1+6\frac{T^{2q}}{q^2}n_4^2\left\|B\right\|^2\left\|B^*\right\|^2 l\right)\frac{T^{2q}}{q^2}Tn_4M_1\widehat{M}_G\left(1+\sup_{t\in J}\mathbb{E}\left\|x(t)\right\|^2\right)$$

which is a contradiction to (8.3). Hence, for some $\rho > 0$, $\Phi(\mathcal{B}_\rho) \subset \mathcal{B}_\rho$.

Step 3. Compactness of Φ.

To prove this, first show that the set $\Phi(\mathcal{B}_\rho)$ is relatively compact in $\mathcal{B}$. Subsequently, one must show that $\Phi(\mathcal{B}_\rho)$ is uniformly bounded. Note that by using the same method as in the above step it can be manifested that the operator Φ is uniformly bounded, i.e.,

$$6n_1\left(1+6\frac{T^{2q}}{q^2}n_4^2\left\|B\right\|^2\left\|B^*\right\|^2 l\right)+36\frac{T^{2q}}{q^2}n_4^2\left\|B\right\|^2\left\|B^*\right\|^2 l\mathbb{E}\left\|x_1\right\|^2$$

$$+6n_2\left(1+6\frac{T^{2q}}{q^2}n_4^2\left\|B\right\|^2\left\|B^*\right\|^2 l\right)$$

$$+6n_3\left(1+6\frac{T^{2q}}{q^2}n_4^2\left\|B\right\|^2\left\|B^*\right\|^2 l\right)+6\frac{T^{2q}}{q^2}n_4\Upsilon\left(1+6\frac{T^{2q}}{q^2}n_4^2\left\|B\right\|^2\left\|B^*\right\|^2 l\right)$$

$$+6\left(1+6\frac{T^{2q}}{q^2}n_4^2\left\|B\right\|^2\left\|B^*\right\|^2 l\right)\frac{T^{2q}}{q^2}Tn_4M_1\widehat{M}_G\left(1+\sup_{t\in J}\mathbb{E}\left\|x(t)\right\|^2\right) < \infty$$

and the set $\Phi(\mathcal{B}_\rho)$ is relatively compact. Finally, one needs to prove that $\Phi(\mathcal{B}_\rho)$ is equicontinuous. For any $x \in \mathcal{B}_\rho$ and $t_1, t_2 \in J$ with $0 < t_1 < t_2 \leq T$, then

$$
\begin{aligned}
\mathbb{E}\left\|\Psi(t_1)-\Psi(t_2)\right\|^2 \le\; & 9\mathbb{E}\left\|[E_{q-p}(At_1^{q-p})-E_{q-p}(At_2^{q-p})]x_0\right\|^2 \\
&+9\mathbb{E}\left\|[At_1^{q-p}E_{q-p,q-p+1}(At_1^{q-p})-At_2^{q-p}E_{q-p,q-p+1}(At_2^{q-p})]x_0\right\|^2 \\
&+9\mathbb{E}\left\|[t_1E_{q-p,2}(At_1^{q-p})-t_2E_{q-p,2}(At_2^{q-p})]x_0'\right\|^2 \\
&+9\mathbb{E}\left\|\int_{t_1}^{t_2}(t_2-s)^{q-1}E_{q-p,q}(A(t_2-s)^{q-p})Bu_x(s)ds\right\|^2 \\
&+9\mathbb{E}\left\|\int_0^{t_1}[(t_1-s)^{q-1}E_{q-p,q}(A(t_1-s)^{q-p})-(t_2-s)^{q-1}E_{q-p,q}(A(t_2-s)^{q-p})]Bu_x(s)ds\right\|^2 \\
&+9\mathbb{E}\left\|\int_{t_1}^{t_2}(t_2-s)^{q-1}E_{q-p,q}(A(t_2-s)^{q-p})v(s)ds\right\|^2 \\
&+9\mathbb{E}\left\|\int_0^{t_1}[(t_1-s)^{q-1}E_{q-p,q}(A(t_1-s)^{q-p})-(t_2-s)^{q-1}E_{q-p,q}(A(t_2-s)^{q-p})]v(s)ds\right\|^2 \\
&+9\mathbb{E}\left\|\int_{t_1}^{t_2}(t_2-s)^{q-1}E_{q-p,q}(A(t_2-s)^{q-p})\left(\int_0^s G(\theta,x(\theta))dw(\theta)\right)ds\right\|^2 \\
&+9\mathbb{E}\left\|\int_0^{t_1}[(t_1-s)^{q-1}E_{q-p,q}(A(t_1-s)^{q-p})-(t_2-s)^{q-1}E_{q-p,q}(A(t_2-s)^{q-p})]\right. \\
&\quad\left.\times\left(\int_0^s G(\theta,x(\theta))dw(\theta)\right)ds\right\|^2 \\
\le\; & 9\mathbb{E}\left\|[E_{q-p}(At_1^{q-p})-E_{q-p}(At_2^{q-p})]x_0\right\|^2 \\
&+9\mathbb{E}\left\|[At_1^{q-p}E_{q-p,q-p+1}(At_1^{q-p})-At_2^{q-p}E_{q-p,q-p+1}(At_2^{q-p})]x_0\right\|^2 \\
&+9\mathbb{E}\left\|[t_1E_{q-p,2}(At_1^{q-p})-t_2E_{q-p,2}(At_2^{q-p})]x_0'\right\|^2 \\
&+9\frac{(t_2-t_1)^q}{q}\int_{t_1}^{t_2}(t_2-s)^{q-1}\left\|E_{q-p,q}(A(t_2-s)^{q-p})\right\|^2\left\|B\right\|^2\mathbb{E}\left\|u_x(s)\right\|^2ds \\
&+9t_1\int_0^{t_1}\left\|[(t_1-s)^{q-1}E_{q-p,q}(A(t_1-s)^{q-p})-(t_2-s)^{q-1}E_{q-p,q}(A(t_2-s)^{q-p})]\right\|^2\left\|B\right\|^2 \\
&\times\mathbb{E}\left\|u_x(s)\right\|^2ds+9\frac{(t_2-t_1)^q}{q}\int_{t_1}^{t_2}(t_2-s)^{q-1}\left\|E_{q-p,q}(A(t_2-s)^{q-p})\right\|^2\varphi_\rho(s)ds \\
&+9t_1\int_0^{t_1}\left\|[(t_1-s)^{q-1}E_{q-p,q}(A(t_1-s)^{q-p})-(t_2-s)^{q-1}E_{q-p,q}(A(t_2-s)^{q-p})]\right\|^2\varphi_\rho(s)ds \\
&+9\frac{(t_2-t_1)^q}{q}\int_{t_1}^{t_2}(t_2-s)^{q-1}\left\|E_{q-p,q}(A(t_2-s)^{q-p})\right\|^2TM_1\widehat{M}_G\left(1+\mathbb{E}\left\|x(s)\right\|^2\right)ds \\
&+9t_1\int_0^{t_1}\left\|[(t_1-s)^{q-1}E_{q-p,q}(A(t_1-s)^{q-p})-(t_2-s)^{q-1}E_{q-p,q}(A(t_2-s)^{q-p})]\right\|^2 \\
&\times TM_1\widehat{M}_G\left(1+\mathbb{E}\left\|x(s)\right\|^2\right)ds.
\end{aligned}
$$

As $t_1 \to t_2$, the right-hand side of the above inequality tends to zero. An application of the Arzela-Ascoli theorem yields that Φ maps $\mathcal{B}_\rho$ into $\mathcal{B}$, i.e., $\Phi:\mathcal{B}_\rho \to \mathcal{P}(\mathcal{B})$ is a compact operator. Thus, $\Phi(\mathcal{B}_\rho)$ is relatively compact.

Step 4. Φ is u.s.c. on $\mathcal{B}_\rho$.

Let $x_n \to x_*$ as $n \to \infty$ and $\Psi_n \to \Psi_*$ as $n \to \infty$. Now derive that $\Psi_* \in \Phi(x_*)$. Since $\Psi_n \in \Phi(x_n)$ means that there exists $v_n \in N_{F,x_n}$ such that

$$\Psi_n(t) = E_{q-p}(At^{q-p})x_0 - At^{q-p}E_{q-p,q-p+1}(At^{q-p})x_0 + tE_{q-p,2}(At^{q-p})x_0' + \int_0^t (t-s)^{q-1}E_{q-p,q}(A(t-s)^{q-p})\left[Bu_{x,n}(s) + v_n(s) + \int_0^s G(\theta,x_n(\theta))dw(\theta)\right]ds. \tag{8.4}$$

One must show that there exists $v_* \in N_{F,x*}$ such that

$$\begin{aligned}\Psi_*(t) &= E_{q-p}(At^{q-p})x_0 - At^{q-p}E_{q-p,q-p+1}(At^{q-p})x_0 + tE_{q-p,2}(At^{q-p})x_0' \\ &\quad + \int_0^t (t-s)^{q-1}E_{q-p,q}(A(t-s)^{q-p})B\Big[B^*E_{q-p,q}(A^*(T-s)^{q-p})\mathbb{W}^{-1} \\ &\quad \times \Big\{\mathbb{E}x_1 - E_{q-p}(AT^{q-p})x_0 + AT^{q-p}E_{q-p,q-p+1}(AT^{q-p})x_0 - TE_{q-p,2}(AT^{q-p})x_0' \\ &\quad - \int_0^T (T-s)^{q-1}E_{q-p,q}(A(T-s)^{q-p})\left(v_*(s) + \int_0^s G(\theta,x_*(\theta))dw(\theta)\right)ds\Big\}\Big](s)ds \\ &\quad + \int_0^t (t-s)^{q-1}E_{q-p,q}(A(t-s)^{q-p})\left[v_*(s) + \int_0^s G(\theta,x_*(\theta))dw(\theta)\right]ds.\end{aligned}$$

Consider the continuous operator defined as

$$\Lambda: L_2(J,\mathbb{R}^n) \to \mathcal{B}, \quad v \mapsto \Lambda(v)(t)$$

such that

$$\begin{aligned}\Lambda(v)(t) &= \int_0^t (t-s)^{q-1}E_{q-p,q}(A(t-s)^{q-p})\Big[v(s) - BB^*E_{q-p,q}(A^*(T-s)^{q-p})\mathbb{W}^{-1} \\ &\quad \times \int_0^T (T-s)^{q-1}E_{q-p,q}(A(T-s)^{q-p})v(s)ds\Big](s)ds.\end{aligned}$$

From Lemma 8.3, it follows that $\Lambda \circ N_F$ is a closed graph operator. Clearly, for each $t \in J$, one can have

$$\Big(\Psi_n(t) - E_{q-p}(At^{q-p})x_0 + At^{q-p}E_{q-p,q-p+1}(At^{q-p})x_0 - tE_{q-p,2}(At^{q-p})x_0' - \int_0^t (t-s)^{q-1}E_{q-p,q}(A(t-s)^{q-p})\left[Bu_{x,n}(s) + v_n(s) + \int_0^s G(\theta,x_n(\theta))dw(\theta)\right]ds\Big) \in \Lambda(N_{F,x_n}).$$

Since $y_n \to y_*$, it follows from Lemma 8.3 that, for some $y_* \in N_{F,x_*}$, then

$$\Big(\Psi_*(t) - E_{q-p}(At^{q-p})x_0 + At^{q-p}E_{q-p,q-p+1}(At^{q-p})x_0 - tE_{q-p,2}(At^{q-p})x_0' \\ - \int_0^t (t-s)^{q-1}E_{q-p,q}(A(t-s)^{q-p})\Big[Bu_{x,*}(s) + v_*(s) + \int_0^s G(\theta, x_*(\theta))dw(\theta)\Big]ds\Big) \in \Lambda(N_{F,x_*}).$$

Clearly, for each $t \in J$, one can set

$$\Big\| \Big(\Psi_n(t) - E_{q-p}(At^{q-p})x_0 + At^{q-p}E_{q-p,q-p+1}(At^{q-p})x_0 - tE_{q-p,2}(At^{q-p})x_0' \\ - \int_0^t (t-s)^{q-1}E_{q-p,q}(A(t-s)^{q-p})\Big[Bu_{x,n}(s) + v_n(s) + \int_0^s G(\theta, x_n(\theta))dw(\theta)\Big]ds\Big) \\ - \Big(\Psi_*(t) - E_{q-p}(At^{q-p})x_0 + At^{q-p}E_{q-p,q-p+1}(At^{q-p})x_0 - tE_{q-p,2}(At^{q-p})x_0' \\ - \int_0^t (t-s)^{q-1}E_{q-p,q}(A(t-s)^{q-p})\Big[Bu_{x,*}(s) + v_*(s) + \int_0^s G(\theta, x_*(\theta))dw(\theta)\Big]ds\Big) \Big\|_{\mathcal{B}}^2 \to 0$$

as $n \to \infty$. From Lemma 8.2, it can be concluded that Φ is u.s.c. As a consequence of Lemma 8.4, it is clear that Φ has a fixed point which is the solution of the system (8.1). □

Theorem 8.3: *(Nonconvex Case) Assume that conditions (H4)–(H5) are satisfied, then the system* (8.1) *has at least one solution on* J*, provided that*

$$4m(t)C_2\left(1 + \frac{T^{2q}}{q^2}n_4^2\|B\|^2\|B^*\|^2 l\right) < 1. \tag{8.5}$$

Proof: Under the assumption (H5) it is easy to see that for each $x \in \mathcal{B}$, the set $S_{F,x}$ is nonempty. Therefore, F has a nonempty measurable selection (by Theorem 8.1). First show that Φ defined in Theorem 8.2 satisfies the assumption of Lemma 8.5. The proof will be given in two steps.

Step 1. $\Phi(x) \in \mathcal{P}_{cl}(\mathcal{B})$ for each $x \in \mathcal{B}$.

Indeed, let $(\Psi_n)n \geq 0 \in \Phi(x)$ such that $\Psi_n \to \Psi$. Then, $\Psi \in \mathcal{B}$ and there exists $v_n \in N_{F,x}$ such that, for each $t \in J$, $\Psi_n(t)$ is defined in (8.4). Using (H5), one can have

$$|v_n(t)| \leq m(t) + m(t)\|x\|^2, \quad n \in \mathbb{N}$$

for a.e. $t \in J$.

The Lebesgue dominated convergence theorem implies that

$$\|v_n - v\|_{L^2} \to 0 \text{ as } n \to \infty.$$

Hence $v \in N_{F,x}$. Then, for each $t \in J$, $\Psi_n(t) \to \Psi(t)$, where

$$\Psi(t) = E_{q-p}(At^{q-p})x_0 - At^{q-p}E_{q-p,q-p+1}(At^{q-p})x_0 + tE_{q-p,2}(At^{q-p})x_0' + \int_0^t (t-s)^{q-1}E_{q-p,q}(A(t-s)^{q-p})\left[Bu_{x,1}(s) + v(s) + \int_0^s G(\theta, x_1(\theta))dw(\theta)\right]ds. \tag{8.6}$$

So, $\Psi \in \Phi(x)$.

Step 2. There exists $\gamma < 1$ such that $H_d(\Phi(x_1), \Phi(x_2)) \leq \gamma \|x_1 - x_2\|_{\mathcal{B}}$ for each $x_1, x_2 \in \mathcal{B}$. Let $x_1, x_2 \in \mathcal{B}$ and $\Psi \in \Phi(x)$. Then, there exists $v \in N_{F,x_1}$ such that $\Psi(t)$ is defined in (8.6). From (H5), it follows that

$$H_d(F(x_1)(t), F(x_2)(t)) \leq m(t)\left[\|x_1(t) - x_2(t)\|^2\right].$$

Hence, there exists $\omega \in N_{F,x_2}$ such that

$$\|v(t) - \omega(t)\|^2 \leq m(t)\left[\|x_1(t) - x_2(t)\|^2\right], \quad t \in J.$$

Consider the map $S: J \to \mathcal{P}(\mathbb{R}^n)$ defined by

$$S(t) = \left\{\omega(t) | \omega: J \to \mathbb{R}^n \text{ is Lebesgue integrable and } \|v(t) - \omega(t)\|^2 \leq m(t)\left[\|x_1(t) - x_2(t)\|^2\right]\right\}.$$

Since the multivalued operator $S(t) \cap F(t, x_2(t))$ is measurable (Proposition 8.1), there exists a function $\overline{v}(t)$ which is a measurable selection for S. So, $\overline{v}(t) \in N_{F,x_2}$, and for each $t \in J$,

$$\|v(t) - \overline{v}(t)\|^2 \leq m(t)\left[\|x_1(t) - x_2(t)\|^2\right].$$

Define the following

$$\overline{\Psi}(t) = E_{q-p}(At^{q-p})x_0 - At^{q-p}E_{q-p,q-p+1}(At^{q-p})x_0 + tE_{q-p,2}(At^{q-p})x_0' + \int_0^t (t-s)^{q-1}E_{q-p,q}(A(t-s)^{q-p})\left[Bu_{x,2}(s) + \overline{v}(s) + \int_0^s G(\theta, x_2(\theta))dw(\theta)\right]ds.$$

Then

$$\mathbb{E}\left\|\Psi(t)-\overline{\Psi}(t)\right\|^2 \leq 4\sum_{k=1}^{2}(\mathbb{J}_k x)(t)+4\frac{T^{2q}}{q^2}n_4^2\left\|B\right\|^2\left\|B^*\right\|^2 l\sum_{k=1}^{2}(\mathbb{J}_k x)(T)$$

$$\leq 4C_2 m(t)\left\|x_1-x_2\right\|^2+6\frac{T^{2q}}{q^2}n_4^2\left\|B\right\|^2\left\|B^*\right\|^2 lC_2 m(t)\left\|x_1-x_2\right\|^2$$

$$\leq 4C_2 m(t)\left\|x_1-x_2\right\|^2\left(1+\frac{T^{2q}}{q^2}n_4^2\left\|B\right\|^2\left\|B^*\right\|^2 l\right)$$

for each $t\in J$. Where

$$(\mathbb{J}_1 x)(t)=\mathbb{E}\left\|\int_0^t (t-s)^{q-1}E_{q-p,q}(A(t-s)^{q-p})[v(s)-\overline{v}(s)]ds\right\|^2$$

$$\leq \frac{T^{2q}}{q^2}\left\|E_{q-p,q}(A(t-\tau)^{q-p})\right\|^2\left\|v(t)-\overline{v}(t)\right\|^2$$

$$\leq \frac{T^{2q}}{q^2}n_4 m(t)\left\|x_1(t)-x_2(t)\right\|^2$$

$$(\mathbb{J}_2 x)(t)=\mathbb{E}\left\|\int_0^t (t-s)^{q-1}E_{q-p,q}(A(t-s)^{q-p})\right.$$

$$\left.\times\left[\int_0^s G(\theta,x_1(\theta))dw(\theta)-\int_0^s G(\theta,x_2(\theta))dw(\theta)\right]ds\right\|^2$$

$$\leq \frac{T^{2q}}{q^2}Tn_4\widehat{M}_G m(t)\left\|x_1(t)-x_2(t)\right\|^2$$

and

$$C_2=\frac{T^{2q}}{q^2}n_4+\frac{T^{2q}}{q^2}Tn_4\widehat{M}_G.$$

Thus

$$\mathbb{E}\left\|\Psi-\overline{\Psi}\right\|_{\mathcal{B}}^2 \leq 4m(t)C_2\left\|x_1-x_2\right\|^2\left(1+\frac{T^{2q}}{q^2}n_4^2\left\|B\right\|^2\left\|B^*\right\|^2 l\right)$$

for each $t\in J$.

From the analogous relation obtained by interchanging the roles of x_1 and x_2, it follows that

$$H_d(\Phi(x_1),\Phi(x_2))\leq 4m(t)C_2\left\|x_1-x_2\right\|^2\left(1+\frac{T^{2q}}{q^2}n_4^2\left\|B\right\|^2\left\|B^*\right\|^2 l\right).$$

By (8.5), Φ is a contraction and thus, by Lemma 8.5, Φ has a fixed point which is the solution to (8.1) on J.

Theorem 8.4: *(Complete controllability) Assume that the hypotheses (H1)–(H5) are satisfied, and the corresponding linear stochastic system is completely controllable on J. Then, the system* (8.1) *is completely controllable.*

Proof: By Theorems 8.2 and 8.3 the operator Φ has a fixed point. So, the control

$$u_x(t) = B^* E_{q-p,q}(A^*(T-t)^{q-p})\mathbb{W}^{-1}\{\mathbb{E}x_1 - E_{q-p}(AT^{q-p})x_0 \\ + AT^{q-p}E_{q-p,q-p+1}(AT^{q-p})x_0 - TE_{q-p,2}(AT^{q-p})x_0' \\ - \int_0^T (T-s)^{q-1}E_{q-p,q}(A(T-s)^{q-p})\left(v(s) + \int_0^s G(\theta, x(\theta))dw(\theta)\right)ds\}.$$

transfers the system (8.1) from x_0 to x_1 at time T. By Definition 8.6, the system is completely controllable on J. □

8.4 DISCUSSION

In this chapter, the nonlinear fractional higher order stochastic integrodifferential inclusion has been considered to investigate the complete controllability in finite dimensional space. Controllability Grammian martrix has been formulated using fractional calculus and the Mittag–Leffler matrix function and utilized to derive the sufficient conditions in stochastic settings to guarantee that the system (8.1) is completely controllable. The main advantage of the proposed technique relies on some hypotheses and fixed point theorems, namely the Bohnenblust–Karlin fixed point theorem for the convex case and the Covitz–Nadler for the nonconvex case for establishing the proposed results. The result can be extended to the infinite dimensional Hilbert space for approximate controllability results.

Further, impulsive fractional differential equations and inclusions have become important in recent years as mathematical models of many phenomena in both physical and social sciences. Upon making some appropriate assumptions on system functions, by adapting the techniques and ideas established in this chapter with suitable modifications, one can easily prove that complete controllability results of impulsive fractional higher order stochastic integrodifferential inclusions are driven by fractional Brownian motion in finite dimensional space.

8.5 CONCLUSION

This chapter has promoted the complete controllability result of fractional higher order stochastic integrodifferential inclusions. Based on fractional calculus, stochastic analysis approach, and suitable fixed point theorems, namely the Bohnenblust–Karlin fixed point theorem for the convex case and the Covitz–Nadler for the nonconvex case, sufficient conditions have been derived for the controllability of fractional higher order stochastic integrodifferential inclusions in finite dimensional space. Many practical problems could be modeled into linear fractional stochastic differential inclusions, hence a suitable controllability Grammian matrix may be constructed to establish the result and an appropriate nonlinear model may also be formulated and one can use proposed technique to derive the controllability results. The result may be extended to infinite dimensional Hilbert space for approximate controllability results by suitable hypotheses on nonlinear functions and appropriate fixed point theorems.

ACKNOWLEDGMENTS

The work of authors are supported by Council of Scientific and Industrial Research (CSIR), Extramural Research Division, Pusa, New Delhi, India under the grant No. 25/(0217)/13/EMR-II.

REFERENCES

Abbas, S., Mouffak, B., 2013. Fractional order Riemann-Liouville integral inclusions with two independent variables and multiple delay. Opuscula Math. 33 (2), 209–222.

Aubin, J.B., Cellina, A., 1984. Differential Inclusions: Set-Valued Maps and Viability Theory. Springer-Verlag, New York.

Azar, A.T., Vaidyanathan, S., Ouannas, A., 2017. Fractional Order Control and Synchronization of Chaotic Systems, Studies in Computational Intelligence. Springer-Verlag, Germany.

Balachandran, K., Kokila, J., 2012. On the controllability of fractional dynamical systems. Int. J. Appl. Math. Comput. Sci. 22 (3), 523–531.

Balasubramaniam, P., 2002. Existence of solutions of functional stochastic differential inclusions. Tamkang J. Math. 33 (1), 25–34.

Balasubramaniam, P., Ntouyas, S., 2006. Controllability for neutral stochastic functional differential inclusions with infinite delay in abstract space. J. Math. Anal. Applicat. 324 (1), 161–176.

Bohnenblust, H.F., Karlin, S., 1950. On a Theorem of Ville, Contributions to the Theory of Games. Princeton University Press, Princeton.

Castaing, C., Valadier, M., 2006. Convex Analysis and Measurable Multifunctions. Springer, New York.

Covitz, H., Nadler, S.B., 1970. Multi-valued contraction mappings in generalized metric spaces. Israel J. Math. 8 (1), 5–11.

Deimling, K., 1992. Multivalued Differential Equations. Walter de Gruyter, New York.

Filippov, A.F., 1988. Differential Equations With Discontinuous Righthand Sides: Control Systems. Springer Science & Business Media, United States.

Gorniewicz, L., 1999. Topological Fixed Point Theory of Multivalued Mappings. Kluwer Academic Publishers, Dordrecht.

Hu, S., Papageorgiou, N., 1997. Handbook of Multivalued Analysis, Volume I: Theory. Kluwer Academic Press, United States.

Hu, S., Papageorgiou, N.S., 2013. Handbook of Multivalued Analysis: Volume II: Applications. Springer Science & Business Media, Dordrecht.

Kexue, J., Jigen, P., 2011. Laplace Transform and fractional differential equations. Appl. Math. Lett. 24 (12), 2019–2023.

Kilbas, A.A., Srivastava, H.M., Trujillo, J.J., 2006. Theory and Applications of Fractional Differential Equations. Elsevier Science Limited, Amsterdam.

Kisielewicz, M., 1991. Differential Inclusions and Optimal Control. Kluwer Academic Publishers, Dordrecht.

Kree, P., 1982. Diffusion equation for multivalued stochastic differential equations. J. Funct. Anal. 49 (1), 73–90.

Lasota, A., Opial, Z., 1965. An application of Kakutani-Ky Fan theorem in the theory of ordinary differential equation. Bulletin De L Academie Polonaise Des Sciences-Serie Des Sciences Mathmatiques Astronomiques Et Physiques 13 (11), 781–786.

Mahmudov, N.I., 2003. Approximate controllability of semilinear deterministic and stochastic evolution equations in abstract spaces. SIAM J. Control Optim. 42 (5), 1604–1622.

Mahmudov, N.I., Denker, A., 2000. On controllability of linear stochastic systems. Int. J. Control 73 (2), 144–151.

Miller, K.S., Ross, B., 1993. An Introduction to the Fractional Calculus and Fractional Differential Equations. Wiley, New York.

Oldham, K., Spanier, J., 1974. The Fractional Calculus: Theory and Applications of Differentiation and Integration to Arbitrary Order. Academic press, New York.

Podlubny, I., 1998. Fractional Differential Equations: An Introduction to Fractional Derivatives, Fractional Differential Equations, to Methods of Their Solution and Some of Their Applications. Academic press, California.

Sabatier, J., Agrawal, O.P., Tenreiro Machado, J.A., 2007. Advances in Fractional Calculus. Springer, Dordrecht.

Sakthivel, R., Ganesh, R., Anthoni, S.M., 2013. Approximate controllability of fractional nonlinear differential inclusions. Appl. Math. Comput. 225, 708–717.

Samko, S.G., Kilbas, A.A., Maricev, I., 1993. Fractional Integrals and Derivatives; Theory and Applications. Gordon and Breach Science Publisher, Amsterdam.

Sathiyaraj, T., Balasubramaniam, P., 2016. Controllability of fractional order stochastic differential inclusions with fractional Brownian motion in finite dimensional space. IEEE/CAA J. Automatica Sinica 3 (4), 400–410.

Wang, J.R., Zhou, Y., 2011. Existence and controllability results for fractional semilinear differential inclusions. Nonlinear Anal. Real World Applicat. 12 (6), 3642–3653.

Chapter 9

Design of Fractional Order Fuzzy Sliding Mode Controller for Nonlinear Complex Systems

Jitendra Kumar[1], Ahmad Taher Azar[2,3], Vineet Kumar[1] and Kamal Pal Singh Rana[1]
[1]Instrumentation and Control Engineering Division, Dwarka, New Delhi, India, [2]Faculty of Computers and Information, Benha University, Benha, Egypt, [3]School of Engineering and Applied Sciences, Nile University, Giza, Egypt

9.1 INTRODUCTION

Rapidly growing complexity of modern engineering systems leads to very high demands on the design and control of nonlinear systems (Azar and Vaidyanathan, 2015a,b,c, 2016; Azar and Zhu, 2015; Meghni et al, 2017a,b, c; Boulkroune et al, 2016a,b; Ghoudelbourk et al., 2016; Azar and Serrano, 2014, 2015a,b,c,d, 2016a,b, 2018; Azar et al., 2017a,b, 2018a,b; Azar 2010a, b, 2012; Mekki et al., 2015; Vaidyanathan & Azar, 2015a,b,c,d, 2016a,b,c,d, e,f,g, 2017a,b,c; Zhu and Azar, 2015; Grassi et al., 2017; Ouannas et al., 2016a,b, 2017a,b,c,d,e,f,g,h,i,j; Singh et al., 2017; Vaidyanathan et al, 2015a, b,c; Wang et al., 2017; Soliman et al., 2017; Tolba et al., 2017).

Researchers and scientists always find an efficient method for controlling nonlinear complex systems. For more than seven decades, linear proportional integral and derivative controllers (PID) have been used to control complex plants but this controller was not able to give satisfactory results for nonlinear, time-varying, uncertain, and complex systems. Several nonlinear classical controllers, such as gain scheduling, model reference adaptive control (MRAC), self-tuning regulator (STR), sliding mode controller (SMC) etc., have been developed to control these types of systems (Åström and Wittenmark, 2008; Khalil, 1996; Utkin, 1977). Out of these control techniques, the sliding mode controller became quite popular among the researchers and scientists because it shows robust behavior. SMC is a type of variable structure controller where the fundamental design is based on the Lyapunov stability theory and due to that it guarantees the

Mathematical Techniques of Fractional Order Systems. DOI: https://doi.org/10.1016/B978-0-12-813592-1.00009-X

bounded input and bounded output (BIBO) stability of overall system. It is a high gain controller where output of system reaches the sliding surface very fast and tries to maintain its position on this surface (Liu and Wang, 2012). This is a type of nonlinear variable order control scheme which is generally designed for nonlinear multiinput–multioutput (MIMO) complex systems. In the present chapter, the design and analysis of the SMC controller, a nonlinear, coupled, MIMO complex system, two-link planar rigid robotic manipulator is considered. Nowadays, robotic manipulators are extensively used in hazardous areas like welding, assembling, manufacturing, painting, etc. in industries. Other applications of robotic manipulators are in the field of automobile industries, robotically assisted surgery, handling of radioactive and biohazardous materials etc. As a manipulator system is a nonlinear, coupled MIMO system where uncertainty can also be realized, it always makes a challenge for control engineers for automatic control purposes. Linear PID controllers fail to give satisfactory results in controlling such types of systems and due to that a robust controller like SMC is always suggested for suitable controlling (Sharma et al., 2014; Azar and Zhu, 2015). Several scientists have suggested classical SMC as well as hybrid of classical SMC with intelligent techniques for controlling the manipulator system. A detailed literature survey for different variants of SMC is presented in the section following.

In this section, a comprehensive literature for controlling of nonlinear, coupled and complex systems by using different variations and modifications of SMC is presented. Starting from classical SMC, the modifications by incorporating soft computing techniques like fuzzy logic (FL), artificial neural networks (ANN), adaptive neuro-fuzzy inference system (ANFIS), etc., to the SMC are presented in this section.

Despite the continuous research in the field of SMC over the last five decades, the significant technical problems such as effects of unmodeled dynamics, uncertainties of the system parameters, chattering, adaptive behavior etc. has attracted researchers and scientists. Out of these problems, SMC offered fast oscillations in the controller output, i.e., chattering which can harm the final control element part of the system. Various technical schemes have been developed and incorporated to the classical SMC to address these complications. An excellent survey on various aspects of SMC has been presented by Yu and Kaynak (2009) where it has been explored that although it has been used for the past half century, enhancements in this area are still required to design the control scheme for nonlinear complex processes. Some current research works explore the use of SMC in different areas like synchronization of chaotic system (Vaidyanathan et al., 2015a; Vaidyanathan and Azar, 2015a,b), control of Furuta pendulum (Azar and Serrano, 2015c), fault tolerance control (Mekki et al., 2015), continuous nonlinear switched systems (Azar and Serrano, 2016a), etc.

To control the manipulator system by using classical SMC techniques, several good works have been reported in the literature (Zhao et al., 2009; Fallaha et al., 2011; Islam and Liu, 2011). Chattering was the common issue in all these research works and to reduce it to a novel fuzzy SMC for robotic manipulator was proposed in Chen et al. (2003) where the signum function is replaced by the boundary layer. It was reported in this work that by arranging the width of the boundary layer using FL technique, an effective reduction in chattering was observed and at the same time a control feature was also maintained. A hybrid structure of SMC with FL technique was also proposed (Roopaei and Jahromi, 2009) to control a class of MIMO nonlinear system in the presence of uncertainties and external disturbances. A series of simulations justified the results that incorporating the FL with SMC reduced the chattering and also the robustness of controller was increased. Extensive works have been reported to this extent to reduce the chattering using FL (Wu and Ham, 1996; Wai, 2007; Sadati and Ghadami, 2008).

Another important field of soft computing technique, ANN is also underpinned with SMC by different scientists and researchers to design the hybrid controller to reduce the effect of chattering (Mu, 2010). It is well known that ANN offers a model free approach to learn from the examples of unknown dynamics. An ANN estimator is used to find the estimated error model (Munoz and Sbarbaro, 2000) to compensate the SMC for reducing the set-point tracking error of discrete nonlinear systems. In another research work performed by Niu et al. (2003), ANN is used to estimate the lumped unmatched uncertainties with time delay states of a dynamic system. A radial basis function neural network (RBFNN) is used (Huang et al., 2003) for better tracking efficacy where a lesser number of parameters has to learn and hence convergence speed is increased. A model free approximator to estimate the adaptive SMC is used which can deal with online learning parameters of matched and unmatched uncertainties. A recurrent neural network (RNN) is used to design the adaptive SMC (Karakasoglu and Sundareshan, 1995) to find the control solution for trajectory tracking task of a robotic manipulator. An RNN with a special learning algorithm for updating the parameters to be learnt is used by Fang et al. (1999). In this work, RNN is constructed to estimate the equivalent control in order to reduce the chattering. It has been shown by authors that incorporation of RNN with SMC enhances convergence due to its feedback mechanism.

Although, the addition of fractional order calculus in SMC control technique improves the overall robustness, chattering in the controller output may increase. To make a proper compromise between robustness and chattering, simultaneous application of fractional order calculus and intelligent techniques are reported. Nowadays, a hybrid structure of classical variable order control schemes and intelligent techniques has been conveyed as a recent trend. Several works are reported in this context which implement the controller by SMC and FL. A novel parameter adjustment scheme was designed by Efe (2008) to improve the robustness of fuzzy SMC by the use of an ANFIS. In this work, fractional order has been used rather than integer order to improve the robustness of controller. The plant is used for study as a two degree of freedom direct-drive robot arm. The obtained result from the order controller is compared to its integer order counterpart, where it is reported that the proposed controller displayed better tracking performance, insensitivity towards disturbances, and a high degree of robustness.

To control two different plants, polar robot and coupled tank system, a novel fractional order fuzzy SMC with proportional derivative surface (FOFSMCPD) is proposed by Delavari et al. (2010a). The discount in signum function is reduced in that work by using the FL when the controller reaches the sliding surface. The parameters of the controller are optimized by using genetic algorithm (GA) which was proposed by J. Holland and his collaborators in the 1960s and 1970s. The performance index is considered as reaching time and on root mean square value of tracking error. Extensive simulation studies revealed that the FOFSMCPD controller outperforms on its integer order counterpart for set point tracking capability and chattering reduction task. In the present chapter, FOFSMCPD controller is considered for a two-link rigid robotic manipulator control task. To show the efficacy of the FOFSMCPD controller, its performance is compared with classical integer order SMCPD (IOSMCPD) controller. The performances of the FOFSMCPD controller are calculated in servo as well as regulatory mode with and without model uncertainty in mass as well as length of both the links.

Followed by the introduction and detailed literature survey presented in Section 9.1, the dynamic mathematical model of the plant is described in Section 9.2. The design and implementation of IOSMCPD and FOFSMCPD controllers are presented in Section 9.3 whereas optimization of controller gains by GA is shown in Section 9.4. The detailed performance evaluation is presented in Section 9.5 and the conclusion of this chapter is given in Section 9.6.

9.2 DYNAMIC MODEL OF MANIPULATOR SYSTEM

In this section, the dynamic model of a two-link planar rigid robotic manipulator system is presented (Craig, 2005). A manipulator diagram is shown in Fig. 9.1 where l_1 and l_2 are considered as length of link-1 and link-2, whereas m_1 and m_2 are taken as mass of link-1 and link-2, respectively. To find the mathematical model of the two-link robotic manipulator system, it has been assumed that masses of both the links are on the tip of respective links. θ_1 and θ_2 are considered as the angular position of link-1 and link-2 in radian whereas τ_1 and τ_2 are taken as input torques to the manipulator system, respectively in N-m. A popular approach, Lagrangian-Euler formula is used to find the mathematical model of the two-link manipulator system. According to this consideration, the kinetic energy of the link will be,

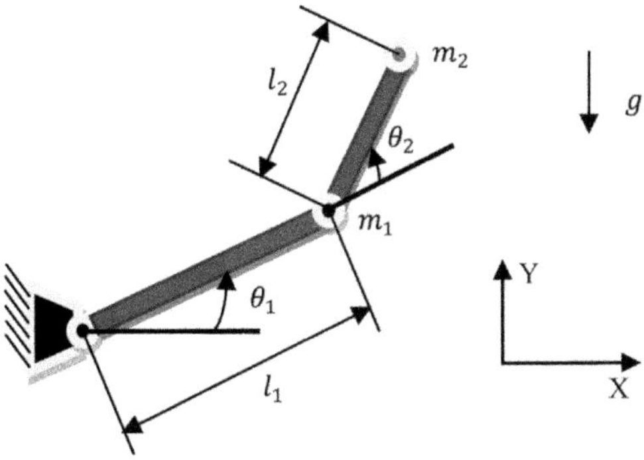

FIGURE 9.1 Two-link planar rigid robotic manipulator diagram.

$$K = \frac{1}{2}mv^2 \tag{9.1}$$

where, v is considered as the linear velocity of the of the link. Now, the linear velocity of link-1 can be expressed as:

$$v_1 = l_1\dot{\theta}_1 \tag{9.2}$$

For link-1, the kinetic energy can be written as:

$$K_1 = \frac{1}{2}m_1v_1^2 \tag{9.3}$$

From (9.2) and (9.3)

$$K_1 = \frac{1}{2}m_1l_1^2(\dot{\theta}_1)^2 \tag{9.4}$$

For link 1, the potential energy can be expressed as:

$$P_1 = m_1l_1g\sin\theta_1 \tag{9.5}$$

To find the linear velocity of link-2, the Cartesian coordinates of the end-point of link-2 are needed. After taking the first derivative, linear velocity of link-2 can be easily obtained. Now, the Cartesian coordinates of the end-point of second link are:

$$x_2 = l_1\cos\theta_1 + l_2\cos(\theta_1 + \theta_2) \tag{9.6}$$

$$y_2 = l_1\sin\theta_1 + l_2\sin(\theta_1 + \theta_2) \tag{9.7}$$

Further, the linear velocity of link-2 can be written as:

$$v_2 = \sqrt{\dot{x}_2{}^2 + \dot{y}_2{}^2} \tag{9.8}$$

From (9.8),

$$v_2^2 = \dot{x}_2{}^2 + \dot{y}_2{}^2 \tag{9.9}$$

From (9.6),

$$\dot{x}_2 = -l_1 \sin\theta_1\dot{\theta}_1 - l_2\sin(\theta_1+\theta_2)(\dot{\theta}_1+\dot{\theta}_2) \tag{9.10}$$

From (9.7),

$$\dot{y}_2 = l_1\cos\theta_1\dot{\theta}_1 + l_2\cos(\theta_1+\theta_2)(\dot{\theta}_1+\dot{\theta}_2) \tag{9.11}$$

Now, from (9.9)–(9.11),

$$\begin{aligned} v_2^2 &= l_1^2S_1^2(\dot{\theta}_1)^2 + l_2^2S_{12}^2(\dot{\theta}_1+\dot{\theta}_2)^2 + 2l_1l_2S_1S_{12}\left((\dot{\theta}_1)^2+\dot{\theta}_1\dot{\theta}_2\right) \\ &+ l_1^2C_1^2(\dot{\theta}_1)^2 + l_2^2C_{12}^2(\dot{\theta}_1+\dot{\theta}_2)^2 + 2l_1l_2C_1C_{12}\left((\dot{\theta}_1)^2+\dot{\theta}_1\dot{\theta}_2\right) \end{aligned} \tag{9.12}$$

To shorten the calculation, some abbreviations are used in (9.12) where, $S_1 = \sin\theta_1$, $C_1 = \cos\theta_1$, $S_2 = \sin\theta_2$, $C_2 = \cos\theta_2$, $S_{12} = \sin(\theta_1+\theta_2)$, and $C_{12} = \cos(\theta_1+\theta_2)$

Further (9.12) can be expressed as:

$$v_2^2 = l_1^2(\dot{\theta}_1)^2 + l_2^2(\dot{\theta}_1+\dot{\theta}_2)^2 + 2l_1l_2C_2\left((\dot{\theta}_1)^2+\dot{\theta}_1\dot{\theta}_2\right) \tag{9.13}$$

The kinetic energy of second link is expressed as:

$$K_2 = \frac{1}{2}m_2v_2^2 \tag{9.14}$$

$$K_2 = \frac{1}{2}m_2\left[l_1^2(\dot{\theta}_1)^2 + l_2^2(\dot{\theta}_1+\dot{\theta}_2)^2 + 2l_1l_2C_2\left((\dot{\theta}_1)^2+\dot{\theta}_1\dot{\theta}_2\right)\right] \tag{9.15}$$

The potential energy of link-2 can be written as:

$$P_2 = m_1l_1gS_1 + m_2l_2gS_{12} \tag{9.16}$$

A scalar function, called the Lagrangian function is defined as the difference between the total kinetic energy and total potential energy of a mechanical system.

$$\text{Lagrangian}, \mathcal{L} = (K-P) \tag{9.17}$$

where, $K = (K_1+K_2)$ is represented as the total kinetic energy, whereas $P = (P_1+P_2)$ is represented as the total potential energy of the two-link manipulator system.

For the present considered system, the Lagrangian can be written as:

$$\begin{aligned} \mathcal{L} &= \frac{1}{2}m_1l_1^2(\dot{\theta}_1)^2 + \frac{1}{2}m_2\left[l_1^2(\dot{\theta}_1)^2 + l_2^2(\dot{\theta}_1+\dot{\theta}_2)^2 + 2l_1l_2C_2\left((\dot{\theta}_1)^2+\dot{\theta}_1\dot{\theta}_2\right)\right] \\ &- m_1l_1gS_1 - m_1l_1gS_1 - m_2l_2gS_{12} \end{aligned} \tag{9.18}$$

The dynamic model of the two-link manipulator system can be obtained by Lagrangian-Euler formula which is expressed as:

$$\tau_i = \frac{d}{dt}\left(\frac{\partial \mathcal{L}}{\partial \dot{\theta}_i}\right) - \frac{\partial \mathcal{L}}{\partial \theta_i} \tag{9.19}$$

where, i represents the number of links present in the manipulator system.

To find the torque for link-1, (9.19) can be expressed as:

$$\tau_1 = \frac{d}{dt}\left(\frac{\partial \mathcal{L}}{\partial \dot{\theta}_1}\right) - \frac{\partial \mathcal{L}}{\partial \theta_1} \tag{9.20}$$

Further, from (9.18),

$$\frac{\partial \mathcal{L}}{\partial \theta_1} = 2m_1gl_1C_1 + m_2gl_2C_{12} \tag{9.21}$$

$$\frac{\partial \mathcal{L}}{\partial \dot{\theta}_1} = m_1l_1^2\dot{\theta}_1 + m_2\left[l_1^2\dot{\theta}_1 + l_2^2\dot{\theta}_1 + l_2^2\dot{\theta}_2 + l_1l_2C_2(2\dot{\theta}_1 + \dot{\theta}_2)\right] \tag{9.22}$$

$$\frac{d}{dt}\left(\frac{\partial \mathcal{L}}{\partial \dot{\theta}_1}\right) = m_1l_1^2\ddot{\theta}_1 + m_2\left[l_1^2\ddot{\theta}_1 + l_2^2\ddot{\theta}_1 + l_2^2\ddot{\theta}_2 - l_1l_2S_2\dot{\theta}_2(2\dot{\theta}_1 + \dot{\theta}_2) + l_1l_2C_2(2\ddot{\theta}_1 + \ddot{\theta}_2)\right] \tag{9.23}$$

Further, from (9.20)–(9.23), the dynamic equation of manipulator for link-1 can be calculated which can be expressed as:

$$\begin{aligned}\tau_1 = {} & l_2^2m_2\left(\frac{d^2\theta_1}{dt^2}\right) + l_2^2m_2\left(\frac{d^2\theta_2}{dt^2}\right) + l_1^2(m_1 + m_2)\left(\frac{d^2\theta_1}{dt^2}\right) \\ & + m_2gl_2\cos(\theta_1 + \theta_2) + (m_1 + m_2)l_1g\cos(\theta_1) - m_2l_1l_2\sin(\theta_2)\left(\frac{d\theta_2}{dt}\right)^2 \\ & - 2m_2l_1l_2\sin(\theta_2)\left(\frac{d\theta_1}{dt}\right)\left(\frac{d\theta_2}{dt}\right) + m_2l_1l_2\cos(\theta_2)\left(2\frac{d^2\theta_1}{dt^2} + \frac{d2\theta_2}{dt^2}\right)\end{aligned} \tag{9.24}$$

To find the torque for link-2, (9.19) can be written as:

$$\tau_2 = \frac{d}{dt}\left(\frac{\partial \mathcal{L}}{\partial \dot{\theta}_2}\right) - \frac{\partial \mathcal{L}}{\partial \theta_2} \tag{9.25}$$

Now, from (9.18),

$$\frac{\partial \mathcal{L}}{\partial \theta_2} = -m_2gl_2C_{12} - 2l_1l_2S_2\left((\dot{\theta}_1)^2 + \dot{\theta}_1\dot{\theta}_2\right) \tag{9.26}$$

$$\frac{\partial \mathcal{L}}{\partial \dot{\theta}_2} = \frac{1}{2}m_2\left[l_2^2(2\dot{\theta}_2 + 2\dot{\theta}_1) + 2l_1l_2C_2\dot{\theta}_1\right] \tag{9.27}$$

$$\frac{d}{dt}\left(\frac{\partial \mathcal{L}}{\partial \dot{\theta}_2}\right) = \frac{1}{2}m_2\left[l_2^2(2\ddot{\theta}_2 + 2\ddot{\theta}_1) + 2l_1l_2(C_2\ddot{\theta}_1 - S_2\dot{\theta}_1\dot{\theta}_2)\right] \tag{9.28}$$

From (9.25)–(9.28), the dynamic equation of manipulator for link-2 can be expressed as:

$$\tau_2 = m_2l_1l_2\sin(\theta_2)\left(\frac{d\theta_1}{dt}\right)^2 + m_2l_1l_2\cos(\theta_2)\left(\frac{d^2\theta_1}{dt^2}\right) + m_2l_2^2\left(\frac{d^2\theta_1}{dt^2} + \frac{d^2\theta_2}{dt^2}\right) + m_2l_2g\cos(\theta_1 + \theta_2) \tag{9.29}$$

The dynamic equation of two-link rigid robotic manipulator system can also be expressed as:

$$\tau = M(\theta)\ddot{\theta} + V(\theta,\dot{\theta}) + G(\theta) \tag{9.30}$$

where $\theta = \begin{bmatrix}\theta_1 & \theta_2\end{bmatrix}^T$ is the angular position of end point of two links.

$$M = \begin{bmatrix} l_1^2(m_1 + m_2) + l_2^2m_2 + 2m_2l_1l_2\cos\theta_2 & m_2l_1l_2\cos\theta_2 + m_2l_2^2 \\ m_2l_1l_2\cos\theta_2 + m_2l_2^2 & m_2l_2^2 \end{bmatrix} \tag{9.31}$$

$$V(\theta,\dot{\theta}) = \begin{bmatrix} -2m_2l_1l_2\sin\theta_2\left(\frac{d\theta_1}{dt}\right)\left(\frac{d\theta_2}{dt}\right) - m_2l_1l_2\sin\theta_2\left(\frac{d\theta_2}{dt}\right)^2 \\ m_2l_1l_2\sin\theta_2\left(\frac{d\theta_1}{dt}\right)^2 \end{bmatrix} \tag{9.32}$$

$$G(\theta) = \begin{bmatrix} (m_1 + m_2)l_1g\cos\theta_1 + m_2gl_2\cos(\theta_1 + \theta_2) \\ m_2gl_2\cos(\theta_1 + \theta_2) \end{bmatrix} \tag{9.33}$$

where $M(\theta)$, $V(\theta,\dot{\theta})$, and $G(\theta)$ are inertia, centripetal, and gravitational matrices respectively.

The parameters used for the manipulator model for simulation is described in Table 9.1 as,

TABLE 9.1 Parameter of Two-Link Manipulator System

Parameter	Symbol	Value	Unit
Length of link-1	l_1	0.8	m
Length of link-2	l_2	0.4	m
Mass of link-1	m_1	0.1	kg
Mass of link-2	m_2	0.1	kg
Acceleration due to gravity	g	9.8	m/s^2

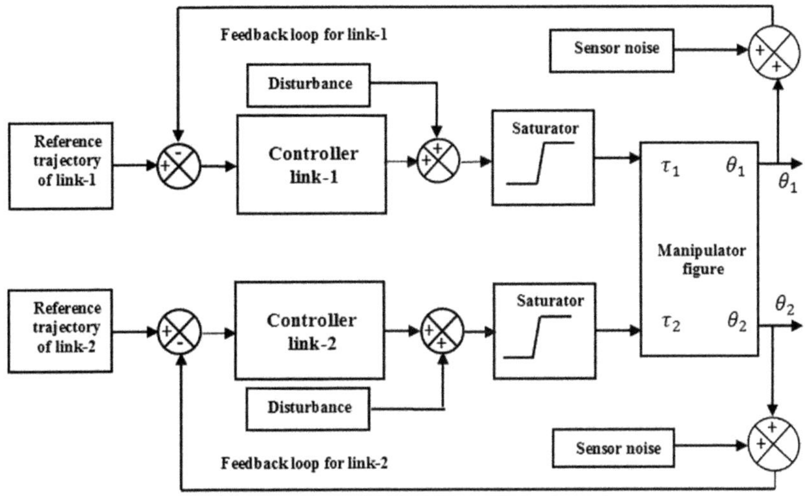

FIGURE 9.2 Basic block diagram of control structure.

9.3 DESIGN AND IMPLEMENTATION OF SLIDING MODE CONTROLLERS

In the present section, the detailed description of IOSMCPD and FOFSMCPD controllers are presented. A two-link manipulator controlled system, consisting of two feedback loops are shown in Fig. 9.2 where disturbance is injected to the controller output.

A saturator is inserted after the controller output to get the appearance of final control element and in this control task, saturation limit is considered as −20 to +20 N-m. It is well known that SMC is a model-based control technique and in the present work Lyapunov stability criteria is used to design it. The main task of the controller is to give the desired output so that the tip of the links should follow the desired trajectory as well as the stability of overall system should be maintained. The detailed design description of the different variants of SMC controllers are presented as follows.

9.4 IOSMCPD CONTROLLER DESIGN

In this section the fundamental design of the IOSMCPD controller with respect to torque is presented. The integer order proportional and derivative sliding surface s for n^{th} order system is considered as,

$$s = \left[\left(\frac{d}{dt} + \lambda \right)^{n-1} e \right] \tag{9.34}$$

For the considered second order ($n = 2$) plant, (9.34) becomes,

$$s = \lambda e + \dot{e} \tag{9.35}$$

Here, λ is a constant and error $e = \theta_r - \theta$, where θ_r is the desired angular position and θ is the current angular position of end points of the links in radian. On differentiation (9.35) becomes,

$$\dot{s} = \lambda \dot{e} + \ddot{e} \tag{9.36}$$

Putting $\ddot{e} = \ddot{\theta}_r - \ddot{\theta}$

$$\dot{s} = \lambda \dot{e} + (\ddot{\theta}_r - \ddot{\theta}) \tag{9.37}$$

The system output is forced to track this surface with the help of a reaching law. The reaching law is to be designed in such a way that it guarantees the stability of the closed loop system. Generally, three reaching laws are widely used and have been reported in the literature namely exponential, constant rate, and power rate laws (Liu and Wang, 2012). In the present work, exponential law as given in (9.38) is utilized. According to this law,

$$\dot{s} = -\epsilon\, sgn(s) - \mathrm{k}s; \epsilon > 0, k > 0 \tag{9.38}$$

where ϵ and k are constants.

As SMC is a high gain switching controller, a fast oscillation phenomenon, called chattering, occurs at the controller output due to the discontinuity in *sgn* function in control action. To resolve this problem, normally saturation function (9.39) is employed rather than *sgn* function (Liu and Wang, 2012).

$$sat(s) = \begin{cases} +1; s > \Delta \\ \dfrac{s}{\Delta}; \ |s| < \Delta \\ -1; s < -\Delta \end{cases} \tag{9.39}$$

where $\Delta(0 < \Delta < 1)$ forms the boundary layer.

Now, from (9.38) and (9.39),

$$\dot{s} = -\epsilon\, sat(s) - \mathrm{k}s; \epsilon > 0, k > 0 \tag{9.40}$$

Equating (9.38) and (9.40),

$$-\epsilon\, sat(s) - \mathrm{k}s = \lambda \dot{e} + (\ddot{\theta}_r - \ddot{\theta}) \tag{9.41}$$

Now substituting $\ddot{\theta}$ from the dynamics of manipulator (9.30), in (9.41) then,

$$-\epsilon\, sat(s) - \mathrm{k}s = \lambda \dot{e} + \left(\ddot{\theta}_r - [M(\theta)]^{-1}\left[\tau - V\left(\theta,\dot{\theta}\right) - G(\theta)\right]\right) \tag{9.42}$$

This can also be expressed as,

$$\tau = V\left(\theta,\dot{\theta}\right) + G(\theta) + [M(\theta)]\left[\epsilon\, sat(s) + \mathrm{k}s + \lambda \dot{e} + \ddot{\theta}_r\right] \tag{9.43}$$

Eq. (9.43) gives the required torques for a two-link manipulator system by an IOSMCPD controller to control the angular positions of end-effector of manipulator links. In this design, two gains, ϵ and k, came from exponential law which uses the manipulator output to track the desired trajectory whereas saturation function is used to reduce the chattering. Therefore, proper variation in ϵ and k can make a suitable compromise between robustness and chattering reduction.

9.5 FOFSMCPD CONTROLLER DESIGN

A generalized integer order sliding surface for second order plant is given in (9.35) which can be expressed as,

$$s = \lambda e + \dot{e} \tag{9.44}$$

It was shown by Efe (2008) and Delavari et al. (2010a) that by inclusion of the fractional order differentiator and integrator instead of the integer order in SMC, the robustness of the controller can be significantly increased. In this context, fractional order calculus has been introduced in the design of SMC to achieve an enhanced level of robustness and the design procedure for the fractional SMC implementation is given below.

After introducing fractional order operator in (9.44) which has been considered from (Efe, 2008; Delavari et al., 2010a,b)

$$s = \lambda e + D^{\alpha} e; 0 < \alpha < 1 \tag{9.45}$$

Eq. (9.45) can be written as,

$$s = \lambda e + D^{\alpha-1} \dot{e} \tag{9.46}$$

On further differentiation (9.46) becomes,

$$\dot{s} = \lambda \dot{e} + D^{\alpha-1} \ddot{e} \tag{9.47}$$

On introducing the fractional order differentiator for the first term in (9.47) it becomes,

$$\dot{s} = \lambda D^{\alpha} e + D^{\alpha-1} \ddot{e} \tag{9.48}$$

In FOFSMCPD controller design, the exponential law (9.40) is utilized so that the system output can be forced to follow this surface. In this regard, comparing (9.40) and (9.48) it becomes,

$$-\epsilon\, sat(s) - \mathrm{k}s = \lambda D^{\alpha} e + D^{\alpha-1}(\ddot{e}) \tag{9.49}$$

Putting $\ddot{e} = \ddot{\theta}_r - \ddot{\theta}$ (9.49) becomes

$$-\epsilon\, sat(s) - \mathrm{k}s = \lambda D^{\alpha} e + D^{\alpha-1}(\ddot{\theta}_r - \ddot{\theta}) \tag{9.50}$$

Substituting $\ddot{\theta}$ from plant dynamic model (9.30) in (9.50), it becomes

$$-\epsilon\, sat(s) - \mathrm{k}s = \lambda D^{\alpha}e + D^{\alpha-1}\left(\ddot{\theta}_r - [M(\theta)]^{-1}\left[\tau - V\left(\theta,\dot{\theta}\right) - G(\theta)\right]\right) \quad (9.51)$$

It can be further simplified as,

$$[M(\theta)]^{-1}\left[\tau - V\left(\theta,\dot{\theta}\right) - G(\theta)\right] = D^{1-\alpha}(\epsilon\, sat(s) + \mathrm{k}s + \lambda D^{\alpha}e) + \ddot{\theta}_r \quad (9.52)$$

$$\tau = [M(\theta)]\left[D^{1-\alpha}(\epsilon\, sat(s) + \mathrm{k}s + \lambda D^{\alpha}e) + \ddot{\theta}_r\right] + V\left(\theta,\dot{\theta}\right) + G(\theta) \quad (9.53)$$

Eq. (9.53) explores the required torques as an FOFSMCPD controller outputs to control the angular positions of end-effector of manipulator system. In the design of FOFSMCPD controller, ϵ and k are the two positive gains of exponential law which direct the manipulator output so as to track the surface, whereas the saturation function reduces the chattering. Therefore, a satisfactory compromise between robustness and chattering can be made after proper variation of these two variables. It is presented in Delavari et al. (2010a) that the concept of an FL control (FLC) can be easily combined with SMC to preserve the advantage of both the techniques. In the present chapter, an FLC is considered to vary the gains ϵ and k with varying surface to reduce chattering. A significant reduction in chattering is obtained which will be explained and presented in Sections 9.5 and 9.6. The detailed design of FLC is presented as follows.

9.5.1 Stability Analysis Using Lyapunov Stability Criteria

The main objective of SMC controller design is the overall guaranteed stability of the feedback control system. According to the Lyapunov stability theorem, the overall system will be stable and will reach the sliding surface if the condition $s\dot{s} < 0$ will be satisfied. For this, the Lyapunov function is considered as:

$$V = \frac{1}{2}s^2 \quad (9.54)$$

$$\dot{V} = s\dot{s} \quad (9.55)$$

Substituting $\dot{s}$ from exponential law (9.40),

$$\dot{V} = \mathrm{s}(-\epsilon\, sat(s) - \mathrm{k}s) \quad \text{where}, \epsilon > 0, k > 0 \quad (9.56)$$

Here, (9.56) governs two cases. In the first case when the sliding surface $\mathrm{s} > 0$ whereas in the second case $s < 0$. After considering the first case:

Case I: When $s > 0$ then $sat(s) > 0$

Let, $sat(s) = \phi$, where, $\phi > 0$.

From Eq. (9.56)

$$\dot{V} = s(-\epsilon\phi - ks) \quad (9.57)$$

$$\dot{V} = -s(\epsilon\phi + ks) \tag{9.58}$$

As, $s > 0$ as well as $\epsilon\phi + ks > 0$

$$\dot{V} = -s(\epsilon\phi + ks) < 0 \tag{9.59}$$

Eq. (9.59) shows that $\dot{V} < 0$ which states that the designed controller for exponential law will be stable for $s > 0$.

Case II: When $s < 0$ then $sat(s) < 0$

Let $sat(s) = -\phi$, where, $\phi > 0$.

From (9.56)

$$\dot{V} = s(\epsilon\phi - ks) \tag{9.60}$$

Let, $\Omega = -s$ such that $\Omega > 0$

Now from (9.59),

$$\dot{V} = -\Omega(\epsilon\phi + \mathrm{k}\Omega) \tag{9.61}$$

$$\text{Since, } \Omega(\epsilon\phi + \mathrm{k}\Omega) > 0, \dot{V} = -\Omega(\epsilon\phi + \mathrm{k}\Omega) < 0 \tag{9.62}$$

Eq. (9.62) shows that $\dot{V} < 0$ which further states that the designed controller for exponential law will be stable for $s < 0$. On the basis of results obtained from (9.59) and (9.62), it can be inferred that the above design of IOFSMCPD and FOFSMCPD controllers will follow the Lyapunov stability theory and plant output will be bounded for bounded input. The block diagram representation of IOFSMCPD and FOFSMCPD controller is shown in Figs. 9.3 and 9.4, respectively.

9.5.2 FLC Design

In the present work, FLC is constituted with SMC in run time for which the values of ϵ and k can make balance between robustness and chattering of the

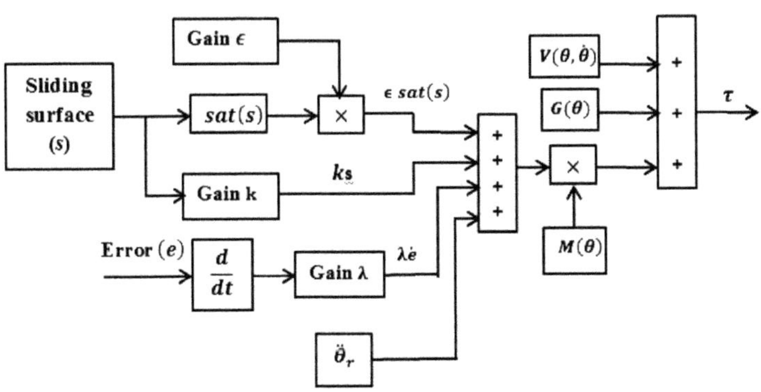

FIGURE 9.3 IOSMCPD controller block diagram.

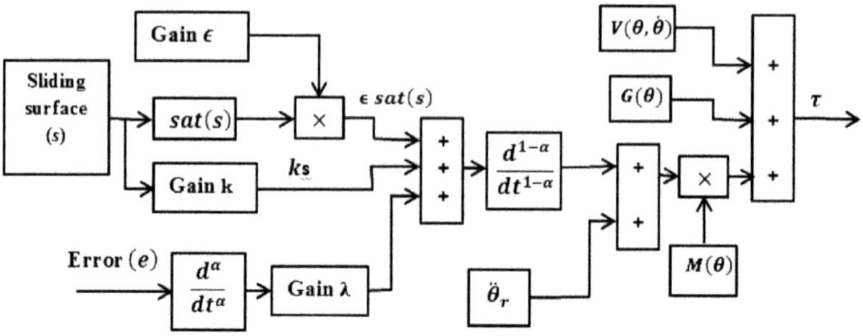

FIGURE 9.4 FOFSMCPD controller block diagram.

TABLE 9.2 Input and Output MFs Terminology

Input MFs		Output MFs	
NB	Negative big	B	Big
NM	Negative medium	M	Medium
NS	Negative small	S	Small
Z	Zero	VS	Very small
PS	Positive small		
PM	Positive medium		
PB	Positive big		

FOFSMCPD controller. A single-input single-output FLC is used where input is used as sliding surface. There are seven triangular input membership functions (MFs) used for input whereas four output MFs are used out of which two are triangular and the remaining two are considered as trapezoidal. The terminology of these MFs is given in Table 9.2 whereas the input and output MFs are shown in Figs. 9.5 and 9.6, respectively. Fuzzy logic block diagram is shown in Fig. 9.7.

In the present work, the fuzzy rules are stated as:

$$R_i\text{:If } s \text{ is } S_i \text{ then o is } O_i$$

where, ϵ and k are defined as $\epsilon = O_i^{*}\ g_\epsilon$ and $k = O_i^{*}\ g_k$. Here g_ϵ and g_k are taken as two different FLC output gains which are obtained from GA while O_i is the output of FLC. Here ϵ and k supplies different values in run-time due to different values of gains g_ϵ and g_k. The used rule base is presented in Table 9.3 which has been considered from Delavari et al. (2010a). The design of rule base signifies that if the value of sliding surface tends to zero, ϵ and k should be minimum, whereas if the value of sliding surface tends to -1 or $+1$, ϵ and k should be maximum. Mamdani type min−max inference

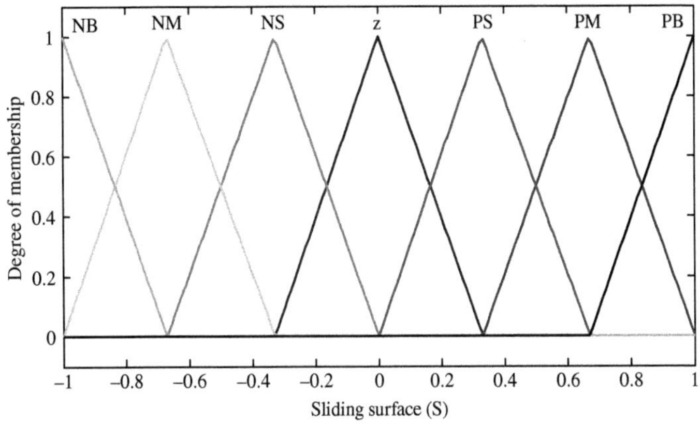

FIGURE 9.5 Input MFs.

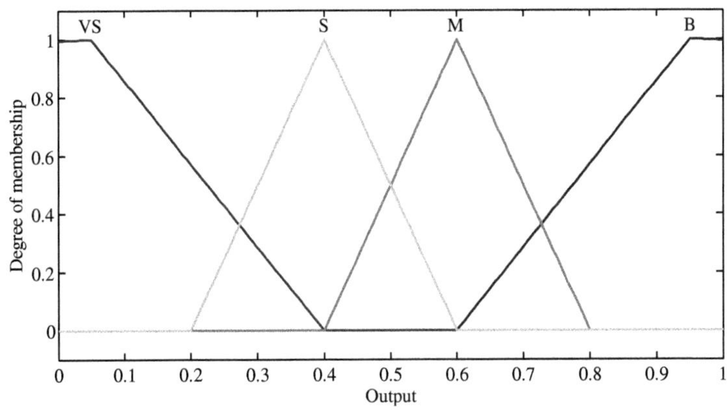

FIGURE 9.6 Output MFs.

TABLE 9.3 Rule Bases for FLC Design

Input (sliding surface)	NB	NM	NS	Z	PS	PM	PB
Output	B	M	S	VS	S	M	B

and the center of gravity defuzzification method have been used in the present work. A V-shape input–output profile is shown in Fig. 9.8 as the resulting control action which implies the rule base design. It can also be seen that for any value of sliding surface, the output of FLC is positive as expected from the definition of exponential law because ϵ and k should be positive as required by this law.

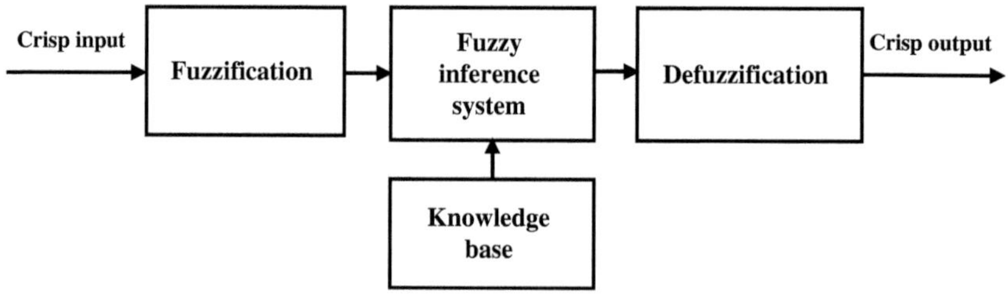

FIGURE 9.7 Fuzzy logic block diagram.

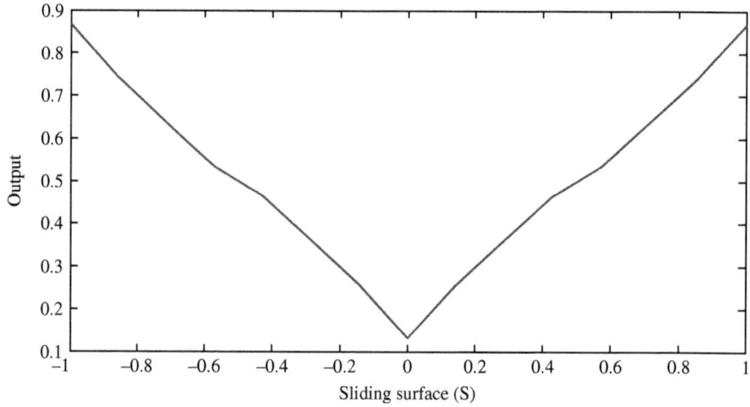

FIGURE 9.8 Input–output surface plot.

9.5.3 Fractional Order Operator Implementation

The combination of fractional order mathematical operators with SMC increases the DOF and as a result accurate solutions are attained which have been used in the field of control in various applications (Sharma et al., 2014, Kumar et al., 2015, 2016; Mishra et al., 2015). A fractional order differentiator and integrator of a function $g(t)$ is represented as, $D^{\alpha}g(t)$ and $D^{-\beta}g(t)$, respectively, where $0<\alpha<1$ and $0<\beta<1$. In the present work, Grünwald–Letnikov (GL) method (9.63) has been used to implement the fractional order operators with a memory size of 100.

$$ {}_aD_t^{\gamma} g(t) = \lim_{h \to 0} \frac{1}{h^{\gamma}} \sum_{i=0}^{[(t-a)/h]} (-1)\binom{\gamma}{i} g(t-ih) \tag{9.63} $$

where D represents the derivative/integrator operator and γ (α, β) is the order of fractional operator. In the present work, α is chosen as the order of differentiator, whose value is found by GA.

9.6 OPTIMIZATION OF CONTROLLER GAINS BY GA

Tuning of the controller parameter plays a vital role in the perfect design of controllers like SMC. A properly tuned controller tries to minimize the objective function of overall system and it assures the process output to track the desired trajectory as well as to reduce the effect of disturbances coming to the system. To tune a variable order controller like SMC, there is not any method that has been specified in the literature survey. In the present work, GA is used to tune the controller parameters.

9.6.1 Description of GA

The idea of GA was developed by John Holland and his coworkers in the 1960s and 1970s. The concept of GA is based on the biological evolution which has been based on the principle of Charles Darwin's theory of natural selection (Yang, 2014). GA has the ability to deal with parallelism as well as complex problems. Following is the procedure to find the optimization of a mathematical expression or a complex control problem which can be also described by a flow chart as shown in Fig. 9.9.

1. Firstly, the algorithm encodes the cost function or objective function which has to be minimized.
2. After that selection criteria will be defined.
3. Further, random population of an individual solution will be created.

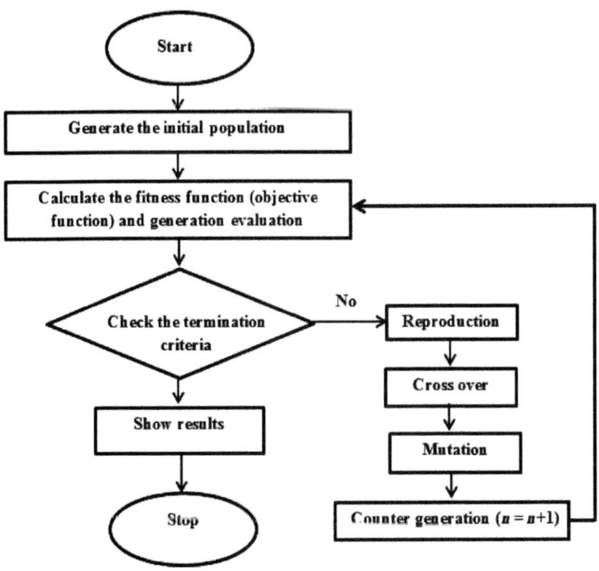

FIGURE 9.9 Flow chart of GA.

4. Objective function of the individuals will be evaluated and after comparing the objective function value, a new population will be generated by using cross-over and mutation.
5. At last, results will be decoded to find the solution of problem.

9.6.2 Tuning of Controller for Trajectory Tracking Task

All the simulations presented in the current chapter have been performed in MATLAB/SIMULINK (R2012a) on a personal computer having Intel core™ i5 processor which is working at 3.33 GHz, 4 GB RAM with a 32-bit operating system. For the ODE solver task, fourth-order Runge–Kutta method was used where the sampling time is considered as 1 ms. Initially optimized gains remain unaltered throughout all the further studies like disturbance rejection and uncertainty analysis. In the present work, the torque restrictions for both links have been considered as [−20, 20] N-m. The reference trajectories (θ_{r1} and θ_{r2}) for link-1 and link-2 have been shown in Eqs. (9.64) and (9.65), respectively, which are expressed as,

$$\theta_{r1} = \{\sin \omega t\} \tag{9.64}$$

$$\theta_{r2} = \{\cos \omega t\} \tag{9.65}$$

where ω is termed as the angular frequency of desired trajectory.

The objective function in the current chapter is taken as:

$$IAE = IAE_1 + IAE_2 \tag{9.66}$$

$$\text{Chatter} = \int [|s_1| + |s_2|]dt \tag{9.67}$$

Here, IAE_1 and IAE_2 represent the integral of absolute errors for link-1 and link-2, respectively. The chatter is defined as the integration of sum of absolute values of surfaces for link-1 and link-2, respectively, as described in Eq. (9.67) where s_1 and s_2 represents the surface of link-1and link-2, respectively. In the present work, the objective function has been taken as the weighted sum of integral of absolute error (*IAE*) and the chatter which is defined in Eq. (9.68).

$$\text{Objective function} = (W_1{}^*IAE) + (W_2{}^*\text{chatter}) \tag{9.68}$$

The aggregate *IAE* is defined as the sum of the *IAE* values of both the links, i.e., $IAE = IAE_1 + IAE_2$. The weightage parameters W_1 and W_2 are chosen as 0.999 and 0.001. These numbers were chosen to give equal weightage to both the terms, *IAE* and chatter, during the optimization. Table 9.4 shows the optimized gains of IOSMCPD and FOFSMCPD controllers where Fig. 9.10 depicts the OBF versus iteration curve for IOSMCPD and FOFSMCPD controllers. The obtained IAE values and objective function values are depicted graphically in Fig. 9.11. It can be easily observed from

TABLE 9.4 Optimized Gain Parameters of IOSMCPD and FOFSMCPD Controllers

Parameters		IOSMCPD		FOFSMCPD	
		Link-1	Link-2	Link-1	Link-2
Normalized fuzzy input gain		–	–	314.42	228.73
Exponential law gains	(g_ε)	311.74	87.60	128.16	58.92
	(g_k)	43.58	220.53	174.46	157.71
Boundary layer (Δ)		0.64	0.64	0.77	0.77
Sliding surface error gain (λ)		38.88	38.88	38.97	38.97
Integer / Fractional order for derivative (α)		1	1	0.89	0.53

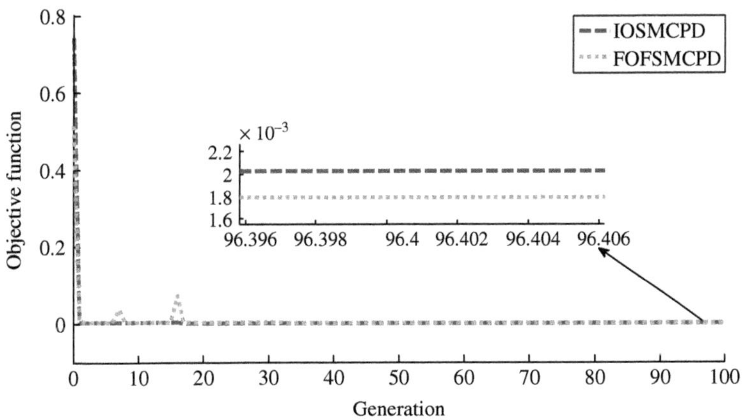

FIGURE 9.10 Objective functions versus iteration curve for IOSMCPD and FOFSMCPD controllers.

this figure that the convergence of the FOFSMCPD controller is better than the IOSMCPD controller. The value of objective functions obtained for the IOSMCPD controller is 0.0020 whereas for the FOFSMCPD controller it is 0.0017 for trajectory tracking task as depicted in Table 9.5. An improvement of 18.18% is obtained in IAE by the FOFSMCPD over the IOSMCPD controller for link-1 whereas link-2 offered the same IAE values by both the controllers. About 15% improvement in objective functions has been obtained by the FOFSMCPD on the IOSMCPD controller. Reference trajectory tracking curves for both links, their controller outputs, and error curves are shown in Figs. 9.12–9.14. A careful observation is obtained from Fig. 9.13 that after taking the value of chatter in the account of objective

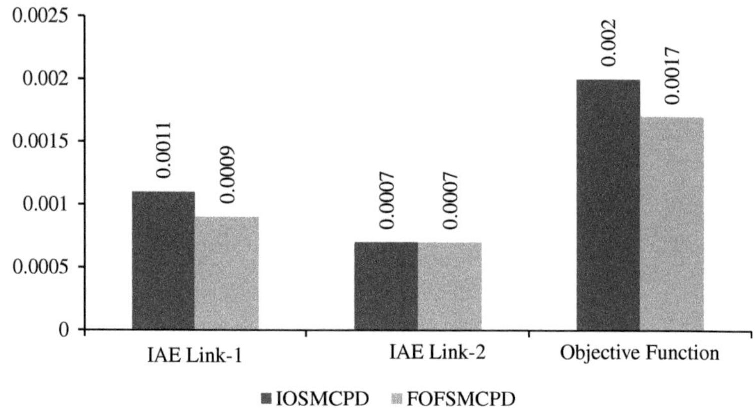

FIGURE 9.11 IAE and OBF values for trajectory tracking.

TABLE 9.5 IAE Values and Objective Functions for Trajectory Tracking Task

	IAE Link-1	IAE Link-2	Objective Function
IOSMCPD	0.0011	0.0007	0.0020
FOFSMCPD	0.0009	0.0007	0.0017

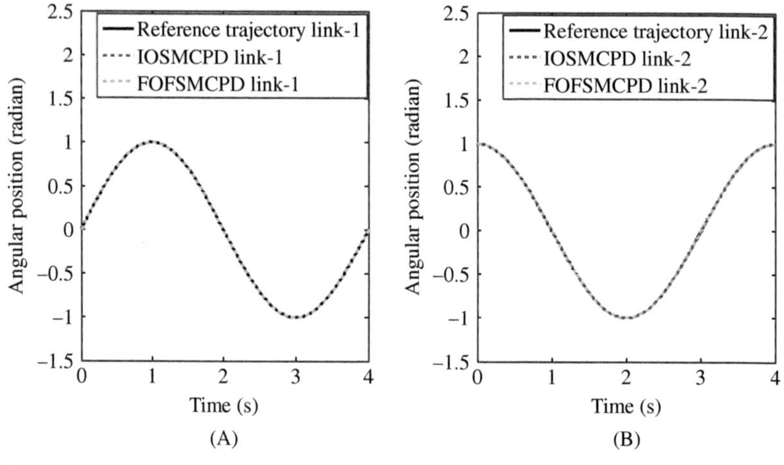

FIGURE 9.12 Trajectory tracking curves (A) link-1; (B) link-2.

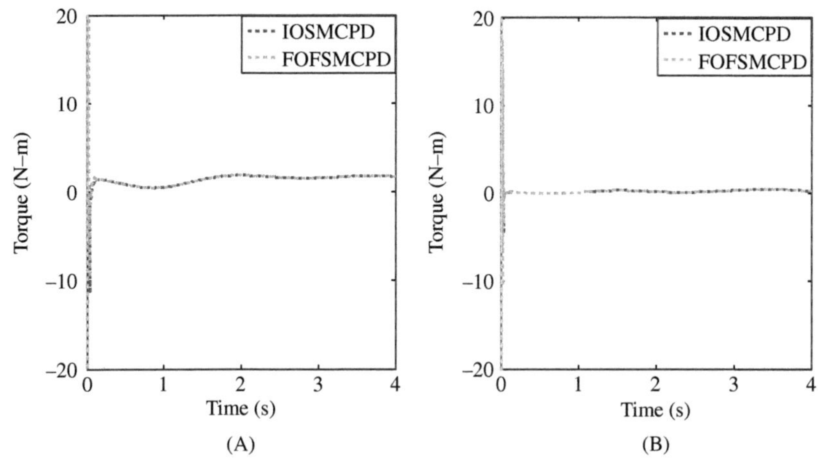

FIGURE 9.13 Controller outputs for trajectory tracking (A) link-1; (B) link-2.

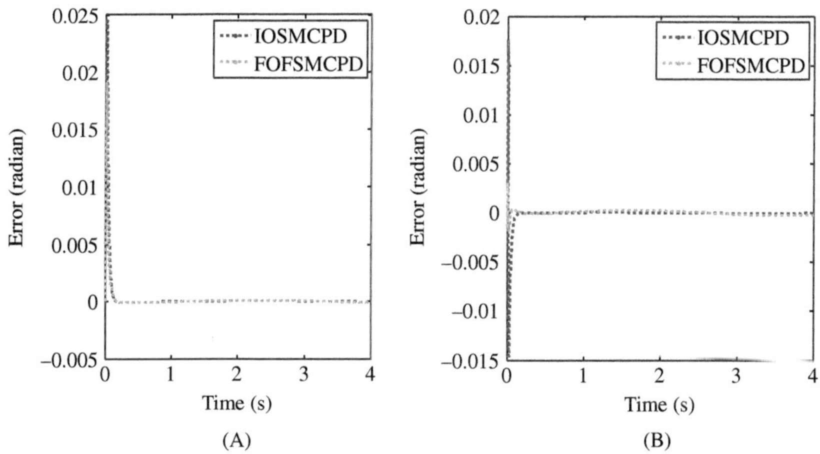

FIGURE 9.14 Error curves for trajectory tracking (A) link-1; (B) link-2.

function, the chattering phenomena is disappeared. $X-Y$ curve of end-point is also presented in Fig. 9.15 where a smooth curve is observed by the FOFSMCPD controller on the IOSMCPD controller.

9.7 DETAILED PERFORMANCE EVALUATION

A robust controller must have the capability to neutralize the effect of disturbances as well as parameter uncertainties. These could occur individually or in combination and the overall tendency would be to degrade the performance from the nominal designed one. Therefore, to further access the performance of the controller's investigations on disturbance rejection and model uncertainty,

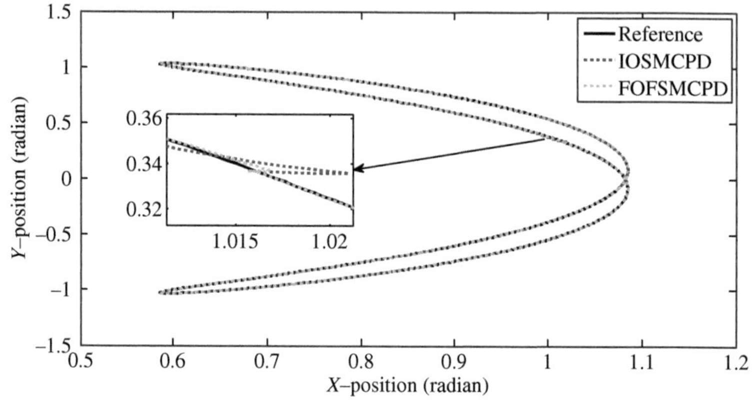

FIGURE 9.15 $X-Y$ curve for trajectory tracking (A) link-1; (B) link-2.

handling capabilities have been carried out and presented in this section. Following are the category wise investigations carried out on IOSMCPD and FOFSMCPD controllers to evaluate the complete performance analysis.

9.7.1 Disturbance Rejection When Disturbance Is Injected at Controller Output

Another task of a robust control scheme is to nullify the effect of unwanted disturbances so as to allow the reference trajectory to be followed smoothly. In the present study, a disturbance signal is injected at the controller output in both the feedback loops at the same time before the saturator. The disturbance signal is given as,

$$d(t) = A[u(t-1) - u(t-3)] \tag{9.69}$$

where, A is the amplitude of disturbances considerd for testing.

According to Eq. (9.69), a pulse signal of amplitude, $A = 0.1$ N-m is considered as the disturbance signal in the present study. A complete analysis is carried out where the obtained IAE values and objective function are depicted in Table 9.6 and same is presented graphically in Fig. 9.16. For this case, a trajectory tracking task, corresponding controller output and error curves are shown in Figs. 9.17–9.19. An $X-Y$ position curve of the endpoint is also shown in Fig. 9.20 where it can be easily observed that the FOFSMCPD controller displays a smoother curve than the IOSMCPD controller. After investigating the obtained values of IAE and objective function, it can be inferred that the FOFSMCPD controller outperforms the IOSMCPD controller. In this study, improvements of 7.69% and 23.80% are obtained in IAE values by the FOFSMCPD over the IOSMCPD controller in link-1 and link-2, respectively. About 18.91% improvement in objective functions are obtained by the FOFSMCPD controller over the IOSMCPD controller.

TABLE 9.6 IAE Values and Objective Functions for Disturbance (at Controller Output) Rejection Task

	IAE Link-1	IAE Link-2	Objective function
IOSMCPD	0.0013	0.0021	0.0037
FOFSMCPD	0.0012	0.0016	0.0030

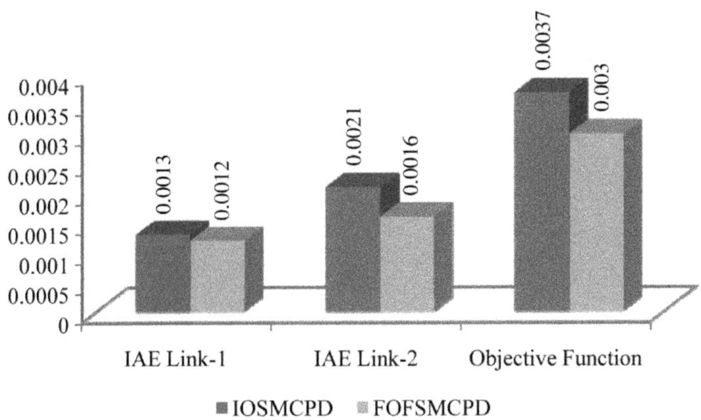

FIGURE 9.16 IAE and objective function values for disturbance (at controller output) rejection study.

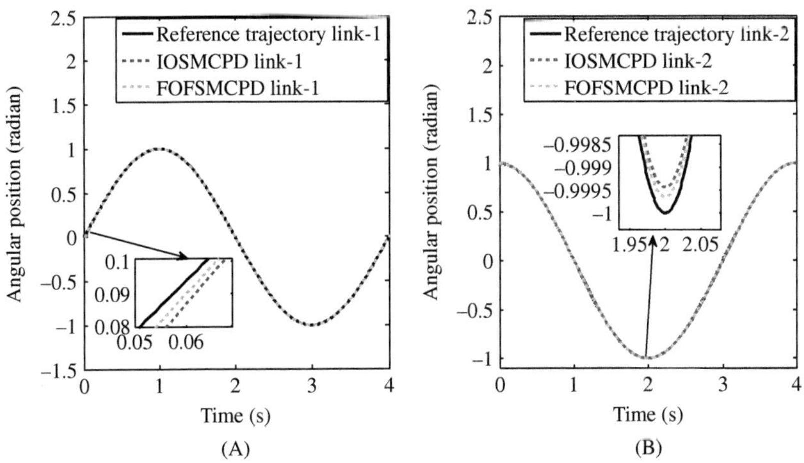

FIGURE 9.17 Trajectory tracking curves for disturbance (at controller output) rejection study (A) link-1; (B) link-2.

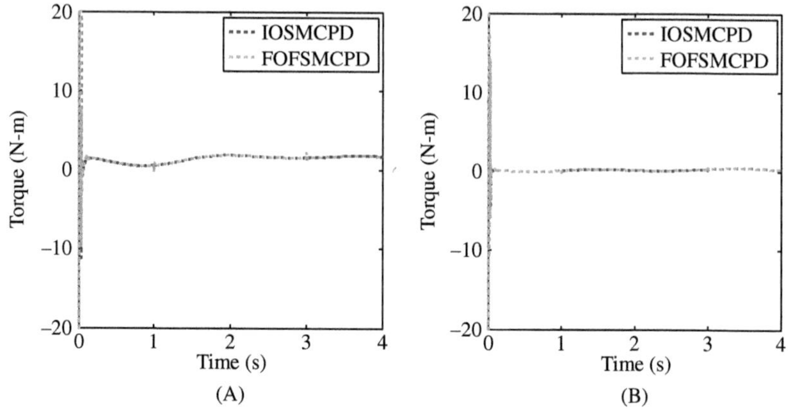

FIGURE 9.18 Corresponding controller output curve (A) link-1; (B) link-2.

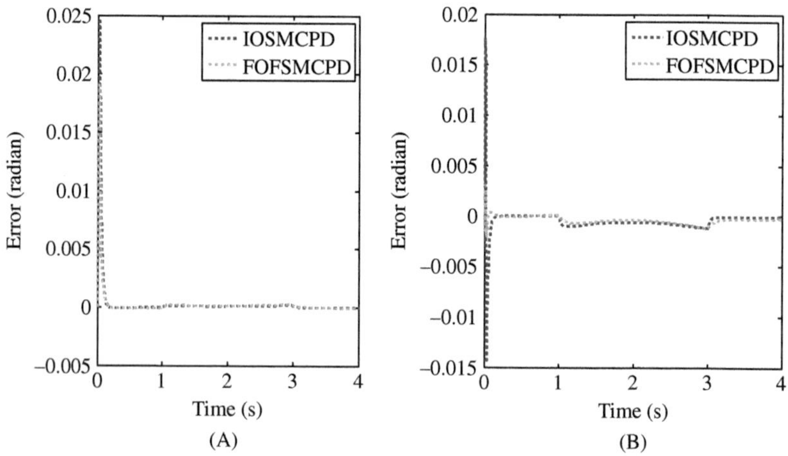

FIGURE 9.19 Corresponding error curve (A) link-1; (B) link-2.

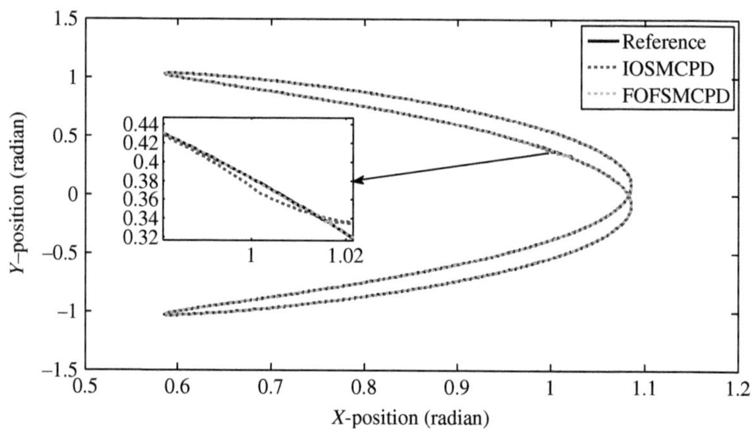

FIGURE 9.20 Corresponding $X-Y$ curve.

9.7.2 Disturbance Rejection When Disturbance Is Injected at Plant Output

Another disturbance signal is considered at the plant output to study the effectiveness of IOSMCPD and FOFSMCPD controllers. In this context, a sinusoidal disturbance signal as given in (9.70) is injected where amplitude $A_d = 0.1$ radian and angular frequency $\omega_d = 0.5$ pi rad/s is considered for disturbance rejection study.

$$\theta_d = \{A_d \sin \omega_d t\} \tag{9.70}$$

Analysis is shown in terms of trajectory tracking task where the obtained values of IAE and objective functions are shown in Table 9.7. The same values are expressed graphically in Fig. 9.21. For this study, a trajectory tracking curve is shown in Fig. 9.22 whereas its corresponding controller output and error curves are shown in Figs. 9.23 and 9.24. An $X-Y$ curve is also shown in Fig. 9.25 which confirms the superiority of the FOFSMCPD controller over the IOSMCPD controller. After investigating the result, it can be inferred that the FOFSMCPD controller shows 9.09% improvement in objective function over the IOSMCPD controller. After comparing the IAE values, 8.16% and 9.09% improvements are obtained by the FOFSMCPD over the IOSMCPD controller in link-1 and link-2, respectively.

TABLE 9.7 IAE Values and Objective Functions for Disturbance (at Plant Output) Rejection Task

	IAE Link-1	IAE Link-2	Objective function
IOSMCPD	0.00098	0.0011	0.0022
FOFSMCPD	0.00090	0.0010	0.0020

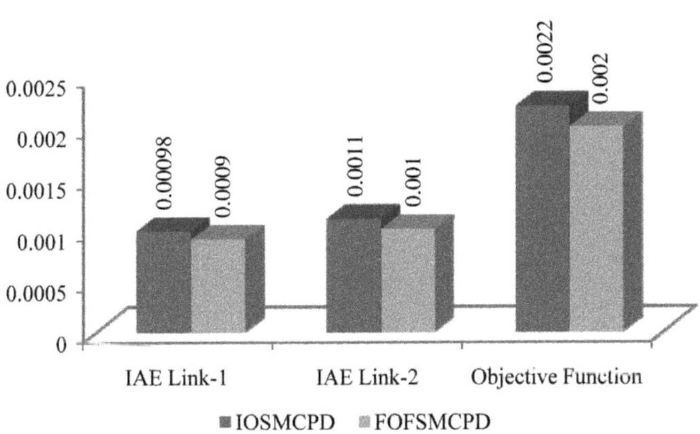

FIGURE 9.21 IAE and objective function values for disturbance (at plant output) rejection study.

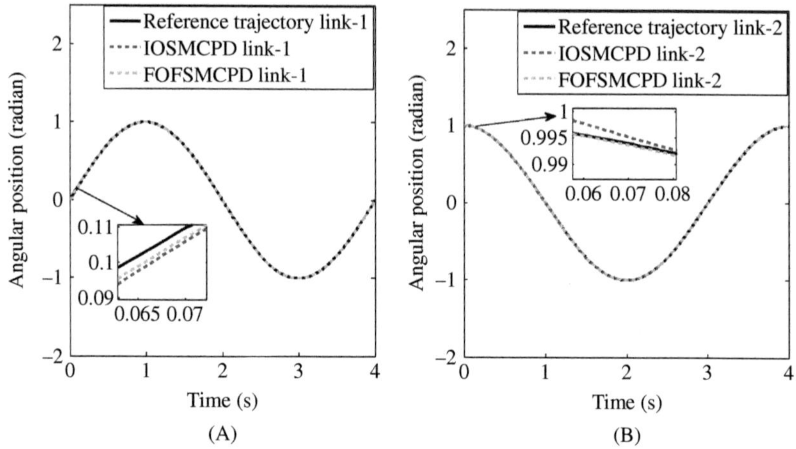

FIGURE 9.22 Trajectory tracking curves for disturbance (at plant output) rejection study (A) link-1; (B) link-2.

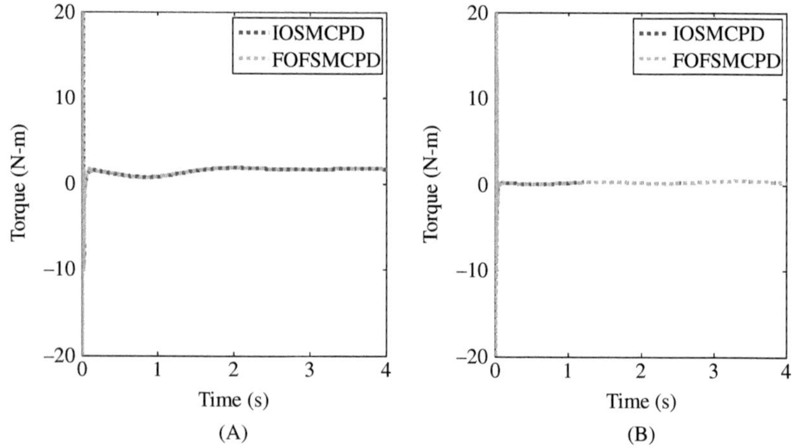

FIGURE 9.23 Corresponding controller output curve (A) link-1; (B) link-2.

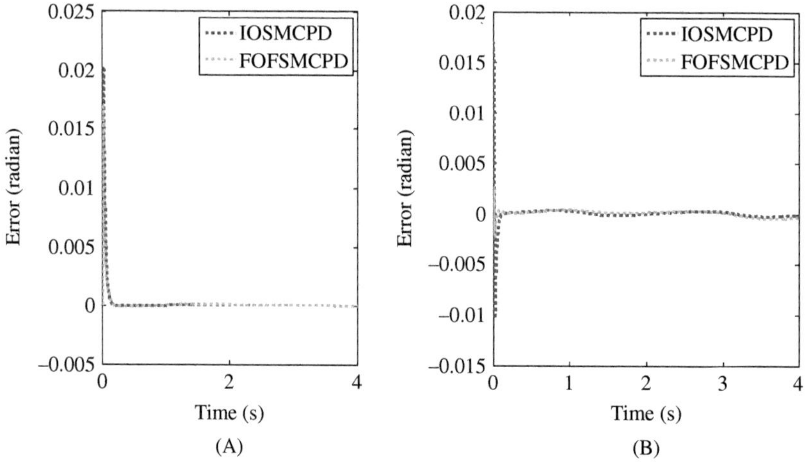

FIGURE 9.24 Corresponding error curve (A) link-1; (B) link-2.

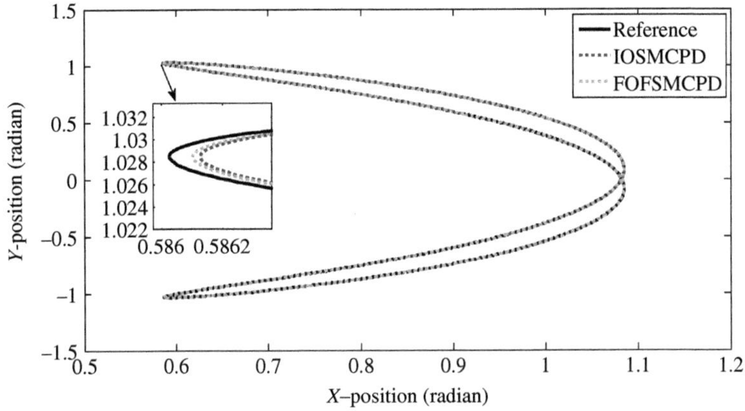

FIGURE 9.25 Corresponding $X-Y$ curve.

TABLE 9.8 IAE Values and Objective Functions for Length Uncertainty in Both Links

	IAE Link-1	IAE Link-2	Objective function
IOSMCPD	0.0015	0.0013	0.0031
FOFSMCPD	0.0013	0.0007	0.0022

9.7.3 Robustness Testing: Uncertainty in Length

To test the robustness of IOSMCPD and FOFSMCPD controllers, the length of both the links of the manipulator system is increased by 10%. Although the controllers have been tuned for the length of both links as 0.8 m for link-1 and 0.4 m for link-2, for robustness testing the length of both links are considered as 0.88 m and 0.44 m for link-1 and link-2, respectively. The obtained values of IAE and objective functions are listed in Table 9.8 where it can be easily observed that a 29.03% improvement in objective function is observed by FOFSMCPD over the IOSMCPD controller. Improvements of 13.33% and 46.15% in IAE values are also obtained by FOFSMCPD over the IOSMCPD controller for link-1 and link-2, respectively.

9.8 ROBUSTNESS TESTING: UNCERTAINTY IN MASS

Another robustness testing is carried out by increasing the mass of both the links by 10%. Although the controllers have been tuned for the mass of both links as 0.1 kg, for robustness testing the mass of both links are considered as 0.11 kg. Table 9.9 indicates the obtained values of IAE and objective

TABLE 9.9 IAE Values and Objective Functions for Mass Uncertainty in Both Links

	IAE Link-1	IAE Link-2	Objective function
IOSMCPD	0.0014	0.0011	0.0027
FOFSMCPD	0.0011	0.0007	0.0021

functions where it can be easily detected that a 22.22% improvement is observed by the FOFSMCPD controller over the IOSMCPD controller. Enhancements of 21.42% and 36.66% are also obtained by the FOFSMCPD over the IOSMCPD controller in IAE values for link-1 and link-2, respectively. On the basis of uncertainty analysis, it can be concluded that the FOFSMCPD controller outperforms the IOSMCPD controller.

9.9 CONCLUSION

In this chapter, integer order sliding mode proportional and derivative (IOSMCPD) controller and fractional order fuzzy sliding mode proportional and derivative (FOFSMCPD) controller are used to control a nonlinear, MIMO, coupled complex system, two-link robotic manipulator. The efficacy of controllers is tested for trajectory tracking task, disturbance rejection, and uncertainty analysis. In a classical sliding mode controller (SMC), chattering is a major problem. This problem is effectively handled by a combination of FL-based intelligent technique and boundary layer technique. Exponential law is used to design the SMC controllers which give the Lyapunov based stability of overall system. The performance index was taken as the weighted sum of integral of absolute error and chatter. The gains of the controllers are tuned by GA. On the basis of the obtained simulated results, it can be concluded that the FOFSMCPD controller outperforms the IOSMCPD controller in all the aspects of performances carried out for evaluation of controllers. As a future extension of this research work, performance of the proposed controller needs to be validated on a real-time hardware system, as well as other variants of the SMC like higher order SMC, Terminal mode SMC may be tried out.

REFERENCES

Åström, K.J., Wittenmark, B., 2008. Adaptive Control. Dover Publications, Mineola, New York.

Azar, A.T., 2010a. Fuzzy Systems. IN-TECH, Vienna, Austria, ISBN 978-953-7619-92-3.

Azar, A.T., 2010b. Adaptive neuro-fuzzy systems. In: Azar, A.T. (Ed.), Fuzzy Systems. IN-TECH, Vienna, Austria, ISBN 978-953-7619-92-3.

Azar, A.T., 2012. Overview of type-2 fuzzy logic systems. Int. J. Fuzzy System Applicat. (IJFSA) 2 (4), 1–28.

Azar, A.T., Serrano, F.E., 2014. Robust IMC-PID tuning for cascade control systems with gain and phase margin specificationsNeur. Comput. Applicat. 25 (5), 983–995Springer . Available from: http://dx.doi.org/10.1007/s00521-014-1560-x.

Azar, A.T., Serrano, F.E., 2015a. Stabilization and control of mechanical systems with backlash. In: Azar, A.T., Vaidyanathan, S. (Eds.), Advanced Intelligent Control Engineering and Automation, Advances in Computational Intelligence and Robotics (ACIR) Book Series. IGI-Global, USA.

Azar, A.T., Serrano, F.E., 2015b. Design and modeling of anti wind up PID controllers. In: Zhu, Q., Azar, A.T. (Eds.), Complex System Modelling and Control Through Intelligent Soft Computations, Studies in Fuzziness and Soft Computing, Vol. 319. Springer-Verlag, Germany, pp. 1–44. Available from: http://dx.doi.org/10.1007/978-3-319-12883-2_1.1.

Azar, A.T., Serrano, F.E., 2015c. Adaptive Sliding mode control of the Furuta pendulum. In: Azar, A.T., Zhu, Q. (Eds.), Advances and Applications in Sliding Mode Control Systems, Studies in Computational Intelligence, Vol. 576. Springer-Verlag GmbH Berlin/Heidelberg, pp. 1–42. Available from: http://dx.doi.org/10.1007/978-3-319-11173-5_1.

Azar, A.T., Serrano, F.E., 2015d. Deadbeat control for multivariable systems with time varying delays. In: Azar, A.T., Vaidyanathan, S. (Eds.), Chaos Modeling and Control Systems Design, Studies in Computational Intelligence, Vol. 581. Springer-Verlag GmbH Berlin/Heidelberg, pp. 97–132. Available from: http://dx.doi.org/10.1007/978-3-319-13132-0_6.

Azar, A.T., Serrano, F.E., 2016a. Robust control for asynchronous switched nonlinear systems with time varying delays. Proceedings of the International Conference on Advanced Intelligent Systems and Informatics 2016, Volume 533, Advances in Intelligent Systems and Computing, pp. 891–899.

Azar, A.T., Serrano, F.E., 2016b. Stabilization of mechanical systems with backlash by PI loop shaping. Int. J. System Dyn. Applicat. (IJSDA) 5 (3), 20–47.

Azar, A.T., Serrano, F.E., 2018. Passivity Based Decoupling of Lagrangian Systems. In: Hassanien A., Shaalan K., Gaber T., Tolba M. (eds) Proceedings of the International Conference on Advanced Intelligent Systems and Informatics 2017. AISI 2017. Advances in Intelligent Systems and Computing, vol 639. Springer, Cham.

Azar, A.T., Vaidyanathan, S., 2015a. Handbook of Research on Advanced Intelligent Control Engineering and Automation. Advances in Computational Intelligence and Robotics (ACIR) Book Series. IGI Global, USA, ISBN 9781466672482.

Azar, A.T., Vaidyanathan, S., 2015b. Computational Intelligence Applications in Modeling and Control. Studies in Computational Intelligence, Vol. 575. Springer-Verlag, Germany, ISBN 978-3-319-11016-5.

Azar, A.T., Vaidyanathan, S., 2015c. Chaos Modeling and Control Systems Design, Studies in Computational Intelligence, Vol. 581. Springer-Verlag, Germany, ISBN 978-3-319-13131-3.

Azar, A.T., Vaidyanathan, S., 2016. Advances in Chaos Theory and Intelligent Control. Studies in Fuzziness and Soft Computing, Vol. 337. Springer-Verlag, Germany, ISBN 978-3-319-30338-3.

Azar, A.T., Zhu, Q., 2015. Advances and Applications in Sliding Mode Control Systems. Studies in Computational Intelligence, Vol. 576. Springer-Verlag, Germany, ISBN 978-3-319-11172-8.

Azar, A.T., Vaidyanathan, S., Ouannas, A., 2017a. Fractional Order Control and Synchronization of Chaotic Systems. Studies in Computational Intelligence, Vol. 688. Springer-Verlag, Germany, ISBN 978-3-319-50248-9.

Azar, A.T., Volos, C., Gerodimos, N.A., Tombras, G.S., Pham, V.T., Radwan, A.G., et al., 2017b. A novel chaotic system without equilibrium: dynamics, synchronization and circuit realization. Complexity vol. 2017, 2017. Available from: https://doi.org/10.1155/2017/7871467. Article ID 7871467, 11 pages.

Azar, A.T., Ouannas, A., Singh, S., 2018a. Control of New Type of Fractional Chaos Synchronization. In: Hassanien A., Shaalan K., Gaber T., Tolba M. (eds) Proceedings of the International Conference on Advanced Intelligent Systems and Informatics 2017. AISI 2017. Advances in Intelligent Systems and Computing, vol 639, pp 47–56, Springer, Cham.

Azar, A.T., Kumar, J., Kumar, V., Rana, K.P.S., 2018b. Control of a Two Link Planar Electrically-Driven Rigid Robotic Manipulator Using Fractional Order SOFC. In: Hassanien A., Shaalan K., Gaber T., Tolba M. (eds) Proceedings of the International Conference on Advanced Intelligent Systems and Informatics 2017. AISI 2017. Advances in Intelligent Systems and Computing, vol 639, pp 57-68, Springer, Cham.

Boulkroune, A., Bouzeriba, A., Bouden, T., Azar, A.T., 2016a. Fuzzy adaptive synchronization of uncertain fractional-order chaotic systems. In: Azar, A.T., Vaidyanathan, S. (Eds.), Advances in Chaos Theory and Intelligent Control. Studies in Fuzziness and Soft Computing, Vol. 337. Springer-Verlag, Germany.

Boulkroune, A., Hamel, S., Azar, A.T., 2016b. Fuzzy control-based function synchronization of unknown chaotic systems with dead-zone input, Advances in Chaos Theory and Intelligent Control. Studies in Fuzziness and Soft Computing, Vol. 337. Springer-Verlag, Germany.

Chen, W.D., Wang, H.T., Tang, D.Z., Wang, H.R., 2003. Fuzzy sliding-mode control based on state for robotic manipulators. In: IEEE International Conference on Machine Learning and Cybernetics, pp. 957-960.

Craig, J.J., 2005. Introduction to Robotics: Mechanics and Control. Pearson/Prentice Hall, Upper Saddle River.

Delavari, H., Ghaderi, R., Ranjbar, A., Momani, S., 2010a. Fuzzy fractional order sliding mode controller for nonlinear systems. Commun. Nonlinear Sci. Numer. Simul. 15 (4), 963–978.

Delavari, H., Ranjbar, A.N., Ghaderi, R., Momani, S., 2010b. Fractional order control of a coupled tank. Nonlinear Dyn. 61 (3), 383–397.

Efe, M.Ö., 2008. Fractional fuzzy adaptive sliding-mode control of a 2-DOF direct-drive robot arm. IEEE Trans. Syst. Man Cybern. Part B 38 (6), 1561–1570.

Fallaha, C.J., Saad, M., Kanaan, H.Y., Al-Haddad, K., 2011. Sliding-mode robot control with exponential reaching law. IEEE Trans. Ind. Electron. 58 (2), 600–610.

Fang, Y., Chow, T.W.S., Li, X.D., 1999. Use of a recurrent neural network in discrete sliding-mode control. IEEE Proc. Control Theory Applicat. 146 (1), 84–90.

Ghoudelbourk, S., Dib, D., Omeiri, A., Azar, A.T., 2016. MPPT Control in wind energy conversion systems and the application of fractional control (PIα) in pitch wind turbine. Int. J. Model. Identif. Control (IJMIC) 26 (2), 140–151.

Grassi, G., Ouannas, A., Azar, A.T., Radwan, A.G., Volos, C., Pham, V.T., et al., 2017. Chaos Synchronisation Of Continuous Systems Via Scalar Signal. The 6th International Conference on Modern Circuits and Systems Technologies (MOCAST), 4-6 May 2017, Thessaloniki Greece.

Huang, S.J., Huang, K.S., Chiou, K.C., 2003. Development and application of a novel radial basis function sliding mode controller. Mechatronics 13 (4), 313–329.

Islam, S., Liu, P.X., 2011. Robust sliding mode control for robot manipulators. IEEE Trans. Ind. Electron. 58 (6), 2444–2453.

Karakasoglu, A., Sundareshan, M.K., 1995. A recurrent neural network based adaptive variable structure model-following control of robotic manipulators. Automatica 31 (10), 1495–1507.

Khalil, H.K., 1996. Nonlinear Control. Prentice Hall, Upper Saddle River, New Jersey.

Kumar, V., Rana, K.P.S., Kumar, J., Mishra, P., Nair, S.S., 2015. A robust fractional order Fuzzy P+ Fuzzy I+ Fuzzy D controller for nonlinear and uncertain system. Int. J. Automat. Comput. 14 (4), 1–15.

Kumar, V., Rana, K.P.S., Kumar, J., Mishra, P., 2016. Self-tuned robust fractional order fuzzy PD controller for uncertain and nonlinear active suspension system. Neur. Comput. Applicat. Available from: https://doi.org/10.1007/s00521-016-2774-x.

Liu, J., Wang, X., 2012. Advanced Sliding Mode Control for Mechanical Systems: Design, Analysis and MATLAB Simulation. Springer, Tsinghua University Press, Beijing.

Meghni, B., Dib, D., Azar, A.T., Ghoudelbourk, S., Saadoun, A., 2017a. Robust Adaptive Supervisory Fractional order Controller For optimal Energy Management in Wind Turbine with Battery Storage. Studies in Computational Intelligence, Vol. 688. Springer-Verlag, Germany, pp. 165–202.

Meghni, B., Dib, D., Azar, A.T., Saadoun, A., 2017b. Effective supervisory controller to extend optimal energy management in hybrid wind turbine under energy and reliability constraints. Int. J. Dyn. Control Springer . Available from: https://doi.org/10.1007/s40435-016-0296-0.

Meghni, B., Dib, D., Azar, A.T., 2017c. A second-order sliding mode and fuzzy logic control to Optimal Energy Management in PMSG Wind Turbine with Battery Storage. Neur. Comput. Applicat. 28 (6), 1417–1434. Available from: https://doi.org/10.1007/s00521-015-2161-z.

Mekki, H., Boukhetala, D., Azar, A.T., 2015. Sliding modes for fault tolerant control. In: Azar, A.T., Zhu, Q. (Eds.), Advances and Applications in Sliding Mode Control systems, Studies in Computational Intelligence book Series, Vol. 576. Springer-Verlag GmbH Berlin/Heidelberg, pp. 407–433. Available from: http://dx.doi.org/10.1007/978-3-319-11173-5_15.

Mishra, P., Kumar, V., Rana, K.P.S., 2015. A fractional order fuzzy PID controller for binary distillation column control. Exp. Syst. Applicat. 42 (22), 8533–8549.

Mu, X., 2010. Fuzzy neural sliding mode control based on genetic algorithm for multi-link robots. In: IEEE Chinese Control and Decision Conference (CCDC), pp. 1766–1770.

Munoz, D., Sbarbaro, D., 2000. An adaptive sliding-mode controller for discrete nonlinear systems. IEEE Trans. Ind. Electron. 43 (3), 574–581.

Niu, Y., Lam, J., Wang, X., Ho, D.W.C., 2003. Sliding-mode control for nonlinear state-delayed systems using neural-network approximation. IEEE Proc. Control Theory Applicat. 150 (3), 233–239.

Ouannas, A., Azar, A.T., Abu-Saris, R., 2016a. A new type of hybrid synchronization between arbitrary hyperchaotic maps. Int. J. Mach. Learn. Cybern. Available from: https://doi.org/10.1007/s13042-016-0566-3.

Ouannas, A., Azar, A.T., Radwan, A.G., 2016b. On Inverse Problem of Generalized Synchronization Between Different Dimensional Integer-Order and Fractional-Order Chaotic Systems. The 28th International Conference on Microelectronics, IEEE, December 17–20, 2016, Cairo, Egypt.

Ouannas, A., Azar, A.T., Ziar, T., Vaidyanathan, S., 2017a. On New Fractional Inverse Matrix Projective Synchronization Schemes. Studies in Computational Intelligence, Vol. 688. Springer-Verlag, Germany, pp. 497–524.

Ouannas, A., Azar, A.T., Ziar, T., Vaidyanathan, S., 2017b. Fractional Inverse Generalized Chaos Synchronization Between Different Dimensional Systems. Studies in Computational Intelligence, Vol. 688. Springer-Verlag, Germany, pp. 525–551.

Ouannas, A., Azar, A.T., Ziar, T., Vaidyanathan, S., 2017c. A New Method To Synchronize Fractional Chaotic Systems With Different Dimensions. Studies in Computational Intelligence, Vol. 688. Springer-Verlag, Germany, pp. 581–611.

Ouannas, A., Azar, A.T., Ziar, T., Radwan, A.G., 2017d. Study On Coexistence of Different Types of Synchronization Between Different dimensional Fractional Chaotic Systems. Studies in Computational Intelligence, Vol. 688. Springer-Verlag, Germany, pp. 637–669.

Ouannas, A., Azar, A.T., Ziar, T., Radwan, A.G., 2017e. Generalized Synchronization of Different Dimensional Integer-order and Fractional Order Chaotic Systems. Studies in Computational Intelligence, Vol. 688. Springer-Verlag, Germany, pp. 671–697.

Ouannas, A., Azar, A.T., Vaidyanathan, S., 2017f. On a simple approach for Q-S synchronization of chaotic dynamical systems in continuous-time. Int. J. Comput. Sci. Math. 8 (1), 20–27.

Ouannas, A., Azar, A.T., Vaidyanathan, S., 2017g. New hybrid synchronization schemes based on coexistence of various types of synchronization between master-slave hyperchaotic systems. Int. J. Comput. Applicat. Technol. 55 (2), 112–120.

Ouannas, A., Azar, A.T., Ziar, T., 2017h. On inverse full state hybrid function projective synchronization for continuous-time chaotic dynamical systems with arbitrary dimensions. Diff. Equat. Dyn. Syst. Available from: https://doi.org/10.1007/s12591-017-0362-x.

Ouannas, A., Azar, A.T., Vaidyanathan, S., 2017i. A robust method for new fractional hybrid chaos synchronization. Math. Methods Appl. Sci. 40 (5), 1804–1812. Available from: https://doi.org/10.1002/mma.4099.

Ouannas, A., Grassi, G., Azar, A.T., Radwan, A.G., Volos, C., Pham, V.T., et al., 2017j. Deadbeat synchronization control in discrete-time chaotic systems. In: The 6th International Conference on Modern Circuits and Systems Technologies (MOCAST), 4-6 May 2017, Thessaloniki Greece.

Roopaei, M., Jahromi, M.Z., 2009. Chattering-free fuzzy sliding mode control in MIMO uncertain systems. Nonlinear Anal. Theory Methods Applicat. 71 (10), 4430–4437.

Sadati, N., Ghadami, R., 2008. Adaptive multi-model sliding mode control of robotic manipulators using soft computing. Neurocomputing 71 (2), 2702–2710.

Sharma, R., Rana, K.P.S., Kumar, V., 2014. Performance analysis of fractional order fuzzy PID controllers applied to a robotic manipulator. Exp. Syst. Applicat. 41 (9), 4274–4289.

Singh, S., Azar, A.T., Ouannas, A., Zhu, Q., Zhang, W., Na, J., 2017. Sliding mode control technique for multi-switching synchronization of chaotic systems. In: 9th International Conference on Modelling, Identification and Control (ICMIC 2017), July 10-12, 2017, Kunming, China.

Soliman, N.S., Said, L.A., Azar, A.T., Madian, A.H., Radwan, A.G., Ouannas, A., 2017. Fractional controllable multi-scroll v-shape attractor with parameters effect. In: The 6th International Conference on Modern Circuits and Systems Technologies (MOCAST), 4-6 May 2017, Thessaloniki Greece.

Tolba, M.F., AbdelAty, A.M., Soliman, N.S., Said, L.A., Madian, A.H., Azar, A.T., et al., 2017. FPGA implementation of two fractional order chaotic systems. Int. J. Electr. Commun. 28 (2017), 162–172.

Utkin, V.I., 1977. Variable structure systems with sliding modes. IEEE Trans. Autom. Control. 22 (2), 212–222.

Vaidyanathan, S., Azar, A.T., 2015a. Anti-synchronization of identical chaotic systems using sliding mode control and an application to vaidyanathan-madhavan chaotic systems. In: Azar, A.T., Zhu, Q. (Eds.), Advances and Applications in Sliding Mode Control systems, Studies in Computational Intelligence book Series, Vol. 576. Springer-Verlag GmbH Berlin/Heidelberg, pp. 527–547. Available from: http://dx.doi.org/10.1007/978-3-319-11173-5_19.

Vaidyanathan, S., Azar, A.T., 2015b. Hybrid synchronization of identical chaotic systems using sliding mode control and an application to vaidyanathan chaotic systems. In: Azar, A.T., Zhu, Q. (Eds.), Advances and Applications in Sliding Mode Control systems, Studies in Computational Intelligence book Series, Vol. 576. Springer-Verlag GmbH Berlin/Heidelberg, pp. 549–569. Available from: http://dx.doi.org/10.1007/978-3-319-11173-5_20.

Vaidyanathan, S., Azar, A.T., 2015c. Analysis, control and synchronization of a nine-term 3-D novel chaotic system. In: Azar, A.T., Vaidyanathan, S. (Eds.), Chaos Modeling and Control Systems Design, Studies in Computational Intelligence, Vol. 581. Springer-Verlag GmbH Berlin/Heidelberg, pp. 3–17. Available from: http://dx.doi.org/10.1007/978-3-319-13132-0_1.

Vaidyanathan, S., Azar, A.T., 2015d. Analysis and control of a 4-D novel hyperchaotic system. In: Azar, A.T., Vaidyanathan, S. (Eds.), Chaos Modeling and Control Systems Design, Studies in Computational Intelligence, Vol. 581. Springer-Verlag GmbH Berlin/Heidelberg, pp. 19–38. Available from: http://dx.doi.org/10.1007/978-3-319-13132-0_2.

Vaidyanathan, S., Azar, A.T., 2016a. Takagi-Sugeno Fuzzy logic controller for liu-chen four-scroll chaotic system. Int. J. Intel. Eng. Infor. 4 (2), 135–150.

Vaidyanathan, S., Azar, A.T., 2016b. Dynamic Analysis, Adaptive Feedback Control and Synchronization of an Eight-Term 3-D Novel Chaotic System with Three Quadratic Nonlinearities. Studies in Fuzziness and Soft Computing, Vol. 337. Springer-Verlag, Germany, pp. 155–178.

Vaidyanathan, S., Azar, A.T., 2016c. Qualitative Study and Adaptive Control of a Novel 4-D Hyperchaotic System with Three Quadratic Nonlinearities. Studies in Fuzziness and Soft Computing, Vol. 337. Springer-Verlag, Germany, pp. 179–202.

Vaidyanathan, S., Azar, A.T., 2016d. A Novel 4-D Four-Wing Chaotic System with Four Quadratic Nonlinearities and its Synchronization via Adaptive Control Method. Advances in Chaos Theory and Intelligent Control. Studies in Fuzziness and Soft Computing, Vol. 337. Springer-Verlag, Germany, pp. 203–224.

Vaidyanathan, S., Azar, A.T., 2016e. Adaptive Control and Synchronization of Halvorsen Circulant Chaotic Systems. Advances in Chaos Theory and Intelligent Control. Studies in Fuzziness and Soft Computing, Vol. 337. Springer-Verlag, Germany, pp. 225–247.

Vaidyanathan, S., Azar, A.T., 2016f. Adaptive Backstepping Control and Synchronization of a Novel 3-D Jerk System with an Exponential Nonlinearity. Advances in Chaos Theory and Intelligent Control. Studies in Fuzziness and Soft Computing, Vol. 337. Springer-Verlag, Germany, pp. 249–274.

Vaidyanathan, S., Azar, A.T., 2016g. Generalized Projective Synchronization of a Novel Hyperchaotic Four-Wing System via Adaptive Control Method. Advances in Chaos Theory and Intelligent Control. Studies in Fuzziness and Soft Computing, Vol. 337. Springer-Verlag, Germany, pp. 275–296.

Vaidyanathan, S., Sampath, S., Azar, A.T., 2015a. Global chaos synchronisation of identical chaotic systems via novel sliding mode control method and its application to Zhu system. Int. J. Model. Identif. Control (IJMIC) 23 (1), 92–100.

Vaidyanathan, S., Azar, A.T., Rajagopal, K., Alexander, P., 2015b. Design and SPICE implementation of a 12-term novel hyperchaotic system and its synchronization via active control (2015). Int. J. Model. Identif. Control (IJMIC) 23 (3), 267–277.

Vaidyanathan, S., Idowu, B.A., Azar, A.T., 2015c. Backstepping controller design for the global chaos synchronization of sprott's jerk systems. In: Azar, A.T., Vaidyanathan, S. (Eds.), Chaos Modeling and Control Systems Design, Studies in Computational Intelligence, Vol. 581. Springer-Verlag GmbH Berlin/Heidelberg, pp. 39–58. Available from: http://dx.doi.org/10.1007/978-3-319-13132-0_3.

Vaidyanathan, S., Azar, A.T., Ouannas, A., 2017a. An Eight-Term 3-D Novel Chaotic System with Three Quadratic Nonlinearities, its Adaptive Feedback Control and Synchronization. Studies in Computational Intelligence, Vol. 688. Springer-Verlag, Germany, pp. 719–746.

Vaidyanathan, S., Zhu, Q., Azar, A.T., 2017b. Adaptive Control of a Novel Nonlinear Double Convection Chaotic System. Studies in Computational Intelligence, Vol. 688. Springer-Verlag, Germany, pp. 357–385.

Vaidyanathan, S., Azar, A.T., Ouannas, A., 2017c. Hyperchaos and Adaptive Control of a Novel Hyperchaotic System with Two Quadratic Nonlinearities. Studies in Computational Intelligence, Vol. 688. Springer-Verlag, Germany, pp. 773–803.

Wai, R.J., 2007. Adaptive fuzzy sliding-mode control for electrical servo drive. IEEE Trans. Ind. Electron. 54 (1), 586–594.

Wang, Z., Volos, C., Kingni, S.T., Azar, A.T., Pham, V.T., 2017. Four-wing attractors in a novel chaotic system with hyperbolic sine nonlinearity. Optik – Int. J. Light Electr. Opt. 131 (2017), 1071–1078.

Wu, J., Ham, T.S., 1996. A sliding-mode approach to fuzzy control design. IEEE Trans. Control Syst. Technol. 4 (2), 141–151.

Yang, X.S., 2014. Nature-Inspired Optimization Algorithms. Elsevier, Jamestown Road, London.

Yu, X., Kaynak, O., 2009. Sliding-mode control with soft computing: a survey. IEEE Trans. Ind. Electron. 56 (9), 3275–3285.

Zhao, D., Li, S., Gao, F., 2009. A new terminal sliding mode control for robotic manipulators. Int. J. Control 82 (10), 1804–1813.

Zhu, Q., Azar, A.T., 2015. Complex system modelling and control through intelligent soft computations, Studies in Fuzziness and Soft Computing, Vol. 319. Springer-Verlag, Germany, ISBN 978-3-319-12882-5.

Chapter 10

Sliding Mode Stabilization and Synchronization of Fractional Order Complex Chaotic and Hyperchaotic Systems

Ahmad Taher Azar[1,2], Fernando E. Serranot[3] and Sundarapandian Vaidyanathan[4]

[1]*Faculty of Computers and Information, Benha University, Benha, Egypt,* [2]*School of Engineering and Applied Sciences, Nile University, Giza, Egypt,* [3]*Central American Technical University (UNITEC), Tegucigalpa, Honduras,* [4]*Vel Tech University, Chennai, Tamil Nadu, India*

10.1 INTRODUCTION

Chaotic systems are dynamical systems that are highly sensitive to initial conditions. This sensitivity is popularly known as the butterfly effect (Rössler, 1976; Lorenz, 1963; Sprott, 1994; Azar and Vaidyanathan, 2016, 2015a,b,c; Zhu and Azar, 2015). The chaos phenomenon was first observed in weather models by Lorenz (1963). The Lyapunov exponent is a measure of the divergence of phase points that are initially very close and can be used to quantify chaotic systems. A positive maximal Lyapunov exponent and phase space compactness are usually taken as defining conditions for a chaotic system. Since the pioneering work by Pecora and Carroll (1990), the chaos synchronization problem has been studied extensively in the literature. Synchronization of chaotic systems is a phenomenon that occurs when two or more chaotic systems are coupled or when a chaotic system drives another chaotic system. Because of the butterfly effect which causes exponential divergence of the trajectories of two identical chaotic systems started with nearly the same initial conditions, the synchronization of chaotic systems is a challenging research problem in the chaos literature (Boulkroune et al., 2016b; Vaidyanathan and Azar, 2015a,b,c,d, 2016a,b,c,d,e,f,g; Wang et al., 2017; Vaidyanathan et al., 2015a,b,c, 2017a,b,c; Ouannas et al., 2017a,b,c, 2016b; Azar et al., 2018b; Grassi et al., 2017; Singh et al., 2017; Azar and

Mathematical Techniques of Fractional Order Systems. DOI: https://doi.org/10.1016/B978-0-12-813592-1.00010-6

Vaidyanathan, 2016; Pham et al., 2017a; Moysis and Azar, 2017; Lamamra et al., 2017). Fractional order complex chaotic and hyperchaotic systems have been recently studied nowadays due to the vast number of physical systems in which this phenomenon is found; systems such as mechanical, electrical, chemical, biological, renewable energy, and control systems (Zhang et al., 2015; Azar et al., 2018a,b, 2017; Ouannas et al., 2016a,b, 2017d,e,f,g, h,i,j; Pham et al., 2017a,b; Tolba et al., 2017a,b; Meghni et al., 2017; Boulkroune et al., 2016a; Ghoudelbourk et al., 2016). For this reason, it's important to design efficient control and synchronization strategies to stabilize these kinds of system to reach the equilibrium point or to synchronize a response system following a reference from a drive system that can be identical or nonidentical.

The stabilization problem for integer order or fractional order complex chaotic system has been recently studied. For example, in El-Sayed et al. (2016), the circuit realization of a fractional order hyperchaotic system is done by designing a suitable control law and studying the effects of the fractional order derivatives. In Zhang et al. (2008) an interesting example in which an analysis of the chaotic and hyperchaotic behavior of a nonautonomous rotational machine is shown where a study of the eigenvalues and the Lyapunov exponents of this mechanical system is done. Something similar is found in Zhang et al. (2015), where a stability analysis of a fractional order nonlinear system is provided considering an order between 0 and 2. Another important work is evinced in Li (2016), where an adaptive tracking control for a fractional order chaotic system is shown considering and not considering the system uncertainties. An interesting example of a fractional order chaotic system is shown in Huang et al. (2014) where a fuzzy state feedback controller is designed, and one important remark of this work is that the state feedback gains are found by linear matrix inequalities (LMIs). A robust controller for a four-wing fractional order hyperchaotic system is shown in Li et al. (2013b) where by using a Lyapunov fractional order stability theorem a state feedback law is found. In Li and Li (2015), an interesting example in which an adaptive integer and fractional order controller is implemented for the stabilization of a fractional order chaotic systems where the Barbalat's Lyapunov-like stability theorem is implemented to derive the proposed controller strategy. Finally in Soukkou et al. (2016), a fractional order hyperchaotic system is stabilized by a generalized prediction based control where the control law is obtained by the Lyapunov and fractional order system stability theorems.

The synchronization of fractional order chaotic and hyperchaotic systems has been reported in the literature. For example, in Abedini et al. (2014), an identical system synchronization using a fractional adaptation law for a 4D Lu hyperchaotic system is shown where the fractional adaptation law is implemented to reduce the convergence of the system parameters. In Su et al. (2016), an interesting example related to the efficient numerical simulation of fractional chaotic systems along with its synchronization is shown where the optimization of numerical algorithm for the simulation of fractional order chaotic systems is shown considering that this reference is important for this study because it's necessary to implement efficient numerical solvers for fractional order chaotic and hyperchaotic complex systems. In Cruz-Ancona and Martnez-Guerra (2017), fractional order controllers for the multisynchronization of fractional order Liouvillian chaotic systems are shown considering that adding diffusive coupling terms in the dynamical controller solves the synchronization problem. Another interesting example can be found in Li et al. (2013a), where fractional order chaotic and hyperchaotic systems are synchronized by a robust control approach. In Xi et al. (2014), Gao et al. (2015), Wang et al. (2014), and Xi et al. (2015), adaptive synchronization controllers for fractional order chaotic and hyperchaotic systems are shown where this approach is important for these studies because the fractional order complex chaotic and hyperchaotic systems are stabilized by an adaptive terminal sliding mode controller.

Fractional complex chaotic and hyperchaotic systems have been recently studied due to the vast number of applications, so the control and synchronization of this kind of systems is of increased importance. For example, in Sun et al. (2016), Matouk (2011), Si et al. (2012), and Wong et al. (2012), different synchronization strategies are shown, but specifically in Si et al. (2012), an interesting approach is shown considering that the system synchronization is done with systems of different fractional orders. In Akbarzadeh-T. et al. (2017), an intelligent synchronization approach is done for implementing a fuzzy type-2 and sliding mode controller to synchronize two chaotic systems. Fractional order hyperchaotic systems synchronization have been studied in Mahmoud (2014), Wang et al. (2014), and Rajagopal et al. (2016), where different synchronization strategies for hyperchaotic systems are shown proving the effectiveness of the proposed approaches. Based on the works of Komurcugil (2012) and Aghababa (2015), the stabilization of a fractional order complex Chen chaotic system is discussed and then a fractional order complex Lorenz hyperchaotic system is shown where the systems are divided between real and imaginary parts in order to design the appropriated controllers to stabilize the systems in their equilibrium points, so the terminal sliding mode controller is implemented. Then the synchronization of nonidentical and identical chaotic and hyperchaotic systems is presented respectively by an adaptive terminal sliding mode controller technique where a fractional order complex chaotic Lorenz system (response system) is synchronized with a fractional order complex chaotic Chen system (drive system) and then a hyperchaotic Lorenz system is synchronized with an identical system.

This chapter is divided in the following sections: in Section 10.2 the fractional order calculus preliminary as well as the chaotic and hyperchaotic problem establishment are shown. In Section 10.3, the stabilization of chaotic and hyperchaotic system is presented. In Section 10.4, the synchronization of chaotic and hyperchaotic systems are shown. In Section 10.5, numerical examples are illustrated. Finally, in Sections 10.6 and 10.7, discussions and conclusion are presented.

10.2 PROBLEM FORMULATION

In this section, some preliminaries of fractional order calculus are shown, specifically, the Riemman–Liouville integration and derivation along with some fractional order calculus properties. Then, the studied chaotic and hyperchaotic systems are analyzed and established because they are used later in the stabilization and synchronization strategies designed in this study. The chaotic systems studied in this chapter are Chen chaotic system and Lorenz chaotic system (Sun et al., 2016; Mahmoud, 2014) and the studied hyperchaotic system is the Lorenz hyperchaotic system (Wang et al., 2014).

10.2.1 Fractional Order Calculus Preliminaries

Consider the function $f(t)$ with the fractional order $\alpha \in (0,1)$ so the fractional order integral is given by Aghababa, 2015):

$$ {}_{t_0}I_t^{\alpha} f(t) = {}_{t_0}D_t^{-\alpha} = \frac{1}{\Gamma(\alpha)} \int_{t_0}^{t} \frac{f(\tau)}{(t-\tau)^{1-\alpha}} d\tau \tag{10.1} $$

where $\Gamma(.)$ is the Gamma function. The Riemann–Liouville fractional derivative is given by (Aghababa, 2015)

$$ {}_{t_0}^{RL}D_t^{\alpha} f(t) = \frac{d^{\alpha} f(t)}{dt^{\alpha}} = \frac{1}{\Gamma(m-\alpha)} \frac{d^m}{dt^m} \int_{t_0}^{t} \frac{f(\tau)}{(t-\tau)^{\alpha-m+1}} d\tau \tag{10.2} $$

where $m-1 < \alpha \le m$ and $m \in \mathbb{N}$. With the following properties (Si et al., 2012)

- If $\alpha = n$ then $D^{\alpha} f(t) = \frac{d^n}{dt^n} f(t)$
- If $\alpha = 0$ then $D^{\alpha} f(t) = f(t)$
- $D^{\alpha}[af(t) + bg(t)] = aD^{\alpha} f(t) + bD^{\alpha} f(t)$
- $D^{\alpha} D^{-\alpha} = D^0 f(t) = f(t)$

10.2.2 Chaotic and Hyperchaotic Systems Used in This Study

Consider the following Chen chaotic system (Mahmoud, 2014; Sun et al., 2016):

$$ \begin{aligned} D^{\alpha} z_1 &= a_1(z_2 - z_1) + u_1 \\ D^{\alpha} z_2 &= (a_3 - a_1) z_1 - z_1 z_3 + a_3 z_2 + u_2 \\ D^{\alpha} z_3 &= \frac{1}{2}(\bar{z}_1 z_2 + z_1 \bar{z}_2) - a_2 z_3 + u_3 \end{aligned} \tag{10.3} $$

where $z_i = z_{ri} + z_{iim} j$, $z_i \in \mathbb{C}$ for $i = 1, 2$, $z_i = z_{ri}$, $z_i \in \mathbb{R}$ for $i = 3$, and finally $\bar{z}_i$ is the complex conjugate. The input variables are $u_i = u_{ri} + u_{iim} j$, $u_i \in \mathbb{C}$ for

$i = 1, 2$, $u_i = u_{ri}$, $u_i \in \mathbb{R}$ for $i = 3$. In order to set the system in chaotic regime the constant values must be $a_1 = 35$, $a_2 = 3$, and $a_3 = 28$ taking into account that the fractional order is $\alpha = 0.95$ and the following initial condition $z(0) = [0.3+0.2j, 0.1+0.3j, -0.1]^T$ so the following phase portraits are obtained for the real and imaginary parts, respectively, as shown in Figs. 10.1 and 10.2.

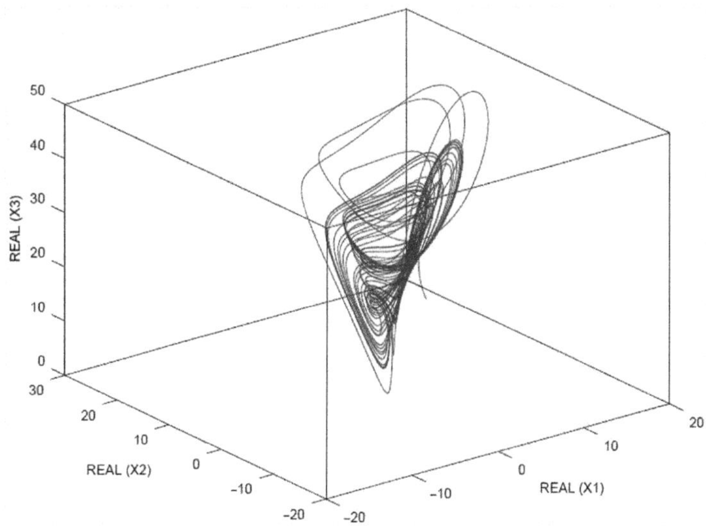

FIGURE 10.1 Phase portrait of the real part of the Chen chaotic system.

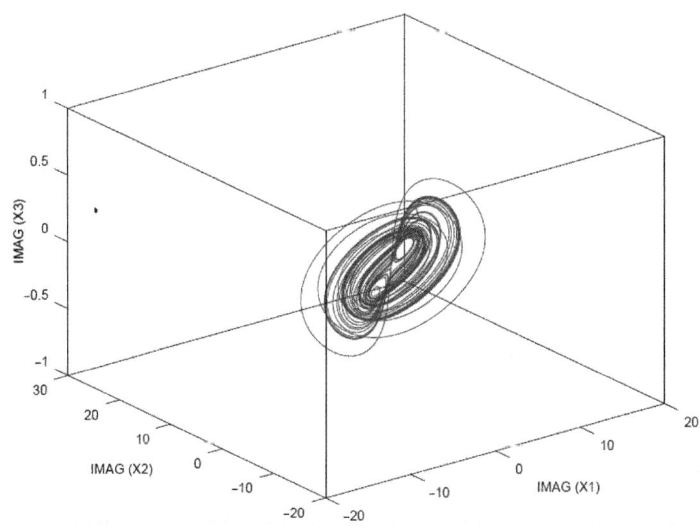

FIGURE 10.2 Phase portrait of the imaginary part of the Chen chaotic system.

Then consider the following Lorenz system (Sun et al., 2016):

$$\begin{aligned}
D^{\alpha}y_1 &= b_1(y_2 - y_1) + u_1 \\
D^{\alpha}y_2 &= b_2y_1 - y_2 - y_1y_3 + u_2 \\
D^{\alpha}y_3 &= -b_3y_3 + \frac{1}{2}(\overline{y}_1y_2 + y_1\overline{y}_2) + u_3
\end{aligned} \tag{10.4}$$

where $y_i = y_{ri} + y_{iim}j$, $y_i \in \mathbb{C}$ for $i = 1, 2$, $y_i = y_{ri}$, $y_i \in \mathbb{R}$ for $i = 3$, and finally $\overline{y}_i$ is the complex conjugate. The input variables are $u_i = u_{ri} + u_{iim}j$, $u_i \in \mathbb{C}$ for $i = 1, 2$, $u_i = u_{ri}$, $u_i \in \mathbb{R}$ for $i = 3$. In order to set the system in chaotic regime the constant values must be $b_1 = 10$, $b_2 = 180$, and $b_3 = 1$ with the initial condition $y(0) = [0.1 + 0.1j, 0.1 + 0.1j, 0.1]^T$ and $\alpha = 0.95$ so the following phase portraits for the real and imaginary parts are shown in Figs. 10.3 and 10.4, respectively.

Finally the following Lorenz hyperchaotic system is given by (Wang et al., 2014)

$$\begin{aligned}
D^{\alpha}y_1 &= c_1(y_2 - y_1) + y_4j + u_1 \\
D^{\alpha}y_2 &= c_3y_1 - y_2 - y_1y_3 + y_4j + u_2 \\
D^{\alpha}y_3 &= \frac{1}{2}(\overline{y}_1y_2 + y_1\overline{y}_2) - c_2y_3 + u_3 \\
D^{\alpha}y_4 &= \frac{1}{2}(\overline{y}_1y_2 + y_1\overline{y}_2) - c_4y_4 + u_4
\end{aligned} \tag{10.5}$$

where $y_i = y_{ri} + y_{iim}j$, $y_i \in \mathbb{C}$ for $i = 1, 2$, $y_i = y_{ri}$, $y_i \in \mathbb{R}$ for $i = 3, 4$, and finally $\overline{y}_i$ is the complex conjugate. The input variables are $u_i = u_{ri} + u_{iim}j$,

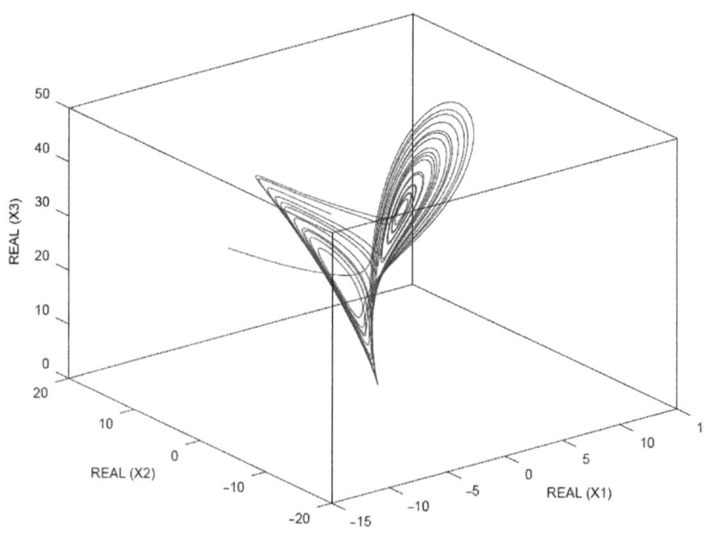

FIGURE 10.3 Phase portrait of the real part of the chaotic Lorenz system.

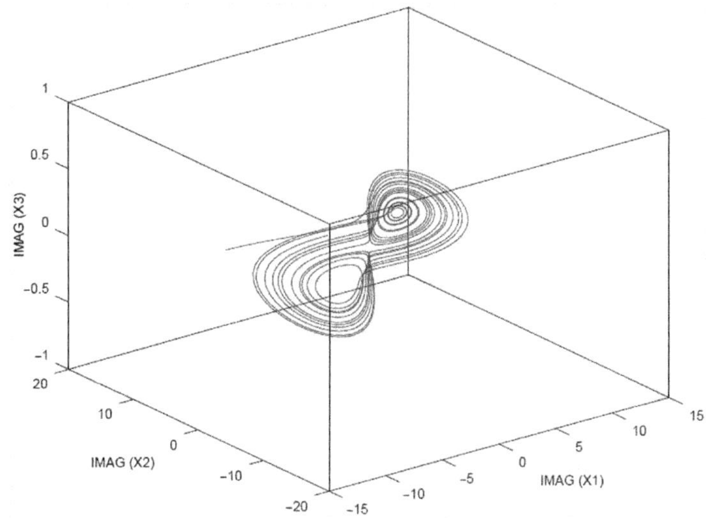

FIGURE 10.4 Phase portrait of the imaginary part of the chaotic Lorenz system.

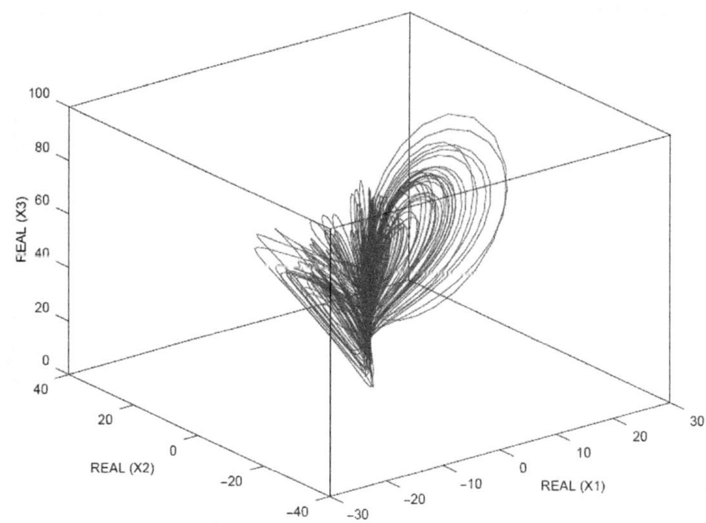

FIGURE 10.5 Phase portrait for the real part of the complex hyperchaotic system.

$u_i \in \mathbb{C}$ for $i = 1, 2$, $u_i = u_{ri}$, $u_i \in \mathbb{R}$ for $i = 3, 4$. Then in order to set the system in hyperchaotic regime the constants must be set as $c_1 = 14$, $c_2 = 5$, $c_3 = 40$, and $c_4 = 13$ with $\alpha = 0.99$ and the following phase portraits are obtained for the real and imaginary parts, respectively, in Figs. 10.5 and 10.6.

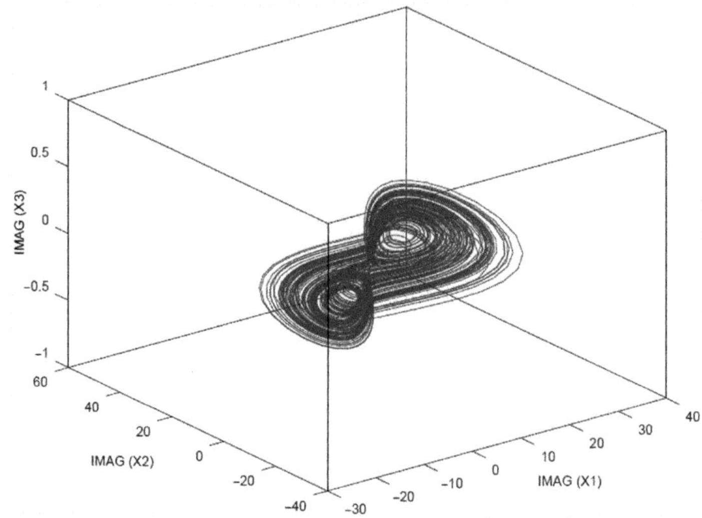

FIGURE 10.6 Phase portrait for the imaginary part of the complex hyperchaotic system.

10.3 TERMINAL SLIDING MODE STABILIZATION OF FRACTIONAL ORDER COMPLEX CHAOTIC SYSTEMS

Before the derivations of the terminal sliding mode controller and synchronization strategies for the fractional order complex chaotic and hyperchaotic system, the systems (10.3)–(10.5) are divided in real and imaginary parts, respectively, as described below:

$$\begin{aligned} D^\alpha z_{r1} &= a_1(z_{r2} - z_{r1}) + u_{r1} \\ D^\alpha z_{1im} &= a_1(z_{2im} - z_{1im}) + u_{1im} \\ D^\alpha z_{r2} &= (a_3 - a_1)z_{r1} - z_{r1}z_{r3} + a_3 z_{r2} + u_{r2} \\ D^\alpha z_{2im} &= (a_3 - a_1)z_{1im} - z_{1im}z_{r3} + a_3 z_{2im} + u_{2im} \\ D^\alpha z_{r3} &= -a_2 z_{r3} + z_{r1}z_{r2} + z_{1im}z_{2im} + u_{r3} \end{aligned} \tag{10.6}$$

$$\begin{aligned} D^\alpha y_{r1} &= b_1(y_{r2} - y_{r1}) + u_{r1} \\ D^\alpha y_{1im} &= b_1(y_{2im} - y_{1im}) + u_{1im} \\ D^\alpha y_{r2} &= b_2 y_{r1} - y_{r2} - y_{r1}y_{r3} + u_{r2} \\ D^\alpha y_{2im} &= b_2 y_{1im} - y_{2im} - y_{1im}y_{r3} - u_{2im} \\ D^\alpha y_{r3} &= -b_3 y_{r3} + y_{r1}y_{r2} + y_{1im}y_{2im} + u_{r3} \end{aligned} \tag{10.7}$$

$$\begin{aligned} D^\alpha y_{r1} &= c_1(y_{r2} - y_{r1}) + u_{r1} \\ D^\alpha y_{1im} &= c_1(y_{2im} - y_{1im}) + y_{r4} + u_{1im} \\ D^\alpha y_{r2} &= c_3 y_{r1} - y_{r2} - y_{r1}y_{r3} + u_{r2} \\ D^\alpha y_{2im} &= c_3 y_{1im} - y_{2im} - y_{1im}y_{r3} + y_{r4} + u_{2im} \\ D^\alpha y_{r3} &= y_{r1}y_{r2} + y_{1im}y_{2im} - c_2 y_{r3} + u_{r3} \\ D^\alpha y_{r4} &= y_{r1}y_{r2} + y_{1im}y_{2im} - \sigma y_{r4} + u_{r4} \end{aligned} \tag{10.8}$$

so for the terminal sliding mode controller design consider the following fractional order dynamical system:

$$D^{\alpha}x(t) = f(x(t)) + U(t) \tag{10.9}$$

where $x(t) \in \mathbb{R}^n$ is the state variable, $f(x(t)) \in \mathbb{R}^n$ is a nonlinear vector, and $U(t) \in \mathbb{R}^n$ is the system input. So, consider the following vector:

$$f(x(t)) = \begin{pmatrix} f_1(x(t)) \\ \vdots \\ f_n(x(t)) \end{pmatrix} \tag{10.10}$$

so the terminal sliding mode controller for the stabilization of system (10.6) and (10.8) is obtained by the following theorem (Aghababa, 2015; Komurcugil, 2012):

Theorem 1: *The following terminal sliding mode control law* $u_i(t) = k_i D^{1-\alpha}(x_i(t) + sign(x_i(t))|x_i(t)|^{\rho}) - f_i(x(t)) - s_i$ *stabilizes the system* (10.9) *in its equilibrium point where* k_i *is the sliding mode gain and* s_i *is the sliding mode surface defined later for* $i = 1, \ldots, n$.

Proof: Consider the following sliding mode surface:

$$s_i(t) = D^{\alpha-1}x_i(t) + k_i D^{-\alpha}(x_i(t) + sign(x_i(t))|x_i(t)|^{\rho}) \tag{10.11}$$

and the following Lyapunov function (Aghababa, 2015):

$$V(t) = \|S(t)\|_1 = \sum_{i=1}^{n} |s_i(t)| \tag{10.12}$$

so deriving (10.12) yields:

$$\dot{V}(t) = \sum_{i=1}^{n} sign(s_i(t))\dot{s}_i \tag{10.13}$$

and implementing the fractional order calculus properties explained in Section 10.2, the following result is obtained:

$$\dot{V}(t) = \sum_{i=1}^{n} sign(s_i(t))[f_i(x(t)) + u_i(t) + k_i D^{1-\alpha}(x_i(t) + sign(x_i(t))|x_i(t)|^{\rho})] \tag{10.14}$$

so by selecting the following control law:

$$u_i(t) = k_i D^{1-\alpha}(x_i(t) + sign(x_i(t))|x_i(t)|^{\rho}) - f_i(x(t)) - s_i(t) \tag{10.15}$$

Then, (10.14) becomes:

$$\dot{V}(t) = -\sum_{i=1}^{n} |s_i(t)| \leq 0 \tag{10.16}$$

and this completes the proof.

10.3.1 Terminal Sliding Mode Control Law for the Chen Chaotic System

The controller for the system (10.6) with control inputs

$$U(t) = [u_{r1}, u_{1im}, u_{r2}, u_{2im}, u_{r3}]^T \tag{10.17}$$

is given by:

$$U(t) = \begin{pmatrix} k_1 D^{1-\alpha}[z_{r1} + sign(z_{r1})|z_{r1}|^{\rho}] - a_1(z_{r2} - z_{r1}) - s_1(t) \\ k_2 D^{1-\alpha}[z_{1im} + sign(z_{1im})|z_{1im}|^{\rho}] - a_1(z_{2im} - z_{1im}) - s_2(t) \\ k_3 D^{1-\alpha}[z_{r2} + sign(z_{r2})|z_{r2}|^{\rho}] - (a_3 - a_1)z_{r1} + z_{r1}z_{r3} - a_3 z_{r2} - s_3(t) \\ k_4 D^{1-\alpha}[z_{2im} + sign(z_{2im})|z_{2im}|^{\rho}] - (a_3 - a_1)z_{1im} + z_{1im}z_{r3} - a_3 z_{2im} - s_4(t) \\ k_5 D^{1-\alpha}[z_{r3} + sign(z_{r3})|z_{r3}|^{\rho}] + a_2 z_{r3} - z_{r1}z_{r2} - z_{1im}z_{2im} - s_5(t) \end{pmatrix} \tag{10.18}$$

10.3.2 Terminal Sliding Mode Control Law for the Lorenz Hyperchaotic System

The control law for the hyperchaotic Lorenz system shown in (10.8) is given in (10.20) with the following control variable:

$$U(t) = [u_{r1}, u_{1im}, u_{r2}, u_{2im}, u_{r3}, u_{r4}]^T; \tag{10.18}$$

so,

$$U(t) = \begin{pmatrix} k_1 D^{1-\alpha}[y_{r1} + sign(y_{r1})|y_{r1}|^{\rho}] - c_1(y_{r2} - y_{r1}) - s_1 \\ k_2 D^{1-\alpha}[y_{1im} + sign(y_{1im})|y_{1im}|^{\rho}] - c_1(y_{2im} - y_{1im}) - y_{r4} - s_2 \\ k_3 D^{1-\alpha}[y_{r2} + sign(y_{r2})|y_{r2}|^{\rho}] - c_3 y_{r1} + y_{r2} + y_{r1}y_{r3} - s_3 \\ k_4 D^{1-\alpha}[y_{2im} + sign(y_{2im})|y_{2im}|^{\rho}] - c_3 y_{1im} + y_{2im} + y_{1im}y_{r3} - y_{r4} - s_4 \\ k_5 D^{1-\alpha}[y_{r3} + sign(y_{r3})|y_{r3}|^{\rho}] - y_{r1}y_{r2} - y_{1im}y_{2im} + c_2 y_{r3} - s_5 \\ k_6 D^{1-\alpha}[y_{r4} + sign(y_{r4})|y_{r4}|^{\rho}] - y_{r1}y_{r2} - y_{1im}y_{2im} + c_4 y_{r4} - s_6 \end{pmatrix} \tag{10.20}$$

10.4 ADAPTIVE TERMINAL SLIDING MODE SYNCHRONIZATION FOR CHAOTIC AND HYPERCHAOTIC SYSTEMS

In this section, the synchronization of nonidentical systems is presented for which the drive system (10.6) is the fractional order complex Chen chaotic system and the response system is the fractional order complex Lorenz chaotic system. Then, the identical synchronization is done with the fractional order complex hyperchaotic system with different initial conditions.

For this purpose consider the following theorem that is valid for the synchronization of chaotic and hyperchaotic systems (identical and nonidentical):

Theorem 2: *The adaptive control law synchronizes chaotic and hyperchaotic systems (identical and nonidentical) with the following control law (Aghababa, 2015; Komurcugil, 2012):*

$$u_i = s_i + f_i(z) - g_i(y) + k_i D^{1-\alpha} sign(e_i)|e_i|^{\rho} \tag{10.21}$$

and the following adaptive gain:

$$\dot{k}_i = \frac{-\,sign(s_i)D^{1-\alpha}e_i}{\Gamma_i} \tag{10.22}$$

where s_i is the sliding surface, $e_i = z_i - y_i$ is the error variable, and Γ_i and k_i are the adaptive law and controller gain constants.

Proof: Consider the following sliding surface (Aghababa, 2015):

$$s_i(t) = D^{\alpha-1}e_i(t) + k_i D^{-\alpha}(e_i(t) + sign(e_i(t))|e_i(t)|^{\rho}) \tag{10.23}$$

and the following Lyapunov function:

$$V(t) = \|S(t)\|_1 + \frac{1}{2}\sum_{i=1}^{n}\Gamma_i k_i^2 = \sum_{i=1}^{n}|s_i| + \frac{1}{2}\sum_{i=1}^{n}\Gamma_i k_i^2 \tag{10.24}$$

so the drive system is defined by:

$$D^{\alpha}Z = f(Z) \tag{10.25}$$

and the response system is given by:

$$D^{\alpha}Y = g(Y) + U \tag{10.26}$$

and the definition of (10.25) and (10.26) are similar to (10.9), defining the error variable as:

$$e_i = z_i - y_i \tag{10.27}$$

Then, the first derivative of (10.24) is given by:

$$\dot{V}(t) = \sum_{i=1}^{n} sign(s_i)\dot{s}_i + \sum_{i=1}^{n}\Gamma_i k_i \dot{k}_i \tag{10.28}$$

so by implementing the properties of fractional calculus and Section 10.2, (10.28) becomes:

$$\dot{V}(t) = \sum_{i=1}^{n} sign(s_i)[f_i(z) - g_i(y) - u_i - s_i + k_i D^{1-\alpha}e_i + k_i D^{1-\alpha} sign(e_i)|e_i|^{\rho}] + \sum_{i=1}^{n}\Gamma_i k_i \dot{k}_i \tag{10.29}$$

so with the following control and adaptive laws the systems are synchronized.

$$\begin{aligned} u_i &= s_i + f_i(z) - g_i(y) + k_i D^{1-\alpha} sign(e_i)|e_i|^\rho \\ \dot{k}_i &= -\frac{sign(s_i) D^{1-\alpha} e_i}{\Gamma_i} \end{aligned} \tag{10.30}$$

so, (10.28) becomes:

$$\dot{V}(t) = -2\sum_{i=1}^{n} |s_i| \leq 0 \tag{10.31}$$

and this completes the proof. It is important to remark that implementing the control and adaptive laws of (10.30), the system (10.7) (response system) is synchronized with system (10.6) (drive system) and system (10.8) is synchronized with the same system but with different initial conditions. This synchronization strategy can be used to synchronize any kind of identical or nonidentical chaotic and hyperchaotic systems.

10.5 NUMERICAL SIMULATION EXAMPLES

In this section, some numerical simulation examples that show the stabilization and synchronization of fractional order complex chaotic and hyperchaotic systems are presented. The parameters and initial conditions for each system are shown in Section 10.2.

10.5.1 Stabilization of Complex Chaotic and Hyperchaotic Systems

In this subsection, the controller strategies for the stabilization of the fractional order complex Chen chaotic system along with the stabilization by a terminal sliding mode control approach of the fractional order complex Lorenz hyperchaotic system are shown and then the results are compared with the outcomes obtained with the controller shown in Li and Li (2015) in order to drive the system variables to the equilibrium point.

10.5.1.1 Stabilization of the Fractional Order Complex Chen Chaotic System

The results obtained in Fig. 10.7A–C show that the controller drives the system variables to the equilibrium point $z = [0+0j, 0+0j, 0]^T$ and in comparison with the outcomes obtained by Li and Li (2015), the variables reach the equilibrium point faster and with less oscillations.

In Fig. 10.8A and B, the real component of the input variables u_1 and u_2 are shown where, as it is noticed, the control variables reach the zero value faster and with less control effort in comparison with the results obtained in Li and Li (2015).

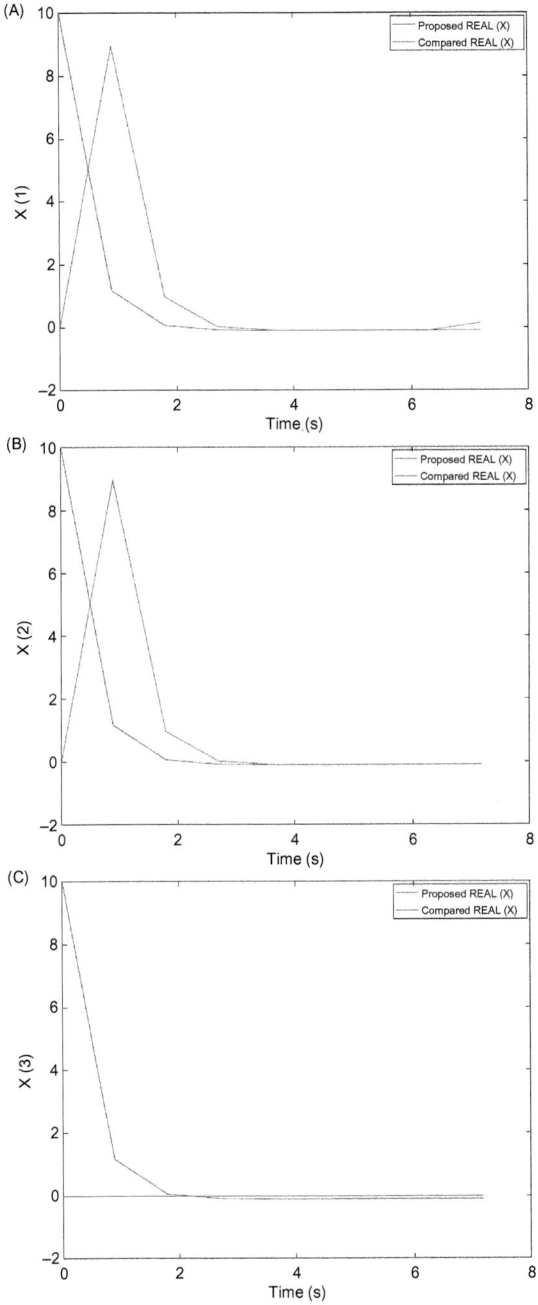

FIGURE 10.7 Real components of (A) variable $x(1)$, (B) variable $x(2)$, (C) variable $x(3)$.

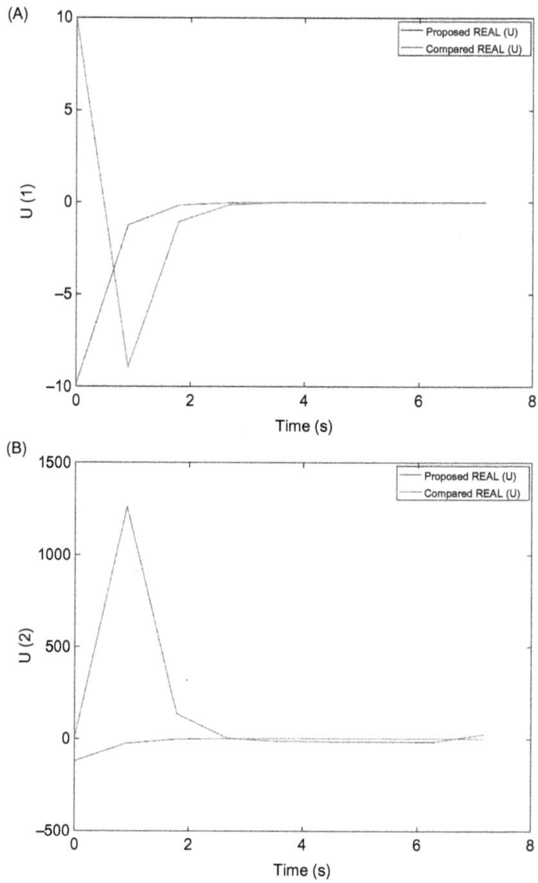

FIGURE 10.8 Real components of the input variables u_1 and u_2.

10.5.1.2 Stabilization of the Fractional Order Complex Lorenz Hyperchaotic System

In Fig. 10.9A and B, the real and imaginary parts of the variable x_1 are shown, where, as can be noticed, there are less oscillations and a faster response in comparison with the outcomes obtained by the approach shown in Li and Li (2015) in order to drive the variables to the equilibrium point.

Then, in Fig. 10.10, the real part of the variable x_2 is shown and as can be noticed there is less overshoot and a faster response in comparison with the outcomes obtained with the strategy shown in Li and Li (2015).

In Fig. 10.11A and B, the real components of the control input variables u_1 and u_2 are shown where the control effort is smaller and a faster response is obtained by the proposed control strategy in comparison with the strategy shown in Li and Li (2015).

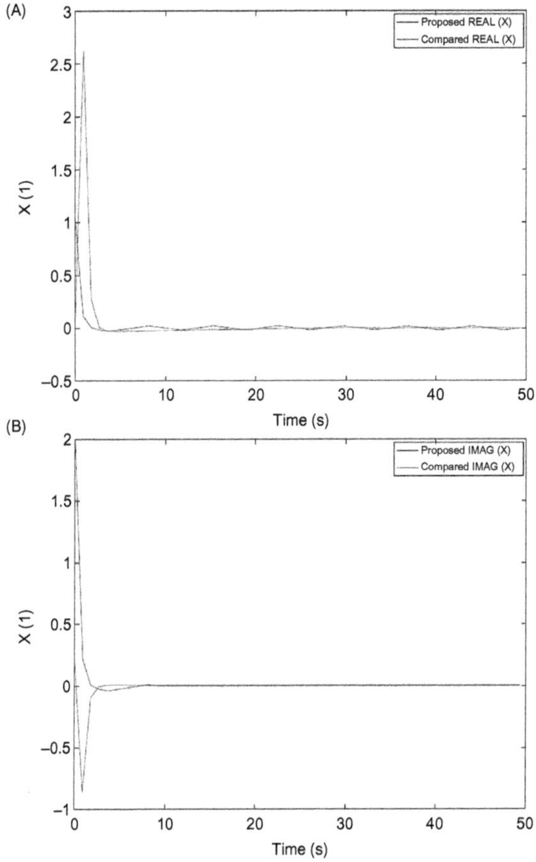

FIGURE 10.9 Real and Imaginary components of the variable $x(1)$.

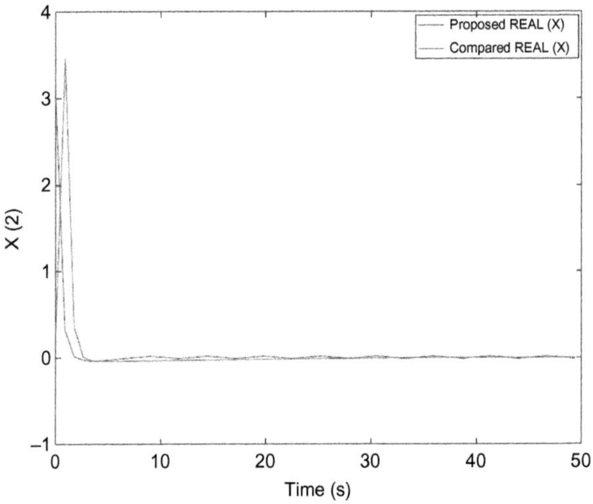

FIGURE 10.10 Real component of the variable $x(2)$.

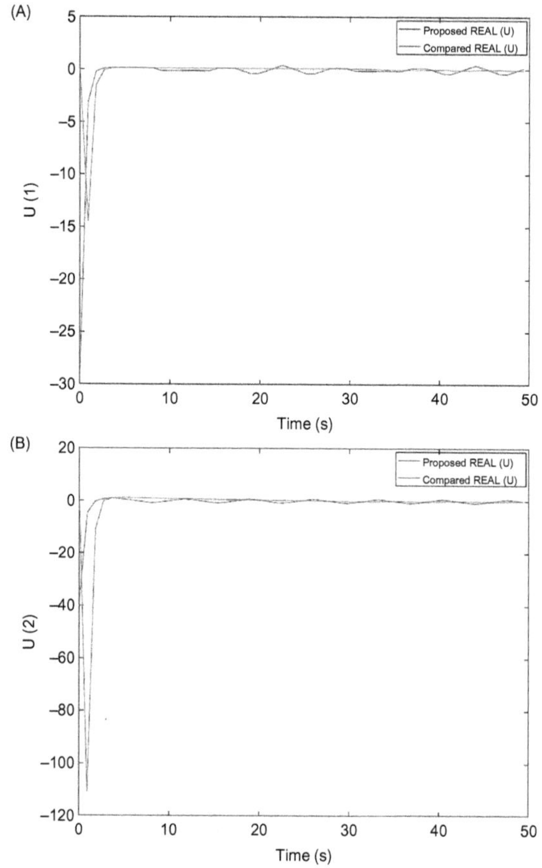

FIGURE 10.11 Real components of the input variables u_1 and u_2.

10.5.2 Synchronization of Fractional Order Complex Chaotic and Hyperchaotic Systems

The synchronization of the fractional order complex chaotic and hyperchaotic system is done by the adaptive terminal sliding mode control strategy. First, a nonidentical stabilization synchronization example is shown where fractional order complex chaotic systems are synchronized (a Chen chaotic system is the drive system and a Lorenz chaotic system is the response system) and then an identical synchronization of two Lorenz hyperchaotic systems are shown.

10.5.2.1 Nonidentical Synchronization of Complex Chaotic Systems

In Figs. 10.12A, B, 10.13A and B, the real and imaginary components of the variables x_1 and x_2 are shown. The real component of the variable x_3 is

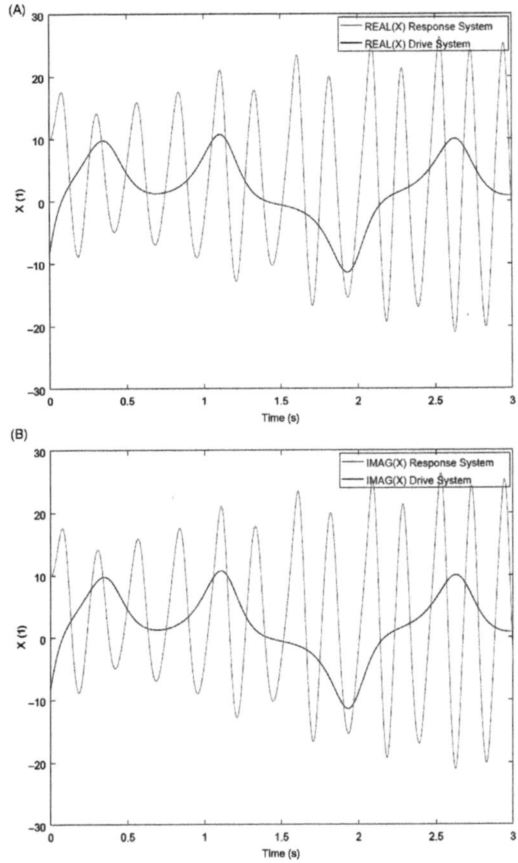

FIGURE 10.12 Real and imaginary components of the variable $x(1)$.

shown in Fig. 10.14. It is noticed that the synchronization between the fractional order complex Chen and Lorenz system is done effectively with a minimum error as corroborated in Figs. 10.15A, B, 10.16A, B, and 10.17.

Then, the control inputs (real and imaginary parts) for the response system are shown in Figs. 10.18A, B, 10.19A, B, and 10.20 where the synchronization is done effectively by this strategy.

10.5.2.2 Identical Synchronization of Fractional Order Complex Hyperchaotic Systems

In this subsection, the identical synchronization of the fractional order complex Lorenz chaotic system with different initial conditions by the adaptive terminal sliding mode controller is presented. In Fig. 10.21A and B the real and imaginary components of the variable x_1 are shown. In Fig. 10.22A–C, the real components of the variables x_2, x_3, and x_4 are shown, respectively.

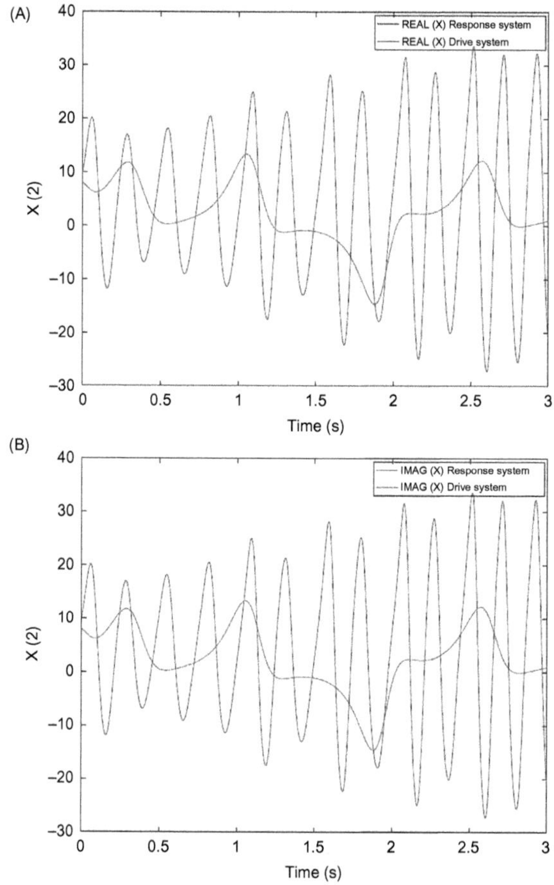

FIGURE 10.13 Real and imaginary components of the variable $x(2)$.

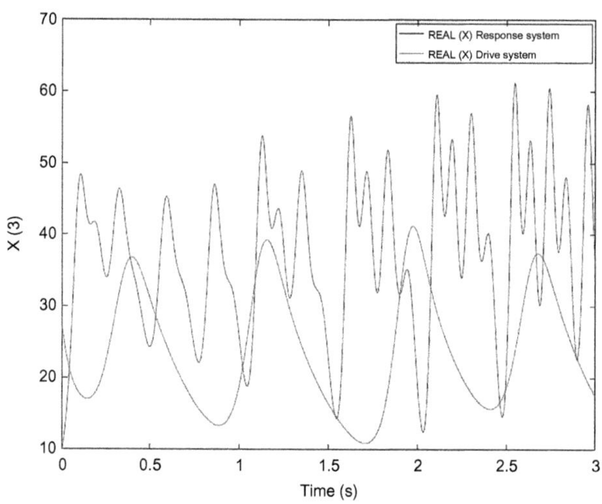

FIGURE 10.14 Real component of the variable $x(3)$.

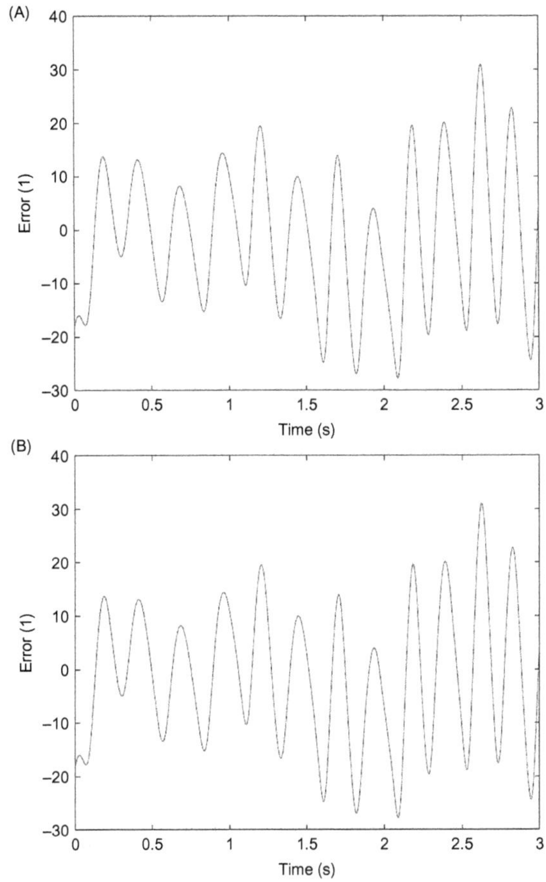

FIGURE 10.15 Real and imaginary components of the error $e(1)$.

As it is noticed these variables are synchronized effectively with a small synchronization error as shown in Figs. 10.23A, B, 10.24A, B, 10.25A, and B, where it is proved the suitability of the performance of the adaptive terminal sliding mode controller.

Finally, in Figs. 10.26a, B, 10.27A, B, 10.28A, and B, the control inputs for the imaginary and real parts are shown where the control effort is necessary for the synchronization of the identical Lorenz systems with different initial conditions.

10.6 DISCUSSION

The results obtained in this chapter provide a suitable contribution for the stabilization and synchronization of fractional order complex chaotic and hyperchaotic systems because similar results have not been reported in the literature. In the case of the stabilization of fractional order chaotic and

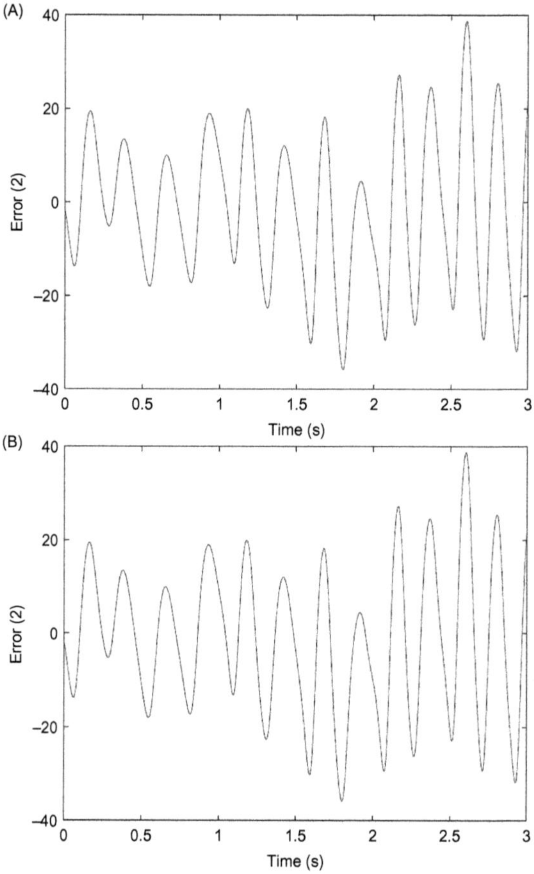

FIGURE 10.16 Real and imaginary components of the error $e(2)$.

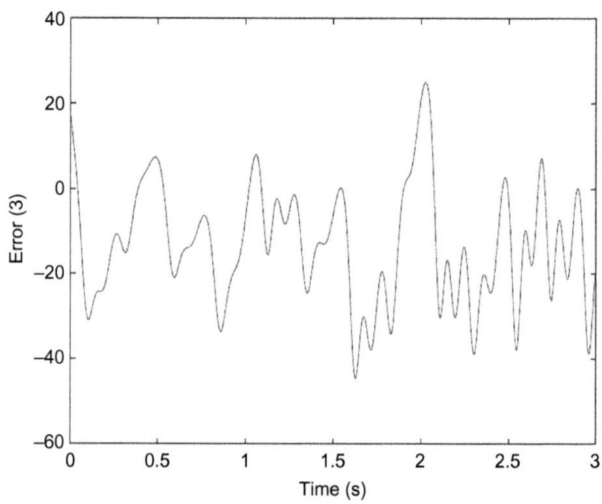

FIGURE 10.17 Real component of the variable $e(3)$.

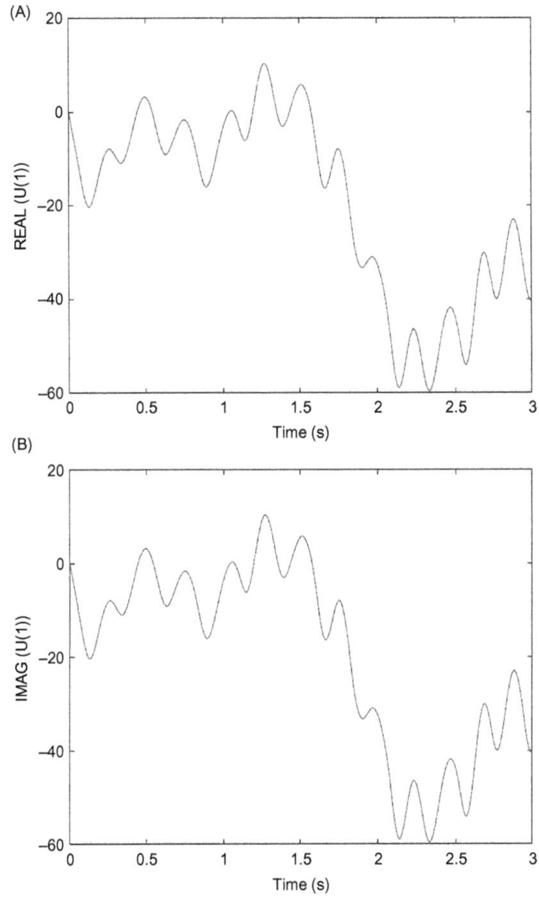

FIGURE 10.18 Real and imaginary components of the input $u(1)$.

hyperchaotic systems, this study provides a deep analysis and theoretical framework in which the terminal sliding mode control strategy stabilizes effectively the two studied systems to reach the equilibrium point. The results obtained by the proposed control strategy provide more accurate and faster outcomes than other control techniques found in literature. In the synchronization case, the sliding mode technique along with its variation have not been studied deeply in recent years considering the capability of this kind of controller to deal with uncertainties. Therefore, a strong theoretical study is provided in this chapter where the synchronization by an adaptive terminal sliding mode controller is done effectively for fractional order complex chaotic and hyperchaotic systems in the identical and nonidentical synchronization cases.

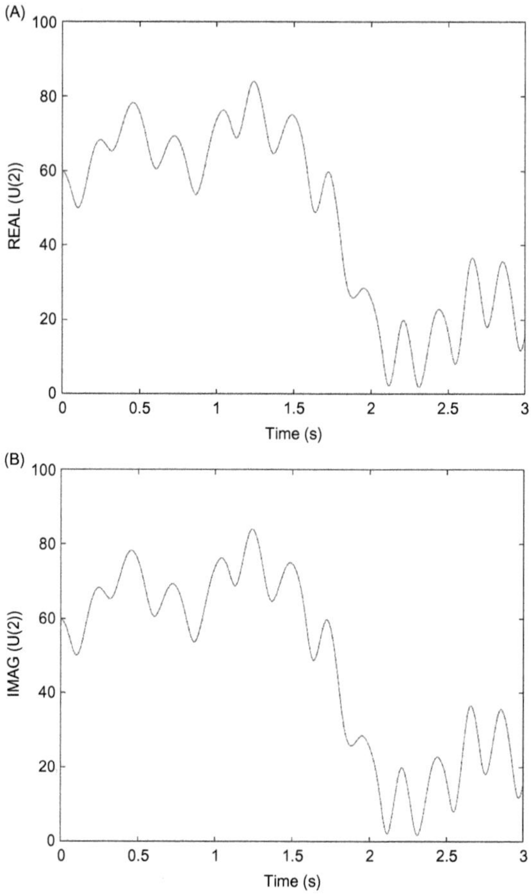

FIGURE 10.19 Real and imaginary components of the input $u(2)$.

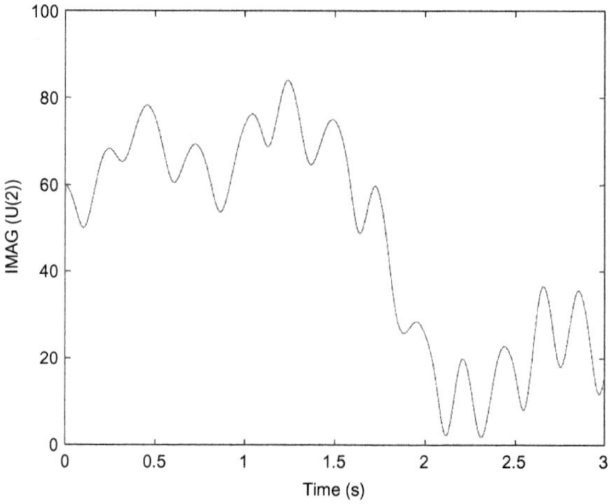

FIGURE 10.20 Real component of the input $u(3)$.

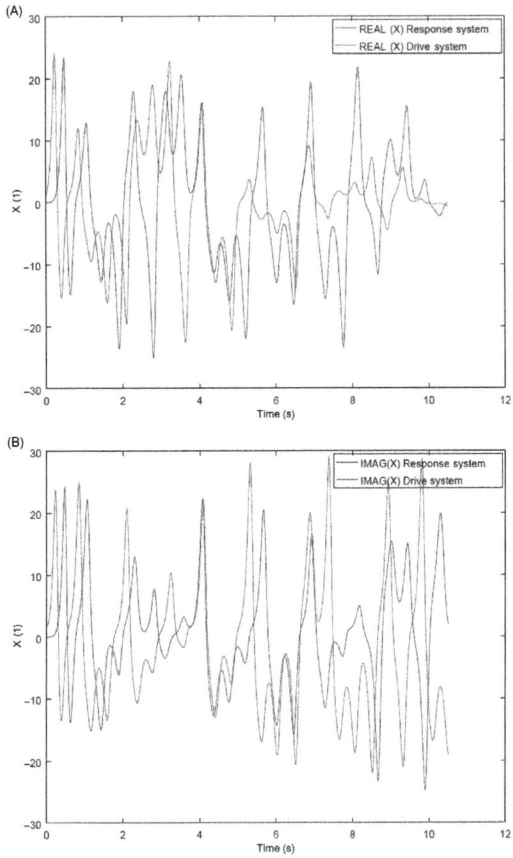

FIGURE 10.21 Real and imaginary components of the variable $x(1)$.

10.7 CONCLUSION

Due to chaos and hyperchaos being a phenomenon found in many kind of physical systems such as mechanical, electrical, chemical, and biological systems, in this book chapter the design of a terminal sliding mode controller along with an adaptive terminal sliding mode technique is proposed for the stabilization and synchronization of fractional order chaotic and hyperchaotic systems. These control and synchronization strategies consist of dividing the real and imaginary parts of the studied systems in order to obtain the desired control and adaptive gain law. Suitable sliding surfaces were designed to derive efficient control and synchronization strategies to obtain the desired system response. As a future direction, the design of sliding mode techniques for these kinds of systems could be considered, such as sliding mode back-stepping control, to provide more contributions in this field.

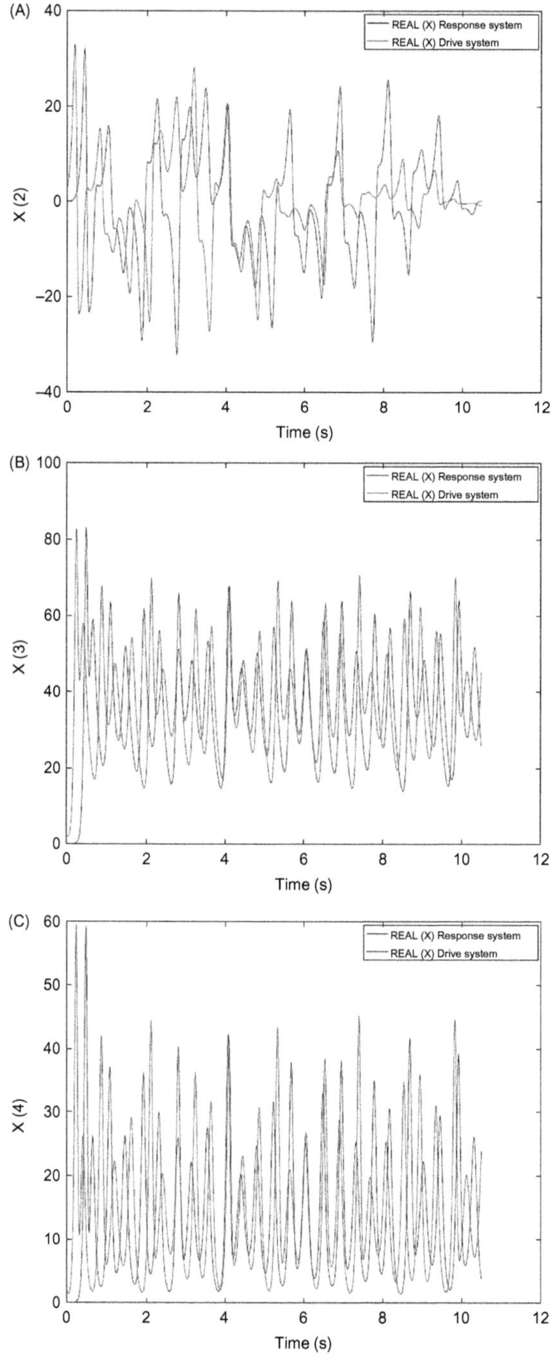

FIGURE 10.22 Real components of (A) variable $x(2)$, (B) variable $x(3)$, (C) variable $x(4)$.

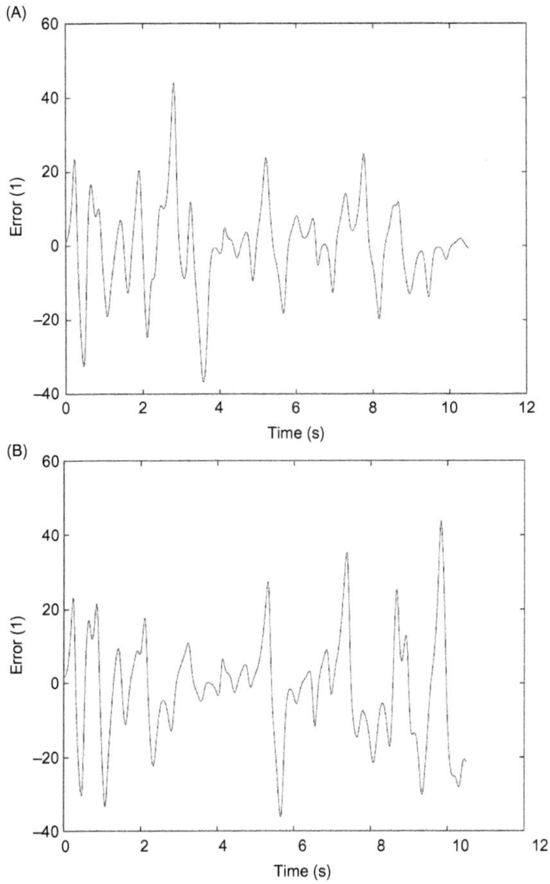

FIGURE 10.23 Real and imaginary components of the error $e(1)$.

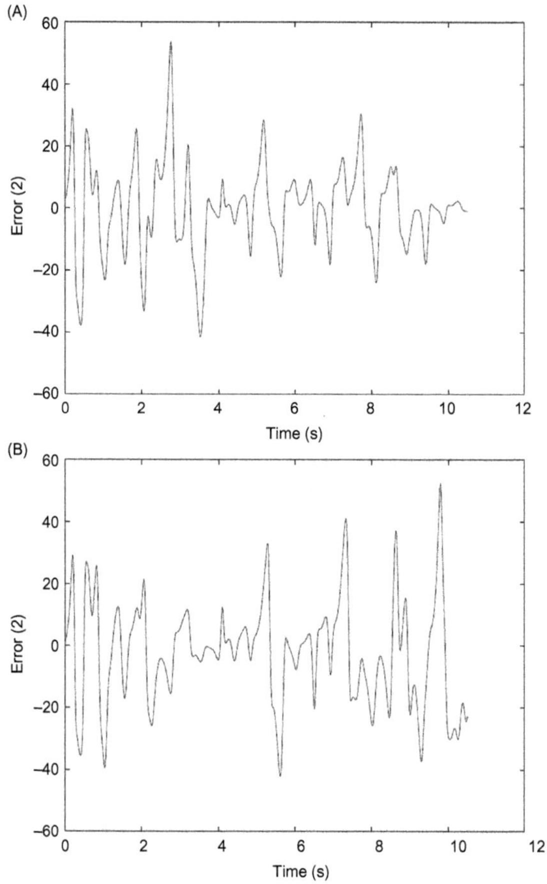

FIGURE 10.24 Real and imaginary components of the error $e(2)$.

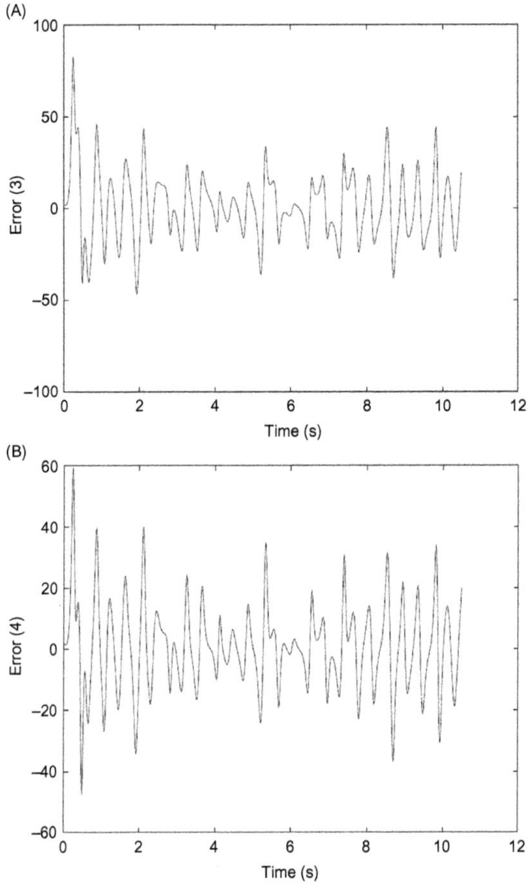

FIGURE 10.25 Real components of the errors $e(3)$ and $e(4)$.

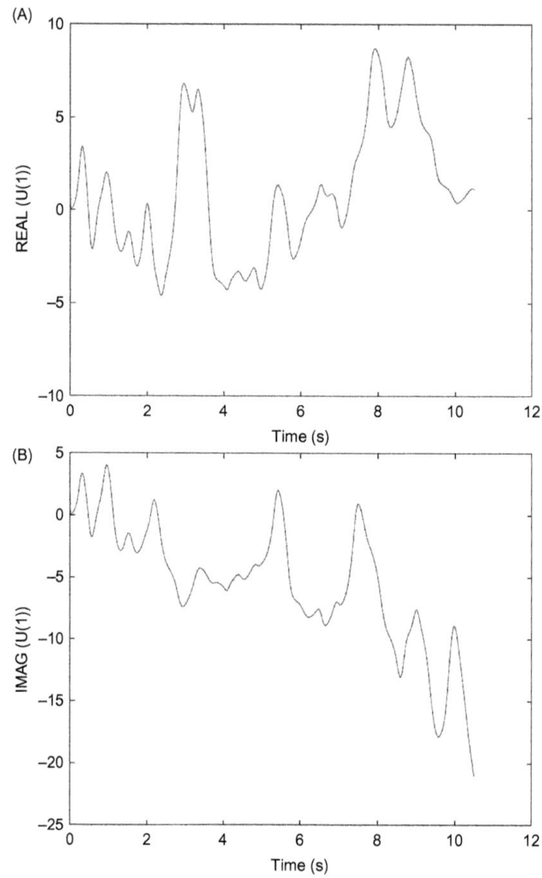

FIGURE 10.26 Real and imaginary components of the input $u(1)$.

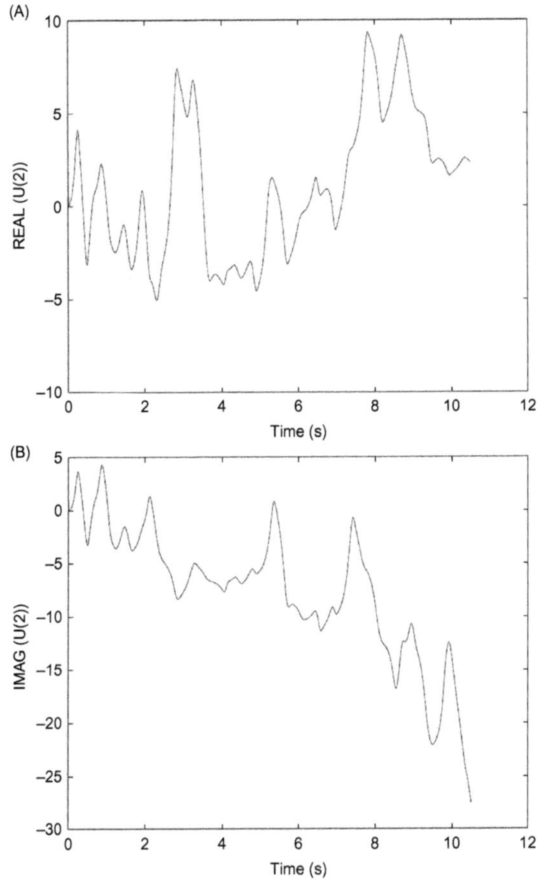

FIGURE 10.27 Real and imaginary components of the input $u(2)$.

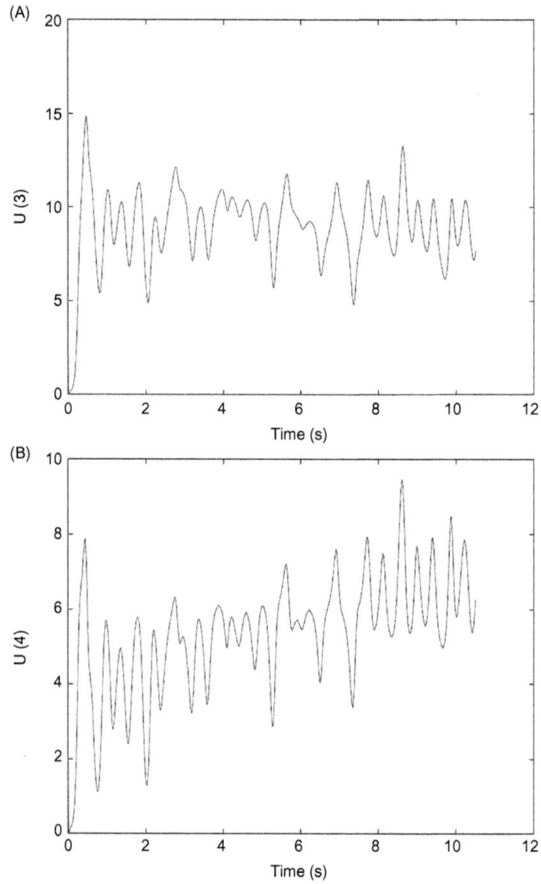

FIGURE 10.28 Real components of the inputs $u(3)$ and $u(4)$.

REFERENCES

Abedini, M., Gomroki, M., Salarieh, H., Meghdari, A., 2014. Identification of 4d 1 hyper-chaotic system using identical systems synchronization and fractional adaptation law. Appl. Math. Model. 38 (19-20), 4652–4661.

Aghababa, M.P., 2015. A fractional sliding mode for finite-time control scheme with application to stabilization of electrostatic and electromechanical transducer. Appl. Math. Model. 39, 6103–6113.

Akbarzadeh-T, M.-R., Hosseini, S., Naghibi-Sistani, M.-B., 2017. Stable indirect adaptive interval type-2 fuzzy sliding-based control and synchronization of two different chaotic systems. Appl. Soft Comput. J. 55 (2017), 576–587.

Azar, A.T., Vaidyanathan, S., 2015a. Chaos Modeling and Control Systems Design. Vol. 581 of Studies inComputational Intelligence. Studies in Computational Intelligence. Springer, Germany.

Azar, A.T., Vaidyanathan, S., 2015b. Computational Intelligence Applications in Modelling and Control. Vol. 575. Studies inComputational Intelligence. Springer, Berlin, Germany.

Azar, A.T., Vaidyanathan, S., 2015c. Handboook of Research on Advanced Intelligent Control Engineering and Automation. IGI Global, New York, USA.

Azar, A.T., Vaidyanathan, S., 2016. Advances in Chaos Theory and Intelligent Control, Vol. 337. Springer, Berlin, Germany.

Azar, A.T., Vaidyanathan, S., Ouannas, A., 2017. Fractional Order Control and Synchronization of Chaotic Systems. Vol. 688. Studies inComputational Intelligence. Springer, Berlin, Germany.

Azar, A.T., Kumar, J., Kumar, V., Rana, K.P.S., 2018a. Control of a Two Link Planar Electrically-Driven Rigid Robotic Manipulator Using Fractional Order SOFC. Springer International Publishing, Cham, pp. 57–68, URL https://doi.org/10.1007/978-3-319-64861-3_6.

Azar, A.T., Ouannas, A., Singh, S., 2018b. Control of New Type of Fractional Chaos Synchronization. Springer International Publishing, Cham, pp. 47–56, URL https://doi.org/10.1007/978-3-319-64861-3_5.

Boulkroune, A., Bouzeriba, A., Bouden, T., Azar, A.T., 2016a. Fuzzy adaptive synchronization of uncertain fractional-order chaotic systems. Advances in Chaos Theory and IntelligentControl. Springer, Berlin, Germany, pp. 681–697.

Boulkroune, A., Hamel, S., Azar, A.T., Vaidyanathan, S., 2016b. Fuzzy control-based function synchronization of unknown chaotic systems with dead-zone input. In: Azar, A.T., Vaidyanathan, S. (Eds.), Advances in Chaos Theory and Intelligent Control. Springer International Publishing, Cham, pp. 699–718.

Cruz-Ancona, C., Martnez-Guerra, R., 2017. Fractional dynamical controllers for generalized multi-synchronization of commensurate fractional order liouvillian chaotic systems. J. Franklin Institute 354 (7), 3054–3096.

El-Sayed, A., Nour, H., Elsaid, A., Matouk, A., Elsonbaty, A., 2016. Dynamical behaviors, circuit realization, chaos control, and synchronization of a new fractional order hyperchaotic system. Appl. Math. Model. 40 (5-6), 3516–3534.

Gao, W., Li, J., Xu, Y., 2015. Adaptive synchronization of uncertain chaotic systems with definite integration scaling function. Optik 126 (19), 1999–2002.

Ghoudelbourk, S., Dib, D., Omeiri, A., Azar, A.T., 2016. Mppt control in wind energy conversion systems and the application of fractional control (piα) in pitch wind turbine. Int. J. Model. Identif. Control 26 (2), 140–151.

Grassi, G., Ouannas, A., Azar, A.T., Radwan, A.G., Volos, C., Pham, V.-T., et al., 2017. Chaos synchronisation of continuous systems via scalar signal. In: 6th International Conference on Modern Circuits and Systems Technologies (MOCAST), pp. 1–4.

Huang, X., Wang, Z., Li, Y., Lu, J., 2014. Design of fuzzy state feedback controller for robust stabilization of uncertain fractional-order chaotic systems. J. Franklin Institute 351 (12), 5480–5493.

Komurcugil, H., 2012. Adaptive terminal sliding-mode control strategy for dc-dc buck converters. ISA Trans. 51, 673–681.

Lamamra, K., Vaidyanathan, S., Azar, A.T., Ben Salah, C., 2017. Chaotic system modelling using a neural network with optimized structure. In: Azar, A.T., Vaidyanathan, S., Ouannas, A. (Eds.), Fractional Order Control and Synchronization of Chaotic Systems. Springer International Publishing, Cham, pp. 833–856.

Li, C., Su, K., Tong, Y., Li, H., 2013a. Robust synchronization for a class of fractional-order chaotic and hyperchaotic systems. Optik 124 (18), 3242–3245.

Li, C., Su, K., Zhang, J., Wei, D., 2013b. Robust control for fractional-order four-wing hyperchaotic system using lmi. Optik 124 (22), 5807–5810.

Li, R., 2016. Adaptive track control for fractional-order chaotic systems with or without uncertainty. Optik 127 (23), 11263–11276.

Li, R., Li, W., 2015. Suppressing chaos for a class of fractional-order chaotic systems by adaptive integer-order and fractional-order feedback control. Optik 126 (21), 2965–2973.

Lorenz, E.N., 1963. Deterministic nonperiodic flow. J. Atmos. Sci. 20, 130–141.

Mahmoud, E., 2014. Generation and suppression of a new hyperchaotic nonlinear model with complex variables. Appl. Math. Model. 38 (17-18), 4445–4459.

Matouk, A., 2011. Chaos, feedback control and synchronization of a fractional-order modified autonomous van der pol-duffing circuit. Commun. Nonlinear Sci. Numer. Simul. 16 (2), 975–986.

Meghni, B., Dib, D., Azar, A.T., Ghoudelbourk, S., Saadoun, A., 2017. Robust adaptive supervisory fractional order controller for optimal energy management in wind turbine with battery storage. In: Azar, A.T., Vaidyanathan, S., Ouannas, A. (Eds.), Fractional Order Control and Synchronization of Chaotic Systems. Springer International Publishing, Cham, pp. 165–202.

Moysis, L., Azar, A.T., 2017. New discrete time 2d chaotic maps. Int. J. System Dynam. Applicat. 6 (1), 77–104.

Ouannas, A., Azar, A.T., Abu-Saris, R., 2016a. A new type of hybrid synchronization between arbitrary hyperchaotic maps. Int. J. Mach. Learn. Cybern.

Ouannas, A., Azar, A.T., Radwan, A.G., 2016b. On inverse problem of generalized synchronization between different dimensional integer-order and fractional-order chaotic systems. In: 2016 28th International Conference on Microelectronics (ICM). pp. 193–196.

Ouannas, A., Azar, A.T., Abu-Saris, R., 2017a. A new type of hybrid synchronization between arbitrary hyperchaotic maps. Int. J. Mach. Learn. Cybern. 8 (6), 1887–1894. URL https://doi.org/10.1007/s13042-016-0566-3.

Ouannas, A., Azar, A.T., Vaidyanathan, S., 2017b. New hybrid synchronization schemes based on coexistence of various types of synchronization between master-slave hyperchaotic systems. Int. J. Comp. Applicat. Technol. 55 (2), 112–120.

Ouannas, A., Azar, A.T., Vaidyanathan, S., 2017c. On a simple approach for q-s synchronization of chaotic dynamical systems in continuous-time. Int. J. Comp. Sci. Math. 8 (1), 20–27.

Ouannas, A., Azar, A.T., Vaidyanathan, S., 2017d. A robust method for new fractional hybrid chaos synchronization. Math. Methods Appl. Sci. 40 (5), 1804–1812. mma.4099. URL https://doi.org/10.1002/mma.4099.

Ouannas, A., Azar, A.T., Ziar, T., 2017e. On inverse full state hybrid function projective synchronization for continuous-time chaotic dynamical systems with arbitrary dimensions. Diff. Eq. Dynam. Systems. Available from: https://doi.org/10.1007/s12591-017-0362-x.

Ouannas, A., Azar, A.T., Ziar, T., Radwan, A.G., 2017f. Generalized synchronization of different dimensional integer-order and fractional order chaotic systems. In: Azar, A.T., Vaidyanathan, S., Ouannas, A. (Eds.), Fractional Order Control and Synchronization of Chaotic Systems. Springer International Publishing, Cham, pp. 671–697.

Ouannas, A., Azar, A.T., Ziar, T., Radwan, A.G., 2017g. A study on coexistence of different types of synchronization between different dimensional fractional chaotic systems. In: Azar, A.T., Vaidyanathan, S., Ouannas, A. (Eds.), Fractional Order Control and Synchronization of Chaotic Systems. Springer International Publishing, Cham, pp. 637–669.

Ouannas, A., Azar, A.T., Ziar, T., Vaidyanathan, S., 2017h. Fractional inverse generalized chaos synchronization between different dimensional systems. In: Azar, A.T., Vaidyanathan, S., Ouannas, A. (Eds.), Fractional Order Control and Synchronization of Chaotic Systems. Springer International Publishing, Cham, pp. 525–551.

Ouannas, A., Azar, A.T., Ziar, T., Vaidyanathan, S., 2017i. A new method to synchronize fractional chaotic systems with different dimensions. In: Azar, A.T., Vaidyanathan, S., Ouannas, A. (Eds.), Fractional Order Control and Synchronization of Chaotic Systems. Springer International Publishing, Cham, pp. 581–611.

Ouannas, A., Azar, A.T., Ziar, T., Vaidyanathan, S., 2017j. On new fractional inverse matrix projective synchronization schemes. In: Azar, A.T., Vaidyanathan, S., Ouannas, A. (Eds.), Fractional Order Control and Synchronization of Chaotic Systems. Springer International Publishing, Cham, pp. 497–524.

Pecora, L., Carroll, T.L., 1990. Synchronization in chaotic systems. Phys. Rev. Lett. 64, 821–824.

Pham, V.-T., Vaidyanathan, S., Volos, C.K., Azar, A.T., Hoang, T.M., Van Yem, V., 2017a. A three-dimensional no-equilibrium chaotic system: analysis, synchronization and its fractional order form. In: Azar, A.T., Vaidyanathan, S., Ouannas, A. (Eds.), Fractional Order Control and Synchronization of Chaotic Systems. Springer International Publishing, Cham, pp. 449–470.

Pham, V.-T., Vaidyanathan, S., Volos, C.K., Jafari, S., Gotthans, T., 2017b. A Three-Dimensional Chaotic System with Square Equilibrium and No-Equilibrium. Springer International Publishing, Cham, pp. 613–635, URL https://doi.org/10.1007/978-3-319-50249-6_21.

Rajagopal, K., Vaidyanathan, S., Karthikeyan, A., Srinivasan, A., 2016. Complex novel 4d memristor hyperchaotic system and its synchronization using adaptive sliding mode control. Alexandria Eng. J.

Rössler, O.E., 1976. An equation for continuous chaos. Phys. Lett. A 57, 397–398.

Si, G., Sun, Z., Zhang, Y., Chen, W., 2012. Projective synchronization of different fractional-order chaotic systems with non-identical orders. Nonlinear Anal. Real World Applicat. 13 (4), 1761–1771.

Singh, S., Azar, A.T., Ouannas, A., Zhu, Q., Zhang, W., Na, J., 2017. Sliding mode control technique for multi-switching synchronization of chaotic systems. In: 9th International Conference on Modelling, Identification and Control (ICMIC 2017), July 10-12, 2017, Kunming, China.

Soukkou, A., Boukabou, A., Leulmi, S., 2016. Design and optimization of generalized prediction-based control scheme to stabilize and synchronize fractional-order hyperchaotic systems. Optik 127 (12), 5070–5077.

Sprott, J., 1994. Some simple chaotic flows. Phys. Rev. E 50, R647–650.

Su, D., Bao, W., Liu, J., Gong, C., 2016. An efficient simulation of the fractional chaotic system and its synchronization. J. Franklin Institute.

Sun, J., Deng, W., Cui, G., Wang, Y., 2016. Real combination synchronization of three fractional-order complex-variable chaotic systems. Optik 127 (23), 11460–11468.

Tolba, M.F., AbdelAty, A.M., Saida, L.A., Elwakil, A.S., Azar, A.T., Madian, A.H., et al., 2017a Fpga realization of caputo and grnwald-letnikov operators. In: 6th International Conference on Modern Circuits and Systems Technologies (MOCAST). pp. 1–4.

Tolba, M.F., AbdelAty, A.M., Soliman, N.S., Said, L.A., Madian, A.H., Azar, A.T., et al., 2017b. Fpga implementation of two fractional order chaotic systems. AEU – Int. J. Electr. Communicat. 78 (Suppl. C), 162–172. URL http://www.sciencedirect.com/science/article/pii/S1434841117303813.

Vaidyanathan, S., Azar, A.T., 2015a. Analysis and control of a 4-d novel hyperchaotic system. In: Azar, A.T., Vaidyanathan, S. (Eds.), Chaos Modeling and Control Systems Design. Vol. 581 of Studies in Computational Intelligence. Springer, Berlin, Germany, pp. 19–38.

Vaidyanathan, S., Azar, A.T., 2015b. Analysis, control and synchronization of a nine-term 3-d novel chaotic system. In: Azar, A.T., Vaidyanathan, S. (Eds.), Chaos Modeling and Control Systems Design. Vol. 581 of Studies in Computational Intelligence. Springer, Berlin, Germany, pp. 3–17.

Vaidyanathan, S., Azar, A.T., 2015c. Anti-synchronization of identical chaotic systems using sliding mode control and an application to vaidyanathan-madhavan chaotic systems. In: Azar, A.T., Zhu, Q. (Eds.), Advances and Applications in Sliding Mode Control Systems. Vol. 576 of Studies in Computational Intelligence. Springer, Berlin, Germany, pp. 527–547.

Vaidyanathan, S., Azar, A.T., 2015d. Hybrid synchronization of identical chaotic systems using sliding mode control and an application to vaidyanathan chaotic systems. In: Azar, A.T., Zhu, Q. (Eds.), Advances and Applications in Sliding Mode Control Systems. Vol. 576 of Studies in Computational Intelligence. Springer, Berlin, Germany, pp. 549–569.

Vaidyanathan, S., Azar, A.T., 2016a. A novel 4-D four-wing chaotic system with four quadratic nonlinearities and its synchronization via adaptive control method. Advances in Chaos Theory and IntelligentControl. Springer, Berlin, Germany, pp. 203–224.

Vaidyanathan, S., Azar, A.T., 2016b. Adaptive backstepping control and synchronization of a novel 3-D jerk system with an exponential nonlinearity. Advances in Chaos Theory and IntelligentControl. Springer, Berlin, Germany, pp. 249–274.

Vaidyanathan, S., Azar, A.T., 2016c. Adaptive control and synchronization of Halvorsen circulant chaotic systems. Advances in Chaos Theory and IntelligentControl. Springer, Berlin, Germany, pp. 225–247.

Vaidyanathan, S., Azar, A.T., 2016d. Dynamic analysis, adaptive feedback control and synchronization of an eight-term 3-D novel chaotic system with three quadratic nonlinearities. Advances in Chaos Theory and IntelligentControl. Springer, Berlin, Germany, pp. 155–178.

Vaidyanathan, S., Azar, A.T., 2016e. Generalized projective synchronization of a novel hyperchaotic four-wing system via adaptive control method. Advances in Chaos Theory and IntelligentControl. Springer, Berlin, Germany, pp. 275–290.

Vaidyanathan, S., Azar, A.T., 2016f. Qualitative study and adaptive control of a novel 4-d hyperchaotic system with three quadratic nonlinearities. In: Azar, A.T., Vaidyanathan, S.

(Eds.), Advances in Chaos Theory and Intelligent Control. Springer International Publishing, Cham, pp. 179–202.

Vaidyanathan, S., Azar, A.T., 2016g. Takagi-Sugeno fuzzy logic controller for Liu-Chen four-scroll chaotic system. Int. J. Intellig. Eng. Inf. 4 (2), 135–150.

Vaidyanathan, S., Azar, A.T., Rajagopal, K., Alexander, P., 2015a. Design and spice implementation of a 12-term novel hyperchaotic system and its synchronisation via active control. Int. J. Model. Identif. Control 23 (3), 267–277.

Vaidyanathan, S., Idowu, B.A., Azar, A.T., 2015b. Backstepping controller design for the global chaos synchronization of sprott's jerk systems. In: Azar, A.T., Vaidyanathan, S. (Eds.), Chaos Modeling and Control Systems Design. Vol. 581 of Studies in Computational Intelligence. Springer, Berlin, Germany, pp. 39–58.

Vaidyanathan, S., Sampath, S., Azar, A.T., 2015c. Global chaos synchronisation of identical chaotic systems via novel sliding mode control method and its application to zhu system. Int. J. Model. Identif. Control 23 (1), 92–100.

Vaidyanathan, S., Azar, A.T., Ouannas, A., 2017a. An eight-term 3-d novel chaotic system with three quadratic nonlinearities, its adaptive feedback control and synchronization. In: Azar, A.T., Vaidyanathan, S., Ouannas, A. (Eds.), Fractional Order Control and Synchronization of Chaotic Systems. Springer International Publishing, Cham, pp. 719–746.

Vaidyanathan, S., Azar, A.T., Ouannas, A., 2017b. Hyperchaos and adaptive control of a novel hyperchaotic system with two quadratic nonlinearities. In: Azar, A.T., Vaidyanathan, S., Ouannas, A. (Eds.), Fractional Order Control and Synchronization of Chaotic Systems. Springer International Publishing, Cham, pp. 773–803.

Vaidyanathan, S., Zhu, Q., Azar, A.T., 2017c. Adaptive control of a novel nonlinear double convection chaotic system. In: Azar, A.T., Vaidyanathan, S., Ouannas, A. (Eds.), Fractional Order Control and Synchronization of Chaotic Systems. Springer International Publishing, Cham, pp. 357–385.

Wang, X.-Y., Zhang, H., Lin, X.-H., 2014. Module-phase synchronization in hyperchaotic complex lorenz system after modified complex projection. Appl. Math. Comput. 232, 91–96.

Wang, Z., Volos, C., Kingni, S.T., Azar, A.T., Pham, V.-T., 2017. Four–wing attractors in a novel chaotic system with hyperbolic sine nonlinearity. Optik 131 (2017), 1071–1078.

Wong, W., Li, H., Leung, S., 2012. Robust synchronization of fractional-order complex dynamical networks with parametric uncertainties. Commun. Nonlinear Sci. Numer. Simul. 17 (12), 4877–4890.

Xi, H., Yu, S., Zhang, R., Xu, L., 2014. Adaptive impulsive synchronization for a class of fractional-order chaotic and hyperchaotic systems. Optik 125 (9), 2036–2040.

Xi, H., Li, Y., Huang, X., 2015. Adaptive function projective combination synchronization of three different fractional-order chaotic systems. Optik 126 (24), 5346–5349.

Zhang, J.-G., Yu, J.-N., Chu, Y.-D., Li, X.-F., Chang, Y.-X., 2008. Bifurcation and chaos of a non-autonomous rotational machine systems. Simul. Model. Practice Theory 16 (10), 1588–1605.

Zhang, R., Tian, G., Yang, S., Cao, H., 2015. Stability analysis of a class of fractional order nonlinear systems with order lying in (0, 2). ISA Trans. 56, 102–110.

Zhu, Q., Azar, A.T., 2015. Complex System Modelling and Control Through Intelligent Soft Computations, Vol. 319. Studies in Fuzziness and Soft Computing, Berlin, Germany.

Chapter 11

Multiswitching Synchronization of Commensurate Fractional Order Hyperchaotic Systems Via Active Control

Shikha Singh[1], Ahmad Taher Azar[2,3], Sundarapandian Vaidyanathan[4], Adel Ouannas[5] and Muzaffar A. Bhat[1]

[1]*Jamia Millia Islamia, New Delhi, India,* [2]*Faculty of Computers and Information, Benha University, Benha, Egypt,* [3]*School of Engineering and Applied Sciences, Nile University, Giza, Egypt,* [4]*Vel Tech University, Chennai, Tamil Nadu, India,* [5]*University of Larbi Tebessi, Tebessa, Algeria*

11.1 INTRODUCTION

Almost three centuries have passed since the emergence of fractional calculus. The major merit of fractional calculus, different from integer calculus, lies in the fact that the fractional order system describes real systems in interdisciplinary fields more elegantly in comparison to the integer order system, as it has memory, and has proven to be a very suitable tool for the description of memory and hereditary properties of various materials and processes (Lyapunov, 1992). During this time many researchers have devoted a lot of time and energy to reveal the investigations on fractional derivatives in detail (Li and Deng, 2007). Among the wide range of studies that have been done in this direction, it has been observed that some of the fractional order differential systems display chaotic behavior (Petras, 2011; El-Sayed et al., 2016; Tolba et al., 2017). Tavazoei and Haeri (2009) suggested two techniques for detecting chaotic properties in a class of fractional order systems. Various researchers have focused to develop the efficient and robust schemes of synchronization and control of chaotic systems. With the evolution of models based on fractional order differential systems the chaos synchronization and chaos control in fractional order chaotic systems are becoming important emerging areas of research. During the studies on the chaos synchronization and chaos control in fractional order chaotic systems it has been observed, that

Mathematical Techniques of Fractional Order Systems. DOI: https://doi.org/10.1016/B978-0-12-813592-1.00011-8

there are lot of differences among the behavior of integer order and fractional order differential systems. The majority of the conclusions based on stabilization for the integer order chaotic systems may not be directly applied to the fractional order chaotic systems. The main hindrance is that the stability regions of fractional order differential systems differ from the integer order systems. It leads to the different stability criteria (Li and Zhang, 2016). The studies on the stability analysis of fractional order chaotic systems are investigated in Wen et al. (2008) and Petras (2008). The history of the subject goes back to the times when Leibnitz in a letter to L-Hospital dated 30th September 1695 raised the following question, "can the meaning of the derivative with integer order be generalized to derivatives with noninteger orders." In modern years plentiful studies and utilizations of systems with fractional order in frequent spaces of engineering and sciences have been presented (Podlubny, 1998; Hilfer, 2000). Fractional calculus can be categorized as applicable mathematics. It has attracted more researchers' interest and has more broad application prospects due to its unique advantages. But until the last 20 years, the fractional order calculus theory was related to practical projects, as it was applied to chaos theory, electromagnetism, signal processing, mechanical engineering, robot control, and so on. In recent years, many dynamical systems with fractional order have been described such as diffusion electromagnetic waves, dielectric polarization, electrode electrode polarization, and viscoelastic systems (Koeller, 1986, 1984; Heaviside, 1970). In comparison with the classical modes with integer order, derivatives with fractional order yield wonderful instruments for the depiction of retention and heritable possessions of different materials and their formation. With the basic text of fractional calculus, it was demonstrated that various dynamical systems with fractional order have chaotic behavior with order less than three. For example, Chuas fractional order circuit (Agarwal et al., 2013), Rössler fractional order system (Li and Chen, 2004), Chen fractional order system (Li and Chen, 2004), Lü fractional order system (Deng and Li, 2005), and modified Duffing fractional order system (Ge and Ou, 2007). Since the work of Pecora and Carroll in 1990 (Pecora and Carroll, 1990), chaos control and chaos synchronization have attracted great interest and received extensive studies in many disciplines such as secure communication, information processing, chemical reaction, and high-performance circuits, etc. Chaos has shown great potential to be useful and brought forth a great fascination. The problems of control of chaos have attracted the attention of researchers and engineers since the early 1990s. Several thousand publications have appeared over the recent decades. In recent years, the control of chaotic and hyperchaotic systems has received great attention due to its potential applications in physics, chemical reactors, biological networks, artificial neural networks, telecommunications, etc. Basically, chaos controlling is the stabilization of an unstable periodic orbit or equilibria by means of tiny perturbations of the system. E. Ott, C. Grebogi, and J. A. Yorke were the first to make the key observation that the infinite number of unstable periodic orbits typically embedded in a chaotic attractor could be taken advantage of for the purpose of achieving control by means of applying only very small perturbations. After making this general point, they illustrated it with a specific method, since called the OGY method (Ott, Grebogi, and Yorke) of achieving stabilization of a chosen unstable periodic orbit. In the OGY method, small, wisely chosen, kicks are applied to the system once per cycle, to maintain it near the desired unstable periodic orbit (Aleksandr et al., 1998). Later on, Several techniques have been devised for chaos control. These may include linear feedback (Becker and Packard, 1994), optimal control (Zhang et al., 2011), adaptive control (Vaidyanathan et al., 2017a,b,c; Vaidyanathan and Azar, 2016c,d,f,g; Khan and Bhat, 2016b), active control (Vaidyanathan et al., 2015a; Khan et al., 2017c), active sliding control (Zhang et al., 2004), passive control (Wang and Liu, 2007), impulsive control (Yang and Chua, 1997), backstepping control (Vaidyanathan et al., 2015b), sliding mode control (Vaidyanathan et al., 2015c), adaptive sliding mode control (Khan et al., 2017a,b) etc. Synchronization of chaos is a phenomenon that may occur when two, or more, dissipative chaotic systems are coupled. Because of the exponential divergence of the nearby trajectories of chaotic system, having two chaotic systems evolving in synchrony might appear surprising. Basically, synchronization of chaos refers to a process wherein two (or many) chaotic systems (either equivalent or nonequivalent) adjust a given property of their motion to a common behavior due to a coupling (Vaidyanathan and Azar, 2016a,b; Azar and Vaidyanathan, 2015a,b,c; Zhu and Azar, 2015; Vaidyanathan and Azar, 2015a,b,c,d; Azar and Zhu, 2015; Vaidyanathan and Azar, 2016h; Boulkroune et al., 2016b; Azar and Vaidyanathan, 2016; Ouannas et al., 2016a,b; Soliman et al., 2017; Ouannas et al., 2015f,g,h,i,j,k; Grassi et al., 2017; Vaidyanathan and Sampath, 2017; Azar et al., 2017b, 2018; Moysis and Azar, 2017; Pham et al., 2017; Lamamra et al., 2017; Ouannas et al., 2017c; Wang et al., 2017; Singh et al., 2017; Munoz-Pacheco et al., 2017; Ouannas et al., 2017a).

In current years more and more attention has been diverted towards the control and synchronization of fractional order chaotic systems (Boulkroune et al., 2016a; Khan and Bhat, 2016a; Pham et al., 2017; Azar et al., 2017a). Various kinds of synchronization phenomenon have been studied, such as complete synchronization (Mahmoud and Mahmoud, 2010), phase synchronization (Rosenblum et al., 1996), generalized synchronization (Ouannas et al., 2017h; Khan et al., 2017c), generalized projective synchronization (Vaidyanathan and Azar, 2016e), lag synchronization (Shahverdiev et al., 2002), antisynchronization (Vaidyanathan and Azar, 2015c), projective synchronization (Mainieri and Rehacek, 1999), modied projective synchronization (Li, 2007), function projective synchronization (Du et al., 2008), modied-function projective synchronization (Du et al., 2009), hybrid synchronization (Ouannas et al., 2016a, 2017b; Vaidyanathan and Azar, 2015d; Ouannas et al., 2017d), and hybrid function projective synchronization (Ouannas et al., 2017e; Khan et al., 2016).

11.2 RECENT WORK

In recent years, some advances have been made with the idea of multiswitching synchronization (Ucar et al., 2008). The different states of the drive system are synchronized with the desired state of the response system in multiswitching synchronization schemes. The relevance of such kinds of synchronization studies to information security is evident in the wide range of possible synchronization directions that exist due to multiswitching synchronization. In spite of these schemes clearly providing improved resistance and anti-attack ability for secure communication, only a few studies of this kind have been reported in the literature (Wang and Sun, 2011; Ajayi et al., 2014; Khan and Bhat, 2017). But this kind of work has been reported in the case of systems having integer order. The problem of multiswitching synchronization of nonidentical fractional order hyperchaotic systems, is an issue to discuss.

Motivated by the above discussion, in this chapter we investigate the multiswitching synchronization between nonidentical fractional order hyperchaotic systems. In this work we design appropriate controllers to synchronize the signals of master fractional order hyperchaotic system with that of the slave fractional order chaotic system in a multiswitching manner via active control technique. The chapter is arranged as follows: in section 11.3, the review and the approximation of fractional operators are described. Section 11.4, describes the problem formulations. In section 11.5, stability of fractional order chaotic systems is examined. In section 11.6, the system description and a brief analysis of fractional order hyperchaotic systems are given. In section 11.7, the main results are discussed. Numerical results are used to verify this technique. Finally in section 11.8, conclusions are drawn.

11.3 THE REVIEW AND THE APPROXIMATION OF A FRACTIONAL OPERATOR

The differintegral operator, designated by ${}_aD_t^\alpha$, is a combined differentiation-integration operator usually found in fractional calculus. For taking both the fractional derivative and the fractional integral this operator is a notation to express them in a single expression and is defined by:

$$ {}_aD_t^\alpha = \begin{cases} \dfrac{d^\alpha}{dt^\alpha}, & :\alpha > 0 \\ 0, & :\alpha = 0 \\ \int_a^t (d\tau)^{-\alpha}, & :\alpha < 0 \end{cases} \tag{11.1} $$

There are so many definitions for fractional derivatives (Podlubny, 1998), but the three most commonly definitions are:

Definition 1: Let $n \in N$ be such that $n-1<\alpha<n$, the Riemann–Liouville definition (Podlubny, 1998) for fractional derivative of order α for any function say $f(t)$ is given as:

$$
\begin{aligned}
{}_aD_t^{\alpha}f(t) &= \frac{d^{\alpha}f(t)}{d(t-a)^{\alpha}} \\
&= \frac{1}{\Gamma(n-\alpha)}\frac{d^n}{dt^n}\int_0^t (t-q)^{n-\alpha-1}f(q)d(q)
\end{aligned}
\tag{11.2}
$$

where n is the first integer such that, $n-1 \leq \alpha < n$ and Γ denotes the Gamma function which is defined as:

$$
\Gamma(x) = \int_0^{\infty} t^{x-1}e^{-t}dt \tag{11.3}
$$

Definition 2: The Grunwald–Letnikov definition (Podlubny, 1998) for derivative of fractional order α, of any function $f(t)$ is defined as under:

$$
\begin{aligned}
{}_aD_t^{\alpha}f(t) &= \frac{d^{\alpha}f(t)}{d(t-a)^{\alpha}} \\
&= \lim_{N\to\infty}\left[\frac{t-a}{N}\right]^{-\alpha}\sum_{j=1}^{N}(-1)^j\binom{\alpha}{j}f\left(t-j\left[\frac{t-a}{N}\right]\right)
\end{aligned}
\tag{11.4}
$$

This definition is considered as simplest and easiest definition to be used in fractional calculus.

Definition 3: Let $f \in C_{-1}^{n}$, $n \in N$. The (left-sided) Caputo definition (Podlubny, 1998) of derivative of fractional order α is defined as:

$$
{}_0D_t^{\alpha} = \begin{cases} \dfrac{1}{\Gamma(n-\alpha)}\displaystyle\int_0^t \frac{f^n(\tau)}{(t-\tau)^{\alpha+1-n}}d\tau, & :n-1<\alpha<n \\ \dfrac{d^nf(t)}{dt^n}, & :\alpha = n \end{cases} \tag{11.5}
$$

where n is the smallest integer, greater than α.

The Grunwald–Letnikov and the Riemann–Liouville definitions are equivalent for the functions $f(t)$ having n continuous derivatives for $t \geq 0$ such that $n-1 \leq \alpha < n$. In terms of Laplace transform, the Riemann–Liouville fractional integral and derivative is as follows:

$$
L\{{}_0D_t^{\alpha}f(t)\} = S^{\alpha}F(s):\alpha \leq 0. \tag{11.6}
$$

$$L\{\,{}_0D_t^\alpha f(t)\} = S^\alpha F(s) - \sum_{k=0}^{n-1} S_0^k D_t^\alpha f(0);\quad n-1<\alpha\leq n\in N \tag{11.7}$$

Unfortunately, the Laplace transform technique used in Riemann–Liouville appears incompatible as it requires the knowledge of the functions $f(t)$ having derivatives at $t=0$ of fractional order (Kilbas et al., 1993). But this kind of difficulty does not occur in the Caputo definition. The Caputo definition for fractional derivative in literature is sometimes named as smooth fractional derivation (Caputo, 1967). With the help of homogeneous initial conditions assumption it is established that the systems with Riemann–Liouville operators are similar to those with Caputo operators (Keil et al., 2012).

In literature, for fractional differential equations the appropriate numerical technique has been developed for those systems which are numerically stable and can be used for all classes of fractional differential equations. To use an improved predictor corrector algorithm we choose the Caputo version for fractional differential equations (Diethelm et al., 2002), in order to get more precise numerical approximation and reduces computational cost. Based on the analytical property of the fractional predictor-corrector algorithm the following differential equation:

$$\begin{cases} D_*^\alpha = g(t,x); 0\leq t\leq T \\ x^k(0) = x_0^k, k=1,2,...,m. \end{cases} \tag{11.8}$$

and the voltera integral equation

$$x(t) = \sum_{k=o}^{m-1} x_0^{(k)}\frac{t^k}{k!} + \frac{1}{\Gamma(\alpha)}\int_0^t \frac{g(\tau,x)}{(t-\tau)^{1-\alpha}}d\tau \tag{11.9}$$

are equivalent. Now set $h=T/N, t_n=nh(n=0,1,2,...,N\in Z^+)$

Eq. (11.6) can be rewritten as follows

$$\begin{aligned} x_h(t_{n+1}) &= \sum_{k=0}^{m-1} x_0^{(k)}\frac{t_{n+1}^k}{k!} + \frac{h^\alpha}{\Gamma(\alpha+2)}g(t_{n+1}, x_h^\theta(t_{n+1})) \\ &+ \frac{h^\alpha}{\Gamma(\alpha+2)}\sum a_{j,n+1}g(t_j, x_h(t_j)) \end{aligned} \tag{11.10}$$

where $x_h^\theta(t_{n+1})$ is the predicted value which is determined by

$$x_h^\theta(t_{n+1}) = \sum_{k=0}^{m-1} x_0^{(k)}\frac{t_{n+1}^k}{k!} + \frac{1}{\Gamma(\alpha)}\sum_{j=0}^{n} b_{j,n+1}g(t_j, x_h(t_j))$$

and

$$
\begin{aligned}
a_{j,n+1} &= n^{\alpha+1} - (n-\alpha)(n+1)^{\alpha+1}; j=0 \\
&= (n-j+2)^{\alpha+1} - (n-j)^{\alpha+1} - 2(n-j+1)^{\alpha+1}; \\
1 \le j \le n & \\
b_{j,n+1} &= \frac{h^{\alpha}}{\alpha}((n-j+1)^{\alpha} - (n-j)^{\alpha})
\end{aligned}
$$

where

$$\max(|x(t_j) - x_h(t_j))|) = O(h^{\theta}), (j = 0, 1, 2, ..., N)$$

is the estimation error and $\theta = \min(2, 1+\alpha)$

11.4 PROBLEM FORMULATION

In this section, we describe the problem formulation to achieve the multi-switching complete synchronization between fractional order hyperchaotic systems by applying active control technique.

Consider the following chaotic system having fractional order α as master system:

$$D^{\alpha}u(t) = Mu(t) + F(u(t)) \tag{11.11}$$

where α represents the fractional order, $u(t) \in \mathbb{R}^{n\times 1}$ is the state vector. $M \in \mathbb{R}^{n\times n}$ is a constant matrix, $Mu(t) \in \mathbb{R}^{n\times 1}$ describes the linear terms and $F(u(t)) \in \mathbb{R}^{n\times 1}$ represents the nonlinear terms in the system (11.11).

Consider the subsequent fractional order system, which acts as slave system as:

$$D^{\alpha}v(t) = Nv(t) + G(v(t)) + \psi_{ij}(u(t), v(t)) \tag{11.12}$$

where $v(t) \in \mathbb{R}^{n\times 1}$ is the state vector. $N \in \mathbb{R}^{n\times n}$ is a constant matrix, $Nv(t) \in \mathbb{R}^{n\times 1}$ describes the linear term, $G(v(t)) \in \mathbb{R}^{n\times 1}$ describes the nonlinear terms in the system (11.12) and $\psi(u(t), v(t)) \in R^{n\times 1}$ is the real feedback controller which is to be designed.

Definition 4: The two systems (11.11) and (11.12) are said to achieve complete synchronization, if

$$\lim_{t\to+\infty} \| e_{ij}(t) \| = \lim_{t\to+\infty} \| v_j(t) - u_i(t) \| = 0$$

where the symbols $\| \bullet \|$ symbolize the matrix norm.

Now the following result is established on the controller of the fractional order chaotic systems, which give the final destination to the problem formulation.

From Eqs. (11.11) and (11.12), the error dynamical system is obtained as

$$D^{\alpha}e_{ij}(t) = Nv_j(t) + G(v_j(t)) + \phi)ij(u,v) - Mu_i(t) - F(u(t)) \qquad (11.13)$$

Our goal is to design a suitable controller $\phi_{ij}(u,v)$, such that the system (11.11) and (11.12) achieve multiswitching complete synchronization in accordance with definition (11.4). In this chapter, we are considering a four-dimensional hyperchaotic system for which the possible switches can be obtained by imposing the conditions on $i, j = 1,2,3,4$. The list of the possible errors for hyperchaotic system whose possible combinations can be used to form the switches is as follows : For $i=j$, we have, $e_{11}, e_{22}, e_{33}, e_{44}$ For $i \neq j$ we have, e_{12}, e_{13}, e_{14}, e_{21}, e_{23}, e_{24}, e_{31}, e_{32}, e_{34}

11.5 STABILITY OF FRACTIONAL ORDER SYSTEMS

Stability of fractional order systems has been thoroughly investigated where necessary and sufficient conditions have been derived in Wolf et al. (1985). The stability region of a linear set of fractional order equations, each of order q, such that $0<q<1$ is shown in Fig. 11.1. An autonomous system is asymptotically stable iff $|\arg \lambda| > \frac{\alpha\pi}{2}$ is satisfied for all eigenvalues λ of matrix A. Also this system is stable iff $|arg\lambda| \geq \frac{\alpha\pi}{2}$ is satisfied for all eigenvalues of a matrix A and those critical eigenvalues which satisfy $|arg\lambda| > \frac{\alpha\pi}{2}$, and have geometric multiplicity one.

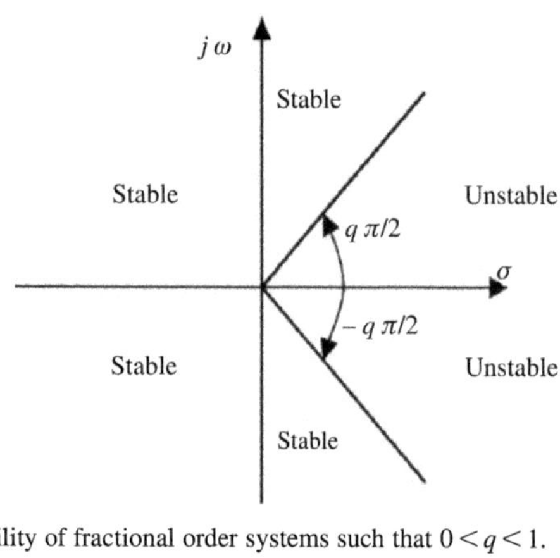

FIGURE 11.1 Stability of fractional order systems such that $0<q<1$.

11.6 SYSTEM DESCRIPTION

The fractional order new hyperchaotic system (Gao et al., 2007) is given by

$$\begin{cases} \dfrac{d^q x_1}{dt^q} = a_1(y_1 - x_1) \\ \dfrac{d^q y_1}{dt^q} = dx_1 - x_1 z_1 + cy_1 - w_1 \\ \dfrac{d^q z_1}{dt^q} = x_1 y_1 - bz_1 \\ \dfrac{d^q w_1}{dt^q} = x_1 + k \end{cases} \tag{11.14}$$

where x_1, y_1, z_1, w_1 are the state variables, q is the fractional order satisfying $0 < q < 1$ and a, b, c, d, k are the parameters.

Applying the discretization scheme, it was found that hyperchaos indeed exists in the new system with fractional order. The parameters are always chosen as $a = 36,\ b = 3,\ c = 28,\ d = 16\ and\ k = 0.5$. It was demonstrated that hyperchaos does exist in the fractional order system with order less than 4. It was found that when $0.72 < q < 1$, the fractional order system (11.14) displays hyperchaotic behaviors. For example, when $q = 0.9$ and $q = 0.72$, hyperchaotic attractors are found and the phase portraits are shown in Figs. 11.2 and 11.3, respectively. We calculated the two largest Lyapunov exponents of this system using the well-known Wolf algorithm. The values of the two largest Lyapunov exponents are $\lambda_1 = 12.3014$ and $\lambda_2 = 0.2318$ when $q = 0.9$. The values of the two largest Lyapunov exponents are $\lambda_1 = 8.2130$ and $\lambda_2 = 0.1015$ when $q = 0.72$. Obviously, the fractional order system (11.14) has hyperchaos. Here we have considered the parameters values $a = 36,\ b = 3,\ c = 28,\ d = -16,\ k = 0.5$ and $q = 0.95$, the phase portraits of the system showing hyperchaotic behavior is shown in Figs. 11.2 and 11.3.

The fractional order hyperchaotic Gao system (Gao et al., 2015) is given by

$$\begin{cases} \dfrac{d^q x_2}{dt^q} = -a_1 x_2 + a_2 y_2 \\ \dfrac{d^q y_2}{dt^q} = a_3 x_2 - x_2 z_2 - y_2 + w_2 \\ \dfrac{d^q z_2}{dt^q} = x_2^2 - a_4(x_2 + z_2) \\ \dfrac{d^q w_2}{dt^q} = -a_5 x_2 \end{cases} \tag{11.15}$$

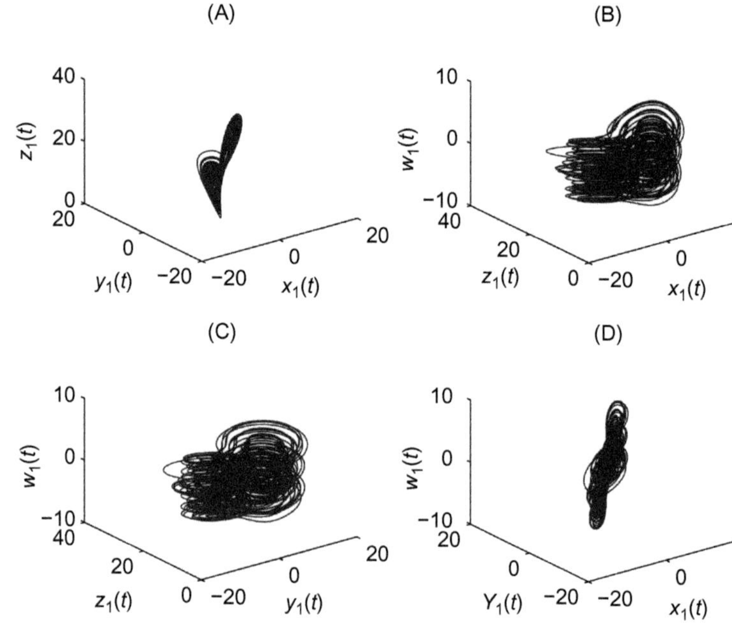

FIGURE 11.2 3D Phase Portrait of the fractional order hyperchaotic Rossler system in
(A) (*x*1, *y*1, *z*1)-space,
(B) (*x*1, *z*1, *w*1)-space,
(C) (*y*1, *z*1, *w*1)-space,
(D) (*x*1, *y*1, *w*1)-space.

where x_2, y_2, z_2, w_2 are the state variables, $q > 0$ is the fractional order and a_1, a_2, a_3, a_4, a_5 are the parameters.

For the parameters values $a_1 = 25$, $a_2 = 60$, $a_3 = 40$, $a_4 = 4$, $a_5 = 5$ and $q = 0.95$ the four Lyapunov exponents of the system (11.15) are calculated as $\gamma_1 = 3.0057$, $\gamma_2 = 0.0304$, $\gamma_3 = -0.1631$, and $\gamma_4 = 46.1578$, respectively.

Therefore, the Kaplan–Yorke dimension of system is

$$\begin{aligned} D &= j + \sum_{i=1}^{j} \frac{\gamma_i}{|\gamma_{j+1}|} \\ &= 3.0622 \end{aligned}$$

where j is the largest integer satisfying $\sum_{i=1}^{j} \gamma_j \geq 0$ and $\sum_{i=1}^{j+1} \gamma_j < 0$. The Lyapunov dimension is fractional, which implies system (11.15) is really a dissipative system. In Figs. 11.4 and 11.5, 3D phase portraits and 2D phase portraits are displayed, it is clear that the system has a double-scroll hyperchaotic attractor.

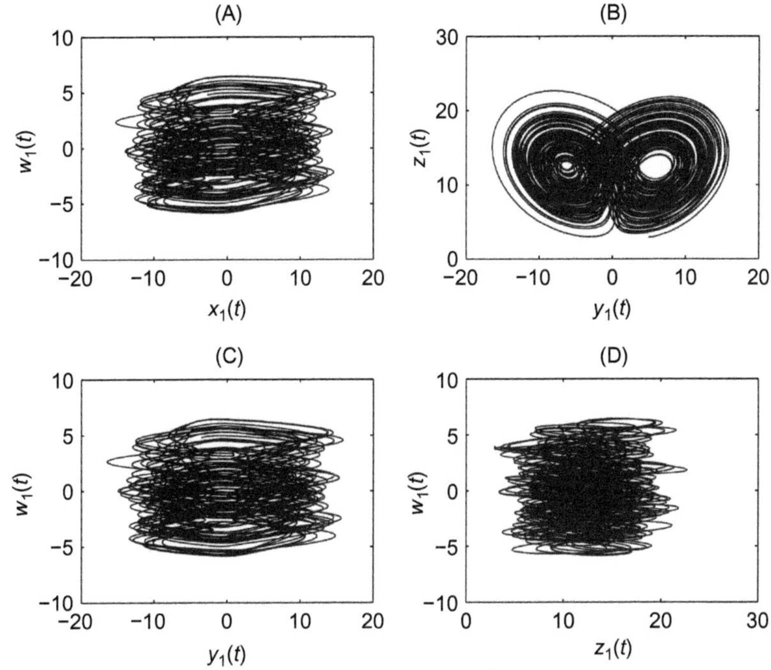

FIGURE 11.3 2D Phase Portrait of the fractional order hyperchaotic Rossler system in
(A) ($x1$, $w1$)-plane,
(B) ($y1$, $z1$)-plane,
(C) ($y1$, $w1$)-plane,
(D) ($z1$, $w1$)-plane.

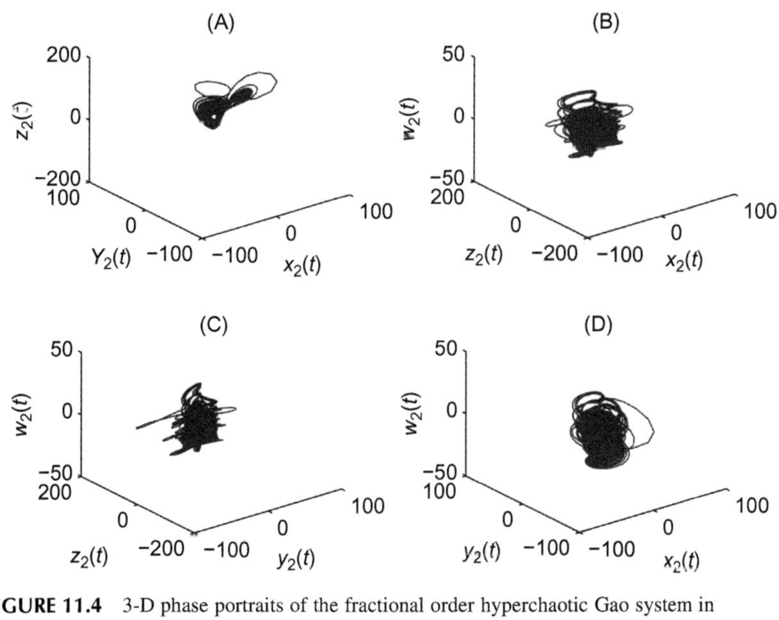

FIGURE 11.4 3-D phase portraits of the fractional order hyperchaotic Gao system in
(A) ($x2$, $y2$, $z2$)-space,
(B) ($x2$, $z2$, $w2$)-space,
(C) ($y2$, $z2$ $w2$)-space,
(D) ($x2$, $y2$, $w2$)-space.

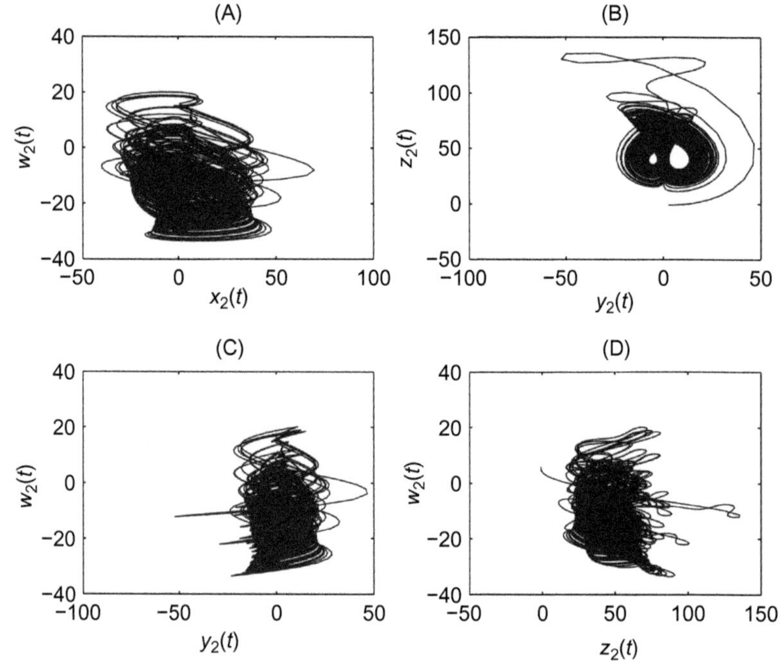

FIGURE 11.5 2-D phase portraits of the fractional order hyperchaotic Gao system in
(A) ($x2$, $w2$)-plane,
(B) ($y2$, z2)-plane,
(C) ($y2$, $w2$)-plane,
(D) ($z2$, $w2$)-plane.

11.7 MULTISWITCHING COMPLETE SYNCHRONIZATION OF FRACTIONAL ORDER HYPERCHAOTIC SYSTEMS

In this section we discuss the multiswitching complete synchronization between the fractional order hyperchaotic systems (11.14) and (11.15). Fractional order chaotic system (11.14) is taken as the drive system.

Fractional order system (11.15) with added controllers is taken as the response system and is represented as :

$$\begin{aligned}
\frac{d^q x_2}{dt^q} &= -\, a_1 x_2 + a_2 y_2 + \psi_1 \\
\frac{d^q y_2}{dt^q} &= a_3 x_2 - x_2 z_2 - y_2 + w_2 + \psi_2 \\
\frac{d^q z_2}{dt^q} &= x_2^2 - a_4 (x_2 + z_2) + \psi_3 \\
\frac{d^q w_2}{dt^q} &= -\, a_5 x_2 + \psi_4
\end{aligned} \tag{11.16}$$

where ψ_1, ψ_2, ψ_3 *and* ψ_4 are the controllers which are to be determined using active control technique. Our goal is to find suitable controllers ψ_1, ψ_2, ψ_3 *and* ψ_4 such that the drive system (11.14) asymptotically synchronizes with response system (11.16).

Out of various possible switches, in this chapter we present results for three randomly selected error state vector combinations.

Let the switching error states be defined as :

$$\begin{cases} e_1 = x_2 - y_1 \\ e_2 = y_2 - z_1 \\ e_3 = z_2 - w_1 \\ e_3 = w_2 - x_1 \end{cases} \quad Switch \quad 1 \tag{11.17}$$

$$\begin{cases} e_1 = x_2 - z_1 \\ e_2 = y_2 - w_1 \\ e_3 = z_2 - x_1 \\ e_3 = w_2 - y_1 \end{cases} \quad Switch \quad 2 \tag{11.18}$$

where we refer to Eqs. (11.17) and (11.18) as switch (11.1) and switch (11.2) respectively.

11.7.1 Switch 1

The error dynamical system for switch 1 is obtained as follows:

$$\begin{cases} \dfrac{d^q e_1}{dt^q} = \dfrac{d^q x_2}{dt^q} - \dfrac{d^q y_1}{dt^q} \\ \dfrac{d^q e_2}{dt^q} = \dfrac{d^q y_2}{dt^q} - \dfrac{d^q z_1}{dt^q} \\ \dfrac{d^q e_3}{dt^q} = \dfrac{d^q z_2}{dt^q} - \dfrac{d^q w_1}{dt^q} \\ \dfrac{d^q e_4}{dt^q} = \dfrac{d^q w_2}{dt^q} - \dfrac{d^q x_1}{dt^q} \end{cases} \tag{11.19}$$

Using Eqs. (11.14), (11.16) and (11.17) the error system (11.19) is obtained as :

$$\begin{cases} \dfrac{d^q e_1}{dt^q} = -a_1 x_2 + a_2 y_2 + \psi_1 - d x_1 + x_1 z_1 - c y_1 + w_1 \\ \dfrac{d^q e_2}{dt^q} = a_3 x_2 - x_2 z_2 - y_2 + w_2 + \psi_2 - x_1 y_1 + b z_1 \\ \dfrac{d^q e_3}{dt^q} = x_2^2 - a_4(x_2 + z_2) + \psi_3 x_1 - k \\ \dfrac{d^q e_4}{dt^q} = -a_5 x_2 + \psi_4 - a_1(y_1 - x_1) \end{cases} \tag{11.20}$$

We define the active control functions $\psi_i(t)$ as follows :

$$\begin{cases} \psi_1 = V_1 + (a_2 + c)y_1 - a_2 y_2 + dx_1 - x_1 z_1 - w_1 \\ \psi_2 = V_2 - a_3 x_2 + x_2 z_2 + (1 - b)z_1 - w_2 + x_1 y_1 \\ \psi_3 = V_3 - x_2^2 + a_4 x_2 + a_4 w_1 + x_1 + k \\ \psi_4 = V_4 + a_5 y_2 + a_1 y_1 - a_1 w_2 \end{cases} \tag{11.21}$$

The terms $V_i(t)$ are linear functions of the error term $e_i(t)$. With the choice of $\psi_i(t)$ given by (11.21), the error system (11.20) becomes

$$\begin{cases} \dfrac{d^q e_1}{dt^q} = -a_2 e_1 + V_1(t) \\ \dfrac{d^q e_2}{dt^q} = -e_2 + V_2(t) \\ \dfrac{d^q e_3}{dt^q} = -a_4 e_3 + V_3(t) \\ \dfrac{d^q e_4}{dt^q} = -a_1 e_4 + V_4(t) \end{cases} \tag{11.22}$$

The control terms $V_i(t)$ are chosen so that the system (11.22) becomes stable. There is not a unique choice for such functions.

We choose

$$\begin{bmatrix} V_1 \\ V_2 \\ V_3 \\ V_4 \end{bmatrix} = A \begin{bmatrix} e_1 \\ e_2 \\ e_3 \\ e_4 \end{bmatrix} \tag{11.23}$$

where A is a 4×4 real matrix, chosen so that all eigenvalues λ_i of the system (11.23) satisfy the condition

$$|arg(\lambda_i)| > \frac{\alpha\pi}{2} \tag{11.24}$$

if we choose

$$A = \begin{bmatrix} a_2 - 61 & 0 & 0 & 0 \\ 0 & -1 & 0 & 0 \\ 0 & 0 & a_4 - 5 & 0 \\ 0 & 0 & 0 & a_1 - 26 \end{bmatrix} \tag{11.25}$$

Then the eigenvalues of the linear system (11.23) are -1, -1, -1, and -1. Hence the condition (11.24) is satisfied for and we get the required synchronization. Numerical simulations for switch 1 are performed to demonstrate the theoretical results. The parameter values for which the system (11.14) and (11.15) exhibits chaotic behavior are taken as $a = 36$, $b = 3$, $c = 28$, $d = -16$, $k = 0.5$, $a_1 = 25$, $a_2 = 60$, $a_3 = 40$, $a_4 = 4, a_5 = 5$. The fractional

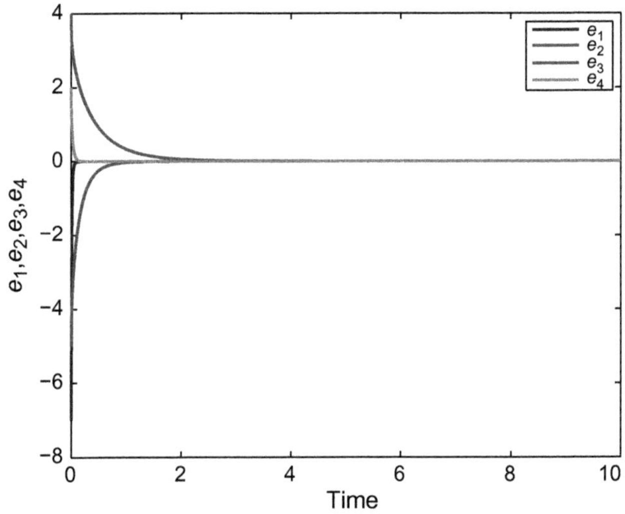

FIGURE 11.6 Synchronization error between states of master and slave systems.

order value for both commensurate fractional order hyperchaotic system is taken as $q = 0.95$ for which the system displays hyperchaotic behavior. The initial states of the master and slave system are arbitrarily chosen as $(x_1(0), y_1(0), z_1(0), w_1(0)) = (4, 5, 3, 4)$ and $(x_2(0), y_2(0), z_2(0), w_2(0)) = (-2, 7, -1, 6)$. The corresponding initial condition for the error states is obtained as $(e_1(0), e_2(0), e_3(0), e_4(0)) = (-7, 0, -5, 2)$. Fig. 11.6 displays the time response of synchronization errors e_1, e_2, e_3, e_4. Figs. 11.7–11.10 illustrate the time response of synchronization states $y_1 \& x_2$, $z_1 \& y_2$, $w_1 \& z_2$, and $x_1 \& w_2$ of the master system (11.14) and slave system (11.16) respectively.

11.7.2 Switch 2

The error dynamical system for switch 2 is obtained as follows:

$$\begin{cases} \dfrac{d^q e_1}{dt^q} = \dfrac{d^q x_2}{dt^q} - \dfrac{d^q z_1}{dt^q} \\ \dfrac{d^q e_2}{dt^q} = \dfrac{d^q y_2}{dt^q} - \dfrac{d^q w_1}{dt^q} \\ \dfrac{d^q e_3}{dt^q} = \dfrac{d^q z_2}{dt^q} - \dfrac{d^q x_1}{dt^q} \\ \dfrac{d^q e_4}{dt^q} = \dfrac{d^q w_2}{dt^q} - \dfrac{d^q y_1}{dt^q} \end{cases} \tag{11.26}$$

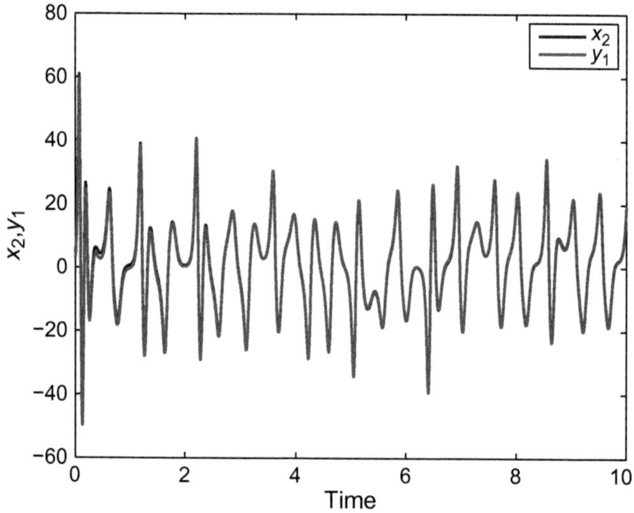

FIGURE 11.7 Time response showing synchronization between states y_1 and x_2.

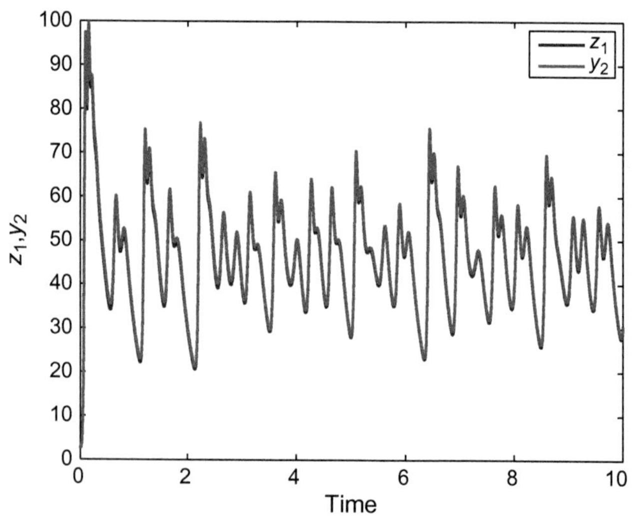

FIGURE 11.8 Time response showing synchronization between states z_1 and y_2.

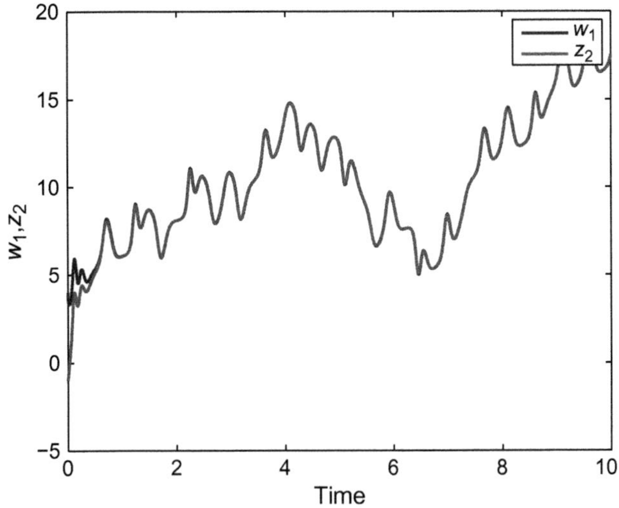

FIGURE 11.9 Time response showing synchronization between states w_1 and z_2.

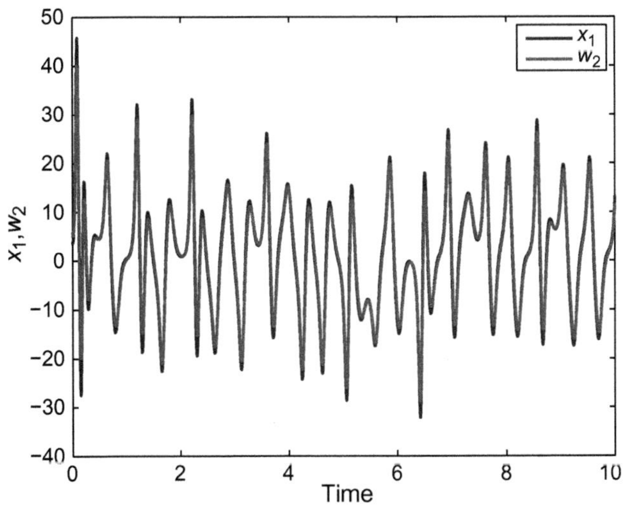

FIGURE 11.10 Time response showing synchronization between states x_1 and w_2.

Using Eqs. (11.14) and (11.16) the error system (11.26) is transformed into the following form:

$$\begin{cases} \dfrac{d^q e_1}{dt^q} = -a_1x_2 + a_2y_2 + \psi_1 - x_1y_1 + bz_1 \\ \dfrac{d^q e_2}{dt^q} = a_3x_2 - x_2z_2 - y_2 + w_2 + \psi_2 - x_1 - k \\ \dfrac{d^q e_3}{dt^q} = x_2^2 - a_4(x_2 + z_2) + \psi_3 - a_1(y_1 - x_1) \\ \dfrac{d^q e_4}{dt^q} = -a_5x_2 + \psi_4 - dx_1 + x_1z_1 - cy_1 + w_1 \end{cases} \tag{11.27}$$

We define the active control functions $\psi_i(t)$ as follows :

$$\begin{cases} \psi_1 = V_1 + a_1z_1 - a_2y_2 + x_1y_1 - bz_1 \\ \psi_2 = V_2 - a_3x_2 + x_2z_2 + w_1 - w_2 + x_1 + k \\ \psi_3 = V_3 - x_2^2 + a_4x_2 + a_4x_1 + a_1y_1 - a_1x_1 \\ \psi_4 = V_4 + a_5x_2 + dx_1 - x_1z_1 + cy_1 - w_1 \end{cases} \tag{11.28}$$

The terms $V_i(t)$ are linear functions of the error term $e_i(t)$. With the choice of $\psi_i(t)$ given by (11.28), the error system (11.27) becomes

$$\begin{cases} \dfrac{d^q e_1}{dt^q} = -a_1e_1 + V_1(t) \\ \dfrac{d^q e_2}{dt^q} = -e_2 + V_2(t) \\ \dfrac{d^q e_3}{dt^q} = -a_4e_3 + V_3(t) \\ \dfrac{d^q e_4}{dt^q} = V_4(t) \end{cases} \tag{11.29}$$

The control terms $V_i(t)$ are chosen so that the system (11.29) becomes stable. There is not a unique choice for such functions.

We choose

$$\begin{bmatrix} V_1 \\ V_2 \\ V_3 \\ V_4 \end{bmatrix} = A \begin{bmatrix} e_1 \\ e_2 \\ e_3 \\ e_4 \end{bmatrix} \tag{11.30}$$

where A is a 4×4 real matrix, chosen so that all eigenvalues λ_i of the system (11.30) satisfy the condition

$$|\arg(\lambda_i)| > \frac{\alpha\pi}{2} \tag{11.31}$$

if we choose

$$A = \begin{bmatrix} a_1 - 26 & 0 & 0 & 0 \\ 0 & -1 & 0 & 0 \\ 0 & 0 & a_4 - 5 & 0 \\ 0 & 0 & 0 & -1 \end{bmatrix} \tag{11.32}$$

Then the eigenvalues of the linear system (11.30) are -1, -1, -1, and -1. Hence the condition (11.31) is satisfied for and we get the required synchronization. Numerical discussions for switch 2 are illustrated to verify the theoretical results. The initial states of the master and slave system are arbitrarily chosen as $(x_1(0),\ y_1(0),\ z_1(0),\ w_1(0)) = (4, 5, 6, 4)$ and $(x_2(0),\ y_2(0),\ z_2(0),\ w_2(0)) = (-2, 3, -1, 6)$. The corresponding initial condition for the error states is obtained as $(e_1(0), e_2(0), e_3(0), e_4(0)) = (-8, -1, -5, 1)$. Fig. 11.11 displays the time response of synchronization errors e_1, e_2, e_3, e_4. Figs. 11.12–11.15 illustrate the time response of synchronization states $z_1\&x_2$, $w_1\&y_2$, $x_1\&z_2$, and $y_1\&w_2$ of the master system (11.14) and slave system (11.16), respectively.

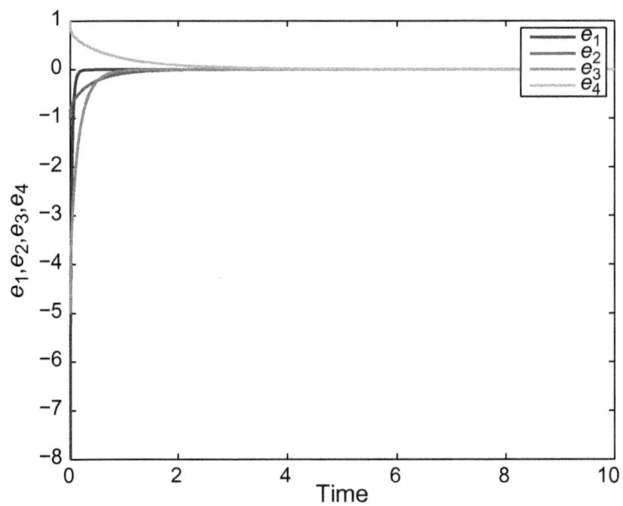

FIGURE 11.11 Synchronization error between states of master and slave systems.

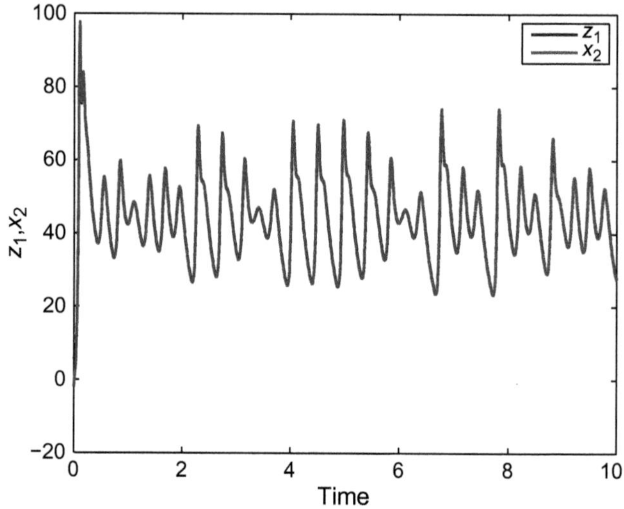

FIGURE 11.12 Time response showing synchronization between states z_1 and x_2.

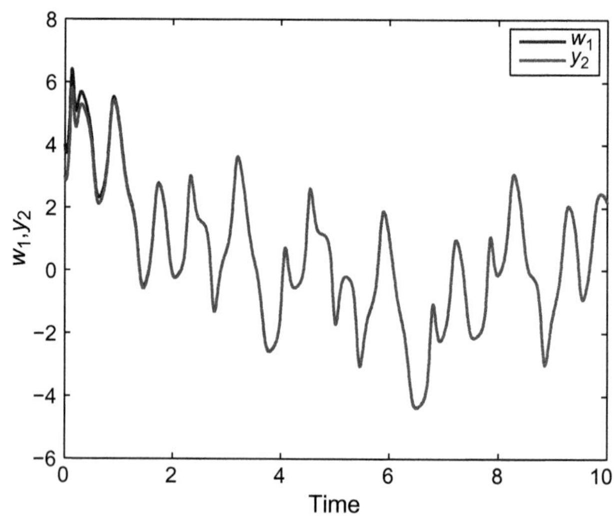

FIGURE 11.13 Time response showing synchronization between states w_1 and y_2.

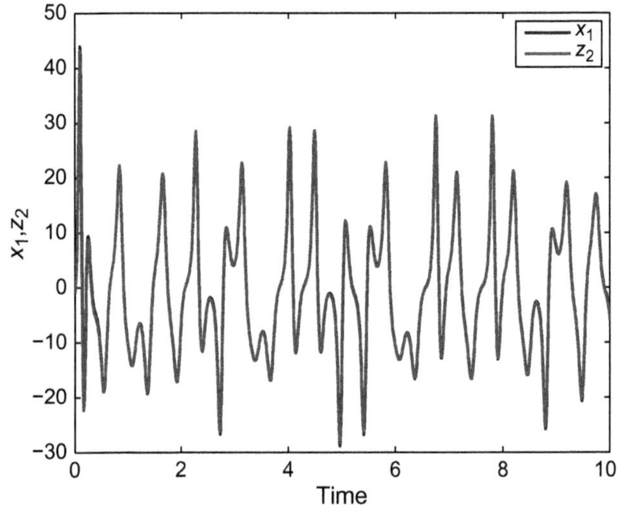

FIGURE 11.14 Time response showing synchronization between states x_1 and z_2.

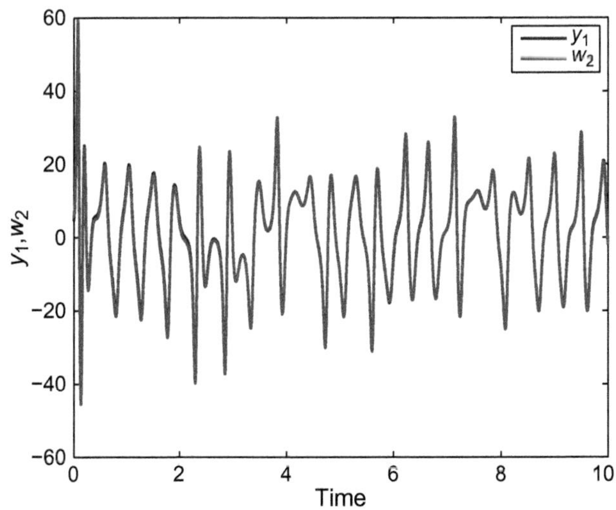

FIGURE 11.15 Time response showing synchronization between states y_1 and w_2.

11.8 CONCLUSION

In this chapter, we investigated multiswitching complete synchronization schemes between nonidentical fractional order hyperchaotic systems. The idea of multiswitching complete synchronization is implemented on two nonidentical hyperchaotic systems. The brief dynamical analysis of hyperchaotic systems is also given. The controllers are obtained using the very well known technique of active control. Based on the stability of fractional order chaotic systems, the stability of fractional order error dynamical system is obtained. Lastly, numerical results are given to confirm the efficiency of the proposed synchronization scheme. Theoretical and numerical results are in excellent agreement.

REFERENCES

Agarwal, R.P., El-Sayed, A.M., Salman, S.M., 2013. Fractional-order chuas system: discretization, bifurcation and chaos. Adv. Diff. Eq. 2013 (1), 320.

Ajayi, A.A., Ojo, S.K., Vincent, E.U., Njah, N.A., 2014. Multiswitching synchronization of a driven hyperchaotic circuit using active backstepping. J. Nonlinear Dynam. 2014.

Aleksandr, L., Pogromsky, A.Y., et al., 1998. Introduction to Control of Oscillations and Chaos., Vol. 35. World Scientific.

Azar, A.T., Vaidyanathan, S., 2015a. Chaos Modeling and Control Systems Design. Vol. 581 of Studies inComputational Intelligence. Studies in Computational Intelligence. Springer, Germany.

Azar, A.T., Vaidyanathan, S., 2015b. Computational Intelligence Applications in Modelling and Control. Vol. 575. Studies inComputational Intelligence. Springer, Berlin, Germany.

Azar, A.T., Vaidyanathan, S., 2015c. Handboook of Research on Advanced Intelligent Control Engineering and Automation. IGI Global, New York, USA, New York, USA.

Azar, A.T., Vaidyanathan, S., 2016. Advances in Chaos Theory and Intelligent Control., Vol. 337. Springer, Berlin, Germany.

Azar, A.T., Zhu, Q., 2015. Advances and Applications in Sliding Mode Control systems. Vol. 576. Studies inComputational Intelligence. Springer, Berlin, Germany.

Azar, A.T., Vaidyanathan, S., Ouannas, A., 2017a. Fractional Order Control and Synchronization of Chaotic Systems. Vol. 688. Studies inComputational Intelligence. Springer, Berlin, Germany.

Azar, A.T., Volos, C., Gerodimos, N.A., Tombras, G.S., Pham, V.-T., Radwan, A.G., et al., 2017b. A novel chaotic system without equilibrium: dynamics, synchronization, and circuit realization. Complexity 2017, (Article ID 7871467), 11 pages.

Azar, A.T., Ouannas, A., Singh, S., 2018. Control of New Type of Fractional Chaos Synchronization. Springer International Publishing, Cham, pp. 47–56, URL https://doi.org/10.1007/978-3-319-64861-3_5.

Becker, G., Packard, A., 1994. Robust performance of linear parametrically varying systems using parametrically-dependent linear feedback. Syst. Control Lett. 23 (3), 205–215.

Boulkroune, A., Bouzeriba, A., Bouden, T., Azar, A.T., 2016a. Fuzzy adaptive synchronization of uncertain fractional-order chaotic systems. Advances in Chaos Theory and Intelligent Control. Springer, Berlin, Germany, pp. 681–697.

Boulkroune, A., Hamel, S., Azar, A.T., Vaidyanathan, S., 2016b. Fuzzy control-based function synchronization of unknown chaotic systems with dead-zone input. In: Azar, A.T., Vaidyanathan, S. (Eds.), Advances in Chaos Theory and Intelligent Control. Springer International Publishing, Cham, pp. 699–718.

Caputo, M., 1967. Linear models of dissipation whose q is almost frequency independentii. Geophys. J. Int. 13 (5), 529–539.

Deng, W., Li, C., 2005. Chaos synchronization of the fractional lü system. Phys. A: Statist. Mech.Applicat. 353, 61–72.

Diethelm, K., Ford, N.J., Freed, A.D., 2002. A predictor-corrector approach for the numerical solution of fractional differential equations. Nonlinear Dynam. 29 (1), 3–22.

Du, H., Zeng, Q., Wang, C., 2008. Function projective synchronization of different chaotic systems with uncertain parameters. Phys. Lett. A 372 (33), 5402–5410.

Du, H., Zeng, Q., Wang, C., 2009. Modified function projective synchronization of chaotic system. Chaos Solitons Fractals 42 (4), 2399–2404.

El-Sayed, A., Nour, H., Elsaid, A., Matouk, A., Elsonbaty, A., 2016. Dynamical behaviors, circuit realization, chaos control, and synchronization of a new fractional order hyperchaotic system. Appl. Math. Model. 40 (5), 3516–3534.

Gao, T., Chen, Z., Yuan, Z., Yu, D., 2007. Adaptive synchronization of a new hyperchaotic system with uncertain parameters. Chaos Solitons Fractals 33 (3), 922–928.

Gao, Y., Liang, C., Wu, Q., Yuan, H., 2015. A new fractional-order hyperchaotic system and its modified projective synchronization. Chaos Solitons Fractals 76, 190–204.

Ge, Z.-M., Ou, C.-Y., 2007. Chaos in a fractional order modified duffing system. Chaos Solitons Fractals 34 (2), 262–291.

Grassi, G., Ouannas, A., Azar, A.T., Radwan, A.G., Volos, C., Pham, V.-T., et al., 2017. Chaos synchronisation of continuous systems via scalar signal. In: 6th International Conference on Modern Circuits and Systems Technologies (MOCAST). pp. 1–4.

Heaviside, O., 1970. Electromagnetic theory (chelsea pc, new york, 1971). Electrical Papers.

Hilfer, R., 2000. Applications of Fractional Calculus in Physics. World Scientific.

Keil, F., Mackens, W., Voß, H., Werther, J., 2012. Scientific Computing in Chemical Engineering II: Computational Fluid Dynamics, Reaction Engineering, and Molecular Properties. Springer Science & Business Media.

Khan, A., Bhat, M.A., 2016a. Hybrid projective synchronization of fractional order chaotic systems with fractional order in the interval (1, 2). Nonlinear Dyn. Syst. Theory 16 (4), 350–365.

Khan, A., Bhat, M.A., 2016b. Hyper-chaotic analysis and adaptive multi-switching synchronization of a novel asymmetric non-linear dynamical system. Int. J. Dynam. Control 1–11.

Khan, A., Bhat, M.A., 2017. Multi-switching combination–combination synchronization of non-identical fractional-order chaotic systems. Math. Methods Appl. Sci.

Khan, A., et al., 2016. Hybrid function projective synchronization of chaotic systems via adaptive control. Int. J. Dynam. Control 1–8.

Khan, A., et al., 2017a. Combination synchronization of genesio time delay chaotic system via robust adaptive sliding mode control. Int. J. Dynam. Control 1–10.

Khan, A., et al., 2017b. Combination synchronization of time-delay chaotic system via robust adaptive sliding mode control. Pramana 88 (6), 91.

Khan, A., et al., 2017c. Increased and reduced order synchronisations between 5d and 6d hyperchaotic systems. Indian J. Ind. Appl. Math. 8 (1), 118–131.

Kilbas, A., Marichev, O., Samko, S., 1993. Fractional Integral and Derivatives (Theory and Applications), 1. Gordon and Breach, Switzerland, p. 1 (993).

Koeller, R., 1984. Applications of fractional calculus to the theory of viscoelasticity. ASME Trans. J. Appl. Mech. 51, 299–307 (ISSN 0021-8936).

Koeller, R., 1986. Polynomial operators, stieltjes convolution, and fractional calculus in hereditary mechanics. Acta Mech. 58 (3), 251–264.

Lamamra, K., Vaidyanathan, S., Azar, A.T., Ben Salah, C., 2017. Chaotic system modelling using a neural network with optimized structure. In: Azar, A.T., Vaidyanathan, S., Ouannas, A. (Eds.), Fractional Order Control and Synchronization of Chaotic Systems. Springer International Publishing, Cham, pp. 833–856.

Li, C., Chen, G., 2004. Chaos and hyperchaos in the fractional-order rössler equations. Phys. A: Statist. Mech. Applicat. 341, 55–61.

Li, C., Deng, W., 2007. Remarks on fractional derivatives. Appl. Math. Comput. 187 (2), 777–784.

Li, C., Zhang, J., 2016. Synchronisation of a fractional-order chaotic system using finite-time input-to-state stability. Int. J. Syst. Sci. 47 (10), 2440–2448.

Li, G.-H., 2007. Modified projective synchronization of chaotic system. Chaos Solitons Fractals 32 (5), 1786–1790.

Lyapunov, A.M., 1992. The general problem of the stability of motion. Int. J. Control 55 (3), 531–534.

Mahmoud, G.M., Mahmoud, E.E., 2010. Complete synchronization of chaotic complex nonlinear systems with uncertain parameters. Nonlinear Dynam. 62 (4), 875–882.

Mainieri, R., Rehacek, J., 1999. Projective synchronization in three-dimensional chaotic systems. Phys. Rev. Lett. 82 (15), 3042.

Moysis, L., Azar, A.T., 2017. New discrete time 2d chaotic maps. Int. J. Syst. Dynam. Applicat. 6 (1), 77–104.

Munoz-Pacheco, J.M., Volos, C., Campos-Canton, E., Azar, A.T., Pham, V.-T., Radwan, A.G., 2017. New trends on modeling, design, and control of chaotic systems 2017 (Article ID 3737681).

Ouannas, A., Azar, A.T., Abu-Saris, R., 2016a. A new type of hybrid synchronization between arbitrary hyperchaotic maps. Int. J. Mach. Learning Cybernetics 1–8.

Ouannas, A., Azar, A.T., Radwan, A.G., 2016b. On inverse problem of generalized synchronization between different dimensional integer-order and fractional-order chaotic systems. In: 2016 28th International Conference on Microelectronics (ICM). pp. 193–196.

Ouannas, A., Azar, A.T., Vaidyanathan, S., 2017a. A new fractional hybrid chaos synchronisation. Int. J. Model. Identif. Control 27 (4), 314–322. URL http://www.inderscienceonline.com/doi/abs/10.1504/IJMIC.2017.084719.

Ouannas, A., Azar, A.T., Vaidyanathan, S., 2017b. New hybrid synchronization schemes based on coexistence of various types of synchronization between master-slave hyperchaotic systems. Int. J. Comput. Applicat. Technol. 55 (2), 112–120.

Ouannas, A., Azar, A.T., Vaidyanathan, S., 2017c. On a simple approach for q-s synchronization of chaotic dynamical systems in continuous-time. Int. J. Comput. Sci. Math. 8 (1), 20–27.

Ouannas, A., Azar, A.T., Vaidyanathan, S., 2017d. A robust method for new fractional hybrid chaos synchronization. Math. Methods Appl. Sci. 40 (5), 1804–1812.

Ouannas, A., Azar, A.T., Ziar, T., 2017e. On inverse full state hybrid function projective synchronization for continuous-time chaotic dynamical systems with arbitrary dimensions. Diff. Eq. Dynam. Syst. 1–14.

Ouannas, A., Azar, A.T., Ziar, T., Radwan, A.G., 2017f. Generalized synchronization of different dimensional integer-order and fractional order chaotic systems. In: Azar, A.T., Vaidyanathan, S., Ouannas, A. (Eds.), Fractional Order Control and Synchronization of Chaotic Systems. Springer International Publishing, Cham, pp. 671–697.

Ouannas, A., Azar, A.T., Ziar, T., Radwan, A.G., 2017g. A study on coexistence of different types of synchronization between different dimensional fractional chaotic systems. In: Azar, A.T., Vaidyanathan, S., Ouannas, A. (Eds.), Fractional Order Control and Synchronization of Chaotic Systems. Springer International Publishing, Cham, pp. 637–669.

Ouannas, A., Azar, A.T., Ziar, T., Vaidyanathan, S., 2017h. Fractional inverse generalized chaos synchronization between different dimensional systems. In: Azar, A.T., Vaidyanathan, S., Ouannas, A. (Eds.), Fractional Order Control and Synchronization of Chaotic Systems. Springer International Publishing, Cham, pp. 525–551.

Ouannas, A., Azar, A.T., Ziar, T., Vaidyanathan, S., 2017i. A new method to synchronize fractional chaotic systems with different dimensions. In: Azar, A.T., Vaidyanathan, S., Ouannas, A. (Eds.), Fractional Order Control and Synchronization of Chaotic Systems. Springer International Publishing, Cham, pp. 581–611.

Ouannas, A., Azar, A.T., Ziar, T., Vaidyanathan, S., 2017j. On new fractional inverse matrix projective synchronization schemes. In: Azar, A.T., Vaidyanathan, S., Ouannas, A. (Eds.), Fractional Order Control and Synchronization of Chaotic Systems. Springer International Publishing, Cham, pp. 497–524.

Ouannas, A., Grassi, G., Azar, A.T., Radwan, A.G., Volos, C., Pham, V.-T., et al., 2017k. Deadbeat synchronization control in discrete-time chaotic systems. In: 6th International Conference on Modern Circuits and Systems Technologies (MOCAST). pp. 1–4.

Pecora, L.M., Carroll, T.L., 1990. Synchronization in chaotic systems. Physical review letters 64 (8), 821.

Petras, I., 2008. Stability of fractional-order systems with rational orders. arXiv preprint arXiv:0811.4102.

Petras, I., 2011. Fractional-order Nonlinear Systems: Modeling, Analysis and Simulation. Springer Science & Business Media.

Pham, V.-T., Vaidyanathan, S., Volos, C.K., Azar, A.T., Hoang, T.M., Van Yem, V., 2017. A three-dimensional no-equilibrium chaotic system: analysis, synchronization and its fractional order form. In: Azar, A.T., Vaidyanathan, S., Ouannas, A. (Eds.), Fractional Order Control and Synchronization of Chaotic Systems. Springer International Publishing, Cham, pp. 449–470.

Podlubny, I., 1998. Fractional Differential Equations: An Introduction to Fractional Derivatives, Fractional Differential Equations, to Methods of Their Solution and Some of Their Applications, Vol. 198. Academic press.

Rosenblum, M.G., Pikovsky, A.S., Kurths, J., 1996. Phase synchronization of chaotic oscillators. Phys. Rev. Lett. 76 (11), 1804.

Shahverdiev, E., Sivaprakasam, S., Shore, K., 2002. Lag synchronization in time-delayed systems. Phys. Lett. A 292 (6), 320–324.

Singh, S., Azar, A.T., Ouannas, A., Zhu, Q., Zhang, W., Na, J., 2017. Sliding mode control technique for multi-switching synchronization of chaotic systems. In: 9th International Conference on Modelling, Identification and Control (ICMIC 2017), July 10-12, 2017, Kunming, China.

Soliman, N.S., Said, L.A., Azar, A.T., Madian, A.H., Radwan, A.G., Ounnas, A., 2017. Fractional controllable multi-scroll v-shape attractor with parameters effect. In: 6th International Conference on Modern Circuits and Systems Technologies (MOCAST). pp. 1–4.

Tavazoei, M.S., Haeri, M., 2009. Describing function based methods for predicting chaos in a class of fractional order differential equations. Nonlinear Dynam. 57 (3), 363–373.

Tolba, M.F., AbdelAty, A.M., Saida, L.A., Elwakil, A.S., Azar, A.T., Madian, A.H., et al., 2017. Fpga realization of caputo and grnwald-letnikov operators. In: 6th International Conference on Modern Circuits and Systems Technologies (MOCAST). pp. 1–4.

Ucar, A., Lonngren, K.E., Bai, E.-W., 2008. Multi-switching synchronization of chaotic systems with active controllers. Chaos Solitons Fractals 38 (1), 254–262.

Vaidyanathan, S., Azar, A.T., 2015a. Analysis and control of a 4-d novel hyperchaotic system. In: Azar, A.T., Vaidyanathan, S. (Eds.), Chaos Modeling and Control Systems Design. Vol. 581 of Studies in Computational Intelligence. Springer, Berlin, Germany, pp. 19–38.

Vaidyanathan, S., Azar, A.T., 2015b. Analysis, control and synchronization of a nine-term 3-d novel chaotic system. In: Azar, A.T., Vaidyanathan, S. (Eds.), Chaos Modeling and Control Systems Design. Vol. 581 of Studies in Computational Intelligence. Springer, Berlin, Germany, pp. 3–17.

Vaidyanathan, S., Azar, A.T., 2015c. Anti-synchronization of identical chaotic systems using sliding mode control and an application to vaidyanathan-madhavan chaotic systems. In: Azar, A. T., Zhu, Q. (Eds.), Advances and Applications in Sliding Mode Control Systems. Vol. 576 of Studies in Computational Intelligence. Springer, Berlin, Germany, pp. 527–547.

Vaidyanathan, S., Azar, A.T., 2015d. Hybrid synchronization of identical chaotic systems using sliding mode control and an application to vaidyanathan chaotic systems. In: Azar, A.T., Zhu, Q. (Eds.), Advances and Applications in Sliding Mode Control Systems. Vol. 576 of Studies in Computational Intelligence. Springer, Berlin, Germany, pp. 549–569.

Vaidyanathan, S., Azar, A.T., 2016a. A novel 4-D four-wing chaotic system with four quadratic nonlinearities and its synchronization via adaptive control method. Advances in Chaos Theory and IntelligentControl. Springer, Berlin, Germany, pp. 203–224.

Vaidyanathan, S., Azar, A.T., 2016b. Adaptive backstepping control and synchronization of a novel 3-D jerk system with an exponential nonlinearity. Advances in Chaos Theory and IntelligentControl. Springer, Berlin, Germany, pp. 249–274.

Vaidyanathan, S., Azar, A.T., 2016c. Adaptive control and synchronization of Halvorsen circulant chaotic systems. Advances in Chaos Theory and IntelligentControl. Springer, Berlin, Germany, pp. 225–247.

Vaidyanathan, S., Azar, A.T., 2016d. Dynamic analysis, adaptive feedback control and synchronization of an eight-term 3-D novel chaotic system with three quadratic nonlinearities. Advances in Chaos Theory and IntelligentControl. Springer, Berlin, Germany, pp. 155–178.

Vaidyanathan, S., Azar, A.T., 2016e. Generalized projective synchronization of a novel hyperchaotic four-wing system via adaptive control method. Advances in Chaos Theory and IntelligentControl. Springer, Berlin, Germany, pp. 275–290.

Vaidyanathan, S., Azar, A.T., 2016f. A novel 4-d four-wing chaotic system with four quadratic nonlinearities and its synchronization via adaptive control method. Advances in Chaos Theory and IntelligentControl. Springer, Berlin, Germany, pp. 203–224.

Vaidyanathan, S., Azar, A.T., 2016g. Qualitative study and adaptive control of a novel 4-d hyperchaotic system with three quadratic nonlinearities. In: Azar, A.T., Vaidyanathan, S. (Eds.), Advances in Chaos Theory and Intelligent Control. Springer International Publishing, Cham, pp. 179–202.

Vaidyanathan, S., Azar, A.T., 2016h. Takagi-Sugeno fuzzy logic controller for Liu-Chen four-scroll chaotic system. Int. J. Intel. Eng. Inf. 4 (2), 135–150.

Vaidyanathan, S., Sampath, S., 2017. Anti-synchronisation of identical chaotic systems via novel sliding control and its application to a novel chaotic system. Int. J. Model. Identif. Control 27 (1), 3–13.

Vaidyanathan, S., Azar, A.T., Rajagopal, K., Alexander, P., 2015a. Design and spice implementation of a 12-term novel hyperchaotic system and its synchronisation via active control. Int. J. Model. Identif. Control 23 (3), 267–277.

Vaidyanathan, S., Idowu, B.A., Azar, A.T., 2015b. Backstepping controller design for the global chaos synchronization of sprott's jerk systems. In: Azar, A.T., Vaidyanathan, S. (Eds.), Chaos Modeling and Control Systems Design. Vol. 581 of Studies in Computational Intelligence. Springer, Berlin, Germany, pp. 39–58.

Vaidyanathan, S., Sampath, S., Azar, A.T., 2015c. Global chaos synchronisation of identical chaotic systems via novel sliding mode control method and its application to zhu system. Int. J. Model. Identif. Control 23 (1), 92–100.

Vaidyanathan, S., Azar, A.T., Ouannas, A., 2017a. An eight-term 3-d novel chaotic system with three quadratic nonlinearities, its adaptive feedback control and synchronization. In: Azar, A.T., Vaidyanathan, S., Ouannas, A. (Eds.), Fractional Order Control and Synchronization of Chaotic Systems. Springer International Publishing, Cham, pp. 719–746.

Vaidyanathan, S., Azar, A.T., Ouannas, A., 2017b. Hyperchaos and adaptive control of a novel hyperchaotic system with two quadratic nonlinearities. In: Azar, A.T., Vaidyanathan, S., Ouannas, A. (Eds.), Fractional Order Control and Synchronization of Chaotic Systems. Springer International Publishing, Cham, pp. 773–803.

Vaidyanathan, S., Zhu, Q., Azar, A.T., 2017c. Adaptive control of a novel nonlinear double convection chaotic system. In: Azar, A.T., Vaidyanathan, S., Ouannas, A. (Eds.), Fractional Order Control and Synchronization of Chaotic Systems. Springer International Publishing, Cham, pp. 357–385.

Wang, F., Liu, C., 2007. Synchronization of unified chaotic system based on passive control. Physica D: Nonlinear Phenomena 225 (1), 55–60.

Wang, X.-Y., Sun, P., 2011. Multi-switching synchronization of chaotic system with adaptive controllers and unknown parameters. Nonlinear Dynam. 63 (4), 599–609.

Wang, Z., Volos, C., Kingni, S.T., Azar, A.T., Pham, V.-T., 2017. Four-wing attractors in a novel chaotic system with hyperbolic sine nonlinearity. Optik – Int. J. Light Electr. Opt. 131, 1071–1078.

Wen, X.-J., Wu, Z.-M., Lu, J.-G., 2008. Stability analysis of a class of nonlinear fractional-order systems. IEEE Trans. Circuits Systems II: Express Briefs 55 (11), 1178–1182.

Wolf, A., Swift, J.B., Swinney, H.L., Vastano, J.A., 1985. Determining lyapunov exponents from a time series. Phys. D: Nonlinear Phen. 16 (3), 285–317.

Yang, T., Chua, L.O., 1997. Impulsive stabilization for control and synchronization of chaotic systems : theory and application to secure communication. IEEE Trans. Circuits Syst. I: Fundam. Theory Applicat. 44 (10), 976–988.

Zhang, H., Ma, X.-K., Liu, W.-Z., 2004. Synchronization of chaotic systems with parametric uncertainty using active sliding mode control. Chaos Solitons Fractals 21 (5), 1249–1257.

Zhang, H., Lewis, F.L., Das, A., 2011. Optimal design for synchronization of cooperative systems: state feedback, observer and output feedback. IEEE Trans. Automatic Control 56 (8), 1948–1952.

Zhu, Q., Azar, A.T., 2015. Complex system modelling and control through intelligent soft computations. Studies in Fuzziness and Soft Computing, Berlin, Germany.

Chapter 12

Dual Combination Synchronization Scheme for Nonidentical Different Dimensional Fractional Order Systems Using Scaling Matrices

Vijay K. Yadav, Mayank Srivastava and Subir Das
Indian Institute of Technology (BHU), Varanasi, Uttar Pradesh, India

12.1 INTRODUCTION

The fractional calculus is a name for the theory of integrals and derivatives of arbitrary real order and also of complex order, which unify and generalize the notions of differintegral orders. Fractional calculus can be used in many modeling and design problems. A dynamical system involving fractional order time derivatives is known as a fractional dynamical system. Introduction of fractional calculus in nonlinear models has rendered a new dimension to the existing problems. Again, due to the nonlocal property of a fractional order differential operator, it takes into account the fact that the future state depends upon the present state as well as all of the history of its previous states. For this realistic property, the fractional order systems are becoming popular. Another reason behind using fractional order derivatives is that these are naturally related to the systems with memory which prevail for most of the physical and scientific system models. The fractional derivative of a function depends on the values of the function over the entire interval. Thus, it is suitable for modeling of the systems with long-range interactions both in space and time. Fractional derivative has the flexibility to allow incorporation of different types of information. The fractional calculus which was in the earlier stage considered as a mathematical curiosity now becomes the object for the extensive development of fractional order partial differential equations for its applications in various physical areas of sciences and engineering. Geometric and physical interpretations of

Mathematical Techniques of Fractional Order Systems. DOI: https://doi.org/10.1016/B978-0-12-813592-1.00012-X

fractional differentiation and fractional integration can be found in Podlubny (1999). Fractional order systems have garnered a lot of interest and appreciation recently due to their ability to provide an exact description of different nonlinear phenomena. The analysis of the existence and uniqueness of solutions for fractional differential equations are more complex than that of classical differential equations due to the nonlocal property and presence of weakly singular kernels. There are many definitions of the fractional derivative; one of the most common definitions is the Caputo definition of fractional derivatives of order α $(0<\alpha<1)$, which can be written as (Podlubny, 1999)

$$\frac{d^{\alpha}f(t)}{dt^{\alpha}} = \begin{cases} \dfrac{1}{\Gamma(n-\alpha)}\displaystyle\int_0^t \frac{f^{(n)}(\tau)}{(t-\tau)^{\alpha-n+1}}\,d\tau, & n-1<\alpha<n \\ \dfrac{d^n f(t)}{dt^n}, & \alpha = n, \end{cases}$$

where n is an integer $n-1\le\alpha<n$.

In the last two decades, fractional dynamics of chaotic systems, being a new field of research, has been reported. Hence, fractional differential equations have been utilized to study dynamical systems in general and applications of chaos in particular. Recently, it has been found that fractional differential equations have many applications in many fields of science like engineering, physics, finance, dielectric polarization, electrode–electrolyte polarization, control systems (Hifer, 2001; Laskin, 2000; Sun et al., 1984; Ichise et al., 1971; Hartley and Lorenzo, 2002; Meghni et al, 2017; Boulkroune et al, 2016a; Ghoudelbourk et al., 2016; Soliman et al., 2017; Tolba et al., 2017a,b), and so on.

Nowadays the applications of dynamical systems have spread to a wide spectrum of disciplines including science, engineering, biology, sociology, etc. During the last few decades the study and analysis of nonlinear dynamical systems have gained enormous popularity due to its important feature of any real-time dynamical system. Chaos is an important phenomenon of dynamical systems which has been comprehensively studied and developed by scientists since the work of Lorenz (1963). A chaotic system has complex dynamical behaviors such as the unpredictability of the long-term future behavior and irregularity. The study of chaotic systems of fractional order is an important topic in nonlinear dynamical systems (Li and Peng, 2004; Hartley et al., 1995; Grigorenko and Grigorenko, 2003; Ouannas et al., 2017g,i,j,k,l; Ouannas et al., 2016a) and it has given the area a new dimension.

Recently lot of researchers have been actively engaged in nonlinear chaotic systems as these systems are rich in dynamics and very much sensitive to initial conditions. Again, due to memory effect and also for non-Markovian and non-Gaussian nature, the fractional calculus has become another area of research. Therefore, chaos and chaos control in fractional order systems have become a hot topic in research. Since there was lack of appropriate mathematical tools, fractional order dynamic systems had not been studied much in the design and practice of control systems during last few decades. In recent years, after emergence of effective methods fractional order systems have become more and more attractive for the systems control community. The TID controller, the fractional PD^{α} controller, the fractional PI^{α} controller, the fractional $PI^{\lambda}D\mu$ controller, and the fractional lead-lag compensator are a few examples of well-known fractional order controllers. In some recent works, it has come into the picture that the fractional order controllers have better disturbance rejection ratios and less sensitivity to plant parameter variations compared to the traditional controllers. At the present time scientists and engineers are using fractional controllers in many practical applications, viz., control of main irrigation canals, lateral and longitudinal control of autonomous vehicles, control of robotic time-delay systems, control of hexapod robots, reducing engine vibrations in automobiles, control of electromechanical systems, and flexible spacecraft attitude control.

The most important achievement in the research of chaos is that chaotic systems can be made to synchronize with each other. Synchronization is a phenomenon that is usually treated as a regime in which two or more coupled or periodic or even chaotic systems exhibit correlated, and sometimes even identical, oscillations. The idea of synchronizing chaotic systems was first introduced by Pecora and Carroll (1990); they showed that it is possible to synchronize chaotic systems through a simple coupling. Chaos synchronization has attracted extensive attentions for its potential applications in physical systems (Lakshmanan and Murali, 1996), chemical systems (Han et al., 1995), neural networks (Wu et al., 2012; Moskalenko et al., 2010), neuron systems (Shuai and Wong, 1998), secure communications (Chai et al., 2012; Chen et al., 2014), ecological system (Blasius et al., 1999), and so on. In recent years, many effective methods have been presented for synchronizing identical and nonidentical chaotic systems. Time-delay feedback approach, active control, adaptive control (Park and Kwon, 2005; Srivastava et al., 2013a,b; Huang et al., 2009), and synchronization via nonlinear feedback control techniques (Grassi et al., 2017; Ouannas et al., 2016b,c; Huang et al., 2004; Chen and Lu, 2002; Singh et al., 2017; Ouannas et al., 2017a,b,c,d,e,h; Boulkroune et al, 2016b; Vaidyanathan et al, 2015a,b,c; Wang et al., 2017; Azar and Zhu, 2015; Azar and Vaidyanathan, 2015a,b,c, 2016; Zhu and Azar, 2015; Vaidyanathan and Azar, 2015a,b,c,d, 2016a,b,c,d,e,f,g, Vaidyanathan et al., 2017a,b,c; Azar et al., 2017a,b,c) have been effectively applied to application of chaos synchronization problem. The conception of synchronization can be extended to complete synchronization, antisynchronization, Projective synchronization, Function projective synchronization, and hybrid synchronization (Yu and Liu, 2003; Liu, 2006; Si et al., 2012; Yadav et al., 2017a; Zhou and Zhu, 2011; Azar et al., 2017a; Ouannas et al., 2017c,d,e), etc.

Dual synchronization is a special circumstance in synchronization in which two different pairs of chaotic systems (two master systems and two slave systems) are synchronized. The dual synchronization of systems plays an important role in many fields, including chaotic secure communication. Recently the dual synchronization of chaotic systems has received less attention. There are only a few results in the literature about the dual synchronization between chaotic systems (Yadav et al., 2016; Agrawal et al., 2012; Ghosh and Chowdhury, 2010; Hassan and Mohammad, 2008). In combination synchronization (Luo et al., 2011; Yadav et al., 2017b; Luo and Wang, 2012), two or more master systems and one slave system are synchronized. This synchronization scheme has advantages over the usual drive response synchronization, such as being able to provide greater security in secure communication. These have motivated the authors to study the dual combination synchronization of fractional order chaotic systems.

Motivated by the above discussions, the aim of this chapter is to study the dual combination synchronization with nonidentical different dimensional fractional order systems using scaling matrices. Firstly, dual combination synchronization is considered, where the four drive systems are fractional order Lu, Qi, Newton–Leipnik, and Volta's systems and the two response systems are the fractional order hyper chaotic Lü and 4D Integral order hyperchaotic systems. Secondly, dual combination synchronization is considered, where the four drive systems are fractional order hyperchaotic Lü, 4D Integral order hyperchaotic, Chen hyperchaotic, and Lorenz hyperchaotic systems and the response systems are fractional order Lu and Qi systems. According to Lyapunov stability theory and active control method, the corresponding controllers are both designed. Finally, several numerical examples are provided to illustrate the obtained results. Numerical simulations demonstrate the effectiveness and feasibility of the method.

The organization of this chapter is as follows. In Section 12.2, problem formulation of the dual combination synchronization scheme of four different chaotic and hyperchaotic systems (master systems), and two hyperchaotic and chaotic systems (response systems) are presented. In Section 12.3, application of the scheme demonstrates numerically the effectiveness of the proposed scheme for dual combination synchronization of among different dimensional fractional order chaotic and hyperchaotic systems. The numerical simulations are presented through Section 12.3 to verify the effectiveness of the proposed method. In Section 12.4, the conclusion of the overall research work is presented.

12.2 PROBLEM FORMULATION

Consider that the first two master systems are taken as

$$D^{\alpha}X_1 = A_1X_1 + F_1(X_1) \tag{12.1}$$

$$D^{\alpha}Y_1 = B_1Y_1 + G_1(Y_1), \tag{12.2}$$

where $A_1 \in R^{m \times m}$, $B_1 \in R^{m \times m}$ are the linear part of the systems, $X_1 = [x_{11}, x_{12}, ..., x_{1m}]^T$ and $Y_1 = [y_{11}, y_{12}, ..., y_{1m}]^T$ are the two state vector of uncoupled master systems (12.1) and (12.2); $F_1:R^m \rightarrow R^m$ and $G_1:R^m \rightarrow R^m$ are the two known real vector valued functions.

Next, another two master systems are considered as

$$D^{\alpha}X_2 = A_2X_2 + F_2(X_2) \tag{12.3}$$

$$D^{\alpha}Y_2 = B_2Y_2 + G_2(Y_2), \tag{12.4}$$

where $A_2 \in R^{m \times m}$, $B_2 \in R^{m \times m}$ are the linear part of the systems, $X_2 = [x_{21}, x_{22}, ..., x_{2m}]^T$ and $Y_2 = [y_{21}, y_{22}, ..., y_{2m}]^T$ are the two state vector of uncoupled master systems (12.3) and (12.4); $F_2:R^m \rightarrow R^m$ and $G_2:R^m \rightarrow R^m$ are the two known real vector valued functions.

Now, the corresponding two slave systems are considered as

$$D^{\alpha}X_3 = A_3X_3 + F_3(X_3) + U_1 \tag{12.5}$$

$$D^{\alpha}Y_3 = B_3Y_3 + G_3(Y_3) + U_2, \tag{12.6}$$

where $X_3 = [x_{31}, x_{32}, ..., x_{3n}]^T$ and $Y_3 = [y_{31}, y_{32}, ..., y_{3n}]^T$ are the two state vector of uncoupled master systems (12.5) and (12.6); $A_3 \in R^{n \times n}$, $B_3 \in R^{n \times n}$ are the linear part of the systems, $F_3:R^n \rightarrow R^n$ and $G_3:R^n \rightarrow R^n$ are the two known real vector valued functions, U_1 and U_2 are the control functions to be designed later.

The error function is defined as

$$e = Z - C(Y + X),$$

where $e = [e_1, e_2]^T$, $X = [X_1, Y_1]^T$, $Y = [X_2, Y_2]^T$, $Z = [X_3, Y_3]^T$ and $C = \begin{bmatrix} C_1 & 0 \\ 0 & C_2 \end{bmatrix}$, then the error function will be

$$e_1 = X_3 - C_1(X_2 + X_1)$$

$$e_2 = Y_3 - C_2(Y_2 + Y_1)$$

The error systems is obtained as

$$\begin{aligned} D^{\alpha}e_1 &= A_3e_1 + A_3C_1(X_2 + X_1) + F_3(X_3) - C_1[A_2X_2 + F_2(X_2) + A_1X_1 + F_1(X_1)] + U_1 \\ D^{\alpha}e_2 &= B_3e_2 + B_3C_2(Y_2 + Y_1) + G_3(Y_3) - C_2[B_2Y_2 + G_2(Y_2) + B_1Y_1 + G_1(Y_1)] + U_2 \end{aligned} \tag{12.7}$$

Theorem 1: (Wang et al., 2010) Consider an autonomous fractional order linear system as

$D^\alpha x = Ax$ with initial condition $x(0) = x_0$, where $x \in R^n$ is a state vector, then the system is asymptotically stable if and only if $|\arg(\lambda_i(A))| > \alpha\pi/2$, $i = 1,\ 2,\ ....,\ n$, where $\arg(\lambda_i(A))$ denotes the argument of the eigenvalues λ_i of A.

Proposition 1: If the control functions are chosen as

$$
\begin{aligned}
U_1 &= C_1[A_2X_2 + F_2(X_2) - A_1X_1 + F_1(X_1)] - A_3C_1(X_2 + X_1) - F_3(X_3) + K_1e_1 \\
U_2 &= C_2[B_2Y_2 + G_2(Y_2) + B_1Y_1 + G_1(Y_1)] - B_3C_2(Y_2 + Y_1) - G_3(Y_3) + K_2e_2,
\end{aligned}
\tag{12.8}
$$

where $K_1 \in R^{n \times n}$, $K_2 \in R^{n \times n}$ are the gain matrices, then the dual combination synchronization will be achieved among considered systems (12.1), (12.2), (12.3), (12.4), (12.5), and (12.6) if and only if all the eigen values λ_i of $A_3 + K_1$ and $B_3 + K_2$satisfy $|\arg(\lambda_i)| > \alpha\pi/2$, where $i = 1,\ 2,\ ....,\ n$.

Proof: : After substituting the values of control function given in Eq. (12.8) into Eq. (12.7), the error system (12.7) is reduced in the following form

$$
\begin{aligned}
D^\alpha e_1 &= (A_3 + K_1)e_1 \\
D^\alpha e_2 &= (B_3 + K_2)e_2
\end{aligned}
\tag{12.9}
$$

In view of theorem 1, it can be concluded that the system (12.9) is asymptotically stable if and only if all the eigen values λ_i of $A_3 + K_1$ and $B_3 + K_2$ satisfy $|\arg(\lambda_i)| > \alpha\pi/2$, where $i = 1,\ 2,\ ....,\ n$.

If $\lim_{t \to \infty} \|e\| = 0$and thus the considered systems are dual combination synchronized and hence this completes the proof.

12.3 APPLICATION OF THE SCHEME

In this section the effectiveness of the proposed scheme for dual combination synchronization among different dimensional fractional order chaotic systems are demonstrated numerically. Here two kinds of cases are taken to discuss dual combination synchronization of chaotic systems: dual combination synchronization with order $n > m$ and dual combination synchronization with order $m > n$.

12.3.1 Dual Combination Synchronization With Order $n > m$

Let us consider the fractional order Lu system (Petras, 2011) and fractional order Qi system (Xiangjun and Yang, 2010) as the first two master systems as

$$\begin{aligned}\frac{d^\alpha x_{11}}{dt^\alpha} &= a_{11}(x_{12} - x_{11})\\ \frac{d^\alpha x_{12}}{dt^\alpha} &= -x_{11}x_{13} + a_{13}x_{12},\\ \frac{d^\alpha x_{13}}{dt^\alpha} &= x_{11}x_{12} - a_{12}x_{13}\end{aligned} \tag{12.10}$$

where $0 \leq \alpha \leq 1$ is the fractional order derivative and a_{11}, a_{12}, a_{13} are system parameters. The phase portraits of system (12.10) at $\alpha = 0.95$ in $x_{11} - x_{12} - x_{13}$ space and $x_{11} - x_{12}$, $x_{11} - x_{13}$, $x_{12} - x_{13}$ planes are shown in Fig. 12.1 for the parameters' values $a_{11} = 36$, $a_{12} = 3$, and $a_{13} = 20$, and initial conditions $(x_{11}(0), x_{12}(0), x_{13}(0)) = (0.2, 0.5, 0.3)$.

$$\begin{aligned}\frac{d^\alpha y_{11}}{dt^\alpha} &= b_{11}(y_{12} - y_{11}) + y_{12}y_{13}\\ \frac{d^\alpha y_{12}}{dt^\alpha} &= b_{13}y_{11} - y_{12} - y_{11}y_{13}\\ \frac{d^\alpha y_{13}}{dt^\alpha} &= -b_{12}y_{13} + y_{11}y_{12}.\end{aligned} \tag{12.11}$$

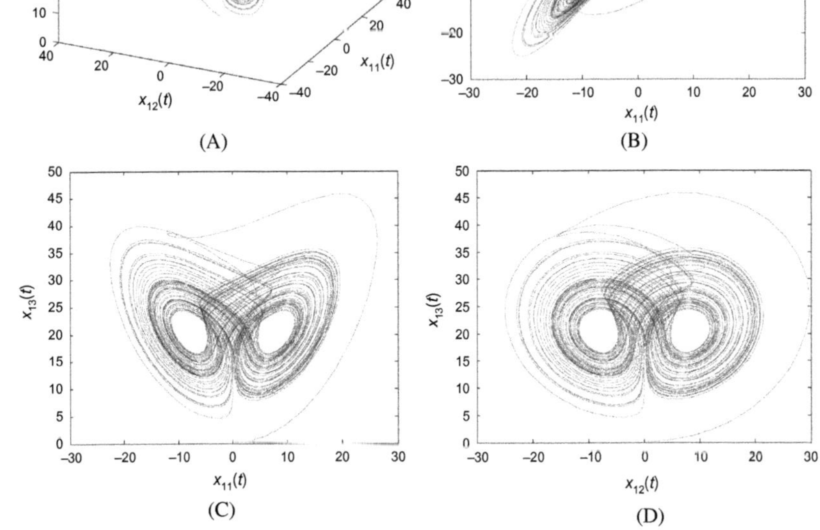

FIGURE 12.1 Phase portraits of Lu system at $\alpha = 0.95$: (A) in $x_{11} - x_{12} - x_{13}$ space, (B) in $x_{11} - x_{12}$ plane, (C) in $x_{11} - x_{13}$ plane, (D) in $x_{12} - x_{13}$ plane.

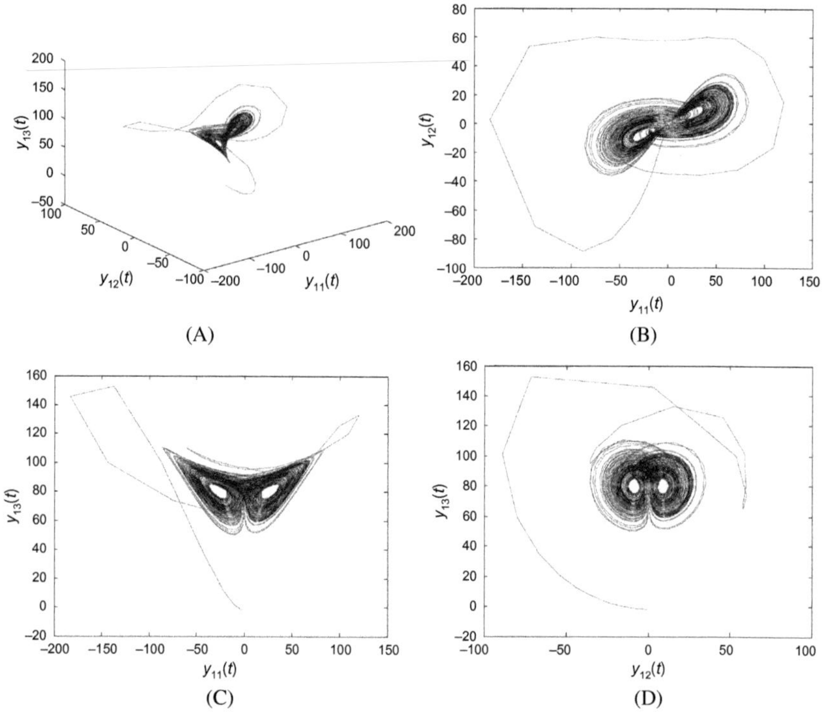

FIGURE 12.2 Phase portraits of Qi system at $\alpha = 0.98$: (A) in $y_{11} - y_{12} - y_{13}$ space, (B) in $y_{11} - y_{12}$ plane, (C) in $y_{11} - y_{13}$ plane, (D) in $y_{12} - y_{13}$ plane.

The phase portrait of system (12.11) at $\alpha = 0.98$ is depicted through Fig. 12.2 for the values of the parameters $b_{11} = 35$, $b_{12} = 8/3$, $b_{13} = 80$ and initial conditions $(y_{11}(0), y_{12}(0), y_{13}(0)) = (-1, -1, -2)$.

The fractional-order Newton–Leipnik system (Sheu et al., 2008) and the fractional-order Volta's System (Petras, 2009) are considered as the other two master systems as

$$\begin{aligned}
\frac{d^{\alpha}x_{21}}{dt^{\alpha}} &= -a_{21}x_{21} + x_{22} + 10x_{22}x_{23} \\
\frac{d^{\alpha}x_{22}}{dt^{\alpha}} &= -x_{21} - 0.4x_{22} + 5x_{21}x_{23} \\
\frac{d^{\alpha}x_{23}}{dt^{\alpha}} &= a_{22}x_{23} - 5x_{21}x_{22},
\end{aligned} \tag{12.12}$$

where $'a_{21}'$ and $'a_{22}'$ are variable parameters. Usually the parameter $'a_{22}'$ is taken in the interval (0, 8.0). The system is ill-behaved outside this interval. As $a_{22} \to 0$, the system shows relatively uninteresting dynamics and for $a_{22} \geq 0.8$ the given system becomes explosive, i.e., the solution diverges to infinity for any initial condition other than the critical points.

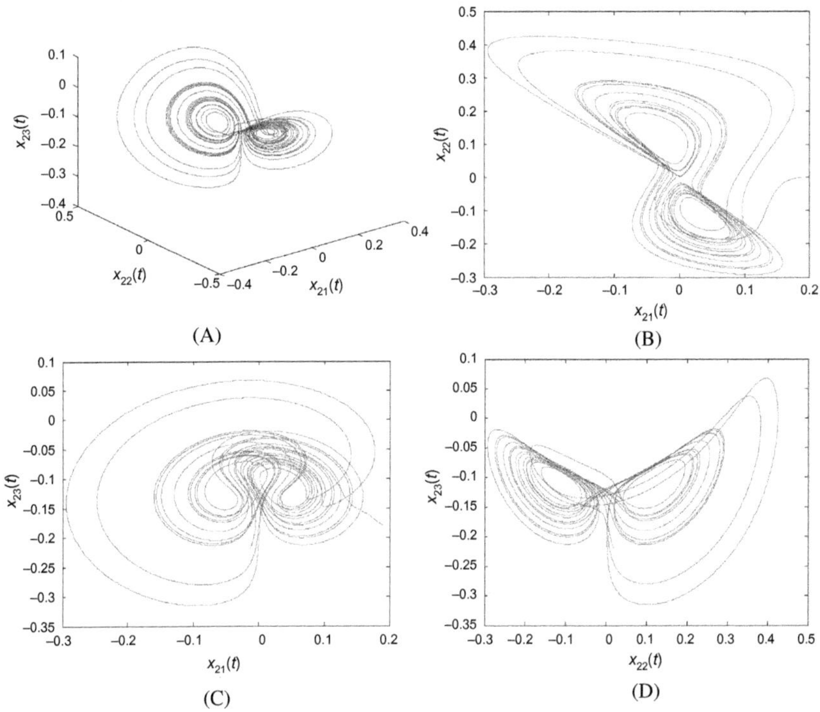

FIGURE 12.3 Phase portraits of Newton–Leipnik system at $\alpha = 0.95$: (A) in $x_{21} - x_{22} - x_{23}$ space, (B) in $x_{21} - x_{22}$ plane, (C) in $x_{21} - x_{23}$ plane, (D) in $x_{22} - x_{23}$ plane.

If the parameters are taken as $a_{21} = 0.4$, $a_{22} = 0.175$, initial condition as $(0.19,\ 0,\ -0.18)$, then at $\alpha = 0.95$, Eq. (12.12) becomes the fractional order Newton–Leipnik chaotic equation. The phase portraits in $x_{21} - x_{22} - x_{23}$ space and $x_{21} - x_{22}$, $x_{21} - x_{23}$, $x_{22} - x_{23}$ planes are shown through the Fig. 12.3A–D respectively.

$$\begin{aligned}
\frac{d^\alpha y_{21}}{dt^\alpha} &= -y_{21} - b_{21}y_{22} - y_{23}y_{22} \\
\frac{d^\alpha y_{22}}{dt^\alpha} &= -y_{22} - b_{22}y_{21} - y_{21}y_{23} \\
\frac{d^\alpha y_{23}}{dt^\alpha} &= b_{23}y_{23} + y_{21}y_{22} + 1,
\end{aligned} \tag{12.13}$$

The chaotic attractors of the system (12.13) are described through Fig. 12.4 at order of derivative $\alpha = 0.99$ for the parameters' values $b_{21} = 19$, $b_{22} = 11$, $b_{23} = 0.73$ and the initial condition is $(8,\ 2,\ 3)$. The phase portraits in $y_{21} - y_{22} - y_{23}$ space and $y_{21} - y_{22}$, $y_{21} - y_{23}$, $y_{22} - y_{23}$ planes are shown through Fig. 12.4A–D respectively.

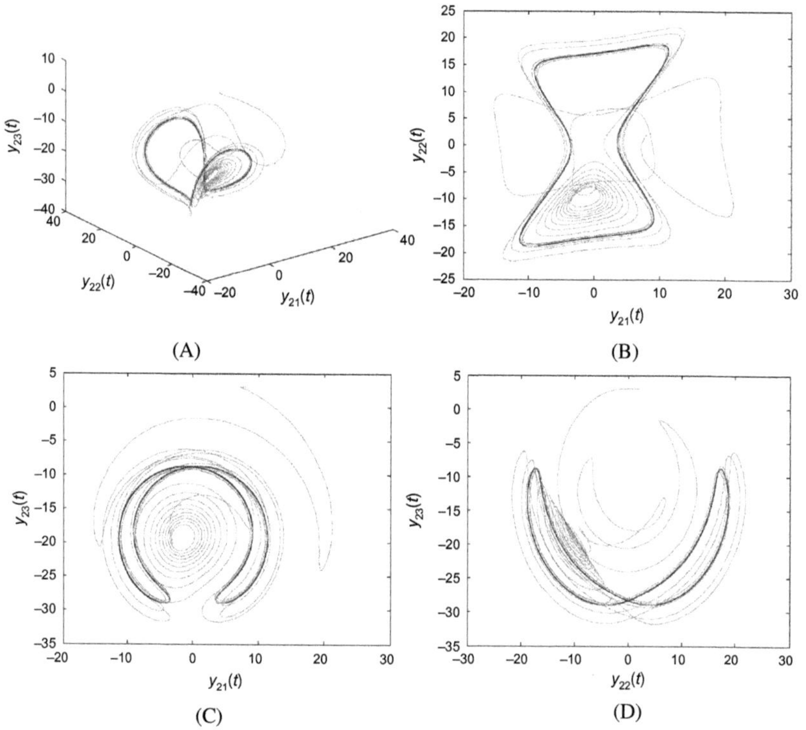

FIGURE 12.4 Phase portraits of Volta's system at $\alpha = 0.99$: (A) in $y_{21} - y_{22} - y_{23}$ space, (B) in $y_{21} - y_{22}$ plane, (C) in $y_{21} - y_{23}$ plane, (D) in $y_{22} - y_{23}$ plane.

The fractional order Lu hyperchaotic system (Pan et al., 2011) and the fractional order 4D Integral order hyperchaotic system (Deng et al., 2009) are taken as the corresponding two response systems as

$$\begin{aligned}
\frac{d^\alpha x_{31}}{dt^\alpha} &= a_{31}(x_{32} - x_{31}) + x_{34} + u_{11} \\
\frac{d^\alpha x_{32}}{dt^\alpha} &= -x_{31}x_{33} + a_{33}x_{32} + u_{12} \\
\frac{d^\alpha x_{33}}{dt^\alpha} &= x_{31}x_{32} - a_{32}x_{33} + u_{13} \\
\frac{d^\alpha x_{34}}{dt^\alpha} &= x_{31}x_{33} + a_{34}x_{34} + u_{14},
\end{aligned} \tag{12.14}$$

where u_{11}, u_{12}, u_{13}, u_{14} are control functions, x_{31}, x_{32}, x_{33}, and x_{34} are states variables and a_{31}, a_{32}, a_{33}, and a_{34} are constant parameters. The phase portraits of (12.14) in $x_{31} - x_{32} - x_{33}$, $x_{31} - x_{32} - x_{34}$, $x_{31} - x_{33} - x_{34}$,

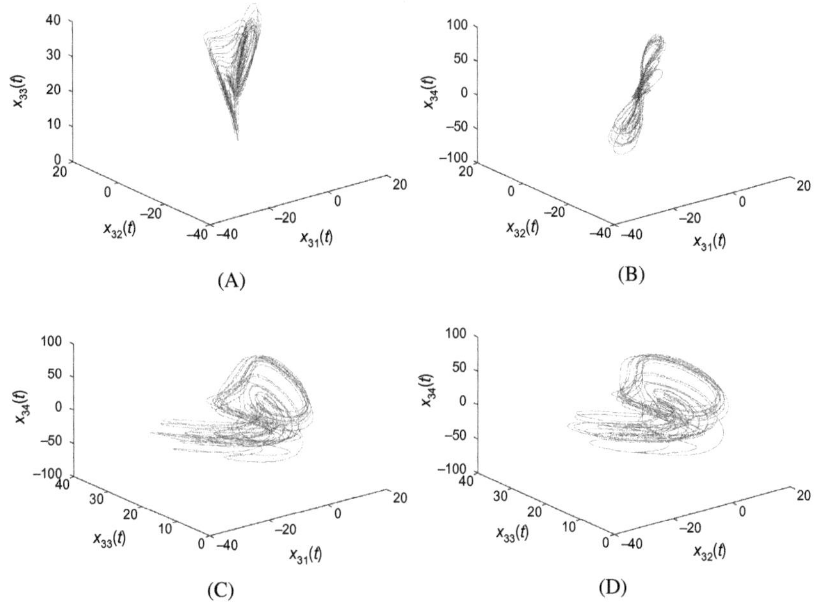

FIGURE 12.5 Phase portraits of Lu hyperchaotic system for $\alpha = 0.95$ (A) in $x_{31} - x_{32} - x_{33}$, space (B) in $x_{31} - x_{32} - x_{34}$ space, (C) in $x_{31} - x_{33} - x_{34}$ space, (D) in $x_{32} - x_{33} - x_{34}$ space.

$x_{32} - x_{33} - x_{34}$ spaces are depicted through Fig. 12.5 for $\alpha = 0.95$ at $a_{31} = 36,\ a_{32} = 3,\ a_{33} = 20$, and $a_{34} = -1$.

$$\begin{aligned}
\frac{d^\alpha y_{31}}{dt^\alpha} &= b_{31}y_{31} - y_{32} + u_{21} \\
\frac{d^\alpha y_{32}}{dt^\alpha} &= y_{31} - y_{32}y_{33}^2 + u_{22} \\
\frac{d^\alpha y_{33}}{dt^\alpha} &= -b_{32}y_{32} - b_{33}y_{33} - b_{34}y_{34} + u_{23} \\
\frac{d^\alpha y_{34}}{dt^\alpha} &= y_{33} + b_{35}y_{34} + u_{24},
\end{aligned} \tag{12.15}$$

where $u_{21},\ u_{22},\ u_{23},\ u_{24}$, are control functions, $y_{31},\ y_{32},\ y_{33}$, and y_{34} are states variables and $b_{31},\ b_{32},\ b_{33},\ b_{34}$, and b_{35} are the constant parameters. The phase portraits of (12.15) in $y_{31} - y_{32} - y_{33},\ y_{31} - y_{32} - y_{34},\ y_{31} - y_{33} - y_{34},\ y_{32} - y_{33} - y_{34}$ spaces are depicted through Fig. 12.6 for $\alpha = 0.95$ at $b_{31} = 0.56,\ b_{32} = 1.0,\ b_{33} = 1.0,\ b_{34} = 6.0$, and $b_{35} = 0.8$.

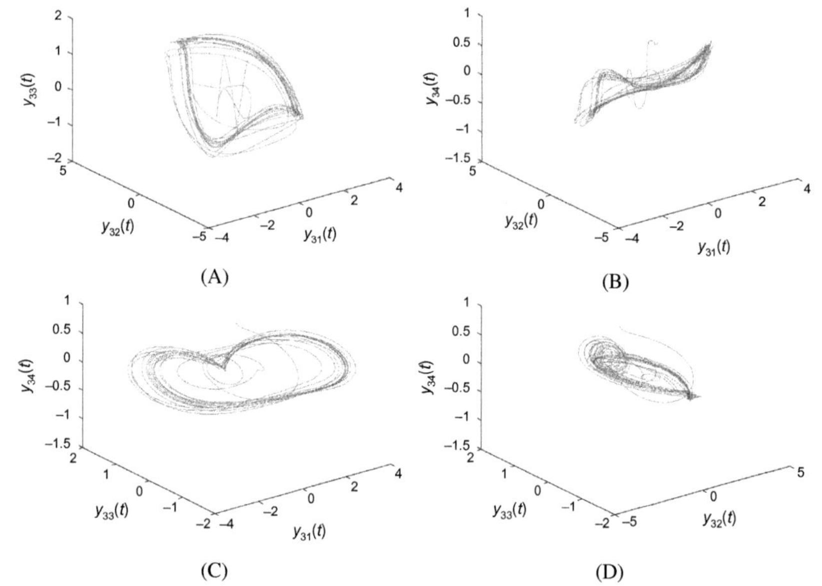

FIGURE 12.6 Phase portraits of 4D Integral order hyperchaotic system for $\alpha = 0.95$ (A) in $y_{31} - y_{32} - y_{33}$, space (B) in $y_{31} - y_{32} - y_{34}$ space, (C) in $y_{31} - y_{33} - y_{34}$ space, (D) in $y_{32} - y_{33} - y_{34}$ space.

Now comparing the systems (12.10)–(12.15) with systems (12.1)–(12.6), we get

$$A_1 = \begin{bmatrix} -a_{11} & a_{11} & 0 \\ 0 & a_{13} & 0 \\ 0 & 0 & -a_{12} \end{bmatrix}, \quad F_1(X_1) = \begin{bmatrix} 0 \\ -x_{11}x_{13} \\ x_{11}x_{12} \end{bmatrix}, \quad B_1 = \begin{bmatrix} -b_{11} & b_{11} & 0 \\ b_{13} & -1 & 0 \\ 0 & 0 & -b_{12} \end{bmatrix}, \quad G_1(Y_1) = \begin{bmatrix} y_{12}y_{13} \\ -y_{11}y_{13} \\ y_{11}y_{12} \end{bmatrix},$$

$$A_2 = \begin{bmatrix} -a_{21} & 1 & 0 \\ -1 & -0.4 & 0 \\ 0 & 0 & a_{22} \end{bmatrix}, \quad F_2(X_2) = \begin{bmatrix} 10x_{22}x_{23} \\ 5x_{21}x_{23} \\ -5x_{21}x_{22} \end{bmatrix}, \quad B_2 = \begin{bmatrix} -1 & b_{21} & 0 \\ -b_{22} & -1 & 0 \\ 0 & 0 & -b_{23} \end{bmatrix}, \quad G_2(Y_2) = \begin{bmatrix} -y_{22}y_{23} \\ -y_{21}y_{23} \\ y_{21}y_{22} + 1 \end{bmatrix},$$

$$A_3 = \begin{bmatrix} -a_{31} & a_{31} & 0 & 1 \\ 0 & a_{33} & 0 & 0 \\ 0 & 0 & -a_{32} & 0 \\ 0 & 0 & 0 & a_{34} \end{bmatrix}, \quad F_3(X_3) = \begin{bmatrix} 0 \\ -x_{31}x_{33} \\ x_{31}x_{32} \\ x_{31}x_{33} \end{bmatrix}, \quad B_3 = \begin{bmatrix} b_{31} & -1 & 0 & 0 \\ 1 & 0 & 0 & 0 \\ 0 & -b_{32} & -b_{33} & -b_{34} \\ 0 & 0 & 1 & b_{35} \end{bmatrix},$$

$$G_3(Y_3) = \begin{bmatrix} 0 \\ -y_{32}y_{33}^2 \\ 0 \\ 0 \end{bmatrix}, \quad U_1 = \begin{bmatrix} u_{11} \\ u_{12} \\ u_{13} \\ u_{14} \end{bmatrix}, \quad U_2 = \begin{bmatrix} u_{21} \\ u_{22} \\ u_{23} \\ u_{24} \end{bmatrix},$$

where $X_1 = [x_{11},\ x_{12},\ x_{13}]$, $Y_1 = [y_{11},\ y_{12},\ y_{13}]$, $X_2 = [x_{21},\ x_{22},\ x_{23}]$, $Y_2 = [y_{21},\ y_{22},\ y_{23}]$, $X_3 = [x_{31},\ x_{32},\ x_{33},\ x_{34}]$, $Y_3 = [y_{31},\ y_{32},\ y_{33},\ y_{34}]$ are the state vectors, $U_1 = [u_{11},\ u_{12},\ u_{13},\ u_{14}]^T$ and $U_2 = [u_{21},\ u_{22},\ u_{23},\ u_{24}]^T$ are the controller to be designed later.

Now choosing the matrix as

$$C_1 = \begin{bmatrix} 1 & 0 & 0 \\ 0 & 1 & 0 \\ 0 & 0 & 1 \\ 0 & 0 & 1 \end{bmatrix}, \quad C_2 = \begin{bmatrix} 1 & 0 & 0 \\ 0 & 1 & 0 \\ 0 & 0 & 1 \\ 0 & 0 & 1 \end{bmatrix},$$

the error functions $e_1 = X_3 - C_1(X_2 + X_1)$, $e_2 = Y_3 - C_2(Y_2 + Y_1)$ can be obtained as

$$\begin{aligned}
e_{11} &= x_{31} - (x_{21} + x_{11}), \\
e_{12} &= x_{32} - (x_{22} + x_{12}), \\
e_{13} &= x_{33} - (x_{23} + x_{13}), \\
e_{14} &= x_{34} - (x_{23} + x_{13}), \\
e_{21} &= y_{31} - (y_{21} + y_{11}), \\
e_{22} &= y_{32} - (y_{22} + y_{12}), \\
e_{23} &= y_{33} - (y_{23} + y_{13}), \\
e_{24} &= y_{34} - (y_{23} + y_{13}),
\end{aligned}$$

where $e_1 = [e_{11},\ e_{12},\ e_{13},\ e_{14}]^T$ and $e_2 = [e_{21},\ e_{22},\ e_{23},\ e_{24}]^T$. In view of preposition 1 the control functions $U_1 = C_1[A_2X_2 + F_2(X_2) - A_1X_1 + F_1(X_1)] - A_3C_1(X_2 + X_1) - F_3(X_3) + K_1e_1$ and $U_2 = C_2[B_2Y_2 + G_2(Y_2) + B_1Y_1 + G_1(Y_1)] - B_3C_2(Y_2 + Y_1) - G_3(Y_3) + K_2e_2$ will be

$$U_1 = \begin{bmatrix} u_{11} \\ u_{12} \\ u_{13} \\ u_{14} \end{bmatrix} = \begin{bmatrix} -a_{21}x_{21} + x_{22} + 10x_{22}x_{23} + a_{11}(x_{12} - x_{11}) + a_{31}(x_{21} + x_{11}) - a_{31}(x_{22} + x_{12}) - x_{23} - x_{13} - a_{31}e_{12} - e_{14} \\ -x_{21} - 0.4x_{22} + 5x_{21}x_{23} - x_{11}x_{13} + a_{13}x_{12} - a_{33}(x_{22} + x_{12}) + x_{31}x_{33} - (a_{33} + 1)e_{12} \\ a_{22}x_{23} - 5x_{21}x_{22} + x_{11}x_{12} - a_{12}x_{13} + a_{32}(x_{23} + x_{13}) - x_{31}x_{32} \\ a_{22}x_{23} - 5x_{21}x_{22} + x_{11}x_{12} - a_{12}x_{13} - a_{34}(x_{23} + x_{13}) - x_{31}x_{33} - (a_{34} + 1)e_{14} \end{bmatrix},$$

$$U_2 = \begin{bmatrix} u_{21} \\ u_{22} \\ u_{23} \\ u_{24} \end{bmatrix} = \begin{bmatrix} -y_{21} - b_{21}y_{22} - y_{23}y_{22} + b_{11}(y_{12} - y_{11}) + y_{12}y_{13} - b_{31}(y_{21} + y_{11}) + y_{22} + y_{12} - (b_{31} + 1)e_{21} + e_{22} \\ -y_{22} - b_{22}y_{21} - y_{21}y_{23} + b_{13}y_{11} - y_{12} - y_{11}y_{13} - y_{21} - y_{11} + y_{32}y_{33}^2 - e_{21} - e_{22} \\ b_{23}y_{23} + y_{21}y_{22} + 1 - b_{12}y_{13} + y_{11}y_{12} + b_{32}(y_{22} + y_{12}) + (b_{33} + b_{34})(y_{23} + y_{13}) + b_{32}e_{22} + b_{34}e_{24} \\ b_{23}y_{23} + y_{21}y_{22} + 1 - b_{12}y_{13} + y_{11}y_{12} - (1 + b_{35})(y_{23} + y_{13}) - e_{23} - (b_{35} + 1)e_{24} \end{bmatrix}.$$

The error systems are given in the form

$$\begin{aligned}
\frac{d^\alpha e_1}{dt^\alpha} &= (A_3 + K_1)e_1 \\
\frac{d^\alpha e_2}{dt^\alpha} &= (B_3 + K_2)e_2.
\end{aligned} \tag{12.16}$$

The preposition 1 confirms that if gain matrices are taken as $K_1 = \begin{bmatrix} 0 & -a_{31} & 0 & -1 \\ 0 & -a_{33} - 1 & 0 & 0 \\ 0 & 0 & 0 & 0 \\ 0 & 0 & 0 & -a_{34} - 1 \end{bmatrix}$, $K_2 = \begin{bmatrix} -b_{31} - 1 & 1 & 0 & 0 \\ -1 & -1 & 0 & 0 \\ 0 & b_{32} & 0 & b_{34} \\ 0 & 0 & -1 & -b_{35} - 1 \end{bmatrix}$, the systems (12.10)–(12.15) realize the dual combination synchronization.

In order to simulate the systems, the earlier values of parameters of the systems are considered and the initial conditions of master systems Lu, Qi, Newton–Leipnik, Volta's chaotic System are taken as $(x_{11}(0),\ x_{12}(0),$

$x_{13}(0)) = (0.2, 0.5, 0.3)$, $(y_{11}(0), y_{12}(0), y_{13}(0)) = (-1, -1, -2)$, $(x_{21}(0), x_{22}(0), x_{23}(0)) = (0.19, 0, -0.18)$, $(y_{21}(0), y_{22}(0), y_{23}(0)) = (8, 2, 3)$ and slave systems Lu hyperchaotic system, 4D Integral order hyperchaotic system are taken as $(x_{31}(0), x_{32}(0), x_{33}(0), x_{34}(0)) = (-10, -14, 12, 10)$ and $(y_{31}(0), y_{32}(0), y_{33}(0), y_{34}(0)) = (1.2, 0.6, 0.8, 0.5)$, respectively. Hence the initial condition of the error system will be $(e_{11}(0), e_{12}(0), e_{13}(0), e_{14}(0), e_{21}(0), e_{22}(0), e_{23}(0), e_{24}(0)) = (-10.39, -14.50, 11.88, 9.88, -5.80, -4, -2, -0.5)$. Fig. 12.7A–H show the chaotic systems are dual combination synchronized and it is also seen from Fig. 12.7I that the error vectors asymptotically converge to zero as time becomes large which implies that dual combination synchronizations among the considered fractional order systems are achieved at $\alpha = 0.98$.

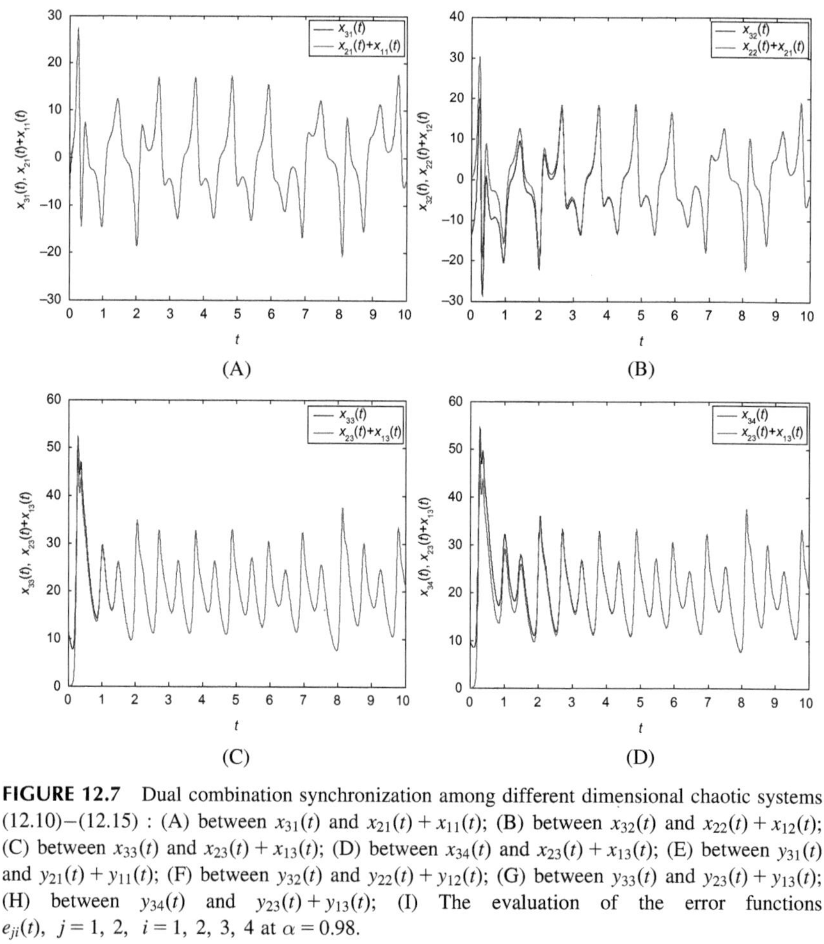

FIGURE 12.7 Dual combination synchronization among different dimensional chaotic systems (12.10)–(12.15) : (A) between $x_{31}(t)$ and $x_{21}(t) + x_{11}(t)$; (B) between $x_{32}(t)$ and $x_{22}(t) + x_{12}(t)$; (C) between $x_{33}(t)$ and $x_{23}(t) + x_{13}(t)$; (D) between $x_{34}(t)$ and $x_{23}(t) + x_{13}(t)$; (E) between $y_{31}(t)$ and $y_{21}(t) + y_{11}(t)$; (F) between $y_{32}(t)$ and $y_{22}(t) + y_{12}(t)$; (G) between $y_{33}(t)$ and $y_{23}(t) + y_{13}(t)$; (H) between $y_{34}(t)$ and $y_{23}(t) + y_{13}(t)$; (I) The evaluation of the error functions $e_{ji}(t)$, $j = 1, 2$, $i = 1, 2, 3, 4$ at $\alpha = 0.98$.

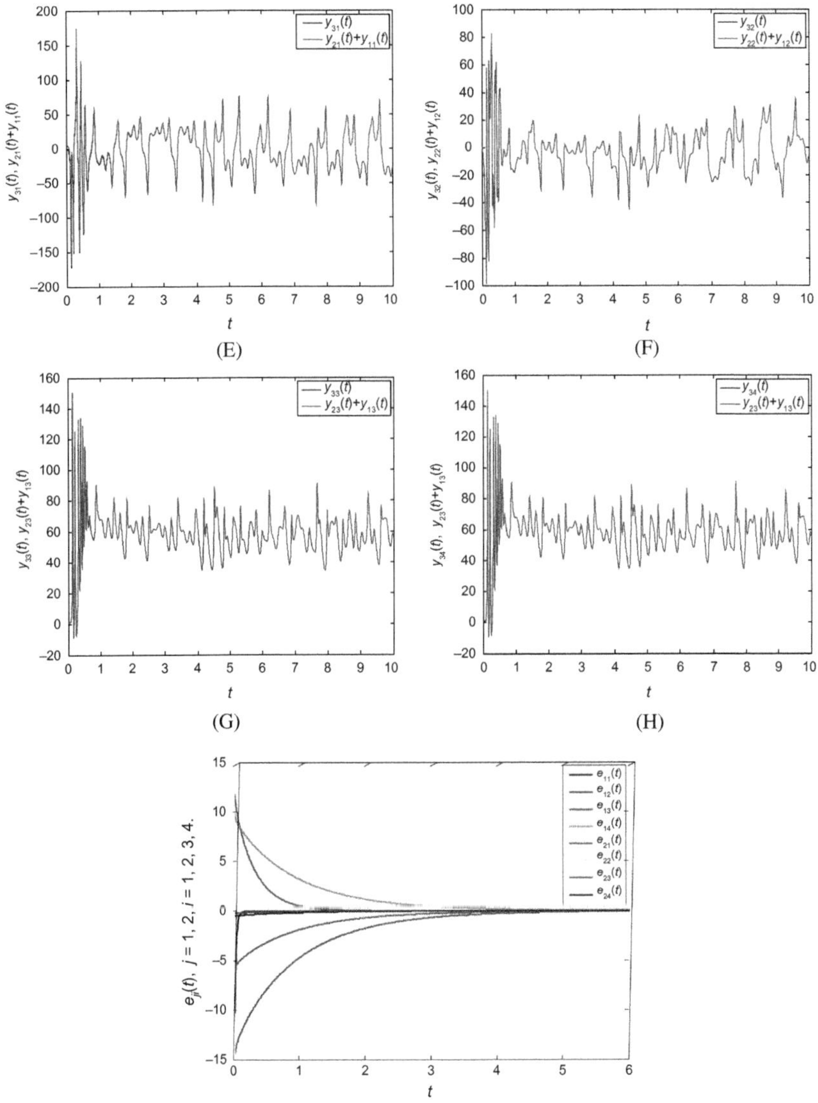

FIGURE 12.7 (Continued).

12.3.2 Dual Combination Synchronization With Order $m > n$

The fractional order Lu hyperchaotic system and the fractional order 4D Integral order hyperchaotic system are taken as the first two master systems as

$$\begin{aligned}
\frac{d^\alpha x_{11}}{dt^\alpha} &= a_{11}(x_{12} - x_{11}) + x_{14} \\
\frac{d^\alpha x_{12}}{dt^\alpha} &= -x_{11}x_{13} + a_{13}x_{12} \\
\frac{d^\alpha x_{13}}{dt^\alpha} &= x_{11}x_{12} - a_{12}x_{13} \\
\frac{d^\alpha x_{14}}{dt^\alpha} &= x_{11}x_{13} + a_{14}x_{14}
\end{aligned} \tag{12.17}$$

and

$$\begin{aligned}
\frac{d^\alpha y_{11}}{dt^\alpha} &= b_{11}y_{11} - y_{12} \\
\frac{d^\alpha y_{12}}{dt^\alpha} &= y_{11} - y_{12}y_{13}^2 \\
\frac{d^\alpha y_{13}}{dt^\alpha} &= -b_{12}y_{12} - b_{13}y_{13} - b_{14}y_{14} \\
\frac{d^\alpha y_{14}}{dt^\alpha} &= y_{13} + b_{15}y_{14},
\end{aligned} \tag{12.18}$$

where x_{1i}, y_{1i} $(i = 1, 2, 3, 4)$ are states variables and a_{1i}, $(i = 1, 2, 3, 4)$, b_{1i} $(i = 1, 2, 3, 4, 5)$ are the constant parameters.

The fractional order Chen hyperchaotic system (Matouk and Elsadany, 2014) and fractional order Lorenz hyperchaotic system (Chen et al., 2011) are considered as

$$\begin{aligned}
\frac{d^\alpha x_{21}}{dt^\alpha} &= a_{21}(x_{22} - x_{21}) + x_{24} \\
\frac{d^\alpha x_{22}}{dt^\alpha} &= a_{22}x_{21} - x_{21}x_{23} + a_{23}x_{22} \\
\frac{d^\alpha x_{23}}{dt^\alpha} &= x_{21}x_{22} - a_{24}x_{23} \\
\frac{d^\alpha x_{24}}{dt^\alpha} &= x_{22}x_{23} + a_{25}x_{24},
\end{aligned} \tag{12.19}$$

where x_{21}, x_{22}, x_{23} and x_{24} are states variables and a_{21}, a_{22}, a_{23}, a_{24}, and a_{25} are constant parameters. The phase portraits of (12.19) in $x_{21} - x_{22} - x_{23}$,

$x_{21}-x_{22}-x_{24}$, $x_{21}-x_{23}-x_{24}$, $x_{22}-x_{23}-x_{24}$ spaces are depicted through Fig. 12.8 for $\alpha = 0.97$ at $a_{21}=35$, $a_{22}=7$, $a_{23}=12$, $a_{24}=3$, $a_{25}=0.3$ and initial condition $(-1,\ -3,\ 2,\ 5)$.

$$\begin{aligned}
\frac{d^{\alpha}y_{21}}{dt^{\alpha}} &= b_{21}(y_{22}-y_{21})+y_{24}\\
\frac{d^{\alpha}y_{22}}{dt^{\alpha}} &= b_{22}y_{21}-y_{22}-y_{21}y_{23}\\
\frac{d^{\alpha}y_{23}}{dt^{\alpha}} &= y_{21}y_{22}-b_{23}y_{23}\\
\frac{d^{\alpha}y_{24}}{dt^{\alpha}} &= -y_{22}y_{23}+b_{24}y_{24},
\end{aligned} \tag{12.20}$$

where b_{21}, b_{22}, b_{23}, and b_{24} are parameters. For the values of parameters $b_{21}=10$, $b_{22}=28$, $b_{23}=8/3$, $b_{24}=-1$and the initial condition $(1.5,\ 3,\ -1,\ 3)$, the system (12.20) exhibits chaotic behavior with commensurate order of derivative $\alpha = 0.98$. The chaotic attractors of the system (12.20) are described through Fig. 12.9. The phase portraits in $y_{21}-y_{22}-y_{23}$, $y_{21}-y_{22}-y_{24}$, $y_{21}-y_{23}-y_{24}$, $y_{22}-y_{23}-y_{24}$ spaces are shown through Fig. 12.9A–D respectively for $\alpha = 0.98$.

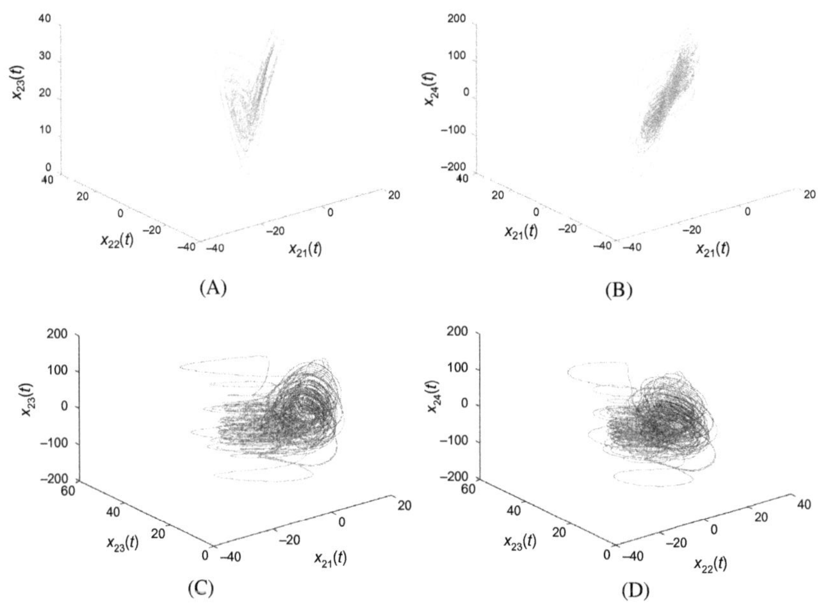

FIGURE 12.8 Phase portraits of Chen hyperchaotic system for $\alpha = 0.97$: (A) in $x_{21}-x_{22}-x_{23}$ space, (B) in $x_{21}-x_{22}-x_{24}$ space, (C) in $x_{21}-x_{23}-x_{24}$ space, (D) in $x_{22}-x_{23}-x_{24}$ space.

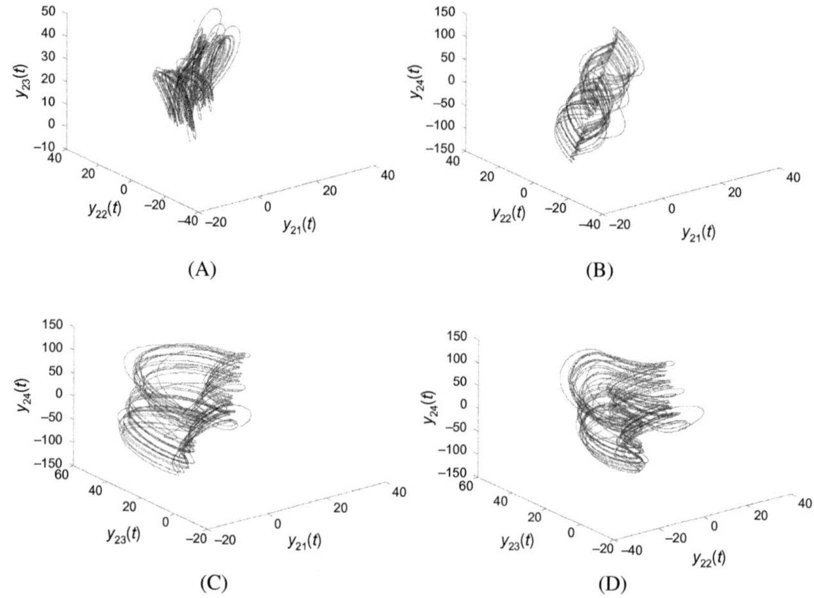

FIGURE 12.9 Phase portraits of Lorenz hyperchaotic system for $\alpha = 0.98$: (A) in $y_{31} - y_{32} - y_{33}$ space, (B) in $y_{31} - y_{32} - y_{34}$ space, (C) in $y_{31} - y_{33} - y_{34}$ space, (D) in $y_{32} - y_{33} - y_{34}$ space.

The fractional order Lu and Qi systems are taken as the corresponding two response systems and are rewritten as

$$\begin{aligned} \frac{d^\alpha x_{31}}{dt^\alpha} &= a_{31}(x_{32} - x_{31}) + u_{11} \\ \frac{d^\alpha x_{32}}{dt^\alpha} &= -x_{31}x_{33} + a_{33}x_{32} + u_{12} \\ \frac{d^\alpha x_{33}}{dt^\alpha} &= x_{31}x_{32} - a_{32}x_{33} + u_{13}, \end{aligned} \tag{12.21}$$

where u_{11}, u_{12}, u_{13} are control functions. The phase portraits of system (12.21) at $\alpha = 0.95$ are shown in Fig. 12.1.

$$\begin{aligned} \frac{d^\alpha y_{31}}{dt^\alpha} &= b_{31}(y_{32} - y_{31}) + y_{32}y_{33} + u_{21} \\ \frac{d^\alpha y_{32}}{dt^\alpha} &= b_{33}y_{31} - y_{32} - y_{31}y_{33} + u_{22} \\ \frac{d^\alpha y_{33}}{dt^\alpha} &= -b_{32}y_{33} + y_{31}y_{32} + u_{23}, \end{aligned} \tag{12.22}$$

where u_{21}, u_{22}, u_{23} are control functions. The phase portraits of system (12.22) at $\alpha = 0.98$ are depicted through Fig. 12.2.

Now again comparing the systems (12.17)–(12.22) with systems (12.1)–(12.6), we get

$$A_1 = \begin{bmatrix} -a_{11} & a_{11} & 0 & 1 \\ 0 & a_{13} & 0 & 0 \\ 0 & 0 & -a_{12} & 0 \\ 0 & 0 & 0 & a_{14} \end{bmatrix}, \quad F_1(X_1) = \begin{bmatrix} 0 \\ -x_{11}x_{13} \\ x_{11}x_{12} \\ x_{11}x_{13} \end{bmatrix}, \quad B_1 = \begin{bmatrix} b_{11} & -1 & 0 & 0 \\ 1 & 0 & 0 & 0 \\ 0 & -b_{12} & -b_{13} & b_{14} \\ 0 & 0 & 1 & b_{15} \end{bmatrix},$$

$$G_1(Y_1) = \begin{bmatrix} 0 \\ -y_{12}y_{13}^2 \\ 0 \\ 0 \end{bmatrix}, \quad A_2 = \begin{bmatrix} -a_{21} & a_{21} & 0 & 1 \\ a_{22} & a_{23} & 0 & 0 \\ 0 & 0 & -a_{24} & 0 \\ 0 & 0 & 0 & a_{25} \end{bmatrix}, \quad F_2(X_2) = \begin{bmatrix} 0 \\ -x_{21}x_{23} \\ x_{21}x_{22} \\ x_{22}x_{23} \end{bmatrix},$$

$$B_2 = \begin{bmatrix} -b_{21} & b_{21} & 0 & 1 \\ b_{22} & -1 & 0 & 0 \\ 0 & 0 & -b_{213} & 0 \\ 0 & 0 & 0 & b_{24} \end{bmatrix}, \quad G_2(Y_2) = \begin{bmatrix} 0 \\ -y_{21}y_{23} \\ y_{21}y_{22} \\ -y_{22}y_{23} \end{bmatrix}, \quad A_3 = \begin{bmatrix} -a_{31} & a_{31} & 0 \\ 0 & a_{33} & 0 \\ 0 & 0 & -a_{32} \end{bmatrix},$$

$$F_3(X_3) = \begin{bmatrix} 0 \\ -x_{31}x_{33} \\ x_{31}x_{32} \end{bmatrix}, \quad B_3 = \begin{bmatrix} -b_{31} & b_{31} & 0 \\ b_{33} & -1 & 0 \\ 0 & 0 & -b_{32} \end{bmatrix},$$

$$G_3(Y_3) = \begin{bmatrix} y_{32}y_{33} \\ -y_{31}y_{33} \\ y_{31}y_{32} \end{bmatrix}, \quad U_1 = \begin{bmatrix} u_{11} \\ u_{12} \\ u_{13} \end{bmatrix}, \quad U_2 = \begin{bmatrix} u_{21} \\ u_{22} \\ u_{23} \end{bmatrix}$$

where $X_1 = [x_{11}, x_{12}, x_{13}, x_{14}]$, $Y_1 = [y_{11}, y_{12}, y_{13}, y_{14}]$, $X_2 = [x_{21}, x_{22}, x_{23}, x_{24}]$, $Y_2 = [y_{21}, y_{22}, y_{23}, y_{24}]$, $X_3 = [x_{31}, x_{32}, x_{33}]$, $Y_3 = [y_{31}, y_{32}, y_{33}]$ are the state vectors, $U_1 = [u_{11}, u_{12}, u_{13}]^T$ and $U_2 = [u_{21}, u_{22}, u_{23}]^T$ are the controller to be designed later.

Now choosing the matrix as

$$C_1 = \begin{bmatrix} 1 & 0 & 0 & 0 \\ 0 & 1 & 0 & 0 \\ 0 & 0 & 1 & 1 \end{bmatrix}, \quad C_2 = \begin{bmatrix} 1 & 0 & 0 & 0 \\ 0 & 1 & 0 & 0 \\ 0 & 0 & 1 & 1 \end{bmatrix},$$

the error functions $e_1 = X_3 - C_1(X_2 + X_1)$, $e_2 = Y_3 - C_2(Y_2 + Y_1)$ can be obtained as

$$\begin{aligned} e_{11} &= x_{31} - (x_{21} + x_{11}), \\ e_{12} &= x_{32} - (x_{22} + x_{12}), \\ e_{13} &= x_{33} - (x_{23} + x_{13}) - (x_{24} + x_{14}), \\ e_{21} &= y_{31} - (y_{21} + y_{11}), \\ e_{22} &= y_{32} - (y_{22} + y_{12}), \\ e_{23} &= y_{33} - (y_{23} + y_{13}) - (y_{24} + y_{14}), \end{aligned}$$

where $e_1 = [e_{11},\ e_{12},\ e_{13}]^T$ and $e_2 = [e_{21},\ e_{22},\ e_{23}]^T$. In view of preposition 1 the control functions $U_1 = C_1[A_2X_2 + F_2(X_2) - A_1X_1 + F_1(X_1)] - A_3C_1(X_2 + X_1) - F_3(X_3) + K_1e_1$ and $U_2 = C_2[B_2Y_2 + G_2(Y_2) + B_1Y_1 + G_1(Y_1)] - B_3C_2(Y_2 + Y_1) - G_3(Y_3) + K_2e_2$ will be

$$U_1 = \begin{bmatrix} u_{11} \\ u_{12} \\ u_{13} \end{bmatrix} = \begin{bmatrix} a_{21}(x_{22}-x_{21})+x_{24}+a_{11}(x_{12}-x_{11})+x_{14}+a_{31}(x_{21}+x_{11})-a_{31}(x_{22}+x_{12})-a_{31}e_{12} \\ a_{22}x_{21}-x_{21}x_{23}+a_{23}x_{22}-x_{11}x_{13}+a_{13}x_{12}-a_{33}(x_{22}+x_{12})+x_{31}x_{33}-(a_{33}+1)e_{12} \\ x_{21}x_{22}-a_{24}x_{23}+x_{11}x_{12}-a_{12}x_{13}+x_{22}x_{23}+a_{25}x_{24}+x_{11}x_{13}+a_{14}x_{14} \\ +a_{32}(x_{23}+x_{13})+a_{32}(x_{24}+x_{14})-x_{31}x_{32} \end{bmatrix},$$

$$U_2 = \begin{bmatrix} u_{21} \\ u_{22} \\ u_{23} \end{bmatrix} = \begin{bmatrix} b_{21}(y_{22}-y_{21})+y_{24}+b_{11}y_{11}-y_{12}+b_{31}(y_{21}+y_{11})-b_{31}(y_{22}+y_{12})-y_{32}y_{33}-b_{31}e_{22} \\ b_{22}y_{21}-y_{22}-y_{21}y_{23}+y_{11}-y_{12}y_{13}^2+b_{33}(y_{21}+y_{11})+(y_{22}+y_{12})+y_{31}y_{33}-b_{33}e_{21} \\ y_{21}y_{22}-b_{23}y_{23}-b_{12}y_{12}-b_{13}y_{13}-b_{14}y_{14}-y_{22}y_{23}+b_{24}y_{24}+y_{13}+b_{15}y_{14} \\ +b_{32}(y_{23}+y_{13})+b_{32}(y_{24}+y_{14})-y_{31}y_{32} \end{bmatrix}.$$

The error systems are given in the form

$$\begin{aligned} \frac{d^\alpha e_1}{dt^\alpha} &= (A_3 + K_1)e_1 \\ \frac{d^\alpha e_2}{dt^\alpha} &= (B_3 + K_2)e_2. \end{aligned} \tag{12.23}$$

The preposition 1 confirms that if we take gain matrices as $K_1 = \begin{bmatrix} 0 & -a_{31} & 0 \\ 0 & -a_{33}-1 & 0 \\ 0 & 0 & 0 \end{bmatrix}$, $K_2 = \begin{bmatrix} 0 & -b_{31} & 0 \\ -b_{33} & 0 & 0 \\ 0 & 0 & 0 \end{bmatrix}$, then the systems (12.17)–(12.22) will be dual combination synchronized.

Fig. 12.10A–G represent the simulation results towards synchronization of the considered chaotic systems considering the earlier values of the parameters of the systems. The initial conditions of master systems Lu, 4D Integral order, Chen, Lorenz hyperchaotic system are taken as $(x_{11}(0),\ x_{12}(0),\ x_{13}(0),\ x_{14}(0)) = (-10,\ -14,\ 12,\ 10)$, $(y_{11}(0),\ y_{12}(0),\ y_{13}(0),\ y_{14}(0)) = (1.2,\ 0.6,\ 0.8,\ 0.5)$, $(x_{21}(0),\ x_{22}(0),\ x_{23}(0),\ x_{24}(0)) = (-1,\ -3,\ 2,\ 5)$, $(y_{21}(0),\ y_{22}(0),\ y_{23}(0),\ y_{24}(0)) = (1.5,\ 3,\ -1,\ 3)$ and slave systems Lu, Qi chaotic systems are taken as $(x_{31}(0),\ x_{32}(0),\ x_{33}(0)) = (0.2,\ 0.5,\ 0.3)$ and $(y_{31}(0),\ y_{32}(0),\ y_{33}(0)) = (-1,\ -1,\ -2)$, respectively. Hence the initial condition of the error system will be $(e_{11}(0),\ e_{12}(0),\ e_{13}(0),\ e_{21}(0),\ e_{22}(0),\ e_{23}(0)) = (11.20,\ 17.50,\ -28.70,\ -3.70,\ -4.60,\ -5.3)$. It is clear from the figures that the error vectors asymptotically converge to zero as time becomes large which justifies the achievement of dual combination synchronization among the considered fractional order systems at order of the derivative $\alpha = 0.98$.

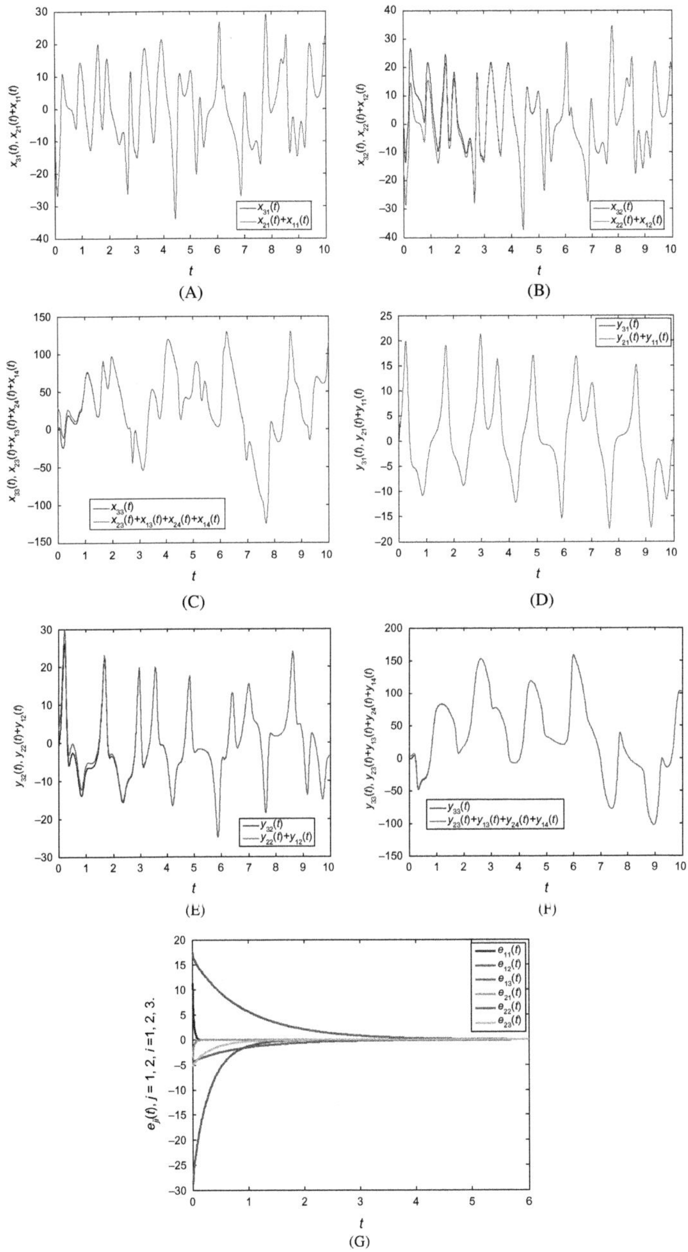

FIGURE 12.10 Dual combination synchronization among different dimensional chaotic systems (12.17)–(12.22) : (A) between $x_{31}(t)$ and $x_{21}(t)+x_{11}(t)$; (B) between $x_{32}(t)$ and $x_{22}(t)+x_{12}(t)$; (C) between $x_{33}(t)$ and $x_{23}(t)+x_{13}(t)+x_{24}(t)+x_{14}(t)$; (D) between $y_{31}(t)$ and $y_{21}(t)+y_{11}(t)$; (E) between $y_{32}(t)$ and $y_{22}(t)+y_{12}(t)$; (F) between $y_{33}(t)$ and $y_{23}(t)+y_{13}(t)+y_{24}(t)+y_{14}(t)$; (G) The evaluation of the error functions $e_{ji}(t)$, $j=1,\ 2,\ i=1,\ 2,\ 3$ at $\alpha=0.98$.

12.4 CONCLUSION

The present article has successfully demonstrated the dual combination synchronization among four fractional order drive systems and two fractional order response systems using scaling matrices separately. Based on the stability analysis, the dual combination synchronization of chaotic systems through controller input parameters on the respective system has been achieved and the components of the error system tend to zero as time becomes large, which helps to find the time required for dual combination synchronization among chaotic systems. Numerical simulation results are given to exhibit the reliability and effectiveness of the proposed dual combination synchronization scheme towards predicting the accuracy of the theory.

REFERENCES

Agrawal, S.K., Srivastava, M., Das, S., 2012. Synchronization of fractional order chaotic systems using active control method. Chaos Solitons Fractals 45 (6), 737–752.

Azar, A.T., Vaidyanathan, S., 2015a. Handbook of research on advanced intelligent control engineering and automation. Advances in Computational Intelligence and Robotics (ACIR) Book Series. IGI Global, USA, ISBN: 9781466672482.

Azar, A.T., Vaidyanathan, S., 2015b. Computational Intelligence applications, Modeling and Control. Studies in Computational Intelligence, Vol. 575. Springer-Verlag, Germany, ISBN: 978-3-319-11016-5.

Azar, A.T., Vaidyanathan, S., 2015c. Chaos Modeling and Control Systems Design, Studies in Computational Intelligence, Vol. 581. Springer-Verlag, Germany, ISBN: 978-3-319-13131-3.

Azar, A.T., Vaidyanathan, S., 2016. Advances in Chaos Theory and Intelligent Control. Studies in Fuzziness and Soft Computing, Vol. 337. Springer-Verlag, Germany, ISBN: 978-3-319-30338-3.

Azar, A.T., Vaidyanathan, S., Ouannas, A., 2017a. Fractional Order Control and Synchronization of Chaotic Systems. Studies in Computational Intelligence, Vol. 688. Springer-Verlag, Germany, ISBN: 978-3-319-50248-9.

Azar, A.T., Volos, C., Gerodimos, N.A., Tombras, G.S., Pham, V.T., Radwan, A.G., et al., 2017b. A novel chaotic system without equilibrium: dynamics, synchronization and circuit realization. Complexity vol. 2017, . Available from: https://doi.org/10.1155/2017/7871467Article ID 7871467, 11pages, 2017.

Azar, A.T., Ouannas, A., Singh, S., 2017c. Control of new type of fractional chaos synchronization, Proceedings of the International Conference on Advanced Intelligent Systems and Informatics 2017, Advances in Intelligent Systems and Computing series, Vol. 639. Springer-Verlag, Germany, pp. 47–56.

Azar, A.T., Zhu, Q., 2015. Advances and Applications in Sliding Mode Control systems. Studies in Computational Intelligence, Vol. 576. Springer-Verlag, Germany, ISBN: 978-3-319-11172-8.

Blasius, B., Huppert, A., Stone, L., 1999. Complex dynamics and phase synchronization in spatially extended ecological systems. Nature 399, 354–359. Available from: https://doi.org/10.1038/20676.

Boulkroune, A., Hamel, S., Azar, A.T., 2016a. Fuzzy control-based function synchronization of unknown chaotic systems with dead-zone input, Advances in Chaos Theory and Intelligent Control. Studies in Fuzziness and Soft Computing, Vol. 337. Springer-Verlag, Germany.

Boulkroune, A., Bouzeriba, A., Bouden, T., Azar, A.T., 2016b. Fuzzy adaptive synchronization of uncertain fractional-order chaotic systems. In: Azar, A.T., Vaidyanathan, S. (Eds.), Advances in Chaos Theory and Intelligent Control. Studies in Fuzziness and Soft Computing, Vol. 337. Springer-Verlag, Germany.

Chai, Y., Chen, L., Wu, R., Sun, J., 2012. Adaptive pinning synchronization in fractional-order complex dynamical networks. Physica A: Stat. Mech. Appl. 391 (22), 5746–5758.

Chen, J., Zeng, Z., Jiang, P., 2014. Global Mittag-Leffler stability and synchronization of memristor-based fractional-order neural networks. Neural Netw. 51, 1–8. Available from: https://doi.org/10.1016/j.neunet.2013.11.016.

Chen, L., Chai, Y., Wu, R., 2011. Lag projective synchronization in fractional-order chaotic (hyperchaotic) systems. Phys. Lett. A 375 (21), 2099–2110.

Chen, S.H., Lu, J., 2002. Parameters identification and synchronization of chaotic systems based upon adaptive control. Phys. Lett. A 299 (4), 353–358.

Deng, H., Li, T., Wang, Q., Li, H., 2009. A fractional-order hyperchaotic system and its synchronization. Chaos Solitons Fractals 41 (2), 962–969.

Ghosh, D., Chowdhury, A.R., 2010. Dual-anticipating, dual and dual-lag synchronization in modulated time-delayed systems. Phys. Lett. A 374 (34), 3425–3436.

Ghoudelbourk, S., Dib, D., Omeiri, A., Azar, A.T., 2016. MPPT Control in wind energy conversion systems and the application of fractional control (PIα) in pitch wind turbine. Int. J. Model. Identif. Control (IJMIC) 26 (2), 140–151.

Grassi G, Ouannas A, Azar AT, Radwan AG, Volos C, Pham VT, Ziar T, Madian A, Kyprianidis IM and Stouboulos IN (2017). Chaos Synchronisation Of Continuous Systems Via Scalar Signal. The 6th International Conference on Modern Circuits and Systems Technologies (MOCAST), 4–6 May 2017, Thessaloniki Greece.

Grigorenko, C.I., Grigorenko, E., 2003. Chaotic dynamics of the fractional Lorenz system. Phys. Rev. Lett. 91 (3), 034101.

Han, S.K., Kerrer, C., Kuramoto, Y., 1995. D-phasing and bursting in coupled neural oscillators. Phys. Rev. Lett. 75 (17), 3190–3193.

Hartley, T.T., Lorenzo, C.F., 2002. Dynamics and control of initialized fractional-order systems. Nonlinear Dyn. 29 (1-4), 201–233.

Hartley, T.T., Lorenzo, C.F., Qammer, H.K., 1995. Chaos in a fractional order Chua's system. IEEE Trans. Circuits Syst. 42 (8), 485–490.

Hassan, S., Mohammad, S., 2008. Dual synchronization of chaotic systems via time-varying gain proportional feedback. Chaos Solitons Fractals 38 (5), 1342–1348.

Hifer, R., 2001. Applications of Fractional Calculus in Physics World Scientific, New Jersey.

Huang, L., Feng, R., Wang, M., 2004. Synchronization of chaotic systems via nonlinear control. Phys. Lett. A 320 (4), 271–275.

Huang, T.W., Li, C.D., Ma, W.X., Huang, H., 2009. Anti-phase and complete-synchronization in coupled Chen system via active control. Int. J. Nonlinear Sci Numer Simulat. 10 (2), 221–226.

Ichise, M., Nagayanagi, Y., Kojima, T., 1971. An analog simulation of non-integer order transfer functions for analysis of electrode processes. J. Electroanal. Chem. Interfacial Electrochem 33 (2), 253–265.

Lakshmanan, M., Murali, K., 1996. Chaos in Nonlinear Oscillators: Controlling and Synchronization. World Scientific, Singapore.

Laskin, N., 2000. Fractional market dynamics. Phys. A 287 (3-4), 482–492.

Li, C.P., Peng, G.J., 2004. Chaos in Chen's system with a fractional order. Chaos Solitons Fract. 22 (2), 443–450.

Liu, W.Q., 2006. Anti-phase synchronization in coupled chaotic oscillators. Phys. Rev. E 73 (5), 57203–57207.

Lorenz, E.N., 1963. Deterministic non-periodic flow. J. Atmos. Sci. 20, 130–141. Available from: https://doi.org/10.1175/1520-0469(1963)020 < 0130:DNF > 2.0.CO;2.

Luo, R.Z., Wang, Y.L., Deng, S.C., 2011. Combination synchronization of three classic chaotic systems using active backstepping design. Chaos 21 (4), 043114.

Luo, R.Z., Wang, Y.L., 2012. Finite-time stochastic combination synchronization of three different chaotic systems and its application in secure communication. Chaos 22 (2), 023109.

Matouk, A.E., Elsadany, A.A., 2014. Achieving synchronization between the fractional-order hyperchaotic Novel and Chen systems via a new nonlinear control technique. Appl. Math. Lett. 29, 30–35.

Meghni, B., Dib, D., Azar, A.T., Ghoudelbourk, S., Saadoun, A., 2017. Robust Adaptive Supervisory Fractional order Controller For optimal Energy Management in Wind Turbine with Battery Storage. Studies in Computational Intelligence, Vol. 688. Springer-Verlag, Germany, pp. 165–202.

Moskalenko, O.I., Koronovskii, A.A., Hramov, A.E., 2010. Generalized synchronization of chaos for secure communication: remarkable stability to noise. Phys. Lett. A 374, 2925–2931.

Ouannas, A., Al-sawalha, M.M., Ziar, T., 2016a. Fractional chaos synchronization schemes for different dimensional systems with non-identical fractional-orders via two scaling matrices. Optik 127 (20), 8410–8418.

Ouannas, A., Azar, A.T., Abu-Saris, R., 2016b. A new type of hybrid synchronization between arbitrary hyperchaotic maps. Int. J. Mach. Learn. Cyber. Available from: https://doi.org/10.1007/s13042-016-0566-3.

Ouannas, A., Azar, A.T., Radwan, A.G., 2016c. On Inverse Problem of Generalized Synchronization Between Different Dimensional Integer-Order and Fractional-Order Chaotic Systems. The 28th International Conference on Microelectronics, IEEE, December 17-20, 2016, Cairo, Egypt.

Ouannas, A., Azar, A.T., Ziar, T., Vaidyanathan, S., 2017a. A New Method To Synchronize Fractional Chaotic Systems With Different Dimensions. Studies in Computational Intelligence, 688. Springer-Verlag, Germany, pp. 581–611.

Ouannas, A., Azar, A.T., Vaidyanathan, S., 2017b. A robust method for new fractional hybrid chaos synchronization. Math. Methods Appl. Sci. 40 (5), 1804–1812. Available from: https://doi.org/10.1002/mma.4099.

Ouannas, A., Azar, A.T., Vaidyanathan, S., 2017c. On a simple approach for Q-S synchronization of chaotic dynamical systems in continuous-time. Int. J. Comput. Sci. Math. 8 (1), 20–27.

Ouannas, A., Azar, A.T., Ziar, T., Radwan, A.G., 2017d. Generalized Synchronization of Different Dimensional Integer-order and Fractional Order Chaotic Systems. Studies in Computational Intelligence, 688. Springer-Verlag, Germany, pp. 671–697.

Ouannas, A., Azar, A.T., Abu-Saris, R., 2017e. A new type of hybrid synchronization between arbitrary hyperchaotic maps. Int. J. Mach. Learn. Cyber. 8 (6), 1887–1894. Available from: https://doi.org/10.1007/s13042-016-0566-3.

Ouannas, A., Azar, A.T., Vaidyanathan, S., 2017g. A new fractional hybrid chaos synchronization. Int. J. Model. Identif. Control 27 (4), 314–322.

Ouannas, A., Azar, A.T., Ziar, T., 2017h. On Inverse Full State Hybrid Function Projective Synchronization for Continuous-time Chaotic Dynamical Systems with Arbitrary Dimensions. Differential Equations and Dynamical Systems. Available from: https://doi.org/10.1007/s12591-017-0362-x.

Ouannas, A., Azar, A.T., Ziar, T., Vaidyanathan, S., 2017i. On New Fractional Inverse Matrix Projective Synchronization Schemes. Studies in Computational Intelligence, Vol. 688. Springer-Verlag, Germany, pp. 497–524.

Ouannas, A., Azar, A.T., Ziar, T., Vaidyanathan, S., 2017j. Fractional Inverse Generalized Chaos Synchronization Between Different Dimensional Systems. Studies in Computational Intelligence, Vol. 688. Springer-Verlag, Germany, pp. 525–551.

Ouannas, A., Azar, A.T., Ziar, T., Radwan, A.G., 2017k. Study On Coexistence of Different Types of Synchronization Between Different dimensional Fractional Chaotic Systems. Studies in Computational Intelligence, Vol. 688. Springer-Verlag, Germany, pp. 637–669.

Ouannas, A., Azar, A.T., Ziar, T., Radwan, A.G., 2017l. Generalized Synchronization of Different Dimensional Integer-order and Fractional Order Chaotic Systems. Studies in Computational Intelligence, Vol. 688. Springer-Verlag, Germany, pp. 671–697.

Pan, L., Zhou, W., Zhou, L., Sun, K., 2011. Chaos synchronization between two different fractional-order hyperchaotic systems. Commun. Nonlinear Sci. Numer. Simulat. 16 (6), 2628–2640.

Park, J.H., Kwon, O.M., 2005. A novel criterion for delayed feedback control of time-delay chaotic systems. Chaos Solitons Fractals 23 (2), 495–501.

Pecora, L.M., Carroll, T.L., 1990. Synchronization in chaotic system. Phys. Rev. Lett. 64 (8), 821–824.

Petras, I., 2009. Chaos in the fractional-order Volta's system: modeling and simulation. Nonlinear Dyn. 57 (1-2), 157–170.

Petras, I., 2011. Fractional-Order Nonlinear Systems. Higher Education Press, Beijing and Springer-Verlag Berlin Heidelberg.

Podlubny, I., 1999. Fractional Differential Equations. Academic Press, New York.

Sheu, L.J., Chen, H.K., Chen, J.H., Tam, L.M., Chen, W.C., Lin, K.T., et al., 2008. Chaos in the Newton–Leipnik system with fractional order. Chaos Solitons Fractals 36 (1), 98–103.

Shuai, J.W., Wong, K.W., 1998. Noise and synchronization in chaotic neural networks. Phys. Rev. E 57 (6), 7002–7007.

Si, G., Sun, Z., Zhang, Y., Chen, W., 2012. Projective synchronization of different fractional-order chaotic systems with non-identical orders. Nonlinear Anal. Real World Appl. 13 (4), 1761–1771.

Singh, S., Azar, A.T., Ouannas, A., Zhu, Q., Zhang, W., Na, J., 2017. Sliding ModeControl Technique for Multi-switching Synchronization of Chaotic Systems. 9th International Conference on Modelling, Identification and Control (ICMIC 2017), July 10-12, 2017, Kunming, China.

Soliman, N.S., Said, L.A., Azar, A.T., Madian, A.H., Radwan, A.G., Ouannas, A., 2017. Fractional Controllable Multi-Scroll V-Shape Attractor with Parameters Effect. The 6th International Conference on Modern Circuits and Systems Technologies (MOCAST), 4-6 May 2017, Thessaloniki Greece.

Srivastava, M., Agrawal, S.K., Das, S., 2013a. Adaptive projective synchronization between different chaotic systems with parametric uncertainties and external disturbances. Pram. J. Phys. 81 (3), 417–437.

Srivastava, M., Das, S., Leung, A.Y.T., 2013b. Hybrid phase synchronization between identical and non-identical three dimensional chaotic systems using active control method. Nonlinear Dynam. 73 (4), 2261–2272.

Sun, H.H., Abdelwahed, A.A., Onaral, B., 1984. Linear approximation for transfer functions with a pole of fractional power. IEEE Trans. Autom. Control 29 (5), 441–444.

Tolba, M.F., AbdelAty, A.M., Said, L.A., Elwakil, A.S., Azar, A.T., Madian, A.H., et al., 2017a. FPGA Realization of Caputo and Grünwald-Letnikov Operators. The 6th International Conference on Modern Circuits and Systems Technologies (MOCAST), 4-6 May 2017 Thessaloniki Greece.

Tolba, M.F., AbdelAty, A.M., Soliman, N.S., Said, L.A., Madian, A.H., Azar, A.T., et al., 2017b. FPGA implementation of two fractional order chaotic systems, Int. J. Electr. Commun., 28. pp. 162–172.

Vaidyanathan, S., Sampath, S., Azar, A.T., 2015a. Global chaos synchronisation of identical chaotic systems via novel sliding mode control method and its application to Zhu system. Int. J. Model. Identif. Control (IJMIC) 23 (1), 92–100.

Vaidyanathan, S., Azar, A.T., Rajagopal, K., Alexander, P., 2015b. Design and SPICE implementation of a 12-term novel hyperchaotic system and its synchronization via active control (2015). Int. J. Model. Identif. Control (IJMIC) 23 (3), 267–277.

Vaidyanathan, S., Idowu, B.A., Azar, A.T., 2015c. Backstepping controller design for the global chaos synchronization of Sprott's Jerk Systems. In: Azar, A.T., Vaidyanathan, S. (Eds.), Chaos Modeling and Control Systems Design, Studies in Computational Intelligence, Vol. 581. Springer-Verlag GmbH, Berlin/Heidelberg, pp. 39–58. Available from: https://doi.org/10.1007/978-3-319-13132-0_3.

Vaidyanathan, S., Azar, A.T., 2015a. Anti-synchronization of identical chaotic systems using sliding mode control and an application to vaidyanathan-madhavan chaotic systems. In: Azar, A.T., Zhu, Q. (Eds.), Advances and Applications in Sliding Mode Control systems, Studies in Computational Intelligence book Series, Vol. 576. Springer-Verlag GmbH, Berlin/Heidelberg, pp. 527–547. Available from: https://doi.org/10.1007/978-3-319-11173-5_19.

Vaidyanathan, S., Azar, A.T., 2015b. Hybrid synchronization of identical chaotic systems using sliding mode control and an application to vaidyanathan chaotic systems. In: Azar, A.T., Zhu, Q. (Eds.), Advances and Applications in Sliding Mode Control systems, Studies in Computational Intelligence book Series, Vol. 576. Springer-Verlag GmbH, Berlin/Heidelberg, pp. 549–569. Available from: https://doi.org/10.1007/978-3-319-11173-5_20.

Vaidyanathan, S., Azar, A.T., 2015c. Analysis, control and synchronization of a nine-term 3-D novel chaotic system. In: Azar, A.T., Vaidyanathan, S. (Eds.), Chaos Modeling and Control Systems Design, Studies in Computational Intelligence, Vol. 581. Springer-Verlag GmbH, Berlin/Heidelberg, pp. 3–17. Available from: https://doi.org/10.1007/978-3-319-13132-0_1.

Vaidyanathan, S., Azar, A.T., 2015d. Analysis and control of a 4-D novel hyperchaotic system. In: Azar, A.T., Vaidyanathan, S. (Eds.), Chaos Modeling and Control Systems Design, Studies in Computational Intelligence, Vol. 581. Springer-Verlag GmbH, Berlin/Heidelberg, pp. 19–38. Available from: https://doi.org/10.1007/978-3-319-13132-0_2.

Vaidyanathan, S., Azar, A.T., 2016a. Takagi-Sugeno Fuzzy Logic Controller for Liu-Chen Four-Scroll Chaotic System. International Journal of Intelligent Engineering Informatics 4 (2), 135–150.

Vaidyanathan, S., Azar, A.T., 2016b. Dynamic Analysis, adaptive feedback control and synchronization of an eight-term 3-D novel chaotic system with three quadratic nonlinearities,

Studies in Fuzziness and Soft Computing, Vol. 337. Springer-Verlag, Germany, pp. 155–178.

Vaidyanathan, S., Azar, A.T., 2016c. Qualitative study and adaptive control of a novel 4-D hyperchaotic system with three quadratic nonlinearities, Studies in Fuzziness and Soft Computing, Vol. 337. Springer-Verlag, Germany, pp. 179–202.

Vaidyanathan, S., Azar, A.T., 2016d. A novel 4-D Four-wing chaotic system with four quadratic nonlinearities and its synchronization via adaptive control method, Advances in Chaos Theory and Intelligent Control. Studies in Fuzziness and Soft Computing, Vol. 337. Springer-Verlag, Germany, pp. 203–224.

Vaidyanathan, S., Azar, A.T., 2016e. Adaptive control and synchronization of halvorsen circulant chaotic systems, Advances in Chaos Theory and Intelligent Control. Studies in Fuzziness and Soft Computing, Vol. 337. Springer-Verlag, Germany, pp. 225–247.

Vaidyanathan, S., Azar, A.T., 2016f. Adaptive backstepping control and synchronization of a novel 3-D Jerk System with an exponential nonlinearity, Advances in Chaos Theory and Intelligent Control. Studies in Fuzziness and Soft Computing, Vol. 337. Springer-Verlag, Germany, pp. 249–274.

Vaidyanathan, S., Azar, A.T., 2016g. Generalized projective synchronization of a novel hyperchaotic four-wing system via adaptive control method, Advances in Chaos Theory and Intelligent Control. Studies in Fuzziness and Soft Computing, Vol. 337. Springer-Verlag, Germany, pp. 275–296.

Vaidyanathan, S., Azar, A.T., Ouannas, A., 2017a. An Eight-Term 3-D Novel Chaotic System with Three Quadratic Nonlinearities, its Adaptive Feedback Control and Synchronization. Studies in Computational Intelligence, Vol. 688. Springer-Verlag, Germany, pp. 719–746.

Vaidyanathan, S., Zhu, Q., Azar, A.T., 2017b. Adaptive control of a novel nonlinear double convection chaotic system, Studies in Computational Intelligence, Vol. 688. Springer-Verlag, Germany, pp. 357–385.

Vaidyanathan, S., Azar, A.T., Ouannas, A., 2017c. Hyperchaos and Adaptive Control of a Novel Hyperchaotic System with Two Quadratic Nonlinearities. Studies in Computational Intelligence, Vol. 688. Springer-Verlag, Germany, pp. 773–803.

Wang, S., Yu, Y., Diao, M., 2010. Hybrid projective synchronization of chaotic fractional order systems with different dimensions. Phys. A 389 (21), 4981–4988.

Wang, Z., Volos, C., Kingni, S.T., Azar, A.T., Pham, V.T., 2017. Four-wing attractors in a novel chaotic system with hyperbolic sine nonlinearity. Optik – Int. J. Light Electr. Optics 131 (2017), 1071–1078.

Wu, X., Wang, H., Lu, H., 2012. Modified generalized projective synchronization of a new fractional-order hyperchaotic system and its application to secure communication. Nonlinear Anal.: Real World Appl. 13 (3), 1441–1450.

Xiangjun, W., Yang, Y., 2010. Chaos in the fractional-order Qi system and its synchronization using active control, International Conference on Intelligent Computing and Cognitive Informatics.

Yadav, V.K., Bhadauria, B.S., Das, S., Singh, A.K., Srivastava, M., 2017a. Stability analysis, chaos control of fractional order chaotic chemical reactor system and its function projective synchronization with parametric uncertainties. Chin. J. Phys. 55 (3), 594–605.

Yadav, V.K., Prasad, G., Som, T., Das, S., 2017b. Combined synchronization of time-delayed chaotic systems with uncertain parameters. Chin. J. Phys. 55 (2), 457–466.

Yadav, V.K., Srikanth, N., Das, S., 2016. Dual function projective synchronization of fractional order complex chaotic systems. Optik 127 (22), 10527–10538.

Yu, H.J., Liu, Y.Z., 2003. Chaotic synchronization based on stability criterion of linear systems. Phys. Lett. A 314 (4), 292–298.

Zhou, P., Zhu, W., 2011. Function projective synchronization for fractional-order chaotic systems. Nonlinear Anal. Real World Appl. 12 (2), 811–816.

Zhu, Q., Azar, A.T., 2015. Complex system modelling and control through intelligent soft computations, Studies in Fuzziness and Soft Computing, Vol. 319. Springer-Verlag, Germany, ISBN: 978-3-319-12882-5.

FURTHER READING

Ouannas, A., Azar, A.T., Vaidyanathan, S., 2017f. On a simple approach for Q-S synchronization of chaotic dynamical systems in continuous-time. Int. J. Comput. Sci. Math. 8 (1), 20–27.

Chapter 13

On the Fractional Order Generalized Discrete Maps

Wafaa S. Sayed[1], Samar M. Ismail[2], Lobna A. Said[3] and Ahmed G. Radwan[1,3]

[1]*Faculty of Engineering, Cairo University, Giza, Egypt,* [2]*Faculty of Information Engineering and Technology (IET), German University in Cairo (GUC), Cairo, Egypt,* [3]*Nanoelectronics Integrated Systems Center (NISC), Nile University, Cairo, Egypt*

13.1 INTRODUCTION

During the last few decades, fractional calculus has become a powerful tool in describing the dynamics of complex systems. It has been emerged in many disciplines such as in control (Luo and Chen, 2012; Rajagopal et al., 2017; Azar et al., 2017a), analog oscillators (Said et al., 2016c,d, 2017), filters (Said et al., 2016a,b), bio-impedance modeling (AboBakr et al., 2017; Freeborn, 2013; Yousri et al., 2017), supercapacitor modeling (Elwakil et al., 2017), stability analysis (Radwan et al., 2009), and chaotic systems (Radwan et al., 2014; Tolba et al., 2017; Ouannas et al., 2017b,a,c).

The Grünwald–Letnikov's (GL) definition for the fractional derivative is described by (Podlubny, 1999):

$$D_x^{\alpha} f(x) = \lim_{h\to 0} \frac{1}{h^{\alpha}} \sum_{n=1}^{\infty} (-1)^i \frac{\Gamma(\alpha+1)}{\Gamma(\alpha-i+1)} f(x-ih), \tag{13.1}$$

where $(m-1) < \alpha < m$, and m is an integer. The GL definition is preferred in discrete applications. The fractional order (FO) parameter α adds extra degrees of freedom which increases the flexibility of any design. Numerical and analytical methods have been developed to study the FO differential equations (El-Sayed et al., 2004). The fractional discrete derivatives were investigated by Atici and Eloe (2009) and Holm (2011), offering a great opportunity to powerfully study the dynamics of discrete systems.

Chaos is an interesting nonlinear phenomenon that has grabbed a great attention in the last three decades (Radwan et al., 2016). Chaotic systems are highly sensitive to initial conditions, implementation and system parameters. A small deviation in any of the system parameters leads to a huge change in

Mathematical Techniques of Fractional Order Systems. DOI: https://doi.org/10.1016/B978-0-12-813592-1.00013-1

its behavior. These significant properties of chaotic systems are highly required in many applications such as: encryption (Ismail et al., 2015, 2017c), biological systems (Moaddy et al., 2012), communication (Chien and Liao, 2005; Dar et al., 2017), chemical reactions (Field and Schneider, 1989), oscillators (Radwan et al., 2003, 2004), and synchronization (Shukla and Sharma, 2017; Azar et al., 2017b; Sayed et al., 2016a; Henein et al., 2016; Sayed et al., 2017b).

Chaotic systems can appear in some continuous forms (Lü et al., 2002), as well as discrete forms (Ausloos and Dirickx, 2006). One of the famous continuous chaotic generators is the butterfly attractor introduced by Lorenz (Lorenz, 1963) which was the inspiration to many researchers to develop other multiscroll versions of this attractor (Elwakil and Özoguz, 2008; Yu et al., 2010; Zidan et al., 2012), The dynamics of several chaotic systems can be more accurately described by FO dynamical equations. Traditionally, the system order must be greater than three to exhibit a chaotic behavior. Yet, that changed with FO systems which were found to exhibit chaotic behavior for total order less than three (Jia et al., 2013; Jia, Lu and Chen, 2006; Soliman et al., 2017). New chaotic behaviors are obtained by generalizing the time derivatives of conventional systems to FO such as FO Lorenz, FO Chen, and FO Lu system (Petras, 2011). Numerical simulation of FO chaotic systems was presented by Petras (2011) along with a collection of the Matlab functions created for the some famous FO chaotic systems.

The best example to study the discrete forms of chaotic systems is the logistic map (Radwan, 2013). It has contributed to the modeling and information processing in many fields, such as theory of business cycle (Pellicer-Lostao and López-Ruiz, 2010), biology (Sutter and Pearl, 1946), chemistry (Malek and Gobal, 2000), and encryption (Ismail et al., 2015, 2017c; Pareek et al., 2006; Radwan et al., 2016). Through the years, modifications have been done to control the chaotic behavior of the logistic map either by changing the coefficients (Sayed et al., 2015b), powers (Radwan, 2013), or generalization into the FO domain (Wu and Baleanu, 2014; El Raheem and Salman, 2014). These modifications add degrees of freedom that give more control and flexibility to design specific maps.

The objective of this chapter is to summarize the generalized one-dimensional discrete maps such as the logistic map and the tent map in the integer order and FO domains. Different generalization techniques are presented to create extended maps. The complete bifurcation diagram is discussed throughout this chapter. Section 13.2 reviews the generalized logistic and tent maps in integer order forms with signed control parameters as well as scaling and shaping parameters. Section 13.3 introduces two forms of discrete FO logistic maps based on GL definition. The system dynamics (fixed points, stability conditions, and bifurcation points) are introduced with several examples for the two maps. Generalized bifurcation diagrams of one of the two studied maps are presented as well as different design examples are investigated with their Lyapunov exponents. Section 13.4 proposes the FO generalized tent map based on one of the two presented discrete FO logistic maps. Section 13.5 summarizes the main contributions of the chapter.

13.2 INTEGER ORDER GENERALIZED DISCRETE MAPS

Discrete one-dimensional chaotic maps are simple iterative or recurrence relations of the form $x_{n+1} = f(x_n)$, which is a function of a single variable and one or few parameters, which are easy to simulate or implement. The conventional equations of three famous discrete one-dimensional maps: the logistic and tent maps and their plots are given in Table 13.1.

Successive values of the iterated variable x_n at discrete time instants n are evaluated using the recurrence relation, starting from a given initial point x_0, to yield a sequence of values called the orbit or discrete time series of the map. The limit of the sequence or its steady state as n approaches infinity may reach a single value, periodic orbit, or theoretically aperiodic orbit (or chaos) depending on the value of the control parameter. A bifurcation diagram is a plot of this steady state solution versus the control parameter(s) of the map. The sudden appearance of a qualitatively different solution for a system as some parameter is varied is called bifurcation and occurs at bifurcation points. The bifurcation diagram and bounded ranges of the parameters and solution of the maps are also given in Table 13.1. For a discrete one-dimensional map $f(x)$, Maximum Lyapunov Exponent (MLE) is given by:

$$\mathrm{MLE} = \lim_{n\to\infty}\left(\frac{1}{n}\sum_{i=0}^{n-1}\ln|f'(x_i)|\right), \tag{13.2}$$

where ln is the natural logarithm. Positive MLE value is an indication that the system exhibits chaotic behavior. The MLE versus the control parameter is shown in Table 13.1.

Novel discrete maps have been proposed in several works, e.g., Ablay (2016), Alpar (2014), Chaves et al. (2016), and Panchuk et al. (2015). Generalized and extended forms of the conventional maps have also been proposed in other research (da Costa et al., 2017; Elhadj and Sprott, 2008; He et al., 2008; Levinsohn et al., 2012; Matthews, 1989; Nejati et al., 2008; Ruan et al., 2004; Vázquez-Medina et al., 2009). Generalizations can make up for the insufficiency of conventional discrete one-dimensional maps for several applications. A set of generalized logistic and tent maps have been recently proposed and utilized in encryption applications. These maps depend mainly on the introduction of extra parameters, which enhance the cryptographic properties of the conventional maps and add extra degrees of freedom. The simplest modification does not include any extra parameters, since it just defines a signed control parameter and considers the whole bifurcation diagram of the map (Sayed et al., 2015b, 2016b, 2017c). The extra parameters

TABLE 13.1 Conventional Logistic and Tent Maps and Their Properties

	Logistic Map	Tent Map
Equation	$x_{n+1} = \lambda x_n(1 - x_n)$	$x_{n+1} = \begin{cases} \mu x_n, & x_n \leq 0.5 \\ \mu(1 - x_n), & x_n \geq 0.5 \end{cases}$
Output range	$x_n \in [0, 1]$	$x_n \in [0, 1]$
Parameter range	$\lambda \in [0, 4]$	$\mu \in [0, 2]$
Time series	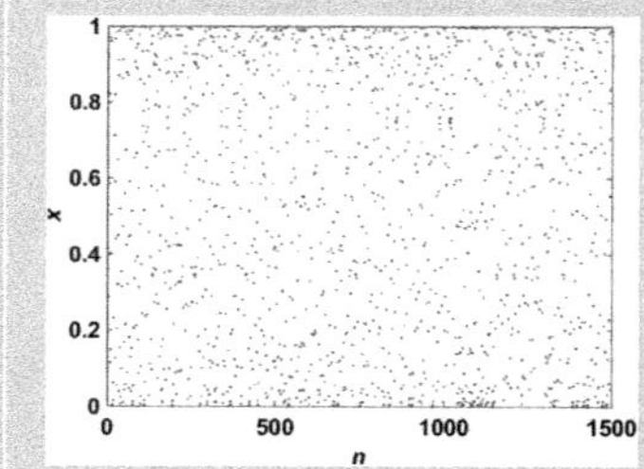	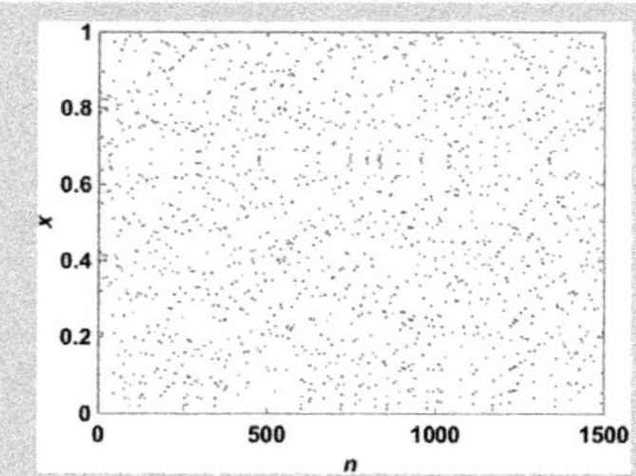
Bifurcation diagram	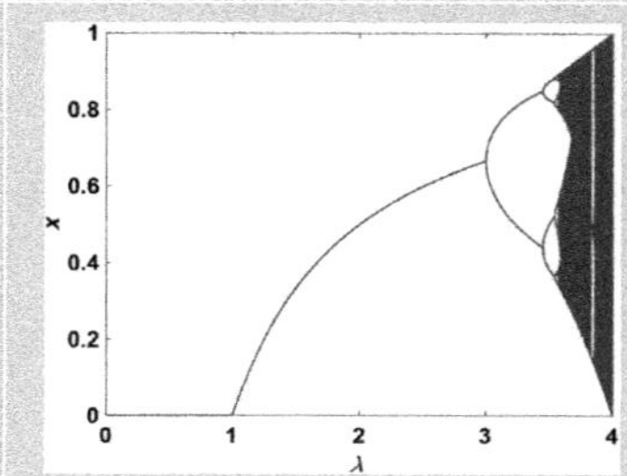	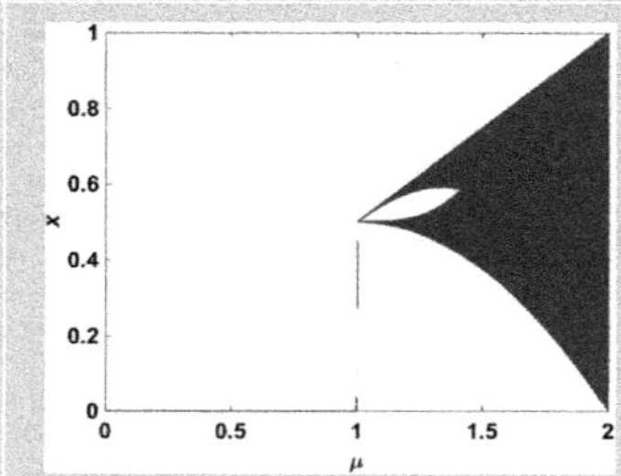

MLE	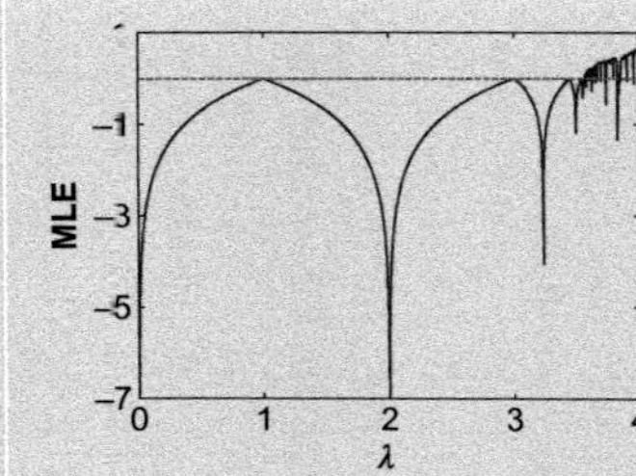	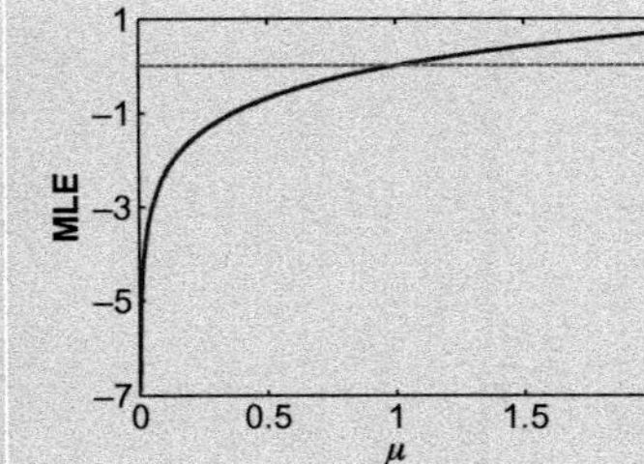

are introduced in the map equations in two different ways. While the parameters introduced using additions and multiplications (Radwan and Abd-El-Hafiz, 2013; Sayed et al., 2015a,b) are called scaling parameters, the parameters introduced using powering (Abd-El-Hafiz et al., 2015; Radwan, 2013; Radwan and Abd-El-Hafiz, 2013) are called shaping parameters. Moreover, a recently introduced map utilizes both ways simultaneously (Sayed et al., 2017a). The following subsections summarize the properties and advantages of these generalized one-dimensional discrete maps.

13.2.1 Maps With Signed Control Parameter

The complete bifurcation diagram of the logistic map is barely mentioned in the well-known chaos textbooks. Yet, in a few researches, it was briefly discussed (Ambadan and Joseph, 2006; Bresten and Jung, 2009; Gutiérrez and Iglesias, 1998; Shimada et al., 2016; Tsuchiya and Yamagishi, 1997) with the claim of having no real-world applications. The use of negatively valued parameters in logistic and tent maps has been recently analyzed by Sayed et al. (2016b, 2015b). The new parameter range provides a controlling capability resulting in a wider output range. The capability of generating chaos from convex maps, in addition to the well-known concave ones, can be utilized in chaos-based communication systems that require multiple modes of operation.

The complete picture of the behavior of the logistic map (Sayed et al., 2015b) is given by the same equation in Table 13.1, but defined for the range $\lambda \in [-2, 4]$ as shown in Fig. 13.1A. Similarly, the complete bifurcation diagram of the tent map (Sayed et al., 2016b) for the range $\mu \in [-2, 2]$ is shown in Fig. 13.1B. For the logistic map, the mathematical analysis required to get the bounds on the parameter λ_{min} and output x_{min-} and x_{max-} of the map shown in Fig. 13.2 is simple and summarized in the following steps. The minimum occurs at the critical point x_c, i.e., $x_{min-} = f(x_c)$ which is solved simultaneously with the two nonlinear equations $x_{\max-} = f(x_{\max-})$ and $x_{\max-} = f(x_{min-})$ to get the three unknowns. Similar analysis can be conducted for the tent map with signed parameter and the other convex maps discussed in this chapter. The left part of the bifurcation diagram, in which the parameter is negative, is called mostly positive map. The name is chosen because the output range has an alternating sign and the wider range exhibits positive values. The ordinary right part, which corresponds to conventional maps, is called positive map. Negative and mostly negative maps were also presented by Sayed et al. (2015b, 2016b).

13.2.2 Maps With Scaling Parameters

The asymmetric nature was added to the logistic and tent maps through the scaling parameters, which control the critical or intersection point. This

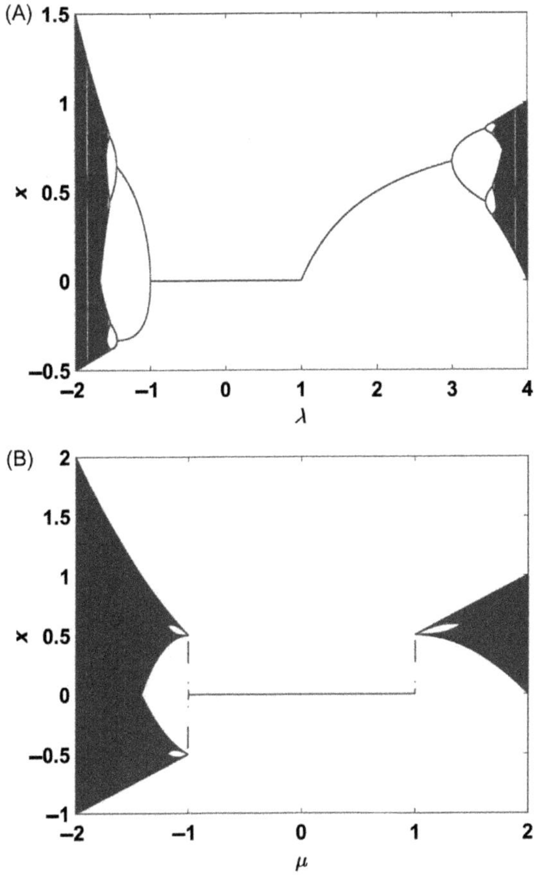

FIGURE 13.1 Bidirectional bifurcation diagram of (A) logistic map and (B) tent map.

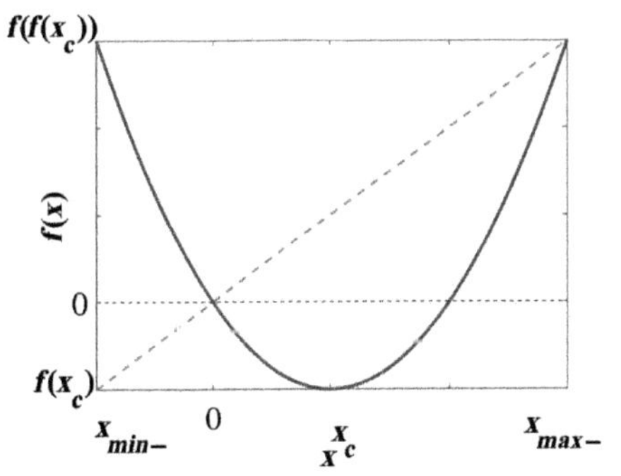

FIGURE 13.2 Methodology of calculating the bounds of the logistic map with negative control parameter.

corresponds to more degrees of freedom required in the coding and communication applications with asymmetric transmission probabilities of binary zeros and ones (Sayed et al., 2015a,b).

13.2.2.1 Generalized Logistic Map With Scaling Parameters

The generalized logistic map with scaling parameters (Sayed et al., 2015b) is given by:

$$f(x) = \lambda x(a - bx), \tag{13.3}$$

Fig. 13.3A and B show the graphs of the map equation in which the output ranges are given in terms of the scaling parameters a and b for $\lambda > 0$ and $\lambda < 0$, respectively.

Fig. 13.4A shows the general schematic of the bidirectional bifurcation diagram of the generalized logistic map with scaling parameters. The figure shows the main bifurcation points and the ranges of the parameter λ and the output x. The effective range of the parameter λ, in which the output is bounded, depends on the scaling parameter a in an inverse proportionality relation. The output range depends on both scaling parameters, where it widens as the value of the parameter a increases and/or the value of the parameter b decreases. These effects of the scaling parameters are better described as special cases, where the map with $a = 1$ is called "vertical

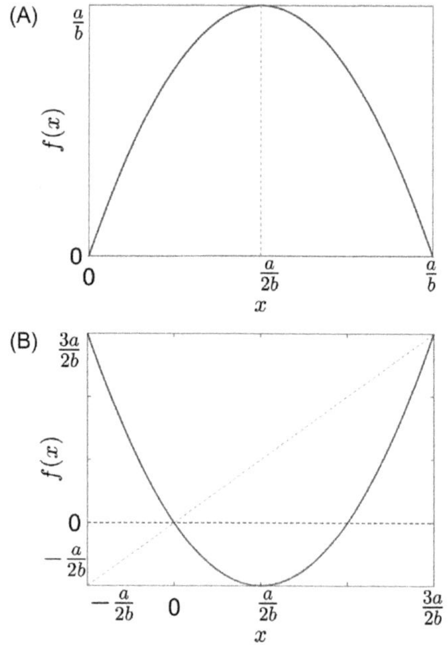

FIGURE 13.3 (A) Positive logistic map and (B) mostly positive logistic map with scaling parameters.

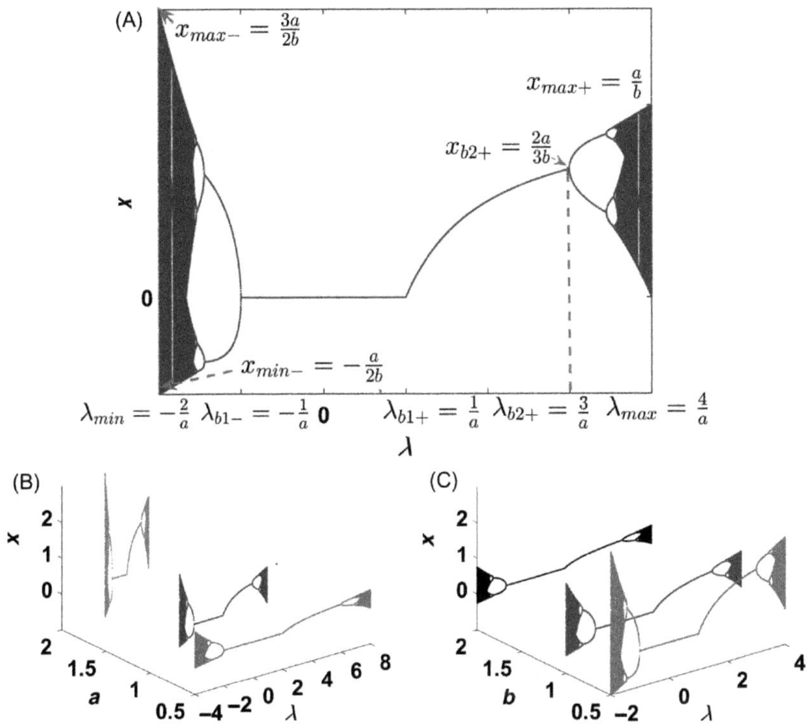

FIGURE 13.4 (A) General bifurcation diagram of the generalized logistic map and 3D snapshots (B) at $b = 1$ and $a = \{0.5, 1, 2\}$ and (C) at $a = 1$ and $b = \{0.5, 1, 2\}$.

scaling map" because only the vertical axis is affected by the parameter b, while the map with $b = 1$ is called "zooming map" because both axes are affected by the parameter a. These effects can be further inferred from the three-dimensional snapshots of bifurcation diagrams against the main system parameter λ for different values of the scaling parameters a and b, which are shown in Fig. 13.4B and C, respectively.

13.2.2.2 Generalized Tent Map With Scaling Parameters

Similarly, the generalized tent map with scaling parameters (Sayed et al., 2015a) is given by:

$$f(x) = \begin{cases} \mu\ x, & x \leq \dfrac{a}{b+1} \\ \mu(a - bx), & x > \dfrac{a}{b+1} \end{cases}. \tag{13.4}$$

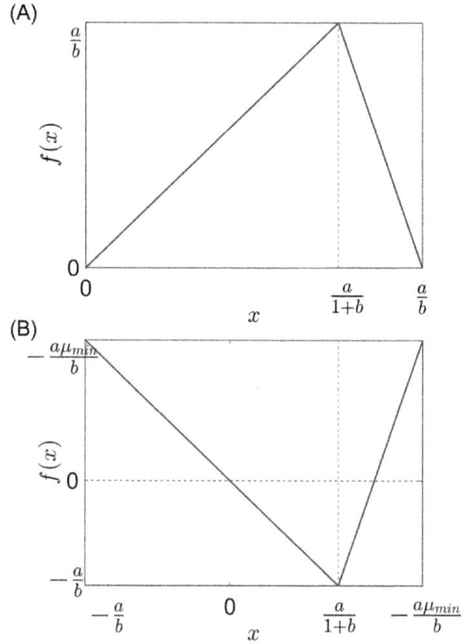

FIGURE 13.5 (A) Positive tent map and (B) mostly positive tent map with scaling parameters, where $\mu_{min} = -\left(1 + \frac{1}{b}\right)$.

Figs. 13.5 and 13.6 show the graphs of the different map versions and their bifurcation diagrams. The diagram changes its shape as the value of the parameter b exceeds 1. The dependence of the range of the output x on the scaling parameters is similar to the generalized logistic map. However, the effective range of the parameter μ depends on the scaling parameter b, instead of a, in an inverse proportionality relation. These effects can be further inferred from the three-dimensional snapshots of bifurcation diagrams shown in Fig. 13.7. The zooming and vertical scaling special cases can be achieved as well, but for the generalized tent map they occur at $a = 1$ and $b = 1$, respectively.

13.2.3 Maps With Shaping Parameters

Effects of other generalized parameters (in the form of arbitrary powers) on the behavior of the logistic map have also been studied in (Radwan, 2013); presenting three special cases. One of these cases is similar to one of the modified maps presented in (Levinsohn et al., 2012), yet, (Levinsohn et al., 2012) confined the powering parameter to positive integers greater than one only. The analyses follow slightly different steps, yet the results are consistent for the parameter ranges which are common in both papers. In addition,

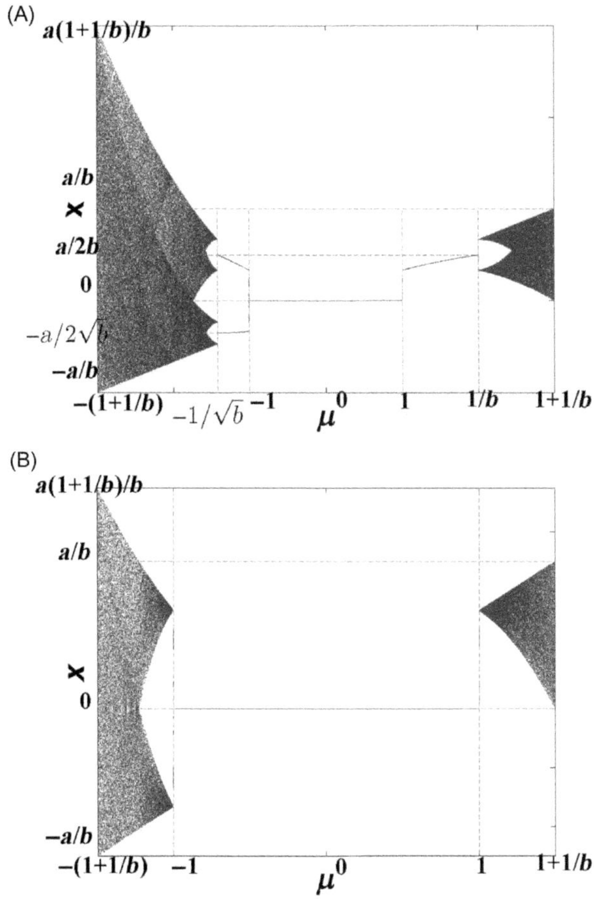

FIGURE 13.6 General schematic of the bidirectional bifurcation diagram of the generalized tent map with scaling parameters a and b in both sides of μ (A) $b<1$ and (B) $b>1$.

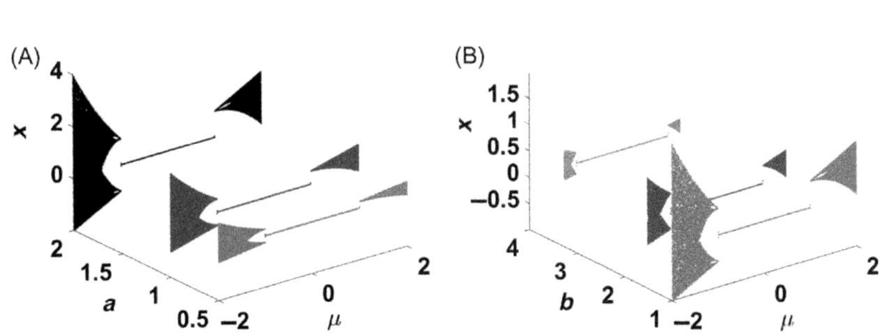

FIGURE 13.7 3D snapshots of the bifurcation diagrams of the generalized tent map (A) at $b=1$ and $a=\{0.5,1,2\}$ and (B) at $a=1$ and $b=\{1,2,4\}$.

Radwan (2013) has studied further properties such as the bifurcation diagram versus the powering parameter, allowing both integer and fractional values of this parameter.

13.2.3.1 Logistic Map With Shaping Parameters

Three different generalizations of the logistic map with arbitrary powers α and β have been proposed by Radwan (2013), which are given by:

$$x_{n+1} = \lambda x_n^{\alpha}(1 - x_n^{\beta}), \tag{13.5}$$

where (α, β) can equal one of three cases (α, α), $(\alpha, 1)$, or $(1, \alpha)$. The bifurcation diagrams of the generic case (α, α) are shown in Fig. 13.8.

13.2.3.2 Tent Map With Shaping Parameters

A similar generalization for the tent map has been introduced by Radwan and Abd-El-Hafiz (2013), which is given by either of the following forms.

$$x_{n+1} = min\left(\mu x_n^{\alpha}, \mu\left(1 - x_n^{\alpha}\right)\right), \qquad \mu, \alpha \in R^{+} \tag{13.6a}$$

$$x_{n+1} = \begin{cases} \mu x_n^{\alpha} & x \leq x_k \\ \mu(1 - x_n^{\alpha}) & x_k < x \end{cases}, \tag{13.6b}$$

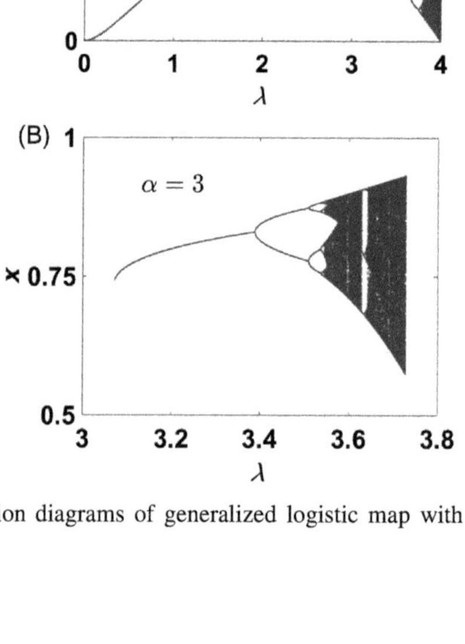

FIGURE 13.8 Bifurcation diagrams of generalized logistic map with arbitrary power for (A) $\alpha < 1$ and (B) $\alpha > 1$.

TABLE 13.2 The Properties of Generalized Tent Map With Arbitrary Power

Property	Value
Intersection point	$x_k = (0.5)^{\frac{1}{\alpha}}$
Fixed points	$x_1^* = 0$, $x_2^* = (1/\mu)^{\frac{1}{\alpha-1}}$, and $x_3^* + \mu x_3^{*\alpha} - \mu = 0$
Bifurcation points	$(\mu_b, x_b) = \left(\left(\frac{1}{\alpha}\right)(\alpha+1)^{1-1/\alpha}, \left(\frac{1}{\alpha+1}\right)^{1/\alpha}\right)$
Bifurcation diagram versus μ	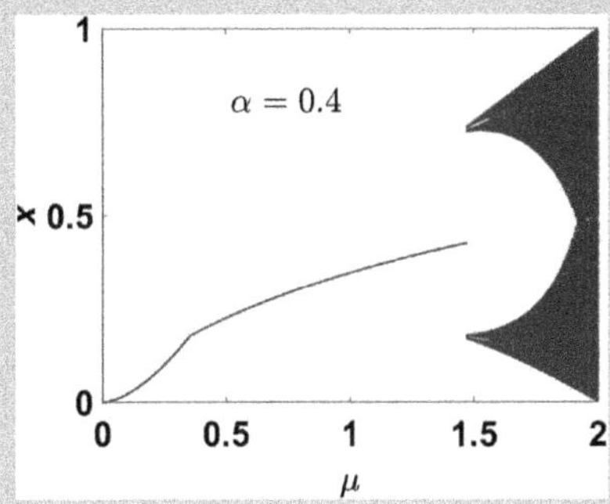

where $x_k = (0.5)^{\frac{1}{\alpha}}$. The properties of the map proposed by Radwan and Abd-El-Hafiz (2013) are summarized in Table 13.2.

13.2.4 Maps With Both Scaling and Shaping Parameters

The idea of the transition map (Sayed et al., 2017a) and its generalizations emerged from the conjugacy between piece-wise linear tent map and quadratic logistic map (Alligood et al., 1996). In addition, both maps exhibit period doubling bifurcation as a route to chaos. The difference in the degree of the relations defining both maps suggests employing powering parameters to get a unified relation. Scaling parameters add extra controllability to the generalized map given by:

$$f(x, r, a, b, \alpha, \beta) = r\,min((\text{sgn}(b)x)^{\alpha}(a-bx)^{\beta}, (\text{sgn}(b)x)^{\beta}(a-bx)^{\alpha}), \qquad (13.7)$$

Fig. 13.9A shows the bifurcation diagram of the generalized map, while Fig. 13.9B shows the variation of the main key-points with the parameter β at fixed values of the other parameters. Mathematical analyses and numerical simulations were conducted by Sayed et al. (2017a) providing general formulae for the main key-points in terms of the map parameters.

13.3 FRACTIONAL ORDER GENERALIZED LOGISTIC MAP

This section combines the generalizations based on signed control parameter and scaling parameters with the FO logistic map. Discrete logistic maps in

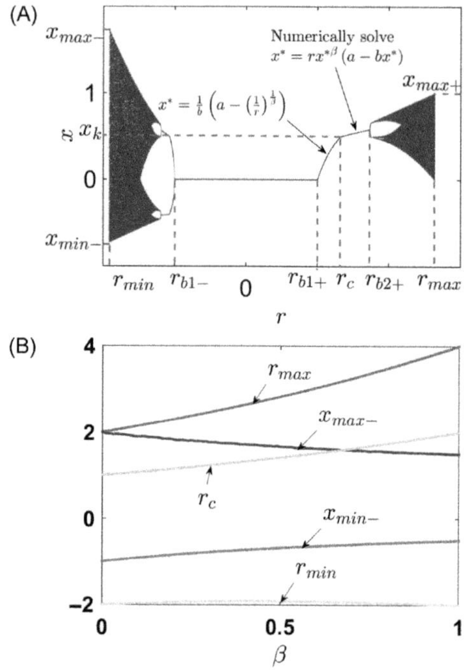

FIGURE 13.9 (A) General bifurcation diagram of the transition map and (B) key-points of bifurcation diagrams versus β at $\alpha = a = b = 1$.

the FO domain were introduced earlier by Wu and Baleanu (2014) and by El Raheem and Salman (2014). They both used the GL definition to reach a FO version out of the conventional logistic map. The recurrence relations of both versions are shown in Table 13.3, with their corresponding bifurcation diagrams. It is obvious that the introduced model in Wu and Baleanu (2014) uses the summation form which implies long memory dependency related to the order ν, while the other model in El Raheem and Salman (2014) is based on a simple approximated discretization scheme. This is the main reason why this work focuses on the latter FO logistic map (El Raheem and Salman, 2014), because of its ease of implementation compared to the other form. Fig. 13.10 shows the complete bifurcation diagram for the positive and negative values of the system parameter for the conventional logistic map in Fig. 13.10A, and FO ones proposed in Wu and Baleanu (2014) and El Raheem and Salman (2014), in Fig. 13.10B and C, respectively.

13.3.1 Dynamics of the Wu Fractional Logistic Map

The FO logistic map introduced by Wu (Wu and Baleanu, 2014) is described in Table 13.3. The intrinsic population growth rate is symbolized by the

TABLE 13.3 Discrete Logistic Equations

Paper	Fractional order logistic map (Wu and Baleanu, 2014)	Fractional order logistic map (El Raheem and Salman, 2014)
Recurrence relation	$x_{n+1} = x(0) + \frac{\mu}{\Gamma(v)}\sum_{j=n-m}^{n}\frac{\Gamma(n-j+v)}{\Gamma(n-j+1)}x_j(1-x_j)$	$x_{n+1} = x_n + \frac{r^{\alpha}}{\Gamma(1+\alpha)}\rho x_n(1-x_n)$
Bifurcation diagram	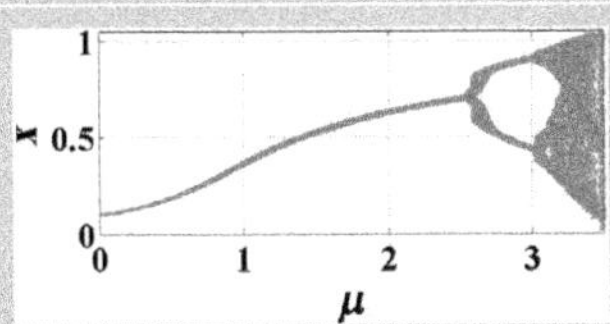	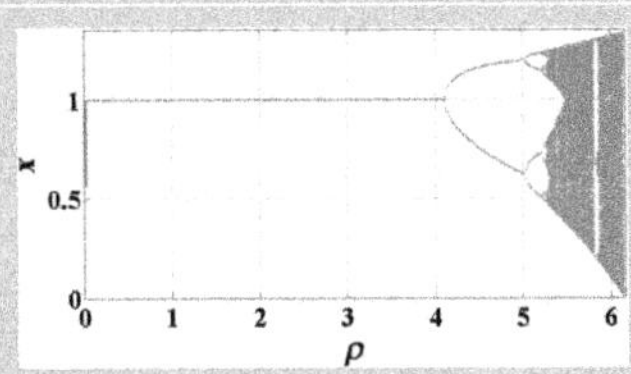
	Ex:- Bifurcation for $v = 0.03$	Ex:- Bifurcation for $(\alpha, r) = (0.6, 0.25)$

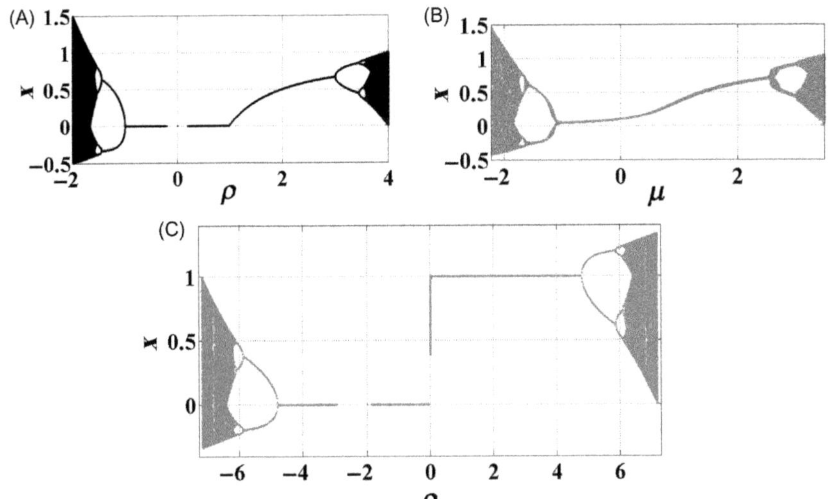

FIGURE 13.10 Complete logistic bifurcation diagrams of the (A) conventional map, (B) fractional order map defined in (Wu and Baleanu, 2014) when $v = 0.02$, and (C) fractional order map defined in (El Raheem and Salman, 2014) when $\alpha = 0.7, r = 0.25$.

parameter μ,while ν is the FO parameter, and $\Gamma(.)$ is the gamma function. The equation is based on a summation form, where all the previous values of the population sizes are added to the memory of the system according to the value of m. This equation is reduced to the famous logistic map if $m = 0$, and the bifurcation diagram in Fig. 13.11A shows that.

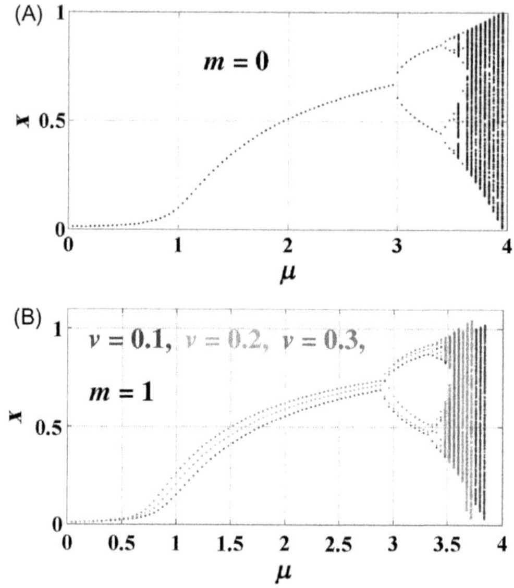

FIGURE 13.11 Bifurcation diagram for (A)$m = 0$, $\nu = 0.5$, and (B) $m = 1$ versus different ν.

For $m = 1$, one delay element will be added to the normal map transforming the map into a 2D-map as follows:

$$x_{n+1} = x(0) + \mu x_n(1 - x_n) + \mu\nu x_{n-1}(1 - x_{n-1}). \tag{13.8}$$

13.3.1.1 Fixed Points Analysis

For the ease of mathematical analysis of this second order map (13.8), it is first converted to a first order one. A new variable is introduced to be $y_{n+1} = x_n$. Thus, the difference equation can be expressed as the following pair:

$$x_{n+1} = x(0) + \mu x_n(1 - x_n) + \mu\nu y_n(1 - y_n), \tag{13.9a}$$

$$y_{n+1} = x_n. \tag{13.9b}$$

The fixed points are calculated by solving the equation $F(x^*, y^*) = (x^*, y^*)$, where the fixed points of this map are:

$$(x^*, y^*) = \left(\frac{1}{2}\left(-k + 1 + \sqrt{(k-1)^2 - 4kx(0)}\right),\right.$$

$$\left.\frac{1}{2}\left(-k + 1 + \sqrt{(k-1)^2 - 4kx(0)}\right)\right), \tag{13.10}$$

where $k = \frac{1}{\mu(1+\nu)}$ and the solution with the minus sign is rejected.

The system behavior changes according to the value of μ, as well as the FO ν. Fixing $\nu = 0.1$, in Table 13.4, the first row shows that the fixed point is stable, and any response will settle to the fixed point value (13.10). Increasing the value of μ, the system starts to develop damped oscillations then settle again to the fixed point value as shown in the second row. As μ increases further in a range $3 < \mu < 3.3$, the fixed point loses its stability and the map starts to bifurcate and oscillates between two values known as orbit 2 or period 2 as

TABLE 13.4 Transient Responses and Phase Plane Portraits of Wu Fractional Order Logistic Map

<table>
<tr><th></th><th>Transient Response</th><th>Phase Plane Portrait</th></tr>
<tr><td>Fixed point</td><td>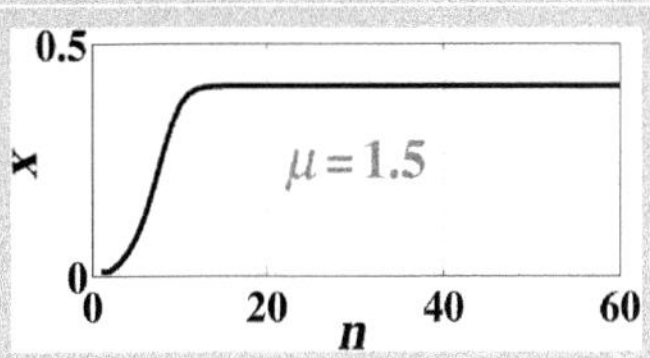
</td><td>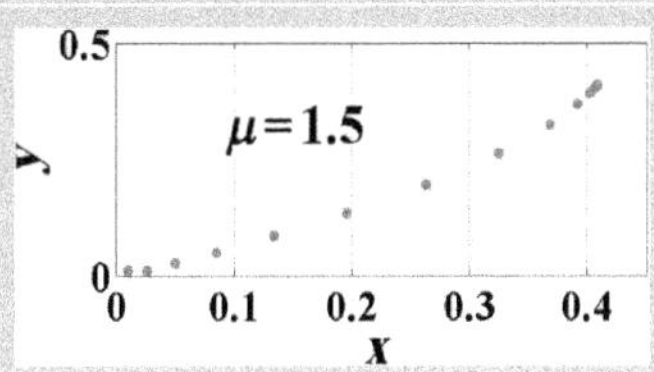
</td></tr>
<tr><td>Damped oscillations</td><td>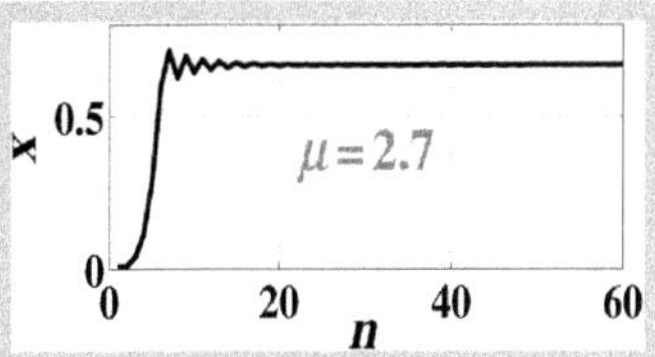
</td><td>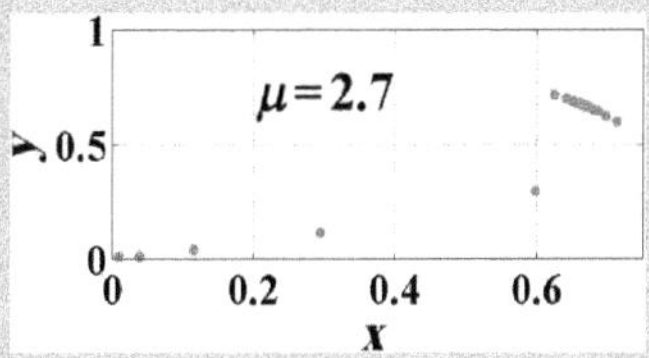
</td></tr>
<tr><td>Period-2</td><td>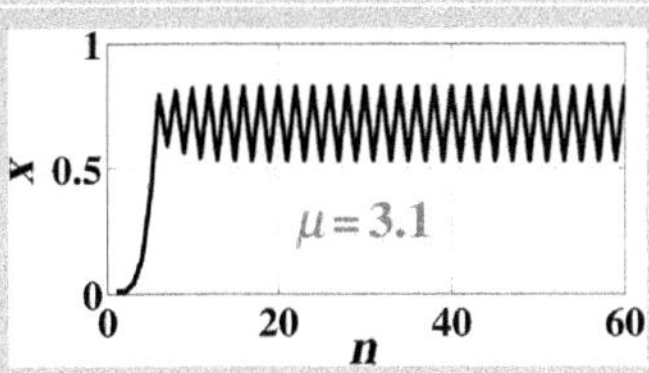
</td><td>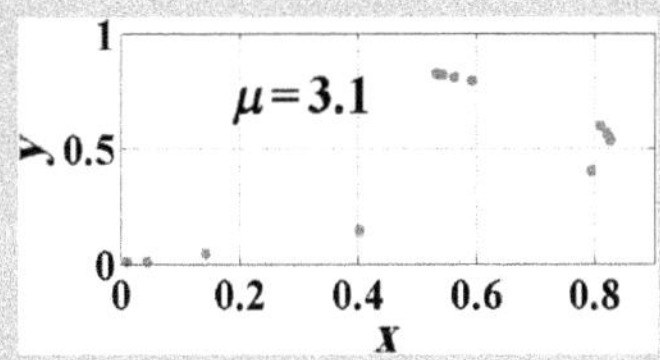
</td></tr>
<tr><td>Period-4</td><td>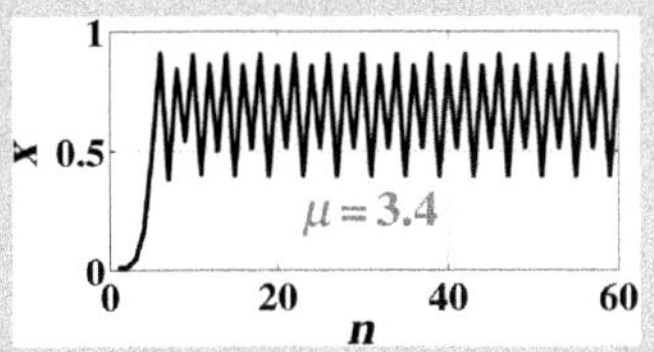
</td><td>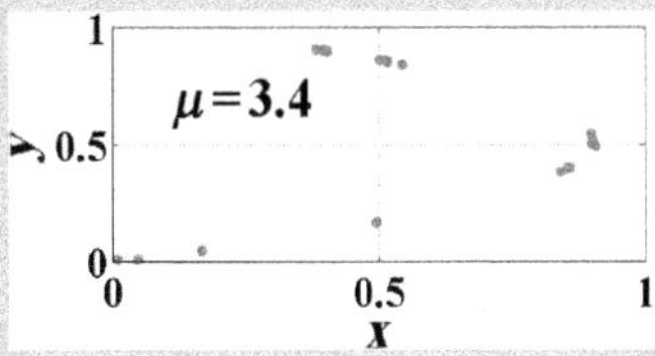
</td></tr>
<tr><td>Chaos</td><td>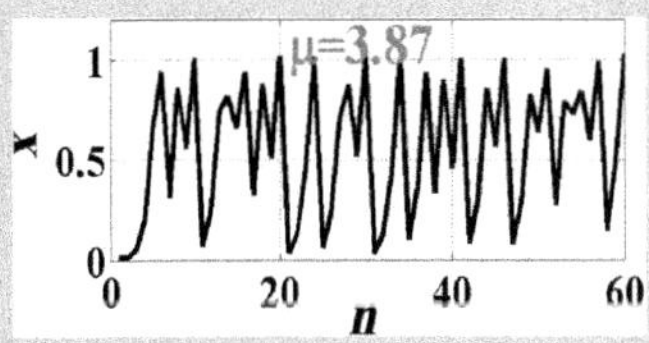
</td><td>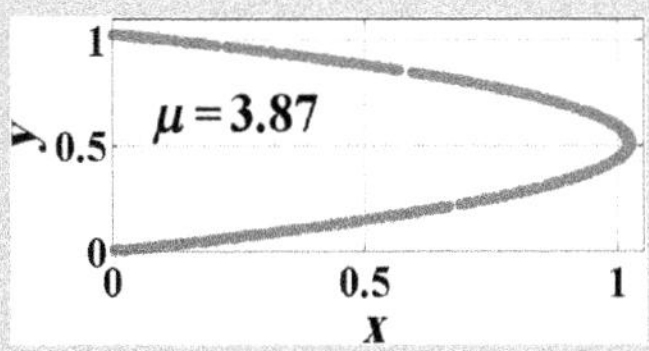
</td></tr>
</table>

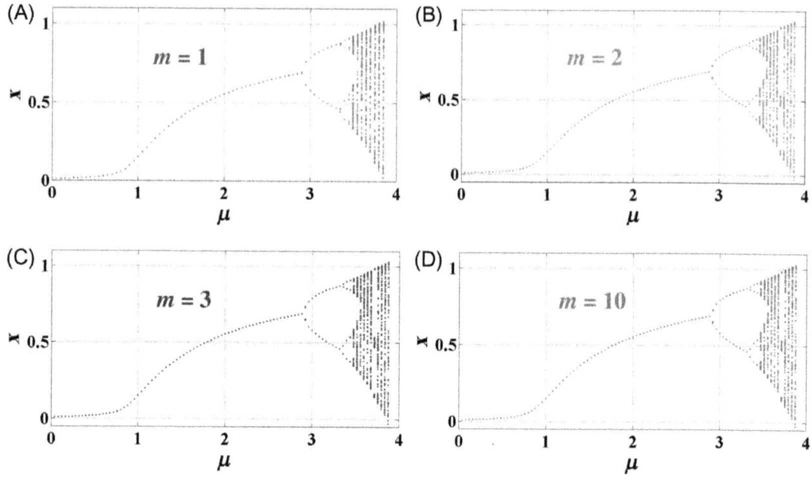

FIGURE 13.12 Bifurcation diagram for $\nu = 0.1$ for (A) $m = 1$, (B) $m = 2$, (C) $m = 3$, and (D) $m = 10$.

shown in the third row of Table 13.4. The fourth row shows that the system will increase its bifurcation to period 4 at $\mu = 3.4$ and with further increase for the parameter μ, bifurcation increases until reaching the chaotic behavior at $\mu = 3.87$. Table 13.4 shows the corresponding phase plane portraits corresponding to each transient response in each range of μ. These different responses are all described by the bifurcation diagram shown in Fig. 13.12A.

The effect of the FO parameter ν is depicted in Fig. 13.11B, for $m = 1$. Changing the value of ν controls the chaotic range of the map. The figure shows that the chaotic behavior is reached at a lower value with increasing ν. Fig. 13.12 presents the effect of adding extra terms from the summation form to the logistic equation, and shows the corresponding bifurcation diagram for each case, e.g., for $m = 2$, $m = 3$,and $m = 10$.

From the previous discussion, it is obvious that the FO logistic map (Wu and Baleanu, 2014) is a bit complicated because of its memory dependence, so the second form of the FO map (El Raheem and Salman, 2014) was adopted for the generalization technique as will be detailed in the next section.

13.3.2 Dynamics of Generalized Discrete Fractional Logistic Map

The proposed generalized fractional logistic equation (Ismail et al., 2017b,a, 2015) is described by:

$$x_{n+1} = x_n + \frac{r^{\alpha}}{\Gamma(1+\alpha)} \rho x_n (a - b x_n) = f(x, r, \rho, \alpha, a, b), \quad (13.11)$$

where a, b are the generalization parameters and $\rho, \alpha \in R^+$. The general fractional logistic map is to be analyzed as a function of r, ρ, α, a and b. The range of ρ, the maximum and the minimum values of the function, x_{max} and x_{min}, respectively, the bifurcation point ρ_b, as well as the value of the function at the bifurcation point x_b are investigated below.

13.3.2.1 Fixed Points and Range of ρ

The fixed points of the map are defined as the points where $x^* = f(x^*, r, \rho, \alpha, a, b)$.

$$x^* = x^* + \frac{r^\alpha}{\Gamma(1+\alpha)} \rho x^* (a - b x_n). \tag{13.12}$$

Therefore, $\rho\, x^*(a - bx^*) = 0$. This results in two fixed points $x_1^* = 0$ and $x_2^* = \frac{a}{b}$. To analyze the complete bifurcation diagram, for both negative and positive values of ρ, the equation is written as follows:

$$x_n + \frac{r^\alpha |\rho|}{\Gamma(1+\alpha)} x_n (a - b x_n) > 0. \tag{13.13}$$

Consider $|k| = \frac{r^\alpha |\rho|}{\Gamma(1+\alpha)}$, and thus $x_n < \frac{a}{b} + \frac{1}{b|k|}$. This results in:

$$x_{max} = \frac{a}{b} + \frac{1}{b|k|}. \tag{13.14}$$

The critical point x_c is the point at which the function has a maximum, it is calculated by equating the derivative of the function $\frac{df(.)}{dx} = 1 + |k|(a - 2bx_c)$ to 0. Thus the value of x_c is obtained:

$$x_c = \frac{a}{2b} + \frac{1}{b|k|} = \frac{x_{max}}{2}. \tag{13.15}$$

It can be seen that $x_c = \frac{x_{max}}{2}$. By substituting by x_c in (13.5), the resultant value should be less than x_{max}; $\frac{a}{2b} + \frac{1}{4b|k|} + \frac{a^2|k|}{4b} < \frac{a}{b} + \frac{1}{b|k|}$.

Therefore, $|k| < \frac{3}{a}$. Substituting by the value of k gives $|\rho| < \frac{3\Gamma(1+\alpha)}{ar^\alpha}$, then ρ can be obtained within a range of $\frac{-3\Gamma(1+\alpha)}{ar^\alpha} < \rho < \frac{3\Gamma(1+\alpha)}{ar^\alpha}$.

$$\rho_{min} = \frac{-3\Gamma(1+\alpha)}{ar^\alpha}. \tag{13.16}$$

$$\rho_{max} = \frac{3\Gamma(1+\alpha)}{ar^\alpha}. \tag{13.17}$$

The maximum value of the function for the positive side bifurcation can be obtained at ρ_{max}:

$$x_{max+} = \frac{a}{b} + \frac{1}{bk} = \frac{4a}{3b}. \tag{13.18}$$

13.3.2.2 Stability Analysis

Stability is studied at each fixed point of the map by finding the first derivative of the function. The fixed points are stable if $|f'(x^*, r, \rho, \alpha, a, b)| < 1$, and a saddle point if$|f'(x^*, r, \rho, \alpha, a, b)| > 1$. The first derivative of the function is given by:

$$f'(x_n, r, |\rho|, \alpha, a, b) = 1 + \frac{r^\alpha}{\Gamma(1+\alpha)}\rho(a - 2bx_n). \tag{13.19}$$

At $|(f'(x^*, r, |\rho_b|, \alpha, a, b)| = 1$, the bifurcation takes place.
At $x_1^* = 0$:
$|1 + \frac{r^\alpha}{\Gamma(1+\alpha)}|\rho_b|a| = 1$. This results in the relation:

$$-2 < \frac{r^\alpha \rho_b a}{\Gamma(1+\alpha)} < 0. \tag{13.20}$$

Hence, for the negative side:

$$\rho_{b-} = \frac{-2\Gamma(1+\alpha)}{ar^\alpha}. \tag{13.21}$$

It is obvious that $\rho_{b-} = \frac{2}{3}\rho_{min}$. Substituting with the value of ρ_{b-}, yields x_{b-}; the function value at the bifurcation point:

$$x_{b-} = f(x_1^*, r, \rho_{b-}, \alpha, a, b) = 0. \tag{13.22}$$

At $x_2^* = \frac{a}{b}$:
$|1 + \frac{r^\alpha}{\Gamma(1+\alpha)}|\rho_b|(a - 2a)| = 1$. This results in the relation:

$$0 < \rho_b < \frac{2\Gamma(1+\alpha)}{ar^\alpha}. \tag{13.23}$$

Therefore, for the positive side:

$$\rho_{b+} = \frac{2\Gamma(1+\alpha)}{ar^\alpha} = \frac{2}{3}\rho_{max}. \tag{13.24}$$

It is obvious that $\rho_{b+} = \frac{2}{3}\rho_{max}$. Substituting with the value of ρ_{b+}, yields the function value at the bifurcation point x_{b+} as:

$$x_{b+} = f(x_2^*, r, \rho_{b+}, \alpha, a, b) = \frac{a}{b} = \frac{3x_{max+}}{4}. \tag{13.25}$$

In conclusion to this analysis, the bifurcation diagram is symmetric with respect to zero. The difference between the minimum value of the function and the maximum value of the function in the positive side is equal to x_{max+}. For the negative side, it is the same difference but shifted downwards, yielding a value of x_{max-} which is equal to x_{b+}, while x_{min} is equal to $\frac{x_{max-}}{3}$.

The relations between the previously derived design parameters are graphically summarized on the complete bifurcation diagram of the proposed

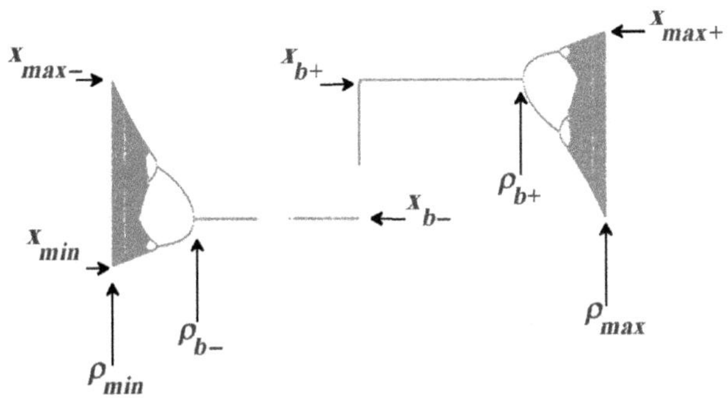

FIGURE 13.13 Complete bifurcation diagram parameters of the fractional order logistic map.

TABLE 13.5 Design Parameters Summary for the Proposed Fractional Logistic Maps

Parameters	Vertical Map	Zooming Map	General Map
ρ_{max}	$\frac{3\Gamma(1+\alpha)}{r^{\alpha}}$	$\frac{3\Gamma(1+\alpha)}{ar^{\alpha}}$	$\frac{3\Gamma(1+\alpha)}{ar^{\alpha}}$
ρ_{min}	$\frac{-3\Gamma(1+\alpha)}{r^{\alpha}}$	$\frac{-3\Gamma(1+\alpha)}{ar^{\alpha}}$	$\frac{-3\Gamma(1+\alpha)}{ar^{\alpha}}$
ρ_{b+}	$\frac{2\Gamma(1+\alpha)}{r^{\alpha}}$	$\frac{2\Gamma(1+\alpha)}{ar^{\alpha}}$	$\frac{2\Gamma(1+\alpha)}{ar^{\alpha}}$
ρ_{b-}	$\frac{-2\Gamma(1+\alpha)}{r^{\alpha}}$	$\frac{-2\Gamma(1+\alpha)}{ar^{\alpha}}$	$\frac{-2\Gamma(1+\alpha)}{ar^{\alpha}}$
x_{max+}	$\frac{4}{3b}$	$\frac{4a}{3}$	$\frac{4a}{3b}$
x_{max-}	$\frac{1}{b} = \frac{3x_{max+}}{4}$	$a = \frac{3x_{max+}}{4}$	$\frac{a}{b} = \frac{3x_{max+}}{4}$
x_{min}	$\frac{-1}{3b}$	$\frac{-a}{3}$	$\frac{-a}{3b} = \frac{-x_{max-}}{3}$
x_{b+}	$\frac{1}{b}$	a	$\frac{a}{b}$
x_{b-}	0	0	0

map in Fig. 13.13. Table 13.5 sums up all the design parameters relations for the proposed generalized FO logistic map, along with two special cases that will be discussed next, the vertical scaling map and the zooming map.

The changes of ρ_{max} and ρ_b versus the parameter a are displayed in Fig. 13.14A, showing an inversely proportionality relation, while the variations of both parameters are shown in Fig. 13.14B versus α, having a direct proportionality with α. Fig.13.14C-D show how x_{max} and x_{min} change versus the parameters b and a, respectively. For the generalized fractional logistic map under study, the effect of the FO parameter α on MLE is depicted in Fig. 13.15.

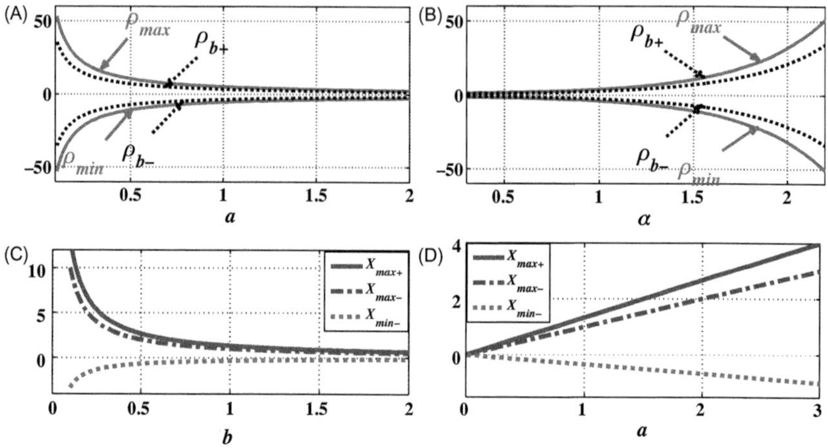

FIGURE 13.14 Changes of ρ_{max}, ρ_{min} and ρ_b (A) versus a, $\alpha = 0.5$, (B) versus $\alpha, a = 3$, changes of x_{max+}, x_{max-} and x_{min-} (C) versus b and (D) versus a.

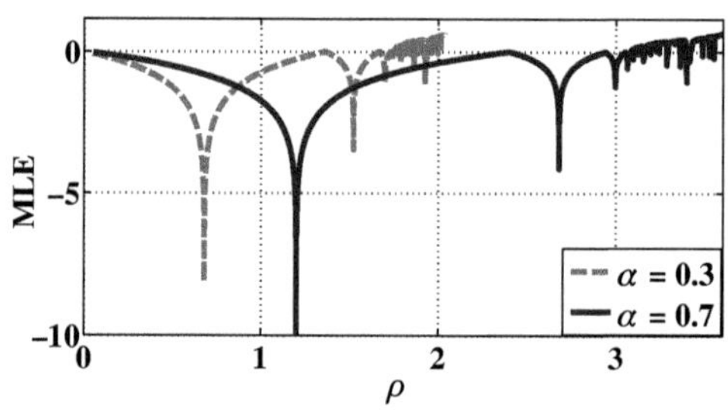

FIGURE 13.15 Lyapunov exponent with different α for $a = b = 2$.

13.3.3 Vertical Scaling and Zooming Maps

Two special maps, vertical scaling and zooming, can be obtained through using the scaling parameters as explained in Section 13.2.2.1. The dynamics analyses equations of both maps are summarized in Table 13.5, while Table 13.6 gives a detailed summary for various simulation analyses for both special maps. The first row of Table 13.6 shows different function iterations, $f^m(x, r, \rho, \alpha, 1, b)$ for the vertical scaling map, and $f^m(x, r, \rho, \alpha, a, 1)$, for the zooming map. The first and the second iterations are shown for the positive values of ρ, while the third and fifth iterations are shown for the negative values of ρ. Table 13.6, second row shows the effect of the FO parameter α, on the chaotic range of both maps. The effect of the generalization parameters (a) or (b) on the chaotic behavior is displayed in the third row of Table 13.6. New bifurcation diagrams versus the FO parameter α are also

TABLE 13.6 Special Maps Simulation Analysis Summary

<table>
<tr><th></th><th>Vertical Scaling Map ($a = 1$)</th><th>Zooming Map ($b = 1$)</th></tr>
<tr><td>Function iterations $\alpha = 0.5$</td><td>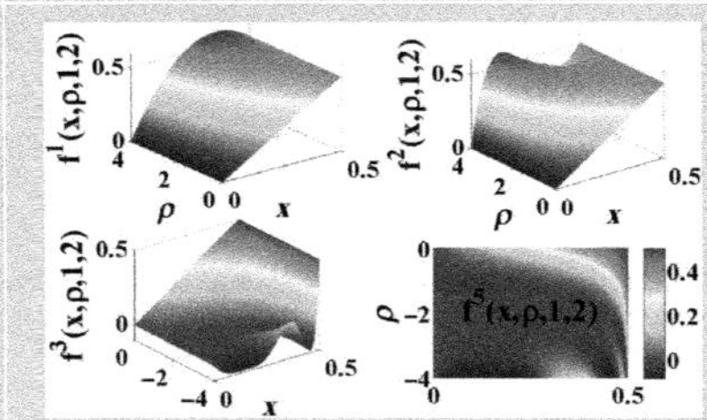
</td><td>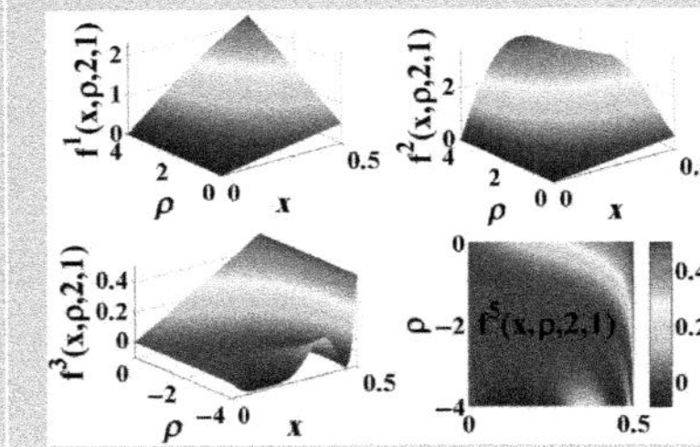
</td></tr>
<tr><td>Effect of α</td><td>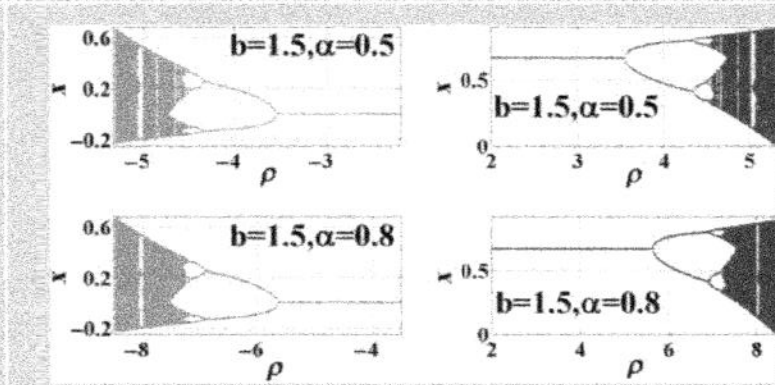
</td><td>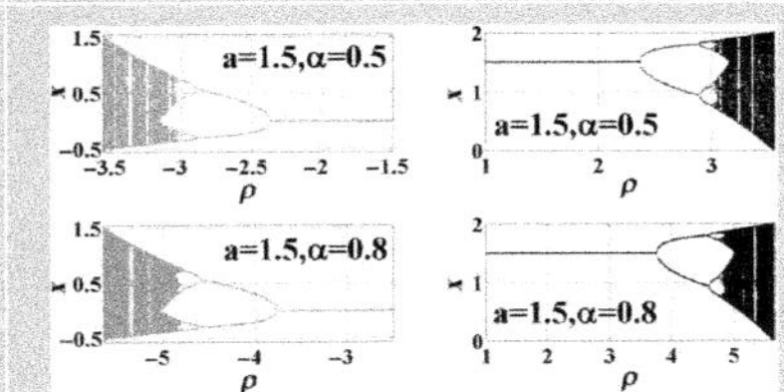
</td></tr>
<tr><td>Effect of generalization parameter $\frac{a}{b}$</td><td>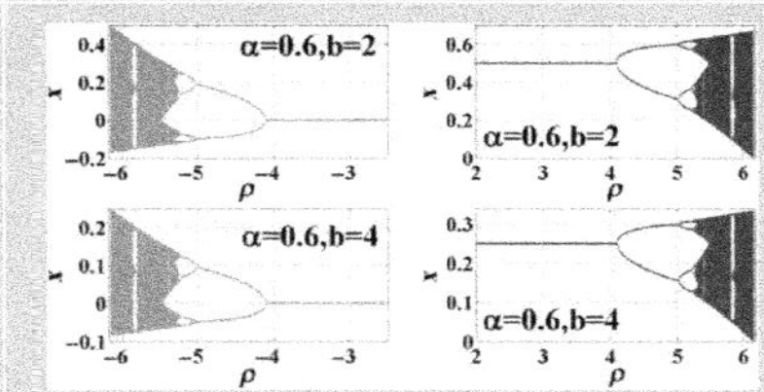
</td><td>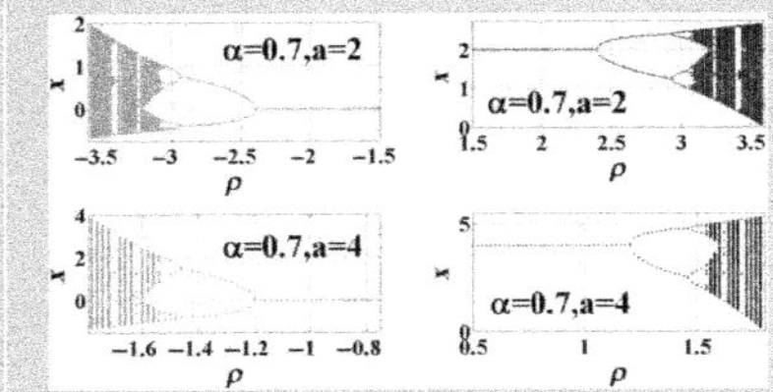
</td></tr>
</table>

(*Continued*)

TABLE 13.6 (Continued)

	Vertical Scaling Map ($a = 1$)	Zooming Map ($b = 1$)
Bifurcation versus α	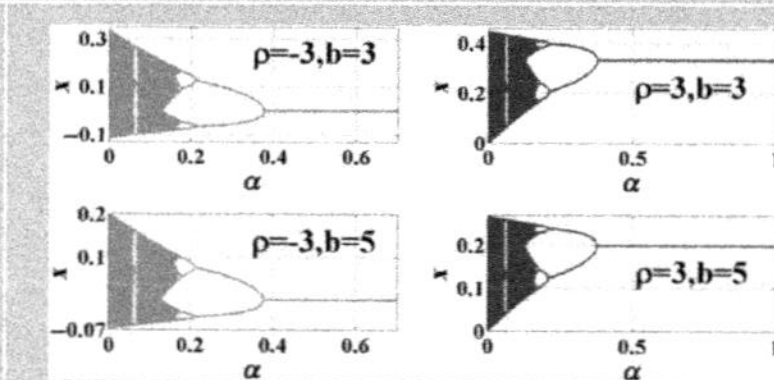	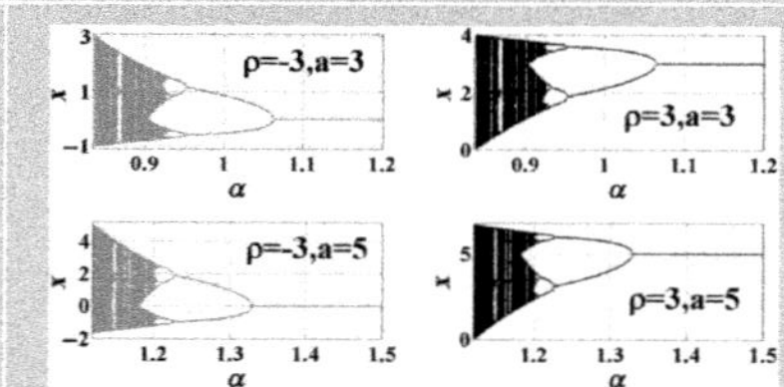
3D Snapshots $\rho-\alpha$	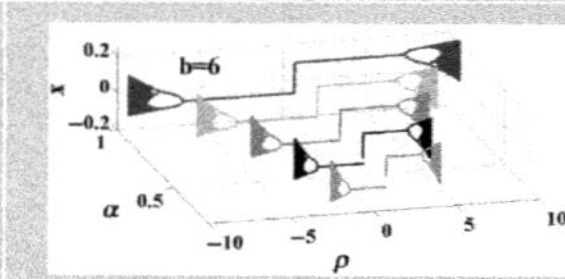	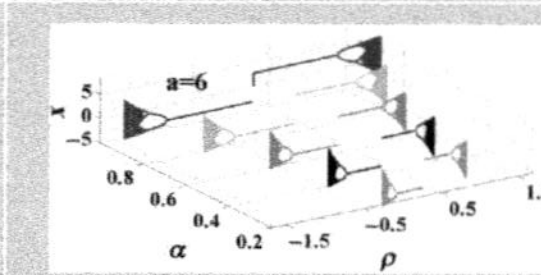
3D-Snapshots $\rho-b$ or $\rho-a$	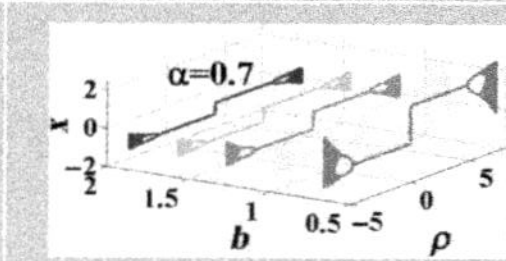	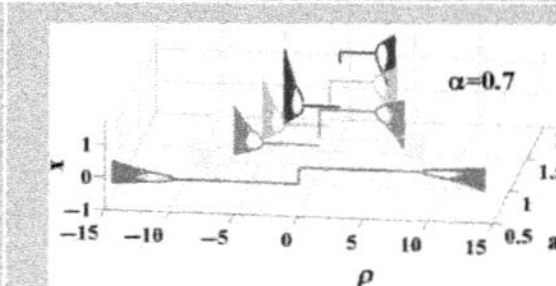

introduced in the fourth row. The last two rows present 3D snapshots of the complete bifurcation diagrams for both maps.

13.3.4 Generalized Fractional Logistic Map Design

Using the generalization parameters (a, b) introduced as well as the FO parameter α, the chaotic range of the map can be fully controlled to achieve a certain bifurcation diagram, allowing different map designs. The function value at which bifurcation takes place x_b at ρ_b can be specified; as well as x_{max} at ρ_{max}; or x_{min}, at ρ_{min}, in both sides positive and negative bifurcations, and hence calculate the values of the generalization parameters (a, b), using the equations derived in Table 13.5, to achieve the specifications required.

Four design cases are introduced and summarized in Table 13.7, with the design specifications, the corresponding bifurcation diagrams, as well as the Lyapunov exponent graph verifying the chaotic region.

13.4 FRACTIONAL ORDER GENERALIZED TENT MAP

This section focuses on the Wu definition (Wu and Baleanu, 2014) of the fractional logistic map and extends it to present the FO tent map depending on the following theorem.

Theorem 1: For the delta fractional difference equation (Chen et al., 2011)

$$^{C}\Delta_a^{\nu}u(t) = f(t+\nu-1, u(t+\nu-1)), \tag{13.26}$$

$$\Delta^k u(a) = u_k, \quad m = [\nu]+1, k = 0, \ldots, m-1, \tag{13.27}$$

the equivalent discrete integral equation can be obtained as

$$u(t) = u_0(t) + \frac{1}{\Gamma(\nu)} \sum_{s=a+m-\nu}^{t-\nu} (t-\sigma(s))^{(\nu-1)} f(s+\nu-1, u(s+\nu-1)), \qquad t \in N_{a+m}, \tag{13.28}$$

where the initial iteration $u_0(t)$ reads

$$u_0(t) = \sum_{k=0}^{m-1} \frac{(t-a)^{(k)}}{k!} \Delta^k u(a). \tag{13.29}$$

TABLE 13.7 Fractional Logistic Map Design Cases With MLE

	Negative Bifurcation		Positive Bifurcation	
	Design I	Design II	Design III	Design IV
	$\rho_{min} = -6$	$\rho b- = -2$	$\rho b = 3$	$\rho_{max} = 6$
	$x_{\min} = -2$	$x_{min} = -4$	$x_{\max} = 6$	$x_b = 3$
	$\alpha = 0.7$	$\alpha = 0.8$	$\alpha = 0.5$	$\alpha = 0.6$
Bifurcation	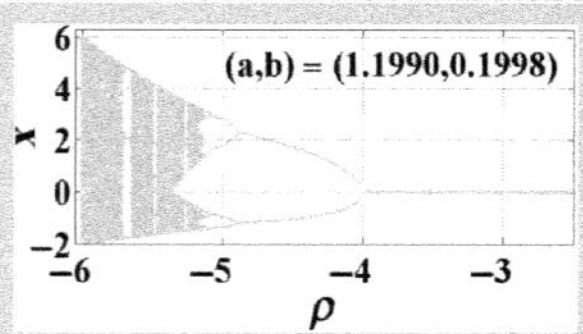	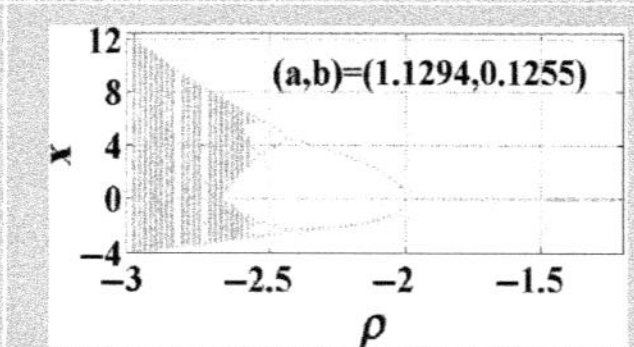	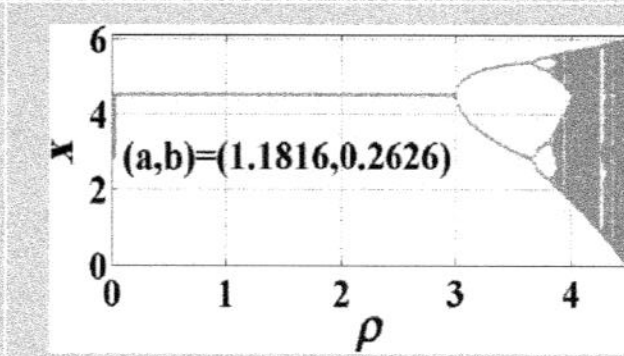	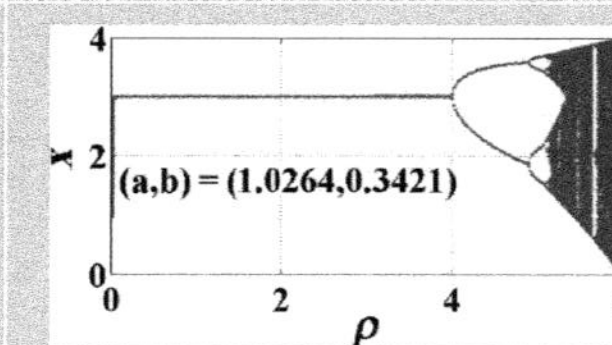
MLE	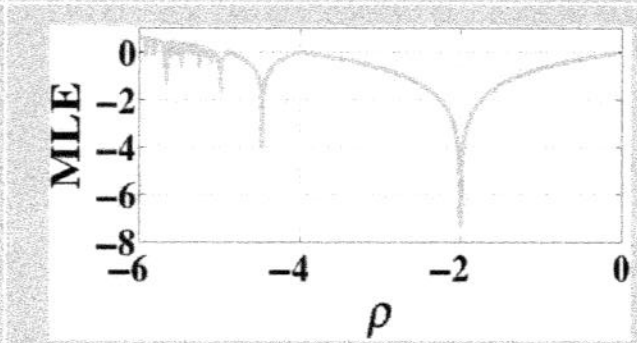	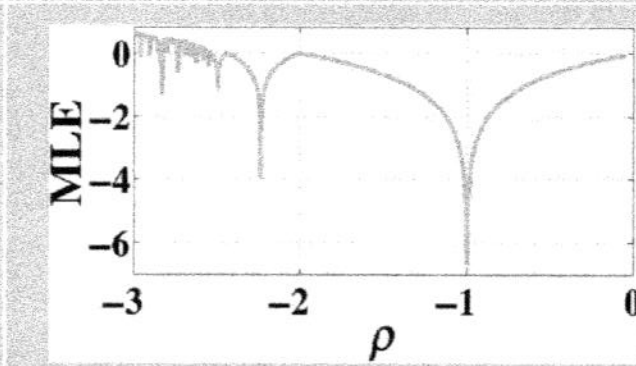	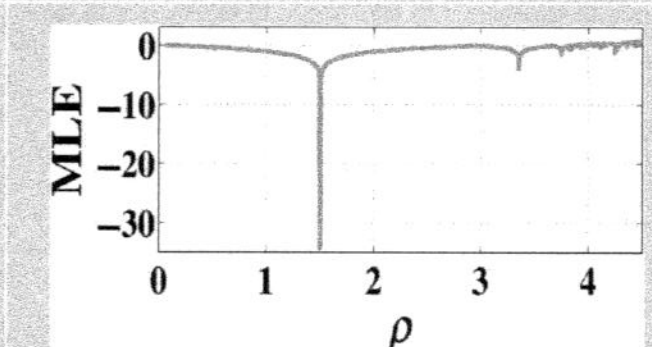	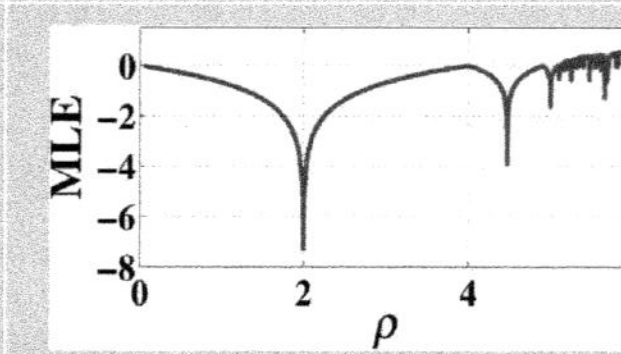

Substituting for f as the generalized tent map with scaling parameters we get the FO generalized tent map according to the Wu definition given by:

$$x_{n+1} = x(0) + \frac{\mu}{\Gamma(\nu)} \sum_{j=n-m}^{n} \frac{\Gamma(n-j+\nu)}{\Gamma(n-j+1)} min\left(x_j, a - bx_j\right). \tag{13.30}$$

13.4.1 Effects of Main System Parameters

The behavior of unity scaling FO tent map at $a = b = 1$ changes according to the value of the system parameter μ, as well as the FO ν. Their effects can be described similar to the FO logistic map. The bifurcation diagram against μ, fixing $\nu = 0.5$, which is shown in Fig. 13.16 describes the responses and

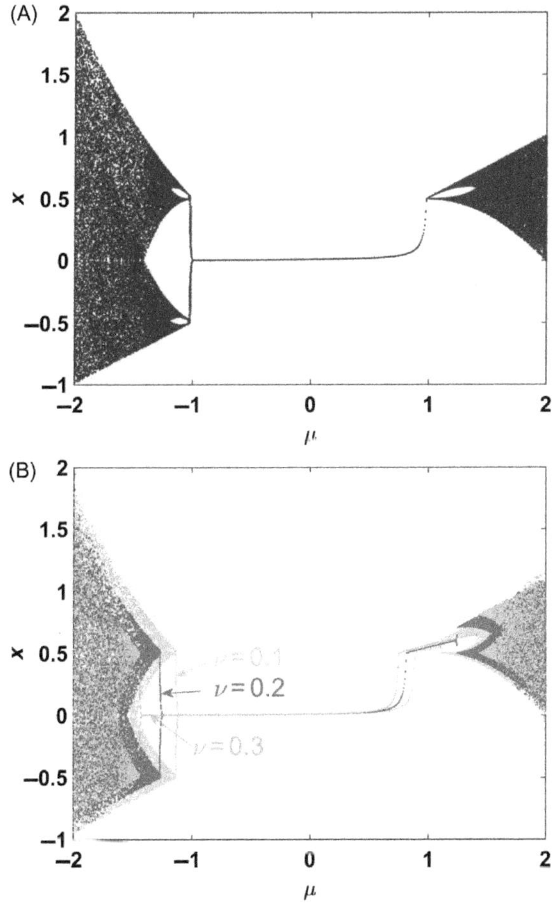

FIGURE 13.16 Bifurcation diagram for (A) $m = 0$, $\nu = 0.5$, and (B) $m = 1$ versus different ν.

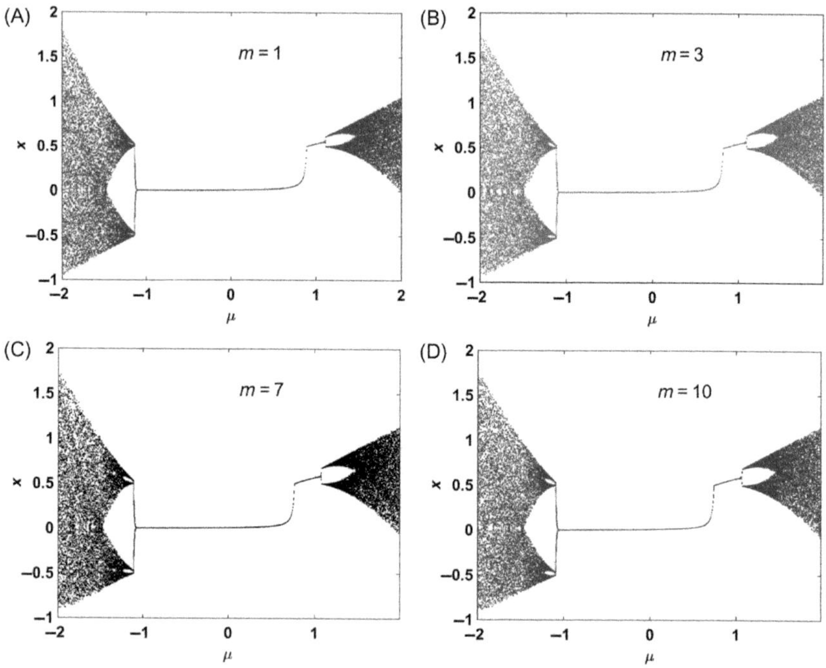

FIGURE 13.17 Bifurcation diagram for $\nu = 0.1$ for (A) $m = 1$, (B) $m = 3$, (C) $m = 7$ and (D) $m = 10$.

the route to chaos. Fig. 13.16 describes the effect of the FO parameter ν, for $m = 1$, where chaos is achieved at a lower value of μ with increasing ν. Fig. 13.17 presents the effect of the parameter m on the bifurcation diagram for $m = 1$, $m = 3$, $m = 7$, and $m = 10$.

13.4.2 Effects of Scaling Parameters

The scaling parameters a and b result in two special maps similar to the FO generalized logistic map, but their effect is reversed. Vertical scaling and zooming maps can be obtained at $b = 1$ and $a = 1$, respectively. The shape of the bifurcation diagram changes as b exceeds 1. Similar results were reported for the integer order generalized tent map with scaling parameters in Section 13.2.2.2. Table 13.8 gives a detailed summary for various simulations for both special maps. The effects of the scaling parameters a or b on the chaotic behavior is illustrated using $m = 2$ and $\nu = 0.3$. Three-dimensional snapshots of the complete bifurcation diagrams are also presented for both maps at different values of the scaling parameters.

TABLE 13.8 Effects of the Scaling Parameters on the Behavior of Fractional Order Generalized Tent Maps

	Vertical Scaling Map ($b = 1$)	Zooming Map ($a = 1$)
Effect of generalization parameter $\frac{a}{b}$	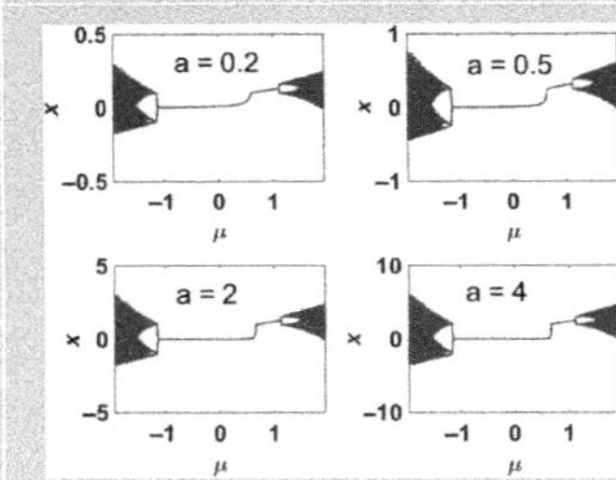	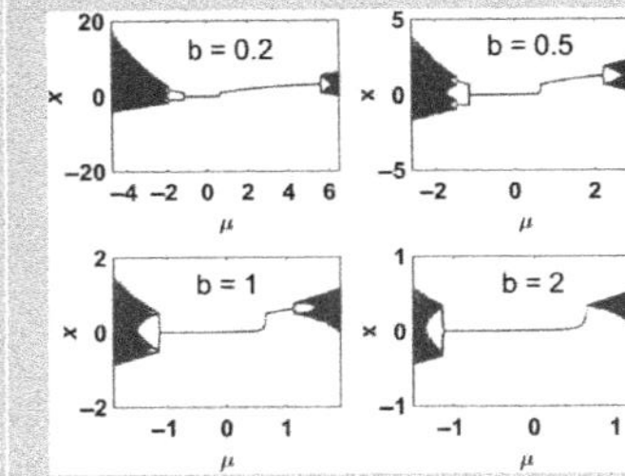
3D-Snapshots $\mu - a$ or $\mu - b$	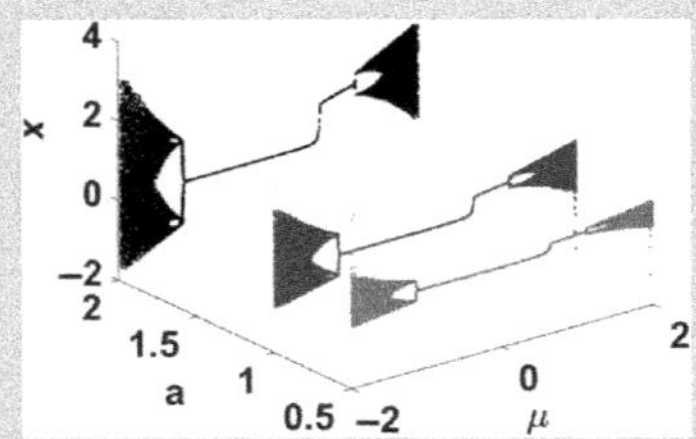	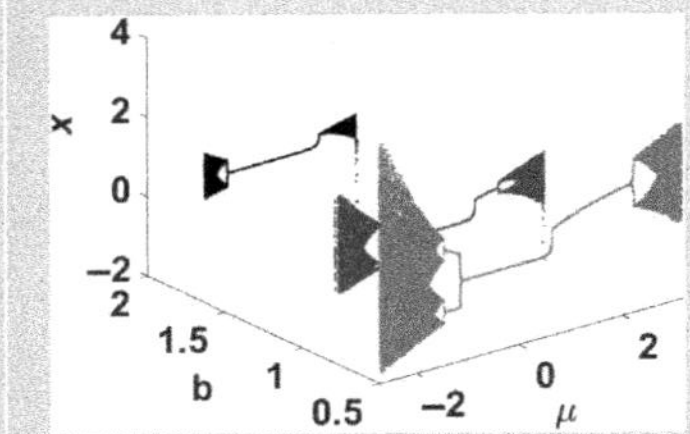

13.5 CONCLUSIONS

In this chapter, generalized forms of two of the most famous one-dimensional discrete maps, the logistic map and the tent map, were presented. The generalizations were applied in both the integer order and the FO domains. Integer order generalized logistic and tent maps with signed control parameter, which results in the complete bifurcation diagram, were reviewed. Extra parameters were added in the form of scaling and/or shaping parameters leading to various types of generalizations. The FO logistic map, based on the GL definition, was presented in two different forms: a truncated form and a summation form with longer memory. The complete bifurcation diagram of the maps with signed control parameter was considered. The effects of the memory parameter and the FO parameter were studied. The behavior of FO generalized logistic map, in the truncated form, with scaling parameters was discussed. Several mathematical analyses of the system dynamics were provided and validated using numerical simulations. Generalized bifurcation diagrams, fixed points, stability conditions, and bifurcation points were provided. In addition, several design examples for two special cases, vertical scaling and zooming maps, were included. The FO generalized tent map, using the summation form, was proposed and its behavior was depicted through various simulations showing the effects of the different parameters.

Similar procedure can be applied for other discrete maps such as the sine map, Gauss map, and Hénon map to obtain generalized forms in both integer order and FO domains. FO generalized discrete maps represent simple, efficient, and secure candidates for pseudo-random number generation in chaotic ciphers. Software implementations and hardware realizations of pseudo-random number generators and encryption schemes based on such maps can be presented. The added FO parameter increases the key space of the encryption scheme resulting in more resistance to brute-force attacks.

REFERENCES

Abd-El-Hafiz, S.K., Radwan, A.G., AbdEl-Haleem, S.H., 2015. Encryption applications of a generalized chaotic map. Appl. Math. Inform. Sci. 9, 3215.

Ablay, G., 2016. Chaotic map construction from common nonlinearities and microcontroller implementations. Int. J. Bifurcation Chaos 26, 1650121.

AboBakr, A., Said, L.A., Madian, A.H., Elwakil, A.S., Radwan, A.G., 2017. Experimental comparison of integer/fractional-order electrical models of plant. AEU-Int. J. Electr. Commun. 80, 1–9.

Alligood, K.T., Sauer, T.D., Yorke, J.A., 1996. Chaos: An Introduction to Dynamical Systems. Springer, New York.

Alpar, O., 2014. Analysis of a new simple one dimensional chaotic map. Nonlinear Dyn. 78, 771–778.

Ambadan, J.T., Joseph, K.B., 2006. Asymmetrical mirror bifurcations in logistic map with a discontinuity at zero, in: National conference on nonlinear systems and dynamics, NCNSD.

Atici, F., Eloe, P., 2009. Initial value problems in discrete fractional calculus. Proc. Am. Math. Soc. 137, 981–989.

Ausloos, M., Dirickx, M., 2006. The Logistic Map and the Route to Chaos: From the Beginnings to Modern Applications. Springer Science & Business Media, New York.

Azar, A.T., Vaidyanathan, S., Ouannas, A., 2017a. Fractional Order Control and Synchronization of Chaotic Systems., Vol. 688. Springer, New York.

Azar, A.T., Volos, C., Gerodimos, N.A., Tombras, G.S., Pham, V.T., Radwan, A.G., et al., 2017b. A novel chaotic system without equilibrium: dynamics, synchronization, and circuit realization. Complexity 2017. Article ID 7871467.

Bresten, C.L., Jung, J.H., 2009. A study on the numerical convergence of the discrete logistic map. Commun. Nonlinear Sci. Numer. Simulat. 14, 3076–3088.

Chaves, D.P., Souza, C.E., Pimentel, C., 2016. A smooth chaotic map with parameterized shape and symmetry. EURASIP J. Adv. Signal Processing 2016, 122.

Chen, F., Luo, X., Zhou, Y., 2011. Existence results for nonlinear fractional difference equation. Adv. Diff. Equ 2011.

Chien, T.I., Liao, T.L., 2005. Design of secure digital communication systems using chaotic modulation, cryptography and chaotic synchronization. Chaos Solitons Fractals 24, 241–255.

da Costa, D.R., Medrano-T, R.O., Leonel, E.D., 2017. Route to chaos and some properties in the boundary crisis of a generalized logistic mapping. Phys. Stat. Mech. Appl.

Dar, M.R., Kant, N.A., Khanday, F.A., 2017. Electronic implementation of fractional-order Newton–Leipnik chaotic system with application to communication. J. Comput. Nonlinear Dyn. 12, 054502.

El Raheem, Z., Salman, S., 2014. On a discretization process of fractional-order logistic differential equation. J. Egypt. Math. Soc. 22, 407–412.

El-Sayed, A., El-Mesiry, A., El-Saka, H., 2004. Numerical solution for multi-term fractional (arbitrary) orders differential equations. Comput. Appl. Math. 23, 33–54.

Elhadj, Z., Sprott, J.C., 2008. The effect of modulating a parameter in the logistic map. Chaos: an Interdisciplinary. J. Nonlinear Sci. 18, 023119.

Elwakil, A.S., Özoguz, S., 2008. A system and circuit for generating "multi-butterflies". Int. J. Bifurcation Chaos 18, 841–844.

Elwakil, A.S., Allagui, A., Freeborn, T., Maundy, B., 2017. Further experimental evidence of the fractional-order energy equation in supercapacitors. AEU-Int. J. Electr. Commun. 78 (8), 209–212.

Field, R.J., Schneider, F., 1989. Oscillating chemical reactions and nonlinear dynamics. J. Chem. Educ. 66, 195.

Freeborn, T.J., 2013. A survey of fractional-order circuit models for biology and biomedicine. IEEE J. Emerg. Selected Topics Circuits Systems 3, 416–424.

Gutiérrez, J.M., Iglesias, A., 1998. Mathematica package for analysis and control of chaos in nonlinear systems. Comput. Phys. 12, 608–619.

He, Y., Zhou, J., Li, C., Yang, J., Li, Q., 2008. A precise chaotic particle swarm optimization algorithm based on improved tent map, in: Fourth International Conference on Natural Computation, 2008. ICNC'08., IEEE. pp. 569–573.

Henein, M.M.R., Sayed, W.S., Radwan., A.G., Abd-El-Hafiz, S.K., 2016. Switched active control synchronization of three fractional order chaotic systems, in: 13th International Conference on Electrical Engineering/Electronics, Computer, Telecommunications and Information Technology.

Holm, M.T., 2011. The laplace transform in discrete fractional calculus. Comput. Math. Applicat. 62, 1591–1601.

Ismail, S.M., Said, L.A., Radwan, A.G., Madian, A.H., Abu-ElYazeed, M.F., Soliman, A.M., 2015. Generalized fractional logistic map suitable for data encryption. 2015 International Conference on Science and Technology (TICST). IEEE, pp. 336–341.

Ismail, S.M., Said, L.A., Rezk, A.A., Radwan, A.G., Madian, A.H., Abu-ElYazeed, M.F., et al., 2017a. Biomedical image encryption based on double-humped and fractional logistic maps. 2017 6th International Conference on Modern Circuits and Systems Technologies (MOCAST). IEEE, pp. 1–4.

Ismail, S.M., Said, L.A., Rezk, A.A., Radwan, A.G., Madian, A.H., Abu-Elyazeed, M.F., et al., 2017b. Generalized fractional logistic map encryption system based on FPGA. {AEU}. Int. J. Electr. Commun. 78, 1–27. Available from: https://doi.org/10.1016/j.aeue.2017.05.010. URL: http://www.sciencedirect.com/science/article/pii/S143484111730376X.

Ismail, S.M., Said, L.A., Rezk, A.A., Radwan, A.G., Madian, A.H., Abu-ElYazeed, M.F., et al., 2017c. Image encryption based on double-humped and delayed logistic maps for biomedical applications. 2017 6th International Conference on Modern Circuits and Systems Technologies (MOCAST). IEEE, pp. 1–4.

Jia, H.Y., Chen, Z.Q., Qi, G.Y., 2013. Topological horseshoe analysis and circuit realization for a fractional-order Lü system. Nonlinear Dynam. 74, 203–212.

Levinsohn, E.A., Mendoza, S.A., Peacock-López, E., 2012. Switching induced complex dynamics in an extended logistic map. Chaos Solitons Fractals 45, 426–432.

Lorenz, E.N., 1963. Deterministic nonperiodic flow. J. Atmos. Sci. 20, 130–141.

Lü, J., Chen, G., Cheng, D., Celikovsky, S., 2002. Bridge the gap between the Lorenz system and the Chen system. Int. J. Bifurcation Chaos 12, 2917–2926.

Lu, J.G., Chen, G., 2006. A note on the fractional-order chen system. Chaos Solitons Fractals 27, 685–688.

Luo, Y., Chen, Y., 2012. Fractional order motion controls. John Wiley & Sons, Chichester.

Malek, K., Gobal, F., 2000. Application of chaotic logistic map for the interpretation of anion-insertion in poly-ortho-aminophenol films. Synth. Metals 113, 167–171. Available from: https://doi.org/10.1016/s0379-6779(00)00194-6. URL: https://doi.org/10.1016%2Fs0379-6779%2800%2900194-6.

Matthews, R., 1989. On the derivation of a "chaotic" encryption algorithm. Cryptologia 13, 29–42.

Moaddy, K., Radwan, A.G., Salama, K.N., Momani, S., Hashim, I., 2012. The fractional-order modeling and synchronization of electrically coupled neuron systems. Comput. Math. Applicat. 64, 3329–3339.

Nejati, H., Beirami, A., Massoud, Y., 2008. A realizable modified tent map for true random number generation. 51st Midwest Symposium on Circuits and Systems MWSCAS. IEEE, pp. 621–624.

Ouannas, A., Azar, A.T., Ziar, T., Radwan, A.G., 2017a. Generalized synchronization of different dimensional integer-order and fractional order chaotic systems. Fractional Order Control and Synchronization of Chaotic Systems. Springer, pp. 671–697.

Ouannas, A., Azar, A.T., Ziar, T., Radwan, A.G., 2017b. A study on coexistence of different types of synchronization between different dimensional fractional chaotic systems. Fractional Order Control and Synchronization of Chaotic Systems. Springer, pp. 637–669.

Ouannas, A., Grassi, G., Azar, A.T., Radwan, A.G., Volos, C., Pham, V.T., et al., 2017c. Dead-beat synchronization control in discrete-time chaotic systems. 2017 6th International Conference on Modern Circuits and Systems Technologies (MOCAST). IEEE, pp. 1–4.

Panchuk, A., Sushko, I., Avrutin, V., 2015. Bifurcation structures in a bimodal piecewise linear map: chaotic dynamics. Int. J. Bifurcation Chaos 25, 1530006.

Pareek, N.K., Patidar, V., Sud, K.K., 2006. Image encryption using chaotic logistic map. Image Vision Comput. 24, 926–934.

Pellicer-Lostao, C., López-Ruiz, R., 2010. A chaotic gas-like model for trading markets. J. Comput. Sci. 1, 24–32.

Petras, I., 2011. Fractional-Order Nonlinear Systems: Modeling, Analysis and Simulation. Springer Science & Business Media, New York.

Podlubny, I., 1999. Fractional Differential Equations: An Introduction to Fractional Derivatives, Fractional Differential Equations, to Methods of Their Solution and Some of Their Applications, Vol. 198. Academic Press, San Diego.

Radwan, A., Soliman, A., El-Sedeek, A., 2004. MOS realization of the modified Lorenz chaotic system. Chaos Solitons Fractals 21, 553–561.

Radwan, A., Moaddy, K., Salama, K.N., Momani, S., Hashim, I., 2014. Control and switching synchronization of fractional order chaotic systems using active control technique. J. Adv. Res. 5, 125–132.

Radwan, A.G., 2013. On some generalized discrete logistic maps. J. Adv. Res. 4, 163–171.

Radwan, A.G., Abd-El-Hafiz, S.K., 2013. Image encryption using generalized tent map. IEEE 20th International Conference on Electronics, Circuits, and Systems (ICECS). IEEE, pp. 653–656.

Radwan, A.G., Soliman, A.M., El-Sedeek, A.L., 2003. MOS realization of the double-scroll-like chaotic equation. IEEE Trans. Circuits Systems I: Fund. Theory Applicat. 50, 285–288.

Radwan, A.G., Soliman, A., Elwakil, A.S., Sedeek, A., 2009. On the stability of linear systems with fractional-order elements. Chaos Solitons Fractals 40, 2317–2328.

Radwan, A.G., AbdElHaleem, S.H., Abd-El-Hafiz, S.K., 2016. Symmetric encryption algorithms using chaotic and non-chaotic generators: a review. J. Adv. Res. 7, 193–208.

Rajagopal, K., Vaidyanathan, S., Karthikeyan, A., Duraisamy, P., 2017. Dynamic analysis and chaos suppression in a fractional order brushless DC motor. Electr. Eng. 99, 721–733.

Ruan, H., Yaz, E.E., Zhai, T., Yaz, Y.I., 2004. A generalization of tent map and its use in EKF based chaotic parameter modulation/demodulation. 43rd IEEE Conference on Decision and Control (CDC), 2004. IEEE, pp. 2071–2075.

Said, L.A., Ismail, S.M., Radwan, A.G., Madian, A.H., El-Yazeed, M.F.A., Soliman, A.M., 2016a. On the optimization of fractional order low-pass filters. Circuits Systems Signal Processing 35, 2017–2039.

Said, L.A., Radwan, A.G., Madian, A.H., Soliman, A.M., 2016b. Fractional-order inverting and non-inverting filters based on CFOA. 39th International Conference on Telecommunications and Signal Processing (TSP). IEEE, pp. 599–602.

Said, L.A., Radwan, A.G., Madian, A.H., Soliman, A.M., 2016c. Fractional order oscillator design based on two-port network. Circuits Systems Signal Processing 35, 3086–3112.

Said, L.A., Radwan, A.G., Madian, A.H., Soliman, A.M., 2016d. Two-port two impedances fractional order oscillators. Microelectr. J. 55, 40–52.

Said, L.A., Radwan, A.G., Madian, A.H., Soliman, A.M., 2017. Three fractional-order-capacitors-based oscillators with controllable phase and frequency. J. Circuits Systems Computers 26, 1750160.

Sayed, W.S., Radwan, A.G., Fahmy, H.A., 2015a. Design of a generalized bidirectional tent map suitable for encryption applications. 11th International Computer Engineering Conference (ICENCO). IEEE, pp. 207–211.

Sayed, W.S., Radwan, A.G., Fahmy, H.A., 2015b. Design of positive, negative, and alternating sign generalized logistic maps. Discrete Dynam. Nat. Soc. 2015.

Sayed, W.S., Radwan., A.G., Abd-El-Hafiz, S.K., 2016a. Generalized synchronization involving a linear combination of fractional-order chaotic systems, in: 13th International Conference on Electrical Engineering/Electronics, Computer, Telecommunications and Information Technology.

Sayed, W.S., Radwan, A.G., Fahmy, H.A., 2016b. Double-sided bifurcations in tent maps: analysis and applications. 3rd International Conference on Advances in Computational Tools for Engineering Applications (ACTEA). IEEE, pp. 207–210.

Sayed, W.S., Fahmy, H.A., Rezk, A.A., Radwan, A.G., 2017a. Generalized smooth transition map between tent and logistic maps. Int. J. Bifurcation Chaos 27, 1730004.

Sayed, W.S., Henein, M.M., Abd-El-Hafiz, S.K., Radwan, A.G., 2017b. Generalized dynamic switched synchronization between combinations of fractional-order chaotic systems. Complexity 2017.

Sayed, W.S., Radwan, A.G., Rezk, A.A., Fahmy, H.A., 2017c. Finite precision logistic map between computational efficiency and accuracy with encryption applications. Complexity 2017.

Shimada, Y., Takagi, E., Ikeguchi, T., 2016. Symmetry of lyapunov exponents in bifurcation structures of one-dimensional maps. Chaos: an Interdisciplinary. J. Nonlinear Sci. 26, 123119.

Shukla, M.K., Sharma, B., 2017. Backstepping based stabilization and synchronization of a class of fractional order chaotic systems. Chaos Solitons Fractals 102, 274–284.

Soliman, N.S., Said, L.A., Azar, A.T., Madian, A.H., Ounnas, A., Radwan, A.G., 2017. Fractional controllable multi-scroll V-shape attractor with parameters effect. 2017 6th International Conference on Modern Circuits and Systems Technologies (MOCAST). IEEE, pp. 1–4.

Sutter, J., Pearl, R., 1946. Introduction to medical biometry and statistics. Population (French Edition) 1, 163. Available from: https://doi.org/10.2307/1524402. URL: https://doi.org/10.2307%2F1524402.

Tolba, M.F., AbdelAty, A.M., Soliman, N.S., Said, L.A., Madian, A.H., Azar, A.T., et al., 2017. FPGA implementation of two fractional order chaotic systems. AEU-Int. J. Electr. Commun. 78, 162–172.

Tsuchiya, T., Yamagishi, D., 1997. The complete bifurcation diagram for the logistic map. Zeitschrift für Naturforschung A 52, 513–516.

Vázquez-Medina, R., Díaz-Méndez, A., del Río-Correa, J.L., López-Hernández, J., 2009. Design of chaotic analog noise generators with logistic map and MOS QT circuits. Chaos Solitons Fractals 40, 1779–1793.

Wu, G.C., Baleanu, D., 2014. Discrete fractional logistic map and its chaos. Nonlinear Dynam. 75, 283–287.

Yousri, D., AbdelAty, A.M., Said, L.A., AboBakr, A., Radwan, A.G., 2017. Biological inspired optimization algorithms for cole-impedance parameters identification. AEU-Int. J. Electr. Commun. 78, 79–89.

Yu, S., Tang, W.K., Lü, J., Chen, G., 2010. Design and implementation of multi-wing butterfly chaotic attractors via Lorenz-type systems. Int. J. Bifurcation Chaos 20, 29–41.

Zidan, M.A., Radwan, A.G., Salama, K.N., 2012. Controllable V-shape multiscroll butterfly attractor: system and circuit implementation. Int. J. Bifurcation Chaos 22, 1250143.

Chapter 14

Applications of Continuous-time Fractional Order Chaotic Systems

Amr M. AbdelAty[1], Ahmad Taher Azar[2,3], Sundarapandi Vaidyanathan[4], Adel Ouannas[5] and Ahmed G. Radwan[6,7]

[1]Engineering Mathematics and Physics Department, Faculty of Engineering, Fayoum University, El Fayoum, Egypt, [2]Faculty of Computers and Information, Benha University, Benha, Egypt, [3]School of Engineering and Applied Sciences, Nile University, Giza, Egypt, [4]Vel Tech University, Chennai, Tamil Nadu, India, [5]University of Larbi Tebessi, Tebessa, Algeria, [6]Faculty of Engineering, Cairo University, Giza, Egypt, [7]Nanoelectronics Integrated Systems Center (NISC), Nile University, Cairo, Egypt

14.1 INTRODUCTION TO CHAOTIC SYSTEMS

In linear systems, a closed form solution of the system response can be easily derived. On the other hand, few nonlinear systems possess this feature and therefore nonlinear systems research relies heavily on computer simulations. It was known that deterministic systems (even nonlinear) are predictable which means that given the initial conditions and the system model, the system response up to any specified time can be easily calculated. However, the introduction of chaos has changed this concept. A system is chaotic when it shows sensitivity to initial conditions which means that two trajectories of the system starting very near to each other will be largely separated after a finite amount of time. In simple words, a chaotic system shows random behavior despite the fact of being deterministic (Parker and Chua, 1987; Azar et al., 2017a; Azar and Vaidyanathan, 2016, 2015a,c). Chaos can also be seen as a bounded unstable dynamic system response that shows sensitivity to initial conditions and has infinite unstable periodic trajectories (Odibat, 2009). The study of chaotic systems gained much attention after Lorenz's seminal work on the subject (Lorenz, 1963). He developed a simplified model for atmospheric convection that consisted of three equations instead of twelve:

$$D^{\alpha}x = \sigma(y - x), \tag{14.1}$$

$$D^{\alpha}y = x(r - z) - y, \tag{14.2}$$

Mathematical Techniques of Fractional Order Systems. DOI: https://doi.org/10.1016/B978-0-12-813592-1.00014-3

$$D^{\alpha}z = xy - bz, \tag{14.3}$$

where x is proportional to the intensity of the convective motion, y is proportional to the temperature difference between the descending and ascending currents and z is proportional to the distortion of the vertical temperature profile from linearity (Lorenz, 1963). The system shows chaotic behavior at parameter values $\sigma = 10$, $b = 8/3$, and $r = 28$. Fig. 14.1 shows the phase diagram of Lorenz system simulated from $t = 0$ to $t = 80$ with initial conditions $x_0 = y_0 = 5$ and $z_0 = 20$. The lines start dark and get lighter as the time increases. Fig. 14.2 shows the simulation of two identical Lorenz responses

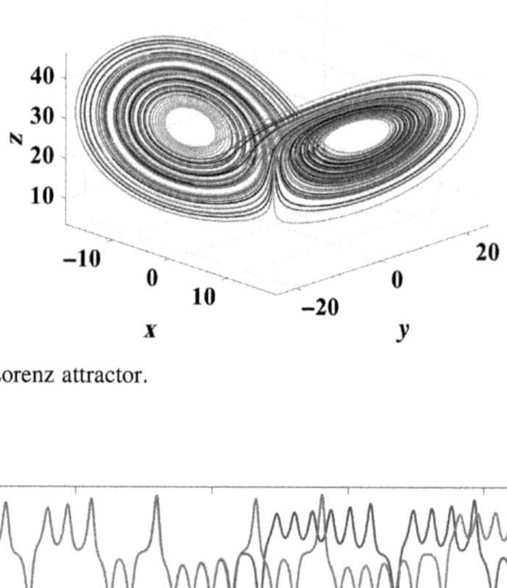

FIGURE 14.1 The Lorenz attractor.

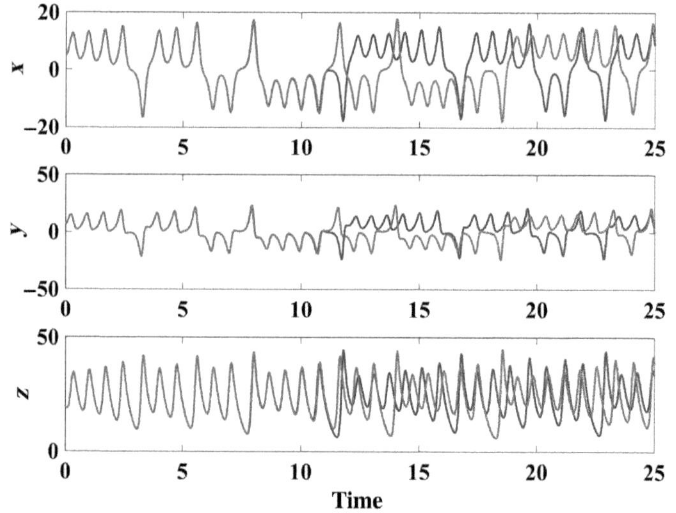

FIGURE 14.2 Lorenz time response comparison for 10^{-4} difference in initial conditions.

having the before mentioned parameter values and initial conditions except that one of them has $x_0 = 5.0001$. It can be seen how the two trajectories separate after nearly 10 time units. After that, many researchers were interested in studying other chaotic system (Rossler, 1976; Arneodo et al., 1981; Chen and Ueta, 1999; Lu and Chen, 2002; Elwakil et al., 2003; Pehlivan et al., 2014; Azar et al., 2017b; Sundarapandian and Pehlivan, 2012; Vaidyanathan and Madhavan, 2013; Vaidyanathan, 2016; Pham et al., 2017a; Wang et al., 2017; Moysis and Azar, 2017).

Lyapunov exponents are one of the quantitative measures of how chaotic the system is. The system is chaotic if it has one positive Lyapunov exponent and hyperchaotic if it has more than one. There are many algorithms to calculate the Lyapunov exponents either from the system equations (model-based) or from time series (Brown et al., 1991; Wolf et al., 1985; Rosenstein et al., 1993). However, sometimes it is sufficient to calculate the maximum Lyapunov exponent (MLE) only to decide whether the system is chaotic or not.

Recently, Leonov and Kuznetsov (2013) have proposed a new classification of chaotic systems: self-excited attractor and hidden attractors. Self-excited attractors have a basin of attraction concentrated about an unstable equilibrium point. From this perspective, most known systems belong to this category, such as Lorenz, Chen, and Rossler systems. On the other hand, systems with line equilibrium, stable equilibrium, or no equilibrium are classified as hidden attractors. Studying these systems is crucial as they play an important role in engineering applications (Pham et al., 2017b). Another important classification of chaotic systems divides them into two main categories: dissipative and conservative. Dissipative systems have a certain range of parameters where chaos behavior exists for most initial conditions while conservative systems have periodic solutions for a wide range of parameter values and initial conditions and exhibit chaos for only a narrow or even certain values of the parameters. Conservative systems can be thought of the exception rather than the rule in the world of chaotic systems (Radwan et al., 2017b, 2003a,b, 2004; Zidan et al., 2012).

Three hundred years ago, fractional calculus was initiated at the same time as integer order calculus. It started out of pure mathematical curiosity in a letter between two of the greatest mathematicians—L'Hopital and Leibniz. The progress was not as fast as its integer order counterpart, but this has changed in the last few decades. Many researchers now use fractional order models to get out of the narrow integer order subspace. Fractional order models are used to represent complex dynamical systems with rich behavior using more compact system description and having more accurate representation. Also, fractional order models are memory dependent due to the integration or weighted summation in their definition which makes them more suitable for describing systems with strong dependency on past states. The three most common definitions of the fractional order differential operator are Grunwald–Letnikov (GL), Riemann–Liouville (RL), and Caputo which are defined as (Kilbas et al., 2006):

$$ {}_aD_t^\alpha f(t) = \lim_{h\to 0} \frac{\sum_{k=0}^{\left[\frac{t-a}{h}\right]} (-1)^k \binom{\alpha}{k} f(t-kh)}{h^\alpha}, t-a=nh, \tag{14.2a} $$

$$ {}_aD_t^\alpha f(t) = \frac{d^m}{dt^m}\left[\frac{1}{\Gamma(m-\alpha)}\int_a^t (t-\tau)^{m-\alpha-1} f(\tau)d\tau\right], \tag{14.2b} $$

$$ {}_a^CD_t^\alpha f(t) = \frac{1}{\Gamma(m-\alpha)}\int_a^t (t-\tau)^{m-\alpha-1} f^{(m)}(\tau)d\tau, \tag{14.2c} $$

respectively, where $\alpha \in R^+$ and $m-1 \leq \alpha \leq m$. If $\alpha < 0$ then the GL definition represents a fractional order integral. Applications of fractional calculus are not limited to: bioengineering (Yousri et al., 2017), control (Haji and Monje, 2017), analog filters (Said et al., 2016a; AbdelAty et al., 2017; Radwan and Fouda, 2013; Soltan et al., 2015, 2014), oscillators (Said et al., 2015a,b, 2016b, 2017, 2014), circuit theory (Radwan and Salama, 2011, 2012; Radwan, 2013; AbdelAty et al., 2016, 2018; Rashad et al., 2017; Radwan et al., 2017a), chemistry (Mizrak and Ozalp, 2017), and image processing (Raghunandan et al., 2017).

Inspired by this new rediscovered fractional order models concept, researchers have proposed either fractional variations of integer order chaotic system or new fractional order systems such as: fractional order Chen system (Asheghan et al., 2011; Wang et al., 2006), fractional order Lorenz system (Grigorenko and Grigorenko, 2003; Xi et al., 2014), FO Rosslers system (Li and Chen, 2004), Coullet system (Shahiri et al., 2012), Liu system (Daftardar-Gejji and Bhalekar, 2010), modied Van der Pol Dufng system (Barbosa et al., 2007), and others (He and Chen, 2017a; Radwan et al., 2014; Tolba et al., 2017; He and Chen, 2017b). However, there are new criteria introduced by Sprott (2011) that any new chaotic system to be proposed should achieve one of the following (Borah and Roy, 2017):

1. The system has to be a model for an important under investigated physical system and give some insight into its dynamics.
2. The system should have unprecedented dynamical behavior.
3. When compared with older systems describing the same dynamical phenomena, the new one should be simpler.

This chapter is an excursion into fractional order continuous-time chaotic systems and their engineering applications. The rest of this chapter is organized as follows: Section 14.2 presents the two common simulations schemes used in literature to solve fractional order differential equations (FDEs); Sections 14.3 and 14.4 discusses selected applications in control and synchronization of fractional order chaotic systems (FOCS); Section 14.5 give examples on how FOCS are used in secure communications and encryption fields; biomedical application are discussed in Section 14.6; while examples of fractional order chaotic motor models are reviewed in Section 14.7; FPGA implementations are discussed in Section 14.8; the concluding remarks and future research directions are presented in Section 14.9.

14.2 SIMULATION OF FRACTIONAL ORDER CHAOTIC SYSTEMS

One of the most common methods used to simulate FOCS is the one based on the GL definition which was described by Monje et al. (2010). Consider the differential equation:

$$D^{\alpha}y(t) + by(t) = u(t), \tag{14.3}$$

which can be rewritten in terms of GL definition as:

$$h^{-\alpha}\sum_{j=0}^{k} w_j^{(\alpha)} y_{k-j} + by_k = q_k, \tag{14.4}$$

where $t_k = kh$, h is the step size and :

$$w_j^{(\alpha)} = (-1)^j \binom{\alpha}{j}. \tag{14.5}$$

The numerical solution is obtained by:

$$y_k = (-by_{k-1} + q_k)h^{\alpha} - \sum_{j=1}^{k} w_j^{(\alpha)} y_{k-j}. \tag{14.6}$$

This method can be easily generalized to a system of FDEs. Consider, e.g., a system with three equations which has the general form:

$$D^{q_1}x = P(x, y, z, t), \tag{14.7a}$$

$$D^{q_2}y = Q(x, y, z, t), \tag{14.7b}$$

$$D^{q_3}z = R(x, y, z, t), \tag{14.7c}$$

where P, Q, and R are, in general, nonlinear functions. The simulation equations are given as (Petráš, 2011):

$$x_{t_k} = P(x(t_{k-1}), y(t_{k-1}), z(t_{k-1}))h^{q_1} - \sum_{j=1}^{m} w_j^{(q_1)} x(t_{k-j}), \tag{14.8a}$$

$$y_{t_k} = Q(x(t_{k-1}), y(t_{k-1}), z(t_{k-1}))h^{q_2} - \sum_{j=1}^{m} w_j^{(q_2)} y(t_{k-j}), \tag{14.8b}$$

$$z_{t_k} = R(x(t_{k-1}), y(t_{k-1}), z(t_{k-1}))h^{q_3} - \sum_{j=1}^{m} w_j^{(q_3)} z(t_{k-j}), \tag{14.8c}$$

where h is the step size. This approach can be thought as a generalization of the Euler method in the integer order domain. For example, the fractional order Liu system is defined as (Daftardar-Gejji and Bhalekar, 2010):

$$D^{q_1}x = -ax - ey^2, \tag{14.9a}$$

$$D^{q_2}y = by - kxz, \tag{14.9b}$$

$$D^{q_3}z = -cz + mxy, \tag{14.9c}$$

Thus, according to (14.8); the simulation equations of Eq. (14.9a–c) is represented as:

$$x_{t_k} = \left(-ax(t_{k-1}) - ey^2(t_{k-1})\right)h^{q_1} - \sum_{j=1}^{k} w_j^{(q_1)} x(t_{k-j}), \tag{14.10a}$$

$$y_{t_k} = (by(t_{k-1}) - kx(t_{k-1})z(t_{k-1}))h^{q_2} - \sum_{j=1}^{k} w_j^{(q_2)} y(t_{k-j}), \tag{14.10b}$$

$$z_{t_k} = (-cz(t_{k-1}) + mx(t_{k-1})y(t_{k-1}))h^{q_3} - \sum_{j=1}^{k} w_j^{(q_3)} z(t_{k-j}), \tag{14.10c}$$

where q_1, q_2, q_3 are the fractional orders, $a = e = 1$, $b = 2.5$, $k = m = 4$ and $c = 5$. It is worth mentioning that this simulation approach is used in many papers and books in literature due to its simplicity specially for chaotic systems (Petráš, 2011). All simulations in this chapter are made using this scheme.

The other method is based on a generalization of Adams–Bashforth–Mounlton numerical algorithm introduced by Diethelm et al. (2002). Consider the initial value problem of Caputo-based FDE:

$$D^q y(t) = f(t, y(t)), 0 \le t \le T, \tag{14.11}$$

$$y^{(k)}(0) = y_0^{(k)}, k = 0, 1, 2, \ldots, m-1, (m = [p]). \tag{14.12}$$

This equation is equivalent to the Volterra integral equation:

$$y(t) = \sum_{k=0}^{\lceil \alpha \rceil - 1} y_0^{(k)} \frac{t^k}{k!} + \frac{1}{\Gamma(\alpha)} \int_0^t (t-\tau)^{\alpha-1} f(\tau, y(\tau)) d\tau. \tag{14.13}$$

The predicted value $y_h^P(t_{n+1})$ is determined by:

$$y_h^P(t_{n+1}) = \sum_{k=0}^{\lceil\alpha\rceil-1} \frac{t_{n+1}^k}{k!} y_0^{(k)} + \frac{1}{\Gamma(\alpha)} \sum_{j=0}^{n} b_{j,n+1} f(t_j, y_h(t_j)), \tag{14.14}$$

where the coefficients $b_{j,n+1}$ are given as:

$$b_{j,n+1} = \frac{h^\alpha}{\alpha}((n+1-j)^\alpha - (n-j)^\alpha). \tag{14.15}$$

Then, the corrector formula is given by:

$$\begin{aligned} y_h(t_{n+1}) = & \sum_{k=0}^{\lceil\alpha\rceil-1} \frac{t_{n+1}^k}{k!} y_0^{(k)} + \frac{h^\alpha}{\Gamma(\alpha+2)} \sum_{j=0}^{n} a_{j,n+1} f(t_j, y_h(t_j)) \\ & + \frac{h^\alpha}{\Gamma(\alpha+2)} f(t_{n+1}, y_j^P(t_{n+1})). \end{aligned}$$

This simulation procedure is more accurate than the GL-based one; however, the simulation is slower due to the extra number of function evaluations. It is worth mentioning that chaotic systems are sensitive to the algorithm and the step size they are simulated with. The smallest changes in the step size may make the system leave the chaotic region and require parameter adjustment to enter it again. So, the algorithm and step size should be considered one of the parameters to be defined when introducing FOCS.

14.3 CONTROL

Chaotic behavior is not always desirable and it is required to apply certain control techniques to avoid this chaotic response and force the system to reach a stable equilibrium point. Control of FOCS is a hot topic now and well-established methods in integer order system are being generalized in order to apply them to fractional order systems (Azar et al., 2018; Vaidyanathan et al., 2017c,a, 2017b, 2015; Vaidyanathan and Sampath, 2017; Meghni et al., 2017a,b; Ghoudelbourk et al., 2016; Vaidyanathan and Azar, 2016a,b,c,d,e,f,g, 2015a,b,c,d,e; Boulkroune et al., 2016a,b; Zhu and Azar, 2015; Azar and Vaidyanathan, 2015b).

14.3.1 Sliding Mode Control

Sliding mode is a nonlinear control method where the state trajectories of the system are forced to be confined in a predefined manifold (the sliding surface) and kept there through switching. The feedback control switches between two continuous control structures based on the position of the system state relative to the sliding surface (Saxena et al., 2017). A very large portion of the papers discussing control of FOCS are based on sliding mode control or one of its variants. This is due to the fact that sliding mode is a well known robust control technique for controlling dynamical systems with model uncertainty.

The controlled fractional order economic system with uncertainties and an external disturbance is given as (Dadras and Momeni, 2010):

$$D^{q_1}x = z + (y - a)x, \tag{14.16a}$$

$$D^{q_2}y = 1 - by - x^2 + \Delta f(x, y, z) - d(t) + u, \tag{14.16b}$$

$$D^{q_3}z = -x - cz, \tag{14.16c}$$

where x is the interest rate, y is the investment demand, z is the price index, u is the control input, and a, b and c are the saving amount, cost per investment, and elasticity of demand, respectively. The uncertainty $|\Delta f(x, y, z)| < \alpha$ and the disturbance $|d(t)| < \beta$ are bounded. The sliding surface was chosen to be:

$$\sigma(t) = D^{q_2 - 1}y(t) + \int_0^t \left(x^2(\tau) + Ky(\tau)\right)d\tau, \tag{14.17}$$

where $K > 0$. The sliding surface must satisfy the two conditions (Dadras and Momeni, 2010):

$$\sigma(t) = 0, \quad \dot{\sigma}(t) = 0. \tag{14.18}$$

Based on Eq. (14.18), the total control law can be written as:

$$u(t) = u_{eq}(t) + u_r(t) = [(b - K)y - 1] + [K_r sign(\sigma)], \tag{14.19}$$

where $u_{eq}(t)$ and $u_r(t)$ are the equivalent and reaching control laws, respectively and K_r is the reaching control gain. It was proved based on the candidate Lyapunov function $V = \frac{1}{2}\sigma^2$ that the closed loop system is globally asymptotically stable if $K_r < -(\alpha + \beta)$ (Dadras and Momeni, 2010). Fig. 14.3 shows the simulation results of the controlled system at $K = 1$, $K_r = -3$ and commensurate order $q = 0.9$. The uncertainty function $\Delta f(x, y, z) = 0.5 \sin(\pi x)\cos(\pi y)\sin(2\pi z)$ and the disturbance function was chosen to be

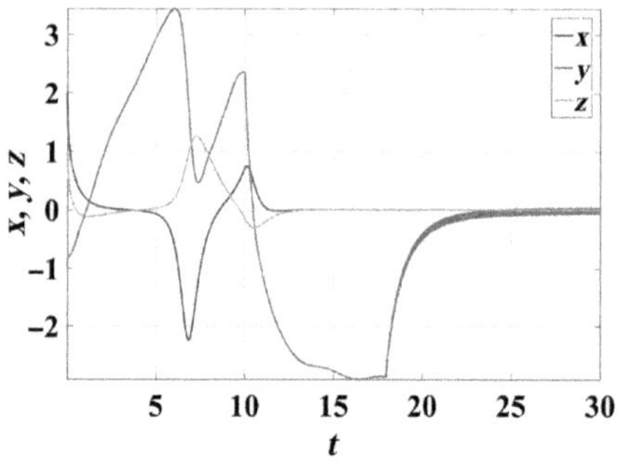

FIGURE 14.3 Sliding mode control example of economic system with external disturbance and model uncertainty.

$d(t) = 0.1\cos(2t)$ which means that $\alpha = 0.5$ and $\beta = 0.1$. The other system parameters were chosen to be: $a = 1.0$, $b = 0.1$, and $c = 1.0$.

Another sliding mode control strategy was studied by Wang et al. (2012) for the same system in Eq. (3.1). Assuming the system can be written in the form:

$$D^q X(t) = AX + Bg(X), \tag{14.20}$$

where $A \in R^{3\times 3}$ and $B \in R^{3\times 2}$ are constant matrices that can be derived from Eq. (14.16a–c). The switching surface was chosen to be:

$$S(t) = D^{q-1}X(t) - \int_0^t (A+K)X(\tau)d\tau, \tag{14.21}$$

where $K \in R^{3\times 3}$ and $S = [S_1, S_2, S_3]^T$. The total control law was derived to be (Wang et al., 2012):

$$u(t) = KX - Bg(X) - (rS + \rho\,\mathrm{sgn}(S)), \tag{14.22}$$

where $\rho, r \geq 0$.

A general sliding mode control for a class of FOCS was derived by Yin et al. (2012, 2013). The systems must have the form:

$$D^{q_1}x = y\cdot f(x,y,z) + z\cdot\phi(x,y,z) - \alpha x, \tag{14.23a}$$

$$D^{q_2}y = g(x,y,z) - \beta y + u(t), \tag{14.23b}$$

$$D^{q_3}z = y\cdot h(x,y,z) - x\cdot\phi(x,y,z) - \gamma z, \tag{14.23c}$$

where u is the control input. Several FOCS can be described by the structure such as: Chen system, Lorenz system, financial system, Lu system, and Liu system. The sliding surface was chosen to be (Yin et al., 2012, 2013):

$$s(t) = D^{q_2-1}y + \int_0^t [x\cdot f(x,y,z) + z\cdot h(x,y,z) + \beta y]d\tau. \tag{14.24}$$

The total control action was derived to be in the form:

$$u(t) = -g(x,y,z) - x\cdot f(x,y,z) - z\cdot h(x,y,z) - K\,\mathrm{sgn}(s). \tag{14.25}$$

This scheme might be considered a generalization of the example in Eq. (14.16a–c).

14.3.2 Other Approaches

Active control of idealized macroeconomic model was proposed by Baskonus et al. (2015). The controlled model is given by the set of equations:

$$D^\alpha x = my + px(d - y^2) + U_1, \tag{14.26a}$$

$$D^\alpha y = -x + cz + U_2, \tag{14.26b}$$

$$D^\alpha z = sx - ry + U_3, \tag{14.26c}$$

where x is the savings of households, y is the gross domestic product (GDP), and z is the foreign capital inflow. The system parameters m, p, d c, s, and r are the marginal propensity, the ratio of capitalized profit, the value of the potential GDP, the between output and capital, the ratio between capital inflow and savings, and the ratio between debt refund and output, respectively. The external active control inputs are U_i, $i=1,2,3$ were designed to ensure that the system is asymptotically stable at one of its three equilibrium points. When the systems parameters are: $m=0.02$, $p=0.4$, $c=50$, $d=1$, $r=0.1$, and $s=10$, the equilibrium points are: $E_0=(0,0,0)$, $E_{1,2}=(\pm 0.024, \pm 2.4, \pm 4.8\times 10^4)$. To stabilize the system at E_0, the control inputs must be:

$$\begin{pmatrix} U_1 \\ U_2 \\ U_3 \end{pmatrix} = \begin{pmatrix} -1.4x+0.98y+0.4xy^2 \\ -y-49.9z \\ -10x-z \end{pmatrix}, \tag{14.27}$$

and to stabilize it at $E_{1,2}$, the control inputs should be:

$$\begin{pmatrix} U_1 \\ U_2 \\ U_3 \end{pmatrix} = \begin{pmatrix} 0.904x+1.02608y\mp 0.002304+0.4xy^2\pm 1.92xy\pm 0.0096y^2 \\ -y-49.9z \\ -10x-z \end{pmatrix}. \tag{14.28}$$

Linear feedback control technique was used by El-Sayed et al. (2016) to achieve stability of a novel hyperchaotic model. The controlled system structure was given as (El-Sayed et al., 2016):

$$D^\alpha x = a_1x+b_1y+c_1xw^2-k_1(x-x^*), \tag{14.29a}$$

$$D^\alpha y = a_2x+b_2y+h_1z+c_2w-k_2(y-y^*), \tag{14.29b}$$

$$D^\alpha z = b_3y-k_3(z-z^*), \tag{14.29c}$$

$$D^\alpha w = a_3x+h_2z-k_4(w-w^*), \tag{14.29d}$$

where $k_i\geq 0$ and $i=1,2,3,4$ are the feedback control gains and (x^*,y^*,z^*,w^*) is an equilibrium point of the system. It was seen that the system is controlled only in the fractional order (El-Sayed et al., 2016). At parameters values: $a_1=5.9$, $c_1=19$, $a_2=7.82$, $b_1=1.36$, $b_2=1.5$, $h_1=8.5$, $c_2=2.7$, $b_3=7.8125$, $a_3=11.6$, $h_2=5.731$, and control gains $k_1=22.2288$, $k_2=k_3=k_4=0$, the system has two symmetric equilibrium points: $E_{1,2}=(\pm 0.1603, 0, \pm 0.3245, \pm 0.5573)$ (El-Sayed et al., 2016). Fig. 14.4 shows the system convergence to these equilibrium points at $\alpha=0.95$ which is practically achieved in under 8 seconds.

14.4 SYNCHRONIZATION

Chaotic synchronization can be defined as the state when two or more chaotic systems adjust their responses or some aspect of them as a result of

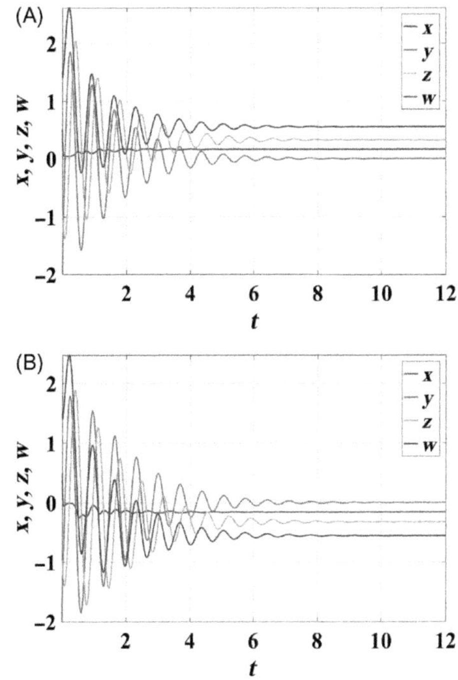

FIGURE 14.4 The feedback controlled system convergence to equilibrium points (A) E_1 and (B) E_2.

coupling or forcing (Stavrinides and Anagnostopoulos, 2013). Pecora and Carroll (1990) were the first to introduce the concept of synchronization in chaotic systems. Two main coupling topologies are found in literature: unidirectional and bidirectional. In unidirectional coupling, two subsystems form a coupled system where one of them is the master or drive system and the other is the slave or response system. The master system response is unconstrained and it drives the slave system. In bidirectional coupling, both systems interact with each other and influence the response of one another. In this configuration, the two systems drive each other into a common manifold and the synchronization is mutual. There are numerous synchronization states in literature such as (Stavrinides and Anagnostopoulos, 2013; Ouannas et al., 2017a,b,c,e,f,g,i,j,k; Ouannas and Grassi, 2016a,b; Ouannas et al., 2016b,c,a, 2017d; Grassi et al., 2017; Soliman et al., 2017; Tolba et al., 2017; Singh et al., 2017):

- Complete synchronization: It is mainly associated with identical systems when there is a complete agreement in trajectories of both systems.
- Generalized synchronization: Used when synchronizing nonidentical systems. The output of the master system is related to the output of the slave system through a given function that holds when both systems are in synchronization Ouannas et al. (2017c).

- Phase synchronization: Happens when the phases of both systems are locked while the dependency of the amplitudes is weak.
- Lag synchronization: When outputs of both systems are identical except for a constant time shift.
- Antisynchronization: Is similar to complete synchronization except the amplitudes are additive inverses to each other (Srivastava et al., 2014).
- Projective synchronization: When both responses are the same except for a constant factor (Jiang et al., 2017; Ouannas et al., 2017i,h).
- Impulsive synchronization: When the coupling is forced to be on/off according to some condition.
- Incomplete synchronization: When the difference between the responses of the synchronized systems occasionally differs from zero.
- Complete desynchronization: It occurs when both systems differ in every detail.

Another important form of synchronization is between two different systems: different dimensions (Ouannas et al., 2017k,f) or even integer and fractional (Ouannas et al., 2017e). The following subsections discuss three examples of synchronization of FOCS.

14.4.1 Synchronization of Fractional Order Modified Van der Pol-Duffing Circuit

The system is given as (Matouk, 2011):

$$D^{\alpha}x = -\nu(x^3 - \mu x - y), \tag{14.30a}$$

$$D^{\alpha}y = x - \gamma y - z, \tag{14.30b}$$

$$D^{\alpha}z = \beta y, \tag{14.30c}$$

where the original integer order system ($\alpha = 1$) exhibits chaotic behavior at parameter values: $\beta = 200$, $\mu = 0.1$, $\nu = 100$ and $\gamma = 1.6$. The system has three equilibrium points: $E_0 = (0, 0, 0)$ and $E_{1,2} = (\pm\mu, 0, \pm\mu)$. The equilibrium point E_0 is a saddle point of index 1 while the other two equilibrium points $E_{1,2}$ are saddle points of index 2. If the eigenvalues at any equilibrium point are given (in general) by the polynomial:

$$P(\lambda) = \lambda^3 + a_1\lambda^2 + a_2\lambda + a_3 = 0, \tag{14.31}$$

then, the discriminant is given as (Matouk, 2011):

$$D(P) = 18a_1a_2a_3 + (a_1a_2)^2 - 4a_3(a_1)^3 - 4(a_2)^3 - 27(a_3)^2. \tag{14.32}$$

Based on the discriminant, the authors derived the fractional Routh–Hurwitz conditions for stability then they proposed two schemes for synchronization of this system: unidirectional linear error feedback coupling and nonlinear control. The master and slave configuration for both schemes can be written as:

$$D^{\alpha}x_1 = -\nu(x_1^3 - \mu x_1 - y_1), D^{\alpha}y_1 = x_1 - \gamma y_1 - z_1, D^{\alpha}z_1 = \beta y_1, \quad (14.33)$$

and:

$$D^{\alpha}x_2 = -\nu(x_2^3 - \mu x_2 - y_2) - u_1, D^{\alpha}y_2 = x_2 - \gamma y_2 - z_2 - u_2, D^{\alpha}z_2 = \beta y_2 - u_3. \quad (14.34)$$

In the case of linear feedback error coupling, the control signals are defined as:

$$u_1 = k_1(x_2 - x_1), u_2 = k_2(y_2 - y_1), u_3 = k_3(z_2 - z_1). \quad (14.35)$$

The synchronization gains were chosen so as to satisfy the conditions given in (Jiang et al., 2003): $k_1 = 280$, $k_2 = 250$, and $k_3 = 100$. For the other synchronization scheme (nonlinear control), the controllers were:

$$u_1 = (k_1 + \mu k_{x_1,x_2})(x_1 - x_2), u_2 = u - 3 = 0, \quad (14.36)$$

where $k_{x_1,x_2} = x_1^2 + x_1x_2 + x_2^2 \geq 0$. The error dynamics of the systems is reduced to (Matouk, 2011):

$$D^{\alpha}e_x = -\nu(-\mu e_x - e_y) + k_1 e_x, D^{\alpha}e_y = e_x - \gamma e_y - e_z, D^{\alpha}e_z = \beta e_y. \quad (14.37)$$

The gain $k_1 = 69.211$ was chosen such that the fractional order error system is locally asymptotically stable and tends to the zero equilibrium point. This means that the master and slave systems can synchronize.

14.4.2 Synchronization of Electrically Coupled Neuron Systems

Fig. 14.5 shows the circuit diagram of single membrane fractional order neuron model. The normalized model is given by (Moaddy et al., 2012):

$$D^{\alpha}X(\tau) = -R\sigma_m X(\tau) - Y(t) + \frac{2\pi aR}{V_R} I_m(\tau), \quad (14.38a)$$

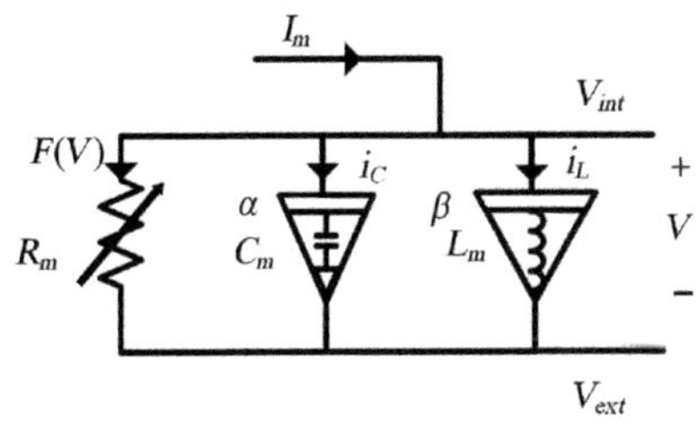

FIGURE 14.5 Fractional order circuit diagram of a single membrane.

$$D^{\beta}Y(\tau) = bX(\tau), \tag{14.38b}$$

where $X(t) = \frac{v(t)}{V_R}, \tau = \frac{t}{RC_m}, Y(t) = \frac{Ri_L(t)}{V_R}$ and single membrane conductance is defined as:

$$\sigma_m = f(v(\tau)) = \left(1 - \frac{v(\tau)}{V_T}\right)\left(1 - \frac{v(\tau)}{V_p}\right), \tag{14.39}$$

where V_T and V_p are the membrane threshold and peak voltages, respectively.

When synchronizing two coupled neurons, their combined system is given by:

$$D^{\alpha}X_1(t) = X_1(t)(X_1(t) - 1)(1 - rX_1(t)) - Y_1(t) + I_0(t) - g(X_1(t) - X_2(t)), \tag{14.40a}$$

$$D^{\beta}Y_1(t) = bX_1(t), \tag{14.40b}$$

$$D^{\alpha}X_2(t) = X_2(t)(X_2(t) - 1)(1 - rX_2(t)) - Y_2(t) + I_0(t) - g(X_2(t) - X_1(t)), \tag{14.40c}$$

$$D^{\beta}Y_2(t) = bX_2(t), \tag{14.40d}$$

where g is the control gain, $r = \frac{V_p}{V_T}$ and $I_0(t)$ is the external stimulation. The dynamical error system is given as:

$$D^{\alpha}e_x = \left((1 + r)(X_1 + X_2) - r(X_1^2 + X_2^2 + X_1X_2)\right)e_x - e_y, \tag{14.41a}$$

$$D^{\alpha}e_y = be_x. \tag{14.69}$$

The authors also discussed the generalized system of N coupled neurons. It can be seen from their simulations that the synchronization is faster as the fractional order decreases and also as the control gain g approaches 0.5 the synchronization is also faster.

14.4.3 Sliding Mode Synchronization of Uncertain Chaotic Systems

Synchronization of a FOCS which involves model uncertainty is a rather difficult task. Hosseinnia et al. (2010) discussed this issue with an example on the Duffing–Holmes system. The master system is given as:

$$D^q x_1 = x_2, \tag{14.42a}$$

$$D^q x_2 = x_1 - \alpha x_2 - x_1^3 + \beta\cos(t), \tag{14.42b}$$

while the slave system is given by:

$$D^q y_1 = y_2, \tag{14.43a}$$

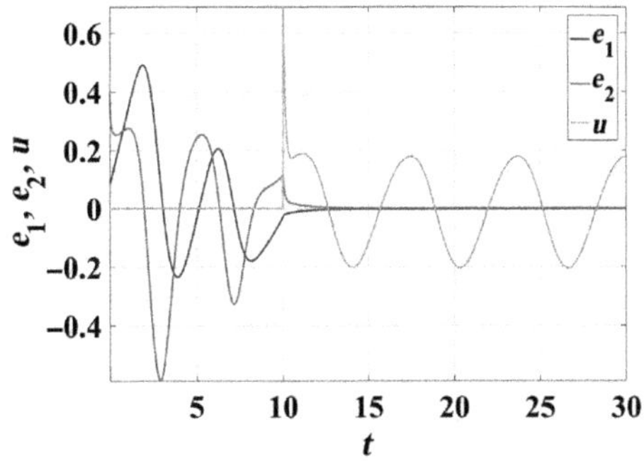

FIGURE 14.6 The synchronization error system response and the control input for Duffing–Holmes system.

$$D^q y_2 = y_1 - \alpha y_2 - y_1^3 + \beta\cos(t) + \Delta f(Y,t) + d(t) + u(t), \qquad (14.43b)$$

where the uncertainty $\Delta f(Y,t) = 0.1\sin(t)\sqrt{y_1^2 + y_2^2}$ and the disturbance $d(t) = 0.1\sin(t)$. The sliding surface was chosen to be:

$$S(t) = c_1(x_1 - y_1) + c_2(x_2 - y_2). \qquad (14.44)$$

The authors proposed sliding mode control action given by:

$$u(t) = \frac{c_1}{c_2}e_2 + e_1 - \alpha x_2 - x_1^3 + \alpha y_2 + y_1^3 - 0.1\sin(t)\left(\sqrt{y_1^2 + y_2 62} + 1\right) + K_s\text{sign}(S(t)), \qquad (14.75)$$

where the parameters were chosen as: $q = 0.9$, $\alpha = 0.25$, $c_1 = c_2 = 1$, and $K_s = 10$. Fig. 14.6 shows the synchronization errors between the master and slave systems where the control action is applied at $t = 10$ seconds. It can be seen that the systems are in synchronization in less than 3 seconds.

14.5 COMMUNICATIONS AND ENCRYPTION

There are two main advantages for applying fractional calculus and fractional order chaotic models in secure communication scheme. Firstly, the nonlinearity and complexity of the system increase which make the system stronger against attacks. Secondly, the fractional order can be thought as a new parameter that changes the system behavior from its integer order counterpart (Luo and Wang, 2013).

14.5.1 Two-Channel Secure Communication

A two-channel secure communication system was introduced in Sheu et al. (2010) based on a fractional order Lorenz system. The proposed system is sensitive to parameter values which makes it suitable for secure communication. The proposed master system at the transmitter side (Sheu et al., 2010):

$$D^{\alpha_1}x_1 = a(x_2 - x_1), \tag{14.46a}$$

$$D^{\alpha_2}x_2 = -x_1x_3 + bx_1 - x_2, \tag{14.46b}$$

$$D^{\alpha_3}x_3 = x_1x_2 - cx_3. \tag{14.46c}$$

where α_1, α_2, and α_3 are the fractional orders and a, b, and c are the system parameters. The signal $x_1(t)$ is chosen to be the synchronization signal used to drive the slave system at the receiver side. The slave system is given by (Sheu et al., 2010):

$$D^{\alpha_1}y_1 = a(y_2 - y_1), \tag{14.47a}$$

$$D^{\alpha_2}y_2 = -x_1y_3 + bx_1 - y_2, \tag{14.47b}$$

$$D^{\alpha_3}y_3 = x_1y_2 - cy_3. \tag{14.47c}$$

It was proved using Laplace transform that the master and slave systems will synchronize. The proposed scheme of the communication system is shown in Fig. 14.7. Two highly nonlinear functions are used to encrypt and decrypt the message. The encryption and decryption operations are given as (Sheu et al., 2010):

$$T_1(t) = x_2^2(t) + (1 + x_2^2(t))S(t), \tag{14.48a}$$

$$S_d(t) = -\frac{y_2^2(t)}{1 + y_2^2(t)} + \frac{T_1(t)}{1 + y_2^2(t)}, \tag{14.48b}$$

where $S(t)$ and $S_d(t)$ are the message signals at the sender and receiver sides respectively. It can be easily proved that when the two systems are synchronized ($x_2 = y_2$) then the received message is the same as the sent one.

As a numerical example, the systems were simulated at the following fractional order, parameters, and initial conditions:

$$(\alpha_1, \alpha_2, \alpha_3) = (0.96, 0.98, 1.1), \tag{14.49a}$$

$$(a, b, c) = (10, 28, 8/3), \tag{14.49b}$$

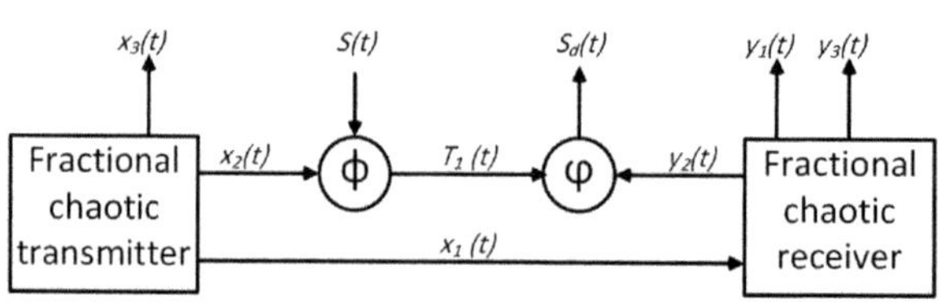

FIGURE 14.7 System block diagram of two channel secure communication system.

$$[x_1(0), x_2(0), x_3(0)] = [-1, -2, 5], \tag{14.49c}$$

$$[y_1(0), y_2(0), y_3(0)] = [1, 2, 1]. \tag{14.49d}$$

Fig. 14.8 shows the synchronization errors between the system states ($e_i = x_i - y_i, i = 1, 2, 3$). Using a plaintext signal $S(t) = 0.05\sin(4\pi t)$, Fig. 14.9A shows comparison between the original and recovered plain texts while Fig. 14.9B shows how a slight mismatch in the parameter α_2 can result in total difference between the recovered and original plain texts.

Another very similar system was proposed by Tang et al. (2012). The system was based on the synchronization of two fractional order Chen systems with time delay. The master and slave systems are given by:

$$D^{\alpha}x_m = a(y_m - x_m), \tag{14.50a}$$

$$D^{\alpha}y_m = (c - a)x_m - x_m z_m + c y_m, \tag{14.50b}$$

$$D^{\alpha}z_m = x_m y_m - b z_m(t - \tau), \tag{14.50c}$$

and,

$$D^{\alpha}x_s = a(y_s - x_s), \tag{14.51a}$$

$$D^{\alpha}y_s = (c - a)x_s - x_s z_s + c y_s + k(y_s - y_m), \tag{14.51b}$$

$$D^{\alpha}z_s = x_s y_s - b z_s(t - \tau), \tag{14.51c}$$

respectively. τ is a constant time delay. It can be seen that the control action is applied to the second state only (y_m) and is proportional to the difference of this state between the master and slave systems. The synchronization convergence was proved using Laplace transform too.

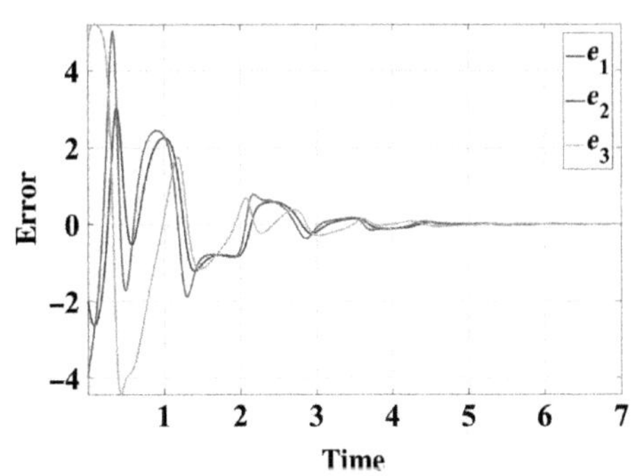

FIGURE 14.8 Synchronization errors between master and slave systems in two channel secure communication.

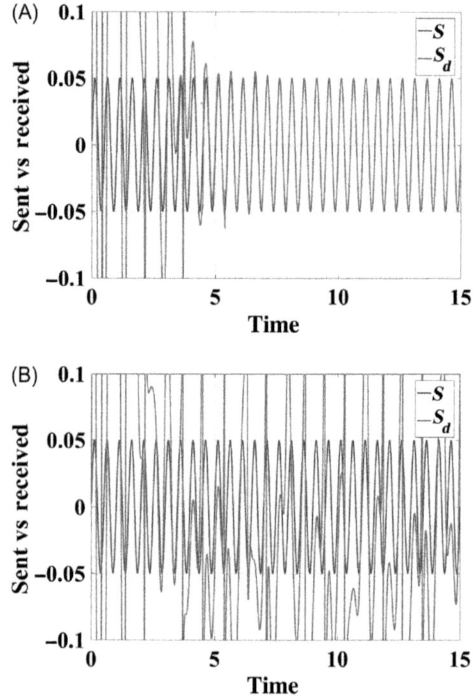

FIGURE 14.9 (A) The sent and received messages at two end of the communication channel. (B) Effect of slight variation in α_2.

Also, the complex version of Chen system was used in a digital secure communication scheme (Luo and Wang, 2013). The complex Chen system is defined as:

$$D^{\alpha_1}x_1 = a_1(x_2 - x_1), \tag{14.52a}$$

$$D^{\alpha_2}x_2 = (a_2 - a_1)x_1 - x_1x_3 + a_2x_2, \tag{14.52b}$$

$$D^{\alpha_3}x_3 = \frac{1}{2}(\bar{x}_1x_2 + x_1\bar{x}_2) - a_3x_3. \tag{14.52c}$$

The variables x_1 and x_2 are complex while the variable x_3 is real. By letting, $x_1 = x'_1 + jx'_2$, $x_2 = x'_3 + jx'_4$, and $x_3 = x'_5$ and separating the real and imaginary parts, the complex Chen system can be written as a real system of five variables: three from the real part and two from the imaginary part. The coupled complex Chen system can be written as:

$$D^{\alpha}x = Ax + f(x, z), \tag{14.53a}$$

$$D^{\alpha}Z = \phi(x, z), \tag{14.53b}$$

$$D^{\alpha}y = Ay + f(y, z) + u, \tag{14.53c}$$

where $x = (x_1, x_2)^T$ and $y = (y_1, y_2)^T$ are the state complex variables vectors and z is the real variable. A hybrid synchronization scheme was proposed as antisynchronization between the real variables and projective synchronization between the imaginary ones. This whole system was used in a switched modulated digital communication scheme. The switching is gained from changing the derivative order. The first master system is the real part and the second master is the imaginary part of x. The modulated signal can be retrieved once the slave systems are in sync with the master. This happens in less than 4 seconds according to the simulations in Luo and Wang (2013).

14.5.2 Substitution Boxes

S-Box is a key component in block cipher as it assures the security of the entire system. It is the only nonlinear component in many encryption algorithms such as IDEA, DES, and AES and most of the attacks on block cipher algorithms are on the S-Box. Two recent papers studied the construction of new 8 bit S-Boxes based on FOCS (Khan and Shah, 2015; Özkaynak et al., 2017). The first one used the fractional order Rössler system as the basis of their construction:

$$D^{\alpha_1}x = -y - x, \tag{14.54a}$$

$$D^{\alpha_2}y = x + \delta y, \tag{14.54b}$$

$$D^{\alpha_3}z = \mu + z(x - \lambda), \tag{14.54c}$$

where δ, μ, and λ are system parameters. The steps for constructing the S-Box are (Khan and Shah, 2015):

1. Solve the system in Eq. (14.54a–c) to generate a time series S of length n.
2. Generate the vector U such that $U = \text{floor}(S \times 256)$.
3. Generate another vector V such that $V = \text{mod}(U, 256)$.
4. Construct the S-Box by choosing the first 256 distinct values in V.

The other paper used the fractional order chaotic Chen system to generate the time series but used it in a different algorithm (Özkaynak et al., 2017). The steps for generating the S-box are (Özkaynak et al., 2017):

1. Generate the time series of the FOC Chen system.
2. Store the three digits after the comma in a vector S.
3. Normalize the digits in S to the range $[0, 255]$ by applying $U = \text{mod}(S, 256)$.
4. The first distinct 256 values are used to construct the table.

There are numerous criterion in literature for the design of a strong S-Box such as: nonlinearity, strict avalanche criterion (SAC), bit independent criterion (BIC), linear and differential approximation property (LAP, DAP). When compared with previous chaotic-based S-Box designs, the constructions in Özkaynak et al. (2017) and Khan and Shah (2015) had higher degrees of nonlinearity.

14.5.3 Image Encryption

FOCS are used as pseudo-random sequence generators in image encryption systems as the fractional orders increase the key space. True color images are stored as 8-bit unsigned integers so the transition from double precision generated from the FOCS simulation to the $[0, 255]$ integer range is an important part of the encryption algorithm.

Dasgupta et al. (2014) used four different FOCS to generate the key. Each system has three variables $[x_1, x_2, x_3]$ and these are designated by additional subscripts for each system: a for FO Lorenz, b for FO Chen, c for FO Liu, and d for FO Volta. The fractional order used in all these systems is $q = 0.95$. The number of pixels in the image is computed and then an equal number of iterations is made for each system variable. The 8-bit unsigned integers are calculated from the system variables as (Dasgupta et al., 2014):

$$X_i = [(|x_i| - \lfloor x_i \rfloor) \times 10^{14}]\text{mod}(256). \tag{14.55}$$

The key space is equal to $(10^{14})^{4\times 6} = 10^{336}$. The key image is generated according to the following Xor operations (Dasgupta et al., 2014):

$$X_R = X_{1a} \oplus X_{3b} \oplus X_{2c} \oplus X_{3d}, \tag{14.56a}$$

$$X_G = X_{2a} \oplus X_{1b} \oplus X_{1c} \oplus X_{2d}, \tag{14.56b}$$

$$X_B = X_{3a} \oplus X_{2b} \oplus X_{3c} \oplus X_{1d}, \tag{14.56c}$$

where X_R, X_G and X_B are the red, green, and blue components of the key image, respectively. Then the original image pixels are Xored with the key image for encryption. For decryption, the encrypted image is then Xored with the key image. The most time-consuming step of this algorithm is the simulation of FOCS as it takes much longer as the number of pixels increase. So, the authors recommended using *C* language for this time-consuming step as well as using the short memory principle to approximate the generated sequence. Fig. 14.10 shows an example encrypted image based on the proposed algorithm in Dasgupta et al. (2014). The image is the standard colored lena image of size 256 pixels.

There are numerous algorithms in literature that depend, in some steps, on FOCS. For example, the one discussed by Radwan et al. (2012) used the fractional order Lorenz system to generate the pseudo-random sequence. One interesting part of this algorithm is the inclusion of a least significant bit

(A) (B) (C)

FIGURE 14.10 Image encryption and decryption examples based on the algorithm in Dasgupta et al. (2014). (A) The original image, (B) the encrypted version and (C) the decrypted version.

(LSB) extractor that extracts the LSB from the x and y state variables and then gives it results to a 3×1 multiplexer which choses the state variable to be used in the Xor process. Another switching scheme was proposed by Hou et al. (2017) where the switching between the three state variables of the FO Chen system is made based on the color level of the pixel as:

$$f(x) = \begin{cases} x_1, 0 \leq m < 85, \\ x_2, 85 \leq m < 170, \\ x_3, 170 \leq m \leq 255, \end{cases} \tag{14.57}$$

where m is integer value at the pixel and the Xor operation is made as: $m_{encrypted} = m \oplus f(x)$. Other notable trials include encryption based on improper FOCS (Zhao et al., 2015) and image encryption using coupled-map lattices (CML) and a FOCS (Wu et al., 2015).

14.6 BIOMEDICAL

14.6.1 Evaluation of Arteriovenous Stenosis in Hemodialysis Patients

A noninvasive strategy was developed by Chen et al. (2013a,b) to evaluate the stage of arteriovenous access (AVA) stenosis and arteriovenous shunt (AVS) stenosis in hemodialysis patients. The authors used a FOCS to monitor the difference in frequency response between the healthy and unhealthy cases.

For AVA, the Burg method was used to estimate the frequency spectra of phonographic signals recorded from the patient and it was found the spectra differed between normal and stenosis cases. Forty-two patients with AVA had received percutaneous transluminal angioplasty (PTA) and participated in this experiment. The Chen–Lee system was adopted to generate the attractor used in AVA stenosis detection. The master and slave systems are defined as:

$$\begin{pmatrix} Dx_1 \\ Dx_2 \\ Dx_3 \end{pmatrix} = \begin{pmatrix} a & -x_3 & 0 \\ x_3 & b & 0 \\ \frac{1}{2}x_2 & 0 & c \end{pmatrix} \begin{pmatrix} x_1 \\ x_2 \\ x_3 \end{pmatrix}, \tag{14.58a}$$

$$\begin{pmatrix} Dy_1 \\ Dy_2 \\ Dy_3 \end{pmatrix} = \begin{pmatrix} a & -y_3 & 0 \\ y_3 & b & 0 \\ \frac{1}{2}y_2 & 0 & c \end{pmatrix} \begin{pmatrix} y_1 \\ y_2 \\ y_3 \end{pmatrix} + \begin{pmatrix} u_1 \\ u_2 \\ u_3 \end{pmatrix}. \tag{14.58b}$$

The error between the master and slave system was modeled using fractional derivatives as (Chen et al., 2013a,b):

$$\begin{pmatrix} D^q e_1 \\ D^q e_2 \\ D^q e_2 \end{pmatrix} = \begin{pmatrix} \frac{a}{\Gamma(2+\alpha)} & 0 & 0 \\ 0 & \frac{b}{\Gamma(2+\alpha)} & 0 \\ 0 & 0 & \frac{c}{\Gamma(2+\alpha)} \end{pmatrix} \begin{pmatrix} e_1^{1+\alpha} \\ e_2^{1+\alpha} \\ e_3^{1+\alpha} \end{pmatrix} + \begin{pmatrix} \frac{e_2 e_3 e_1^{\alpha}}{\Gamma(1+\alpha)} \\ \frac{(e_3)^2 e_2^{1+\alpha}}{a\Gamma(2+\alpha)} \\ 0 \end{pmatrix}, \tag{14.59}$$

where $q = 1 - \alpha$ and $0 < q < 1$. The discrete version of error system is denoted by : $e_1[i]$, $e_2[i]$ and $e_3[i]$. The dynamic error equations in the discrete case are defined by:

$$\Phi_1[i] = \frac{a}{\Gamma(2+\alpha)}(e_1[i])^{1+\alpha} - \frac{e_2[i]e_3[i]e_1^{\alpha}[i]}{\Gamma(1+\alpha)}, \tag{14.60a}$$

$$\Phi_2[i] = \left(\frac{b}{\Gamma(2+\alpha)} + \frac{(e_3[i]^2}{a\Gamma(2+\alpha)}\right) e_2^{1+\alpha}[i], \tag{14.60b}$$

$$\Phi_3[i] = \frac{c}{\Gamma(2+\alpha)} e_3^{1+\alpha}[i]. \tag{14.60c}$$

The performance index was defined as (Chen et al., 2013a,b):

$$\Psi = \sqrt{\phi_1^2 + \phi_2^2 + \phi_3^2}, \tag{14.61}$$

where $\phi_j = \max(\Phi_j[i])$, $\Phi_j[i] \in \mathbb{R}^{n-2}$, $j = 1, 2, 3$, and n is the number of samples. This index was used to evaluate the degree of AVA stenosis and surgical improvement rate (SIR) after the PTA treatment. The same procedure was applied to AVS on 13 patients who also received PTA treatment whenever needed. It was noticed that the phase trajectories of the system in Eq. (14.60a–c) became larger as the stenosis became more deteriorated.

14.6.2 Generalized Chaotic Susceptible Infected Recovered Epidemic Model

Mathematical modeling of epidemiological spread of diseases help predict and control the outbreak of an epidemic. Epidemiological models are in fact population models where the rate of growth of each group of population (susceptible or infected) is given by a differential equation that describes the interaction between the groups and the external factors if they exist. Ansari et al. (2015) proposed a modified SIR model:

$$D^{\alpha}S = rS\left(1 - \frac{S}{k}\right) - \frac{\beta SI}{1 + aS}, \tag{14.62a}$$

$$D^{\alpha}I = \frac{\beta IZ}{1 + aZ} - \mu I - \gamma I, \tag{14.62b}$$

$$D^{\alpha}Z = \frac{1}{T}(S - Z), \tag{14.62c}$$

where S, R, and Z are the densities of the susceptible and infected within the population and the information factor, respectively. r is the intrinsic growth rate of susceptible, k is the carrying capacity of susceptible, a is the saturation factor that measures the inhibitory effect, β is the transmission or contact rate, γ is the rate of recovery from infection, and μ is the death rates. Lastly, T is the average delay of the collected information on the disease. The system has three equilibrium points: the trivial equilibrium point $E_1 = (0, 0, 0)$, the disease free point $E_2 = (k, 0, k)$, and the endemic point at $E_3 = (\hat{S}, \hat{I}, \hat{Z})$ where:

$$\hat{S} = \frac{\mu + \gamma}{\beta - a(\mu + \gamma)}, \hat{I} = \frac{r}{\beta k}(1 + a\hat{S})(k - \hat{S}), \hat{Z} = \hat{S}. \tag{14.63}$$

The authors proposed a synchronization scheme based on an active control method. The master system is defined as (Ansari et al., 2015):

$$D^{\alpha}S_m = rS_m\left(1 - \frac{S_m}{k}\right) - \frac{\beta S_m I_m}{1 + aS_m}, \tag{14.64a}$$

$$D^{\alpha}I_m = \frac{\beta I_m Z_m}{1 + aZ_m} - \mu I_m - \gamma I_m, \tag{14.64b}$$

$$D^{\alpha}Z_m = \frac{1}{T}(S_m - Z_m), \tag{14.64c}$$

while the slave system is given by:

$$D^{\alpha}S_s = rS_s\left(1 - \frac{S_s}{k}\right) - \frac{\beta S_s I_s}{1 + aS_s} + u_1(t), \tag{14.65a}$$

$$D^{\alpha}I_s = \frac{\beta I_s Z_s}{1 + aZ_s} - \mu I_s - \gamma I_s + u_2(t), \tag{14.65b}$$

$$D^{\alpha}Z_s = \frac{1}{T}(S_s - Z_s) + u_3(t). \tag{14.65c}$$

The control inputs were chosen to be:

$$u_1(t) = V_1(t) + \frac{r}{k}(S_s^2 - S_m^2) + \frac{\beta S_s I_s}{1 + aS_s} - \frac{\beta S_m I_m}{1 + aS_m}, \tag{14.66a}$$

$$u_2(t) = V_2(t) - \frac{\beta I_s Z_s}{1 + aZ_s} + \frac{\beta I_m Z_m}{1 + aZ_m}, \tag{14.66b}$$

$$u_3(t) = V_3(t), \tag{14.66c}$$

where:

$$\begin{pmatrix} V_1(t) \\ V_2(t) \\ V_3(t) \end{pmatrix} = \begin{pmatrix} -(r+1) & 0 & 0 \\ 0 & (\mu + + \gamma - 1) & 0 \\ \dfrac{-1}{T} & 0 & \dfrac{1}{T} - 1 \end{pmatrix} \begin{pmatrix} S_m - S_s \\ I_m - I_s \\ Z_m - Z_s \end{pmatrix}. \tag{14.67}$$

Fig. 14.11 shows the simulation results of the this synchronization system where the control action is applied at $t = 5$ seconds. The simulation parameters are: $r = 2$, $k = 5$, $a = 0.01$, $\beta = 0.5$, $\mu = 0.3$, $\gamma = 0.2$, and $T = 0.85$. The fractional order $alpha = 0.9$. It was observed that the synchronization time increases as the order approaches the integer case ($\alpha = 1.0$).

14.7 MOTORS

Chaotic behavior is highly undesirable in motors as it causes many problems, such as low frequency oscillations in the current, torque ripples, and even motor collapse. Hence, identifying the range of parameters at which the motor exhibit chaos is important for motor protection. Once the motor enters the chaotic region, chaos control must be applied to prevent system collapse. This section discusses two fractional order nonlinear motor models introduced in literature: fractional order Brushless DC Motor model (BLDCM) and fractional order Permanent Magnet Synchronous Motor model (PMSM).

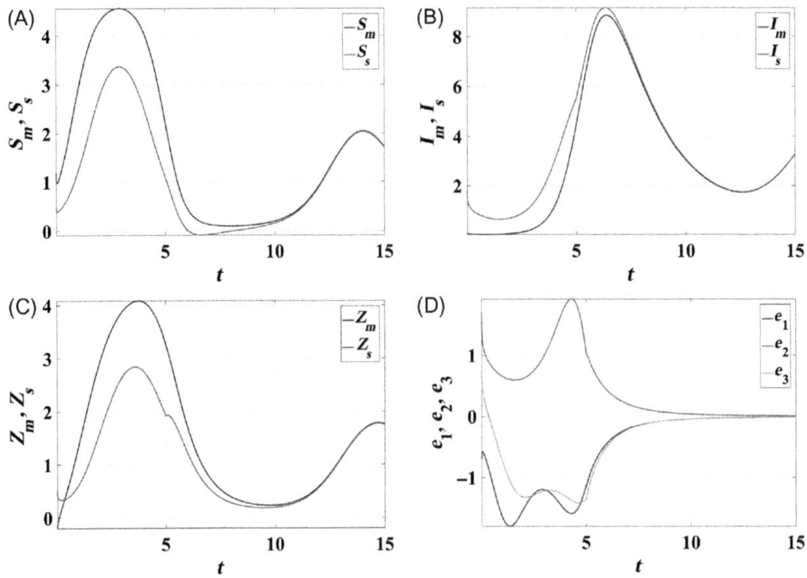

FIGURE 14.11 Synchronization results for the *SIZ* epidemic model for $\alpha = 0.9$ and control action applied at $t = 5$ seconds.

14.7.1 Brushless DC Motor

The fractional order model of BLDCM was first introduced by Zhou et al. (2015) as a generalization of the model by Wei et al. (2014). The model is given as (Zhou et al., 2015):

$$D^q x_d = -0.875x_d + x_q x_a, \tag{14.68a}$$

$$D^q x_q = -x_q - x_d x_a + 55x_a, \tag{14.68b}$$

$$D^q x_a = 4(x_q - x_a), \tag{14.68c}$$

where $q \in [0, 1]$ is the fractional order, D^q denotes the Caputo fractional derivative, x_d is the direct axis current, x_q is the quadrature axis current, and x_a is the motor angular velocity. Fig. 14.12 shows the system in Eq. (14.68a–c) at $q = 0.97$. This system has three equilibrium points: $E_1 = (0, 0, 0)$, $E_{2,3} = (54, \pm 6.87386, \pm 6.87386)$.

The controlled version of the system can be written as (Zhou et al., 2015):

$$D^q x_d = -0.875x_d + x_q x_a, \tag{14.69a}$$

$$D^q x_q = -x_q - x_d x_a + 55x_a + u(x_a), \tag{14.69b}$$

$$D^q x_a = 4(x_q - x_a) + u(x_q), \tag{14.69c}$$

where $u(x_a)$ and $u(x_q)$ are the control inputs. This control system is single input, so the inputs $u(x_a)$ and $u(x_q)$ are interlocked. Using the Generalized Gronwall inequality and Laplace transform of the Mittag–Leffler function, the system was proven to converge to the equilibrium point E_1 at some conditions on $u(x_a)$ and $u(x_q)$. The first case is at $u(x_a) = (m - 55)x_a$, $u(x_q) = 0$, and $m < 1$. The second case is at $u(x_q) = (n - 4)x_q$, $u(x_a) = 0$, and $n < 4/55$. Fig. 14.13 shows the time responses on the system at the first and second

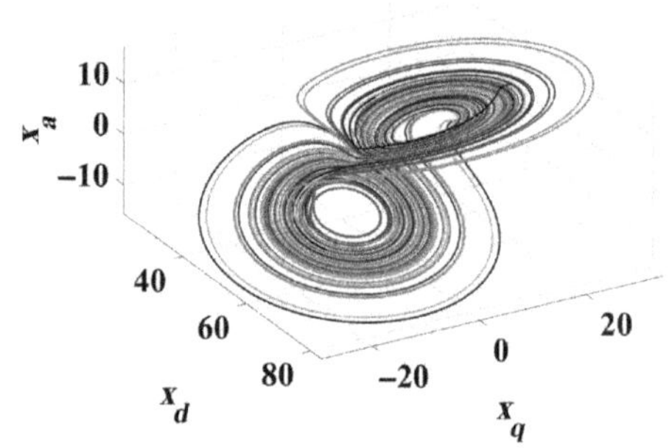

FIGURE 14.12 The BLDCM system at $q = 0.97$.

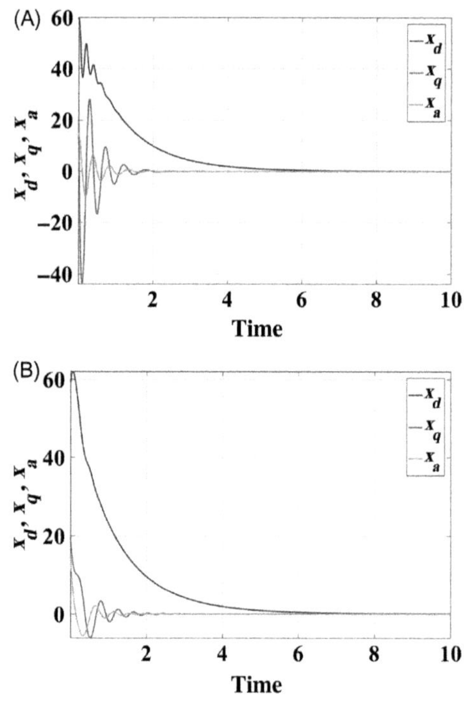

FIGURE 14.13 The controlled BLDCM system at (A) the first case, $m = -10$ and (B) the second case, $n = -5$.

case. The parameter m was chosen to be $m = -10$ in the first case while the parameter n was chosen to be $n = -5$ at the second case. It can be seen that the system reaches the equilibrium point in less 10 seconds. The initial conditions are $x_d = 60$ Amp., $x_q = 20$ Amp. and $x_a = 15$ rad/sec. It can be seen that the state variables x_a and x_q exhibit oscillatory decay while the variable x_d decays without oscillations. Also, the further the value of the variable from the equilibrium point, the longer the time it takes to stabilize. Moreover, oscillations from the first case have larger overshoots than in the second case.

Later on, the authors proposed a synchronization scheme for the same system (Shen and Zhou, 2016). Consider the system in Eq. (14.68a–c) to be the master system, so the slave system can be written as (Shen and Zhou, 2016):

$$D^q y_d = -0.875 y_d + y_q y_a + u_d, \tag{14.70a}$$

$$D^q y_q = -y_q - y_d y_a + 55 y_a + u_q, \tag{14.70b}$$

$$D^q y_a = 4(y_q - y_a) + u_a, \tag{14.70c}$$

where u_d, u_q, and u_a are the feedback controller inputs given by:

$$\begin{bmatrix} u_d \\ u_q \\ u_a \end{bmatrix} = \begin{bmatrix} -x_q \\ x_d + k - 55 - \gamma \\ 0 \end{bmatrix} (y_a - x_a), \tag{14.71}$$

where k is a real number and was chosen to be equal to either one of the three cases: $2k_q k_a$, k_q, and k_a. It was proved that chaos synchronization can be achieved if $|k_q| < 1$ and $|k_a| < \sqrt{\gamma}$. Fig. 14.14 shows the synchronization errors between the master and slave systems at $\gamma = 4$ and $k = -2$. It is worth noting that the synchronization is achieved in less than 5 seconds.

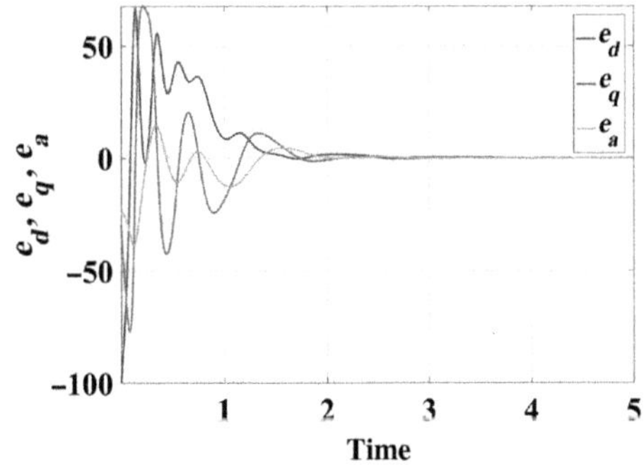

FIGURE 14.14 Synchronization errors of the BLDCM system at $q = 0.97$.

14.7.2 Permanent Magnet Synchronous Motor

Generalization of the integer model of PMSM to the fractional order domain was discussed by Chun-Lai et al. (2012). The dimensionless form of the system is given by (Chun-Lai et al., 2012):

$$D^{\alpha_1} i_d = -i_d + \omega i_q, \tag{14.72a}$$

$$D^{\alpha_2} i_q = -i_q - \omega i_d + \gamma\omega, \tag{14.72b}$$

$$D^{\alpha_3} \omega = \sigma(i_q - \omega), \tag{14.72c}$$

where i_d and i_q are the stator currents, ω is the angular velocity, and $\alpha_i, i = 1, 2, 3$ are the fractional orders. The system has three equilibrium points: $E_1 = (0, 0, 0)$ and $E_{2,3} = (\gamma - 1, \pm\sqrt{\gamma - 1}, \pm\sqrt{\gamma - 1})$. The authors then proposed an adaptive control strategy where the controlled system has the form (Chun-Lai et al., 2012):

$$D^{\alpha_1} i_d = -i_d + \omega i_q - d_1, \tag{14.73a}$$

$$D^{\alpha_2} i_q = -i_q - \omega i_d + \gamma\omega + d_2 - \rho(i_q - i_{q^*}), \tag{14.73b}$$

$$D^{\alpha_3} \omega = \sigma(i_q - \omega) + d_3, \tag{14.73c}$$

$$D^{\alpha_j} \rho = \eta(i_q - i_{q^*})^2, \tag{14.73d}$$

where d_i is a random disturbance such that $|d_1| \leq 0.5$, $|d_2| \leq 1$, and $|d_3| \leq 0.3$ and i_d^* is a constant. The variable ρ is part of the adaptive feedback control scheme. Fig. 14.15 shows the simulation result of the controlled system with and without external disturbance where the control action was applied at $t = 5$ seconds. The simulation parameters are: $\alpha_1 = 0.98$, $\alpha_2 = 1.0$, $\alpha_3 = 0.99$, $\alpha_j = 0.5$, $i_d(0) = 2.5$, $i_q(0) = 3$, and $\omega = 1$. The control parameter i_q^* is set to be equal to 7.0 to force the system to the equilibrium point E_2. The simulations show how the system is robust against external disturbances.

14.8 FPGA IMPLEMENTATIONS

There are two main approaches for FPGA implementation of FOCS in literature: HDL coder-based (Rajagopal et al., 2017; Shah et al., 2017; Rana et al., 2016), and designing from scratch (Tolba et al., 2017). Although the former approach is easier, it gives the designer less involvement in the optimization of the generated HDL code. The latter approach is rather tedious but it permits the designer to fully control the details of each HDL block in the system. The next subsections discuss examples from both approaches.

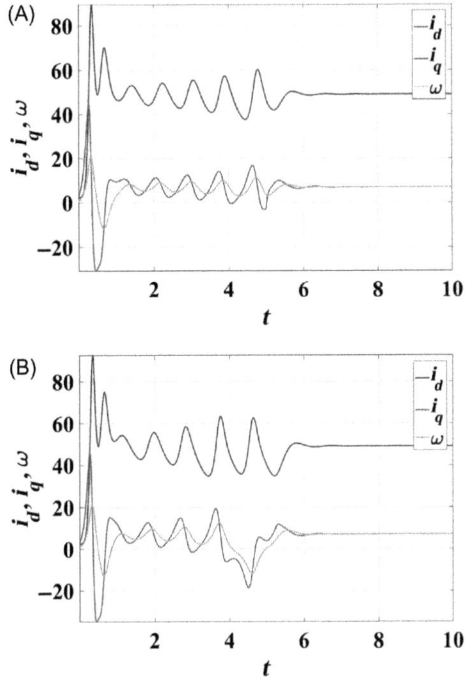

FIGURE 14.15 The controlled PMSM system at (A) without disturbance and (B) with disturbance.

14.8.1 HDL Coder-based Implementation of FOCS

Several papers in literature have utilized the functionalities of MATLAB HDL coder to implement FOCS. The main steps of such design are (Shah et al., 2017):

1. The system is modeled in Simulink using the HDL Coder toolbox where the designers have to use some of the 200 supported blocks in this toolbox.
2. The model is then simulated to obtain time domain responses of the state variables.
3. The HDL Coder is used to generate VHDL code, then the code is synthesized in Quartus or Xilinix.
4. The code is simulated using ModelSim.
5. The compiled code is then transferred to the FPGA.
6. The time response from the FPGA is then compared with the MATLAB simulated version generated earlier.

Shah et al. (2017) used this procedure to implement five different FOCS: Lorenz, Chen, Lü, Arneodo, and Lorenz Hyperchaotic system. The implementation based on the fractional version of the Tustin operator with the generating function is defined as (Barbosa and Machado, 2006; Vinagre et al., 2003):

$$(w(z^{-1}))^{\pm\alpha} = \left(\frac{2}{T}\frac{1-z^{-1}}{1+z^{-1}}\right)^{\pm\alpha}, \tag{14.74}$$

where T is the sampling time and $0<\alpha<1$. The expression in the right hand side of Eq. (14.74) can be evaluated either by power series expansion (PSE) or continued fraction expansion (CFE). Shah et al. (2017) utilized the latter method due to the fact that the CFE has wider range on convergence than PSE. For example, the Arneodo system was implemented using the set of equations:

$$\frac{N_x(z^{-1})}{D_x(z^{-1})}x(k) = y(k-1), \tag{14.75a}$$

$$\frac{N_y(z^{-1})}{D_y(z^{-1})}y(k) = z(k-1), \tag{14.75b}$$

$$\frac{N_z(z^{-1})}{D_z(z^{-1})}z(k) = -\beta_1 x(k) - \beta_2 y(k) - \beta_3 z(k-1) + \beta_4 x^3(k), \tag{14.75c}$$

where N_x, D_x, N_y, D_y, N_z, and D_z are the polynomials of the CFE approximation. The system exhibits chaotic behavior with the parameters: $\beta_1 = -5.5$, $\beta_2 = 3.5$, $\beta_3 = 0.8$, $\beta_4 = -1.0$, and commensurate order $\alpha > 0.86$. The implementation is made on DE2-115 board that utilizes a Cyclone IV (EP4CE115F29C7N) FPGA chip. The state variables are represented by 32-bits fixed point representation. This implementation utilized: 57% of the available (300) 9-bit multipliers, 5% of the available combinational functions, and only 1% of the available logic registers. The MLE of the generated Arneodo system was calculated to be 0.2614.

14.8.2 Grünwald–Letnikov Based FPGA Implementation of FOCS

In order to implement Eq. (14.8a–c) on FPGA, only a limited number of past values of the variables can be used. However, as the size of this window approaches ∞, the solution gets closer to the exact according to the short memory principle (Podlubny, 1998). Based on this assumptions, the fractional order multiscroll attractor can be simulated by (Tolba et al., 2017):

$$x_{t_k} = (x(t_{k-1}) - y(t_{k-1}))h^{q_1} - \sum_{j=1}^{L} w_j^{(q_1)} x(t_{k-j}), \tag{14.76a}$$

$$y_{t_k} = (sign(x(t_{k-1}))[1 - mz(t_{k-1}) + G(z(t_{k-1}))])h^{q_2} - \sum_{j=1}^{L} w_j^{(q_2)} y(t_{k-j}), \tag{14.76b}$$

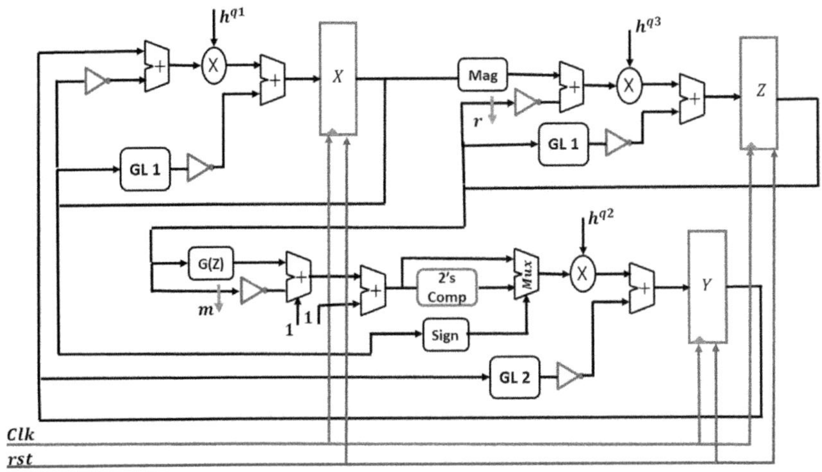

FIGURE 14.16 FPGA block diagram for FO multiscroll attractor (Tolba et al., 2017).

$$z_{t_k} = (|x(t_{k-1})| - rz(t_{k-1}))h^{q_3} - \sum_{j=1}^{L} w_j^{(q_3)} z(t_{k-j}), \quad (14.76c)$$

$$G(z) = \begin{pmatrix} 0 & z < s_0 \\ d_1 & s_0 < z < s_1 \\ \vdots & \\ d_{N-1} & z < s_{N-1} \end{pmatrix}, \quad (14.76d)$$

where m and r are constants and $G(z)$ is a staircase piecewise linear function. q_i, $i = 1, 2, 3$ are the fractional orders of the system and L is the window size of past states. Fig. 14.16 shows the block diagram for FPGA implementation of the system in Eq. (14.76a–d). The GL block represents the last summation term in each equation. It consists of two parts: a look up table (LUT) and a dot product of row and column vectors. The system was implemented on Xilinx FPGA Virtex 5 XC5VLX50T. It is quite interesting how the usual system parameters in addition to the step size h and windows size L can affect the behavior of the scrolls and also their corresponding MLE. This make the FOCS even more tunable (Tolba et al., 2017).

14.9 CONCLUSION

This chapter has made a quick overview of the applications of FOCS in various scientific and engineering fields. The study of chaos continues to be an exciting part of studying nonlinear dynamical system. It can be seen that in some applications chaotic behavior is an advantage while in others it has to be avoided. The addition of fractional derivatives in the description of chaotic systems has increased its complexity and flexibility. This is due to the nonlocality of these models and their memory dependency which have made the realization more complex and the models more representative. The fractional orders are extra parameters that can be used to increase the range of system parameters at which the system shows chaotic response.

Most of the applications of FOCS have some sort of chaos control or synchronization in their structure. Chaos control is used when chaotic behavior is undesirable and needs to be annihilated or diminished while synchronization is preferred when two systems needs to follow some aspects of each other. The use of fractional models in either the model or the control or both allowed more control over the convergence time. There are still unexplored areas in nonlinear control of FOCS like back-stepping control.

The area of secure communications benefits from FOCS in two main ways: the structure of the communication system itself and the encryption of the transferred data. For example, in the first case, both chaotic systems at the sender and receiver ends need to be in sync for successful transmission of the message. In the second case, FOCS are used as pseudo-random number generators which can be used in either block chiper or stream cipher. The possibilities are unlimited in this area and the output algorithms have to be put to standardized tests to check their reliability. Biomedical applications exploited the tools of control and synchronization and it can really benefit from faster control algorithms that could help to avoid disasters like in epidemiological modeling or making faster diagnosis.

Fractional order models of motors are nonlinear and the chaotic range of operations needs to be avoided as it causes damage and total failure of the system. If a change in system parameters occurs and the system moves into the chaotic range, chaos control needs to act fast to stabilize the system back to its safe operating conditions. So, fast identification of chaotic response is an important topic to help this purpose. FPGA realizations of FOCS shorten the gap between the theoretical study and market demand. Also, FOCS are very sensitive to their implementation which allows for even more tuning to achieve higher MLE and other desirable characteristics. Also, there will be always a trade-off between generating an accurate response and system complexity in FPGA design.

REFERENCES

AbdelAty, A.M., Radwan, A.G., Ahmed, W.A., Faied, M., 2016. Charging and discharging rc $_{\alpha}$ circuit under riemann-liouville and caputo fractional derivatives. In: Electrical Engineering/ Electronics, Computer, Telecommunications and Information Technology (ECTI-CON), 2016 13th International Conference on (pp. 1–4). IEEE.

AbdelAty, A.M., Soltan, A., Ahmed, W.A., Radwan, A.G., 2017. On the analysis and design of fractional-order chebyshev complex filter. Circuits Systems Signal Processing 2017, 1–24.

AbdelAty, A.M., Radwan, A.G., Elwakil, A.S., Psychalinos, C., 2018. Transient and steady-state response of a fractional-order dynamic pv model under different loads. J. Circuits Systems Computers 27 (02), 1850023.

Ansari, S.P., Agrawal, S.K., Das, S., 2015. Stability analysis of fractional-order generalized chaotic susceptible–infected–recovered epidemic model and its synchronization using active control method. Pramana 84 (1), 23–32.

Arneodo, A., Coullet, P., Tresser, C., 1981. Possible new strange attractors with spiral structure. Commun. Math. Phys. 79 (4), 573–579.

Asheghan, M.M., Beheshti, M.T.H., Tavazoei, M.S., 2011. Robust synchronization of perturbed chen's fractional-order chaotic systems. Commun. Nonlinear Sci. Numer. Simulat. 16 (2), 1044–1051.

Azar, A.T., Vaidyanathan, S., 2015a. Chaos Modeling and Control Systems Design volume 581 of Studies in Computational Intelligence. Studies in Computational Intelligence, Springer, Germany.

Azar, A.T., Vaidyanathan, S., 2015b. Computational Intelligence Applications in Modelling and Control, Vol. 575. Studies in Computational Intelligence, Springer, Berlin, Germany.

Azar, A.T., Vaidyanathan, S., 2015c. Handbook of Research on Advanced Intelligent Control Engineering and Automation. IGI Global, New York, NY.

Azar, A.T., Vaidyanathan, S., 2016. Advances in Chaos Theory and Intelligent Control, Vol. 337. Springer, Berlin, Germany.

Azar, A.T., Vaidyanathan, S., Ouannas, A., 2017a. Fractional Order Control and Synchronization of Chaotic Systems, Vol. 688. Studies in Computational Intelligence, Springer, Berlin, Germany.

Azar, A.T., Volos, C., Gerodimos, N.A., Tombras, G.S., Pham, V.-T., Radwan, A.G., et al., 2017b. A novel chaotic system without equilibrium: dynamics, synchronization, and circuit realization. Complexity 2017, (Article ID7871467), 11 pages.

Azar, A.T., Ouannas, A., Singh, S., 2018. Control of new type of fractional chaos synchronization. In: Hassanien, A.E., Shaalan, K., Gaber, T., Tolba, M.F. (Eds.), Proceedings of the International Conference on Advanced Intelligent Systems and Informatics 2017. Springer International Publishing, Cham, pp. 47–56.

Barbosa, R.S., Machado, J.T., 2006. Implementation of discrete-time fractional-order controllers based on ls approximations. Acta Polytech. Hung. 3 (4), 5–22.

Barbosa, R.S., Machado, J.T., Vinagre, B., Calderon, A., 2007. Analysis of the van der pol oscillator containing derivatives of fractional order. J. Vibrat. Control 13 (9-10), 1291–1301.

Baskonus, H.M., Mekkaoui, T., Hammouch, Z., Bulut, H., 2015. Active control of a chaotic fractional order economic system. Entropy 17 (8), 5771–5783.

Borah, M., Roy, B.K., 2017. Can fractional-order coexisting attractors undergo a rotational phenomenon?. ISA Trans. Available from: http://dx.doi.org/10.1016/j.isatra.2017.02.007.

Boulkroune, A., Bouzeriba, A., Bouden, T., Azar, A.T., 2016a. Fuzzy adaptive synchronization of uncertain fractional-order chaotic systems. Advances in Chaos Theory and IntelligentControl. Springer, Berlin, Germany, pp. 681–697.

Boulkroune, A., Hamel, S., Azar, A.T., Vaidyanathan, S., 2016b. Fuzzy control-based function synchronization of unknown chaotic systems with dead-zone input. In: Azar, A.T., Vaidyanathan, S. (Eds.), Advances in Chaos Theory and Intelligent Control. Springer International Publishing, Cham, pp. 699–718.

Brown, R., Bryant, P., Abarbanel, H.D., 1991. Computing the lyapunov spectrum of a dynamical system from an observed time series. Phys. Rev. A 43 (6), 2787.

Chen, G., Ueta, T., 1999. Yet another chaotic attractor. Int. J. Bifurcation Chaos 09 (07), 1465–1466.

Chen, W.-L., Chen, T., Lin, C.-H., Chen, P.-J., Kan, C. D., 2013a. Phonographic signal with a fractional-order chaotic system: a novel and simple algorithm for analyzing residual arteriovenous access stenosis. Med. Biol. Eng. Comput. 51 (9), 1011–1019.

Chen, W.L., Lin, C.H., Kan, C.D., Chen, T., 2013b. The feasibility of application phonoangiography with fractional order chaotic system in arteriovenous shunt stenosis detection. In 1st International Conference on Orange Technologies (ICOT) (pp. 123–126).

Chun-Lai, L., Si-Min, Y., Xiao-Shu, L., 2012. Fractional-order permanent magnet synchronous motor and its adaptive chaotic control. Chin. Phys. B 21 (10), 100506.

Dadras, S., Momeni, H.R., 2010. Control of a fractional-order economical system via sliding mode. Phys. A: Statist. Mech. Applicat. 389 (12), 2434–2442.

Daftardar-Gejji, V., Bhalekar, S., 2010. Chaos in fractional ordered liu system. Comput. Math. Applicat. 59 (3), 1117–1127.

Dasgupta, T., Paral, P., Bhattacharya, S., 2014. Colour image encryption based on multiple fractional order chaotic systems. In Proceedings of The 2014 International Conference on Control, Instrumentation, Energy and Communication (CIEC) (pp. 583–587).

Diethelm, K., Ford, N.J., Freed, A.D., 2002. A predictor-corrector approach for the numerical solution of fractional differential equations. Nonlinear Dynam. 29 (1), 3–22.

El-Sayed, A., Nour, H., Elsaid, A., Matouk, A., Elsonbaty, A., 2016. Dynamical behaviors, circuit realization, chaos control, and synchronization of a new fractional order hyperchaotic system. Appl. Math. Modelling 40 (5), 3516–3534.

Elwakil, A.S., Zouz, S., Kennedy, M.P., 2003. A four-wing butterfly attractor from a fully autonomous system. Int. J. Bifurcation Chaos 13 (10), 3093–3098.

Ghoudelbourk, S., Dib, D., Omeiri, A., Azar, A.T., 2016. Mppt control in wind energy conversion systems and the application of fractional control (pi α) in pitch wind turbine. Int. J. Modelling Identif. Control 26 (2), 140–151.

Grassi, G., Ouannas, A., Azar, A.T., Radwan, A.G., Volos, C., Pham, V.-T., et al., 2017. Chaos synchronisation of continuous systems via scalar signal. In 6th International Conference on Modern Circuits and Systems Technologies (MOCAST) (pp. 1–4).

Grigorenko, I., Grigorenko, E., 2003. Chaotic dynamics of the fractional lorenz system. Phys. Rev. Lett. 91, 3.

Haji, V.H., Monje, C.A., 2017. Fractional-order PID control of a chopper-fed DC motor drive using a novel firefly algorithm with dynamic control mechanism. Soft Computing .

He, J.-M., Chen, F.-Q., 2017a. A new fractional order hyperchaotic rabinovich system and its dynamical behaviors. Int. J. Non Linear Mech. 95, 73–81.

He, J.-M., Chen, F.-Q., 2017b. A new fractional order hyperchaotic rabinovich system and its dynamical behaviors. Int. J. Non Linear Mech. 95, 73–81.

Hosseinnia, S., Ghaderi, R., Ranjbar, A., Mahmoudian, M., Momani, S., 2010. Sliding mode synchronization of an uncertain fractional order chaotic system. Comput. Math. Applicat. 59 (5), 1637–1643. Fractional Differentiation and Its Applications.

Hou, J., Xi, R., Liu, P., Liu, T., 2017. The switching fractional order chaotic system and its application to image encryption. IEEE/CAA J. Automat. Sinica 4 (2), 381–388.

Jiang, G.-P., Tang, W.K.-S., Chen, G., 2003. A simple global synchronization criterion for coupled chaotic systems. Chaos Solitons Fractals 15 (5), 925–935.

Jiang, C.-M., Liu, S.-T., Zhang, F.-F., 2017. Complex modified projective synchronization for fractional-order chaotic complex systems. Int. J. Automat. Computing .

Khan, M., Shah, T., 2015. An efficient construction of substitution box with fractional chaotic system. Signal Image Video Processing 9 (6), 1335–1338.

Kilbas, A., Srivastava, H.M., Trujillo, J., 2006. Theory and Applications of Fractional Differential Equations. Elsevier, New York, NY.

Leonov, G.A., Kuznetsov, N.V., 2013. Hidden attractors in dynamical systems. from hidden oscillations in hilbert-kolmogorov, aizerman, and kalman problems to hidden chaotic attractor in chua circuits. Int. J. Bifurcation and Chaos 23 (01), 1330002.

Li, C., Chen, G., 2004. Chaos and hyperchaos in the fractional-order rssler equations. Phys. A: Statist. Mech. Applicat. 341, 55–61.

Lorenz, E.N., 1963. Deterministic nonperiodic flow. J. Atmos. Sci. 20 (2), 130–141.

Lu, J., Chen, G., 2002. A new chaotic attractor coined. Int. J. Bifurcation and Chaos 12 (03), 659–661.

Luo, C., Wang, X., 2013. Chaos generated from the fractional-order complex chen system and its application to digital secure communication. Int. J. Modern Phys. C 24 (04), 1350025.

Matouk, A., 2011. Chaos, feedback control and synchronization of a fractional-order modified autonomous van der polduffing circuit. Commun. Nonlinear Sci. Numer. Simulat. 16 (2), 975–986.

Meghni, B., Dib, D., Azar, A.T., 2017a. A second-order sliding mode and fuzzy logic control to optimal energy management in wind turbine with battery storage. Neural Comput. Applicat. 28 (6), 14171434.

Meghni, B., Dib, D., Azar, A.T., Ghoudelbourk, S., Saadoun, A., 2017b. Robust adaptive supervisory fractional order controller for optimal energy management in wind turbine with battery storage. In: Azar, A.T., Vaidyanathan, S., Ouannas, A. (Eds.), Fractional Order Control and Synchronization of Chaotic Systems. Springer International Publishing, Cham, pp. 165–202.

Mizrak, O.O., Ozalp, N., 2017. Fractional analog of a chemical system inspired by braess' paradox. Computat. Appl. Math.

Moaddy, K., Radwan, A., Salama, K., Momani, S., Hashim, I., 2012. The fractional-order modeling and synchronization of electrically coupled neuron systems. Comput. Math. Applicat. 64 (10), 3329–3339. Advances in FDE, III.

Monje, C.A., Chen, Y., Vinagre, B.M., Xue, D., Feliu-Batlle, V., 2010. Fractional-order Systems and Controls: Fundamentals and Applications. Springer Science & Business Media, New York.

Moysis, L., Azar, A.T., 2017. New discrete time 2d chaotic maps. Int. J. System Dynam. Applicat. 6 (1), 77–104.

Odibat, Z.M., 2009. Adaptive feedback control and synchronization of non-identical chaotic fractional order systems. Nonlinear Dynam. 60 (4), 479–487.

Ouannas, A., Grassi, G., 2016a. Inverse full state hybrid projective synchronization for chaotic maps with different dimensions. Chin. Phys. B 25 (9), 090503.

Ouannas, A., Grassi, G., 2016b. A new approach to study the coexistence of some synchronization types between chaotic maps with different dimensions. Nonlinear Dynam. 86 (2), 1319–1328.

Ouannas, A., Al-sawalha, M.M., Ziar, T., 2016a. Fractional chaos synchronization schemes for different dimensional systems with non-identical fractional-orders via two scaling matrices. Optik – Int. J. Light Electron Optics 127 (20), 8410–8418.

Ouannas, A., Azar, A.T., Abu-Saris, R., 2016b. A new type of hybrid synchronization between arbitrary hyperchaotic maps. Int. J. Mach. Learn. Cybern. 1–8.

Ouannas, A., Azar, A.T., Vaidyanathan, S., 2016c. A robust method for new fractional hybrid chaos synchronization. Math. Methods Appl. Sci. (pp. n/a–n/a).

Ouannas, A., Azar, A.T., Vaidyanathan, S., 2017a. New hybrid synchronization schemes based on coexistence of various types of synchronization between master-slave hyperchaotic systems. Int. J. Comput. Appl. Technol. 55 (2), 112–120.

Ouannas, A., Azar, A.T., Vaidyanathan, S., 2017b. On a simple approach for q-s synchronization of chaotic dynamical systems in continuous-time. Int. J. Comput. Sci. Math. 8 (1), 20–27.

Ouannas, A., Azar, A.T., Vaidyanathan, S., 2017c. A robust method for new fractional hybrid chaos synchronization. Math. Methods Appl. Sci. 40 (5), 1804–1812.

Ouannas, A., Azar, A.T., Ziar, T., 2017d. On inverse full state hybrid function projective synchronization for continuous-time chaotic dynamical systems with arbitrary dimensions. Different. Equat. Dynam. Systems .

Ouannas, A., Azar, A.T., Ziar, T., Radwan, A.G., 2017e. Generalized synchronization of different dimensional integer-order and fractional order chaotic systems. Fractional Order Control and Synchronization of Chaotic Systems. Springer, New York, pp. 671–697.

Ouannas, A., Azar, A.T., Ziar, T., Radwan, A.G., 2017f. A study on coexistence of different types of synchronization between different dimensional fractional chaotic systems. Fractional Order Control and Synchronization of Chaotic Systems. Springer, New York, pp. 637–669.

Ouannas, A., Azar, A.T., Ziar, T., Vaidyanathan, S., 2017g. Fractional inverse generalized chaos synchronization between different dimensional systems. In: Azar, A.T., Vaidyanathan, S., Ouannas, A. (Eds.), Fractional Order Control and Synchronization of Chaotic Systems. Springer International Publishing, Cham, pp. 525–551.

Ouannas, A., Azar, A.T., Ziar, T., Vaidyanathan, S., 2017h. Fractional inverse generalized chaos synchronization between different dimensional systems. Fractional Order Control and Synchronization of Chaotic Systems. Springer, New York, pp. 525–551.

Ouannas, A., Azar, A.T., Ziar, T., Vaidyanathan, S., 2017i. On new fractional inverse matrix projective synchronization schemes. Fractional Order Control and Synchronization of Chaotic Systems. Springer, New York, pp. 497–524.

Ouannas, A., Grassi, G., Azar, A.T., Radwan, A.G., Volos, C., Pham, V.-T., et al., 2017j. Deadbeat synchronization control in discrete-time chaotic systems. In 6th International Conference on Modern Circuits and Systems Technologies (MOCAST) (pp. 1–4).

Ouannas, A., Ziar, T., Azar, A.T., Vaidyanathan, S., 2017k. A new method to synchronize fractional chaotic systems with different dimensions. Fractional Order Control and Synchronization of Chaotic Systems. Springer, New York, pp. 581–611.

Özkaynak, F., Çelik, V., Özer, A.B., 2017. A new s-box construction method based on the fractional-order chaotic chen system. Signal Image Video Processing 11 (4), 659–664.

Parker, T., Chua, L., 1987. Chaos: a tutorial for engineers. Proc. IEEE 75 (8), 982–1008.

Pecora, L.M., Carroll, T.L., 1990. Synchronization in chaotic systems. Phys. Rev. Lett. 64 (8), 821–824.

Pehlivan, I., Moroz, I.M., Vaidyanathan, S., 2014. Analysis, synchronization and circuit design of a novel butterfly attractor. J. Sound Vib. 333 (20), 5077–5096.

Petráš, I., 2011. Fractional-Order Nonlinear Systems. Springer Berlin Heidelberg, New York.

Pham, V.-T., Vaidyanathan, S., Volos, C.K., Azar, A.T. (2017a). Dynamics, synchronization and fractional order form of a chaotic system without equilibrium. In Nonlinear Systems: Design, Applications and Analysis.

Pham, V.-T., Vaidyanathan, S., Volos, C.K., Azar, A.T., Hoang, T.M., Yem, V.V., 2017b. A three-dimensional no-equilibrium chaotic system: analysis, synchronization and its fractional order form. Fractional Order Control and Synchronization of Chaotic Systems. Springer International Publishing, Cham, pp. 449–470.

Podlubny, I., 1998. Fractional Differential Equations: An Introduction to Fractional Derivatives, Fractional Differential Equations, to Methods of Their Solution and Some of Their Applications, Vol. 198. Academic Press, San Diego.

Radwan, A., Soliman, A., El-Sedeek, A., 2004. Mos realization of the modified lorenz chaotic system. Chaos Solitons Fractals 21 (3), 553–561.

Radwan, A., Moaddy, K., Salama, K., Momani, S., Hashim, I., 2014. Control and switching synchronization of fractional order chaotic systems using active control technique. J. Adv. Res. 5 (1), 125–132.

Radwan, A.G., 2013. Resonance and quality factor of the fractional *rlc* fractional circuit. IEEE J. Emerg. Selected Topics Circuits Systems 3 (3), 377–385.

Radwan, A.G., Fouda, M.E., 2013. Optimization of fractional-order RLC filters. Circuits Systems Signal Processing 32 (5), 2097–2118.

Radwan, A.G., Salama, K.N., 2011. Passive and active elements using fractional $L_\beta C_\alpha$ circuit. IEEE Trans. Circuits Systems I: Regular Papers 58 (10), 2388–2397.

Radwan, A.G., Salama, K.N., 2012. Fractional-order RC and RL circuits. Circuits Systems Signal Processing 31 (6), 1901–1915.

Radwan, A.G., Soliman, A.M., El-Sedeek, A.-L., 2003a. An inductorless cmos realization of chuas circuit. Chaos Solitons Fractals 18 (1), 149–158.

Radwan, A.G., Soliman, A.M., El-Sedeek, A.-L., 2003b. Mos realization of the double-scroll-like chaotic equation. IEEE Trans. Circuits Systems I: Fundament. Theory Applicat. 50 (2), 285–288.

Radwan, A.G., Abd-El-Hafiz, S.K., AbdElHaleem, S.H., 2012. Image encryption in the fractional-order domain. In 2012 International Conference on Engineering and Technology (ICET) (pp. 1–6).

Radwan, A.G., Emira, A.A., AbdelAty, A.M., Azar, A.T., 2017a. Modeling and analysis of fractional order dc-dc converter. ISA Trans.

Radwan, A.G., Sayed, W.S., Abd-El-Hafiz, S.K., 2017b. Control and synchronization of fractional-order chaotic systems. Fractional Order Control and Synchronization of Chaotic Systems. Springer International Publishing, Cham, pp. 325–355.

Raghunandan, K.S., Shivakumara, P., Jalab, H.A., Ibrahim, R.W., Kumar, G.H., Pal, U., et al., 2017. Riesz fractional based model for enhancing license plate detection and recognition. IEEE Trans. Circuits Systems Video Technol. (99), 1–1.

Rajagopal, K., Karthikeyan, A., Duraisamy, P., 2017. Hyperchaotic chameleon: fractional order FPGA implementation. Complexity 2017.

Rana, K., Kumar, V., Mittra, N., Pramanik, N., 2016. Implementation of fractional order integrator/differentiator on field programmable gate array. Alexandria Eng. J. 55 (2), 1765–1773.

Rashad, S.H., Hamed, E.M., Fouda, M.E., AbdelAty, A.M., Said, L.A., Radwan, A.G., 2017. On the analysis of current-controlled fractional-order memristor emulator. In Modern Circuits and Systems Technologies (MOCAST), 2017 6th International Conference on (pp. 1–4). IEEE.

Rosenstein, M.T., Collins, J.J., De Luca, C.J., 1993. A practical method for calculating largest lyapunov exponents from small data sets. Phys. D: Nonlinear Phenomena 65 (1-2), 117–134.

Rossler, O., 1976. An equation for continuous chaos. Phys. Lett. A 57 (5), 397–398.

Said, L.A., Madian, A.H., Radwan, A.G., Soliman, A.M., 2014. Fractional order oscillator with independent control of phase and frequency. In Electronic Design (ICED), 2014 2nd International Conference on (pp. 224–229). IEEE.

Said, L.A., Radwan, A.G., Madian, A.H., Soliman, A.M., 2015a. Fractional order oscillator design based on two-port network. Circuits Systems Signal Processing 35 (9), 3086–3112.

Said, L.A., Radwan, A.G., Madian, A.H., Soliman, A.M., 2015b. Fractional order oscillators with single non-zero transmission matrix element. In Science and Technology (TICST), 2015 International Conference on (pp. 56–60). IEEE.

Said, L.A., Ismail, S.M., Radwan, A.G., Madian, A.H., El-Yazeed, M.F.A., Soliman, A.M., 2016a. On the optimization of fractional order low-pass filters. Circuits Systems Signal Processing 35 (6), 2017–2039.

Said, L.A., Radwan, A.G., Madian, A.H., Soliman, A.M., 2016b. Two-port two impedances fractional order oscillators. Microelectron. J. 55, 40–52.

Said, L.A., Radwan, A.G., Madian, A.H., Soliman, A.M., 2017. Three fractional-order-capacitors-based oscillators with controllable phase and frequency. J. Circuits Systems Computers 26 (10), 1750160.

Saxena, A., Tandon, A., Saxena, A., Rana, K.P.S., Kumar, V., 2017. On the terminal full order sliding mode control of uncertain chaotic systems. In: Azar, A.T., Vaidyanathan, S., Ouannas, A. (Eds.), Fractional Order Control and Synchronization of Chaotic Systems. Springer International Publishing, Cham, pp. 387–430.

Shah, D.K., Chaurasiya, R.B., Vyawahare, V.A., Pichhode, K., Patil, M.D., 2017. FPGA implementation of fractional-order chaotic systems. AEU – Int. J. Electr. Communicat. 78, 245–257.

Shahiri, T., Ranjbar, A., Ghaderi, R., Karami, M., Hosseinnia, S., 2012. Adaptive backstepping chaos synchronization of fractional order coullet systems with mismatched parameters. arXiv preprint arXiv *1206.2026*.

Shen, S., Zhou, P., 2016. Synchronization of the fractional-order brushless dc motors chaotic system. J. Control Sci. Eng. 2016.

Sheu, L.J., Chen, W.C., Chen, Y.C., Weng, W.T., 2010. A two-channel secure communication using fractional chaotic systems. World Acad. Sci. Eng. Technol. 65, 1057–1061.

Singh, S., Azar, A.T., Ouannas, A., Zhu, Q., Zhan, W., Na, J., 2017. Sliding mode control technique for multi-switching synchronization of chaotic systems. In 9th International Conference on Modelling, Identification and Control (ICMIC 2017), July 10-12, 2017, Kunming, China.

Soliman, N.S., Said, L.A., Azar, A.T., Madian, A.H., Radwan, A.G., Ounnas, A., 2017. Fractional controllable multi-scroll v-shape attractor with parameters effect. In 6th International Conference on Modern Circuits and Systems Technologies (MOCAST) (pp. 1–4).

Soltan, A., Radwan, A.G., Soliman, A.M., 2014. Ccii based fractional filters of different orders. J. Adv. Res. 5 (2), 157–164.

Soltan, A., Radwan, A.G., Soliman, A.M., 2015. Fractional order sallen-key and khn filters: stability and poles allocation. Circuits Systems Signal Processing 34 (5), 1461–1480.

Sprott, J.C., 2011. A proposed standard for the publication of new chaotic systems. Int. J. Bifurcation and Chaos 21 (09), 2391–2394.

Srivastava, M., Ansari, S.P., Agrawal, S.K., Das, S., Leung, A.Y.T., 2014. Anti-synchronization between identical and non-identical fractional-order chaotic systems using active control method. Nonlinear Dynam. 76 (2), 905–914.

Stavrinides, S.G., Anagnostopoulos, A.N., 2013. The route from synchronization to desynchronization of chaotic operating circuits and systems. Understanding Complex Systems. Springer Berlin Heidelberg, New York, pp. 229–275.

Sundarapandian, V., Pehlivan, I., 2012. Analysis, control, synchronization, and circuit design of a novel chaotic system. Math. Computer Modelling 55 (7), 1904–1915.

Tang, J., Zou, C., Zhao, L., 2012. Chaos synchronization of fractional order time-delay chen system and its application in secure communication. J. Convergence Inf. Technol. 7 (2), 124–131.

Tolba, M.F., AbdelAty, A.M., Soliman, N.S., Said, L.A., Madian, A.H., Azar, A.T., et al., 2017. {FPGA} implementation of two fractional order chaotic systems. AEU – Int. J. Electr. Communicat. 78, 162–172.

Vaidyanathan, S., 2016. A novel highly hyperchaotic system and its adaptive control. Advances and Applications in Chaotic Systems. Springer International Publishing, Cham, pp. 513–535.

Vaidyanathan, S., Azar, A.T., 2015a. Analysis and control of a 4-d novel hyperchaotic system. In: Azar, A.T., Vaidyanathan, S. (Eds.), Chaos Modeling and Control Systems Design. Springer volume 581 of Studies in Computational Intelligence, Berlin, Germany, pp. 19–38.

Vaidyanathan, S., Azar, A.T., 2015b. Analysis and control of a 4-d novel hyperchaotic system. In: Azar, A.T., Vaidyanathan, S. (Eds.), Chaos Modeling and Control Systems Design. Springer volume 581 of Studies in Computational Intelligence, Berlin, Germany, pp. 19–38.

Vaidyanathan, S., Azar, A.T., 2015c. Analysis, control and synchronization of a nine-term 3-d novel chaotic system. In: Azar, A.T., Vaidyanathan, S. (Eds.), Chaos Modeling and Control Systems Design. Springer volume 581 of Studies in Computational Intelligence, Berlin, Germany, pp. 3–17.

Vaidyanathan, S., Azar, A.T., 2015d. Anti-synchronization of identical chaotic systems using sliding mode control and an application to vaidyanathan-madhavan chaotic systems. In: Azar, A. T., Zhu, Q. (Eds.), Advances and Applications in Sliding Mode Control Systems. Springer volume 576 of Studies in Computational Intelligence, Berlin, Germany, pp. 527–547.

Vaidyanathan, S., Azar, A.T., 2015e. Hybrid synchronization of identical chaotic systems using sliding mode control and an application to vaidyanathan chaotic systems. In: Azar, A.T., Zhu, Q. (Eds.), Advances and Applications in Sliding Mode Control Systems. Springer volume 576 of Studies in Computational Intelligence, Berlin, Germany, pp. 549–569.

Vaidyanathan, S., Azar, A.T., 2016a. A novel 4-D four-wing chaotic system with four quadratic nonlinearities and its synchronization via adaptive control method. Advances in Chaos Theory and IntelligentControl. Springer, Berlin, Germany, pp. 203–224.

Vaidyanathan, S., Azar, A.T., 2016b. Adaptive backstepping control and synchronization of a novel 3-D jerk system with an exponential nonlinearity. Advances in Chaos Theory and IntelligentControl. Springer, Berlin, Germany, pp. 249–274.

Vaidyanathan, S., Azar, A.T., 2016c. Adaptive control and synchronization of Halvorsen circulant chaotic systems. Advances in Chaos Theory and IntelligentControl. Springer, Berlin, Germany, pp. 225–247.

Vaidyanathan, S., Azar, A.T., 2016d. Dynamic analysis, adaptive feedback control and synchronization of an eight-term 3-D novel chaotic system with three quadratic nonlinearities. Advances in Chaos Theory and IntelligentControl. Springer, Berlin, Germany, pp. 155–178.

Vaidyanathan, S., Azar, A.T., 2016e. Generalized projective synchronization of a novel hyperchaotic four-wing system via adaptive control method. Advances in Chaos Theory and IntelligentControl. Springer, Berlin, Germany, pp. 275–290.

Vaidyanathan, S., Azar, A.T., 2016f. Qualitative study and adaptive control of a novel 4-d hyperchaotic system with three quadratic nonlinearities. In: Azar, A.T., Vaidyanathan, S. (Eds.), Advances in Chaos Theory and Intelligent Control. Springer International Publishing, Cham, pp. 179–202.

Vaidyanathan, S., Azar, A.T., 2016g. Takagi-Sugeno fuzzy logic controller for Liu-Chen four-scroll chaotic system. Int. J. Intelligent Eng. Inf. 4 (2), 135–150.

Vaidyanathan, S., Madhavan, K., 2013. Analysis, adaptive control and synchronization of a seven-term novel 3-d chaotic system. Int. J. Control Theory Applicat. 6 (2), 121–137.

Vaidyanathan, S., Sampath, S., 2017. Anti-synchronisation of identical chaotic systems via novel sliding control and its application to a novel chaotic system. Int. J. Modelling Identif. Control 27 (1), 3–13.

Vaidyanathan, S., Idowu, B.A., Azar, A.T., 2015. Backstepping controller design for the global chaos synchronization of sprott's jerk systems. In: Azar, A.T., Vaidyanathan, S. (Eds.), Chaos Modeling and Control Systems Design. Springer volume 581 of Studies in Computational Intelligence, Berlin, Germany, pp. 39–58.

Vaidyanathan, S., Azar, A.T., Ouannas, A., 2017a. An eight-term 3-d novel chaotic system with three quadratic nonlinearities, its adaptive feedback control and synchronization. In: Azar, A.T., Vaidyanathan, S., Ouannas, A. (Eds.), Fractional Order Control and Synchronization of Chaotic Systems. Springer International Publishing, Cham, pp. 719–746.

Vaidyanathan, S., Azar, A.T., Ouannas, A., 2017b. Hyperchaos and adaptive control of a novel hyperchaotic system with two quadratic nonlinearities. In: Azar, A.T., Vaidyanathan, S., Ouannas, A. (Eds.), Fractional Order Control and Synchronization of Chaotic Systems. Springer International Publishing, Cham, pp. 773–803.

Vaidyanathan, S., Zhu, Q., Azar, A.T., 2017c. Adaptive control of a novel nonlinear double convection chaotic system. In: Azar, A.T., Vaidyanathan, S., Ouannas, A. (Eds.), Fractional Order Control and Synchronization of Chaotic Systems. Springer International Publishing, Cham, pp. 357–385.

Vinagre, B.M., Chen, Y.Q., Petr, I., 2003. Two direct tustin discretization methods for fractional-order differentiator/integrator. J. Franklin Institute 340 (5), 349–362.

Wang, J., Xiong, X., Zhang, Y., 2006. Extending synchronization scheme to chaotic fractional-order chen systems. Phys. A: Statist. Mech. Applicat. 370 (2), 279–285.

Wang, Z., Huang, X., Shen, H., 2012. Control of an uncertain fractional order economic system via adaptive sliding mode. Neurocomputing 83, 83–88.

Wang, Z., Volos, C., Kingni, S.T., Azar, A.T., Pham, V.-T., 2017. Four-wing attractors in a novel chaotic system with hyperbolic sine nonlinearity. Optik – Int. J. Light Electron Optics 131, 1071–1078.

Wei, D.Q., Wan, L., Luo, X.S., Zeng, S.Y., Zhang, B., 2014. Global exponential stabilization for chaotic brushless dc motors with a single input. Nonlinear Dynam. 77 (1), 209–212.

Wolf, A., Swift, J.B., Swinney, H.L., Vastano, J.A., 1985. Determining lyapunov exponents from a time series. Phys. D: Nonlinear Phenomena 16 (3), 285–317.

Wu, X., Li, Y., Kurths, J., 2015. A new color image encryption scheme using cml and a fractional-order chaotic system. PLoS One 10 (3), e0119660.

Xi, H., Li, Y., Huang, X., 2014. Generation and nonlinear dynamical analyses of fractional-order memristor-based lorenz systems. Entropy 16 (12), 6240–6253.

Yin, C., ming Zhong, S., fan Chen, W., 2012. Design of sliding mode controller for a class of fractional-order chaotic systems. Commun. Nonlinear Sci. Numer. Simulat. 17 (1), 356–366.

Yin, C., Dadras, S., ming Zhong, S., Chen, Y., 2013. Control of a novel class of fractional-order chaotic systems via adaptive sliding mode control approach. Appl. Math. Modelling 37 (4), 2469–2483.

Yousri, D., AbdelAty, A.M., Said, L.A., AboBakr, A., Radwan, A.G., 2017. Biological inspired optimization algorithms for cole-impedance parameters identification. AEU – Int. J. Electr. Communicat. 78, 79–89.

Zhao, J., Wang, S., Chang, Y., Li, X., 2015. A novel image encryption scheme based on an improper fractional-order chaotic system. Nonlinear Dynam. 80 (4), 1721–1729.

Zhou, P., Bai, R.-j, Zheng, J.-m, 2015. Stabilization of a fractional-order chaotic brushless dc motor via a single input. Nonlinear Dynam. 82 (1), 519–525.

Zhu, Q., Azar, A.T., 2015. Complex System Modelling and Control Through Intelligent Soft Computations, Vol. 319. Studies in Fuzziness and Soft Computing, Berlin, Germany.

Zidan, M.A., Radwan, A.G., Salama, K.N., 2012. Controllable v-shape multiscroll butterfly attractor: system and circuit implementation. Int. J. Bifurcation and Chaos 22 (06), 1250143.

Chapter 15

Multiswitching Synchronization Between Chaotic Fractional Order Systems of Different Dimensions

Samuel T. Ogunjo[1], Kayode S. Ojo[2] and Ibiyinka A. Fuwape[1,3]
[1]*Federal University of Technology, Akure, Nigeria,* [2]*University of Lagos, Lagos, Nigeria,*
[3]*Michael and Cecilia Ibru University, Ughelli, Nigeria*

15.1 INTRODUCTION

Chaos is defined as the aperiodic long-term behavior in a deterministic system that exhibits sensitive dependence on initial conditions (Strogatz, 2000; Azar et al., 2017b; Pham et al., 2017; Wang et al., 2017; Azar and Vaidyanathan, 2015a,c, 2016). Chaotic systems are characterized by one or more positive Lyapunov exponents. The study of chaotic systems have evolved from the first discovered Lorenz system into maps, partial differential equations, delayed differential equations, fractional order systems, and time series analyses (Ogunjo et al., 2017).

Due to the sensitivity to initial conditions, the existing belief before the work of Pecora and Carroll (1990) was that the trajectory of two chaotic systems cannot converge. Boccaletti et al. (2002) defined synchronization of chaotic systems as a process wherein two (or many) chaotic systems (either equivalent or nonequivalent) adjust a given property of their motion to a common behavior due to a coupling or to a forcing. This implies that given two chaotic systems (identical or nonidentical): the drive $y_i(t)$ and the response $x_i(t)$. The trajectory of the two systems will differ due to sensitivity to initial condition. However, if suitable control $u_i(t)$ is added to $x_i(t)$, the trajectory of $x_i(t)$ can be made to coincide with that of $y_i(t)$ such that

$$\lim_{t \to \infty} ||y_i(t) - x)i(t)|| = 0, \ \forall t \geq 0$$

Mathematical Techniques of Fractional Order Systems. DOI: https://doi.org/10.1016/B978-0-12-813592-1.00015-5

Practical applications of synchronization can be found in communication systems where it is used for encoding of information. It is also useful in understanding of biological phenomena and robotics (Ojo and Ogunjo, 2012).

Different types of synchronization such as generalized synchronization, lag synchronization, complete synchronization, phase synchronization, and projective synchronization have been introduced (Vaidyanathan et al., 2015a, b,c; Azar and Vaidyanathan, 2015b; Zhu and Azar, 2015; Vaidyanathan and Azar, 2015a,b,c,d, 2016a,b,c,d,e,f; Boulkroune et al., 2016a,b). Furthermore, different techniques for synchronization have been proposed. These include: Open Plus Close loop (OPCL), active control, direct method, backstepping, etc. Synchronization of chaotic systems began with identical systems. It has evolved to include synchronization between different systems, increased and reduced order systems, compound synchronization, combination–combination synchronization and many other forms (Vincent et al., 2015; Ojo et al., 2013b, 2014a,b). Switching synchronization of chaotic systems is one in which the different states of the response system are synchronized with the desired state of the drive system (Uar et al., 2008; Zheng, 2016; Ajayi et al., 2014). Vincent et al. (2015) opined that the synchronization of different states of the slave system synchronizing with desired states of the master in the master–slave configuration will improve security of information transmission through synchronization.

From the Grünwald–Letnikov definition of fractional order systems, the fractional order derivative of order α can be written as (Petras, 2011)

$$D_t^{\alpha} f(t) = \lim_{h \to 0} \frac{1}{h^{\alpha}} \sum_{j=0}^{\infty} (-1)^j \begin{pmatrix} \alpha \\ j \end{pmatrix} f(t - jh) \tag{15.1}$$

where the binomial coefficients can be written in terms of the Gamma function as

$$\begin{pmatrix} \alpha \\ j \end{pmatrix} = \frac{\Gamma(\alpha + 1)}{\Gamma(j + 1)\Gamma(\alpha - j + 1)}$$

The Riemann–Liouville definition of fractional derivative is given as

$$D_t^{-\alpha} f(t) = \frac{1}{\Gamma(n - \alpha)} \frac{d^n}{dt^n} \int_a^t \frac{f(\tau)}{(t - \tau)^{\alpha+1}} d\tau \tag{15.2}$$

The Caputo fractional derivatives can be written as

$$D_t^{\alpha} f(t) = \frac{1}{\Gamma(n - \alpha)} \int_a^t \frac{f^{(n)}(\tau)}{(t - \tau)^{\alpha-n+1}} d\tau, \; n - 1 < \alpha < n \tag{15.3}$$

If the order of fractional order are the same ($\alpha_1 = \alpha_2 = \alpha_3 = \alpha_n = \alpha$), the system is called commensurate; otherwise, it is called incommensurate

(Golmankhaneh et al., 2015). Different concepts such as population, anomalous diffusion, lithium ion battery model, viscoelasticity, and control systems have been found to be better described by fractional calculus (Rivero et al., 2011; Jiang et al., 2017; Azar et al., 2018a; Meghni et al., 2017a,b,c; Tolba et al., 2017; Soliman et al., 2017).

Several studies have been conducted on the synchronization of fractional order systems, multiswitching synchronization, and different order systems (Azar et al., 2017a, 2018b; Singh et al., 2017; Ouannas et al., 2016, 2017a,b,c,d, e,f,g,h; Grassi et al., 2017; Ouannas et al., 2017i). This current work aims to implement multiswitching synchronization of two fractional order systems with different dimensions. Related works to this research are stated in Section 15.2, a description of the chaotic systems used is made in Section 15.3, the design of controllers will be carried out in Section 15.4, the results obtained are explained in Section 15.5, and conclusions are drawn in Section 15.6.

15.2 RELATED WORK

The synchronization of chaotic systems with different dimensions has been studied for integer order systems. Some of the studies include increased order synchronization of two systems (Ogunjo, 2013; Ojo et al., 2014c), hybrid function projective combination synchronization (Ojo et al., 2014b), reduced order function projective combination synchronization (Ojo et al., 2014a), backstepping fuzzy adaptive control (Wang and Fan, 2015), and experimental designs (Adelakun et al., 2017) (Fig. 15.1).

Fractional order synchronization has been carried out using adaptive control (Hajipour and Aminabadi, 2016), generalized synchronization (Lu, 2005; Ge and Ou, 2008), multiswitching synchronization (Vincent et al., 2015), active sliding mode controller (Tavazoei and Haeri, 2008), nonidentical chaotic fractional order systems (Golmankhaneh et al., 2015), inverse matrix projective synchronization Ouannas et al. (2017c), and a new matrix scaling method for multidimensional synchronization.

15.3 SYSTEM DESCRIPTION

15.3.1 Duffing Oscillator

One of the most widely studied dynamical systems is the Duffing oscillator. It is given as

$$\ddot{y} + b\dot{y} + \frac{dV(y)}{dy} = g(f, \omega, t) \tag{15.4}$$

where the variable y is the displacement from the equilibrium position, f, b, g, and ω are the forcing strength, damping parameters, periodic driving force, and the angular frequency, respectively Ogunjo (2013).

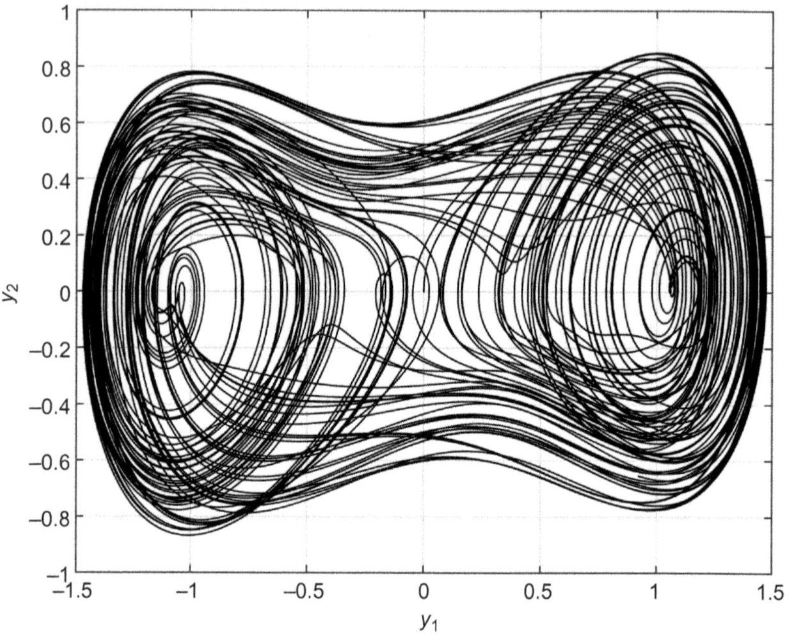

FIGURE 15.1 Chaotic attractor of the Duffing system with $q=0.98$, $\alpha=0.25$, $\beta=0.35$, and $\omega=1$.

Eq. (15.4) can be written as

$$\begin{aligned}\dot{y}_1 &= y_2\\ \dot{y}_2 &= -by_2-\alpha y_1-\beta x_1^3+f\cos\omega t\end{aligned}\tag{15.5}$$

The fractional order Duffing oscillator is given as

$$\begin{aligned}\frac{d^{q_1}y_1}{dt^{q_1}} &= y_2\\ \frac{d^{q_2}y_2}{dt^{q_2}} &= -by_2-\alpha y_1-\beta x_1^3+f\cos\omega t\end{aligned}\tag{15.6}$$

Nonlinear dynamical analysis of the fractional order Duffing system has been carried out by Li et al. (2015) and He and Luo (2012). Applications of the Duffing equation include detection of water treeing in XLPE power cable (Wu et al., 2013) and extracting the features of early mechanical failure signal (Hu and Wen, 2003). Synchronization of the Duffing oscillator has been investigated using active sliding mode controller (Tavazoei and Haeri, 2008), a modified Duffing system with excited parameters (Ge and Ou, 2008), and via active control (He and Luo, 2012).

15.3.2 Arenodo System

Arneodo et al. (1982) proposed a third order differential equation for the thermohaline convection given as

$$\begin{aligned} \frac{dx_1}{dt} &= x_2 \\ \frac{dx_2}{dt} &= x_3 \\ \frac{dx_3}{dt} &= -\beta_1 x_1 - \beta_2 x_2 - \beta_3 x_3 + \beta_4 x_1^3 \end{aligned} \tag{15.7}$$

where β_1, β_2, β_3, and β_4 are constant parameters. The integer order form of the Arneodo system was found to have three unstable equilibrium points and positive Lyapunov exponent (Motallebzadeh et al., 2009).

The fractional order Arneodo system was introduced by Lu (2005) and described as

$$\begin{aligned} \frac{d^{q_1} x_1}{dt} &= x_2 \\ \frac{d^{q_2} x_2}{dt} &= x_3 \\ \frac{d^{q_3} x_3}{dt} &= -\beta_1 x_1 - \beta_2 x_2 - \beta_3 x_3 + \beta_4 x_1^3 \end{aligned} \tag{15.8}$$

where q is the fractional order satisfying $0 < q \leq 1$. The fractional order Arneodo was found to have a maximum Lyapunov exponent of 0.22 when $\beta_3 = 0.4$ and $q = 0.9$ (Lu, 2005). The phase portrait of the fractional order Arneodo system is shown in Fig. 15.2. In this study, $\beta_1 = -5.5$, $\beta_2 = 3.5$, $\beta_3 = 0.4$, $\beta_4 = -1$, $q_1 = q_2 = q_3 = 0.9$.

The Arenodo chaotic system has been the subject of different research, such as control of chaos in the integer order Arneodo system (Motallebzadeh et al., 2009), synchronization of the fractional order Arenodo system (Hajipour and Aminabadi, 2016; Lu, 2005), backstepping fuzzy adaptive control synchronization (Wang and Fan, 2015), and FPGA implementation (Shah et al., 2017).

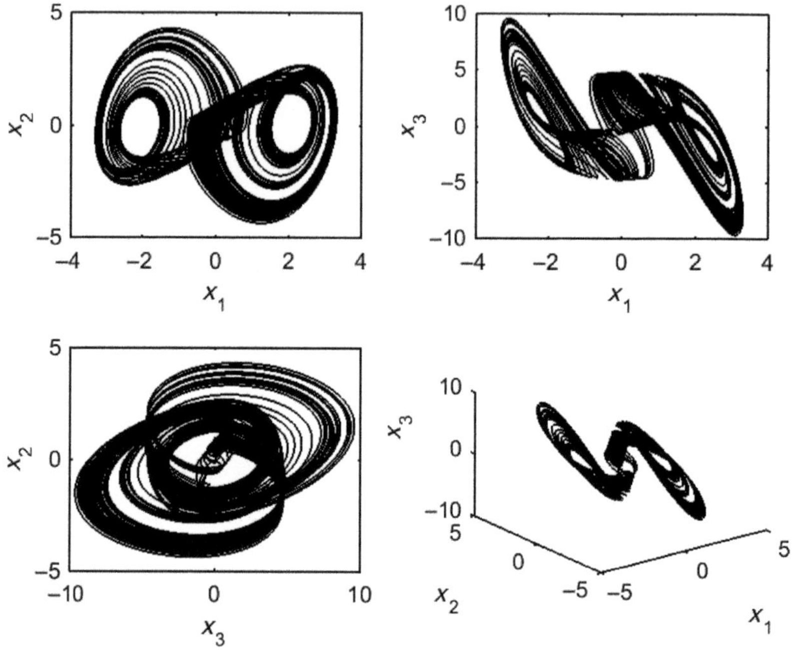

FIGURE 15.2 Chaotic attractor in different planes for the Arneodo system with $q = 0.9$ and $\beta_3 = 0.4$.

15.4 DESIGN OF CONTROLLERS

Let the drive system be the Arneodo system

$$\begin{aligned}\frac{d^q x_1}{dt^q} &= x_2 \\ \frac{d^q x_2}{dt^q} &= x_3 \\ \frac{d^q x_3}{dt^q} &= -\beta_1 x_1 - \beta_2 x_2 - \beta_3 x_3 + \beta_4 x_1^3\end{aligned} \tag{15.9}$$

And the response system defined as the driven Duffing oscillator

$$\begin{aligned}\frac{d^q y_1}{dt^q} &= y_2 + u_1(t) \\ \frac{d^q y_2}{dt^q} &= y_1 - y_1^3 - \alpha y_2 + f\cos\omega t + u_2(t)\end{aligned} \tag{15.10}$$

where $u_1(t)$ and $u_2(t)$ are controllers to be designed.

There are six possible switching combinations that could exist for this drive-response system given in Eqs. (15.9) and (15.10). They are:

$$\begin{aligned} e_{11} &= y_1 - x_1 \\ e_{12} &= y_2 - x_2 \end{aligned} \qquad \text{Switch 1} \tag{15.11}$$

$$\begin{aligned} e_{21} &= y_1 - x_1 \\ e_{22} &= y_2 - x_3 \end{aligned} \qquad \text{Switch 2} \tag{15.12}$$

$$\begin{aligned} e_{31} &= y_1 - x_2 \\ e_{32} &= y_2 - x_1 \end{aligned} \qquad \text{Switch 3} \tag{15.13}$$

$$\begin{aligned} e_{41} &= y_1 - x_2 \\ e_{42} &= y_2 - x_3 \end{aligned} \qquad \text{Switch 4} \tag{15.14}$$

$$\begin{aligned} e_{51} &= y_1 - x_3 \\ e_{52} &= y_2 - x_1 \end{aligned} \qquad \text{Switch 5} \tag{15.15}$$

$$\begin{aligned} e_{61} &= y_1 - x_3 \\ e_{62} &= y_2 - x_2 \end{aligned} \qquad \text{Switch 6} \tag{15.16}$$

15.4.1 Case 1

For Switch 1 defined by Eq. (15.11), the time derivative of the errors is given by

$$\begin{aligned} \frac{d^q e_{11}}{dt^q} &= y_2 + u_{11}(t) - x_2 \\ &= e_{12} + u_1 t \\ \frac{d^q e_{12}}{dt^q} &= y_1 - y_1^3 - \alpha y_2 + f\cos\omega t + u_2(t) - x_3 \\ &= e_{11} + x_1 - y_1^3 - \alpha(e_{12} + x_2) + f\cos\omega t + u_{12}(t) - x_3 \\ &= e_{12} - \alpha e_{12} + f_{12}(x, y) + u_2(t) \end{aligned}$$

where $f_{12}(x, y)$ are the nonlinear terms in e_{11} and e_{12} given as

$$f_{12}(x, y) = x_1 - y_1^3 - \alpha x_2 + f\cos\omega t - x_3$$

Theorem 1: *If the control function* $u_1(t)$ *and* $u_2(t)$ *are chosen such that*

$$\begin{aligned} u_1(t) &= k_1 e_{11} + e_{12} \\ u_2(t) &= -f_{12}(x, y) + e_{11} + (k_2 - \alpha)e_{12} \end{aligned} \tag{15.17}$$

then the drive system 15.9 *will achieve multiswitching synchronization with the response system* 15.10

Proof: The method of active control is employed to prove theorem 1. This is predicated on its simplicity, efficiency, and fast convergence (Ojo et al., 2013a; Ogunjo et al., 2017)

The time derivative of the errors can also be written as

$$\begin{pmatrix} \dfrac{d^q e_{11}}{dt^q} \\ \dfrac{d^q e_{12}}{dt^q} \end{pmatrix} = \begin{pmatrix} 0 & 1 \\ 1 & -\alpha \end{pmatrix}\begin{pmatrix} e_{11} \\ e_{12} \end{pmatrix} + \begin{pmatrix} 0 \\ f_{12}(x,y) \end{pmatrix} + \begin{pmatrix} u_1(t) \\ u_2(t) \end{pmatrix}$$

In order to eliminate the nonlinear term in e_1 and e_2, we define the active control input $u_1(t)$ and $u_2(t)$ as

$$\begin{aligned} u_1(t) &= V_{11}(t) \\ u_2(t) &= e_{11} - \alpha e_{12} - f_{12}(x,y) + V_{12}(t) \end{aligned}$$

which leads to

$$\begin{aligned} \frac{d^q e_{11}}{dt^q} &= e_{12} + V_{11}(t) \\ \frac{d^q e_{12}}{dt^q} &= e_1 1 - \alpha e_{12} + V_{12}(t) \end{aligned} \tag{15.18}$$

The synchronization error (Eq. 15.18) is a linear system with active control inputs $V_{11}(t)$ and $V_{12}(t)$. The feedback control which would stabilize the system so that e_{11} and e_{12} converge to zero at time $t \to \infty$ are to be designed. As a result, we chose

$$\begin{pmatrix} V_{11}(t) \\ V_{12}(t) \end{pmatrix} = A \begin{pmatrix} e_{11} \\ e_{12} \end{pmatrix}$$

where A is a 2×2 constant matrix. In order to make the closed-loop system stable, the matrix A should be chosen such that the eigenvalues λ_i of A satisfies

$$|\arg(\lambda_i)| > 0.5\pi\alpha, \; i = 1, 2, \ldots \tag{15.19}$$

There are varieties of choices for choosing matrix A. A good choice of A is

$$A = \begin{pmatrix} k_1 & 1 \\ 1 & (k_2 - \alpha) \end{pmatrix} \tag{15.20}$$

Matrix A satisfies the condition 15.19 for $k_1, \; k_2 > 0$. Therefore, multi-switching synchronization is achieved.

15.4.2 Case 2

For Switch 2 defined by Eq. (15.12), the time derivative of the errors is given by

$$\frac{d^q e_{21}}{dt^q} = y_2 - x_2 + u_1(t)$$
$$= e_{22} + x_3 - x_2 + u_1(t)$$
$$\frac{d^q e_{22}}{dt^q} = y_1 - y_1^3 - \alpha y_2 + f\cos\omega t + u_2(t) - \beta_1 x_1 + \beta_2 x_2 + \beta_3 x_3 - \beta_4 x_1^3$$
$$= e_{21} - \alpha e_{22} - \alpha x_3 + x_1 + f_{21}(x,y) + u_2$$

where f_{21} are nonliear terms in e_{11} and e_{12} given as

$$f_{21} = -y_1^3 + f\cos\omega t - \beta_1 x_1 + \beta_2 x_2 + \beta_3 x_3 - \beta_4 x_1^3$$

Theorem 2: *If the control function $u_1(t)$ and $u_2(t)$ are chosen such that*

$$\begin{aligned} u_1(t) &= -x_3 + x_2 + k_1 e_{21} + e_{22} \\ u_2(t) &= \alpha x_3 - x_1 - f_{21}(x,y) + e_{21} + (k_2 - \alpha e_{22}) \end{aligned} \tag{15.21}$$

then the drive system 15.9 *will achieve multiswitching synchronization with the response system* 15.10

Proof: The method of active control is employed to prove theorem 2.

We redefine the controller in order to eliminate nonlinear terms in e_{21} and e_{22}, so,

$$\begin{aligned} u_1(t) &= -x_3 + x_2 + V_{21}(t) \\ u_2(t) &= \alpha x_3 - x_1 - f_{21}(x,y) + V_{22}(t) \end{aligned}$$

where $V_{21}(t)$ and $V_{22}(t)$ are virtual control functions to be determined.

Following the same procedure as in Switch 1, we have

$$\begin{pmatrix} V_{11}(t) \\ V_{12}(t) \end{pmatrix} = A \begin{pmatrix} e_{11} \\ e_{12} \end{pmatrix}$$

Matrix B is chosen as

$$B = \begin{pmatrix} k_1 & 1 \\ 1 & (k_2 - \alpha) \end{pmatrix} \tag{15.22}$$

This yields

$$V_{21} = k_1 e_{21} + e_{22} \tag{15.23}$$

$$V_{22} = e_{21} + (k_2 - \alpha) e_{22} \tag{15.24}$$

This function results in control functions given by Eq. (15.21) to give multiswitching synchronization with switch 2.

15.4.3 Case 3

For Switch 3 defined by Eq. (15.13), the time derivative of the errors is given by

$$\frac{d^q e_{31}}{dt^q} = y_2 + u_1(t)$$
$$= e_{32} + x_1 - x_3 + u_1(t)$$
$$\frac{d^q e_{32}}{dt^q} = y_1 - y_1^3 - \alpha y_2 + f\cos\omega t + u_2 - x_2$$
$$= e_{31} - \alpha e_{32} + f_{31}(x, y) + u_2$$

where f_{13} are nonliear terms in e_{11} and e_{12} given as

$$f_{13} = -y_1^3 - \alpha x_1 + f\cos\omega t$$

Theorem 3: *If the control function* $u_1(t)$ *and* $u_2(t)$ *are chosen such that*

$$\begin{aligned} u_1(t) &= -x_1 + x_3 + k_1 e_{31} + e_{32} \\ u_2(t) &= -f_{13}(x, y) + e_{31} + (k_2 - \alpha)e_{32} \end{aligned} \tag{15.25}$$

then the drive system 15.9 *will achieve multiswitching synchronization with the response system* 15.10

Proof: The controller is defined as

$$u_1(t) = V_{31}(t) - x_1 + x_3$$
$$u_2(t) = -f_{13}(x, y) + V_{32}$$

where $V_{31}(t)$ and $V_{32}(t)$ are virtual control functions to be determined.

Following the same procedure as in Switch 1, we have

$$\begin{pmatrix} V_{31}(t) \\ V_{32}(t) \end{pmatrix} = C\begin{pmatrix} e_{31} \\ e_{32} \end{pmatrix}$$

Matrix C is chosen as

$$C = \begin{pmatrix} k_1 & 1 \\ 1 & (k_2 - \alpha) \end{pmatrix} \tag{15.26}$$

This yields

$$V_{31} = k_1 e_{31} + e_{32} \tag{15.27}$$

$$V_{32} = e_{31} + (k_2 - \alpha)e_{32} \tag{15.28}$$

This function results in control functions given by Eq. (15.25) to give multiswitching synchronization with switch 3.

15.4.4 Case 4

For Switch 4 defined by Eq. (15.14), the time derivative of the errors is given by

$$\begin{aligned}\frac{d^q e_{41}}{dt^q} &= y_2 - x_3 + u_1(t)\\ &= e_{42} + u_1(t)\\ \frac{d^q e_{42}}{dt^q} &= y_1 - y_1^3 - \alpha y_2 + f\cos\omega t + u_2(t) + \beta_1 x_1 + \beta_2 x_2 + \beta_3 x_3 - \beta_4 x_1^3\\ &= e_{41} - \alpha e_{42} - f_{41}(x,y) + u_2\end{aligned}$$

where f_{41} are nonliear terms in e_{11} and e_{12} given as

$$f_{41} = x_2 - \alpha x_3 - y_1^3 + f\cos\omega t + \beta_1 x_1 + \beta_2 x_2 + \beta_3 x_3 - \beta_4 x_1^3$$

Theorem 4: *If the control function* $u_1(t)$ *and* $u_2(t)$ *are chosen such that*

$$\begin{aligned}u_1(t) &= k_1 e_{41} + e_{42}\\ u_2(t) &= -f_{41}(x,y) + e_{41} + (k_2 - \alpha)e_{42}\end{aligned} \tag{15.29}$$

then the drive system 15.9 *will achieve multiswitching synchronization with the response system* 15.10

Proof: We redefine $u_1(t)$ and $u_2(t)$ to eliminate all the nonlinear terms in e_{41} and e_{42}

$$\begin{aligned}u_1(t) &= V_{41}(t)\\ u_2(t) &= -f_{41}(x,y) + V_{42}(t)\end{aligned}$$

$$\begin{pmatrix} V_{41}(t)\\ V_{42}(t)\end{pmatrix} = D\begin{pmatrix} e_{41}\\ e_{42}\end{pmatrix}$$

where D is a 2×2 constant matrix. In order to make the closed-loop system stable, the matrix A should be chosen such that the eigenvalues λ_i of D satisfies

$$|\arg(\lambda_i)| > 0.5\pi\alpha,\ i = 1, 2, \ldots.$$

We chose D as

$$D = \begin{pmatrix} k_1 & 1 \\ 1 & (k_2 - \alpha) \end{pmatrix} \tag{15.30}$$

Matrix D satisfies the condition 15.19 for $k_1,\ k_2 > 0$. Therefore, multi-switching synchronization is achieved.

15.4.5 Case 5

For Switch 5 defined by Eq. (15.15), the time derivative of the errors is given by

$$\frac{d^q e_{51}}{dt^q} = y_2 + \beta_1 x_1 + \beta x_2 + \beta x_3 - \beta x_1^3 + u_1(t)$$

$$= e_{51} + x_1 + \beta_1 x_1 + \beta x_2 + \beta x_3 - \beta x_1^3 + u_1(t)$$
$$= e_{51} + f_{51}(x, y) + u_1$$

where f_{51} is defined as:

$$f_{51} = x_1 + \beta_1 x_1 + \beta_2 x_2 + \beta_3 x_3 - \beta_4 x_1^3$$
$$\frac{d^q e_{52}}{dt^q} = y_1 - y_1^3 - \alpha y_2 + f\cos\omega t + u_2(t) - x_2$$

$$= e_{51} - x_3 - y_1^3 - \alpha x_1 + f\cos\omega t - x_2 - \alpha e_{52} + u_2$$

where f_{52} are nonlinear terms in e_{11} and e_{12} given as

$$f_{52} = x_3 - y_1^3 - \alpha x_1 + f\cos\omega t - x_2$$
$$\frac{d^q e_{52}}{dt^q} = e_{51} - \alpha e_{52} + f_{52}(x, y) + u_2$$

Theorem 5: *If the control function $u_1(t)$ and $u_2(t)$ are chosen such that*

$$\begin{aligned} u_1(t) &= -f_{51}(x, y) + k_1 e_{51} + e_{52} \\ &= -f_{52}(x, y) + e_{51} + (k_2 - \alpha) e_{52} \end{aligned} \tag{15.31}$$

then the drive system 15.9 *will achieve multiswitching synchronization with the response system* 15.10

Proof: The method of active control is employed to prove theorem 5.

In order to eliminate the nonlinear term in e_1 and e_2, we define the active control input $u_1(t)$ and $u_2(t)$ as

$$u_1(t) = -f_{51}(x, y) + V_{51}(t)$$
$$u_2(t) = -f_{52}(x, y) + V_{52}(t)$$

The synchronization error is a linear system with active control inputs $V_{51}(t)$ and $V_{52}(t)$. The feedback control which would stabilize the system so that e_{51} and e_{52} converge to zero at time $t \to \infty$ are to be designed. As a result, we chose

$$\begin{pmatrix} V_{51}(t) \\ V_{52}(t) \end{pmatrix} = E \begin{pmatrix} e_{51} \\ e_{52} \end{pmatrix}$$

where E is a 2×2 constant matrix. In order to make the closed-loop system stable, the matrix E should be chosen such that the eigenvalues λ_i of E satisfies

$$|\arg(\lambda_i)| > 0.5\pi\alpha, \; i = 1, 2, \ldots. \tag{15.32}$$

There are varieties of choices for choosing matrix E. A good choice of E is

$$E = \begin{pmatrix} k_1 & 1 \\ 1 & (k_2 - \alpha) \end{pmatrix} \tag{15.33}$$

Matrix E satisfies the condition (15.35) for $k_1, \; k_2 > 0$. Therefore, multiswitching synchronization is achieved.

15.4.6 Case 6

For Switch 6 defined by Eq. (15.16), the time derivative of the errors is given by

$$\begin{aligned} \frac{d^q e_{61}}{dt^q} &= y_2 - \beta_1 x_1 + \beta_2 x_2 + \beta_3 x_3 - \beta_4 x_1^3 + u_1(t) \\ &= e_{62} - f_{61}(x, y) + u_1(t) \\ \frac{d^q e_{62}}{dt^q} &= y_1 \quad y_1^3 - \alpha y_2 + f\cos\omega t + u_2(t) - x_3 \\ &= e_{61} - \alpha e_{62} + f_{62}(x, y) + u_2 \end{aligned}$$

where f_{62} are nonlinear terms in e_{11} and e_{12} given as

$$\begin{aligned} f_{61} &= x_2 + \beta_1 x_1 + \beta_2 x_2 + \beta_3 x_3 - \beta_4 x_1^3 \\ f_{62} &= -\alpha x_3 - y_1^3 + f\cos\omega t \end{aligned}$$

Theorem 6: *If the control function $u_1(t)$ and $u_2(t)$ are chosen such that*

$$\begin{aligned} u_1(t) &= k_1 e_{11} + e_{12} \\ u_2(t) &= -f_{12}(x, y) + e_{11} + (k_2 - \alpha) e_{12} \end{aligned} \tag{15.34}$$

then the drive system 15.9 *will achieve multiswitching synchronization with the response system* 15.10

Proof: The method of active control is employed to prove theorem 6.

In order to eliminate the nonlinear term in e_61 and e_62, we define the active control input $u_1(t)$ and $u_2(t)$ as

$$u_1(t) = -f_{61}(x,y) + V_{61}(t)$$
$$u_2(t) = -f_{62}(x,y) + V_{62}(t)$$

The synchronization error is a linear system with active control inputs $V_{61}(t)$ and $V_{62}(t)$. The feedback control which would stabilize the system so that e_{61} and e_{62} converge to zero at time $t\to\infty$ are to be designed. As a result, we chose

$$\begin{pmatrix} V_{61}(t) \\ V_{62}(t) \end{pmatrix} = F\begin{pmatrix} e_{61} \\ e_{62} \end{pmatrix}$$

where F is a 2×2 constant matrix. In order to make the closed-loop system stable, the matrix F should be chosen such that the eigenvalues λ_i of F satisfies

$$|\arg(\lambda_i)| > 0.5\pi\alpha,\ i = 1,2,\ldots. \tag{15.35}$$

There are varieties of choices for choosing matrix F. A good choice of F is

$$F = \begin{pmatrix} k_1 & 1 \\ 1 & (k_2-\alpha) \end{pmatrix} \tag{15.36}$$

Matrix F satisfies the condition (15.35) for $k_1,\ k_2 > 0$. Therefore, multi-switching synchronization is achieved.

15.5 NUMERICAL SIMULATION OF RESULTS

To verify the effectiveness of the proposed controllers for the synchronization of the drive and response systems, solutions of fractional order systems were implemented using algorithms by Petras (2011). According to Petras (2011), the general numerical solution of the fractional differential equation

$${}_aD_t^q y(t) = f(y(t),t) \tag{15.37}$$

can be expressed as

$$y(t_k) = f(y(t_k),t_k)h^q - \sum_{j=v}^{k} c_j^{(q)} y(t_{t-j}) \tag{15.38}$$

where $c_i^{(q)}$ is given as

$$c_0^{(q)} = 1$$
$$c_j^{(q)} = \left(1 - \frac{1+q}{j}c_{j-1}^q\right) \tag{15.39}$$

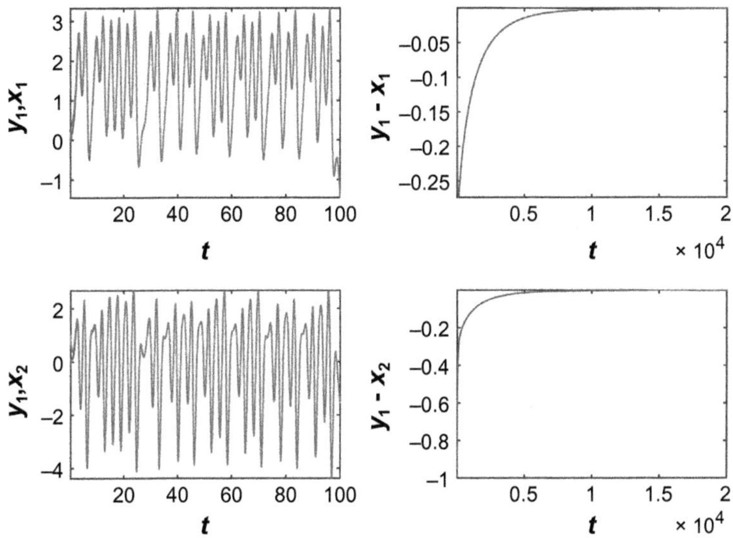

FIGURE 15.3 Realization of synchronization between drive and response using switch 1 (Eq. 15.11).

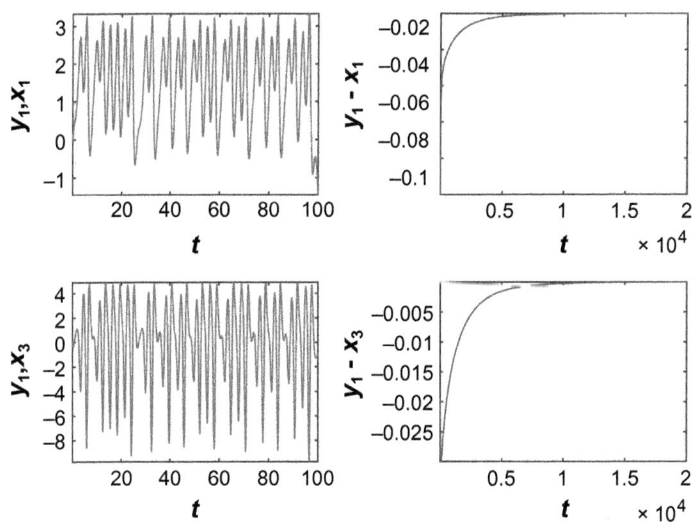

FIGURE 15.4 Realization of synchronization between drive and response using switch 2 (Eq. 15.12).

The initial conditions $x_i(1,2,3)$ and $y_i(1,2)$ were taken as $(0.1,1,0.02)$ and $(0,0)$, respectively. The order of both the drive and response were taken to be 0.9 with an integration time step of 0.005. The system parameters were chosen as $\beta_1 = -5.5$, $\beta_2 = 3.5$, $\beta_3 = 0.4$, $\beta_4 = -1$, $b = 0.15$, $\alpha = 1$, $\omega = 1$, $f = 0.3$, and $\beta = 1$ to ensure that both the drive and response systems are chaotic. The results obtained are shown in Figs. 15.3–15.8 for switch 11–16

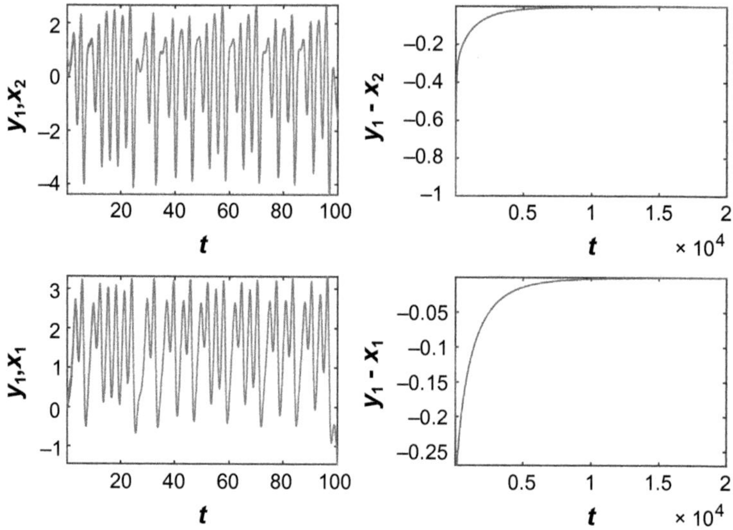

FIGURE 15.5 Realization of synchronization between drive and response using switch 3 (Eq. 15.13).

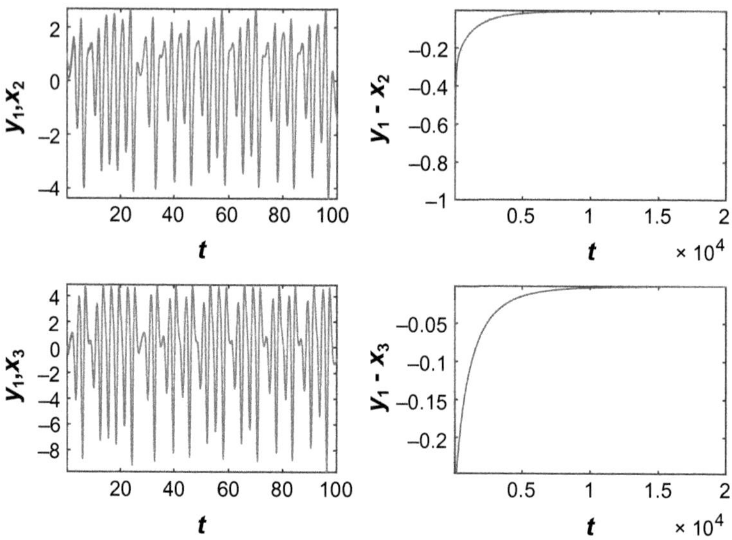

FIGURE 15.6 Realization of synchronization between drive and response using switch 4 (Eq. 15.14).

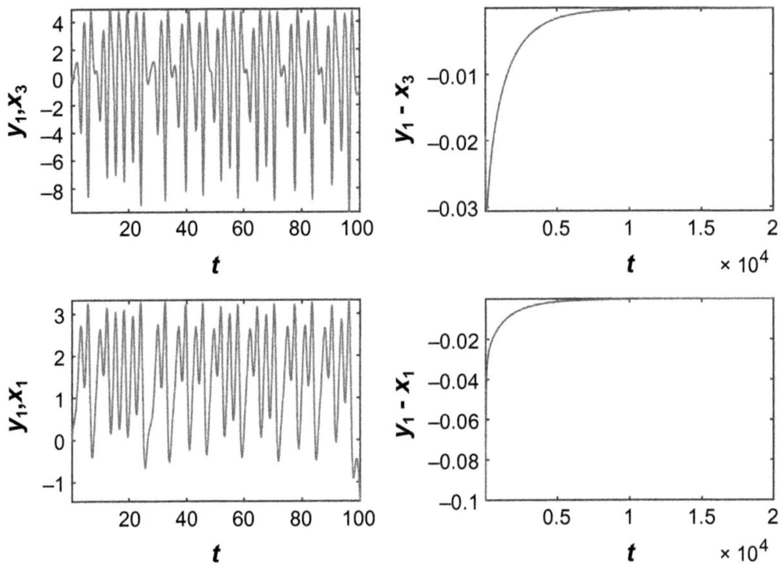

FIGURE 15.7 Realization of synchronization between drive and response using switch 5 (Eq. 15.15).

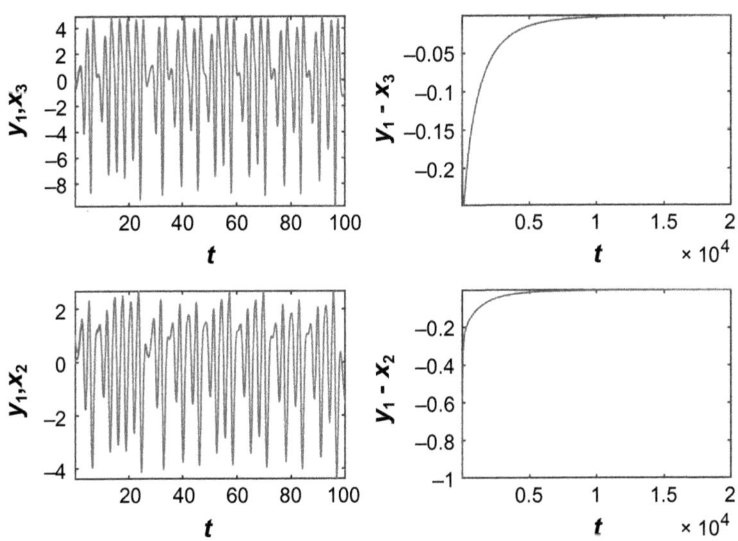

FIGURE 15.8 Realization of synchronization between drive and response using switch 5 (Eq. 15.16).

respectively. From the results presented, it could be observed that synchronization was achieved by the convergence of the trajectories of the systems and error state variables to zero. Fast and exponential convergence of the error could be observed in all the different switches. The proposed controllers, by virtue of the convergence, could be said to be effective. By applying the proposed controllers in this work added complexity, which translates to better security will be introduced into the communication system. The multiswitching component in this work can be used as a prearranged key in a secure communication system.

15.6 CONCLUSION

The multiswitching synchronization of two fractional order chaotic systems with different dimensions have been implemented using the method of active controls. Six possible switches were proposed and controllers designed for each of the six switches. Numerical simulations were carried out to test the effectiveness of the proposed controllers. From the results presented, the controllers were found to be effective by the convergence of the error dynamics to zero.

However, there is the opportunity for further studies. Practical implementation of the proposed controllers using electronic simulation is hereby recommended. Also recommended is the implementation of multiswitching combination and combinaiton–combination synchronization of fractional order system with different dimensions. In particular, multiswitching synchronization of two-, three-, and four-dimensional fractional order systems is proposed.

REFERENCES

Adelakun, O., Ogunjo, S., Fuwape, I., 2017. Dynamics of delayed memristive systems in combination chaotic circuits. Advances in Memristors, Memristive Devices and Systems. Springer International Publishing, pp. 477–492. Available from: https://link.springer.com/chapter/10.1007%2F978-3-319-51724-7_19.

Ajayi, A.A., Ojo, S.K., Vincent, E.U., Njah, N.A., 2014. Multiswitching synchronization of a driven hyperchaotic circuit using active backstepping. J. Nonlin. Dynam. 2014. Available from: http://dx.doi.org/10.1155/2014/918586.

Arneodo, A., Coullet, P., Spiegel, E.A., 1982. Chaos in a finite macroscopic system. Phys. Lett. A 92 (8), 369–373.

Azar, A.T., Vaidyanathan, S., 2015a. Chaos Modeling and Control Systems Design. Vol. 581 of Studies inComputational Intelligence. Studies in Computational Intelligence. Springer, Germany.

Azar, A.T., Vaidyanathan, S., 2015b. Computational Intelligence Applications in Modelling and Control. Vol. 575. Studies inComputational Intelligence. Springer, Berlin, Germany.

Azar, A.T., Vaidyanathan, S., 2015c. Handboook of Research on Advanced Intelligent Control Engineering and Automation. IGI Global, New York, USA.

Azar, A.T., Vaidyanathan, S., 2016. Advances in Chaos Theory and Intelligent Control, Vol. 337. Springer, Berlin, Germany.

Azar, A.T., Vaidyanathan, S., Ouannas, A., 2017a. Fractional Order Control and Synchronization of Chaotic Systems. Vol. 688. Studies inComputational Intelligence. Springer, Berlin, Germany.

Azar, A.T., Volos, C., Gerodimos, N.A., Tombras, G.S., Pham, V.-T., Radwan, A.G., et al., 2017b. A novel chaotic system without equilibrium: dynamics, synchronization, and circuit realization. Complexity 2017 (Article ID 7871467), 11 pages.

Azar, A.T., Kumar, J., Kumar, V., Rana, K.P.S., 2018a. Control of a Two Link Planar Electrically-Driven Rigid Robotic Manipulator Using Fractional Order SOFC. Springer International Publishing, Cham, pp. 57−68. Available from: https://doi.org/10.1007/978-3-319-64861-3_6.

Azar, A.T., Ouannas, A., Singh, S., 2018b. Control of New Type of Fractional Chaos Synchronization. Springer International Publishing, Cham, pp. 47−56. Available from: https://doi.org/10.1007/978-3-319-64861-3_5.

Boccaletti, S., Kurths, J., Osipov, G., Valladares, D.L., Zhou, C.S., 2002. The synchronization of chaotic systems. Phys. Rep. 366, 1−101. Available from: https://doi.org/10.1016/S0370-1573(02)00137-0.

Boulkroune, A., Bouzeriba, A., Bouden, T., Azar, A.T., 2016a. Fuzzy adaptive synchronization of uncertain fractional-order chaotic systems. Advances in Chaos Theory and IntelligentControl. Springer, Berlin, Germany, pp. 681−697.

Boulkroune, A., Hamel, S., Azar, A.T., Vaidyanathan, S., 2016b. Fuzzy control-based function synchronization of unknown chaotic systems with dead-zone input. In: Azar, A.T., Vaidyanathan, S. (Eds.), Advances in Chaos Theory and Intelligent Control. Springer International Publishing, Cham, pp. 699−718.

Ge, Z.-M., Ou, C.-Y., 2008. Chaos synchronization of fractional order modified duffing systems with parameters excited by a chaotic signal. Chaos Solitons Fractals 35 (4), 705−717. Available from: http://www.sciencedirect.com/science/article/pii/S096007790600484X.

Golmankhaneh, A.K., Arefi, R., Baleanu, D., 2015. Synchronization in a nonidentical fractional order of a proposed modified system. J. Vibrat. Control 21 (6), 1154−1161.

Grassi, G., Ouannas, A., Azar, A.T., Radwan, A.G., Volos, C., Pham, V.-T., et al., 2017. Chaos synchronisation of continuous systems via scalar signal. In: 6th International Conference on Modern Circuits and Systems Technologies (MOCAST), pp. 1−4.

Hajipour, A., Aminabadi, S.S., 2016. Synchronization of chaotic arneodo system of incommensurate fractional order with unknown parameters using adaptive method. Optik-Int. J. Light Electron Optics 127 (19), 7704−7709.

He, G.-T., Luo, M.-K., 2012. Dynamic behavior of fractional order duffing chaotic system and its synchronization via singly active control. Appl. Math. Mech. 33 (5), 567−582. Available from: http://dx.doi.org/10.1007/s10483-012-1571-6.

Hu, N., Wen, X., 2003. The application of duffing oscillator in characteristic signal detection of early fault. J. Sound Vib. 268 (5), 917−931. Available from: http://www.sciencedirect.com/science/article/pii/S0022460X03000026.

Jiang, Y., Xia, B., Zhao, X., Nguyen, T., Mi, C., de Callafon, R.A., 2017. Data-based fractional differential models for non-linear dynamic modeling of a lithium-ion battery. Energy 135, 171−181. Available from: http://www.sciencedirect.com/science/article/pii/S0360544217311064.

Li, Z., Chen, D., Zhu, J., Liu, Y., 2015. Nonlinear dynamics of fractional order duffing system. Chaos Solitons Fractals 81, 111−116.

Lu, J.G., 2005. Chaotic dynamics and synchronization of fractional-order arneodoâ€™s systems. Chaos Solitons Fractals 26 (4), 1125–1133.

Meghni, B., Dib, D., Azar, A.T., 2017a. A second-order sliding mode and fuzzy logic control to optimal energy management in wind turbine with battery storage. Neural Comput. Applicat. 28 (6), 1417–1434.

Meghni, B., Dib, D., Azar, A.T., Ghoudelbourk, S., Saadoun, A., 2017b. Robust adaptive supervisory fractional order controller for optimal energy management in wind turbine with battery storage. In: Azar, A.T., Vaidyanathan, S., Ouannas, A. (Eds.), Fractional Order Control and Synchronization of Chaotic Systems. Springer International Publishing, Cham, pp. 165–202.

Meghni, B., Dib, D., Azar, A.T., Saadoun, A., 2017c. Effective supervisory controller to extend optimal energy management in hybrid wind turbine under energy and reliability constraints. Int. J. Dynam. Control 1–15.

Motallebzadeh, F., Motallebzadeh, F., Dadras, S., Ozgoli, S., 2009. Controlling chaos in arneodo system. In: Control and Automation, 2009. MED'09. 17th Mediterranean Conference on. IEEE, pp. 314–319.

Ogunjo, S., Ojo, K., Fuwape, I., 2017. Comparison of three different synchronization schemes for fractional chaotic systems. Fractional Order Control and Synchronization of Chaotic Systems. Springer International Publishing, pp. 471–495. Available from: https://doi.org/10.1007/978-3-319-50249-6_16.

Ogunjo, S.T., 2013. Increased and reduced order synchronization of 2d and 3d dynamical systems. Int. J. Nonlin. Sci. 16 (2), 105–112.

Ojo, K., Ogunjo, S., 2012. Synchronization of 4d rabinovich hyperchaotic system for secure communication. J. Niger. Assoc. Math. Phys. 21, 35–40.

Ojo, K., Njah, A., Ogunjo, S., 2013a. Comparison of backstepping and modified active control in projective synchronization of chaos in an extended bonhoffer van der pol oscillator. Pramana 80 (5), 825–835. Available from: https://doi.org/10.1007/s12043-013-0526-3.

Ojo, K., Ogunjo, S., Williams, O., 2013b. Mixed tracking and projective synchronization of 5d hyperchaotic system using active control. Cybern. Phys. 2 (1), 31–36.

Ojo, K., Njah, A., Ogunjo, S., Olusola, O., 2014a. Reduced order function projective combination synchronization of three josephson junctions using backstepping technique. Nonlin. Dynam. System Theory 14 (2), 119.

Ojo, K., Njah, A.N., Ogunjo, S.T., Olusola, O.I., et al., 2014b. Reduced order hybrid function projective combination synchronization of three josephson junctions. Arch. Control Sci. 24 (1), 99–113. Available from: https://doi.org/10.2478/acsc-2014-0007.

Ojo, K.S., Ogunjo, S.T., Njah, A.N., Fuwape, I.A., 2014c. Increased-order generalized synchronization of chaotic and hyperchaotic systems. Pramana 1–13. Available from: https://doi.org/10.1007/s12043-014-0835-1.

Ouannas, A., Azar, A.T., Abu-Saris, R., 2016. A new type of hybrid synchronization between arbitrary hyperchaotic maps. Int. J. Mach. Learn. Cybern. 1–8.

Ouannas, A., Azar, A.T., Vaidyanathan, S., 2017a. New hybrid synchronization schemes based on coexistence of various types of synchronization between master-slave hyperchaotic systems. Int. J. Computer Applications in Technology 55 (2), 112–120.

Ouannas, A., Azar, A.T., Vaidyanathan, S., 2017b. On a simple approach for q-s synchronization of chaotic dynamical systems in continuous-time. Int. J. Comput. Sci. Math. 8 (1), 20–27.

Ouannas, A., Azar, A.T., Ziar, T., 2017c. On inverse full state hybrid function projective synchronization for continuous-time chaotic dynamical systems with arbitrary dimensions. Diff. Eq. Dynam. Systems. Available from: https://doi.org/10.1007/s12591-017-0362-x.

Ouannas, A., Azar, A.T., Ziar, T., Radwan, A.G., 2017d. Generalized synchronization of different dimensional integer-order and fractional order chaotic systems. In: Azar, A.T., Vaidyanathan, S., Ouannas, A. (Eds.), Fractional Order Control and Synchronization of Chaotic Systems. Springer International Publishing, Cham, pp. 671–697.

Ouannas, A., Azar, A.T., Ziar, T., Radwan, A.G., 2017e. A study on coexistence of different types of synchronization between different dimensional fractional chaotic systems. In: Azar, A.T., Vaidyanathan, S., Ouannas, A. (Eds.), Fractional Order Control and Synchronization of Chaotic Systems. Springer International Publishing, Cham, pp. 637–669.

Ouannas, A., Azar, A.T., Ziar, T., Vaidyanathan, S., 2017f. Fractional inverse generalized chaos synchronization between different dimensional systems. In: Azar, A.T., Vaidyanathan, S., Ouannas, A. (Eds.), Fractional Order Control and Synchronization of Chaotic Systems. Springer International Publishing, Cham, pp. 525–551.

Ouannas, A., Azar, A.T., Ziar, T., Vaidyanathan, S., 2017g. A new method to synchronize fractional chaotic systems with different dimensions. In: Azar, A.T., Vaidyanathan, S., Ouannas, A. (Eds.), Fractional Order Control and Synchronization of Chaotic Systems. Springer International Publishing, Cham, pp. 581–611.

Ouannas, A., Azar, A.T., Ziar, T., Vaidyanathan, S., 2017h. On new fractional inverse matrix projective synchronization schemes. In: Azar, A.T., Vaidyanathan, S., Ouannas, A. (Eds.), Fractional Order Control and Synchronization of Chaotic Systems. Springer International Publishing, Cham, pp. 497–524.

Ouannas, A., Grassi, G., Azar, A.T., Radwan, A.G., Volos, C., Pham, V.-T., et al., 2017i . Deadbeat synchronization control in discrete-time chaotic systems. In: 6th International Conference on Modern Circuits and Systems Technologies (MOCAST), pp. 1–4.

Pecora, L.M., Carroll, T.L., 1990. Synchronization in chaotic systems. Phys. Rev. Lett. 64, 821–824. Available from: https://link.aps.org/doi/10.1103/PhysRevLett.64.821.

Petras, I., 2011. Fractional-Order Nonlinear Systems: Modeling, Analysis and Simulation. Springer Science & Business Media, New York.

Pham, V.-T., Vaidyanathan, S., Volos, C.K., Azar, A.T., Hoang, T.M., Van Yem, V., 2017. A three-dimensional no-equilibrium chaotic system: analysis, synchronization and its fractional order form. In: Azar, A.T., Vaidyanathan, S., Ouannas, A. (Eds.), Fractional Order Control and Synchronization of Chaotic Systems. Springer International Publishing, Cham, pp. 449–470.

Rivero, M., Trujillo, J.J., Vzquez, L., Velasco, M.P., 2011. Fractional dynamics of populations. Appl. Math. Comput. 218 (3), 1089–1095. special Issue in Honour of Hari M. Srivastava on his 70th Birth Anniversary. http://www.sciencedirect.com/science/article/pii/S0096300311003717.

Shah, D.K., Chaurasiya, R.B., Vyawahare, V.A., Pichhode, K., Patil, M.D., 2017. {FPGA} implementation of fractional-order chaotic systems. {AEU} – Int. J. Electr. Communicat. 78, 245–257. Available from: http://www.sciencedirect.com/science/article/pii/S1434841117303643.

Singh, S., Azar, A.T., Ouannas, A., Zhu, Q., Zhang, W., Na, J., 2017. Sliding mode control technique for multi-switching synchronization of chaotic systems. In: 9th International Conference on Modelling, Identification and Control (ICMIC 2017), July 10-12, 2017, Kunming, China.

Soliman, N.S., Said, L.A., Azar, A.T., Madian, A.H., Radwan, A.G., Ounnas, A., 2017. Fractional controllable multi-scroll v-shape attractor with parameters effect. In: 6th International Conference on Modern Circuits and Systems Technologies (MOCAST). pp. 1–4.

Strogatz, S.H., 2000. Nonlinear Dynamics and Chaos: Applications to Physics, Biology, Chemistry, and Engineering: With Applications to Physics, Biology, Chemistry and Engineering.

Tavazoei, M.S., Haeri, M., 2008. Synchronization of chaotic fractional-order systems via active sliding mode controller. Phys. A: Statist. Mech. Its Applicat 387 (1), 57–70. Available from: http://www.sciencedirect.com/science/article/pii/S0378437107008849.

Tolba, M.F., AbdelAty, A.M., Saida, L.A., Elwakil, A.S., Azar, A.T., Madian, A.H., et al., 2017. Fpga realization of caputo and grnwald-letnikov operators. In: 6th International Conference on Modern Circuits and Systems Technologies (MOCAST). pp. 1–4.

Uar, A., Lonngren, K.E., Bai, E.-W., 2008. Multi-switching synchronization of chaotic systems with active controllers. Chaos Solitons Fractals 38 (1), 254–262. Available from: http://www.sciencedirect.com/science/article/pii/S0960077906010484.

Vaidyanathan, S., Azar, A.T., 2015a. Analysis and control of a 4-d novel hyperchaotic system. In: Azar, A.T., Vaidyanathan, S. (Eds.), Chaos Modeling and Control Systems Design. Vol. 581 of Studies in Computational Intelligence. Springer, Berlin, Germany, pp. 19–38.

Vaidyanathan, S., Azar, A.T., 2015b. Analysis, control and synchronization of a nine-term 3-d novel chaotic system. In: Azar, A.T., Vaidyanathan, S. (Eds.), Chaos Modeling and Control Systems Design. Vol. 581 of Studies in Computational Intelligence. Springer, Berlin, Germany, pp. 3–17.

Vaidyanathan, S., Azar, A.T., 2015c. Anti-synchronization of identical chaotic systems using sliding mode control and an application to vaidyanathan-madhavan chaotic systems. In: Azar, A. T., Zhu, Q. (Eds.), Advances and Applications in Sliding Mode Control Systems. Vol. 576 of Studies in Computational Intelligence. Springer, Berlin, Germany, pp. 527–547.

Vaidyanathan, S., Azar, A.T., 2015d. Hybrid synchronization of identical chaotic systems using sliding mode control and an application to vaidyanathan chaotic systems. In: Azar, A.T., Zhu, Q. (Eds.), Advances and Applications in Sliding Mode Control Systems. Vol. 576 of Studies in Computational Intelligence. Springer, Berlin, Germany, pp. 549–569.

Vaidyanathan, S., Azar, A.T., 2016a. A novel 4-D four-wing chaotic system with four quadratic nonlinearities and its synchronization via adaptive control method. Advances in Chaos Theory and IntelligentControl. Springer, Berlin, Germany, pp. 203–224.

Vaidyanathan, S., Azar, A.T., 2016b. Adaptive backstepping control and synchronization of a novel 3-D jerk system with an exponential nonlinearity. Advances in Chaos Theory and IntelligentControl. Springer, Berlin, Germany, pp. 249–274.

Vaidyanathan, S., Azar, A.T., 2016c. Adaptive control and synchronization of Halvorsen circulant chaotic systems. Advances in Chaos Theory and IntelligentControl. Springer, Berlin, Germany, pp. 225–247.

Vaidyanathan, S., Azar, A.T., 2016d. Dynamic analysis, adaptive feedback control and synchronization of an eight-term 3-D novel chaotic system with three quadratic nonlinearities. Advances in Chaos Theory and IntelligentControl. Springer, Berlin, Germany, pp. 155–178.

Vaidyanathan, S., Azar, A.T., 2016e. Qualitative study and adaptive control of a novel 4-d hyperchaotic system with three quadratic nonlinearities. In: Azar, A.T., Vaidyanathan, S. (Eds.), Advances in Chaos Theory and Intelligent Control. Springer International Publishing, Cham, pp. 179–202.

Vaidyanathan, S., Azar, A.T., 2016f. Takagi-Sugeno fuzzy logic controller for Liu-Chen four-scroll chaotic system. Int. J. Intellig. Eng. Inf. 4 (2), 135–150.

Vaidyanathan, S., Azar, A.T., Rajagopal, K., Alexander, P., 2015a. Design and spice implementation of a 12-term novel hyperchaotic system and its synchronisation via active control. Int. J. Model. Identif. Control 23 (3), 267–277.

Vaidyanathan, S., Idowu, B.A., Azar, A.T., 2015b. Backstepping controller design for the global chaos synchronization of sprott's jerk systems. In: Azar, A.T., Vaidyanathan, S. (Eds.), Chaos Modeling and Control Systems Design. Vol. 581 of Studies in Computational Intelligence. Springer, Berlin, Germany, pp. 39–58.

Vaidyanathan, S., Sampath, S., Azar, A.T., 2015c. Global chaos synchronisation of identical chaotic systems via novel sliding mode control method and its application to zhu system. Int. J. Model. Identif. Control 23 (1), 92–100.

Vincent, U.E., Saseyi, A.O., McClintock, P.V.E., 2015. Multi-switching combination synchronization of chaotic systems. Nonlin. Dynam. 80 (1), 845–854. Available from: http://dx.doi.org/10.1007/s11071-015-1910-y.

Wang, W., Fan, Y., 2015. Synchronization of arneodo chaotic system via backstepping fuzzy adaptive control. Optik-Int. J. Light Electr. Opt. 126 (20), 2679–2683.

Wang, Z., Volos, C., Kingni, S.T., Azar, A.T., Pham, V.-T., 2017. Four-wing attractors in a novel chaotic system with hyperbolic sine nonlinearity. Optik-Int. J. Light Electr. Opt. 131, 1071–1078.

Wu, Y., Deng, D., Ma, S., 2013. Application of duffing oscillator from big cycle motion to chaotic motion in the detection of water treeing in xlpe power cable. In: 2013 IEEE International Conference on Signal Processing, Communication and Computing (ICSPCC 2013), pp. 1–4.

Zheng, S., 2016. Multi-switching combination synchronization of three different chaotic systems via nonlinear control. Optik-Int. J. Light Electr. Opt. 127 (21), 10247–10258. Available from: http://www.sciencedirect.com/science/article/pii/S0030402616308774.

Zhu, Q., Azar, A.T., 2015. Complex System Modelling and Control Through Intelligent Soft Computations, Vol. 319. Studies in Fuzziness and Soft Computing, Berlin, Germany.

Chapter 16

Dynamics, Synchronization and Fractional Order Form of a Chaotic System With Infinite Equilibria

Viet–Thanh Pham[1,2], Gokul P.M.[2], Tomasz Kapitaniak[2], Christos Volos[3] and Ahmad Taher Azar[4,5]

[1]Hanoi University of Science and Technology, Hanoi, Vietnam, [2]Lodz University of Technology, Lodz, Poland, [3]Aristotle University of Thessaloniki, Thessaloniki, Greece, [4]Faculty of Computers and Information, Benha University, Benha, Egypt, [5]School of Engineering and Applied Sciences, Nile University, Giza, Egypt

16.1 INTRODUCTION

In the past decades, the chaotic phenomena in nature, environmental science, physics, economics, and especially in engineering have attracted the interest of the research community (Lorenz, 1963; Strogatz, 1994; Sprott, 2003; Chen and Yu, 2003; Azar et al., 2017a,b; Azar and Vaidyanathan, 2015a,b,c, 2016; Zhu and Azar, 2015; Pham et al., 2017c; Vaidyanathan et al., 2017a,b,c; Pham et al., 2017a; Moysis and Azar, 2017; Lamamra et al., 2017; Ouannas et al., 2017b). After the studies of Lorenz's system (Lorenz, 1963) and Rössler's system (Rössler, 1976), a great number of research works on chaotic systems have been reported (Chen and Ueta, 1999; Sprott, 1994, 2010). The literature on chaos has highlighted several special chaotic systems such as memristor-based systems (Wu et al., 2016; Wu and Wang, 2016), systems with multiwing butterfly chaotic attractors (Yu et al., 2010; Wang et al., 2017), chaotic flow with a continuously adjustable attractor dimension (Munmuangsaen et al., 2015), systems with multiscroll chaotic oscillator (Lin et al., 2015; Soliman et al., 2017), chaotic systems with heart-shaped attractors (Wu et al., 2016), chaotic systems with hyperbolic sine nonlinearity (Wang et al., 2017), chaotic systems with different families of hidden attractors (Pham et al., 2016d), hyperchaotic systems without equilibrium (Wang et al., 2012), or chaotic systems with time delay (Valli et al., 2014; Ahmad et al., 2016). Moreover, there is a large volume of

Mathematical Techniques of Fractional Order Systems. DOI: https://doi.org/10.1016/B978-0-12-813592-1.00016-7

published studies describing the role of chaos in practical applications (Yalcin et al., 2004; Volos et al., 2012, 2013; Boulkroune et al., 2016a,b). Chaos and nonlinear dynamics have been used in the design of chaotic masking communication applications (Cicek et al., 2016; Tolba et al., 2017), image encryption (Min et al., 2015), radar design (Liu et al., 2007; Esmaeili-Najafabadi et al., 2017), and secure communications (Mata-Machuca et al., 2012).

Equilibrium points play a critical role in the discovery of chaotic systems (Lü and Chen, 2006; Wang and Chen, 2012, 2013). Previous research has established that there is the presence of unstable equilibria in conventional chaotic systems, in which their chaos can be verified with the Shilnikov criterion (Shilnikov, 1965; Shilnikov et al., 1998). Recent developments in the field of chaos have led to a renewed interest in the number of equilibrium points in chaotic systems (Wang and Chen, 2013).

Although a considerable amount of literature has been published on chaotic systems with countable equilibrium, there are some works related to systems with uncountable equilibrium (Jafari and Sprott, 2013). Nine chaotic flows with a line equilibrium were introduced by Jafari and Sprott (Jafari and Sprott, 2013). Li and Sprott found chaotic flows with a single nonquadratic term, which have one or more lines of equilibrium points (Li and Sprott, 2014a). Li et al. proposed a method to control the amplitude of the variables in chaotic systems (Li et al., 2015a). The method could generate infinite lines of equilibrium points. Chen and Yang presented a Lorenz-type hyperchaotic system with a curve of equilibria (Chen and Yang, 2015). Bistability was investigated in a hyperchaotic system with a line equilibrium (Li et al., 2014a). Memristive system with a line of equilibria was reported in Li et al. (2014b) while a 4D memristive circuit with infinitely many stable equilibria was introduced in Li et al. (2015b). Zhou and Yang studied hyperchaos, chaos, and horseshoe in a 4D nonlinear system with an infinite number of equilibrium points (Zhou and Yang, 2014). A new class of chaotic systems with circular equilibrium was found by using a search program (Gotthans and Petržela, 2015). Gotthans et al. designed simple chaotic flow with circle and square equilibrium (Gotthans et al., 2016). A chaotic system with equilibria located on the rounded square loop was implemented by electronic components (Pham et al., 2016a). In addition, a chaotic system with different shapes of equilibria was constructed (Pham et al., 2016c). So far, however, there has been little discussion about chaotic systems with an open curve of equilibrium points (Pham et al., 2016b; Barati et al., 2016).

The aim of this chapter is to investigate a 3D autonomous chaotic system, which has an infinite number of equilibrium points. In the next section, we introduce the description of the system and discover its dynamics via various tools of nonlinear theory. Antisynchronization of new chaotic systems with infinite equilibria is studied in Section 16.3. Section 16.4 introduces the corresponding fractional form of the 3D system with infinite equilibria. Finally, conclusion r are drawn in Section 16.5.

16.2 MODEL AND DYNAMICS OF THE SYSTEM WITH AN INFINITE NUMBER OF EQUILIBRIUM POINTS

Recently, Gotthans et al. have proposed an approach for investigating new chaotic systems with an infinte number of equilibrium (Gotthans and Petržela, 2015; Gotthans et al., 2016). By constructing general models with expected features and applying a systematical search routine, authors have introduced chaotic flows with circle and square equilibrium (Jafari et al., 2013).

Based on the effective approach of Gotthans et al., in this work we consider a general form given by

$$\begin{cases} \dot{x} = -z \\ \dot{y} = xz^2 + a\mathrm{sgn}(z) \\ \dot{z} = f_1(x,y) + zf_2(x,y,z) \end{cases} \tag{16.1}$$

in which state variables are x, y, and z and a is a positive parameter. In the general form, two nonlinear functions are denoted as $f_1(x,y)$ and $f_2(x,y,z)$. It is noted that the signum function sgn(z) has been used in known systems because it can be conveniently implemented using an operational amplifier (Piper and Sprott, 2010). The signum function is defined by:

$$\mathrm{sgn}(z) = \begin{cases} -1, & x<0 \\ 0, & x=0 \\ 1, & x>0 \end{cases} \tag{16.2}$$

It is trivial to find the equilibrium points of general form (16.1) by solving the three following equations:

$$-z=0 \tag{16.3}$$

$$xz^2 + a\mathrm{sgn}(z) = 0 \tag{16.4}$$

$$f_1(x,y) + zf_2(x,y,z) = 0 \tag{16.5}$$

From Eq. (16.3) we have $z=0$. Thus, Eq. (16.4) is correct for all x. By substituting Eq. (16.3) into Eq. (16.5), we get

$$f_1(x,y) = 0. \tag{16.6}$$

It means that the equilibrium points of general form (16.1) are located on the curve (16.6). In this chapter we select the nonlinear function $f_1(x,y)$ as:

$$f_1(x,y) = x - be^y \tag{16.7}$$

where b is a positive parameter. The nonlinear function $f_2(x,y,z)$ is chosen as follows:

$$\begin{aligned} f_2(x,y,z) = {} & a_1x + a_2y + a_3z + a_4xy + a_5xz + a_6yz \\ & + a_7x^2 + a_8y^2 + a_9z^2 + a_{10} \end{aligned} \tag{16.8}$$

in which 10 parameters are $a_i(i=1,\ldots,10)$).

We have applied the systematic search procedure (Jafari et al., 2013) into proposed general model (16.1) in order to find chaotic cases. A simple case has been found for

$$\begin{cases} a_1 = a_2 = a_3 = a_4 = a_5 = a_6 = a_7 = a_{10} = 0 \\ a_8 = c \\ a_9 = -1 \end{cases} \tag{16.9}$$

In other words, we have a new three-dimensional system

$$\begin{cases} \dot{x} = -z \\ \dot{y} = xz^2 + a\operatorname{sgn}(z) \\ \dot{z} = x - be^y + z(cy^2 - z^2) \end{cases} \tag{16.10}$$

in which three state variables are x, y, and z. It is noted that in system (16.10) three positive parameters are a, b and c $(a, b, c > 0)$.

The equilibrium points $E(x^*, y^*, 0)$ of system (16.10) are located on a curve described by

$$x^* = be^{y^*} \tag{16.11}$$

The curve of equilibrium points is illustrated in Fig. 16.1.

It is noted that system (16.10) is different from common chaotic systems, which have a countable number of equilibrium points.

It is interesting that chaos has been observed in systems with infinite equilibria (16.10). For example, Figs. 16.2 and 16.3 display chaotic behaviors of system (16.10) for $a = 0.1$, $b = 0.1$, $c = 1$ and initial conditions $(x(0), y(0), z(0)) = (0.1, 0.1, 0.1)$. By applying the algorithm in Wolf et al. (1985), we get Lyapunov exponents of system with infinite equilibria (16.10):

$$L_1 = 0.0668, \quad L_2 = 0, \quad L_3 = -0.5771. \tag{16.12}$$

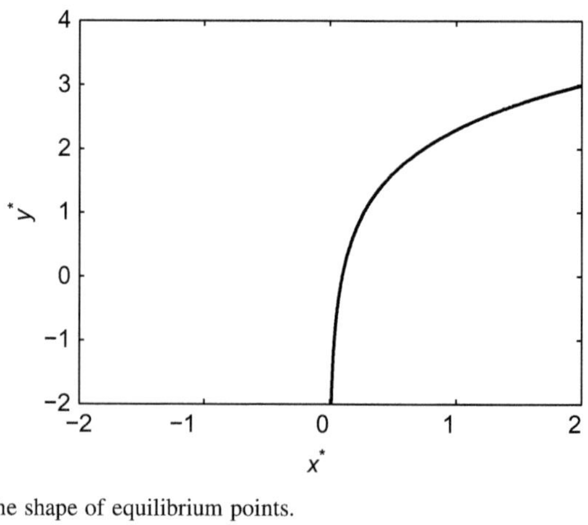

FIGURE 16.1 The shape of equilibrium points.

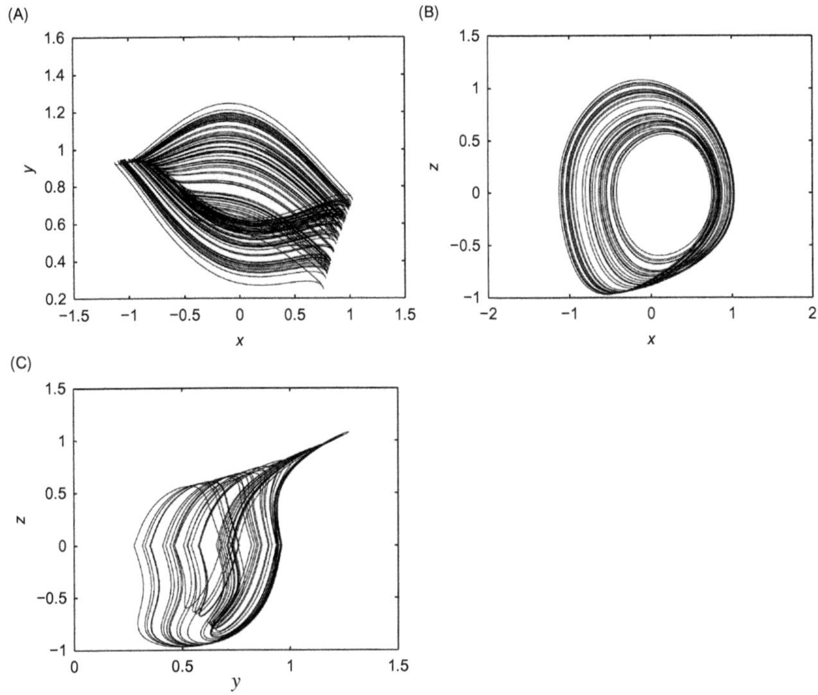

FIGURE 16.2 Chaotic attractors of system with infinite equilibria (16.10) in (A) $x-y$ plane, (B) $x-z$ plane, and (C) $y-z$ plane for $a=0.1$, $b=0.1$, $c=1$ and initial conditions $(x(0), y(0), z(0)) = (0.1, 0.1, 0.1)$.

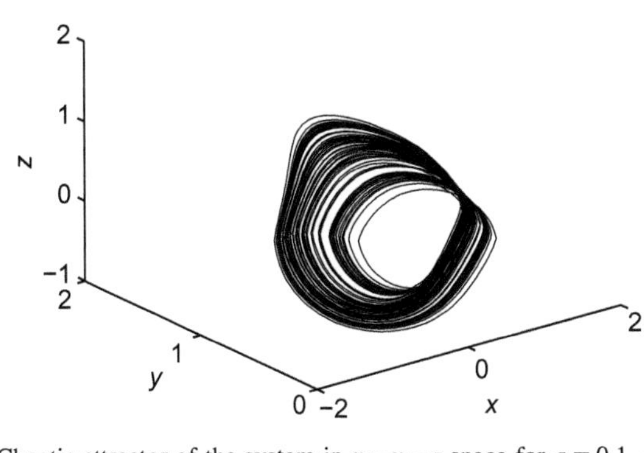

FIGURE 16.3 Chaotic attractor of the system in $x-y-z$ space for $a=0.1$, $b=0.1$, $c=1$ and initial conditions $(x(0), y(0), z(0)) = (0.1, 0.1, 0.1)$.

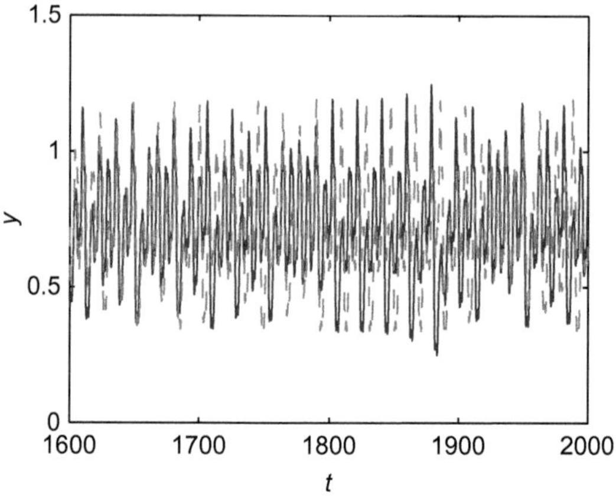

FIGURE 16.4 Presentation of the variable y in system with infinite equilibria (16.10) for two different initial conditions $(x(0), y(0), z(0)) = (0.1, 0.1, 0.1)$ (blue solid) and $(x(0), y(0), z(0)) = (0.1, 0.101, 0.1)$ (red dashed).

In addition, Kaplan–York dimension of the system is calculated by:

$$D_{KY} = 2 + \frac{L_1 + L_2}{|L_3|} = 2.1158. \tag{16.13}$$

As has been known, a typical feature of a chaotic system is the sensitive dependence on initial conditions. Fig. 16.4 illustrates such a feature of a system with infinite equilibria (16.10). We have changed initial conditions slightly from $(x(0), y(0), z(0)) = (0.1, 0.1, 0.1)$ to $(x(0), y(0), z(0)) = (0.1, 0.101, 0.1)$. As can be seen in Fig. 16.4, a small change of initial conditions leads to large differences in the system's state.

Moreover, we have also verified the chaoticity of a system (16.10) with infinite equilibria by using Poincaré maps and frequency spectra. Poincaré maps of system (16.10) are presented in Fig. 16.5. Fig. 16.6 shows the frequency spectra of system (16.10). It is easy to see from Figs. 16.5 and 16.6 that the system (16.10) with an infinite number of equilibrium points is chaotic.

We have investigated the dynamics of system (16.10) with infinite equilibria by changing the value of the bifurcation parameter b from 0.06 to 0.12. Figs. 16.7 and 16.8 present the bifurcation diagram and the diagram of maximal Lyapunov exponents (MLEs) of the system, respectively. From Figs. 16.7 and 16.8, obviously we can see the presence of a period-doubling route to chaos when increasing the value of the parameter b. For example, system (16.10) with infinite equilibria displays period-1 oscillation for $b = 0.07$ (Fig. 16.9A), period-2 oscillation for $b = 0.08$ (Fig. 16.9B), and period-4 oscillation for $b = 0.092$ (Fig. 16.9C). Chaotic oscillations can be observed for $b > 0.934$.

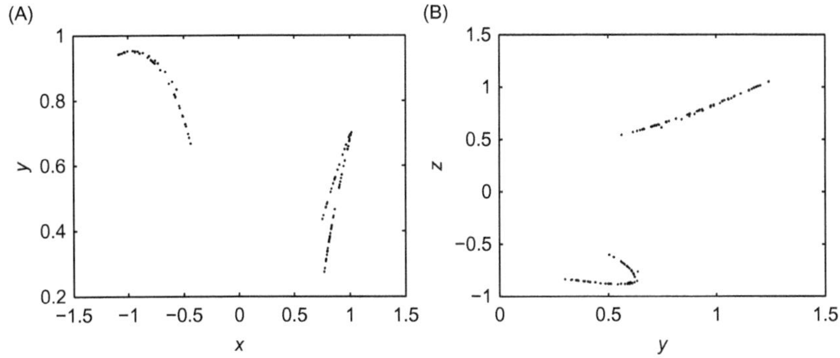

FIGURE 16.5 Poincaré map of system with infinite equilibria (16.10) in (A) $x-y$ plane, and (B) $y-z$ plane.

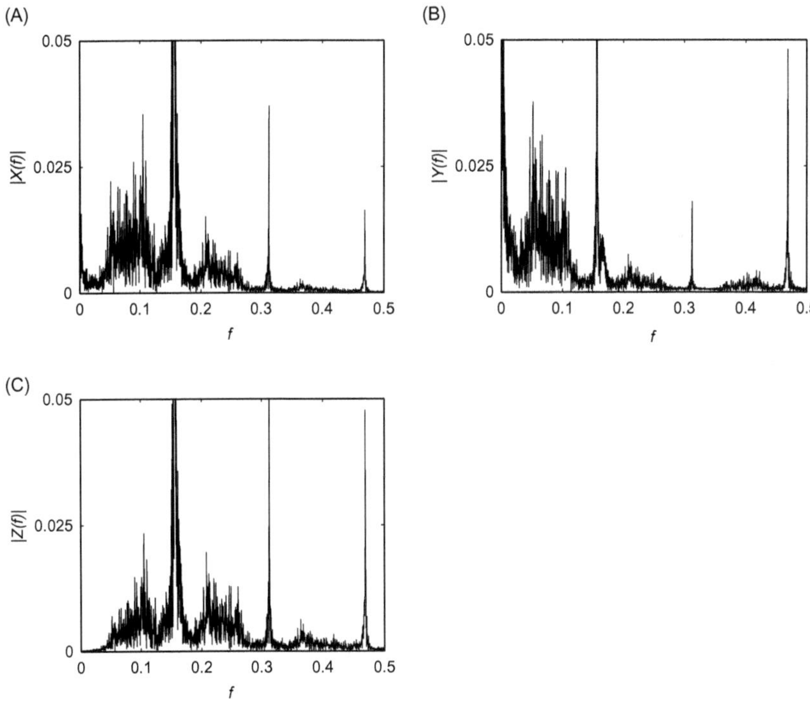

FIGURE 16.6 The frequency spectra of the system with infinite equilibria (A) single-sided amplitude spectrum of x, (B) single-sided amplitude spectrum of y, and (C) single-sided amplitude spectrum of z.

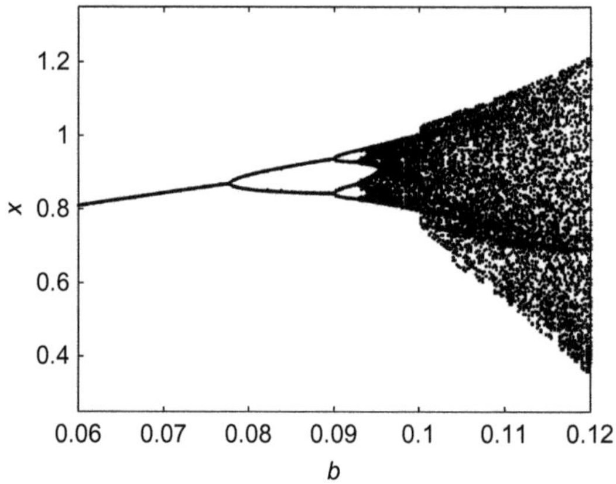

FIGURE 16.7 Bifurcation diagram of system with infinite equilibria (16.10) for $a = 0.1$, $c = 1$ when changing the value of the parameter b from 0.06 to 0.12.

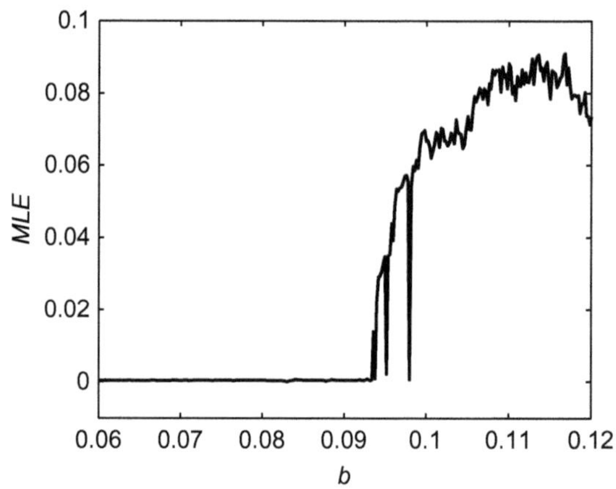

FIGURE 16.8 Maximum Lyapunov exponents of system with infinite equilibria (16.10) for $a = 0.1$, $c = 1$ and $b \in [0.06, 0.12]$.

Researchers have shown an increased interest in multistability (Hens et al., 2015; Li and Sprott, 2014b). It is now well-established from a variety of studies that multistability leads to different qualitative behavior in a given nonlinear dynamical system for the same parameter values (Li and Sprott, 2014b). A considerable amount of literature has been published on multistability. Multistability was reported in different areas such as rigid body motion with linear feedback control (Leipnik and Newton, 1981), laser diode with optical feedback (Masoller, 1994), power system model (Vaithianathan

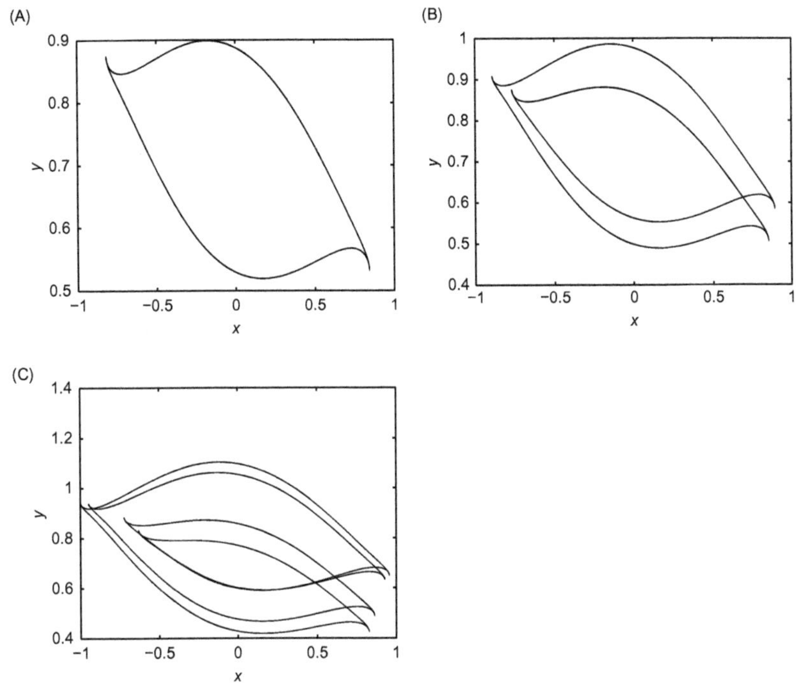

FIGURE 16.9 Three cases of limit cycles in system with infinite equilibria (16.10) for (A) $b = 0.07$, (B) $b = 0.08$, and (C) $b = 0.092$.

and Veijun, 1999), food chain (Upadhyay, 2003), biological system (Cushing et al., 2007), or Koper–Gaspard model of electrochemical oscillatory reactions (Massoudi et al., 2010). It is noting that systems with coexisting attractors were potential candidates for modeling memory in neural networks (Zeng and Zheng, 2012; Zeng et al., 2010). In addition, multistability in chaotic oscillators has received tremendous research efforts in the literature (Pivka et al., 1994; Kengne et al., 2014a,b,c; Kengne, 2015). Interestingly, we have found that a system with infinite equilibria (16.10) is a multistable system. Figs. 16.10 and 16.11 reported bifurcation diagrams of system with infinite equilibria (16.10) for $b = 0.1$, $c = 1$, and different initial conditions. Figs. 16.10 and 16.11 indicate the presence of multistability in system (16.10). For instance, coexistence of chaotic attractor and limit cycle attractor is illustrated in Fig. 16.12.

16.3 ANTISYNCHRONIZATION OF TWO IDENTICAL SYSTEMS WITHOUT EQUILIBRIUM

Synchronization is an interesting phenomenon which was reported in different systems and has a broad range of applications (Holstein-Rathlou et al.,

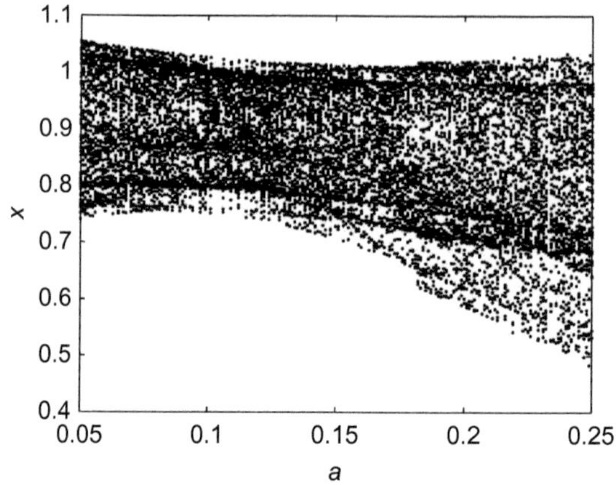

FIGURE 16.10 Bifurcation diagram of system infinite equilibria (16.10) for $b = 0.1$, $c = 1$, and initial conditions $(x(0), y(0), z(0)) = (0.1, 0.1, 0.1)$ when changing the value of the parameter a from 0.05 to 0.25.

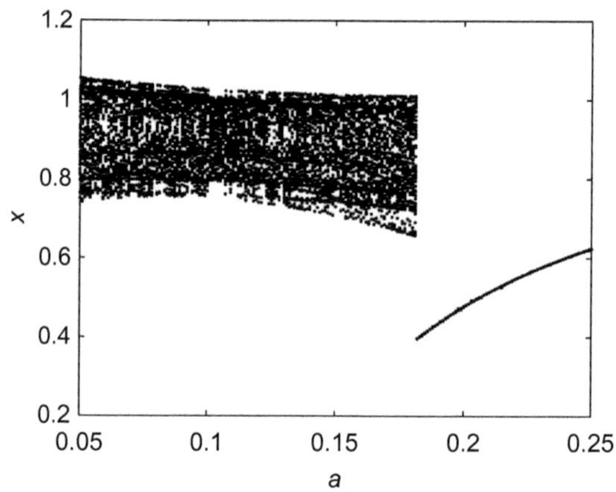

FIGURE 16.11 Bifurcation diagram of system infinite equilibria (16.10) for $b = 0.1$, $c = 1$, and initial conditions $(x(0), y(0), z(0)) = (-0.1, 0.1, 0.1)$ when changing the value of the parameter a from 0.05 to 0.25.

2001; Yanchuk et al., 2001; Mosekilde et al., 2002). Especially, after the study of Pecora and Carrol about synchronization in chaotic systems (Pecora and Carroll, 1990), a great deal of work related to chaos synchronization has attracted the interest of the research community (Vaidyanathan and Pehlivan, 2012; Zhao et al., 2011; Jeong et al., 2013; Yang, 2013; Trejo-Guerra et al., 2013; Pai, 2014). The possibility of the synchronization of chaotic systems plays a vital role in practical applications (Boccaletti et al., 2002; Fortuna and Frasca, 2007; Abdullah, 2013; Sun et al., 2015; Banerjee, 2010; Mata-Machuca et al., 2012; Volos et al., 2013; Aguilar-Lopez et al., 2014;

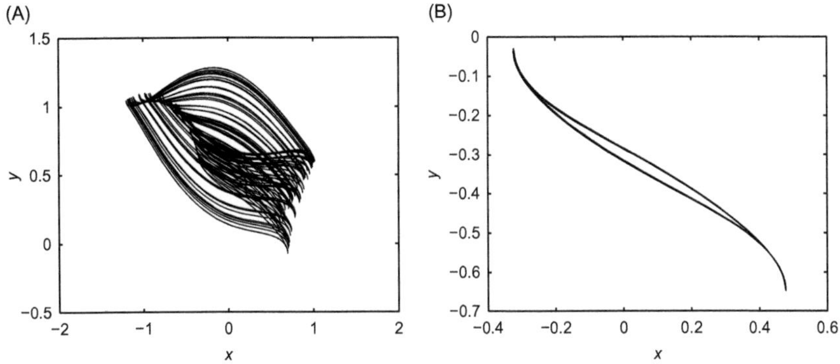

FIGURE 16.12 Coexisting attractors of system with infinite equilibria (16.10) in $x-y$ plane for $a=0.2$, $b=0.1$, $c=1$: (A) initial conditions $(x(0), y(0), z(0)) = (0.1, 0.1, 0.1)$ and (B) initial conditions $(x(0), y(0), z(0)) = (-0.1, 0.1, 0.1)$.

Cicek et al., 2016). Therefore, numerous chaos synchronization schemes have been investigated such as complete synchronization, phase synchronization, lag synchronization, generalized synchronization, antisynchronization, antiphase synchronization, protective synchronization, and so on (Vaidyanathan and Azar, 2015a,b,c,d; Vaidyanathan and Azar, 2016a,b,c,d,e, f,g; Vaidyanathan et al., 2015a,b,c; Ouannas et al., 2016; Azar et al., 2018b; Ouannas et al., 2017a,c; Grassi et al., 2017; Singh et al., 2017).

In this section, we investigate antisynchronization between two new no-equilibrium systems with infinite equilibria. Such two systems are called master and slave systems. Antisynchronization is an interesting scheme in which two systems can be synchronized in amplitude, but with opposite signs (Vaidyanathan and Azar, 2015c; Pham et al., 2017a,b).

Firstly, we consider the master system with infinite equilibria. The master system is described by

$$\begin{cases} \dot{x}_1 = -z_1 \\ \dot{y}_1 = x_1 z_1^2 + a\,\mathrm{sgn}(z_1) \\ \dot{z}_1 = x_1 - be^{y_1} + cy_1^2 z_1 - z_1^3 \end{cases} \tag{16.14}$$

where three state variables are x_1, y_1, z_1. In a master system with infinite equilibria (16.14), the unknown system parameters are a, b and c.

The slave system with infinite equilibria is given in the following form:

$$\begin{cases} \dot{x}_2 = -z_2 + u_x \\ \dot{y}_2 = x_2 z_2^2 + a\,\mathrm{sgn}(z_2) + u_y \\ \dot{z}_2 = x_2 - be^{y_2} + cy_2^2 z_2 - z_2^3 + u_z \end{cases} \tag{16.15}$$

in which x_2, y_2, z_2 are three variables and $\mathrm{u} = [u_x, u_y, u_z]^T$ is an adaptive control.

The state errors of the antisynchronization are defined by

$$\begin{cases} e_x = x_2 + x_1, \\ e_y = y_2 + y_1, \\ e_z = z_2 + z_1. \end{cases} \tag{16.16}$$

It is trivial to see that the state error dynamics of the antisynchronization are given by:

$$\begin{cases} \dot{e}_x = \dot{x}_2 + \dot{x}_1, \\ \dot{e}_y = \dot{y}_2 + \dot{y}_1, \\ \dot{e}_z = \dot{z}_2 + \dot{z}_1. \end{cases} \tag{16.17}$$

By taking the difference between the unknown system parameters (a, b, c) and the estimation of the unknown parameters ($\hat{a}$, $\hat{b}$, $\hat{c}$), the parameter estimation errors are calculated as:

$$\begin{cases} e_a = a - \hat{a}, \\ e_b = b - \hat{b}, \\ e_c = c - \hat{c}, \end{cases} \tag{16.18}$$

Similarly, we get the parameter estimation error dynamics by differentiating Eq. (16.18). Thus the parameter estimation error dynamics are:

$$\begin{cases} \dot{e}_a = -\dot{\hat{a}}, \\ \dot{e}_b = -\dot{\hat{b}}, \\ \dot{e}_c = -\dot{\hat{c}}, \end{cases} \tag{16.19}$$

The main aim of this section is to design an adaptive control to antisynchronize the slave no-equilibrium system (16.15) and the master no-equilibrium system (16.14). Therefore, the adaptive control has been constructed as follows:

$$\begin{cases} u_x = e_z - k_x e_x \\ u_y = -x_1 z_1^2 - x_2 z_2^2 - \hat{a}(\mathrm{sgn}(z_1) + \mathrm{sgn}(z_2)) - k_y e_y \\ u_z = -e_x + \hat{b}(e^{y_1} + e^{y_2}) - \hat{c}(y_1^2 z_1 + y_2^2 z_2) + z_1^3 + z_2^3 - k_z e_z \end{cases} \tag{16.20}$$

It is noted that in (16.20), three positive gain constants are k_x, k_y, k_z. In addition, we have designed the following parameter update law:

$$\begin{cases} \dot{\hat{a}} = (\mathrm{sgn}(z_1) + \mathrm{sgn}(z_2))e_y \\ \dot{\hat{b}} = -(e^{y_1} + e^{y_2})e_z \\ \dot{\hat{c}} = (y_1^2 z_1 + y_2^2 z_2)e_z \end{cases} \tag{16.21}$$

The 3D chaotic systems with infinite equilibrium (16.14) and (16.15) with unknown system parameters are globally and exponentially antisynchronized by using the adaptive controller (16.20) and the parameter update law (16.21), where k_x, k_y, k_z are positive gain constants.

In order to confirm the antisynchronization of the slave system (16.15) and the master system (16.14) when applying the designed adaptive control (16.20), the Lyapunov function is selected as follows:

$$V(e_x, e_y, e_z, e_a, e_b, e_c) = \frac{1}{2}(e_x^2 + e_y^2 + e_z^2 + e_a^2 + e_b^2 + e_c^2). \tag{16.22}$$

From Eq. (16.22), we obtain the differentiation of V:

$$\dot{V} = e_x\dot{e}_x + e_y\dot{e}_y + e_z\dot{e}_z + e_a\dot{e}_a + e_b\dot{e}_b + e_c\dot{e}_c. \tag{16.23}$$

It is simple to verify that by combining (16.14), (16.15), and (16.20), synchronization error dynamics are rewritten by:

$$\begin{cases} \dot{e}_x = -k_x e_x \\ \dot{e}_y = (\text{sgn}(z_1) + \text{sgn}(z_2))e_a - k_y e_y \\ \dot{e}_z = -(e^{y_1} + e^{y_2})e_b + (y_1^2 z_1 + y_2^2 z_2)e_c - k_z e_z \end{cases} \tag{16.24}$$

Similarly, we get the differentiation of the Lyapunov function by substituting Eqs. (16.19) and (16.24) into Eq. (16.23):

$$\dot{V} = -k_x e_x^2 - k_y e_y^2 - k_z e_z^2. \tag{16.25}$$

It is simple to verify that the differentiation of V is a negative semidefinite function. Therefore, according to Barbalat's lemma (Khalil, 2002), we have $e_x \to 0$, $e_y \to 0$, and $e_z \to 0$ exponentially as $t \to \infty$. In other words, the antisynchronization between the slave system with infinite equilibria and the master system with infinite equilibria is achieved.

An example is presented to illustrate the correction of the proposed antisynchronization scheme. In this example, the parameter values of the master system and the slave system are selected as:

$$a = 0.1, \quad b = 0.1, \quad c = 1. \tag{16.26}$$

We assume that the initial states of the master system with infinite equilibria are taken as:

$$x_1(0) = 0.1, \quad y_1(0) = 0.1, \quad z_1(0) = 0.1. \tag{16.27}$$

We take the following initial states for the slave system with infinite equilibria:

$$x_2(0) = 0.1, \quad y_2(0) = 0.2, \quad z_2(0) = 0.3. \tag{16.28}$$

In this example, the positive gain constants are given by:

$$k_x = 6, \quad k_y = 6, \quad k_z = 6, \tag{16.29}$$

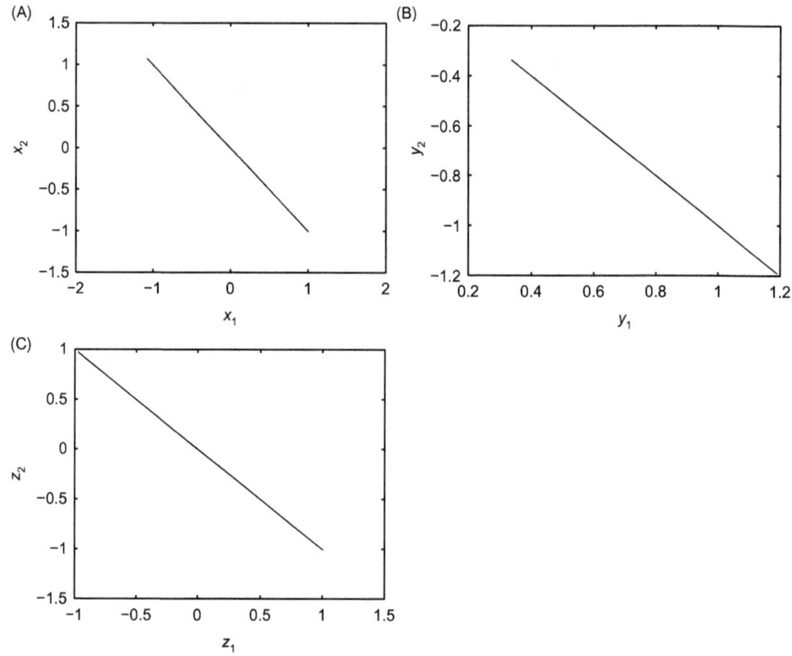

FIGURE 16.13 Phase portraits show the antisynchronization of the master system with infinite equilibria and the slave system with infinite equilibria (A) x_1 vs. x_2, (B) y_1 vs. y_2 and (C) z_1 vs. z_2.

while the initial condition of the parameter estimate is chosen as:

$$\hat{a}(0) = 0.2, \quad \hat{b}(0) = 0.2, \quad \hat{c}(0) = 1.2. \tag{16.30}$$

The classical fourth order Runge–Kutta method has been used for numerical results which are shown in Figs. 16.13 and 16.14. Fig. 16.13 displays the phase portraits of states in the master and slave systems with infinite equilibria. The time series of the master and slave systems are presented in Fig. 16.14. As can be seen from Figs. 16.13 and 16.14, the antisynchronization of the slave system with infinite equilibria and the master system with infinite equilibria has been achieved.

16.4 FRACTIONAL ORDER FORM OF THE SYSTEM WITH INFINITE EQUILIBRIA

Fractional calculus has been studied since the 17th century and has been applied in numerous fields such as physics, electrical circuit, chemical engineering, control systems, electromagnetic theory, etc (Heaviside, 1971;

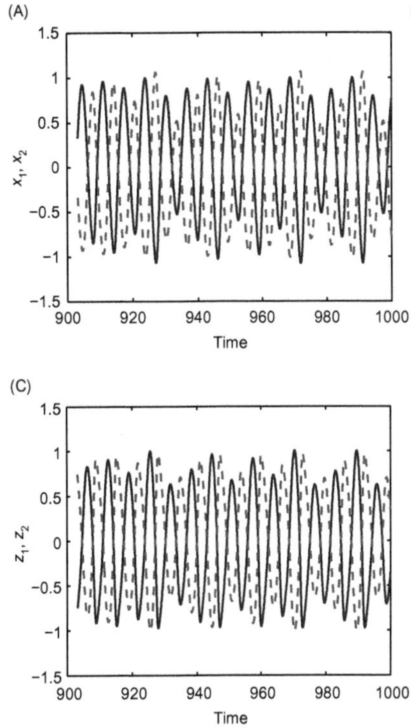

FIGURE 16.14 Time series of the master system with infinite equilibria (blue solid) and the slave system with infinite equilibria (red dashed): (A) x_1 and x_2, (B) y_1 and y_2 and (C) z_1 and z_2.

Sun et al., 1984; Bagley and Calico, 1991; Westerlund and Ekstam, 1994; Jenson and Jeffreys, 1997; Hifer, 2001; Maundy et al., 2010; Elwakil, 2010; Meghni et al., 2017a,b; Azar et al., 2018a; Ghoudelbourk et al., 2016; Freeborn et al., 2013). Chaotic behavior has been observed in various fractional order systems (Li and Peng, 2004; Deng and Li, 2005; Lu and Chen, 2006; Li et al., 2006; Ouannas et al., 2007, 2017d,e,f,g,h,i). Moreover, the complexity and the advantages of fractional order systems have been used in different engineering applications ranging from image encryption, authenticated encryption scheme, to affine cipher using date of birth (Muthukumar and Balasubramaniam, 2013; Muthukumar et al., 2014a,b; Xu et al., 2014). Therefore, in this section we will consider the fractional order form of the introduced system with infinite equilibria

The literature on fractional calculus has highlighted several definitions of the fractional order derivative. However it is worth noting that three common definitions are Grunwald–Letnikov, Riemann–Liouville, and Caputo definitions (Podlubny, 1999; Diethelm, 2010; Monje et al., 1999; Petras, 2011). In this section, the Caputo definition is utilized. The Caputo definition is described by

$$ {}_0D_t^q f(t) = \frac{1}{\Gamma(m-q)} \int_0^t \frac{f^{(m)}(\tau)}{(t-\tau)^{q+1-m}} d\tau, \quad m-1 < q < m. \tag{16.31} $$

In the Caputo definition (16.31), m is the first integer which is not less than $q(m = \lceil q \rceil)$ and Γ is the Gamma function:

$$ \Gamma(z) = \int_0^\infty t^{z-1} e^{-t} dt. \tag{16.32} $$

When considering the effect of fractional order derivative on the introduced system with infinite equilibria (16.10), we concentrate on its fractional order form given by:

$$ \begin{cases} D^q x = -z \\ D^q y = xz^2 + a\mathrm{sgn}(z) \\ D^q z = x - be^y + z(cy^2 - z^2) \end{cases} \tag{16.33} $$

where the three state variables are x, y, and z while q is the fractional order $(0 < q < 1)$. It is noted that the three positive parameters are a, b, c $(a, b, c > 0)$. From (16.33), it is trivial to verify that fractional order system (16.33) has an infinite number of equilibrium points.

For investigating fractional order system (16.33), we have applied Adams–Bashforth–Moulton algorithm (Diethelm and Ford, 2002; Diethelm et al., 2004). As a result, fractional order system (16.33) can be rewritten in the following form:

$$ \begin{cases} x_h(t_{n+1}) = x(0) + \dfrac{h^q}{\Gamma(q+2)}(-z^{(p)}(t_{n+1})) + \dfrac{h^q}{\Gamma(q+2)} \displaystyle\sum_{j=0}^{n} \alpha_{j,n+1}(-z(t_j)) \\ y_h(t_{n+1}) = y(0) + \dfrac{h^q}{\Gamma(q+2)} \left(x^{(p)}(t_{n+1})(z^{(p)}(t_{n+1}))^2 + a\mathrm{sgn}(z^{(p)}(t_{n+1}))\right) \\ \qquad + \dfrac{h^q}{\Gamma(q+2)} \displaystyle\sum_{j=0}^{n} \alpha_{j,n+1}\left(x(t_j)(z(t_j))^2 + a\mathrm{sgn}(z(t_j))\right) \\ z_h(t_{n+1}) = z(0) + \dfrac{h^q}{\Gamma(q+2)}\left(x^{(p)}(t_{n+1}) - be^{y^{(p)}(t_{n+1})}\right) \\ \qquad + \dfrac{h^q}{\Gamma(q+2)}(c(y^{(p)}(t_{n+1}))^2 z^{(p)}(t_{n+1}) - (z^{(p)}(t_{n+1}))^3) \\ \qquad + \dfrac{h^q}{\Gamma(q+2)} \displaystyle\sum_{j=0}^{n} \alpha_{j,n+1}(x(t_j) - be^{y(t_j)} + c(y(t_j))^2 z(t_j) - (z(t_j))^3) \end{cases} \tag{16.34} $$

Here the predicted variables are given by:

$$\begin{cases} x^{(p)}(t_{n+1}) = x(0) + \dfrac{1}{\Gamma(q)}\sum_{j=0}^{n}\beta_{j,n+1}(-z(t_j)) \\ y^{(p)}(t_{n+1}) = y(0) + \dfrac{1}{\Gamma(q)}\sum_{j=0}^{n}\beta_{j,n+1}(x(t_j)(z(t_j))^2 + a\operatorname{sgn}(z(t_j)))) \\ z^{(p)}(t_{n+1}) = z(0) + \dfrac{1}{\Gamma(q)}\sum_{j=0}^{n}\beta_{j,n+1}(x(t_j) - be^{y(t_j)} + c(y(t_j))^2 z(t_j) - (z(t_j))^3) \end{cases} \tag{16.35}$$

It is noted that in (16.35), $\alpha_{j,n+1}$ and $\beta_{j,n+1}$ are described as follows

$$\alpha_{j,n+1} = \begin{cases} n^{q+1} - (n-q)(n+1)^q & j=0 \\ (n-j+2)^{q+1} + (n-j)^{q+1} - 2(n-j+1)^{q+1} & 1 \le j \le n \\ 1 & j=n+1 \end{cases} \tag{16.36}$$

and

$$\beta_{j,n+1} = \frac{h^q}{q}((n+1-j)^q - (n-j)^q) \quad 0 \le j \le n \tag{16.37}$$

It is interesting that for $q = 0.99$, fractional order system (16.33) generates chaotic behavior as shown Fig. 16.15. In order to verify the chaoticity of fractional order system infinite equilibria (16.33), its largest Lyapunov exponent has been calculated by applying the practical method in (Rosenstein et al., 1993). The largest Lyapunov exponent of the fractional order system (16.33) for $q = 0.99$ is 0.1298.

Previous research has established that the "0−1" test is an efficient test for confirming chaos. Gottwald and Melbourne (Gottwald and Melbourne, 2004, 2009) have introduced and developed the "0−1" test, in which Gottwald and Melbourne constructed a random walk-type process from the data and investigated the variance of the random walk scales with time (Cafagna and Grassi, 2008; Gottwald and Melbourne, 2009). Thus, the test is useful and has advantages, for example, the test does not require phase space reconstruction (Cafagna and Grassi, 2008; Gottwald and Melbourne, 2009). We have also used the "0−1" test to confirm the chaos of fractional order system (16.33) with infinite equilibria for $q = 0.99$.

For implementing the test, a discrete map $\phi(n)$ from the fractional order system has been considered. Two functions $p(n)$, $q(n)$ are defined in the following forms:

$$p(n) = \sum_{j=1}^{n} \phi(j)\cos(\theta(j)) \tag{16.38}$$

and

$$q(n) = \sum_{j=1}^{n} \phi(j)\sin(\theta(j)) \tag{16.39}$$

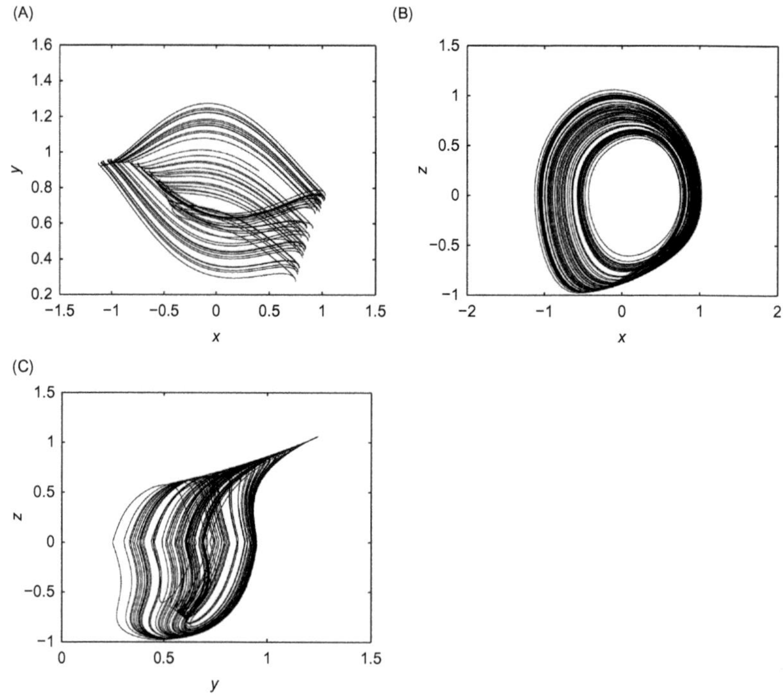

FIGURE 16.15 Phase portrait of fractional order system with infinite equilibria (16.33) in (A) $x-y$ plane, (B) $x-z$ plane, and (C) $y-z$ plane for $q=0.99$, $a=0.1$, $b=0.1$, $c=1$ and initial conditions $(x(0), y(0), z(0))=(0.1, 0.1, 0.1)$.

It is worth noting that $\theta(j)$ is described by

$$\theta(j)=jc+\sum_{i=1}^{j}\phi(i) \tag{16.40}$$

for $c\in(0,\pi)$. From Eq. (16.38) and Eq. (16.39), the mean square displacement $M(n)$ is calculated as follows:

$$M(n)=\frac{1}{N}\sum_{j=1}^{N}((p(j+n)-p(j))^2+(q(j+n)-q(j))^2) \tag{16.41}$$

Therefore, we get the asymptotic growth rate

$$K=\lim_{n\to\infty}\frac{\log M(n)}{\log(n)} \tag{16.42}$$

The calculated value of the asymptotic growth rate of fractional order system (16.33) for $q=0.99$ is $K=0.9354$. As a result, fractional order system (16.33) is chaotic because K is close to 1.

When the fractional derivative order decreases, fractional order system (16.33) displays a point attractor, for example at $q = 0.98$ (see Fig. 16.16). Moreover, it is clearly seen in Fig. 16.17 that the trajectories of fractional order system (16.33) converge to a point.

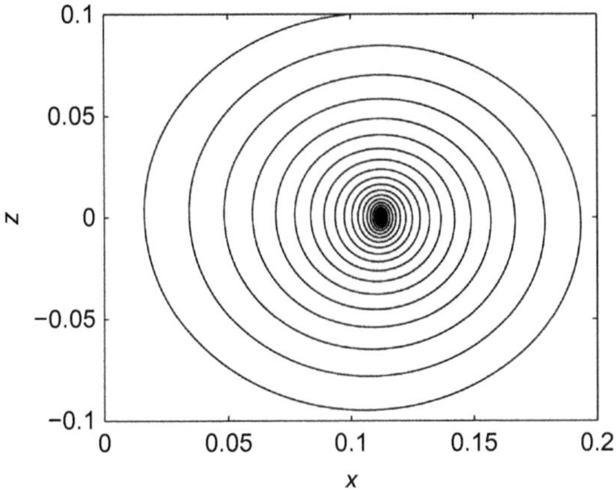

FIGURE 16.16 Point attractor of fractional order system with infinite equilibria (16.33) in $x-z$ plane for $q = 0.98$, $a = 0.1$, $b = 0.1$, $c = 1$ and initial conditions $(x(0), y(0), z(0)) = (0.1, 0.1, 0.1)$.

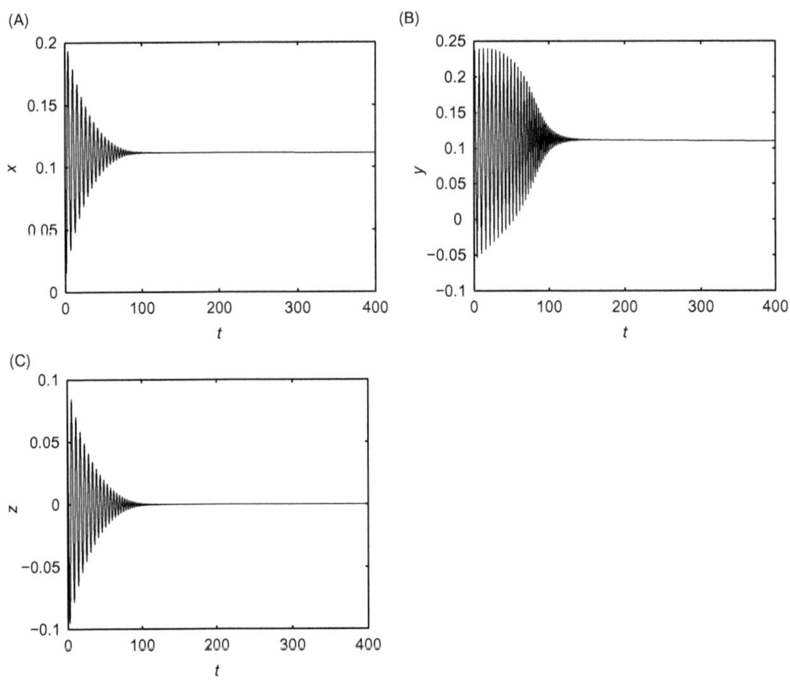

FIGURE 16.17 The time series fractional order system with infinite equilibria (16.33) (A) $x(t)$ plane, (B) $y(t)$ plane, and (C) $z(t)$ for $q = 0.98$, $a = 0.1$, $b = 0.1$, $c = 1$ and initial conditions $(x(0), y(0), z(0)) = (0.1, 0.1, 0.1)$.

16.5 CONCLUSIONS

In this chapter, a chaotic system with seven terms has been studied. The system is different from a conventional system because it includes signum and exponential nonlinear terms. It is interesting that the system is special because there is the presence of an infinite number of equilibrium points. By using phase portraits, Poincaré map, bifurcation diagram, and Lyapunov exponents, we have investigated the system's dynamics. We have observed chaos and coexisting attractors in such a system. In addition, we have proposed an adaptive control for antisynchronization of chaotic systems with infinite equilibria. Fractional order form of the system with infinite equilibria can exhibit chaotic and point attractors. Practical applications of the new system will be discovered in our future works.

REFERENCES

Abdullah, A., 2013. Synchronization and secure communication of uncertain chaotic systems based on full order and reduced order output–affine observers. Appl. Math. Comput. 219, 10000–10011.

Aguilar-Lopez, R., Martinez-Guerra, R., Perez-Pinacho, C., 2014. Nonlinear observer for synchronization of chaotic systems with application to secure data transmission. Eur. Phys. J. Special Topics 223, 1541–1548.

Ahmad, I., Saaban, A.B., Ibrahim, A.B., Shahzad, M., Al-sawalha, M.M., 2016. Reduced order synchronization of time–delay chaotic systems with known and unknown parameters. Optik 127 (13), 5506–5514.

Azar, A.T., Vaidyanathan, S., 2015a. Chaos Modeling and Control Systems Design. Vol. 581 of Studies in Computational Intelligence. Studies in Computational Intelligence. Springer, Germany.

Azar, A.T., Vaidyanathan, S., 2015b. Computational Intelligence Applications in Modelling and Control. Vol. 575. Studies in Computational Intelligence. Springer, Berlin, Germany.

Azar, A.T., Vaidyanathan, S., 2015c. Handboook of Research on Advanced Intelligent Control Engineering and Automation. IGI Global, New York, USA.

Azar, A.T., Vaidyanathan, S., 2016. Advances in Chaos Theory and Intelligent Control, Vol. 337. Springer, Berlin, Germany.

Azar, A.T., Kumar, J., Kumar, V., Rana, K.P.S., 2018a. Control of a Two Link Planar Electrically-Driven Rigid Robotic Manipulator Using Fractional Order SOFC. Springer International Publishing, Cham, pp. 57–68. Available from: https://doi.org/10.1007/978-3-319-64861-3_6.

Azar, A.T., Ouannas, A., Singh, S., 2018b. Control of New Type of Fractional Chaos Synchronization. Springer International Publishing, Cham, pp. 47–56. Available from: https://doi.org/10.1007/978-3-319-64861-3_5.

Azar, A.T., Vaidyanathan, S., Ouannas, A., 2017a. Fractional Order Control and Synchronization of Chaotic Systems. Vol. 688. Studies in Computational Intelligence. Springer, Berlin, Germany.

Azar, A.T., Volos, C., Gerodimos, N.A., Tombras, G.S., Pham, V.-T., Radwan, A.G., et al., 2017b. A novel chaotic system without equilibrium: dynamics, synchronization, and circuit realization. Complexity 2017, (Article ID 7871467), 11 pages.

Bagley, R.L., Calico, R.A., 1991. Fractional order state equations for the control of visco–elastically damped structers. J. Guide Conteol Dyn. 14, 304–311.

Banerjee, S., 2010. Chaos Synchronization and Cryptography for Secure Communication. IGI Global, USA.

Barati, K., Jafari, S., Sprott, J.C., Pham, V.-T., 2016. Simple chaotic flows with a curve of equilibria. Int. J. Bifurcation Chaos 26 (12), 1630034.

Boccaletti, S., Kurths, J., Osipov, G., Valladares, D.L., Zhou, C.S., 2002. The synchronization of chaotic systems. Phys. Rep. 366, 1–101.

Boulkroune, A., Bouzeriba, A., Bouden, T., Azar, A.T., 2016a. Fuzzy adaptive synchronization of uncertain fractional-order chaotic systems. Advances in Chaos Theory and Intelligent Control. Springer, Berlin, Germany, pp. 681–697.

Boulkroune, A., Hamel, S., Azar, A.T., Vaidyanathan, S., 2016b. Fuzzy control-based function synchronization of unknown chaotic systems with dead-zone input. In: Azar, A.T., Vaidyanathan, S. (Eds.), Advances in Chaos Theory and Intelligent Control. Springer International Publishing, Cham, pp. 699–718.

Cafagna, D., Grassi, G., 2008. Fractional order Chua's circuit: time–domain analysis, bifurcation, chaotic behaviour and test for chaos. Int. J. Bifurcation Chaos 18, 615–639.

Chen, G., Yu, X., 2003. Chaos Control: Theory and Applications. Springer, Berlin.

Chen, G.R., Ueta, T., 1999. Yet another chaotic attractor. Int. J. Bifurcation and Chaos 9, 1465–1466.

Chen, Y., Yang, Q., 2015. A new Lorenz–type hyperchaotic system with a curve of equilibria. Math. Comput. Simul. 112 (2015), 40–55.

Cicek, S., Ferikoglu, A., Pehlivan, I., 2016. A new 3D chaotic system: dynamical analysis, electronic circuit design, active control synchronization and chaotic masking communication application. Optik 127 (8), 4024–4030.

Cushing, J.M., Henson, S.M., Blackburn, C.C., 2007. Multiple mixed attractors in a competition model. J. Biol. Dyn. 1, 347–362.

Deng, W.H., Li, C.P., 2005. Chaos synchronization of the fractional Lü system. Physica A 353, 61–72.

Diethelm, K., 2010. The Analysis of Fractional Differential Equations, An Application–Oriented Exposition Using Differential Operators of Caputo Type. Springer, Berlin.

Diethelm, K., Ford, N.J., 2002. Analysis of fractional differential equations. J. Math. Anal. Appl. 265, 229–248.

Diethelm, K., Ford, N.J., Freed, A.D., 2004. Detailed error analysis for a fractional Adams method. Numer. Algorithms 36, 31–52.

Elwakil, A., 2010. Fractional order circuits and systems: an emerging interdisciplinary research area. IEEE Circuits Syst. Mag. 10, 40–50.

Esmaeili-Najafabadi, H., Ataei, M., Sabahi, M.F., 2017. Designing sequence with minimum PSL using Chebyshev distance and its application for chaotic MIMO radar waveform design. IEEE. Trans. Signal Process. 65 (3), 690–704.

Fortuna, L., Frasca, M., 2007. Experimental synchronization of single–transistor–based chaotic circuits. Chaos 17, 043118–1–5.

Freeborn, T.J., Maundy, B., Elwakil, A.S., 2013. Measurement of supercapacitor fractional order model parameters from voltage–excited step response. IEEE J. Emerging Sel. Top. Circuits Syst. 3, 367–376.

Ghoudelbourk, S., Dib, D., Omeiri, A., Azar, A.T., 2016. Mppt control in wind energy conversion systems and the application of fractional control (piα) in pitch wind turbine. Int. J. Model. Identif. Control 26 (2), 140–151.

Gotthans, T., Petržela, J., 2015. New class of chaotic systems with circular equilibrium. Nonlinear Dyns. 81 (3), 1143–1149.

Gotthans, T., Sportt, J.C., Petržela, J., 2016. Simple chaotic flow with circle and square equilibrium. Int. J. Bifurcation and Chaos 26 (8), 1650137.

Gottwald, G.A., Melbourne, I., 2004. A new test for chaos in deterministic systems. Proc R. Soc. London A: Math. Phys. Eng. Sci. 460, 603–611.

Gottwald, G.A., Melbourne, I., 2009. On the implementation of the 0–1 test for chaos. SIAM J. Appl. Dyn. Syst. 8, 129–145.

Grassi, G., Ouannas, A., Azar, A.T., Radwan, A.G., Volos, C., Pham, V.-T., et al., 2017. Chaos synchronisation of continuous systems via scalar signal. In: 6th International Conference on Modern Circuits and Systems Technologies (MOCAST). pp. 1–4.

Heaviside, O., 1971. Electromagnetic Theory. Academic Press, New York, USA.

Hens, C., Dana, S.K., Feudel, U., 2015. Extreme multistability: attractors manipulation and robustness. Chaos 25 (5), 053112.

Hifer, R., 2001. Applications of Fractional Calculus in Physics. World Scientific, Singapore.

Holstein-Rathlou, N.H., Yip, K.P., Sosnovtseva, O.V., Mosekilde, E., 2001. Synchronization phenomena in nephron–nephron interaction. Physica D 1q 417–426.

Jafari, S., Sprott, J.C., 2013. Simple chaotic flows with a line equilibrium. Chaos Solitons Fractals 57, 79–84.

Jafari, S., Sprott, J., Golpayegani, S.M.R.H., 2013. Elementary quadratic chaotic flows with no equilibria. Phys. Lett. A 377, 699–702.

Jenson, V.G., Jeffreys, G.V., 1997. Mathematical Methods in Chemical Enginerring. Academic Press, New York, USA.

Jeong, S.C., Ji, D.H., Park, J.H., Won, S.C., 2013. Adaptive synchronization for uncertain chaotic neural networks with mixed time delays using fuzzy disturbance observer. Appl. Math. Comput. 219, 5984–5995.

Kengne, J., 2015. Coexistence of chaos with hyperchaos, period?-3 doubling bifurcation, and transient chaos in the hyperchaotic oscillator with gyrators. Int. J. Bifurcat. Chaos 25 (4), 1550052.

Kengne, J., Chedjou, J.C., Fozin, T.F., Kyamakya, K., Kenne, G., 2014a. On the analysis of semiconductor diode based chaotic and hyperchaotic chaotic generators – a case study. Nonlinear Dyn. 77, 373–386.

Kengne, J., Chedjou, J.C., Kom, M., Kyamakya, K., Tamba, V.K., 2014b. Regular oscillations, chaos, and multistability in a system of two coupled van der pol oscillators: numerical and experimental studies. Nonlinear Dyn. 76, 1119–1132.

Kengne, J., Njitacke, Z.T., Fotsin, H.B., 2014c. Dynamical analysis of a simple autonomous jerk system with multiple attractors. Nonlinear Dyn. 83 (1–2), 751–765.

Khalil, H.K., 2002. Nonlinear Systems, third ed. Prentice Hall, New Jersey, USA.

Lamamra, K., Vaidyanathan, S., Azar, A.T., Ben Salah, C., 2017. Chaotic system modelling using a neural network with optimized structure. In: Azar, A.T., Vaidyanathan, S., Ouannas, A. (Eds.), Fractional Order Control and Synchronization of Chaotic Systems. Springer International Publishing, Cham, pp. 833–856.

Leipnik, R.B., Newton, T.A., 1981. Double strange attractors in rigid body motion with linear feedback control. Phys. Lett. A 86, 63–87.

Li, C., Peng, G., 2004. Chaos in Chen's system with a fractional order. Chaos Solitons Fractals 22, 443–450.

Li, C., Sprott, J.C., 2014a. Chaotic flows with a single nonquadratic term. Phys. Lett. A 378 (3), 178–183.

Li, C., Sprott, J.C., 2014b. Finding coexisting attractors using amplitude control. Nonlinear Dyn. 78, 2059–2064.

Li, C., Sportt, J.C., Thio, W., 2014a. Bistability in a hyperchaotic system with a line equilibrium. J. Exp. Theoretical Phys. 118 (3), 494–500.

Li, Q., Hu, S., Tang, S., Zeng, G., 2014b. Hyperchaos and horseshoe in a 4D memristive system with a line of equilibria and its implementation. Int. J. Cir. Theory Applicat. 42, 1172–1188.

Li, C., Sprott, J.C., Yuan, Z., Li, H., 2015a. Constructing chaotic systems with total amplitude control. Int. J. Bifurcation Chaos 25 (10), 1530025.

Li, Q., Zeng, H., Li, J., 2015b. Hyperchaos in a 4D memristive circuit with infinitely many stable equilibria. Nonlinear Dyn. 79 (4), 2295–2308.

Li, C.P., Deng, W.H., Xu, D., 2006. Chaos synchronization of the Chua system with a fractional order. Physica A 360, 171–185.

Lin, Y., Wang, C., He, H., 2015. A simple multi–scroll chaotic oscillator employing CCIIs. Optik 126 (7–8), 824–827.

Liu, Z., Zhu, X., Hu, W., Jiang, F., 2007. Principles of chaotic signal radar. Int. J. Bifurcation Chaos 93, 1735–1739.

Lorenz, E.N., 1963. Deterministic nonperiodic flow. J. Atmos. Sci. 20, 130–141.

Lu, J.G., Chen, G., 2006. A note on the fractional order Chen system. Chaos Solitons Fractals 27, 685–688.

Lü, J.H., Chen, G.R., 2006. Generating multiscroll chaotic attractors: theories, methods and applications. Int. J. Bif. Chaos 16, 775–858.

Masoller, C., 1994. Coexistence of attractors in a laser diode with optical feedback from a large external cavity. Phys. Rev. A 50, 2569–2578.

Massoudi, A., Mahjani, M.G., Jafarian, M., 2010. Multiple attractors in Koper–Gaspard model of electrochemical. J. Electroanal. Chem. 647, 74–86.

Mata-Machuca, J., Martinez-Guerra, R., Aguilar-Lopez, R., Aguilar-Ibanez, C., 2012. A chaotic system in synchronization and secure communications. Commun. Nonlinear Sci. Numer. Simul. 17, 1706–1713.

Maundy, B., Elwakil, A., Gift, S., 2010. On a multivibrator that employs a fractional capacitor. Analog Integr. Circuits Signal Process. 62, 99–103.

Meghni, B., Dib, D., Azar, A.T., 2017a. A second-order sliding mode and fuzzy logic control to optimal energy management in wind turbine with battery storage. Neural Comput. Applicat. 28 (6), 1417–1434.

Meghni, B., Dib, D., Azar, A.T., Ghoudelbourk, S., Saadoun, A., 2017b. Robust adaptive supervisory fractional order controller for optimal energy management in wind turbine with battery storage. In: Azar, A.T., Vaidyanathan, S., Ouannas, A. (Eds.), Fractional Order Control and Synchronization of Chaotic Systems. Springer International Publishing, Cham, pp. 165–202.

Min, L., Yang, X., Chen, G., Wang, D., 2015. Some polynomial chaotic maps without equilibria and an application to image encryption with avalanche effects. Int. J. Bifurcation Chaos 25 (9), 1550124.

Monje, C.A., Chen, Y.Q., Vinagre, B.M., Xue, D., Feliu, V., 1999. Fractional Differential Equations. Academic Press, New York.

Mosekilde, E., Maistrenko, Y., Postnov, D., 2002. Chaotic Synchronization: Applications to Living Systems. World Scientific, Singapore.

Moysis, L., Azar, A.T., 2017. New discrete time 2d chaotic maps. Int J. Syst. Dyn. Applicat. 6 (1), 77–104.

Munmuangsaen, B., Sprott, J.C., Thio, W.J.-C., Buscarino, A., Fortuna, L., 2015. A simple chaotic flow with a continuously adjustable attractor dimension. Int. J. Bifurcation Chaos 25 (12), 1530036.

Muthukumar, P., Balasubramaniam, P., 2013. Feedback synchronization of the fractional order reverse butterfly–shaped chaotic system and its application to digital cryptography. Nonlinear Dyn. 74, 1169–1181.

Muthukumar, P., Balasubramaniam, P., Ratnavelu, K., 2014a. Fast projective synchronization of fractional order chaotic and reverse chaotic systems with its application to an affine cipher using date of birth (DOB). Nonlinear Dyn. 80, 1883–1897.

Muthukumar, P., Balasubramaniam, P., Ratnavelu, K., 2014b. Synchronization of a novel fractional order stretch–twist–fold (STF) flow chaotic system and its application to a new authenticated encryption scheme (AES). Nonlinear Dyn. 77, 1547–1559.

Ouannas, A., Azar, A.T., Vaidyanathan, S., 2007. A robust method for new fractional hybrid chaos synchronization. Math. Meth. Appl. Sci. 40, 1804–1812.

Ouannas, A., Azar, A.T., Radwan, A.G., Dec 2016. On inverse problem of generalized synchronization between different dimensional integer-order and fractional-order chaotic systems. In: 2016 28th International Conference on Microelectronics (ICM). pp. 193–196.

Ouannas, A., Azar, A.T., Vaidyanathan, S., 2017a. New hybrid synchronization schemes based on coexistence of various types of synchronization between master-slave hyperchaotic systems. Int. J. Computer Applicat. Technol. 55 (2), 112–120.

Ouannas, A., Azar, A.T., Vaidyanathan, S., 2017b. On a simple approach for q-s synchronization of chaotic dynamical systems in continuous-time. Int. J. Computing Sci. Math. 8 (1), 20–27.

Ouannas, A., Azar, A.T., Ziar, T., 2017c. On inverse full state hybrid function projective synchronization for continuous-time chaotic dynamical systems with arbitrary dimensions. Different. Eq. Dyn. Syst. Available from: https://doi.org/10.1007/s12591-017-0362-x.

Ouannas, A., Azar, A.T., Ziar, T., Radwan, A.G., 2017d. Generalized synchronization of different dimensional integer-order and fractional order chaotic systems. In: Azar, A.T., Vaidyanathan, S., Ouannas, A. (Eds.), Fractional Order Control and Synchronization of Chaotic Systems. Springer International Publishing, Cham, pp. 671–697.

Ouannas, A., Azar, A.T., Ziar, T., Radwan, A.G., 2017e. A study on coexistence of different types of synchronization between different dimensional fractional chaotic systems. In: Azar, A.T., Vaidyanathan, S., Ouannas, A. (Eds.), Fractional Order Control and Synchronization of Chaotic Systems. Springer International Publishing, Cham, pp. 637–669.

Ouannas, A., Azar, A.T., Ziar, T., Vaidyanathan, S., 2017f. Fractional inverse generalized chaos synchronization between different dimensional systems. In: Azar, A.T., Vaidyanathan, S., Ouannas, A. (Eds.), Fractional Order Control and Synchronization of Chaotic Systems. Springer International Publishing, Cham, pp. 525–551.

Ouannas, A., Azar, A.T., Ziar, T., Vaidyanathan, S., 2017g. A new method to synchronize fractional chaotic systems with different dimensions. In: Azar, A.T., Vaidyanathan, S., Ouannas, A. (Eds.), Fractional Order Control and Synchronization of Chaotic Systems. Springer International Publishing, Cham, pp. 581–611.

Ouannas, A., Azar, A.T., Ziar, T., Vaidyanathan, S., 2017h. On new fractional inverse matrix projective synchronization schemes. In: Azar, A.T., Vaidyanathan, S., Ouannas, A. (Eds.), Fractional Order Control and Synchronization of Chaotic Systems. Springer International Publishing, Cham, pp. 497–524.

Ouannas, A., Grassi, G., Azar, A.T., Radwan, A.G., Volos, C., Pham, V.-T., et al., 2017i. Deadbeat synchronization control in discrete-time chaotic systems. In: 6th International Conference on Modern Circuits and Systems Technologies (MOCAST), pp. 1–4.

Pai, M.-C., 2014. Global synchronization of uncertain chaotic systems via discrete−time sliding mode control. Appl. Math. Comput. 227, 663−671.

Pecora, L., Carroll, T.L., 1990. Synchronization in chaotic systems. Phys. Rev. Lett. 64, 821−824.

Petras, I., 2011. Fractional-Order Nonlinear Systems, Modeling, Analysis and Simulation. Higher Education Press and Springer, Beijing and Berlin.

Pham, V.-T., Jafari, S., Volos, C., Giakoumis, A., Vaidyanathan, S., Kapitaniak, T., 2016a. A chaotic system with equilibria located on the rounded square loop and its circuit implementation. IEEE Trans. Circuits Syst.-II: Express Briefs 63 (9), 878−882.

Pham, V.-T., Jafari, S., Volos, C., Vaidyanathan, S., Kapitaniak, T., 2016b. A chaotic system with infinite equilibria located on a piecewise linear curve. Optik 127 (20), 9111−9117.

Pham, V.T., Jafari, S., Wang, X., Ma, J., 2016c. A chaotic system with different shapes of equilibria. Int. J. Bifurcation Chaos 26 (4), 1650069.

Pham, V.-T., Volos, C., Jafari, S., Vaidyanathan, S., Kapitaniak, T., Wang, X., 2016d. A chaotic systems with different families of hidden attractors. Int. J. Bifurcation Chaos 25 (8), 1650139.

Pham, V.-T., Vaidyanathan, S., Volos, C.K., Azar, A.T., Hoang, T.M., Van Yem, V., 2017a. A three-dimensional no-equilibrium chaotic system: analysis, synchronization and its fractional order form. In: Azar, A.T., Vaidyanathan, S., Ouannas, A. (Eds.), Fractional Order Control and Synchronization of Chaotic Systems. Springer International Publishing, Cham, pp. 449−470.

Pham, V.-T., Vaidyanathan, S., Volos, C.K., Jafari, S., Gotthans, T., 2017b. A Three-Dimensional Chaotic System with Square Equilibrium and No-Equilibrium. Springer International Publishing, Cham, pp. 613−635. Available from: https://doi.org/10.1007/978-3-319-50249-6_21.

Pham, V.-T., Volos, C., Kapitaniak, T., 2017c. Systems with Hidden Attractors: From Theory to Realization in Circuits. Springer, Germany.

Piper, J.R., Sprott, J.C., 2010. Simple autonomous chaotic circuits. IEEE Trans. Circuits Syst.-II: Express Briefs 57, 730−734.

Pivka, L., Wu, C.W., Huang, A., 1994. Chua's oscillator: a compendium of chaotic phenomena. J. Frankl. Inst. 331, 705−741.

Podlubny, I., 1999. Fractional Differential Equations. Academic Press, New York.

Rosenstein, M.T., Collins, J.J., Luca, C.J.D., 1993. A practical method for calculating largest Lyapunov exponents from small data sets. Physica D 65, 117.

Rössler, O.E., 1976. An equation for continuous chaos. Phys. Lett. A 57, 397−398.

Shilnikov, L., Shilnikov, A., Turaev, D., Chua, L., 1998. Methods of Qualitative Theory in Nonlinear Dynamics. World Scientific, Singapore.

Shilnikov, L.P., 1965. A case of the existence of a countable number of periodic motions. Sov. Math. Docklady 6, 163−166.

Singh, S., Azar, A.T., Ouannas, A., Zhu, Q., Zhang, W., Na, J., 2017. Sliding mode control technique for multi-switching synchronization of chaotic systems. In: 9th International Conference on Modelling, Identification and Control (ICMIC 2017), July 10−12, 2017, Kunming, China.

Soliman, N.S., Said, L.A., Azar, A.T., Madian, A.H., Radwan, A.G., Ounnas, A., 2017. Fractional controllable multi-scroll v-shape attractor with parameters effect. In: 6th International Conference on Modern Circuits and Systems Technologies (MOCAST), pp. 1−4.

Sprott, J., 1994. Some simple chaotic flows. Phys. Rev. E 50, R647−650.

Sprott, J.C., 2003. Chaos and Times−Series Analysis. Oxford University Press, Oxford.

Sprott, J.C., 2010. Elegant Chaos Algebraically Simple Chaotic Flows. World Scientific, Singapore.

Strogatz, S.H., 1994. Nonlinear Dynamics and Chaos: With Applications to Physics, Biology, Chemistry, and Engineering. Perseus Books, Massachusetts, US.

Sun, H., Abdelwahed, A., Onaral, B., 1984. Linear approximation for transfer function with a pole of fractional order. IEEE Trans. Autom. Control 29, 441–444.

Sun, J., Guo, J., Yang, C., Zheng, A., Zhang, X., 2015. Adaptive generalized hybrid function projective disclocated synchronization of new four–dimensional uncertain chaotic systems. Appl. Math. Comput. 252 (2015), 304–314.

Tolba, M.F., AbdelAty, A.M., Saida, L.A., Elwakil, A.S., Azar, A.T., Madian, A.H., et al., 2017. Fpga realization of caputo and grnwald-letnikov operators. In: 6th International Conference on Modern Circuits and Systems Technologies (MOCAST, pp. 1–4.

Trejo-Guerra, R., Tlelo-Cuautle, E., Carbajal-Gomez, V.H., Rodriguez-Gomez, G., 2013. A survey on the integrated design of chaotic oscillators. Appl. Math. Comput. 219, 5113–5122.

Upadhyay, R.K., 2003. Multiple attractors and crisis route to chaos in a model of food–chain. Chaos Solitons Fractals 16, 737–747.

Vaidyanathan, S., Azar, A.T., 2015a. Analysis and control of a 4-d novel hyperchaotic system. In: Azar, A.T., Vaidyanathan, S. (Eds.), Chaos Modeling and Control Systems Design. Vol. 581 of Studies in Computational Intelligence. Springer, Berlin, Germany, pp. 19–38.

Vaidyanathan, S., Azar, A.T., 2015b. Analysis, control and synchronization of a nine-term 3-d novel chaotic system. In: Azar, A.T., Vaidyanathan, S. (Eds.), Chaos Modeling and Control Systems Design. Vol. 581 of Studies in Computational Intelligence. Springer, Berlin, Germany, pp. 3–17.

Vaidyanathan, S., Azar, A.T., 2015c. Anti-synchronization of identical chaotic systems using sliding mode control and an application to vaidyanathan-madhavan chaotic systems. In: Azar, A. T., Zhu, Q. (Eds.), Advances and Applications in Sliding Mode Control Systems. Vol. 576 of Studies in Computational Intelligence. Springer, Berlin, Germany, pp. 527–547.

Vaidyanathan, S., Azar, A.T., 2015d. Hybrid synchronization of identical chaotic systems using sliding mode control and an application to vaidyanathan chaotic systems. In: Azar, A.T., Zhu, Q. (Eds.), Advances and Applications in Sliding Mode Control Systems. Vol. 576 of Studies in Computational Intelligence. Springer, Berlin, Germany, pp. 549–569.

Vaidyanathan, S., Azar, A.T., 2016a. A novel 4-D four-wing chaotic system with four quadratic nonlinearities and its synchronization via adaptive control method. Advances in Chaos Theory and Intelligent Control. Springer, Berlin, Germany, pp. 203–224.

Vaidyanathan, S., Azar, A.T., 2016b. Adaptive backstepping control and synchronization of a novel 3-D jerk system with an exponential nonlinearity. Advances in Chaos Theory and Intelligent Control. Springer, Berlin, Germany, pp. 249–274.

Vaidyanathan, S., Azar, A.T., 2016c. Adaptive control and synchronization of Halvorsen circulant chaotic systems. Advances in Chaos Theory and Intelligent Control. Springer, Berlin, Germany, pp. 225–247.

Vaidyanathan, S., Azar, A.T., 2016d. Dynamic analysis, adaptive feedback control and synchronization of an eight-term 3-D novel chaotic system with three quadratic nonlinearities. Advances in Chaos Theory and Intelligent Control. Springer, Berlin, Germany, pp. 155–178.

Vaidyanathan, S., Azar, A.T., 2016e. Generalized projective synchronization of a novel hyperchaotic four-wing system via adaptive control method. Advances in Chaos Theory and Intelligent Control. Springer, Berlin, Germany, pp. 275–290.

Vaidyanathan, S., Azar, A.T., 2016f. Qualitative study and adaptive control of a novel 4-d hyperchaotic system with three quadratic nonlinearities. In: Azar, A.T., Vaidyanathan, S. (Eds.), Advances in Chaos Theory and Intelligent Control. Springer International Publishing, Cham, pp. 179–202.

Vaidyanathan, S., Azar, A.T., 2016g. Takagi-Sugeno fuzzy logic controller for Liu-Chen four-scroll chaotic system. International Journal of Intelligent Engineering Informatics 4 (2), 135–150.

Vaidyanathan, S., Pehlivan, I., 2012. Analysis, control, synchronization, and circuit design of a novel chaotic system. Math. Comp. Modelling 55, 1904–1915.

Vaidyanathan, S., Azar, A.T., Ouannas, A., 2017a. An eight-term 3-d novel chaotic system with three quadratic nonlinearities, its adaptive feedback control and synchronization. In: Azar, A.T., Vaidyanathan, S., Ouannas, A. (Eds.), Fractional Order Control and Synchronization of Chaotic Systems. Springer International Publishing, Cham, pp. 719–746.

Vaidyanathan, S., Azar, A.T., Ouannas, A., 2017b. Hyperchaos and adaptive control of a novel hyperchaotic system with two quadratic nonlinearities. In: Azar, A.T., Vaidyanathan, S., Ouannas, A. (Eds.), Fractional Order Control and Synchronization of Chaotic Systems. Springer International Publishing, Cham, pp. 773–803.

Vaidyanathan, S., Zhu, Q., Azar, A.T., 2017c. Adaptive control of a novel nonlinear double convection chaotic system. In: Azar, A.T., Vaidyanathan, S., Ouannas, A. (Eds.), Fractional Order Control and Synchronization of Chaotic Systems. Springer International Publishing, Cham, pp. 357–385.

Vaidyanathan, S., Azar, A.T., Rajagopal, K., Alexander, P., 2015a. Design and spice implementation of a 12-term novel hyperchaotic system and its synchronisation via active control. Int. J. Model. Identif. Control 23 (3), 267–277.

Vaidyanathan, S., Idowu, B.A., Azar, A.T., 2015b. Backstepping controller design for the global chaos synchronization of sprott's jerk systems. In: Azar, A.T., Vaidyanathan, S. (Eds.), Chaos Modeling and Control Systems Design. Vol. 581 of Studies in Computational Intelligence. Springer, Berlin, Germany, pp. 39–58.

Vaidyanathan, S., Sampath, S., Azar, A.T., 2015c. Global chaos synchronisation of identical chaotic systems via novel sliding mode control method and its application to zhu system. Int. J. Model. Identif. Control 23 (1), 92–100.

Vaithianathan, V., Veijun, J., 1999. Coexistence of four different attractors in a fundamental power system model. IEEE Trans. Circuits Syst. I 46, 405–409.

Valli, D., Muthuswamy, B., Banerjee, S., Ariffin, M.R.K., Wahad, A.W.A., Ganesan, K., et al., 2014. Synchronization in coupled ikeda delay systems experimental observations using field programmable gate arrays. Eur. Phys. J. Special Topics 223, 1465–1479.

Volos, C.K., Kyprianidis, I.M., Stouboulos, I.N., 2012. A chaotic path planning generator for autonomous mobile robots. Robot. Auto. Systems 60, 651–656.

Volos, C.K., Kyprianidis, I.M., Stouboulos, I.N., 2013. Image encryption process based on chaotic synchronization phenomena. Signal Processing 93, 1328–1340.

Wang, X., Chen, G., 2012. A chaotic system with only one stable equilibrium. Commun. Nonlinear Sci. Numer. Simul. 17, 1264–1272.

Wang, X., Chen, G., 2013. Constructing a chaotic system with any number of equilibria. Nonlinear Dyn. 71, 429–436.

Wang, Z., Chen, S., Ochola, E.O., Sun, Y., 2012. A hyperchaotic system without equilibrium. Nonlinear Dyn. 69, 531–537.

Wang, Z., Volos, C., Kingni, S.T., Azar, A.T., Pham, V.-T., 2017. Four–wing attractors in a novel chaotic system with hyperbolic sine nonlinearity. Optik 131 (2017), 1071–1078.

Westerlund, S., Ekstam, L., 1994. Capacitor theory. IEEE Trans. Dielectr. Electr. Insul. 1, 826–839.

Wolf, A., Swift, J.B., Swinney, H.L., Vastano, J.A., 1985. Determining Lyapunov exponents from a time series. Phys. D 16, 285–317.

Wu, J., Wang, L., Chen, G., Duan, S., 2016. A memristive chaotic system with heart–shaped attractors and its implementation. Chaos Solitons Fractals 92 (2016), 20–29.

Wu, R., Wang, C., 2016. A new simple chaotic circuit based on memristor. Int. J. Bifurcation Chaos 26 (9), 1650145.

Xu, Y., Wang, H., Li, Y., Pei, B., 2014. Image encryption based on synchronization of fractional chaotic systems. Commun. Nonlinear Sci. Numer. Simul. 19, 3735–3744.

Yalcin, M.E., Suykens, J.A.K., Vandewalle, J., 2004. True random bit generation from a double–scroll attractor. IEEE Trans. Circuits Syst. I, Regular Papers 51, 1395–1404.

Yanchuk, S., Maistrenko, Y., Mosekilde, E., 2001. Loss of synchronization in coupled rössler systems. Phys. D 154, 26–42.

Yang, C.-C., 2013. One input control of exponential synchronization for a four–dimensional chaotic system. Appl. Math. Comput. 219, 5152–5161.

Yu, S., Tang, W.K.S., Lu, J., Chen, G., 2010. Design and implementation of multi–wing butterfly chaotic attractors via Lorenz–type systems. Int. J. Bifurcation Chaos 20, 29–41.

Zeng, Z., Zheng, W., 2012. Multistability of neural networks with time–varying delays and concave–convex characteristic. IEEE Trans. Neural Netw. Learn. Syst. 23, 293–305.

Zeng, Z., Huang, T., Zheng, W., 2010. Multistability of recurrent networks with time–varying delays and the piecewise linear activation function. IEEE Trans. Neural Netw. 21, 1371–1377.

Zhao, X., Li, Z., Li, S., 2011. Synchronization of a chaotic finance system. Appl. Math. Comput. 217, 6031–6039.

Zhou, P., Yang, F., 2014. Hyperchaos, chaos, and horseshoe in a 4D nonlinear system with an infinite number of equilibrium points. Nonlinear Dyn. 76, 473–480.

Zhu, Q., Azar, A.T., 2015. Complex System Modelling and Control Through Intelligent Soft Computations. Vol. 319. Studies in Fuzziness and Soft Computing, Berlin, Germany.

Chapter 17

Chaotic Properties of Various Types of Hidden Attractors in Integer and Fractional Order Domains

Wafaa S. Sayed[1], Ahmed G. Radwan[1,2] and Salwa K. Abd-El-Hafiz[1]
[1]*Faculty of Engineering, Cairo University, Giza, Egypt,* [2]*Nanoelectronics Integrated Systems Center, (NISC) Nile University, Cairo, Egypt*

17.1 INTRODUCTION

In the recent decades, chaotic systems have been analyzed and implemented on electronic platforms (Radwan et al., 2004, 2007; Radwan, 2013a; Radwan et al., 2003; Sayed et al., 2017d; Zidan et al., 2012). They are characterized by the sensitivity to initial conditions, which is a required property for many applications such as chemical reactions (Han et al., 1995), biological systems (Moaddy et al., 2012; Strogatz, 2014), circuit theory (Radwan, 2012, 2013b; Radwan et al., 2008a,b), electronics (Schöll, 2001), control and synchronization (Azar and Vaidyanathan, 2015, 2016; Azar et al., 2017a; Radwan et al., 2013, 2014a; Sayed et al., 2017b, 2016), secure communication (Chien and Liao, 2005; Frey, 1993; Lau and Tse, 2003), and cryptography (Abd-El-Hafiz et al., 2016, 2015, 2014; Abdelhaleem et al., 2014; Barakat et al., 2013; Kocarev and Lian, 2011; Radwan and Abd-El-Hafiz, 2013, 2014; Radwan et al., 2014b, 2015b; Sayed et al., 2017a, 2015a,b).

Meanwhile, fractional calculus has flourished and found its way to real-world applications in various fields including viscoelasticity, robotics, feedback amplifiers, electrical circuits, control theory, electro analytical chemistry, fractional multipoles, electromagnetics, bioengineering, and image encryption (AbdelAty et al., 2017; Caponetto, 2010; Fouda and Radwan, 2015; Fouda et al., 2016; Goreno and Mainardi, 1997; Ismail et al., 2015; Magin, 2006; Petras, 2011; Psychalinos et al., 2016; Radwan et al., 2012, 2015a, 2016, 2011a,b; Radwan and Fouda, 2013; Semary et al., 2016; Shamim et al., 2011; Soltan et al., 2012, 2015, 2017). Fractional calculus is

Mathematical Techniques of Fractional Order Systems. DOI: https://doi.org/10.1016/B978-0-12-813592-1.00017-9

more suitable for modeling the continuous nonstandard behaviors of nature due to the flexibility offered by the extra degrees of freedom and including memory effects. Recently, most of the chaotic dynamical systems based on integer order calculus have been extended into the fractional order domain to fit the experimental data much more precisely than the integer order modeling (AboBakr et al., 2017). This came in association with the advances in numerical methods for solving fractional order systems and their electronic implementations (Caponetto, 2010; Semary et al., 2017; Tolba et al., 2017).

Continuous flows expressed in terms of ordinary differential equations can have numerous types of posttransient solution(s). An attractor is defined as the set of points approached by the orbit as the number of iterations increase to infinity representing its long-term behavior. For a continuous system of differential equations, the equilibrium points are defined to be those points at which all time derivatives equal zero. The linear stability of each of the obtained points can be determined by calculating the eigenvalues of the linearized Jacobian matrix (Sprott, 1994) (Routh–Hurwitz criterion). Specifically, if all eigenvalues have negative real parts, then the system is stable near the equilibrium point. If any eigenvalue has a positive real part, then the point is unstable. If the matrix has at least one eigenvalue with positive real part, at least one with negative real part, and no eigenvalues with zero real part, then the point is called a saddle (Alligood et al., 1996).

Many previous works focused on coming up with novel continuous chaotic systems, especially those which are algebraically simple and hence suitable for hardware realization (Sprott, 1994, 2000, 2007; Sayed et al., 2017c). Most familiar examples of chaotic flows occur in such self-excited systems having one or more saddle points (Alligood et al., 1996). A characteristic of a saddle equilibrium point is that the trajectories starting at some of the initial values converge to it while the others diverge from it. If most of the initial values move away or diverge from the saddle point, then it is unstable. While the set of initial values that converge to the saddle are called the stable manifold of the saddle, those which diverge are called the unstable manifold. Such saddle points allow homoclinic and heteroclinic orbits and locating strange attractors by choosing an initial condition on the unstable manifold in the vicinity of the saddle point (Jafari et al., 2015).

Several researches in the last few years have presented hidden attractors with no equilibria (Jafari et al., 2013), one stable equilibrium (Molaie et al., 2013), and a line of equilibrium points (Jafari and Sprott, 2013). Consequently, hidden attractors have also been reported in memristive systems (Pham et al., 2017a) and hyperchaotic systems (Li and Sprott, 2014; Vaidyanathan et al., 2015). Applications such as circuit realization and synchronization of hidden attractors have been presented (Azar et al., 2017b; Pham et al., 2017b,c). Familiar self-excited attractors have been extended to the fractional order domain in several works (Petras, 2011; Radwan et al., 2014a). Other fewer researches focused on hidden attractors in the fractional order domain (Danca, 2017; Henein et al., 2016; Radwan et al., 2017; Volos et al., 2017). Studying the effect of the different parameters on the behavior of systems of differential equations and reporting the parameter values at which these systems exhibit chaos represents a rich research field. For systems with hidden attractors, such a study becomes rather harder than the self-excited case, especially in the fractional order domain.

This chapter presents an overview of a set of various hidden attractors (Jafari et al., 2013, 2015; Pham et al., 2015, 2014). With respect to the type of equilibrium, hidden attractors are classified into three main categories: rare flows with no equilibria, rare flows with a line of equilibrium points, and rare flows with one stable equilibrium. Another classification is concerned with the type of nonlinear terms that appear in the equations of the system. The properties of these hidden attractors are reviewed for integer order derivatives. In addition, their behavior is explored for other parameter ranges and in the fractional order domain. Section 17.2 reviews the numerical methods used for solving the systems and generating the time series of their responses for integer order and fractional order derivatives. Section 17.3 provides an overview about the systems including their equations, the specified parameter values and initial conditions, and the corresponding responses in the form of attractor diagrams and the corresponding Lyapunov exponent values. While Section 17.4 focuses on a set of systems and explores their sensitivity to parameters variations, Section 17.5 presents their sensitivity to fractional orders. Section 17.6 summarizes the contributions of the chapter.

17.2 NUMERICAL SOLUTION OF INTEGER AND FRACTIONAL ORDER DIFFERENTIAL EQUATIONS

A system of ordinary first order differential equations can be solved numerically through well-established techniques. One of the most famous and widely used solvers is the fourth order Runge Kutta method (RK-4), where a MATLAB implementation can be found in Mueller (2011). MATLAB also provides the function ode45 (Senan, 2012) as a solver. Several software packages can also be used to study the behavior of systems of equations and their chaotic properties and will be mentioned in Sections 17.3 and 17.4.

Finding robust and stable numerical and analytical methods for solving fractional differential equations has been an active research topic. These methods include the fractional difference method, the Adomian decomposition method, the homotopy-perturbation method, the variational iteration method and the Adams–Bashforth–Moulton method. Recently, the Non-Standard Finite Difference (NSFD) scheme (Mickens, 2000, 2005) has been applied for the numerical solutions of fractional differential equations (Hussian et al., 2008). The scheme has been developed as an alternative method for solving a wide range of problems whose mathematical models involve algebraic, differential, biological models and chaotic systems. The definition of Grünwald–Letnikov derivative has been used in numerical analysis to discretize the fractional differential equations. The technique has many advantages over the classical techniques, and provides an efficient numerical solution.

The Caputo fractional derivative (Gorenflo and Mainardi, 1997) of order α is defined as:

$$D^{\alpha}f(t) = \frac{d^{\alpha}f(t)}{dt^{\alpha}} = \begin{cases} \dfrac{1}{\Gamma(m-\alpha)}\displaystyle\int_0^t \frac{f^m(\tau)}{(t-\tau)^{\alpha-m+1}}d\tau & m-1<\alpha<m \\ \dfrac{d^m}{dt^m}f(t) & \alpha=m \end{cases}, \tag{17.1}$$

where m is the first integer greater than α and $\Gamma(.)$ is the gamma function defined by:

$$\Gamma(z) = \int_0^{\infty} e^{-t}t^{z-1}dt, \quad \Gamma(z+1) = z\Gamma(z). \tag{17.2}$$

Consider the fractional order differential equation

$$D^{\alpha}x(t) = f(t,x). \tag{17.3}$$

Grünwald–Letnikov method of approximation (Hussian et al., 2008) is defined as follows:

$$D^{\alpha}x(t) = \lim_{h\to 0} h^{-\alpha}\sum_{j=0}^{t/h}(-1)^j\binom{\alpha}{j}x(t-jh), \tag{17.4}$$

where h is the step size. This equation can be discretized as follows:

$$\sum_{j=0}^{n+1} c_j^{\alpha}x(t-jh) = f(t_n, x(t_n)), \quad j=1,2,3,\ldots \tag{17.5}$$

where $t_n = nh$ and c_j^{α} are the Grünwald–Letnikov coefficients defined as:

$$c_j^{\alpha} = \left(1-\frac{1+\alpha}{j}\right)c_{j-1}^{\alpha}, \quad j=1,2,3,\ldots, \quad c_0^{\alpha} = h^{-\alpha}. \tag{17.6}$$

The NSFD discretization technique is based on replacing the step size h by a function $\phi(h)$ (Hussian et al., 2008; Moaddy et al., 2012) and applying it with (17.5) to solve (17.3).

The same algebraic manipulation can be applied to a system of three fractional order differential equations

$$\begin{aligned} D^{\alpha}x &= f(x,y,z),\\ D^{\beta}y &= g(x,y,z),\\ D^{\gamma}z &= h(x,y,z), \end{aligned} \tag{17.7}$$

where $0 < \alpha, \beta, \gamma \leq 1$, to obtain the corresponding solutions. Discretized solutions to the systems could be obtained using (17.5) and NSFD. Nonlinear terms including the same state variable that is being calculated are replaced with the aid of the nonlocal discrete representations. For example, in the equation of $D^{\beta}y$, the following rules are used for replacement:

$$y^2 \approx y_n y_{n+1}, \quad xy \approx 2x_{n+1}y_n - x_{n+1}y_{n+1}, \quad \text{and} \quad zy \approx 2z_n y_n - z_n y_{n+1}. \tag{17.8}$$

The relations used for solving the systems will be given in Section 17.5.

17.3 SURVEY OF SOME SYSTEMS WITH HIDDEN ATTRACTORS

Fig. 17.1 shows the categorization of the hidden attractors, which are included in the selected papers (Jafari et al., 2013, 2015; Pham et al., 2015, 2014), from the viewpoints of type of equilibrium points and nonlinearity. The selected systems are reviewed in this section by placing each of them in the category to which it belongs with the aid of Tables 17.1 and 17.2.

The history of hidden attractors dates back to the early works by J.C. Sprott (1994) about two decades ago, which uncovered some simple chaotic flows. The case of a system of differential equations with no equilibria that exhibits chaotic behavior was first reported in the system which he called case A, however, it was not analyzed in this sense. On the other hand, the classification of such chaotic flows into flows with no equilibria, a stable equilibrium point, or line equilibria is a more recent topic and their handling in this sense of classification or grouping started around 5 years ago (Wang and Chen, 2012). Hidden attractors allow unexpected and potentially disastrous responses to perturbations in a structure like a bridge or an

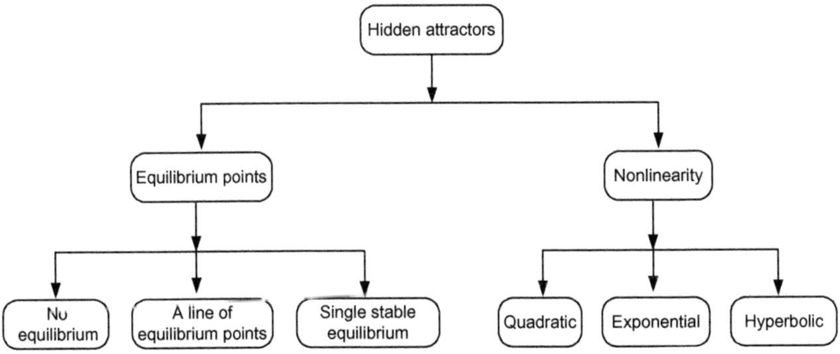

FIGURE 17.1 Categories of the reviewed hidden attractors.

TABLE 17.1 Hidden Attractors With No Equilibria and Quadratic Nonlinearities

	Equations	Attractor	Properties & Main Characteristics
(1)	$\dot{x} = y$ $\dot{y} = -x - zy$ $\dot{z} = y^2 - a$	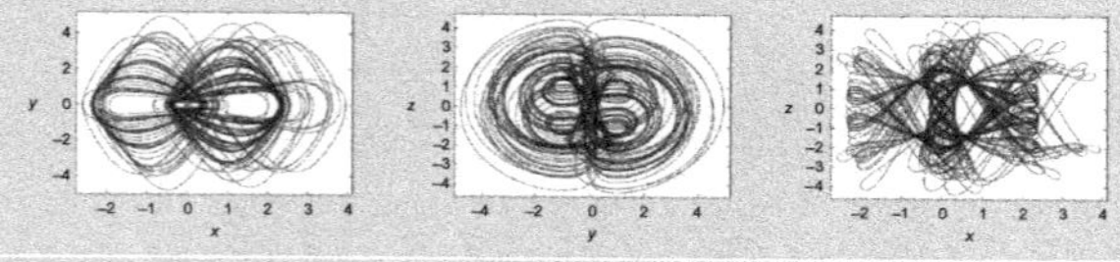	ICs: (0,5,0) $a = 1.0$ LEs:(0.0138, 0, − 0.0138) • A conservative system.
(2)	$\dot{x} = -y$ $\dot{y} = x + z$ $\dot{z} = 2y^2 + xz - a$	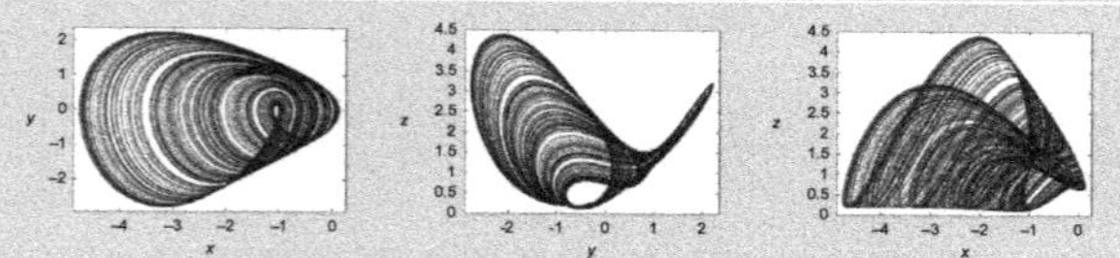	ICs: (0, 0.4, 1) $a = 0.35$ LEs:(0.0776, 0, − 1.5008)
(3)	$\dot{x} = y$ $\dot{y} = z$ $\dot{z} = -y + 0.1x^2 + 1.1xz + a$	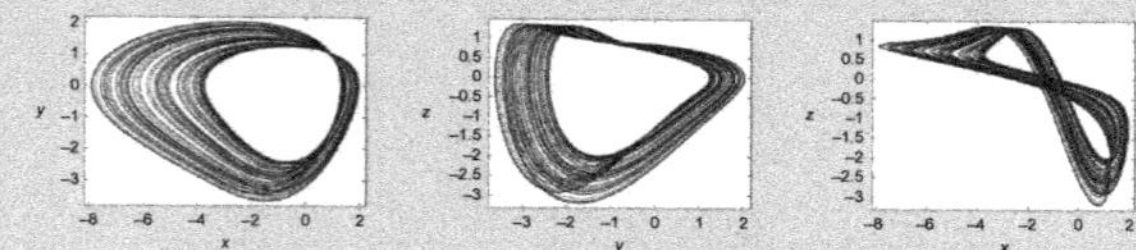	ICs: (1,1,-1) $a = 1.0$ LEs: (0.0522,0,-2.6585)
(4)	$\dot{x} = -0.1y + a$ $\dot{y} = x + z$ $\dot{z} = xz - 3y$	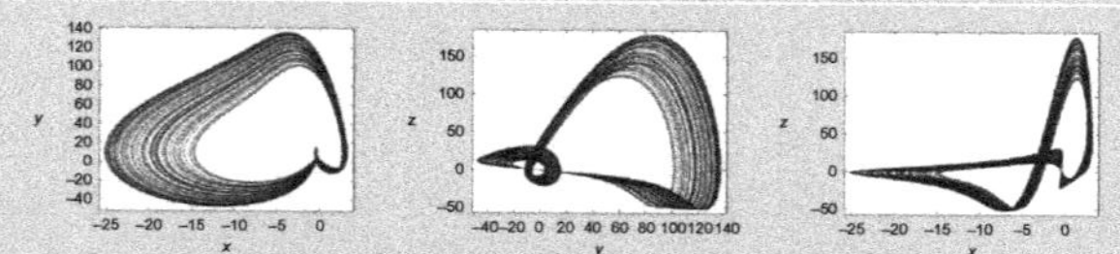	ICs:(− 8.2, 0, − 5) $a = 1.0$ LEs:(0.0235, 0, − 8.480)
(5)	$\dot{x} = 2y$ $\dot{y} = -2x - z$ $\dot{z} = -y^2 + z^2 + a$	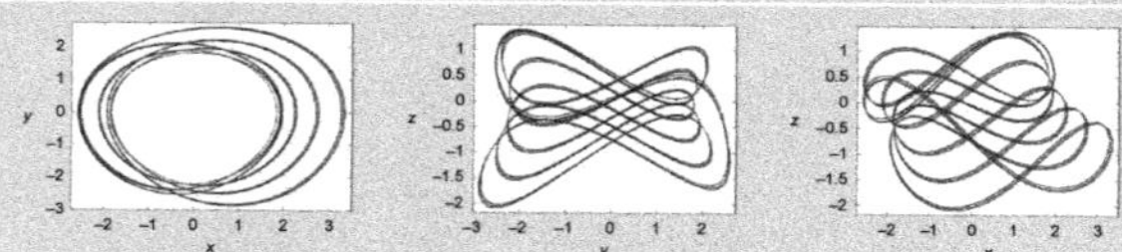	ICs:(0.98, 1.8, − 0.7) $a = 2.0$ LEs:(0.0168, 0, − 0.3622)

(6)	$\dot{x} = y$ $\dot{y} = z$ $\dot{z} = -y - xz - yz - a$	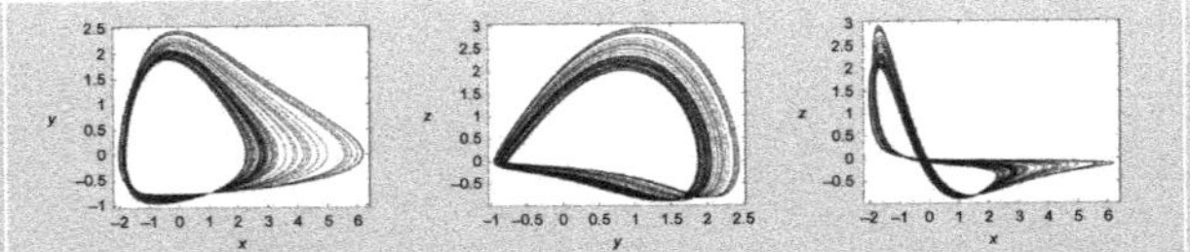	ICs:$(0, 3, -0.1)$ $a = 0.75$ LEs:$(0.0280, 0, -3.4341)$
(7)	$\dot{x} = y$ $\dot{y} = -x + z$ $\dot{z} = -0.8x^2 + z^2 + a$	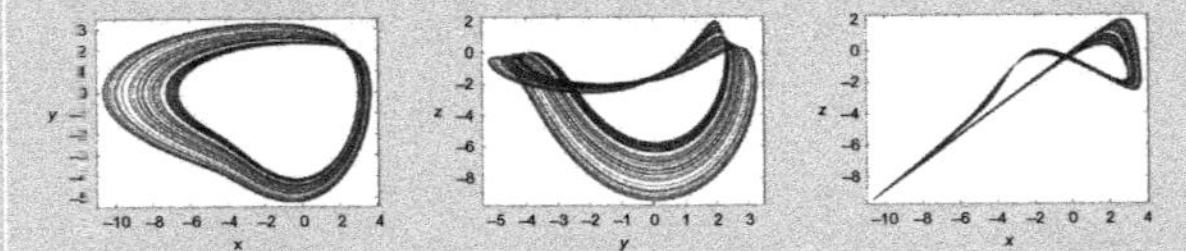	ICs:$(0, 2.3, 0)$ $a = 2.0$ LEs:$(0.0252, 0, -6.8524)$
(8)	$\dot{x} = y$ $\dot{y} = -x - yz$ $\dot{z} = xy + 0.5x^2 - a$	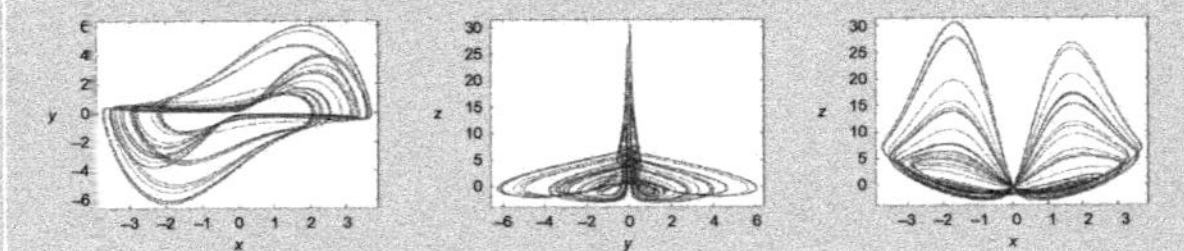	ICs:$(0, 0.1, 0)$ $a = 1.3$ LEs:$(0.0314, 0, -10.2108)$
(9)	$\dot{x} = y$ $\dot{y} = -x - yz$ $\dot{z} = -xz + 7x^2 - a$	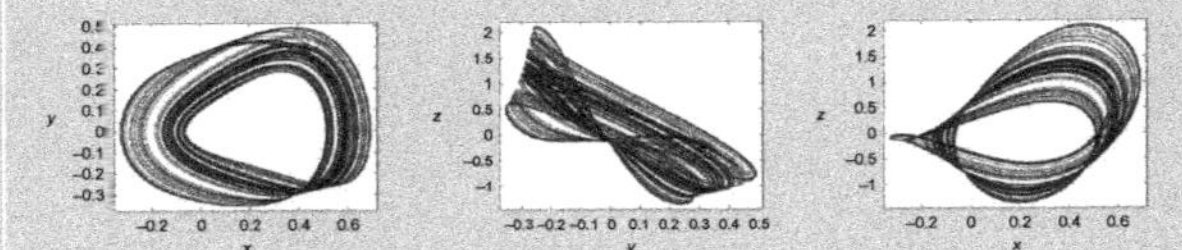	ICs:$(0.5, 0, 0)$ $a = 0.55$ LEs:$(0.0504, 0, -0.3264)$
(10)	$\dot{x} = z$ $\dot{y} = z - y$ $\dot{z} = -0.9y - xy + xz + a$	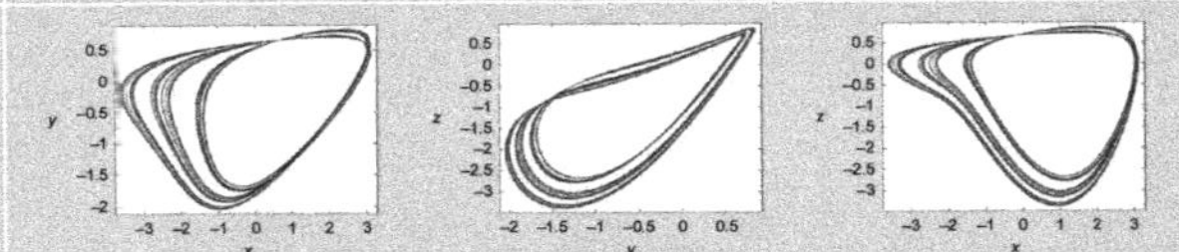	ICs:$(1, 0.7, 0.8)$ $a = 0.6$ LEs:$(0.0061, 0, -1.3002)$
(11)	$\dot{x} = y$ $\dot{y} = -x + z$ $\dot{z} = z - 2xy - 1.8xz - a$	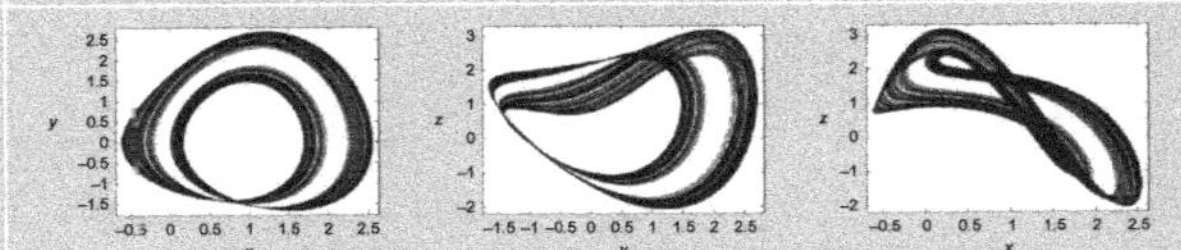	ICs:$(0, 1.6, 3)$ $a = 1.0$ LEs:$(0.0706, 0, -0.6456)$

(*Continued*)

TABLE 17.1 (Continued)

	Equations	Attractor	Properties & Main Characteristics
(12)	$\dot{x} = z$ $\dot{y} = x - y$ $\dot{z} = -4x^2 + 8xy + yz + a$		ICs:(0.5, 0, − 1) $a = 0.1$ LEs:(0.0654, 0, − 2.0398)
(13)	$\dot{x} = -y$ $\dot{y} = x + z$ $\dot{z} = xy + xz + 0.2yz - a$	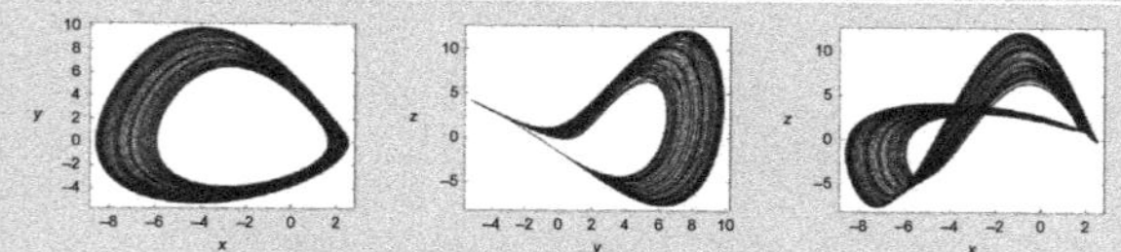	ICs:(2.5, 0, 0) $a = 0.4$ LEs: (0.1028, 0, -2.1282)
(14)	$\dot{x} = y$ $\dot{y} = z$ $\dot{z} = x^2 - y^2 + 2xz + yz + a$	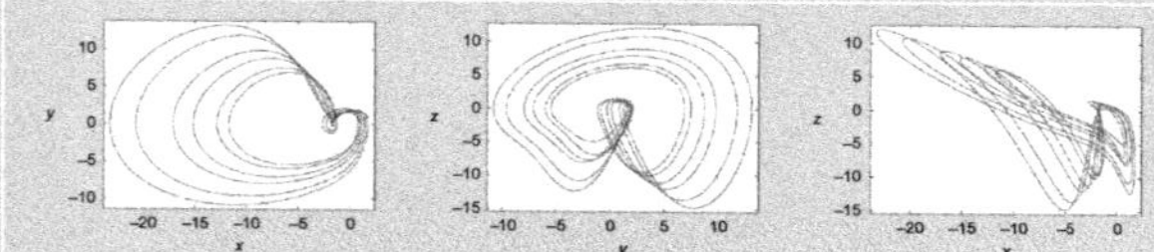	ICs:(1, 0, − 4) $a = 1.0$ LEs:(0.0532, 0, − 11.8580)
(15)	$\dot{x} = y$ $\dot{y} = z$ $\dot{z} = x^2 - y^2 + xy + 0.4xz + a$	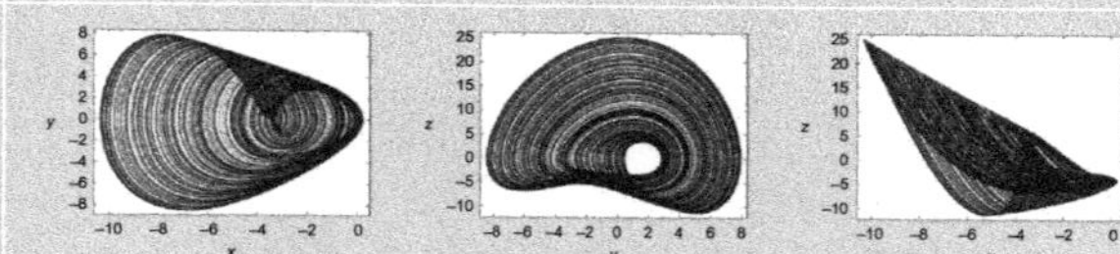	ICs:(0, 1, − 4.9) $a = 1.0$ LEs:(0.1101, 0, − 1.3879)
(16)	$\dot{x} = -0.8x - 0.5y^2 + xz + a$ $\dot{y} = -0.8y - 0.5z^2 + yx + a$ $\dot{z} = -0.8z - 0.5x^2 + zy + a$	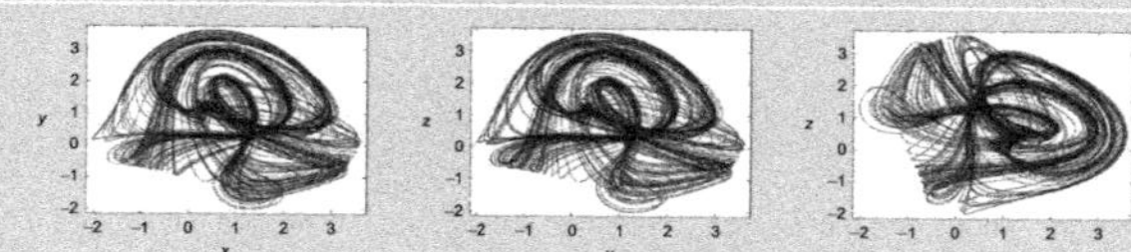	ICs:(0, 1, − 1) $a = 1.0$ LEs:(0.0607, 0, − 0.1883)

(17)	$\dot{x} = -y - z^2 + 2.3xy + a$ $\dot{y} = -z - x^2 + 2.3yz + a$ $\dot{z} = -x - y^2 + 2.3zx + a$	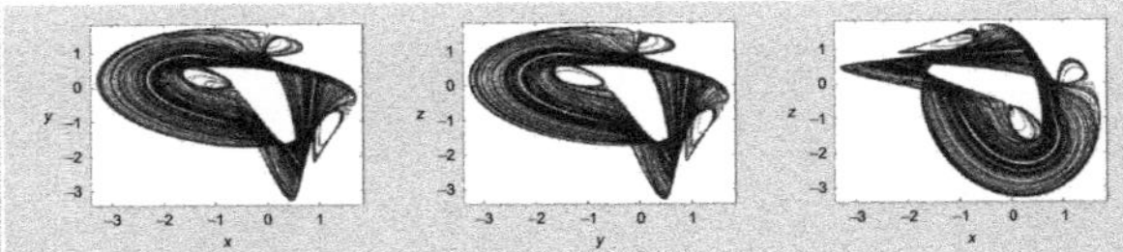	ICs:(1, − 1, 0) $a = 2.0$ LEs:(0.2257, 0, − 1.7477)
(18)	$\dot{x} = -y$ $\dot{y} = x + z$ $\dot{z} = 2y^2 + xz - 0.35$	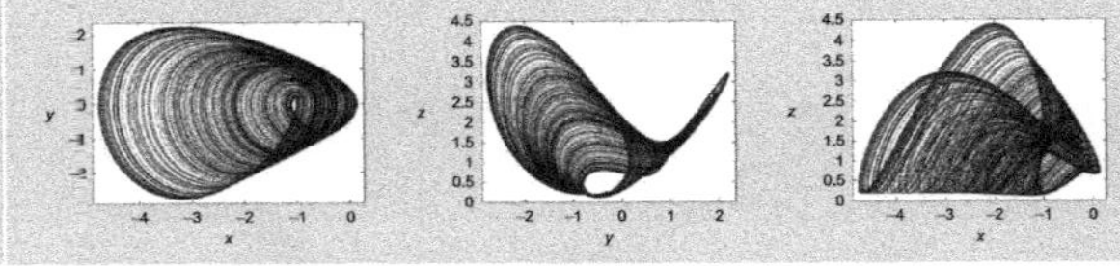	ICs:(− 1.6, 0.82, 1.9) LEs:(0.0776, 0, − 1.5008)

TABLE 17.2 Further Hidden Attractors of Various Types

	Equations	Attractor	Properties & main characteristics
(19)	$\dot{x} = y$ $\dot{y} = z$ $\dot{z} = -x - 0.6y - 2z + z^2 - 0.4xy$	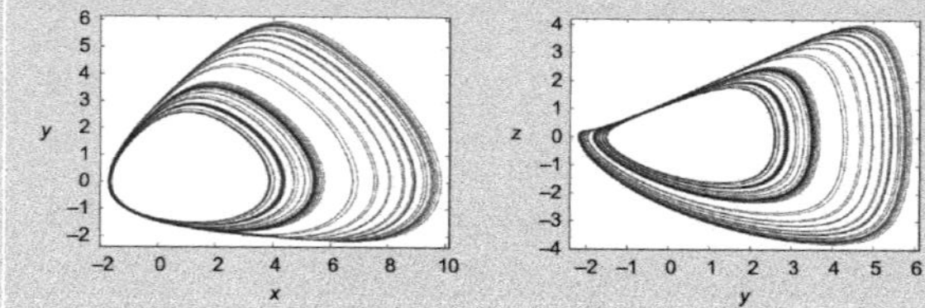	ICs:$(4, -2, 0)$ LEs:$(0.0377, 0, -2.0377)$ • Quadratic nonlinearities. • A single stable equilibrium point at (0,0,0).
(20)	$\dot{x} = y$ $\dot{y} = -x + yz$ $\dot{z} = -x - 15xy - xz$	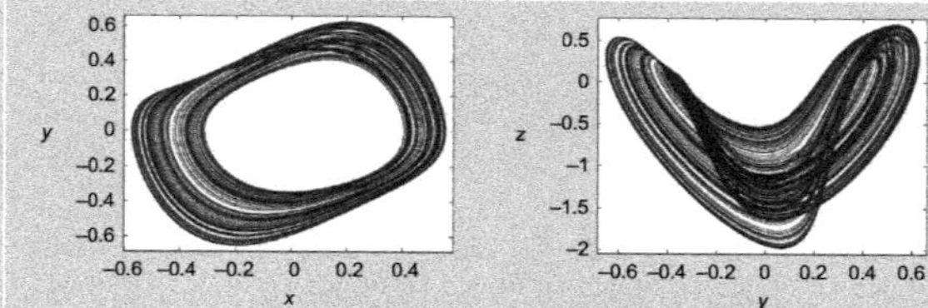	ICs:$(0, 0.5, 0.5)$ LEs:$(0.0717, 0, -0.5232)$ • Quadratic nonlinearities. • The z-axis is the line equilibrium of this system, i.e., $(0, 0, z)$ with no other equilibria.
(21)	$\dot{x} = y$ $\dot{y} = -x + yz$ $\dot{z} = z + ax^2 - y^2 - b$	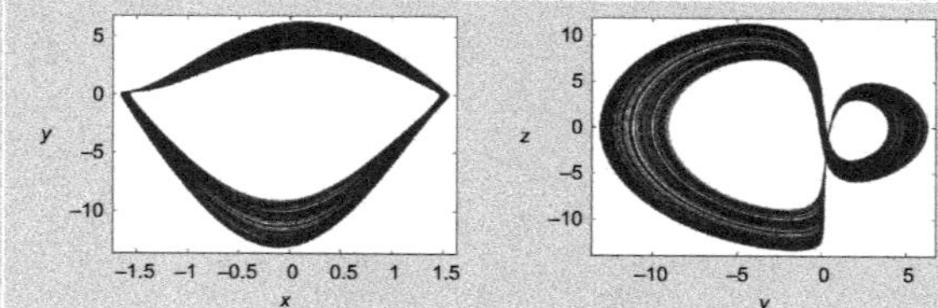	ICs:$(0, 3.9, 0.7)$ Parameter(s):$a = 8.888, b = 4$ • Quadratic nonlinearities.

(22)	$\dot{x} = -x + 1.6\tan(x) + 2\tanh(y) + \tanh(z)$ $\dot{y} = -y + aw^2\tanh(x) + 1.5\tanh(y)$ $\dot{z} = -z + 3\tanh(x) - 2\tanh(y) + \tanh(z) + b$ $\dot{w} = \tanh(x)$	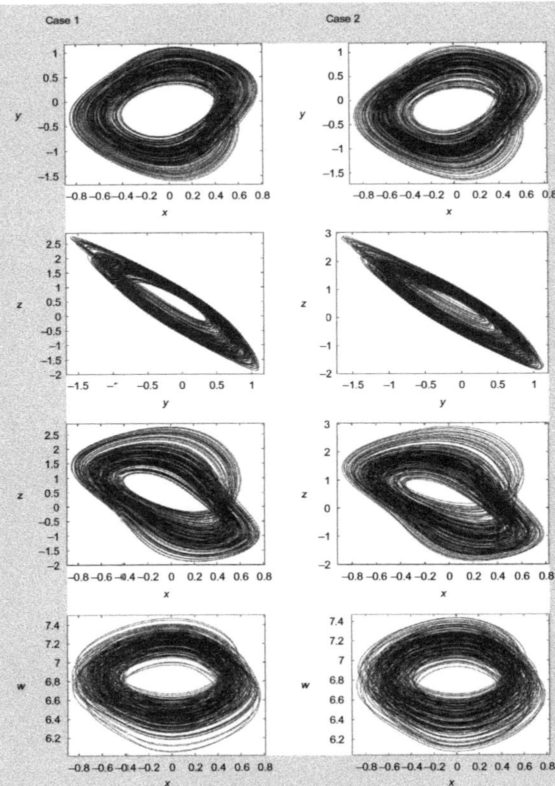	ICs:(0, 0.01, 0.01, 0) • Hyperbolic and quadratic nonlinearities. $a = -0.05$ Case 1: Parameter(s):$b = 0$ • Line equilibrium along w-axis. LEs:(0.029, 0.0088, 0, −0.1151) Case 2: Parameter(s):$b = -0.001$ • No equilibria. LEs:(0.0291, 0.0098, 0, −0.1152)

(*Continued*)

TABLE 17.2 (Continued)

	Equations	Attractor	Properties & main characteristics
(23)	$\dot{x} = a(y - x)$ $\dot{y} = -by + xz + cw$ $\dot{z} = d - e^{xy}$ $\dot{w} = -mx$	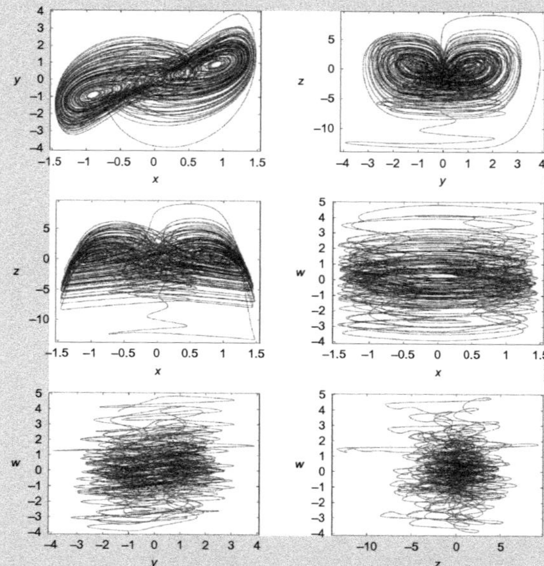	ICs:$(-1.6, 0.9, 1, 0.1)$ Parameter(s):$a = 0.87,\ b = 0.5,\ c = 0.37,\ d = 2.5,\ m = 0.5$ LEs:$(0.1357, 0.0275, 0, -1.5326)$ • No equilibria. • Exponential and quadratic nonlinearities.

airplane wing. Moreover, hidden attractors appear in mechanical systems, phase-locked loops, drilling systems, or electronic circuits. Due to its importance in engineering applications, the subject attracted the interest of many other researchers, such as Leonov and Kuznetsov (Kuznetsov and Leonov, 2014; Leonov and Kuznetsov, 2013), to investigate not only hidden attractors but also hidden oscillations. It is worth mentioning that hidden attractors/oscillations have been discovered for some conventional continuous systems at certain combinations of parameters and initial conditions.

The clear classification of the newly proposed systems flourished only a few years ago. Jafari et al. (2013) provided 17 elementary three-dimensional chaotic flows with quadratic nonlinearities that have the unusual feature of lacking any equilibrium points. The main objective was to find the algebraically simplest cases, which cannot be further reduced by the removal of terms without losing chaotic behavior. The search method depended on including at least one constant term in the system of equations that prevents the presence of equilibrium points. Three methods have been used to generate more examples preserving simplicity: (1) adding a constant term a to a set of systems known to be nonhyperbolic, i.e., the equilibria have eigenvalues with zero real part; (2) investigating other cases in which the equilibrium points are algebraically shown to be imaginary; and (3) adding a constant to each of the derivatives in known chaotic systems. The resulting systems are listed as (1)–(17) in Table 17.1. Table 17.2 shows additional hidden attractors of various types. For instance, Jafari et al. (2015) discussed examples on different groups of hidden attractors with no equilibria (18), single stable equilibrium (19), line of equilibrium points (20) in addition to presenting another new chaotic system containing hidden attractor (21). Although the equilibrium is an unstable node, the attractor appears to be hidden because it cannot be found by starting with initial conditions in the vicinity of this equilibrium. In addition, the range of parameters corresponding to bounded solutions is relatively small.

Pham et al. (2014) presented a hidden hyperchaotic attractor in a novel simple memristive neural network. A model of new simple Hopfield memristive neural network is introduced, system (22), with the presence of an input bias current. The network consists of only three neurons and one memristive synaptic weight and results in a fourth-order model. The hidden attractor can be detected in two different cases: with no equilibrium points and with a line of equilibrium points. Another no equilibrium chaotic system, but with an exponential nonlinearity, has been presented in 2015 by Pham et al. (2015). The proposed fourth-order dynamic system with no equilibria, system (23), exhibits a hyperchaotic attractor for certain values of the parameters.

Tables 17.1 and 17.2 provide a summary of the systems presented in the selected set of papers, their equations, and a combination of parameters and initial conditions which are known to produce chaos. The attractor diagrams have been plotted using Economics and Finance (E&F) chaos software (Diks et al., 2008) at the values of parameters and initial conditions specified in the tables.

One of the most widely used indicators of chaotic behavior is a positive Maximum Lyapunov Exponent (MLE). Lyapunov Exponents (LEs) measure the sensitivity to initial conditions through the exponential divergence of nearby trajectories. Lyapunov exponent values are included in the tables, when given in the reviewed papers, at the specified parameters and initial conditions. The values satisfy the two conditions for chaos production in dissipative chaotic systems. First, the summation of LEs is less than zero. Second, the MLE is positive, which accounts for chaotic behavior.

It is worth mentioning that some continuous systems that were studied in the set of papers belong to the category of conservative chaotic systems such as system (1) in Table 17.1. Dissipative systems category, to which most of the studied systems belong, usually exhibit chaos for most initial conditions in a specified range of parameters. On the other hand, the summation of LEs in a conservative system is zero. It exhibits periodic and quasi-periodic solutions for most values of parameters and initial conditions, and can exhibit chaos for special values only. Consequently, dissipative systems usually appear in most applications of chaos theory such as chaos-based communication, physical, and financial modeling. It should be noted that conservative systems have another different set of applications where they are useful to study the development of chaos in some kinds of systems.

Table 17.1 focuses on hidden attractors with no equilibrium and quadratic nonlinearities. Systems (2)–(17) approach chaos through a succession of period-doubling limit cycles, with an attractor dimension slightly greater than 2.0. System (4) is the simplest among them as it has only six terms and a single quadratic nonlinearity. Systems (16) and (17) are circulant systems, i.e., they are symmetric with respect to a cyclic rotation of the variables x, y and z. In addition, they exhibit the largest Kaplan–Yorke dimension, $D_{KY} = 2.3224$, among the other systems with a wide range of the parameter that produces chaos.

Table 17.2 presents more systems from different categories and their characteristic properties. System (20) is hidden since only a tiny portion of the uncountably many unstable points on the equilibrium line intersect the basin of the chaotic attractor. System (21) presents a special type of hidden attractors in which the only equilibrium is unstable. However, it is considered hidden because it has a small parameter space and the strange attractor cannot be found by starting with initial conditions in the vicinity of the equilibrium. Systems (22) and (23) are examples of higher dimensional systems with different types of nonlinearities.

17.4 SENSITIVITY TO PARAMETER VARIATIONS

This section provides some extra results and simulations for a selected set of the systems summarized in Tables 17.1 and 17.2. A simulation-based procedure for deciding the type of a system's response through plotting the MLE against system parameters is discussed for integer order. The calculations are carried out with the aid of E&F chaos software (Diks et al., 2008), Lyapunov Exponent Toolbox (LET) (Siu, 1998), and the Lyapunov exponent calculation function of a MATLAB-based program for Dynamical System investigation (MATDS) (Govorukhin, 2004b,a). Calculations of Lyapunov exponents have been carried out for $10,000$ iterations up to accuracy of four decimal places to ensure reaching a posttransient value.

The procedure makes use of the calculated Lyapunov exponents in determining approximate ranges of parameters that produce chaos. The flows of most dynamical systems with respect to parameter variations exhibit the following pattern of different types of solutions: stable or fixed $\rightarrow$ periodic $\rightarrow$ quasi-periodic $\rightarrow$ chaotic $\rightarrow$ unstable or divergent, whether in the direction of increasing the parameter or vice versa. To have a dissipative dynamical system, the values of all Lyapunov exponents should sum to a negative number. Lyapunov exponents, arranged in descending order and denoted by $(\lambda_1, \lambda_2, \lambda_3)$, can be used to determine the type of the response as follows (assuming three dimensional phase space for simplicity):

- $(+, 0, -)$: Chaos or strange attractor.
- $(-, -, -)$: Fixed point.
- $(0, -, -)$: Limit cycle.
- $(0, 0, -)$: Torus.

17.4.1 A Conservative Hidden Attractor With No Equilibria and Quadratic Nonlinearities: System (1)

System (1) is studied as a sample from the set of third-order systems (1) to (17) with no equilibria and quadratic nonlinearities. It is also chosen as an example of conservative systems. Fig. 17.2A shows that the range $0 < a < 1.7$ can exhibit chaotic behavior with various positive values of MLE. Fig. 17.2B shows the time evolution of Lyapunov exponents at $a = 1$, where the system exhibits one positive MLE accounting for chaotic behavior. The MLE value approaches the value given in Table 17.1 as time advances. As the value of a increases, MLE exhibits a small positive value near to zero and the corresponding phase portrait shows a quasi-periodic solution, for instance, at $a = 10$ as shown in Fig. 17.2C. The three LEs are $(0, 0, -)$ at $a = 10$ which is consistent with the obtained phase portrait of a torus. The sum of LEs tends to zero at $a = 1$ and $a = 10$ as shown in Fig. 17.2B and D, respectively, indicating that it is a conservative system.

The effect of initial conditions is a very important factor in conservative systems. Varying the initial condition can drift the solution from chaotic behavior at the same values of the parameters. For example, Fig. 17.2E shows a completely different quasi-periodic torus starting at initial condition $(0, 1, 0)$ instead of $(0, 5, 0)$ that was used in Jafari et al. (2013) and yielded chaotic behavior.

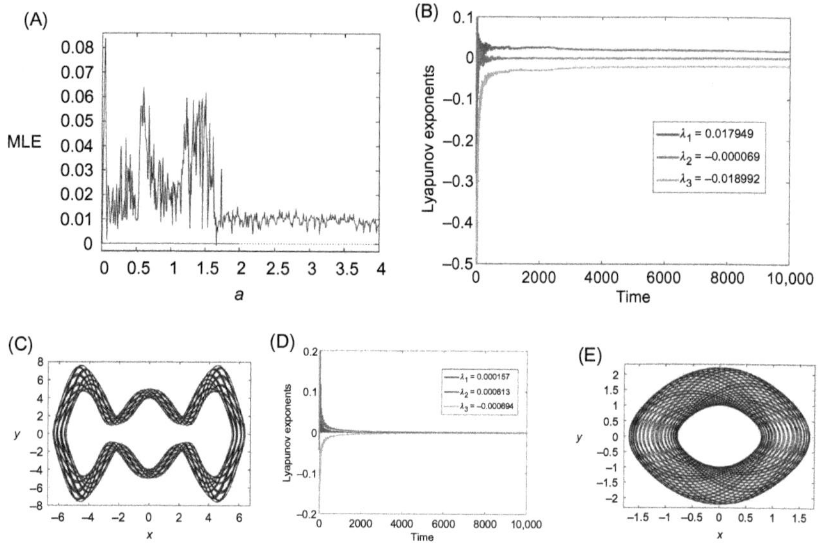

FIGURE 17.2 (A) MLE versus the parameter a for system (1). (B) Time evolution of Lyapunov exponents at $a = 1$. (C) Phase portrait at $a = 10$. (D) Time evolution of Lyapunov exponents at $a = 10$. (E) Phase portrait at $a = 1$ and an initial point (0,1,0).

17.4.2 A Dissipative Hidden Attractor With No Equilibria and Quadratic Nonlinearities: System (2)

Fig. 17.3A shows that the system exhibits positive MLE in the range $0.07 < a < 0.45$. The value $a = 0.35$ given in Jafari et al. (2013) is shown to satisfy the conditions on Lyapunov exponents values for chaotic behavior as shown in Fig. 17.3B. For values of a slightly greater than 0.45, the response starts to diverge, while for a slightly less than 0.07, periodic responses start to appear as shown in Fig. 17.3C. The $(0, -, -)$ values for the three Lyapunov exponents shown in Fig. 17.3D indicate a limit cycle.

17.4.3 A Dissipative Hidden Attractor With No Equilibria and Quadratic Nonlinearities: System (16)

This system is characterized by having a wide range of the parameter a that can produce chaotic solution since it has a large Kaplan–Yorke dimension as mentioned in Section 17.3. Although the value of MLE can decrease, become negative, or diverge at certain values of a, it may increase again and exhibit a positive value accounting for chaotic behavior as shown in Fig. 17.4. Consequently, it is not straightforward to identify the range of a with positive MLE for this system using the explained procedure. Fig. 17.4C illustrates the chaotic behavior exhibited at $a = 5$ given that Jafari et al. (2013) used $a = 1$.

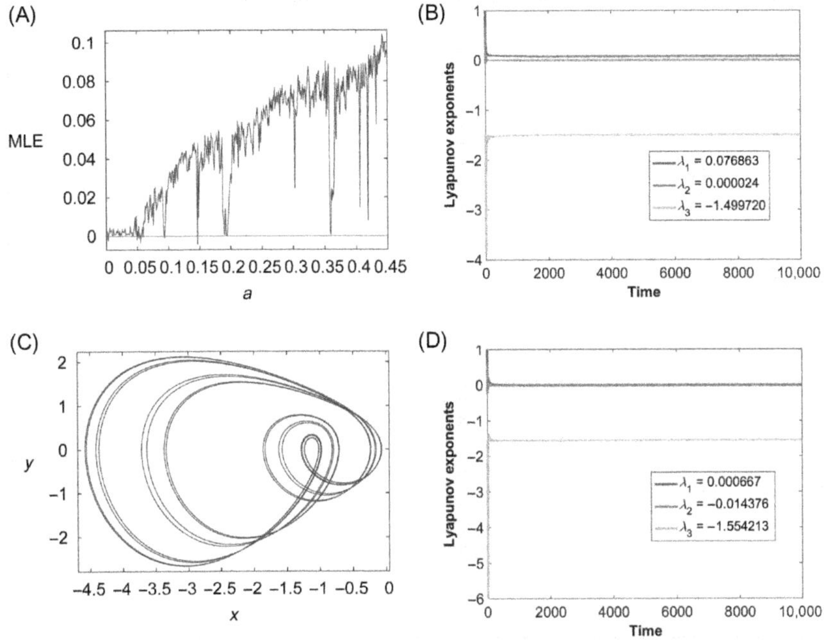

FIGURE 17.3 (A) MLE versus the parameter a for system (2). (B) Time evolution of Lyapunov exponents at $a = 0.35$. (C) Phase portrait at $a = 0.05$. (D) the corresponding time evolution of LEs.

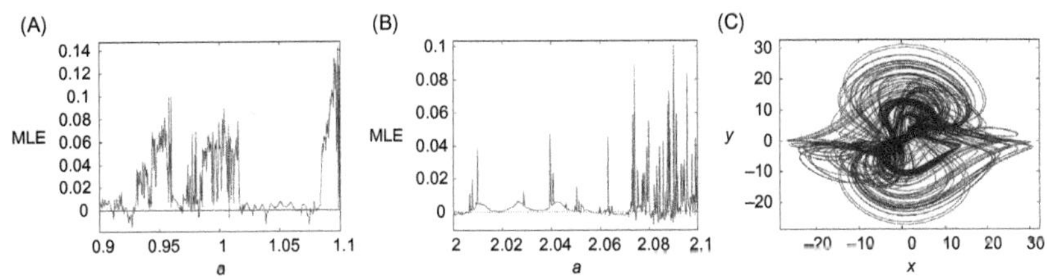

FIGURE 17.4 MLE versus a in different ranges for system (16). (A) $0.9 < a < 1.1$; (B) $2 < a < 2.1$; and (C) Phase portrait at $a = 5$.

17.4.4 A Dissipative Hidden Attractor With Single Stable Equilibrium Point: System (19)

The equations of the system was given in Jafari et al. (2015) with constant coefficients instead of parameters. The coefficients 0.4 and 0.6 that appear in the system's equations in Table 17.2 are replaced by parameters a and b, respectively. Fig. 17.5 shows that the ranges of both coefficients that correspond to positive MLE values are limited.

17.5 SENSITIVITY TO FRACTIONAL ORDERS

In this section, we discuss the sensitivity of the selected systems to fractional orders. Numerical simulations are used to identify when they generate

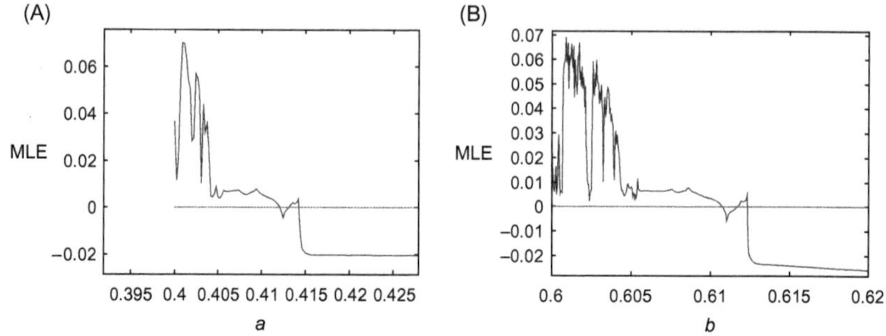

FIGURE 17.5 MLE for system (19) (A) versus a (a parameter instead of the coefficient 0.4) and (B) versus b (a parameter instead of 0.6).

TABLE 17.3 Discretized Solutions of the Fractional Order Extensions of the Selected Systems

System (1)	**System (2)**
$x_{n+1} = \frac{-\sum_{j=1}^{n+1} c_j^{\alpha} x_{n+1-j} + y_n}{c_0^{\alpha}}$	$x_{n+1} = \frac{-\sum_{j=1}^{n+1} c_j^{\alpha} x_{n+1-j} - y_n}{c_0^{\alpha}}$
$y_{n+1} = \frac{-\sum_{j=1}^{n+1} c_j^{\beta} y_{n+1-j} - x_{n+1} - 2z_n y_n}{c_0^{\beta} - z_n}$	$y_{n+1} = \frac{-\sum_{j=1}^{n+1} c_j^{\beta} y_{n+1-j} + x_n + z_n}{c_0^{\beta}}$
$z_{n+1} = \frac{-\sum_{j=1}^{n+1} c_j^{\gamma} z_{n+1-j} + y_{n+1}^2 - a}{c_0^{\gamma}}$	$z_{n+1} = \frac{-\sum_{j=1}^{n+1} c_j^{\gamma} z_{n+1-j} + 2y_n y_{n+1} + 2x_{n+1} z_n - a}{c_0^{\gamma} + x_{n+1}}$
System (16)	**System (20)**
$x_{n+1} = \frac{-\sum_{j=1}^{n+1} c_j^{\alpha} x_{n+1-j} - 0.8x_n - 0.5y_n^2 + 2x_n z_n + a}{c_0^{\alpha} + z_n}$	$x_{n+1} = \frac{-\sum_{j=1}^{n+1} c_j^{\alpha} x_{n+1-j} + y_n}{c_0^{\alpha}}$
$y_{n+1} = \frac{-\sum_{j=1}^{n+1} c_j^{\beta} y_{n+1-j} - 0.8y_n - 0.5z_n^2 + 2x_{n+1} y_n + a}{c_0^{\beta} + x_{n+1}}$	$y_{n+1} = \frac{-\sum_{j=1}^{n+1} c_j^{\beta} y_{n+1-j} - x_n + 2y_n z_n}{c_0^{\beta} + z_n}$
$z_{n+1} = \frac{-\sum_{j=1}^{n+1} c_j^{\gamma} z_{n+1-j} - 0.8z_n - 0.5x_n^2 + 2y_{n+1} z_n + a}{c_0^{\gamma} + y_{n+1}}$	$z_{n+1} = \frac{-\sum_{j=1}^{n+1} c_j^{\gamma} z_{n+1-j} - x_n - 2x_{n+1} z_n - 15x_n y_n}{c_0^{\gamma} - x_{n+1}}$

periodic or chaotic responses. In addition, we compare the shape of their attractors in integer order and fractional order. For simplicity, the three fractional orders in the system of fractional differential Eq. (17.7) are assumed to be equal, i.e., in this section $\alpha = \beta = \gamma$ and the unified fractional-order is denoted by α. NSFD with $\phi(h) = 1 - e^{-h}$ is used to solve the systems of fractional order differential equations using the formulas given in Table 17.3. A time step of 0.005 and a total simulation time of 200 are used, except where stated otherwise.

Tables 17.4–17.7 show the posttransient time series of the three phase space dimensions x, y and z as well as the posttransient attractor diagram,

TABLE 17.4 System (1) Responses Versus the Fractional Order α at Parameter Value $a = 1.2$

$\alpha = 0.9$	$\alpha = 0.94$	$\alpha = 0.99$	$\alpha = 1$
Periodic	Periodic	Quasi-periodic	Chaotic
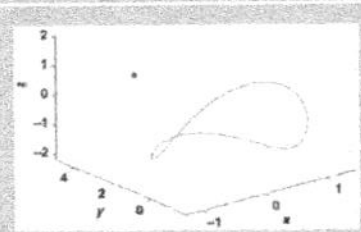	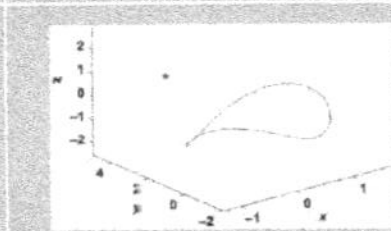	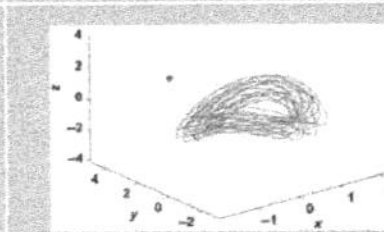	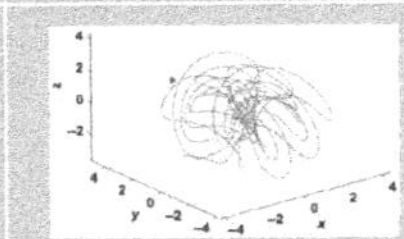
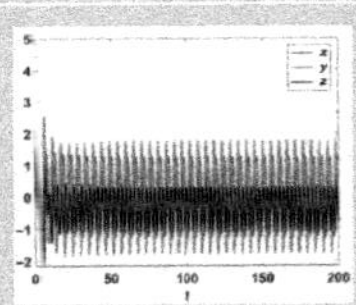	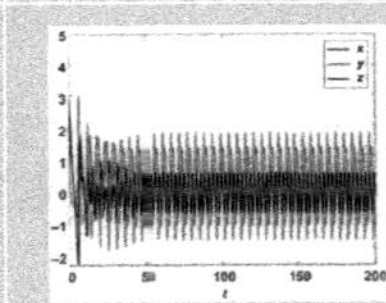	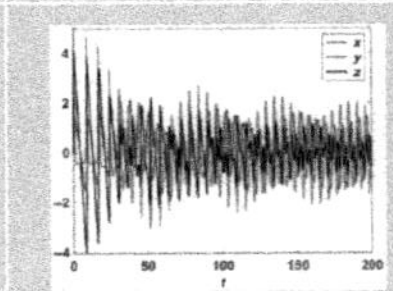	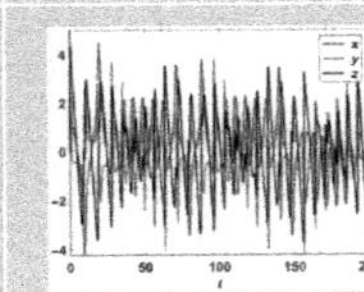

TABLE 17.5 System (2) Responses Versus the Fractional Order α at Parameter Value $a = 0.4$

$\alpha = 0.96$	$\alpha = 0.98$	$\alpha = 1$
Divergent	Periodic	Chaotic
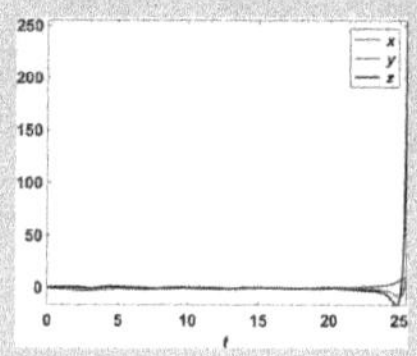	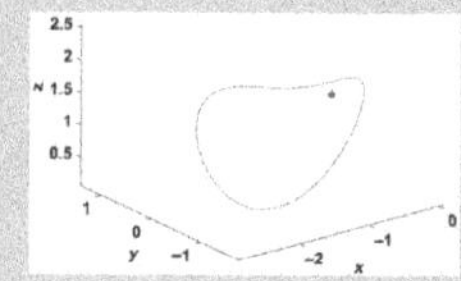 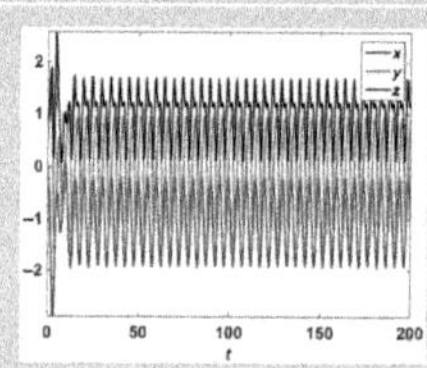	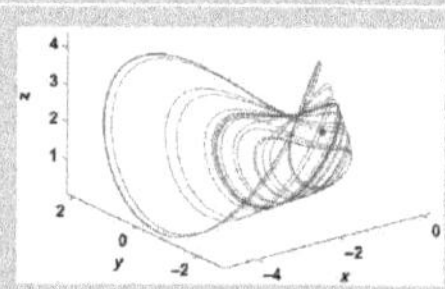 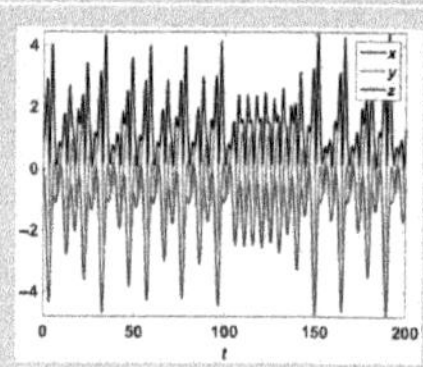

TABLE 17.6 System (16) Responses Versus the Fractional Order α at Parameter Value $a = 1$

$\alpha = 0.79$	$\alpha = 0.95$	$\alpha = 0.98$	$\alpha = 1$
Divergent	Periodic	Quasi-Periodic	Chaotic
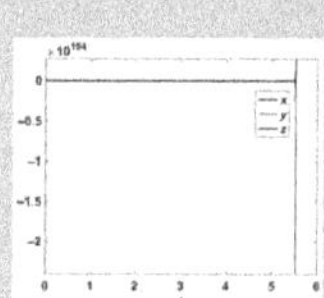	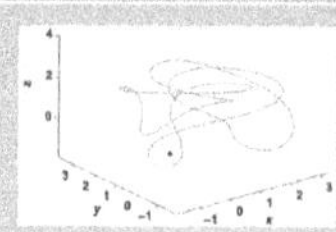 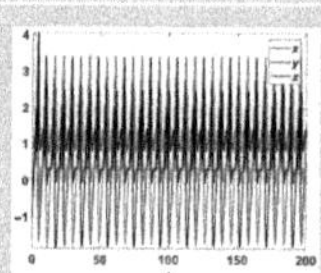	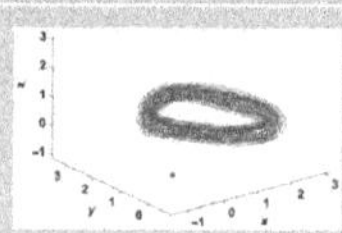 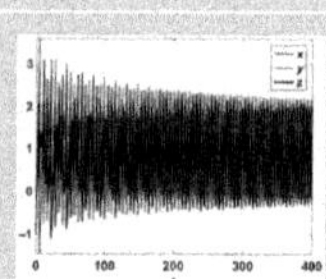	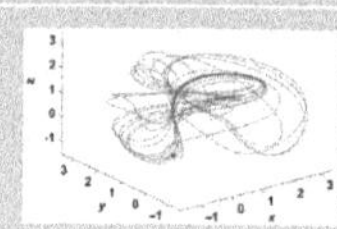 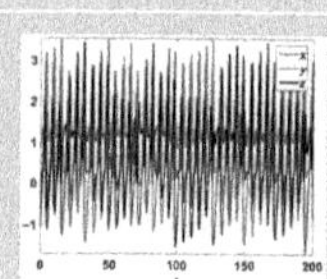

with the initial point marked in red, illustrating the obtained type of solution (periodic or chaotic) for different values of the fractional order.

Sensitivities of some other systems to parameter variations and fractional orders were tested. For example, system (4) is characterized by high sensitivity to parameter variation, where it diverges for values of a slightly lower or higher than 1.0. It is similarly sensitive to variations in the fractional order α and diverges for values slightly lower than 1.0.

TABLE 17.7 System (20) Responses Versus the Fractional Order α

$\alpha = 0.9$	$\alpha = 0.99$	$\alpha = 1$
Stable	Periodic	Chaotic
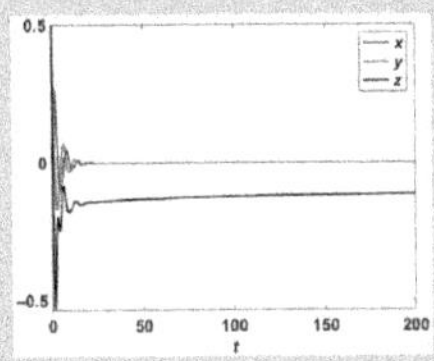	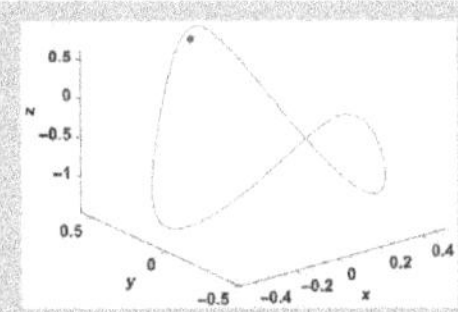 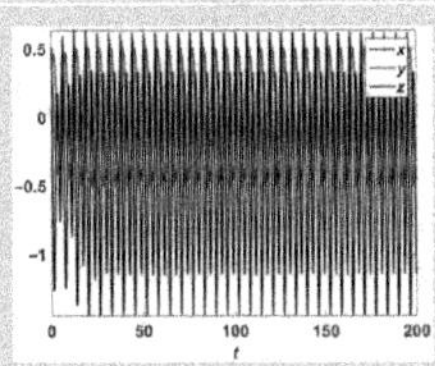	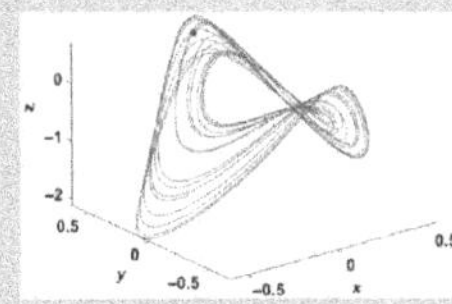 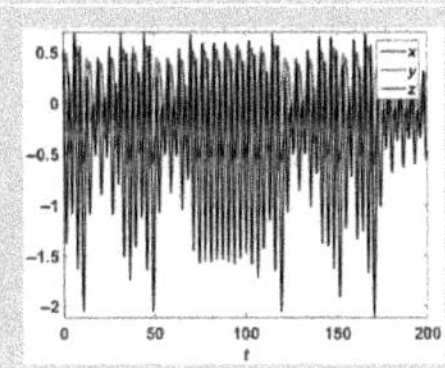

17.6 CONCLUSIONS

This chapter has presented a review of various types of continuous chaotic systems with hidden attractors, which were introduced in several recent papers. Hidden attractors with no equilibria, with a line of equilibrium points, and with a single stable equilibrium point have been reviewed. The nonlinear terms that appear in the equations of the systems included quadratic, exponential, and hyperbolic nonlinearities. Both dissipative and conservative systems were included as well. The resulting behaviors of these systems have been validated using suitable software packages and their properties have been summarized. The attractor diagrams have been plotted at the given parameter values and initial conditions and were shown to be chaotic. Lyapunov exponents values of the dissipative systems sum to a negative value with a positive MLE. Further simulations have been conducted for several selected systems in a range of parameters around the values specified in the papers. The sensitivity of those selected systems to parameter variations is asserted through plotting the MLE against the parameter. Intervals of chaotic behavior, rather than single values for the parameters, have been found to exhibit positive MLE. In addition, extensions of some systems to the fractional order domain have been presented discussing their sensitivity to the fractional order parameter.

REFERENCES

AbdelAty, A.M., Soltan, A., Ahmed, W.A., Radwan, A.G., 2017. On the analysis and design of fractional-order chebyshev complex filter. Circuits Systems Signal Processing 1–24.

Abd-El-Hafiz, S.K., Radwan, A.G., AbdelHaleem, S.H., Barakat, M.L., 2014. A fractal-based image encryption system. IET Image Proc. 8, 742–752.

Abd-El-Hafiz, S.K., Radwan, A.G., AbdEl-Haleem, S.H., 2015. Encryption applications of a generalized chaotic map. Appl. Math. Inf. Sci. 9, 3215.

Abd-El-Hafiz, S.K., AbdElHaleem, S.H., Radwan, A.G., 2016. Novel permutation measures for image encryption algorithms. Opt. Lasers Eng. 85, 72–83.

Abdelhaleem, S.H., Radwan, A.G., Abd-El-Hafiz, S.K., 2014. A chess-based chaotic block cipher, in: IEEE 12th International New Circuits and Systems Conference (NEWCAS), 2014, IEEE. pp. 405–408.

AboBakr, A., Said, L.A., Madian, A.H., Elwakil, A.S., Radwan, A.G., 2017. Experimental comparison of integer/fractional-order electrical models of plant. AEU-Int. J. Electr. Commun.

Alligood, K.T., Sauer, T.D., Yorke, J.A., 1996. Chaos: An Introduction to Dynamical Systems. Springer, New York.

Azar, A.T., Vaidyanathan, S., 2015. Chaos Modeling and Control Systems Design. Volume 581 of Studies in Computational Intelligence. Springer, New York.

Azar, A.T., Vaidyanathan, S., 2016. Advances in Chaos Theory and Intelligent Control. Volume 337 of Studies in Fuzziness and Soft Computing. Springer, New York.

Azar, A.T., Vaidyanathan, S., Ouannas, A., 2017a. Fractional Order Control and Synchronization of chaotic systems. Volume 688 of Studies in Computational Intelligence. Springer, New York.

Azar, A.T., Volos, C., Gerodimos, N.A., Tombras, G.S., Pham, V.T., Radwan, A.G., et al., 2017b. A novel chaotic system without equilibrium: dynamics, synchronization, and circuit realization. Complexity 2017.

Barakat, M.L., Mansingka, A.S., Radwan, A.G., Salama, K.N., 2013. Generalized hardware post-processing technique for chaos-based pseudorandom number generators. ETRI J. 35, 448–458.

Caponetto, R., 2010. Fractional Order Systems: Modeling and Control Applications, Vol. 72. World Scientific, Hackensack, NJ.

Chien, T.I., Liao, T.L., 2005. Design of secure digital communication systems using chaotic modulation, cryptography and chaotic synchronization. Chaos Solitons Fractals 24, 241–255.

Danca, M.F., 2017. Hidden chaotic attractors in fractional-order systems. Nonlinear Dynam. 1–10.

Diks, C., Hommes, C., Panchenko, V., Van Der Weide, R., 2008. E&F Chaos: a user friendly software package for nonlinear economic dynamics. Computat. Econ. 32, 221–244.

Fouda, M.E., Radwan, A.G., 2015. Fractional-order memristor response under dc and periodic signals. Circuits Systems Signal Processing 34, 961–970.

Fouda, M.E., Soltan, A., Radwan, A.G., Soliman, A.M., 2016. Fractional-order multi-phase oscillators design and analysis suitable for higher-order PSK applications. Analog Integ. Circuits Signal Processing 87, 301–312.

Frey, D.R., 1993. Chaotic digital encoding: an approach to secure communication. Circuits and Systems II: Analog and Digital Signal Processing, IEEE Transactions on 40, 660–666.

Gorenflo, R., Mainardi, F., 1997. Fractional Calculus. Springer, New York.

Govorukhin, V.N., 2004a. Calculation lyapunov exponents for ODE. MATLAB Central File Exchange, file ID 4628.

Govorukhin, V.N., 2004b. MATDS: MATLAB based program for dynamical systems investigation. http://www.math.rsu.ru/mexmat/kvm/matds/.

Han, S.K., Kurrer, C., Kuramoto, Y., 1995. Dephasing and bursting in coupled neural oscillators. Phys. Rev. Lett. 75, 3190.

Henein, M.M.R., Sayed, W.S., Radwan., A.G., Abd-El-Hafiz, S.K., 2016. Switched active control synchronization of three fractional order chaotic systems, in: 13th International Conference on Electrical Engineering/Electronics, Computer, Telecommunications and Information Technology.

Hussian, G., Alnaser, M., Momani, S., 2008. Non-standard discretization of fractional differential equations, in: Proceeding of 8th Seminar of Differential Equations and Dynamical Systems in, Isfahan, Iran.

Ismail, S.M., Said, L.A., Radwan, A.G., Madian, A.H., Abu-ElYazeed, M.F., Soliman, A.M., 2015. Generalized fractional logistic map suitable for data encryption, in: Science and Technology (TICST), 2015 International Conference on, IEEE. pp. 336–341.

Jafari, S., Sprott, J., 2013. Simple chaotic flows with a line equilibrium. Chaos Solitons Fractals 57, 79–84.

Jafari, S., Sprott, J.C., Golpayegani, S.M.R.H., 2013. Elementary quadratic chaotic flows with no equilibria. Phys. Lett. A 377, 699–702.

Jafari, S., Sprott, J.C., Nazarimehr, F., 2015. Recent new examples of hidden attractors. Eur. Phys. J.

Kocarev, L., Lian, S., 2011. Chaos-based Cryptography: Theory, Algorithms and Applications, Vol. 354. Springer, New York.

Kuznetsov, N.V., Leonov, G.A., 2014. Hidden attractors in dynamical systems: systems with no equilibria, multistability and coexisting attractors. IFAC World Congress 19, 5445–5454.

Lau, F., Tse, C.K., 2003. Chaos-based Digital Communication Systems. Springer, New York.

Leonov, G.A., Kuznetsov, N.V., 2013. Hidden attractors in dynamical systems. from hidden oscillations in Hilbert–Kolmogorov, Aizerman, and Kalman problems to hidden chaotic attractor in Chua circuits. Int. J. Bifurcat. Chaos 23, 1330002.

Li, C., Sprott, J.C., 2014. Coexisting hidden attractors in a 4-D simplified lorenz system. Int. J. Bifurcat. Chaos 24, 1450034.

Magin, R.L., 2006. Fractional Calculus in Bioengineering. Begell House Redding, Redding.

Mickens, R.E., 2000. Applications of Nonstandard Finite Difference Schemes. World Scientific, Hackensack, NJ.

Mickens, R.E., 2005. Advances in the Applications of Nonstandard Finite Difference Schemes. World Scientific, Hackensack, NJ.

Moaddy, K., Radwan, A.G., Salama, K.N., Momani, S., Hashim, I., 2012. The fractional-order modeling and synchronization of electrically coupled neuron systems. Comput. Math. Applicat. 64, 3329–3339.

Molaie, M., Jafari, S., Sprott, J.C., Golpayegani, S.M.R.H., 2013. Simple chaotic flows with one stable equilibrium. Int. J. Bifurcation Chaos 23, 1350188.

Mueller, S.T., 2011. Numerical method and step size variation in the lorenz equations. http://ir.library.oregonstate.edu/xmlui/bitstream/handle/1957/25245/S.T.Mueller_Thesis.pdf.

Petras, I., 2011. Fractional-order Nonlinear Systems: Modeling, Analysis and Simulation. Springer Science & Business Media, New York.

Pham, V.T., Volos, C., Jafari, S., Wang, X., Vaidyanathan, S., 2014. Hidden hyperchaotic attractor in a novel simple memristive neural network. Optoelectr. Adv. Mater. Rapid Commun. 8, 1157–1163.

Pham, V.T., Vaidyanathan, S., Volos, C.K., Jafari, S., 2015. Hidden attractors in a chaotic system with an exponential nonlinear term. Eur. Phys. J. Spec. Topics 224, 1507–1517.

Pham, V.T., Vaidyanathan, S., Volos, C., Tlelo-Cuautle, E., Tahir, F.R., 2017a. A memristive system with hidden attractors and its engineering application. Advances in Memristors, Memristive Devices and Systems. Springer, New York, pp. 81–99.

Pham, V.T., Volos, C., Kapitaniak, T., 2017b. Synchronization of systems with hidden attractors. Systems with Hidden Attractors. Springer, New York, pp. 65–77.

Pham, V.T., Volos, C., Kapitaniak, T., 2017c. Systems with hidden attractors: from theory to realization in circuits.

Psychalinos, C., Elwakil, A.S., Radwan, A.G., Biswas, K., 2016. Guest editorial: fractional-order circuits and systems: theory, design, and applications. Circuits Systems Signal Processing 35, 1807–1813.

Radwan, A., 2012. Stability analysis of the fractional-order $RL_\beta C_\alpha$ circuit. J. Fract. Calc. Appl 3, 1–15.

Radwan, A., Soliman, A., El-Sedeek, A., 2004. MOS realization of the modified Lorenz chaotic system. Chaos Solitons Fractals 21, 553–561.

Radwan, A., Soliman, A.M., Elwakil, A.S., 2007. 1-D digitally-controlled multiscroll chaos generator. Int. J. Bifurcation Chaos 17, 227–242.

Radwan, A., Moaddy, K., Hashim, I., 2013. Amplitude modulation and synchronization of fractional-order memristor-based Chua's circuit. Abstract and Applied Analysis. Hindawi Publishing Corporation.

Radwan, A., Moaddy, K., Salama, K.N., Momani, S., Hashim, I., 2014a. Control and switching synchronization of fractional order chaotic systems using active control technique. J. Adv. Res. 5, 125–132.

Radwan, A.G., Abd-El-Hafiz, S.K., AbdElHaleem, S.H., 2014b. An image encryption system based on generalized discrete maps, in: IEEE International Conference on Electronics, Circuits and Systems (ICECS), 21st, IEEE. pp. 283–286.

Radwan, A.G., 2013a. On some generalized discrete logistic maps. J. Adv. Res. 4, 163–171.

Radwan, A.G., 2013b. Resonance and quality factor of the fractional circuit. IEEE J. Emerg. Selected Topics Circuits Systems 3, 377–385.

Radwan, A.G., Abd-El-Hafiz, S.K., 2013. Image encryption using generalized tent map, in: IEEE 20th International Conference on Electronics, Circuits, and Systems (ICECS), IEEE. pp. 653–656.

Radwan, A.G., Abd-El-Hafiz, S.K., 2014. The effect of multi-scrolls distribution on image encryption, in: 21st IEEE International Conference on Electronics, Circuits and Systems (ICECS), 2014, IEEE. pp. 435–438.

Radwan, A.G., Fouda, M.E., 2013. Optimization of fractional-order RLC filters. Circuits Systems Signal Processing 32, 2097–2118.

Radwan, A.G., Soliman, A.M., El-Sedeek, A.L., 2003. An inductorless CMOS realization of Chuas circuit. Chaos Solitons Fractals 18, 149–158.

Radwan, A.G., Elwakil, A.S., Soliman, A.M., 2008a. Fractional-order sinusoidal oscillators: design procedure and practical examples. IEEE Transactions on Circuits and Systems I: Regular Papers 55, 2051–2063.

Radwan, A.G., Soliman, A.M., Elwakil, A.S., 2008b. First-order filters generalized to the fractional domain. J. Circuits Systems Computers 17, 55–66.

Radwan, A.G., Moaddy, K., Momani, S., 2011a. Stability and non-standard finite difference method of the generalized Chuas circuit. Comput. Math. Applicat. 62, 961–970.

Radwan, A.G., Shamim, A., Salama, K.N., 2011b. Theory of fractional order elements based impedance matching networks. Microwave Wireless Components Letters, IEEE 21, 120–122.

Radwan, A.G., Abd-El-Hafiz, S.K., AbdElHaleem, S.H., 2012. Image encryption in the fractional-order domain, in: International Conference on Engineering and Technology (ICET), 2012, IEEE. pp. 1–6.

Radwan, A.G., Abd-El-Hafiz, S.K., AbdElHaleem, S.H., 2015a. Image encryption based on fractional-order chaotic generators, in: 2015 International Symposium on Nonlinear Theory and its Applications NOLTA2015, Kowloon, Hong Kong, China, December 1-4, 2015, IEEE. pp. 688–691.

Radwan, A.G., AbdElHaleem, S.H., Abd-El-Hafiz, S.K., 2015b. Symmetric encryption algorithms using chaotic and non-chaotic generators: a review. J. Adv. Res. Available from: http://dx.doi.org/10.1016/j.jare.2015.07.002.

Radwan, A.G., Maundy, B.J., Elwakil, A.S., 2016. Fractional-order oscillators. oscillator circuits: frontiers in design. Anal. Applicat. 32, 25.

Radwan, A.G., Sayed, W.S., Abd-El-Hafiz, S.K., 2017. Control and synchronization of fractional-order chaotic systems. Fractional Order Control and Synchronization of Chaotic Systems. Springer, New York, pp. 325–355.

Sayed, W.S., Radwan, A.G., Fahmy, H.A., 2015a. Design of a generalized bidirectional tent map suitable for encryption applications, in: 11th International Computer Engineering Conference (ICENCO), IEEE. pp. 207–211.

Sayed, W.S., Radwan, A.G., Fahmy, H.A., 2015b. Design of positive, negative, and alternating sign generalized logistic maps. Discr. Dynam. Nat. Soc. 2015.

Sayed, W.S., Radwan, A.G., Abd-El-Hafiz, S.K., 2016. Generalized synchronization involving a linear combination of fractional-order chaotic systems, in: 13th International Conference on Electrical Engineering/Electronics, Computer, Telecommunications and Information Technology.

Sayed, W.S., Fahmy, H.A., Rezk, A.A., Radwan, A.G., 2017a. Generalized smooth transition map between tent and logistic maps. Int. J. Bifurcation Chaos 27, 1730004.

Sayed, W.S., Henein, M.M., Abd-El-Hafiz, S.K., Radwan, A.G., 2017b. Generalized dynamic switched synchronization between combinations of fractional-order chaotic systems. Complexity 2017.

Sayed, W.S., Radwan, A.G., Fahmy, H.A., 2017c. Chaotic systems based on jerk equation and discrete maps with scaling parameters, in: 6th International Conference on Modern Circuits and Systems Technologies (MOCAST), IEEE. pp. 1–4.

Sayed, W.S., Radwan, A.G., Rezk, A.A., Fahmy, H.A., 2017d. Finite precision logistic map between computational efficiency and accuracy with encryption applications. Complexity 2017.

Schöll, E., 2001. Nonlinear Spatio-Temporal Dynamics and Chaos in Semiconductors, Vol. 10. Cambridge University Press, Cambridge.

Semary, M.S., Radwan, A.G., Hassan, H.N., 2016. Fundamentals of fractional-order lti circuits and systems: number of poles, stability, time and frequency responses. Int. J.Circuit Theory Applicat. 44, 2114–2133.

Semary, M.S., Hassan, H.N., Radwan, A.G., 2017. Controlled picard method for solving nonlinear fractional reaction–diffusion models in porous catalysts. Chem. Eng. Commun. 204, 635–647.

Senan, N.A.F., 2012. A brief introduction to using ode45 in matlab.

Shamim, A., Radwan, A.G., Salama, K.N., 2011. Fractional Smith chart theory. Microwave Wireless Compon Lett IEEE 21, 117–119.

Siu, S., 1998. Lyapunov exponent toolbox. MATLAB Central File Exchange, file ID 233.

Soltan, A., Radwan, A.G., Soliman, A.M., 2012. Fractional order filter with two fractional elements of dependant orders. Microelectron. J. 43, 818–827.

Soltan, A., Radwan, A.G., Soliman, A.M., 2015. Fractional order Sallen–Key and KHN filters: stability and poles allocation. Circuits Systems Signal Processing 34, 1461–1480.

Soltan, A., Soliman, A.M., Radwan, A.G., 2017. Fractional-order impedance transformation based on three port mutators. AEU-Int. J. Electr. Commun.

Sprott, J.C., 1994. Some simple chaotic flows. Phys. Rev. E 50, R647.

Sprott, J.C., 2000. A new class of chaotic circuit. Phys. Lett. A 266, 19–23.

Sprott, J.C., 2007. A simple chaotic delay differential equation. Phys. Lett. A 366, 397–402.

Strogatz, S.H., 2014. Nonlinear Dynamics and Chaos: With Applications to Physics, Biology, Chemistry, and Engineering. Westview Press.

Tolba, M.F., AbdelAty, A.M., Soliman, N.S., Said, L.A., Madian, A.H., Azar, A.T., et al., 2017. FPGA implementation of two fractional order chaotic systems. AEU-Int. J. Electr. Commun. 78, 162–172.

Vaidyanathan, S., Pham, V.T., Volos, C., 2015. A 5-D hyperchaotic Rikitake dynamo system with hidden attractors. Eur. Phys. J. Special Topics 224, 1575–1592.

Volos, C., Pham, V.T., Zambrano-Serrano, E., Munoz-Pacheco, J., Vaidyanathan, S., Tlelo-Cuautle, E., 2017. Analysis of a 4-D hyperchaotic fractional-order memristive system with hidden attractors. Advances in Memristors, Memristive Devices and Systems. Springer, New York, pp. 207–235.

Wang, X., Chen, G., 2012. A chaotic system with only one stable equilibrium. Commun. Nonlinear Sci. Numer. Simulat. 17, 1264–1272.

Zidan, M.A., Radwan, A.G., Salama, K.N., 2012. Controllable V-shape multiscroll butterfly attractor: system and circuit implementation. Int. J. Bifurcation Chaos 22, 1250143.

Chapter 18

Parameters Identification of Fractional Order Permanent Magnet Synchronous Motor Models Using Chaotic Meta-Heuristic Algorithms

Dalia Yousri, Dalia Allam and Magdy Eteiba
Fayoum University, Fayoum, Egypt

18.1 INTRODUCTION

Nowadays, fractional calculus is considered as an effective mathematical tool when dealing with noninteger orders derivatives and integrals. It has became a hot topic in recent research, where the physical behavior of different systems can be mimicked as fractional order differential equations (Miller and Ross, 1993; Debnath, 2003, 2004). As the integer order differential equation is just a special case of the fractional one, many researchers prefer to use the fractional calculus in modeling of many systems to enhance their design flexibility by providing extra degrees of freedom (Debnath, 2003). Moreover, the solution of fractional order differential equations can be accomplished by several numerical methods (Diethelm, 2010; Petráš, 2011). That's why the fractional calculus has been employed in different fields such as electrical circuit models (Gu et al., 2016; Khateb et al., 2016), control systems (Azar et al., 2017; Azar and Vaidyanathan, 2015a,b, 2016; Boulkroune et al., 2016a,b; Zamani et al., 2016; Xu et al., 2016; Zhu and Azar, 2015), and electric drives (Rajagopal et al., 2016; Zhou et al., 2015), etc.

The permanent magnet synchronous motor (PMSM) is widely used in different industrial applications due to its simple structure, its high efficiency, and low manufacturing cost. However, the PMSM exhibits an undesirable behavior which is known as a chaotic behavior at certain ranges of the load disturbance and of the system parameters (Li et al., 2002; Liu et al., 2011;

Mathematical Techniques of Fractional Order Systems. DOI: https://doi.org/10.1016/B978-0-12-813592-1.00018-0

Rahimi et al., 2016). This chaotic behavior may cause torque ripples, instability of the motor speed, and current low-frequency oscillations which may damage the motor (Li et al., 2002; Sun et al., 2016. Therefore, there is a necessity for controlling this behavior to modify the motor behavior and to prevent its damage. The control process of this unexpected behavior requires an accurate description of the physical behavior of the motor (Li and Wu, 2016; Li et al., 2012) and an accurate estimation method of the motor parameters (Zribi et al., 2009; Ataei et al., 2010; Yu et al., 2011).

Currently, the nonlinear differential equations of the PMSM are modeled as fractional order equations to simulate the nature of the motor accurately. Where the fractional order calculus provides a profound insight into the physical processes of the motor and requires an infinite number of terms is the memory of all past events (Xue et al., 2015), while the integer order calculus is considered as a local operator. Hence, the PMSM commensurate and the incommensurate fractional order models have been introduced and they are approved as more accurate and more flexible models than the integer ones (Li and Wu, 2016; Li et al., 2012; Xue et al., 2015). The chaotic systems such as PMSM are complex and nonlinear systems that are effected by the initial conditions. Consequently, an irregular and unpredictable behavior results. Estimation of states of this nonlinear system to control its behavior is difficult because it's hard to define the parameters of such a system. Therefore, the unknown parameters identification of the PMSM fractional order models is a very crucial issue to control the chaotic behavior in the motor and to protect it from damage. This motivates the authors to search for efficient techniques to identify the models parameters accurately.

Only a few numerical techniques have been published to extract the fractional order PMSM models parameters (Li et al., 2012; Xue et al., 2015) at certain operating conditions. Therefore, there is a real need for efficient and a simple parameter estimation methods to identify these parameters accurately.

Recently, meta-heuristic optimization techniques have become a promising method in the identification of the models parameters in different fields (Alam et al., 2015; Allam et al., 2016; Yousri et al., 2017). These algorithms haven't been applied in extracting the unknown parameters of these models yet, despite their ability to optimize such a problem simply and simultaneously without any assumptions. This fact has motivated the authors to efficiently estimate the PMSM models parameters using the meta-heuristic algorithms for the first time.

Integration between the chaos maps and the nature-inspired algorithms to adjust some of their parameters is becoming a new research approach (Gandomi and Yang, 2014). This approach improves the exploration and/or the exploitation ability of the original techniques based on the randomization effectiveness of the chaos theory. Consequently, the accuracy and the speed of convergence of the developed algorithms are improved as in the cases of

Chaotic Particle Swarm Optimization (Liu et al., 2005), Chaotic Bee Colony Optimization (Alatas, 2010b), Chaotic Harmony Search (Alatas, 2010a), and Chaotic Bat Optimization Algorithm (Rezaee Jordehi, 2015; Gandomi and Yang, 2014).

In this chapter, two main targets are accomplished. The first one is the utilization of commensurate and incommensurate fractional order differential equations for modeling of the PMSM to achieve a better description of the motor physical behavior. The second target is introducing novel developed meta-heuristic optimization algorithms to identify the parameters of commensurate and incommensurate fractional order PMSM models corresponding to the chaotic behavior in the motor. The developed applied algorithms such as Chaotic Grey Wolf Optimizer and the Chaotic Grasshopper Optimization Algorithms, in addition to their original versions which are called Grey Wolf Optimizer and Grasshopper Optimizer, are proposed for the first time. The comparison between the results of the introduced developed algorithms and their original versions is carried out to recommend the more suitable optimization technique to estimate the models parameters efficiently. The final outcome clarifies that the chaotic algorithms extract the models parameters accurately with less error and less execution time compared with the original ones.

The rest of the manuscript is organized as follows: Section 18.2 presents the PMSM motor models. The problem formulation is described in section 18.3. Section 18.4 introduces an overview for the meta-heuristic algorithm. Simulation and results are investigated in section 18.5. Section 18.6 presents the conclusion.

18.2 PMSM MODELS

In general, the PMSM model can be described as a system of nonlinear differential equations (Liu et al., 2008). These equations can be established as fractional order differential equations with noninteger derivative orders that may have fractional values lower or greater than 1 (Li et al., 2012).

18.2.1 Fractional order PMSM

In the fractional order model of PMSM, the system of differential equations are formulated with noninteger derivative orders. This feature enhances the flexibility of the PMSM model design by providing extra degrees of freedom in the model (Xue et al., 2015; Li et al., 2012). In Xue et al. (2015), the authors propose a commensurate fractional order PMSM model where all derivative orders have equal values. However, in Li et al. (2012), the incommensurate fractional order PMSM model is introduced with the derivative orders of different values. Generally the system of the differential equations can be written as follows 18.1 (Li et al., 2012)

$$\begin{aligned}
\frac{d^{q_1} i_d}{dt} &= -\, i_d + \omega i_q + u_d \\
\frac{d^{q_2} i_q}{dt} &= -\, i_d - \omega i_q + \gamma\omega + u_q \\
\frac{d^{q_3} \omega}{dt} &= \sigma(i_q - \omega) - T_L
\end{aligned} \tag{18.1}$$

where q_i (i = 1,2,3) are the noninteger derivative orders, i_d and i_q are the stator current in d, q axis, respectively. ω is the rotor angular velocity. u_d and u_q are the stator voltage in the d, q axis. σ, γ are the dimensionless operating parameters of the system. T_L is the load torque.

If the external inputs of the system are equal to 0 where $T_L = u_d = u_q = 0$, the previous system of equations will be rewritten as follows 18.2 (Li et al., 2012).

$$\begin{aligned}
\frac{d^{q_1} i_d}{dt} &= -\, i_d + \omega i_q \\
\frac{d^{q_2} i_q}{dt} &= -\, i_d - \omega i_q + \gamma\omega \\
\frac{d^{q_3} \omega}{dt} &= \sigma(i_q - \omega)
\end{aligned} \tag{18.2}$$

The chaotic behavior in the PMSM occurs when the parameters of the commensurate fractional order PMSM model σ, γ are equal to 10, 100, respectively, and $q_1 = q_2 = q_3 = q = 0.95$ as mentioned in Xue et al. (2015). While, with respect to the data reported in Li et al. (2012), the chaotic behavior in the PMSM occurs when the parameters of the incommensurate fractional order PMSM model σ, γ, q_1, q_2, and q_3 are equal to 4, 50, 0.99, 1, and 0.98, respectively. The obtained chaotic behaviors from the system 18.2 at the two cases of commensurate and incommensurate fractional order models with the initial conditions $[i_d, i_q, \omega] = [2.5, 3, 1]$ are plotted in Fig. 18.1A and B, respectively

18.3 PROBLEM FORMULATION

Practically, the dimensionless unknown parameters of the commensurate and the incommensurate fractional order PMSM models are (σ, γ, $q_1 = q_2 = q_3 =$ q) and (σ, γ, q_1, q_2, q_3), respectively. These parameters are necessary to be identified accurately because at these parameters the PMSM exhibits unexpected performance which is known as a chaotic behavior. Hence, the main target of the introduced optimization techniques is to extract these parameters via comparison between the measured response of the system and the estimated one (Tofighi et al., 2013).

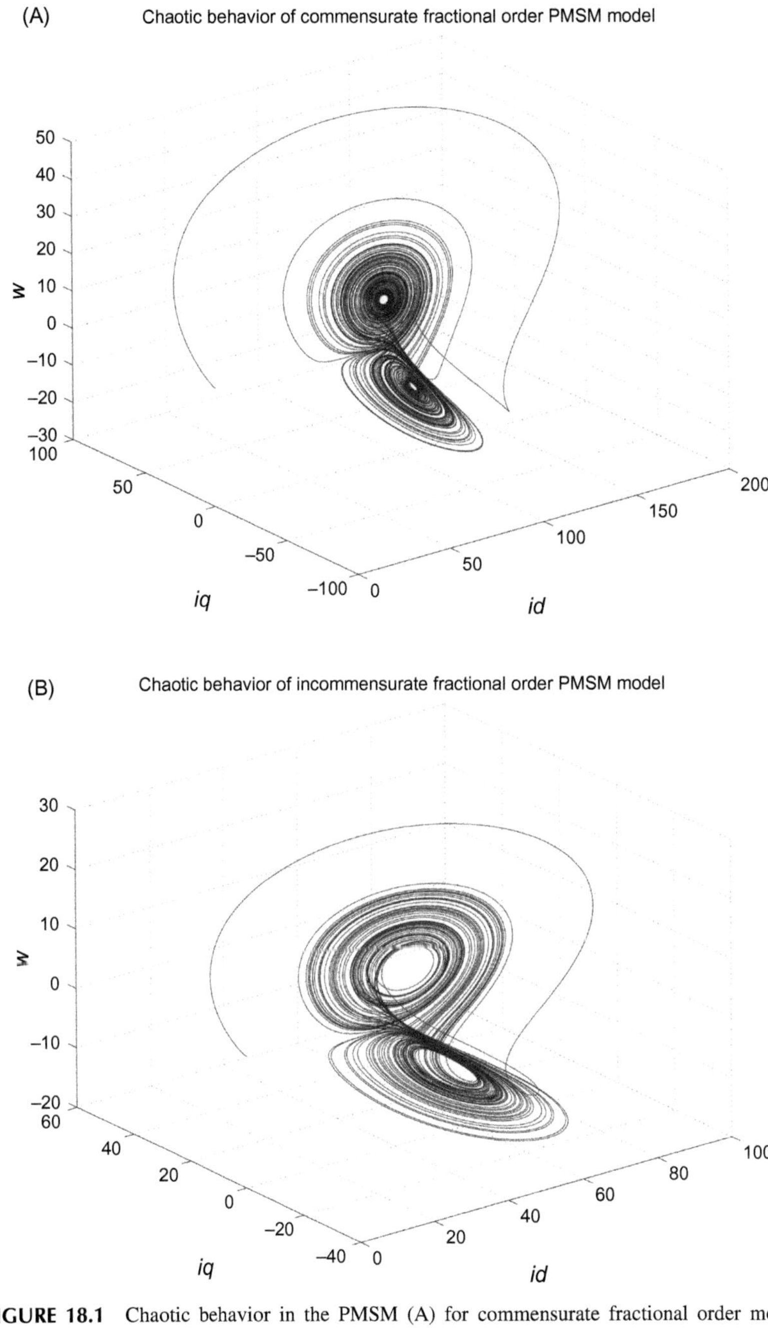

FIGURE 18.1 Chaotic behavior in the PMSM (A) for commensurate fractional order model and (B) for incommensurate fractional order model.

The nonlinear equations of the PMSM are generally described as follows 18.3.

$$D_t^q X(t) = f(X_t, X_0, \theta) \tag{18.3}$$

where X denotes the state vector and X_0 is the initial state vector. $\theta = (\theta_1, \theta_2, \ldots, \theta_d)^T$ is a set of original parameters. q is the noninteger derivative orders.

Similarly, the nonlinear equation of the PMSM with the estimated parameters can be written as follows 18.4

$$D_t^q \hat{X}(t) = f(\hat{X}_t, X_0, \hat{\theta}) \tag{18.4}$$

where $\hat{X}$ is the estimated state vector, $\hat{\theta} = (\hat{\theta}_1, \hat{\theta}_2, \ldots, \hat{\theta}_d)^T$ denotes a set of the extracted parameters and q is the noninteger derivative orders.

The proposed nature-inspired optimization algorithms search for the global unknown parameters of the models based on a fitness function which is the Mean Square Error (MSE) between the known and the estimated system as in 18.5.

$$MSE = \frac{1}{k} \sum_{i=1}^{k} |X(i) - \hat{X}(i)|^2 \tag{18.5}$$

where K is the number of samples.

18.4 PROPOSED CHAOTIC OPTIMIZATION TECHNIQUES OVERVIEW

Developing the meta-heuristic algorithms by mixing the chaos maps with them to tune some of their parameters is becoming a new research trend (Gandomi and Yang, 2014). In this chapter, 10 different chaos maps are integrated with the Grey Wolf Optimizer (GWO) and Grasshopper Optimization algorithm (GOA) to improve the performance of the original algorithms.

18.4.1 Chaos Maps

Almost all the meta-heuristic optimization algorithms are random-based techniques. This randomization is satisfied by using uniform or Gaussian distribution. Nowadays, a recent approach has been introduced to substitute this randomness with chaos maps to capitalize from the better properties of the chaos maps randomization (Alatas, 2010a, Rezaee Jordehi, 2015; Gandomi and Yang, 2014). In this approach, integrating the properties of chaos with the original algorithms assists the standard algorithms to converge for the global solution rapidly and accurately, especially in the case of multimodal functions problems (Gandomi and Yang, 2014; Saremi et al., 2014; Rezaee Jordehi, 2015; Emary and Zawbaa, 2016). In this manuscript, 10 different one-dimensional chaos maps are employed to tune some parameters of the original versions of the GWO and the GOA techniques as a result set of Chaotic GWO and Chaotic GOA are generated. The used chaos maps are listed in Table 18.1. Their distribution values are equal to the number of maximum iterations with an initial value equal to 0.7 as illustrated in Fig. 18.2.

18.4.2 Chaotic Grey Wolf Optimization Algorithm

18.4.2.1 Grey Wolf Optimization Algorithm Overview (GWO)

GWO technique is developed by Mirjalili et al. (2014) to mimic the hunting process of grey wolves in nature. The leader of the wolves named alpha (α) is responsible for the hunting process and is helped by beta (β) and delta (δ) wolves, However the omega (ω) wolves follow these three wolves.

The hunting mechanism is summarized and mathematically modeled as in the following equations:

1. **Searching for prey (Exploration phase)**: In this stage the wolves walk stochastically from each other looking for their preys. This step is modeled as in Eqs. (18.6)–(18.7).

$$\vec{D} = |\vec{C}.\vec{X}_{rand}(l) - \vec{X}(t)| \tag{18.6}$$

$$\vec{X}(t+1) = \vec{X}_{rand}(t) - \vec{A}.\vec{D} \tag{18.7}$$

where l is the current iteration, A and C are coefficient vectors, $||$ is the absolute value, $\vec{X}_{rand}$ is a random position vector, and $\vec{X}$ is the current position vector. Vector A is utilized with a random values > 1 or < -1 to force the search agent to diverge from the prey (Mirjalili et al., 2014).

2. **Encircling prey**: Once, the alpha, beta, and delta wolves detect the nearest position for the prey location, the other wolves follow this location as in Eqs. (18.8)–(18.9).

$$\vec{D} = |\vec{C}.\vec{X}_p(t) - \vec{X}(t)| \tag{18.8}$$

$$\vec{X}(t+1) = \vec{X}_p(t) - \vec{A}.\vec{D} \tag{18.9}$$

$$\vec{A} = 2\vec{a}.\vec{r} - \vec{a} \tag{18.10}$$

$$\vec{C} = 2.\vec{r} \tag{18.11}$$

Where $\vec{X}_p$ is the position vector of the prey, and $\vec{X}$ is the positions of the grey wolf. The vectors A and C can be defined by Eqs. (18.10)–(18.12). The vector a is linearly decreased from 2 to 0 over the course of iterations and r is random vector in [0,1] (Mirjalili et al. (2014). The vector C has random values in [0, 2].

TABLE 18.1 Chaos Maps Functions

Number	Function Name	Chaotic Map	Range
1	Chebyshev (Gandomi and Yang, 2014)	$x_{i+1} = \cos(i\cos^{-1}(x_i))$	(−1,1)
2	Circl (Saremi et al., 2014)	$x_{i+1} = mod\left(x_i + b - \left(\frac{a}{2\pi}\right)\sin(2\pi x_k, 1)\right), a = 0.5 \text{ and } b = 0.2$	(0,1)
3	Gauss/mouse (Gandomi and Yang, 2014)	$x_{i+1} = \begin{cases} 1 & x_i = 0 \\ \frac{1}{mod(x_i, 1)} & \text{Otherwise} \end{cases}$	(0,1)
4	Iterative (Gandomi and Yang, 2014)	$x_{i+1} = \sin\left(\frac{a\pi}{x_i}\right), a = 0.7$	(0,1)
5	Logistic (Gandomi and Yang, 2014)	$x_{i+1} = ax_i(1 - x_i), a = 4$	(0,1)
6	Piecewise (Gandomi and Yang, 2014)	$x_{i+1} = \begin{cases} \frac{x_i}{P} & 0 <= x_i < P \\ \frac{x_i - P}{0.5 - P} & P <= x_i < 0.5 \\ \frac{1 - P - x_i}{0.5 - P} & 0.5 <= x_i < 1 - P \\ \frac{1 - x_i}{P} & 1 - P <= x_i < 1 \end{cases}, P = 0.4$	(0,1)
7	Sine (Gandomi and Yang, 2014)	$x_{i+1} = \frac{a}{4} sin(\pi x_i), a = 4$	(0,1)
8	Singer (Gandomi and Yang, 2014)	$x_{i+1} = \mu(7.86x_i - 23.31x_i^2 + 28.75x_i^3 - 13.302875x_i^4), \mu = 1.07$	(0,1)
9	Sinusoidal (Gandomi and Yang, 2014)	$x_{i+1} = ax_i^2 \sin(\pi x_i), a = 2.3$	(0,1)
10	Tent (Gandomi and Yang, 2014)	$x_{i+1} = \begin{cases} \frac{x_i}{0.7} & x_i < 0.7 \\ \frac{10}{3}(1 - x_i) & x_i >= 0.7 \end{cases}$	(0,1)

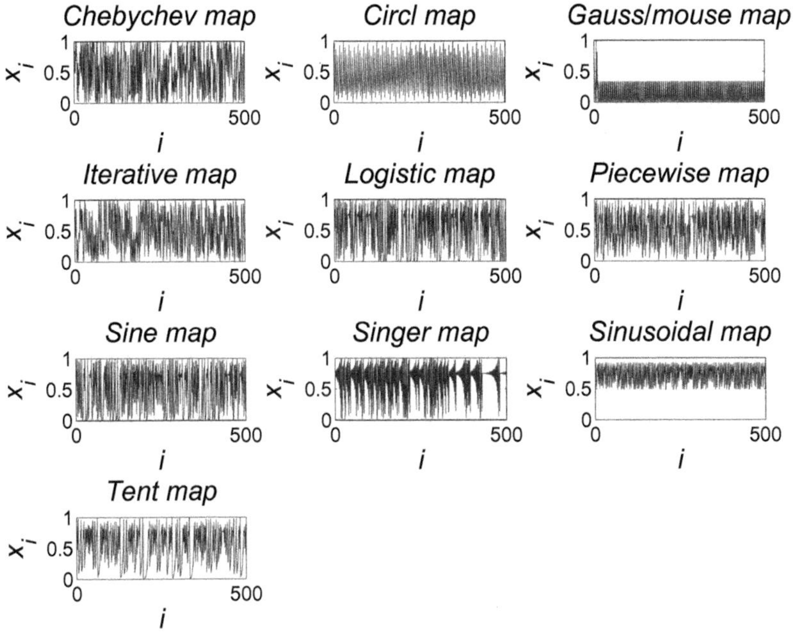

FIGURE 18.2 Distribution values of chaotic maps.

3. **Hunting prey**: In this stage, the wolves have a certain knowledge about the prey position and all the wolves walk toward it hunting it. This behavior can be formulated mathematically as follows in Eqs. (18.12)–(18.14) (Mirjalili et al., 2014):

$$\begin{aligned}\vec{D}_\alpha &= |\vec{C}_1.\vec{X}_\alpha - \vec{X}| \\ \vec{D}_\beta &= |\vec{C}_2.\vec{X}_\beta - \vec{X}| \\ \vec{D}_\delta &= |\vec{C}_3.\vec{X}_\delta - \vec{X}|\end{aligned} \tag{18.12}$$

$$\begin{aligned}\vec{X}_1 &= \vec{X}_\alpha - \vec{A}_1.\vec{D}_\alpha \\ \vec{X}_2 &= \vec{X}_\beta - \vec{A}_2.\vec{D}_\beta \\ \vec{X}_3 &= \vec{X}_\delta - \vec{A}_3.\vec{D}_\delta\end{aligned} \tag{18.13}$$

$$\vec{X}(t+1) = \frac{\vec{X}_1 + \vec{X}_2 + \vec{X}_3}{3} \tag{18.14}$$

4. **Attacking prey (Exploitation phase)**: For the last stage, the wolves attack the prey. This process is done through decreasing parameter a linearly from 2 to 0, consequently the fluctuation range of A is decreased (Mirjalili et al., 2014). The flowchart of the GWO technique is in Fig. 18.3

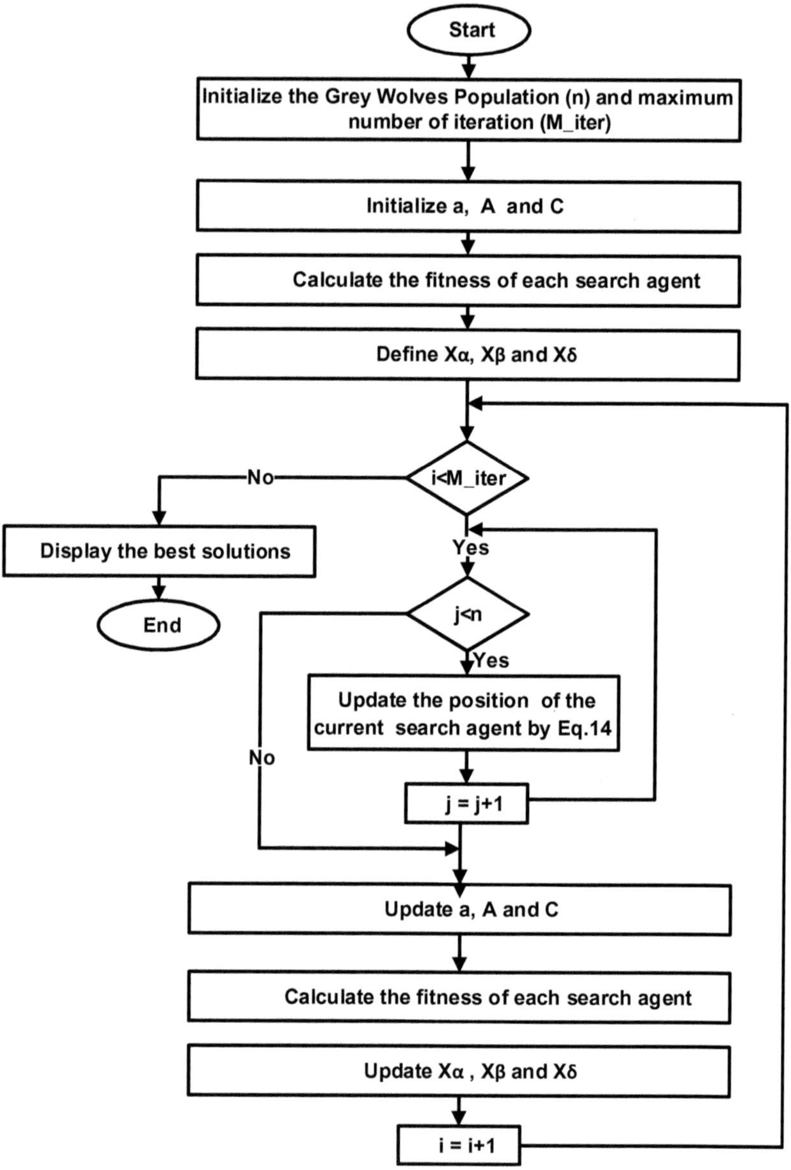

FIGURE 18.3 GWO technique flowchart.

18.4.2.2 Chaotic Grey Wolf Optimizer Strategy (CGWO)

From the previous controlling equations of the hunting mechanism of wolves, it's illustrated that there are two essential parameters A and C that are trade-offs between the exploration and exploitation phases.

In GWO, parameter A changes according to the lineally decreasing of vector a from 2 to 0. While in CGWO, the vector a is decreasing from the same range chaotically, consequently A will be changed based on the chaos maps as illustrated in Eq. (18.15).

$$a = \left(a_i - i.\frac{(a_f - a_i)}{I} \right).Chaos(i); \tag{18.15}$$

where a_i, a_f are the initial and final values of the parameters a, they are adjusted as 2 and 0.00001 respectively. i, I are the current iteration and the maximum number of iterations respectively. $Ch\vec{a}os$ is the chaotic map.

Additionally, parameters C in the GWO has random values between 0 and 2 with the uniform distribution. However, in CGWO, the parameter C has a random values in the same interval over the course of iterations with respect to the chaos maps. Therefore, the chaos maps are normalized to lie in the same ranges [0 2] as follows in Eqs. (18.16)–(18.23)

$$Norm\,\vec{C}haos = \frac{Ch\vec{a}os.(d-e)}{b-a} + e \tag{18.16}$$

$$C = Norm\,\vec{C}haos(i) \tag{18.17}$$

where $[a\ b]$ are the minimum and maximum ranges of chaotic maps. $[e\ d]$ are the range of the normalization. The $[e\ d]$ interval is selected as [0.00001 2]. $Norm\vec{C}haos$ is the normalized chaos map.

18.4.3 Chaotic Grasshopper Optimization Algorithm

18.4.3.1 Grasshopper Optimization Algorithm Overview

GOA algorithm has been recently introduced in Saremi et al. (2017) to mimic the behavior of the grasshopper swarms in nature. The grasshoppers are found in two phases of life. In the nymph phase, the movement of grasshopper is slow with small steps, however the long range and the abrupt movement is the main feature of the swarm in the adulthood phase (Saremi et al., 2017). In both of these phases, the main target of the grasshoppers is searching for the sources of food (Saremi et al., 2017). These features can be mathematically modeled to generate GOA technique (Saremi et al., 2017) as in Eqs. (18.18)–(18.19):

$$X_i = S_i + G_i + A_i \tag{18.18}$$

where X_i is the i-th grasshopper position, $S_i = \sum_{j=1, j\neq i}^{N} s(d_{ij})\widehat{d_{ij}}$ is the social interaction between grasshoppers, G_i is the gravity force, and A_i is the wind advection (Saremi et al., 2017). $d_{ij} = |x_j - x_i|$ is the distance between i-th and j-th grasshopper and $\widehat{d_{ij}} = \frac{|x_j - x_i|}{d_{ij}}$ is the unit vector of the distance. The function $s = fe^{\frac{-r}{l}} - e^{-r}$ defines social forces between grasshoppers. f shows the intensity of attraction and l indicates the attractive length scale (Saremi et al., 2017).

Eq. 18.18 is considered to be a controlled equation of the GOA algorithm. However, this mathematical equation cannot be employed directly to solve the optimization problems, because the grasshoppers quickly arrive at the local zone. Therefore, the controlled equation Eq. 18.18 of GOA algorithm will be modified as in the following Eq. 18.19 (Saremi et al., 2017).

$$X_i^d = c\left(\sum_{j=1, j\neq i}^{N} c\frac{ub_d - lb_d}{2} s(|x_j - x_i|)\frac{|x_j - x_i|}{d_{ij}}\right) + \widehat{T_d} \tag{18.19}$$

where ub_d and lb_d are the upper and lower boundaries in the D th dimension. T_d is the value best solution obtained so far, and c is a decreasing coefficient from 1 to zero over the course of iteration by this relation $\left(c_{max} - j\left(\frac{c_{max} - c_{min}}{J}\right)\right)$, j is the current iteration, and J is the maximum number of iterations. c_{max}, c_{min} are 1 and 0.00001, respectively (Saremi et al., 2017).

The first term of Eq. 18.19, $c\left(\sum_{j=1, j\neq i}^{N} c\frac{ub_d - lb_d}{2} s(|x_j - x_i|)\frac{|x_j - x_i|}{d_{ij}}\right)$ simulates the interaction of grasshoppers in nature, whereas, the second term, T_d, implements their tendency to move towards the source of food (Saremi et al., 2017). Accordingly, characteristics of grasshoppers in natural have been implemented in this Eq. 18.19 (Saremi et al., 2017). The flowchart of the GOA algorithm is shown in Fig. 18.4.

18.4.3.2 Chaotic Grasshopper Optimizer Strategy (CGOA)

As illustrated in Eq. (18.19), there are three important parameters c, f, and l that control the behavior of the GOA technique. The parameter c is the most essential one that helps in balancing between exploration and exploitation phases, c value linearly decreases from 1 to 0 over the course of iterations. While, both f and l are responsible for managing the social interaction between the grasshoppers to prevent them from sticking in local minima. In GOA, f and l values are selected as 0.5 and 1.5 from the intervals [0 1] and [1 2], respectively. While, in the developed CGOA technique, the values of c, f, and l are changed based on the chaos maps distribution as illustrated in the following Eqs. (18.20)–(18.21). Where c will chaotically deceased from

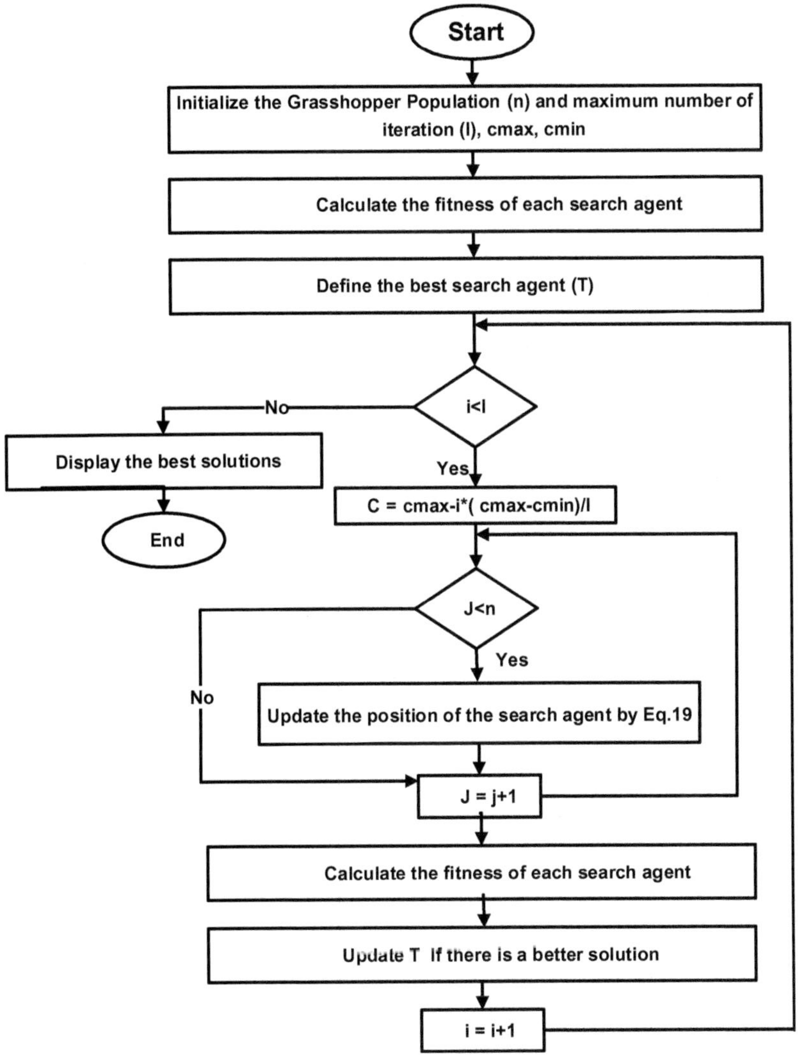

FIGURE 18.4 GOA algorithm flowchart.

1 to 0 over the course of iterations to improve the fluctuation of c as in Eq. (18.20).

$$c = \left(c_i - i.\frac{(c_f - c_i)}{I}\right).Chaos(i); \tag{18.20}$$

where c_i, c_f are the initial and final values of the parameters c, they are adjusted as 1 and 0.00001 respectively. i, I are the current iteration and the maximum number of iterations respectively. $Ch\vec{a}os$ is the chaotic map.

The f and l values are changed abruptly with the chaos maps over the course of iteration, which may help the trapped grasshoppers to release themselves from local minima. Therefore, the chaos maps are normalized to lie in the same ranges as follows in Eq. (18.21).

$$NormChaos\vec{} = \frac{Ch\vec{a}os.(d-e)}{b-a} + e \tag{18.21}$$

$$l = Norm\vec{C}haos(i) \tag{18.22}$$

$$f = Norm\vec{C}haos(i) \tag{18.23}$$

where $[a\ b]$ are the range of chaotic maps. $[e\ d]$ are the range of the normalization. The $[e\ d]$ interval is selected as [0.3 0.7] and [1.3 1.7] for f and l, respectively. $Norm\vec{C}haos$ is the normalized chaos map.

18.5 SIMULATION AND RESULTS

The parameters of the commensurate and incommensurate fractional order PMSM models are extracted using the GWO, GOA, CGWO, and CGOA techniques. Statistical analysis is performed to compare the results of the proposed algorithms. This analysis includes the mean value and the Standard deviation (STD) of the identified models parameters in addition to the STD of the fitness function (MSE). Moreover, the mean and the STD of the absolute Percentage Error (APE) between the identified parameters and the reported global values in (Li et al., 2012; Xue et al., 2015) are calculated. Both of the STD and mean absolute percentage error (MAPE) can be calculated as follows:

Standard Deviation (STD)

$$STD = \frac{1}{N}\sum_{i=1}^{N}\left(Z(i) - \overline{Z}(i)\right) \tag{18.24}$$

where N is the number of runs, Z is the extracted values such as estimated parameters or MSE or APE at each run, while $\overline{Z}$ is the mean value of these extracted values over the total number of runs.

Mean absolute percentage error (MAPE) of the estimated parameters is calculated as follows:

$$MAPE = \frac{100}{N}\sum_{i=1}^{N}\left|\frac{Y(i) - \hat{Y}(i)}{Y}\right| \tag{18.25}$$

where Y is the global value of the model parameters and $\hat{Y}$ is the estimated parameters values.

18.5.1 Commensurate Fractional Order Permanent Magnet Synchronous Motor Model

Based on the reported data in Xue et al. (2015), the PMSM model displays a chaotic behavior when the commensurate fractional order model parameters σ, γ, and $q_1 = q_2 = q_3 = q$ are 10, 100, and 0.95, respectively, and the initial condition of $[i_d, i_q, \omega]$ is equal to [2.5, 3, 1]. The length of the samples vector is equal to 100 samples with sampling steps equal to 0.001 s. The proposed algorithms are applied to extract the corresponding parameters of the chaotic behavior in the motor. For each algorithm the population size is selected to be 30, the number of iterations is 200 and the range of the lower and upper boundaries of the problem variables are adjusted as follows $5 \leq \sigma \leq 15$, $80 \leq \gamma \leq 120$, and $0.9 \leq q \leq 1$. The comparison between the results of the introduced techniques is carried out to illustrate which algorithm is the more adequate one to optimize this problem. To verify the results, 20 independent runs are established for each algorithm.

Mean and STD values of the identified parameters of the commensurate fractional order model as well as the best and STD values of the fitness function over 20 runs are reported in Table 18.2. The listed data clarifies that the CGWO achieves more accurate and consistent results than that of GWO and similarly the results of CGOA is better than that of GOA technique. Therefore, it's obvious that the randomization of the chaos maps affects the performance of the algorithms where the accuracy and consistency of the applied algorithms are enhanced. Moreover it's clarified that the results of the CGWO are the most precise with respect to the other employed techniques.

To investigate the accuracy of the introduced techniques, the mean and the STD values of the APE of the estimated parameters are calculated as in Table 18.3. The mean and STD values clarify that the CGWO achieves the least deviation from the global values of the model parameters. While, CGOA has the second rank in the accuracy of the extracted parameters followed by GWO, and GOA has the last rank. Accordingly, the chaos maps improve the exploitation and the exploration phases of the original proposed algorithms.

Additionally, the mean convergence curves of the employed algorithms with different maps over 20 runs are plotted in Fig. 18.5. The figures show that CGWO with many types of built in chaos maps converges to the least value of the fitness function faster than the other algorithms. While the CGOA technique with the logistic, piecewise, and sinusoidal maps reaches the least value of the fitness function faster than that of the CGWO and the other algorithms. Therefore, the CGWO and CGOA are recommended to extract the parameters of the commensurate fractional order PMSM model.

TABLE 18.2 Mean and STD the Estimated Parameters as well as the Best and STD of the MSE for the Commensurate Fractional Order PMSM Model by the Proposed Algorithms Over 20 Runs

Algorithms		σ(*mean* ± *STD*)	γ(*mean* ± *STD*)	*q*(*mean* ± *STD*)	Best MSE ± STD
CGWO	GWO	10.0014 ± 0.0115	99.9857 ± 0.0438	$0.9500 \pm 6.4341 \times 10^{-5}$	$1.1614 \times 10^{-4} \pm 1.6430 \times 10^{-2}$
	CGWO1	$10.0001 \pm 2.7939 \times 10^{-4}$	$99.9990 \pm 9.9788 \times 10^{-4}$	$0.9500 \pm 9.1110 \times 10^{-7}$	$2.7590 \times 10^{-7} \pm 8.7827 \times 10^{-6}$
	CGWO2	$10.0000 \pm 13.8855 \times 10^{-4}$	$99.9976 \pm 37.2056 \times 10^{-4}$	$0.9500 \pm 4.5710 \times 10^{-6}$	$2.7772 \times 10^{-6} \pm 2.3953 \times 10^{-4}$
	CGWO3	$10.0008 \pm 15.4403 \times 10^{-4}$	$99.9956 \pm 70.2156 \times 10^{-4}$	$0.9500 \pm 5.5271 \times 10^{-6}$	$5.1488 \times 10^{-6} \pm 3.5955 \times 10^{-4}$
	CGWO4	$10.0001 \pm 5.4193 \times 10^{-4}$	$99.9991 \pm 17.8904 \times 10^{-4}$	$0.9499 \pm 2.2393 \times 10^{-7}$	$3.3078 \times 10^{-7} \pm 4.5198 \times 10^{-5}$
	CGWO5	$10.0001 \pm 2.7388 \times 10^{-4}$	$99.9991 \pm 20.4359 \times 10^{-4}$	$0.9500 \pm 8.0010 \times 10^{-7}$	$7.5099 \times 10^{-8} \pm 2.6439 \times 10^{-5}$
	CGWO6	$10.0002 \pm 13.0202 \times 10^{-4}$	$99.9971 \pm 57.3652 \times 10^{-4}$	$0.9499 \pm 1.0068 \times 10^{-5}$	$1.8660 \times 10^{-6} \pm 3.5760 \times -4$
	CGWO7	$10.0000 \pm 5.1355 \times 10^{-4}$	$99.9991 \pm 15.7978 \times 10^{-4}$	$0.9500 \pm 1.6009 \times 10^{-6}$	$1.5642 \times 10^{-6} \pm 2.5750 \times 10^{-5}$
	CGWO8	$10.0003 \pm 37.0185 \times 10^{-4}$	$100 \pm 0 \times 10^{-4}$	$0.9500 \pm 1.0930 \times 10^{-5}$	$9.4343 \times 10^{-6} \pm 1.8487 \times 10^{-3}$
	CGWO9	$9.9998 \pm 31.1781 \times 10^{-4}$	$99.9959 \pm 83.1832 \times 10^{-4}$	$0.9499 \pm 3.1190 \times 10^{-5}$	$5.8418 \times 10^{-5} \pm 2.1504 \times 10^{-3}$
	CGWO10	$10.0001 \pm 13.5541 \times 10^{-4}$	$99.9963 \pm 92.7834 \times 10^{-4}$	$0.9499 \pm 9.2365 \times 10^{-6}$	$6.5734 \times 10^{-6} \pm 4.1403 \times 10^{-4}$

CGOA	GOA	9.6782 ± 0.4540	99.9733 ± 0.11766	0.9449 ± 0.00752	$1.0562 \times 10^{-7} \pm 7.1343 \times 10^{+1}$
	CGOA1	$9.9998 \pm 71.0088 \times 10^{-4}$	$99.9992 \pm 34.4903 \times 10^{-4}$	$0.9499 \pm 1.0588 \times 10^{-4}$	$1.3327 \times 10^{-13} \pm 5.1116 \times 10^{-2}$
	CGOA2	9.9512 ± 0.15209	$99.9999 \pm 0.6550 \times 10^{-4}$	$0.9492 \pm 23.2833 \times 10^{-4}$	$1.7100 \times 10^{-12} \pm 1.4445 \times 10^{+1}$
	CGOA3	10.5917 ± 1.1598	96.2670 ± 3.0602	$0.95034 \pm 96.7812 \times 10^{-4}$	$9.3310 \times 10^{-1} \pm 1.2573 \times 10^{+2}$
	CGOA4	$9.8897 \pm 0.27399 \times 10^{-4}$	$99.9907 \pm 278.8525 \times 10^{-4}$	$0.9482 \pm 44.5497 \times 10^{-4}$	$1.3202 \times 10^{-7} \pm 3.8657 \times 10^{+1}$
	CGOA5	$9.9999 \pm 0.8924 \times 10^{-4}$	$99.9999 \pm 0.3209 \times 10^{-4}$	$0.9499 \pm 0.0135 \times 10^{-4}$	$1.7935 \times 10^{-14} \pm 8.5292 \times 10^{-6}$
	CGOA6	$10.000 \pm 7.6465 \times 10^{-4}$	$99.9998 \pm 4.6410 \times 10^{-4}$	$0.9500 \pm 0.1120 \times 10^{-4}$	$7.2870 \times 10^{-11} \pm 5.2240 \times 10^{-4}$
	CGOA7	$9.9979 \pm 71.0527 \times 10^{-4}$	$99.9999 \pm 0.5700 \times 10^{-4}$	$0.9499 \pm 1.0806 \times 10^{-4}$	$8.1750 \times 10^{-15} \pm 6.9663 \times 10^{-2}$
	CGOA8	$9.9771 \pm 870.1828 \times 10^{-4}$	$99.9999 \pm 0.2336 \times 10^{-4}$	$0.9496 \pm 13.0466 \times 10^{-4}$	$3.9726 \times 10^{-10} \pm 6.7899 \times 10^{0}$
	CGOA9	$9.9994 \pm 37.6001 \times 10^{-4}$	$99.9993 \pm 15.7982 \times 10^{-4}$	$0.9499 \pm 0.5629 \times 10^{-4}$	$4.8692 \times 10^{-10} \pm 9.8623 \times 10^{-2}$
	CGOA10	10.2136 ± 0.95425	99.5875 ± 1.84309800	$0.9519 \pm 89.0933 \times 10^{-4}$	$9.1468 \times 10^{-11} \pm 1.0431 \times 10^{+2}$

TABLE 18.3 Mean and STD of the Absolute Percentage Error Between the Values of the Estimated Parameters and the Global Ones for the Commensurate Fractional order PMSM Model Over 20 Runs

Algorithms		$\sigma(MAPE \pm STD(APE))$	$\gamma(MAPE \pm STD(APE))$	$q(MAPE \pm STD(APE))$
CGWO	GWO	$8.7521 \times 10^{-2} \pm 7.3856 \times 10^{-2}$	$1.4243 \times 10^{-2} \pm 4.3842 \times 10^{-2}$	$4.9290 \times 10^{-3} \pm 4.5742 \times 10^{-3}$
	CGWO1	$2.4967 \times 10^{-3} \pm 2.1672 \times 10^{-3}$	$9.4693 \times 10^{-4} \pm 9.9788 \times 10^{-4}$	$7.7566 \times 10^{-5} \pm 7.8056 \times 10^{-5}$
	CGWO2	$1.0192 \times 10^{-2} \pm 9.1613 \times 10^{-3}$	$2.3169 \times 10^{-3} \pm 3.7206 \times 10^{-3}$	$3.9155 \times 10^{-4} \pm 2.6604 \times 10^{-4}$
	CGWO3	$1.2077 \times 10^{-2} \pm 1.2454 \times 10^{-2}$	$4.3651 \times 10^{-3} \pm 7.0216 \times 10^{-3}$	$4.7166 \times 10^{-4} \pm 3.5781 \times 10^{-4}$
	CGWO4	$4.2379 \times 10^{-3} \pm 3.3962 \times 10^{-3}$	$8.1508 \times 10^{-4} \pm 1.7890 \times 10^{-3}$	$2.4595 \times 10^{-4} \pm 1.8062 \times 10^{-4}$
	CGWO5	$1.5798 \times 10^{-3} \pm 2.4933 \times 10^{-3}$	$8.4301 \times 10^{-4} \pm 2.0436 \times 10^{-3}$	$6.1924 \times 10^{-5} \pm 5.8693 \times 10^{-5}$
	CGWO6	$9.9918 \times 10^{-3} \pm 8.4723 \times 10^{-3}$	$2.8561 \times 10^{-3} \pm 5.7365 \times 10^{-3}$	$8.6609 \times 10^{-4} \pm 7.0172 \times 10^{-4}$
	CGWO7	$3.9461 \times 10^{-3} \pm 3.1931 \times 10^{-3}$	$8.7701 \times 10^{-4} \pm 1.5798 \times 10^{-3}$	$1.3488 \times 10^{-4} \pm 1.1276 \times 10^{-4}$
	CGWO8	$2.6469 \times 10^{-2} \pm 2.5459 \times 10^{-2}$	$0.0000 \times 10^{0} \pm 0.0000 \times 10^{0}$	$1.6058 \times 10^{-3} \pm 1.1972 \times 10^{-3}$
	CGWO9	$2.4887 \times 10^{-2} \pm 1.7981 \times 10^{-2}$	$4.0308 \times 10^{-3} \pm 8.3183 \times 10^{-3}$	$2.5994 \times 10^{-3} \pm 2.0653 \times 10^{-3}$
	CGWO10	$1.0702 \times 10^{-2} \pm 8.1948 \times 10^{-3}$	$3.6260 \times 10^{-3} \pm 9.2783 \times 10^{-3}$	$8.1045 \times 10^{-4} \pm 5.9733 \times 10^{-4}$

CGOA	GOA	$3.4349 \times 10^{0} \pm 4.3695 \times 10^{0}$	$2.6684 \times 10^{-2} \pm 1.1766 \times 10^{-1}$	$5.8187 \times 10^{-1} \pm 7.5646 \times 10^{-1}$
	CGOA1	$2.1876 \times {}^{-2} \pm 6.7398 \times 10^{-2}$	$7.7391 \times 10^{-4} \pm 3.4490 \times 10^{-3}$	$3.4255 \times 10^{-3} \pm 1.0585 \times 10^{-2}$
	CGOA2	$4.8851 \times 10^{-1} \pm 1.5208 \times 10^{0}$	$2.3102 \times 10^{-5} \pm 6.5505 \times 10^{-5}$	$7.8536 \times 10^{-2} \pm 2.4505 \times 10^{-1}$
	CGOA3	$9.9446 \times 10^{0} \pm 8.2034 \times 10^{0}$	$5.7330 \times 10^{0} \pm 4.0603 \times 10^{0}$	$6.9948 \times 10^{-1} \pm 7.2401 \times 10^{-1}$
	CGOA4	$1.3076 \times 10^{0} \pm 2.6433 \times 10^{0}$	$9.2280 \times 10^{-3} \pm 2.7885 \times 10^{-2}$	$2.1541 \times 10^{-1} \pm 4.5390 \times 10^{-1}$
	CGOA5	$3.8269 \times 10^{-4} \pm 8.1746 \times 10^{-4}$	$8.9856 \times 10^{-6} \pm 3.2094 \times 10^{-5}$	$6.0969 \times 10^{-5} \pm 1.3062 \times 10^{-4}$
	CGOA6	$3.9139 \times 10^{-3} \pm 6.5167 \times 10^{-3}$	$1.6389 \times 10^{-4} \pm 4.6411 \times 10^{-4}$	$6.1015 \times 10^{-4} \pm 1.0003 \times 10^{-3}$
	CGOA7	$2.1648 \times 10^{-2} \pm 7.0830 \times 10^{-2}$	$2.7020 \times 10^{-5} \pm 5.7010 \times 10^{-5}$	$3.4685 \times 10^{-3} \pm 1.1342 \times 10^{-2}$
	CGOA8	$2.2858 \times 10^{-1} \pm 8.7008 \times 10^{-1}$	$1.2594 \times 10^{-5} \pm 2.3363 \times 10^{-5}$	$3.6078 \times 10^{-2} \pm 1.3732 \times 10^{-1}$
	CGOA9	$2.2737 \times 10^{-2} \pm 2.9950 \times 10^{-2}$	$6.5632 \times 10^{-4} \pm 1.5798 \times 10^{-3}$	$3.5858 \times 10^{-3} \pm 4.7399 \times 10^{-3}$
	CGOA10	$2.1384 \times 10^{0} \pm 9.5421 \times 10^{0}$	$4.1241 \times 10^{-1} \pm 1.8431 \times 10^{0}$	$2.1042 \times 10^{-1} \pm 9.3774 \times 10^{-1}$

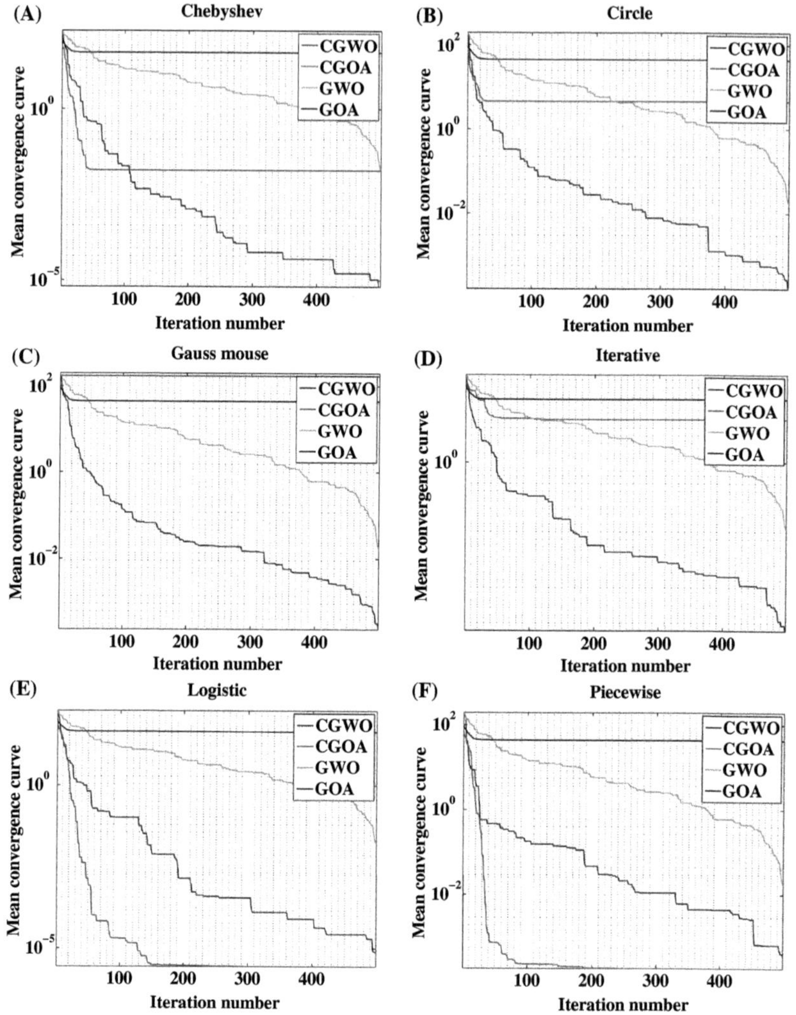

FIGURE 18.5 Mean convergence curves of the introduced algorithms for the commensurate fractional order PMSM model.

18.5.2 Incommensurate Fractional Order Permanent Magnet Synchronous Motor Model

In this section the GWO, GOA, CGWO, and CGOA algorithms are tested to extract the incommensurate fractional order PMSM model parameters depending on the data reported in (Li et al., 2012). Where the PMSM system 18.2 displays a chaotic behavior when the values of the model parameters σ, γ, q_1, q_2, and q_3 are 4, 50, 0.99, 1, and 0.98, respectively, and the initial condition of $[i_d, i_q, \omega]$ is equal to [2.5, 3, 1]. The length of the samples vector is

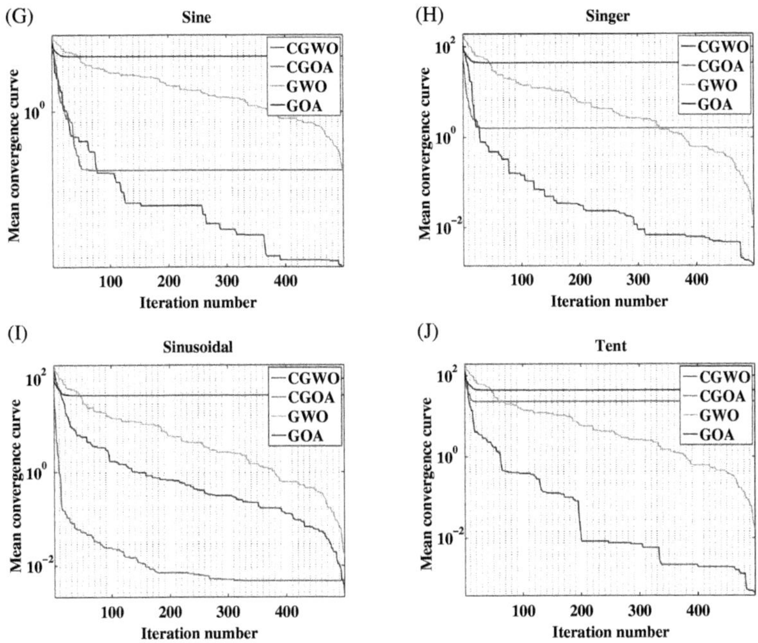

FIGURE 18.5 (Continued).

equal to 100 samples with sampling step equals to 0.001 s. For each algorithm the population size is selected to be 50 search agent, the number of iterations is 500, the range of the lower and upper boundaries of the problem variables are adjusted as follows $6 \le \sigma \le 8$ and $40 \le \gamma \le 60$ and $0.9 \le q_1, q_2$ and $q_3 \le 1$. The comparison between the results of the applied techniques is accomplished over 20 independent runs to recommend the more suitable algorithm for this problem.

To discuss the accuracy, the consistency, and the speed of convergence of the utilized optimization algorithms in the incommensurate fractional order model parameters estimation, the mean and the STD of the estimated model parameters are listed in Table 18.4. Moreover, the mean and the STD values of the APE of the estimated parameters over 20 runs are reported in Table 18.5. In addition, the mean convergence curves of the applied algorithms with different maps over 20 runs are drawn in Fig. 18.6.

From Tables 18.4 and 18.5, it's observed that CGWO algorithm achieves more accurate and consistent results than CGOA and outperforms the results of the standard algorithms (GWO and GOA). The mean and the STD of the identified parameters show that CGWO has a better performance than the other techniques. Moreover, the best value of the fitness function and its STD of CGWO and CGOA are better than that of GWO and GOA algorithms. Additionally, the mean and STD of the APE of the estimated

TABLE 18.4 Mean and STD of the Estimated Parameters as well as the Best and STD of MSE for the Incommensurate Fractional Order PMSM Model by the Introduced Algorithms

Algorithms		σ	γ	q_1	q_2	q_3	Best MSE ± STD
CGWO	GWO	$4.0000 \pm 8.1707 \times 10^{-3}$	$49.9849 \pm 2.1914 \times 10^{-2}$	$0.9903 \pm 7.8789 \times 10^{-4}$	$0.9994 \pm 7.0836 \times 10^{-4}$	$0.9805 \pm 7.6211 \times 10^{-4}$	$1.4319 \times 10^{-3} \pm 4.2572 \times 10^{-3}$
	CGWO1	$3.9998 \pm 2.4728 \times 10^{-3}$	$49.9970 \pm 3.3582 \times 10^{-3}$	$0.9900 \pm 1.1483 \times 10^{-4}$	$0.9999 \pm 2.0036 \times 10^{-4}$	$0.9801 \pm 1.6906 \times 10^{-4}$	$3.3456 \times 10^{-6} \pm 4.6202 \times 10^{-4}$
	CGWO2	$4.0008 \pm 1.4495 \times 10^{-3}$	$49.9932 \pm 7.1498 \times 10^{-3}$	$0.9901 \pm 1.8338 \times 10^{-4}$	$0.9999 \pm 1.4353 \times 10^{-4}$	$0.9802 \pm 1.8442 \times 10^{-4}$	$3.6685 \times 10^{-6} \pm 3.3285 \times 10^{-4}$
	CGWO3	$4.0008 \pm 1.1213 \times 10^{-3}$	$49.9968 \pm 3.3464 \times 10^{-3}$	$0.9901 \pm 9.3944 \times 10^{-5}$	$1.0000 \pm 5.3116 \times 10^{-5}$	$0.9801 \pm 5.6697 \times 10^{-5}$	$7.5097 \times 10^{-6} \pm 6.9201 \times 10^{-5}$
	CGWO4	$4.0002 \pm 2.3156 \times 10^{-3}$	$49.9954 \pm 5.2149 \times 10^{-3}$	$0.9901 \pm 2.0821 \times 10^{-4}$	$0.9999 \pm 1.8251 \times 10^{-4}$	$0.9801 \pm 1.6220 \times 10^{-4}$	$1.2839 \times 10^{-5} \pm 4.0259 \times 10^{-4}$
	CGWO5	$3.9993 \pm 2.0633 \times 10^{-3}$	$49.9984 \pm 2.6288 \times 10^{-3}$	$0.9900 \pm 1.3018 \times 10^{-4}$	$0.9999 \pm 1.9032 \times 10^{-4}$	$0.9801 \pm 2.0745 \times 10^{-4}$	$1.1375 \times 10^{-5} \pm 3.4494 \times 10^{-4}$
	CGWO6	$4.0010 \pm 3.1370 \times 10^{-3}$	$49.9983 \pm 4.2481 \times 10^{-3}$	$0.9902 \pm 3.4169 \times 10^{-4}$	$0.9999 \pm 1.1072 \times 10^{-4}$	$0.9801 \pm 1.1813 \times 10^{-4}$	$3.2565 \times 10^{-5} \pm 6.5649 \times 10^{-4}$
	CGWO7	$3.9997 \pm 1.3020 \times 10^{-3}$	$49.9991 \pm 2.2369 \times 10^{-3}$	$0.9900 \pm 9.9040 \times 10^{-5}$	$1.0000 \pm 9.2788 \times 10^{-5}$	$0.9800 \pm 8.7591 \times 10^{-5}$	$3.9396 \times 10^{-6} \pm 1.6357 \times 10^{-4}$
	CGWO8	$4.0002 \pm 3.0967 \times 10^{-3}$	$49.9951 \pm 9.9718 \times 10^{-3}$	$0.9901 \pm 4.1441 \times 10^{-4}$	$0.9999 \pm 2.5243 \times 10^{-4}$	$0.9801 \pm 2.9379 \times 10^{-4}$	$3.0929 \times 10^{-4} \pm 1.1959 \times 10^{-3}$
	CGWO9	$4.0008 \pm 4.4419 \times 10^{-3}$	$49.9899 \pm 1.7294 \times 10^{-3}$	$0.9902 \pm 5.3244 \times 10^{-4}$	$0.9998 \pm 2.1632 \times 10^{-4}$	$0.9802 \pm 2.6908 \times 10^{-4}$	$5.4552 \times 10^{-4} \pm 1.4507 \times 10^{-3}$
	CGWO10	$4.0004 \pm 2.6115 \times 10^{-3}$	$49.9946 \pm 6.9844 \times 10^{-3}$	$0.9901 \pm 2.4591 \times 10^{-4}$	$0.9999 \pm 1.1578 \times 10^{-4}$	$0.9801 \pm 1.3404 \times 10^{-4}$	$1.8528 \times 10^{-5} \pm 5.3643 \times 10^{-4}$
CGOA	GOA	$3.9801 \pm 1.8670 \times 10^{-2}$	$49.9575 \pm 5.7709 \times 10^{-2}$	$0.9877 \pm 7.9707 \times 10^{-3}$	$0.9961 \pm 5.1493 \times 10^{-3}$	$0.9792 \pm 1.3558 \times 10^{-2}$	$3.8324 \times 10^{-6} \pm 7.2794 \times 10^{0}$

	CGOA1	$3.9889 \pm 2.0956 \times 10^{-2}$	$49.9676 \pm 6.5564 \times 10^{-2}$	$0.9905 \pm 2.1158 \times 10^{-3}$	$0.9971 \pm 5.6567 \times 10^{-3}$	$0.9826 \pm 6.6274 \times 10^{-3}$	$1.4388 \times 10^{-9} \pm 3.0567 \times 10^{-1}$
	CGOA2	$3.9952 \pm 1.0664 \times 10^{-2}$	$49.9865 \pm 3.1030 \times 10^{-2}$	$0.9904 \pm 8.5858 \times 10^{-4}$	$0.9987 \pm 2.8308 \times 10^{-3}$	$0.9813 \pm 3.0696 \times 10^{-3}$	$2.4653 \times 10^{-10} \pm 6.7161 \times 10^{-2}$
	CGOA3	$3.8764 \pm 1.8057 \times 10^{-1}$	$47.6184 \pm 1.9015 \times 10^{0}$	$0.9521 \pm 2.2832 \times 10^{-2}$	$0.9821 \pm 1.4166 \times 10^{-2}$	$0.9789 \pm 1.6511 \times 10^{-2}$	$3.5396 \times 10^{0} \pm 3.1049 \times 10^{1}$
	CGOA4	$3.9856 \pm 1.9079 \times 10^{-2}$	$49.9600 \pm 5.8421 \times 10^{-2}$	$0.9904 \pm 2.3407 \times 10^{-3}$	$0.9964 \pm 5.2179 \times 10^{-3}$	$0.9827 \pm 6.7281 \times 10^{-3}$	$4.3041 \times 10^{-6} \pm 2.5448 \times 10^{-1}$
	CGOA5	$3.9958 \pm 6.8685 \times 10^{-3}$	$49.9885 \pm 1.9382 \times 10^{-2}$	$0.9903 \pm 5.5553 \times 10^{-4}$	$0.9989 \pm 1.8174 \times 10^{-3}$	$0.9812 \pm 1.9713 \times 10^{-3}$	$1.8725 \times 10^{-9} \pm 2.5906 \times 10^{-2}$
	CGOA6	$3.9920 \pm 1.3571 \times 10^{-2}$	$49.9770 \pm 4.0197 \times 10^{-2}$	$0.9906 \pm 1.0904 \times 10^{-3}$	$0.9979 \pm 3.6087 \times 10^{-3}$	$0.9823 \pm 3.9156 \times 10^{-3}$	$2.1158 \times 10^{-10} \pm 1.0539 \times 10^{-1}$
	CGOA7	$3.9964 \pm 5.8943 \times 10^{-3}$	$49.9917 \pm 1.6473 \times 10^{-2}$	$0.9899 \pm 1.2458 \times 10^{-3}$	$0.9992 \pm 1.5622 \times 10^{-3}$	$0.9803 \pm 2.5394 \times 10^{-3}$	$5.5318 \times 10^{-8} \pm 2.0905 \times 10^{-1}$
	CGOA8	$3.9934 \pm 1.0259 \times 10^{-2}$	$49.9812 \pm 2.9697 \times 10^{-2}$	$0.9905 \pm 8.2587 \times 10^{-4}$	$0.9982 \pm 2.7237 \times 10^{-3}$	$0.9819 \pm 2.9503 \times 10^{-3}$	$5.8679 \times 10^{-9} \pm 5.9284 \times 10^{-2}$
	CGOA9	$3.9959 \pm 5.9267 \times 10^{-3}$	$49.9887 \pm 1.6025 \times 10^{-2}$	$0.9902 \pm 5.9900 \times 10^{-4}$	$0.9989 \pm 1.5559 \times 10^{-3}$	$0.9809 \pm 1.8700 \times 10^{-3}$	$1.0394 \times 10^{-5} \pm 1.9723 \times 10^{-2}$
	CGOA10	$3.9886 \pm 1.3104 \times 10^{-2}$	$49.9676 \pm 3.7899 \times 10^{-2}$	$0.9909 \pm 1.0551 \times 10^{-3}$	$0.9970 \pm 3.4751 \times 10^{-3}$	$0.9832 \pm 3.7701 \times 10^{-3}$	$3.6319 \times 10^{-6} \pm 7.6663 \times 10^{-2}$

TABLE 18.5 Mean and STD of the Absolute Percentage Error Between the Values of the Estimated Parameters and the Global Ones for the Incommensurate Fractional Order PMSM Model Over 20 Runs

Algorithms		$\sigma(MAPE \pm STD(APE))$	$\gamma(MAPE \pm STD(APE))$	$q_1(MAPE \pm STD(APE))$	$q_2(MAPE \pm STD(APE))$	$q_3(MAPE \pm STD(APE))$
CGWO	GWO	$1.5695 \times 10^{-1} \pm 1.2568 \times 10^{-1}$	$3.0080 \times 10^{-2} \pm 4.3827 \times 10^{-2}$	$7.1541 \times 10^{-2} \pm 4.7972 \times 10^{-2}$	$5.3743 \times 10^{-2} \pm 7.0836 \times 10^{-2}$	$7.1254 \times 10^{-2} \pm 6.7203 \times 10^{-2}$
	CGWO1	$4.2511 \times 10^{-2} \pm 4.4180 \times 10^{-2}$	$5.9663 \times 10^{-3} \pm 6.7164 \times 10^{-3}$	$9.5431 \times 10^{-3} \pm 7.4928 \times 10^{-3}$	$1.1066 \times 10^{-2} \pm 2.0036 \times 10^{-2}$	$1.2201 \times 10^{-2} \pm 1.7144 \times 10^{-2}$
	CGWO2	$3.1973 \times 10^{-2} \pm 2.4948 \times 10^{-2}$	$1.3624 \times 10^{-2} \pm 1.4300 \times 10^{-2}$	$1.9073 \times 10^{-2} \pm 1.3813 \times 10^{-2}$	$1.2408 \times 10^{-2} \pm 1.4353 \times 10^{-2}$	$1.7399 \times 10^{-2} \pm 1.7463 \times 10^{-2}$
	CGWO3	$2.8045 \times 10^{-2} \pm 1.8479 \times 10^{-2}$	$6.3881 \times 10^{-3} \pm 6.6929 \times 10^{-3}$	$1.1654 \times 10^{-2} \pm 7.4158 \times 10^{-3}$	$4.5158 \times 10^{-3} \pm 5.3116 \times 10^{-3}$	$7.0132 \times 10^{-3} \pm 5.5953 \times 10^{-3}$
	CGWO4	$4.4391 \times 10^{-2} \pm 3.6083 \times 10^{-2}$	$9.1857 \times 10^{-3} \pm 1.0430 \times 10^{-2}$	$1.7789 \times 10^{-2} \pm 1.3999 \times 10^{-2}$	$1.4223 \times 10^{-2} \pm 1.8251 \times 10^{-2}$	$1.5250 \times 10^{-2} \pm 1.6422 \times 10^{-2}$
	CGWO5	$4.1684 \times 10^{-2} \pm 3.4525 \times 10^{-2}$	$3.1344 \times 10^{-3} \pm 5.2576 \times 10^{-3}$	$9.4292 \times 10^{-3} \pm 9.0888 \times 10^{-3}$	$1.2867 \times 10^{-2} \pm 1.9032 \times 10^{-2}$	$1.5149 \times 10^{-2} \pm 2.0241 \times 10^{-2}$
	CGWO6	$6.0873 \times 10^{-2} \pm 5.4391 \times 10^{-2}$	$3.4190 \times 10^{-3} \pm 8.4963 \times 10^{-3}$	$3.1403 \times 10^{-2} \pm 2.5027 \times 10^{-2}$	$8.5472 \times 10^{-3} \pm 1.1072 \times 10^{-2}$	$1.1976 \times 10^{-2} \pm 1.0560 \times 10^{-2}$
	CGWO7	$2.0985 \times 10^{-2} \pm 2.5889 \times 10^{-2}$	$1.8211 \times 10^{-3} \pm 4.4738 \times 10^{-3}$	$7.8333 \times 10^{-3} \pm 5.9588 \times 10^{-3}$	$4.1059 \times 10^{-3} \pm 9.2788 \times 10^{-3}$	$5.8632 \times 10^{-3} \pm 8.1920 \times 10^{-3}$
	CGWO8	$5.4449 \times 10^{-2} \pm 5.3868 \times 10^{-2}$	$9.7713 \times 10^{-3} \pm 1.9944 \times 10^{-2}$	$3.1949 \times 10^{-2} \pm 2.6994 \times 10^{-2}$	$1.3050 \times 10^{-2} \pm 2.5243 \times 10^{-2}$	$2.0095 \times 10^{-2} \pm 2.5507 \times 10^{-2}$
	CGWO9	$9.2809 \times 10^{-2} \pm 6.0998 \times 10^{-2}$	$2.0250 \times 10^{-2} \pm 3.4589 \times 10^{-2}$	$4.7200 \times 10^{-2} \pm 2.8463 \times 10^{-2}$	$1.5529 \times 10^{-2} \pm 2.1632 \times 10^{-2}$	$2.5442 \times 10^{-2} \pm 2.4888 \times 10^{-2}$
	CGWO10	$4.6417 \times 10^{-2} \pm 4.5600 \times 10^{-2}$	$1.0737 \times 10^{-2} \pm 1.3969 \times 10^{-2}$	$1.6091 \times 10^{-2} \pm 1.9468 \times 10^{-2}$	$5.5866 \times 10^{-3} \pm 1.1578 \times 10^{-2}$	$8.5636 \times 10^{-3} \pm 1.2770 \times 10^{-2}$

CGOA	GOA	$4.9677 \times 10^{-1} \pm 4.6675 \times 10^{-1}$	$8.5077 \times 10^{-2} \pm 1.1542 \times 10^{-1}$	$4.6506 \times 10^{-1} \pm 6.9014 \times 10^{-1}$	$3.8669 \times 10^{-1} \pm 5.1493 \times 10^{-1}$	$9.3986 \times 10^{-1} \pm 9.9599 \times 10^{-1}$
	CGOA1	$2.7832 \times 10^{-1} \pm 5.2389 \times 10^{-1}$	$6.4776 \times 10^{-2} \pm 1.3113 \times 10^{-1}$	$1.2042 \times 10^{-1} \pm 1.8312 \times 10^{-1}$	$2.8731 \times 10^{-1} \pm 5.6567 \times 10^{-1}$	$3.7217 \times 10^{-1} \pm 6.2022 \times 10^{-1}$
	CGOA2	$1.1892 \times 10^{-1} \pm 2.6661 \times 10^{-1}$	$2.6972 \times 10^{-2} \pm 6.2060 \times 10^{-2}$	$3.9228 \times 10^{-2} \pm 8.5618 \times 10^{-2}$	$1.2543 \times 10^{-1} \pm 2.8308 \times 10^{-1}$	$1.4002 \times 10^{-1} \pm 3.1136 \times 10^{-1}$
	CGOA3	$4.7023 \times 10^{0} \pm 2.6730 \times 10^{0}$	$4.7633 \times 10^{0} \pm 3.8030 \times 10^{0}$	$3.8252 \times 10^{0} \pm 2.3063 \times 10^{0}$	$1.7902 \times 10^{0} \pm 1.4166 \times 10^{0}$	$1.3797 \times 10^{0} \pm 9.2066 \times 10^{-1}$
	CGOA4	$3.6018 \times 10^{-1} \pm 4.7699 \times 10^{-1}$	$8.0054 \times 10^{-2} \pm 1.1684 \times 10^{-1}$	$1.8236 \times 10^{-1} \pm 1.4970 \times 10^{-1}$	$3.6242 \times 10^{-1} \pm 5.2179 \times 10^{-1}$	$5.1963 \times 10^{-1} \pm 5.1952 \times 10^{-1}$
	CGOA5	$1.0410 \times 10^{-1} \pm 1.7171 \times 10^{-1}$	$2.3035 \times 10^{-2} \pm 3.8764 \times 10^{-2}$	$3.4455 \times 10^{-2} \pm 5.4656 \times 10^{-2}$	$1.0960 \times 10^{-1} \pm 1.8174 \times 10^{-1}$	$1.2224 \times 10^{-1} \pm 1.9867 \times 10^{-1}$
	CGOA6	$2.0065 \times 10^{-1} \pm 3.3926 \times 10^{-1}$	$4.6014 \times 10^{-2} \pm 8.0394 \times 10^{-2}$	$6.5573 \times 10^{-2} \pm 1.0913 \times 10^{-1}$	$2.1265 \times 10^{-1} \pm 3.6087 \times 10^{-1}$	$2.3556 \times 10^{-1} \pm 3.9785 \times 10^{-1}$
	CGOA7	$8.9675 \times 10^{-2} \pm 1.4736 \times 10^{-1}$	$1.6536 \times 10^{-2} \pm 3.2946 \times 10^{-2}$	$6.0895 \times 10^{-2} \pm 1.1003 \times 10^{-1}$	$7.9203 \times 10^{-2} \pm 1.5622 \times 10^{-1}$	$1.4342 \times 10^{-1} \pm 2.1556 \times 10^{-1}$
	CGOA8	$1.6629 \times 10^{-1} \pm 2.5644 \times 10^{-1}$	$3.7570 \times 10^{-2} \pm 5.9393 \times 10^{-2}$	$5.4248 \times 10^{-2} \pm 8.2381 \times 10^{-2}$	$1.7583 \times 10^{-1} \pm 2.7237 \times 10^{-1}$	$1.9458 \times 10^{-1} \pm 2.9930 \times 10^{-1}$
	CGOA9	$1.0416 \times 10^{-1} \pm 1.4684 \times 10^{-1}$	$2.2543 \times 10^{-2} \pm 3.2049 \times 10^{-2}$	$4.5540 \times 10^{-2} \pm 4.3219 \times 10^{-2}$	$1.0590 \times 10^{-1} \pm 1.5559 \times 10^{-1}$	$1.3948 \times 10^{-1} \pm 1.5865 \times 10^{-1}$
	CGOA10	$2.8565 \times 10^{-1} \pm 3.2759 \times 10^{-1}$	$5.4729 \times 10^{-2} \pm 7.5797 \times 10^{-2}$	$9.2645 \times 10^{-2} \pm 1.0468 \times 10^{-1}$	$3.0181 \times 10^{-1} \pm 3.4751 \times 10^{-1}$	$3.3376 \times 10^{-1} \pm 3.8148 \times 10^{-1}$

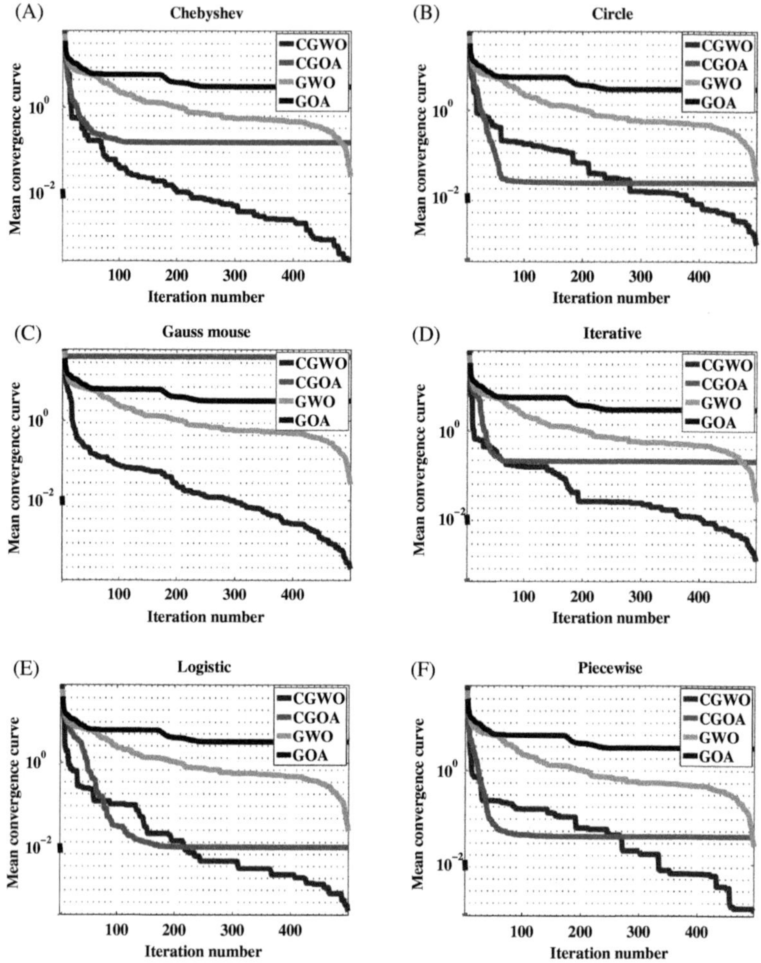

FIGURE 18.6 Mean convergence curves of the proposed algorithms for the incommensurate fractional order PMSM model.

parameters obtained by CGWO is less than that of the other techniques. The previous statistical analysis prove that the integration between the chaos maps and the standard algorithms improves the accuracy and the efficiency of the parameters identification of the incommensurate fractional order PMSM model.

From Fig. 18.5, the mean convergence curves of the utilized algorithms show that at almost chaos maps, the CGWO achieves the fastest convergence to the least value of the fitness function compared with the CGOA, GWO, and GOA techniques. While in the case of the sinusoidal map, the CGOA reaches the least value of the fitness function than that of the CGWO and the

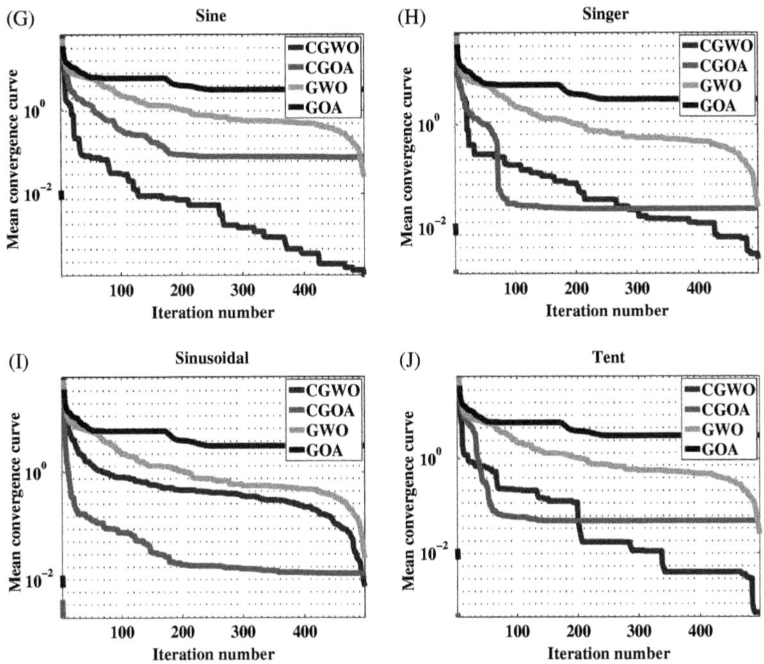

FIGURE 18.6 (Continued).

other algorithms. Consequently, the proposed chaotic biologically inspired optimizers are the more suitable techniques for identifying the parameters of the incommensurate fractional order PMSM model, especially the CGWO with almost of chaos maps and CGOA with sinusoidal map.

18.6 CONCLUSION

The commensurate and incommensurate fractional order PMSM models have been recently proposed to provide more flexibility for the modeling of the motor. Moreover, they provide a deeper vision of the physical behavior of the motor. Therefore, an accurate estimation of the corresponding parameters of these models to the chaotic behavior in the motor is the main target of this work. For this purpose, both of the original meta-heuristic algorithms and the modified ones through integration with chaos maps are introduced. The proposed algorithms such as Grey Wolf Optimizer, Grasshopper Optimizer, Chaotic Grey Wolf Optimizer, and the Chaotic Grasshopper Optimization Algorithm were tested and evaluated to recommend the most suitable one for this application. The main finding is that the Chaotic Grey Wolf Optimizer with almost chaos maps and the Chaotic Grasshopper Optimization Algorithm with the sinusoidal map are the more efficient techniques for the parameters estimation of the commensurate and incommensurate fractional order models. Consequently, with the aid of these techniques, an accurate and fast prediction for the chaotic behavior parameters in the motor is achieved and that may help in controlling the system and may protect the motor from damage.

REFERENCES

Alam, D.F., Yousri, D.A., Eteiba, M.B., 2015. Flower pollination algorithm based solar PV parameter estimation. Energy Convers. Manage. 101, 410–422.

Alatas, B., 2010a. Chaotic harmony search algorithms. Appl. Math. Comput. 216 (9), 2687–2699. Available from: https://doi.org/10.1016/j.amc.2010.03.114.

Alatas, B., 2010b. Expert Systems with Applications Chaotic bee colony algorithms for global numerical optimization. Exp. Syst. Applicat. 37 (8), 5682–5687. Available from: https://doi.org/10.1016/j.eswa.2010.02.042.

Allam, D., Yousri, D.A., Eteiba, M.B., 2016. Parameters extraction of the three diode model for the multi-crystalline solar cell / module using Moth-Flame Optimization Algorithm. Energy Convers. Manage. 123, 535–548. Available from: https://doi.org/10.1016/j.enconman.2016.06.052.

Ataei, M., Kiyoumarsi, A., Ghorbani, B., 2010. Control of chaos in permanent magnet synchronous motor by using optimal lyapunov exponents placement. Phys. Lett. A 374 (41), 4226–4230. Available from: https://doi.org/10.1016%2Fj.physleta.2010.08.047.

Chaos Modeling and Control Systems Design. In: Azar, A.T., Vaidyanathan, S. (Eds.), Springer International Publishing. Available from: https://doi.org/10.1007%2F978-3-319-13132-0.

Azar, A.T., Vaidyanathan, S. (Eds.), 2015b. Computational Intelligence Applications in Modeling and Control. Springer International Publishing. Available from: https://doi.org/10.1007%F2978-3-319-11017-2.

Azar, A.T., Vaidyanathan, S. (Eds.), 2016. Advances in Chaos Theory and Intelligent Control. Springer International Publishing. Available from: https://doi.org/10.1007%F978-3-319-30340-6.

Azar, A.T., Vaidyanathan, S., Ouannas, A. (Eds.), 2017. Fractional Order Control and Synchronization of Chaotic Systems. Springer International Publishing. Available from: https://doi.org/10.1007%F978-3-319-50249-6.

Boulkroune, A., Bouzeriba, A., Bouden, T., Azar, A.T., 2016a. Fuzzy adaptive synchronization of uncertain fractional-order chaotic systems. Advances in Chaos Theory and Intelligent Control. Springer International Publishing, pp. 681–697. Available from: https://doi.org/10.1007%2F978-3-319-30340-6_28.

Boulkroune, A., Hamel, S., Azar, A.T., Vaidyanathan, S., 2016b. Fuzzy control-based function synchronization of unknown chaotic systems with dead-zone input. Advances in Chaos Theory and Intelligent Control. Springer International Publishing, pp. 699–718. Available from: https://doi.org/10.1007%2F978-3-319-30340-6_29.

Debnath, L., 2003. Recent applications of fractional calculus to science and engineering. Int. J. Math. Sci. 2003 (54), 3413–3442. Available from: http://www.hindawi.com/journals/ijmms/2003/753601/abs/.

Debnath, L., 2004. A brief historical introduction to fractional calculus. Int. J. Math. Edu. Sci. Tech. 35 (4), 487–501. Available from: http://www.tandfonline.com/doi/abs/10.1080/0020739041000168657l.

Diethelm, K., 2010. The analysis of fractional differential equations: an application-oriented exposition using differential operators of Caputo type. http://www.amazon.com/Analysis-Fractional-Differential-Equations-Application-Oriented/dp/3642145736/ref = pd{_}sim{_}b{_}7.

Emary, E., Zawbaa, H.M., July 2016. Impact of chaos functions on modern swarm optimizers. PLoS One 11 (7), e0158738. Available from: https://doi.org/10.1371%2Fjournal.pone.0158738.

Gandomi, A.H., Yang, X.S., 2014. Chaotic bat algorithm. J. Comp. Sci. 5 (2), 224–232.

Gu, W., Yu, Y., Hu, W., 2016. Parameter estimation of unknown fractional-order memristor-based chaotic systems by a hybrid artificial bee colony algorithm combined with differential evolution. Nonlinear Dyn. 84 (2), 779–795. Available from: https://doi.org/10.1007/s11071-015-2527-x.

Khateb, F., Kubánek, D., Tsirimokou, G., Psychalinos, C., 2016. Fractional-order filters based on low-voltage DDCCs. Microelectr. J. 50, 50–59. Available from: http://linkinghub.elsevier.com/retrieve/pii/S0026269216000331.

Li, C.-L., Wu, L., 2016. Optik Sliding mode control for synchronization of fractional permanent magnet synchronous motors with finite time. Optik - Int. J. Light Electr. Opt. 127 (6), 3329–3332. Available from: https://doi.org/10.1016/j.ijleo.2015.12.102.

Li, C.-L., Yu, S.-M., Luo, X.-S., 2012. Fractional-order permanent magnet synchronous motor and its adaptive chaotic control. Chin. Phys. B 21 (10), 100506. Available from: https://doi.org/10.1088%2F1674-1056%2F21%2F10%2F100506.

Li, Z., Park, J.B., Joo, Y.H., Zhang, B., Chen, G., 2002. Bifurcations and chaos in a permanent-magnet synchronous motor. IEEE Trans. Circ. Sys. I: Fund. Theory Applicat. 49 (3), 383–387. Available from: https://doi.org/10.1109%2F81.989176.

Liu, B., Wang, L., Jin, Y.-H., Tang, F., Huang, D.-X., 2005. Improved particle swarm optimization combined with chaos. Chaos Solitons Fractals 25 (5), 1261–1271. Available from: https://doi.org/10.1016%2Fj.chaos.2004.11.095.

Liu, K., Zhang, Q., Chen, J., Zhu, Z.Q., Zhang, J., May 2011. Online multiparameter estimation of nonsalient-pole PM synchronous machines with temperature variation tracking. IEEE Trans. Ind. Electron. 58 (5), 1776–1788. Available from: https://doi.org/10.1109%2Ftie.2010.2054055.

Liu, L., Liu, W., Cartes, D.A., 2008. Particle swarm optimization-based parameter identification applied to permanent magnet synchronous motors. Eng. Applicat. Artificial Intelligence 21 (7), 1092–1100.

Miller, K.S., Ross, B., 1993. An introduction to the fractional calculus and fractional differential equations. John Wiley & Sons, New York.

Mirjalili, S., Mirjalili, S.M., Lewis, A., 2014. Grey wolf optimizer. Adv. Eng. Software 69, 46–61. Available from: https://doi.org/10.1016%2Fj.advengsoft.2013.12.007.

Petráš, I., 2011. Fractional-Order Chaotic Systems. Springer, Berlin Heidelberg, pp. 103–184. Available from: https://doi.org/10.1007%2F978-3-642-18101-6_5.

Rahimi, A., Bavafa, F., Aghababaei, S., Khooban, M.H., Naghavi, S.V., 2016. The online parameter identification of chaotic behaviour in permanent magnet synchronous motor by self-adaptive learning bat-inspired algorithm. Int. J. Elect. Power Energy Syst. 78, 285–291. Available from: https://doi.org/10.1016%2Fj.1jepes.2015.11.084.

Rajagopal, K., Vaidhyanathan, S., Karthikeyan, A., Duraisamy, P., 2016. Dynamic analysis and chaos suppression in a fractional order brushless DC motor. Elect. Eng. Available from: https://doi.org/10.1007/s00202-016-0444-8.

Rezaee Jordehi, A., 2015. Chaotic bat swarm optimisation (CBSO). Appl. Soft Comput. J. 26, 523–530.

Saremi, S., Mirjalili, S., Lewis, A., 2014. Biogeography-based optimisation with chaos. Neural Comput. Appl. 25 (5), 1077–1097. Available from: https://doi.org/10.1007%2Fs00521-014-1597-x.

Saremi, S., Mirjalili, S., Lewis, A., 2017. Grasshopper optimisation algorithm: Theory and application. Adv. Eng. Softw. 105, 30–47. Available from: https://doi.org/10.1016%2Fj.advengsoft.2017.01.004.

Sun, Y., Wu, X., Bai, L., Wei, Z., Sun, G., 2016. Finite-time synchronization control and parameter identification of uncertain permanent magnet synchronous motor. Neurocomputing 207, 511–518. Available from: https://doi.org/10.1016%2Fj.neucom.2016.05.036.

Tofighi, E.M., Mahdizadeh, A., Feyzi, M.R., 2013. Real-time estimation and tracking of parameters in permanent magnet synchronous motor using a modified two-stage particle swarm optimization algorithm. 2013 IEEE International Symposium on Sensorless Control for Electrical Drives and Predictive Control of Electrical Drives and Power Electronics (SLED/PRECEDE) (October), 1–7. http://ieeexplore.ieee.org/lpdocs/epic03/wrapper.htm?arnumber = 6684496.

Xu, Y., Zhou, J., Xue, X., Fu, W., Zhu, W., Li, C., 2016. An adaptively fast fuzzy fractional order PID control for pumped storage hydro unit using improved gravitational search algorithm. Energy Convers. Manage. 111, 67–78. Available from: https://doi.org/10.1016/j.enconman.2015.12.049.

Xue, W., Li, Y., Cang, S., Jia, H., Wang, Z., Xue, W., et al., 2015. Chaotic behavior and circuit implementation of a fractional-order permanent magnet synchronous motor model. J. Franklin Inst. Available from: https://doi.org/10.1016/j.jfranklin.2015.05.025.

Yousri, D., AbdelAty, A.M., Said, L.A., AboBakr, A., Radwan, A.G., 2017. Biological inspired optimization algorithms for cole-impedance parameters identification. {AEU} – Int. J. Electr. Commun. 78, 79–89. Available from: http://www.sciencedirect.com/science/article/pii/S143484111730376X.

Yu, J., Chen, B., Yu, H., Gao, J., 2011. Nonlinear Analysis : Real World Applications Adaptive fuzzy tracking control for the chaotic permanent magnet synchronous motor drive system via backstepping. Nonlinear Anal. Real World Appl. 12 (1), 671–681. Available from: https://doi.org/10.1016/j.nonrwa.2010.07.009.

Zamani, A., Barakati, S.M., Youso, S., 2016. Design of a fractional order PID controller using GBMO algorithm for load frequency control with governor saturation consideration.

Zhou, P., Bai, R.-J., Zheng, J.-M., 2015. Stabilization of a fractional-order chaotic brushless DC motor via a single input. Nonlinear Dyn. 82 (1–2), 519–525. Available from: http://link.springer.com/10.1007/s11071-015-2172-4.

Zhu, Q., Azar, A.T. (Eds.), 2015. Complex System Modelling and Control Through Intelligent Soft Computations. Springer International Publishing. Available from: https://doi.org/10.1007%2F978-3-319-12883-2.

Zribi, M., Oteafy, A., Smaoui, N., 2009. Chaos, Solitons and Fractals Controlling chaos in the permanent magnet synchronous motor 41, 1266–1276.

Chapter 19

Control and Synchronization of a Fractional Order Hyperchaotic System via Backstepping and Active Backstepping Approach

Manoj K. Shukla[1,2], Bharat B. Sharma[1] and Ahmad Taher Azar[3,4]

[1]National Institute of Technology, Hamirpur, Himachal Pradesh, India, [2]Lovely Professional University, Punjab, India, [3]Faculty of Computers and Information, Benha University, Benha, Egypt, [4]School of Engineering and Applied Sciences, Nile University, Giza, Egypt

19.1 INTRODUCTION

Many physical phenomena have an intrinsic fractional order description and so fractional order calculus is necessary to replicate their input–output characteristics. Fractional calculus finds wide application in different unrelated topics such as: transmission line theory, chemical analysis of aqueous solutions, viscoelasticity, electromagnetic wave, quantum mechanical calculations, signal processing, robotics, bioengineering, control systems, etc. The incorporation of two degrees of freedom from the use of a fractional order integrator and differentiator provides a greater degree of flexibility and hence makes it possible to further improve the performance of traditional PID controllers. Fractional order calculus allows us to describe and model a real system, more accurately than the classical integer methods, which further enables us to analyze them in a better way and design better controllers. Fractional order chaotic system can be helpful in random number generation and more secure communication. The additional parameters of differintegral orders give more flexibility to the designer. Fractional derivatives provide an excellent instrument for the description of memory and hereditary properties of various materials and processes.

19.1.1 Introduction to Chaotic Systems

Nonlinear systems are very interesting to engineers, physicists, and mathematicians because most real physical systems are inherently nonlinear in

Mathematical Techniques of Fractional Order Systems. DOI: https://doi.org/10.1016/B978-0-12-813592-1.00019-2

nature (Krstic et al., 1995). Nonlinear equations are difficult to be solved by analytical methods and give rise to interesting phenomena such as bifurcation and chaos. Even simple nonlinear (or piecewise linear) dynamical systems can exhibit a completely unpredictable behavior, the so-called deterministic chaos. Chaos theory has been so surprising because chaos can also be found within trivial systems. To be called "chaotic," a system should also show sensitive dependence on initial conditions, in the sense that neighboring orbits separate exponentially fast, on average. Chaos is an aperiodic long-term behavior in a deterministic system that exhibits sensitive dependence on initial conditions (Strogatz, 2014).

There are a number of areas in which chaos finds its applications such as in lasers, biological systems, chemical reactors, power converters, etc. However, chaos is not always desirable, sometimes it has to be regulated to avoid the deteriorating effect on the system (Azar and Vaidyanathan, 2015a, b,c, 2016; Azar and Zhu, 2015; Azar et al., 2017a,b,c). The control and stabilization of chaotic systems have gained enough attention in recent years (Vaidyanathan and Azar, 2015a,b,c,d, 2016a,b,c,d,e,f,g; Azar et al., 2017b; Zhu and Azar, 2015; Lamamra et al., 2017; Wang et al., 2017). Control of a chaotic system can be useful in two ways: to minimize the chaotic behavior; and to bring the system out of chaotic behavior or to increase the extent of chaoticity in nonlinear dynamical systems. Control of chaos by different approaches has been obtained in a variety of systems (Boulkroune et al, 2016a,b; Singh et al., 2017) including weather phenomenon (Lorenz, 1963), turbulent fluids (Lesieur, 2012), oscillating chemical reactions (Field and Schneider, 1989), magneto-elastic oscillators (Feeny et al., 2001), cardiac tissues (Chialvo et al., 1990), etc. Chaos synchronization, on the other hand, occurs when a chaotic system (master system) drives another identical or nonidentical chaotic system (slave system) such that the output of the slave system follows the output of master system, asymptotically. After the pioneering work of Ott et al. (1990) and Pecora et al. (1997), chaos synchronization and control has become a topic of great interest (Boulkroune et al., 2016a,b; Yang et al., 1998; Yassen, 2005; Astakhov et al., 1997; Ouannas et al., 2016a,b, 2017a,b,c,d,e,f,g,h,i,j). Various methods have been proposed until now to achieve the synchronization and stabilization of chaotic systems, including parametric perturbation (Astakhov et al., 1997), sliding mode control (Vaidyanathan et al., 2015a; Vaidyanathan and Azar, 2015a, b), active control Vaidyanathan et al., 2015b), adaptive control (Zhang et al., 2006; Kebriaei and Javad Yazdanpanah, 2010; Grassi et al., 2017; Vaidyanathan and Azar, 2016c,d,e,g; Vaidyanathan et al., 2017b,c), variable structure control, backstepping control (Wang and Ge, 2001; Vaidyanathan et al., 2015c; Vaidyanathan and Azar, 2016f), observer based synchronization (Sharma and Kar, 2011), contraction theory based approach (Sharma and Kar, 2009b), intelligent control (Zhu and Azar, 2015), and so on. Chaos synchronization finds its application in secure communication

(Sharma and Kar, 2009a), robotics (Nakamura and Sekiguchi, 2001), biological implants (Mormann et al., 2000), etc.

19.1.2 Introduction to Fractional Calculus

The story of fractional calculus begins with a question asked by L. Hospital, some 300 years ago, to Leibniz regarding his publication for the nth derivative of a function, i.e., what would the result be if $n = 1/2$. The answer given by Leibniz is now well-known to the research fraternity, i.e., "an apparent paradox from which 1-day useful consequences will be drawn." Fractional calculus is basically a generalization of integer order differentiation and n-fold integration to arbitrary order. Due to the unavailability of solution techniques, fractional calculus has not been explored much for almost 300 years, although some pioneers contributed significantly during the 19th century towards the development of fractional calculus (Samko et al., 1993). Furthermore, a number of researchers have also helped in the advancement of this area (e.g., Hardy and Beier, 1994; Oldham and Spanier, 1974; Srivastava and Owa, 1989; Samko et al., 1993; Miller and Ross, 1993; Podlubny, 1998). In view of contributions made by various researchers, in recent years, various methods for approximation of the fractional derivative and integral came into existence, which further enabled us to use it in wide areas such as bioengineering (West, 2007), groundwater flow problem (Atangana and Bildik, 2013; Atangana and Vermeulen, 2014), diffusion of heat (Angulo et al., 2000), signal processing (Tseng, 2007), robotics (da Graça Marcos et al., 2008), quantum theory (Laskin, 2002), electrical engineering (Carlson and Halijak, 1964; Meghni et al., 2017a,b,c; Soliman et al., 2017; Tolba et al., 2017), etc. Fractional calculus finds wide applicability in control systems too (Tenreiro Machado et al., 2010; Tabari and Kamyad, 2013; Mishra and Chandra, 2014). The application revolves around two pillars: one is of fractional order systems and the other is fractional order control. It has been established that fractional calculus gives a more realistic modeling of systems and the best performance is achieved when a fractional order controller is employed for a fractional order system (Podlubny, 1999). Also, it is confirmed now that the best fractional order controller can be better than the best integer order controller. Enough research has been done on fractional order PID controller and its tuning that fractional order PID controllers (Podlubny, 1994; Petras, 2011) are now being used in real-time applications as their performance is better than the traditional ones.

19.1.3 Introduction to Fractional Order Chaotic Systems

Chaotic systems can also be modeled more accurately by noninteger order differential equations and hence are called fractional order chaotic systems. It is known that chaotic behavior is exhibited by the systems, for system order greater than three, but recently, researchers have established that some fractional order differential systems exhibit chaotic behavior for total order less than three, such as fractional order Chen system (Wang et al., 2006; Li and Peng, 2004), fractional order Lorenz system (Grigorenko and Grigorenko, 2003; Yu et al., 2009). Various other fractional order chaotic systems have also been studied in recent years such as, Chua's circuit (Hartley et al., 1995), Rössler's system (Li and Chen, 2004), Lu system (Lu, 2006; Ouannas et al., 2017a,b,c,d,e,i), etc. Some fractional order hyperchaotic systems have also been analyzed and presented in literature (Yang and Liu, 2013; Liu and Lu, 2010), such as hyperchaotic Rössler's system (Zhang et al., 2005), hyperchaotic Henon map (Shukla and Sharma, 2017e), Lorenz–Stenflo system (Wang et al., 2014), etc. Various techniques mentioned above have also been utilized for control of various classes of chaotic systems such as sliding mode control (Chen et al., 2012; Aghababa, 2012), active control (Shukla and Sharma, 2015; Azar et al., 2017a) etc. Synchronization of integer order systems has been studied extensively and nonlinear control methodologies used for such systems are also extended to synchronize fractional order systems, (Pham et al., 2017; Asheghan et al., 2011; Mohadeszadeh and Delavari, 2017; Wang and Zhang, 2006; Agrawal et al., 2012; Borah et al., 2016).

Backstepping control developed by Krstic et al. (1995) is one of the most popular nonlinear techniques of controller design and has been used intensively for the control and synchronization of integer order nonlinear systems. It employs a systematic approach based on Lyapunov theory where the uncertainty can be linearly parameterized and guarantees global stability of systems. The state variables are treated as independent controllers for subsystems and subsequently each step leads to an updated control law for the following step. The controller algorithm for each step is chosen such that the corresponding Lyapunov function is satisfied which further guarantees the stability of each subsystem. A similar approach can be used for controller design for fractional order nonlinear systems provided there exists some methodology to exploit the existing Lyapunov stability analysis. Extension of Lyapunov theory to fractional order nonlinear systems along with development of Mittag–Leffler stability concept is presented in Li et al. (2010). Use of these stability concepts for design of nonlinear feedback controllers for fractional order nonlinear systems is the topic of interest and it has been exploited in various research works (Wei et al., 2016; Zhao et al., 2016; Ding et al., 2015; Shukla and Sharma, 2017a,b,c,d,e; Wang et al., 2006; Shukla and Sharma, 2018).

Active backstepping approach (Zhang et al., 2005) is a combination of active control and backstepping approach. Unlike backstepping strategy, it involves design of multiple controllers but the step-wise procedure is derived from backstepping approach design methods to control chaos. For a wider class of strict-feedback system, backstepping design guarantees global asymptotic stabilities, tracking, and transient performance. Active control technique is not only employed for strict-feedback systems, but gives the flexibility to control a wide variety of nonlinear systems including chaotic systems.

A number of state feedback controller design algorithms have been proposed in the literature. For example, in Yuan et al. (2013), Chen et al. (2012), Yin et al. (2013, 2012), traditional Lyapunov stability has been used to analyze the system stability, whereas in this work, the stability analysis and consequently the stabilizing and synchronizing controller design is carried out by combining fractional order extension of Lyapunov direct method and Mittag–Leffler stability which offers a more realistic approach for investigating the system stability. The proposed controller enables the states to converge in a smaller time, in comparison with Chen et al. (2012) and Yin et al. (2012). On the other hand, the strategy presented in the current manuscript gives a freedom of choosing systems in which nonlinearity can be there with any state.

Application of fractional order extension of Lyapunov stability and Mittag–Leffler stability for the backstepping based control of fractional order chaotic systems is a less explored area. Also, most of the state feedback controllers use traditional Lyapunov stability concept. This is a motivation to design controllers for fractional order systems using fractional order extension of Lyapunov stability approach. In the previous works (Shukla and Sharma, 2017a,c,d), stabilization of a different class of three-dimensional fractional order chaotic system was addressed. Here, in present work, application of Mittag–Leffler stability criterion along with backstepping control approach is proposed to address the control and synchronization of a hyperchaotic system. The main contribution of this work is application of backstepping strategy and active backstepping strategy to design controllers for stabilization and synchronization of fractional order hyperchaotic Lorenz–Stenflo system. Systematic derivation of stabilizing controller, while exploiting Mittag–Leffler and Lyapunov stability results, is presented initially and later on these results are extended to meet the two-system synchronization in master–slave configuration for addressed hyperchaotic systems.

This chapter is organized as follows. Section 19.2 explains the fundamentals of fractional calculus and stability of fractional order chaotic systems. System description, control, and synchronization strategies are given in Section 19.3. Section 19.4 deals with design of controllers for stabilization and synchronization of two example systems. Detailed numerical simulations results are presented in Section 19.5. The work is concluded in Section 19.6.

19.2 PRELIMINARIES OF FRACTIONAL CALCULUS

Fractional calculus is a generalization of integer order calculus to noninteger order fundamental operator ${}_aD_t^q$ where a and t are the limits of operation and

q is the order which can be any real number. For the purpose of handling fractional calculus, the integro-differential operator is defined as:

$$ {}_aD_t^q = \begin{cases} \dfrac{d^q}{dt^q} & q>0 \\ 1 & q=0 \\ \int_a^t (d\tau)^{-q} & q<0 \end{cases} \tag{19.1} $$

The available literature mainly highlights three prominent definitions which are used for general fractional order derivative or integral and these are Grunwald–Letnikov (GL), Riemann–Liouville (RL), and the Caputo definition. These popular definitions are elaborated as follows:

A. *Grunwald–Letnikov Definition*

$$ {}_a^{GL}D_t^q f(t) = \lim_{h\to 0} \frac{1}{h^q} \sum_{j=0}^{\infty} (-1)^j \binom{q}{j} f(t-jh) \tag{19.2} $$

B. *Riemann–Liouville Definition*

For integral:

$$ \mathcal{J}^q f(t) \triangleq \frac{1}{\Gamma(q)} \int_0^t (t-\tau)^{q-1} f(\tau) d\tau \tag{19.3} $$

For Derivative:

$$ {}_a^{RL}D_t^q f(t) = \frac{1}{\Gamma(m-q)} \frac{d^m}{dt^m} \int_a^t \frac{f(\tau)}{(t-\tau)^{q-m+1}} d\tau $$

where, $\Gamma(m) = (m-1)!$, $t>0, q \in \mathbb{R}^+ m-1<q<m$

C. *Caputo Definition (Derivative)*

$$ {}_a^{C}D_t^q f(t) = \frac{1}{\Gamma(m-q)} \int_0^t \frac{f^m(\tau)}{(t-\tau)^{q-m+1}} f(\tau) d\tau \tag{19.4} $$

In the case of fractional order control, the above three definitions are used commonly. GL definition gives slightly inaccurate results during the initial phase of the simulation. RL definition is basically for fractional order integrations and it cannot be directly used for fractional order differentiation. The most suitable definition out of the abovementioned definitions is the Caputo definition as in this case the initial conditions for fractional order differential equations (FODE) are similar to those for integer order differential equations. Under the homogenous initial conditions, the RL and the Caputo derivatives are equivalent. The Caputo definition is related to RL definition as,

$$ {}_a^{RL}D_t^q f(t) = {}_a^{C}D_t^q f(t) + \sum_{k=0}^{m-1} \frac{(t-\tau)^{k-q}}{\Gamma(k-q+1)} f^k(a) $$

The initial conditions are defined as, $f^k(a)=0,(k=0,1,\ldots,n-1)$ and $m-1<q<m$.

Integer order derivatives and integrals can be very clearly interpreted geometrically and physically. But, the physical and geometric significance of fractional calculus is not straightforward and is a topic of research. With the help of the concept of transformed time and projection of integration on different axes, Podlubny (2002) has made an effort to explain the physical and geometrical significance of fractional calculus. Some basic Podlubny (2002) properties of fractional calculus are given as follows:

1. For $\alpha=n$, where n is integer, the operation $D^\alpha f(t)$ gives the same result as classical differentiation of integer order n.
2. For $\alpha=0$ the operation $D^\alpha f(t)$ is the identity operator: $D^0 f(t)=f(t)$.
3. Fractional differentiation and fractional integration are linear operation:
$$D^\alpha(\lambda f(t)+\mu g(t))=\lambda D^\alpha f(t)+\mu D^\alpha g(t).$$
4. The additive index law $D^\alpha D^\beta f(t)=D^\beta D^\alpha f(t)=D^{\alpha+\beta}f(t)$ holds under some reasonable constraints on the function $f(t)$.
5. The Leibniz's rule for fractional differentiation is given as:
$$D^\alpha(\phi(t)f(t))=\sum_{k=0}^{\infty}\binom{\alpha}{k}\phi^{(k)}(t)D^{\alpha-k}f(t).$$
6. The Laplace transform of fractional order derivative is defined as:
$$\mathcal{L}\left[{}_aD_t^a f(t)\right]=s^\alpha\mathcal{L}[f(t)]-\sum_{k=1}^{n-1}s^k\left[{}_aD_t^{a-k-1}f(t)\right]_{t=0},\quad n-1<\alpha<n.$$

19.2.1 Numerical Solution of Fractional Order Differential Equations

A number of approximation techniques have been proposed in literature for the solution of FODEs. One of the common numerical methods for the solution of nonlinear FODEs has been derived from the GL definition (Vinagre et al., 2003; Dorcak, 1994; Petras, 2011). The explicit numerical approximation of the qth order derivative at the points kh $(k=1,2,\ldots)$ is given as:

$$_{k-L_m/h}D_{kh}^q f(t)\approx h^{-q}\sum_{j=0}^{k}(-1)^j\binom{q}{j}f(t_{k-j})=h^{-q}\sum_{j=0}^{k}c_j^{(q)}f(t_{k-j})\qquad(19.5)$$

where, $L_m=$ 'memory length', $h=$ time step of calculation. $c_j^{(q)}$ $(j=0,1,\ldots,k)$ are the binomial coefficients and are given as

$$c_0^{(q)}=1,c_j^{(q)}=\left(1-\frac{1+q}{j}\right)c_{j-1}^{(q)}\qquad(19.6)$$

Employing the above two definitions, the general numerical solution of the nonlinear FODE of form,

$$ {}_aD_t^q y(t) = f(y(t), t) $$

can be expressed by using following description:

$$ y(t_k) = f(y(t_k), t_k)h^q - \sum_{j=1}^{k} c_j^{(q)} y(t_{k-j}) \tag{19.7} $$

In the present work, the above description has been used for solving FODE and also for the simulation of the proposed results on computer.

19.2.2 Stability of Fractional Order Nonlinear Dynamical System

Owing to increasing application of fractional order nonlinear dynamical systems, their stability analysis has attracted a great amount of interest from the research community. Although, the stability analysis is not as simple as it is for their linear counterpart, still some development has been made in this regard. Matignon (1996; Petras, 2011) has shown up that one cannot use exponential stability to analyze asymptotic stability of fractional order systems. A definition given by Tavazoei and Haeri which uses matrix properties is given below.

Definition 1: (Tavazoei and Haeri, 2007): (**For commensurate order systems**): The equilibrium points are asymptotically stable for $q_1 = q_2 = \ldots = q_n = q$ if all the eigen values $\lambda_i (i = 1, 2, \ldots, n)$ of the Jacobian matrix $J = \partial f / \partial x$, where $f = [f_1, f_2, \ldots f_n]^T$, evaluated at equilibrium E^*, satisfy the condition

$$ \left|arg(eig(J))\right| = \left|arg(\lambda_i)\right| > q\frac{\pi}{2}, i = 1, 2 \ldots, n $$

Definition 2: (Tavazoei and Haeri, 2008): (**For incommensurate order systems**): When incommensurate fractional order system is considered $q_1 \neq q_2 \neq \ldots \neq q_n$ and suppose that m is the LCM of the denominators u_i's of q_i's, where $q_i = \frac{v_i}{u_i}$, $v_i, u_i \in \mathbb{Z}^+$, for $i = 1, 2, \ldots, n$ and set $\gamma = 1/m$. The system is asymptotically stable if

$$ \left|arg(\lambda)\right| > \gamma\frac{\pi}{2} $$

For all the roots λ of the following equation:

$$ det(diag([\lambda^{mq_1} \lambda^{mq_2} \ldots \lambda^{mq_n}]) - J) = 0 $$

Mittag–Leffler stability and extension of Lyapunov direct method for fractional order nonlinear systems is given by Li et al. (2010). Mittag–Leffler stability is given by following theorem.

Theorem 1: (Li et al., 2010) The solution of following fractional order non-autonomous system

$$D_t^{\alpha} x(t) = f(t, x) \tag{19.8}$$

is said to be Mittag–Leffler stable if

$$\|x(t)\| \leq \{m[x(t_0)]E_{\alpha}(-\lambda(t-t_0)^{\alpha})\}^{b} \tag{19.9}$$

where t_0 is the initial time, $\alpha \in (0, 1)$, $\lambda > 0$, $b > 0$, $m(0) = 0$, $m(x) \geq 0$ and $m(x)$ is locally lipschitz on $x \in \mathbb{B} \in \mathbb{R}^n$ with Lipschitz constant m_0.

In the above theorem, a function E_{α} has been used. This function is called Mittag–Leffler function (Mathai and Haubold, 2008) and has significant importance in fractional calculus. It was proposed by Agarwal and Humbert in 1953. Mittag–Leffler function in two parameter form can be defined as:

$$E_{\alpha,\beta}(z) = \sum_{k=0}^{\infty} \frac{z^k}{\Gamma(\alpha k + \beta)}$$

where, $z \in \mathbb{C}, \alpha, \beta \in \mathbb{C}$ and $\Re(\alpha) > 0$

Also for $\alpha = \beta = 1$, the function reduces to conventional exponential function, $E_{1,1}(z) = e^z$

For $\beta = 1$, Mittag–Leffler function can be defined in one parameter form as follows:

$$E_{\alpha,1}(z) = \sum_{k-0}^{\infty} \frac{z^k}{\Gamma(\alpha k + 1)} \equiv E_{\alpha}(z)$$

A number of researchers have tried to extend Lyapunov stability criterion to fractional order systems. The following theorem can be stated for the same.

Theorem 2: (Li et al., 2010) Let $x = 0$ be an equilibrium point for the non-autonomous fractional order system (19.8). Assume that there exists a Lyapunov function $V(t, x(t))$ and class $\mathcal{K}$ functions $\gamma_i (i = 1, 2, 3)$ satisfying

$$\gamma_1 \|x\| \leq V(t, x(t)) \leq \gamma_2 \|x\| \tag{19.10}$$

$$D_t^{\alpha} V(t, x(t)) \leq -\gamma_3 \|x\| \tag{19.11}$$

where $t \geq 0$, $\alpha \in (0, 1)$. Then $x = 0$ is Mittag–Leffler stable, asymptotically.

To use the above theorems in our problem, the following lemma and theorem are described here.

Lemma 1: (Ding et al., 2015) **Power law for fractional order derivative:** Let $x(t) \in \mathbb{R}$ be a real continuously differentiable function. Then for any $p = 2^n$, $n \in N$,

$$D_t^{\alpha} x^p(t) \leq p x^{p-1}(t) D_t^{\alpha} x(t) \tag{19.12}$$

where $0 < \alpha < 1$ is the fractional order.

From the above lemma, the following corollary is concluded:

Corollary 1: (Aguila-Camacho et al., 2014) Let $x(t) \in \mathbb{R}$ *be* a continuous and derivable function. Then, for any time t

$$\frac{1}{2} D_t^{\alpha} x^2(t) \leq x(t) D_t^{\alpha} x(t), \forall \alpha \in (0, 1) \tag{19.13}$$

Theorem 3: (Ding et al., 2015) The fractional order system (19.8), with control law $u = \beta(x)$ *is* stable if for a $p = 2^n$, $n \in N$,

$$x^{p-1} D_t^{\alpha} x(t) = x^{p-1} f(x, \beta(x)) \leq 0 \tag{19.14}$$

and the system with $u = \beta(x)$ is asymptotically Mittag–Leffler stable if $x^{p-1} f(x, \beta(x)) < 0$

The conclusion can be summarized now in the form of a theorem as given below:

Theorem 4: (Ding et al., 2015) The stability of the system (19.8) with a controller defined as $u = \beta(x)$ is guaranteed by the Lyapunov function $V(t, x(t))$ *a*nd its fractional order derivative if following condition is met, where γ is class $\mathcal{K}$ function.

$$D_t^{\alpha} V(t, x(t)) \leq -\gamma g(x, \beta(x)),$$

also if

$$D_t^{\alpha} V(t, x(t)) < -\gamma g(x, \beta(x)),$$

then it can be concluded that, the corresponding controller u stabilizes the system globally and asymptotically.

19.3 SYSTEM DESCRIPTION

The fractional order hyperchaotic system which is under focus in this chapter is Lorenz–Stenflo system. A hyperchaotic system generally exhibits chaotic behavior with at least two positive Lyapunov exponents. Initially, let us have a brief description of traditional (integer order) Lorenz–Stenflo system. The integer order version of this system was first proposed by Stenflo (1996) in

1996. It is basically an extended version of the famous Lorenz system. Further investigations and analysis were made in Zhou et al. (1997). The integer order Lorenz–Stenflo system can be described by following expressions.

$$\begin{aligned}\dot{x}(t) &= ay(t) - ax(t) + dw(t)\\ \dot{y}(t) &= cx(t) - y(t) - x(t)z(t)\\ \dot{z}(t) &= x(t)y(t) - bz(t)\\ \dot{w}(t) &= -x(t) - aw(t)\end{aligned} \tag{19.15}$$

where x, y, z, and w are the system states and a, b, c, and d are the system parameters. The hyperchaotic behavior of the above system is shown in Fig. 19.1. The system parameters are taken as $a=1$, $b=0.5$, $c=26$, and $d=1.5$. The simulation has been carried out for 200 seconds for the initial conditions $(x(0), y(0), z(0), w(0)) = (-0.592, 0.04, 0.72, 0.5)$.

The dynamic analysis of fractional order version of the hyperchaotic Lorenz–Stenflo system was first discussed in Wang et al. (2014). System dynamics of fractional order Lorenz–Stenflo system are as follows:

$$\begin{aligned}D_t^{q_1}x(t) &= ay(t) - ax(t) + dw(t)\\ D_t^{q_2}y(t) &= cx(t) - y(t) - x(t)z(t)\\ D_t^{q_3}z(t) &= x(t)y(t) - bz(t)\\ D_t^{(q_4)}w(t) &= -x(t) - aw(t)\end{aligned} \tag{19.16}$$

where q_1, q_2, q_3, and q_4 are the orders of the fractional derivatives. The hyperchaotic behavior of the above system is depicted in Fig. 19.2 for the system parameters as $a=1$, $b=0.5$, $c=26$, and $d=1.5$. The order of derivatives is taken as $q_1=q_2=q_3=q_4=0.98$ and the simulation time is 200 seconds. The initial conditions are chosen as $(x(0), y(0), z(0), w(0)) = (1, 0.2, 0.2, -0.2)$.

19.4 CONTROL AND SYNCHRONIZATION VIA BACKSTEPPING TECHNIQUE

As the dynamics of the fractional order Lorenz–Stenflo system given in (19.16), are not in strict-feedback form, a simple transformation has to be carried out. For, $w=x_1$, $x=x_2$, $y=x_3$, $z=x_4$ and $q_1=q_2=q_3=q_4=q$, the system is transformed to:

$$\begin{aligned}D_t^q x_1 &= -x_2 - ax_1\\ D_t^q x_2 &= ax_3 - ax_2 + dx_1\\ D_t^q x_3 &= cx_2 - x_2x_4 - x_3 + u\\ D_t^q x_4 &= x_2x_3 - bx_4\end{aligned} \tag{19.17}$$

where, u is the controller to be designed. As the above system is in the desired form, two different strategies can be applied for control and synchronization. In this section, the backstepping strategy is presented and in the next section active backstepping technique will be described.

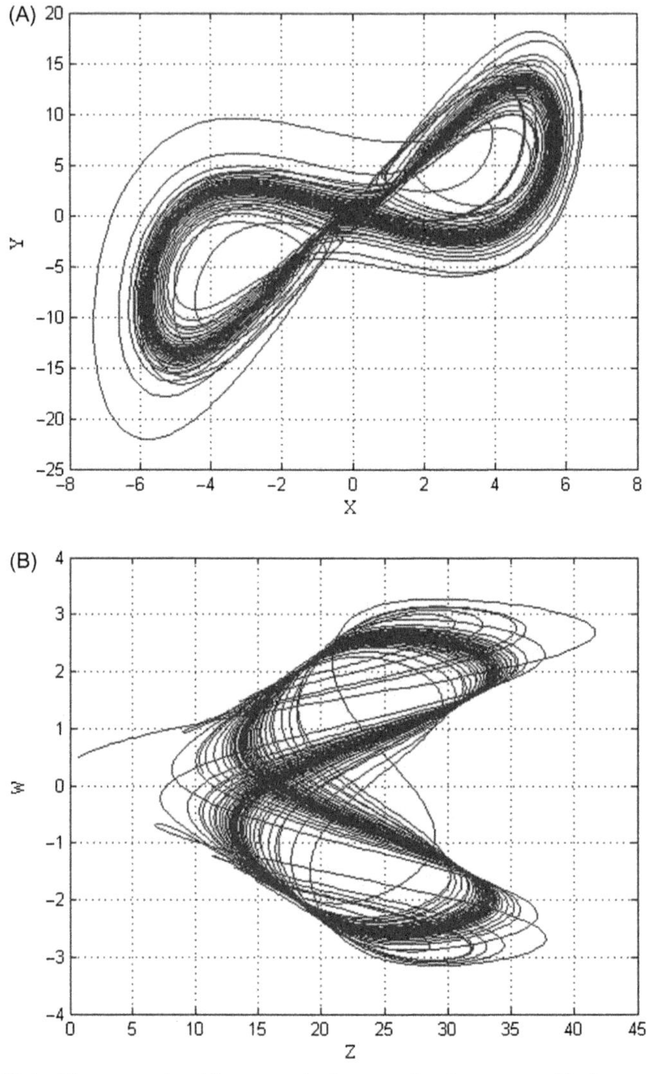

FIGURE 19.1 Phase portrait of integer order Lorenz–Stenflo system: (A) for states x and y, (B) for states z and w, (C) for states y and z, (D) for states z, w, and x.

19.4.1 Controller Design for Stabilization

For the system (19.17), let $z_1 = x_1$ and $z_2 = x_2 - \alpha_1$, where α_1 is the virtual controller for the first subsystem. It gives,

$$D^q z_1 = -(z_2 + \alpha_1) - a x_1 \tag{19.18}$$

The Lyapunov function can be chosen as below for the subsystem in (19.18),

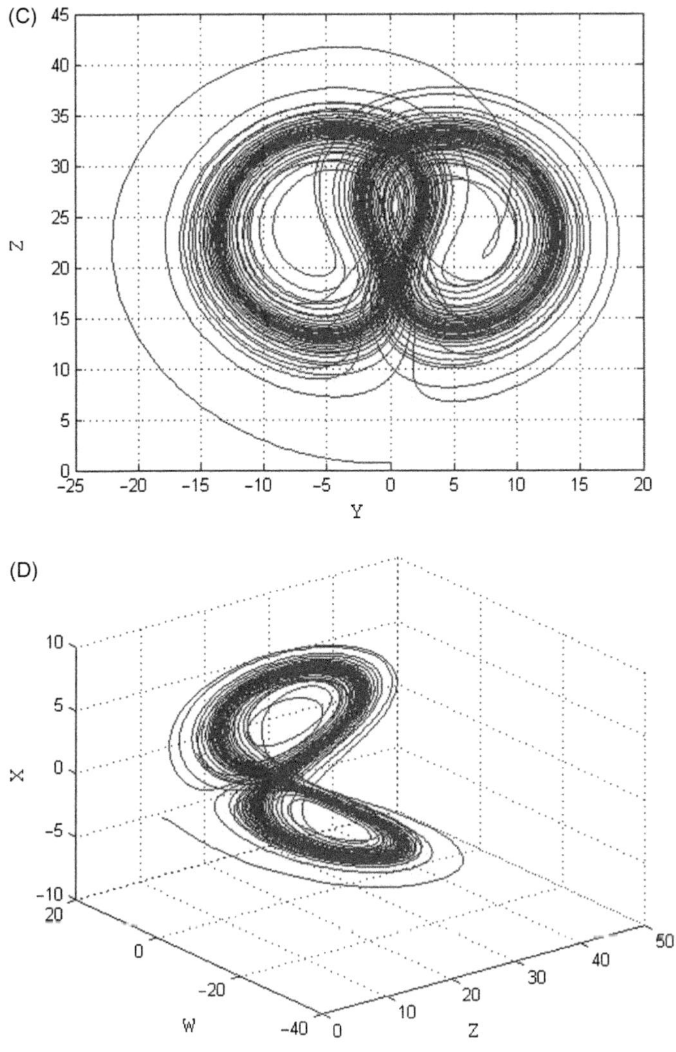

FIGURE 19.1 (Continued).

$$V_1 = \frac{1}{2} z_1^2$$

The fractional order derivative of V_1 is expressed as

$$D^q V_1 \leq z_1(-(z_2 + \alpha_1) - a x_1)$$

With the choice of the virtual controller as: $\alpha_1 = 0$, the above function is reduced to,

$$D^q V_1 \leq -a z_1^2 - z_1 z_2$$

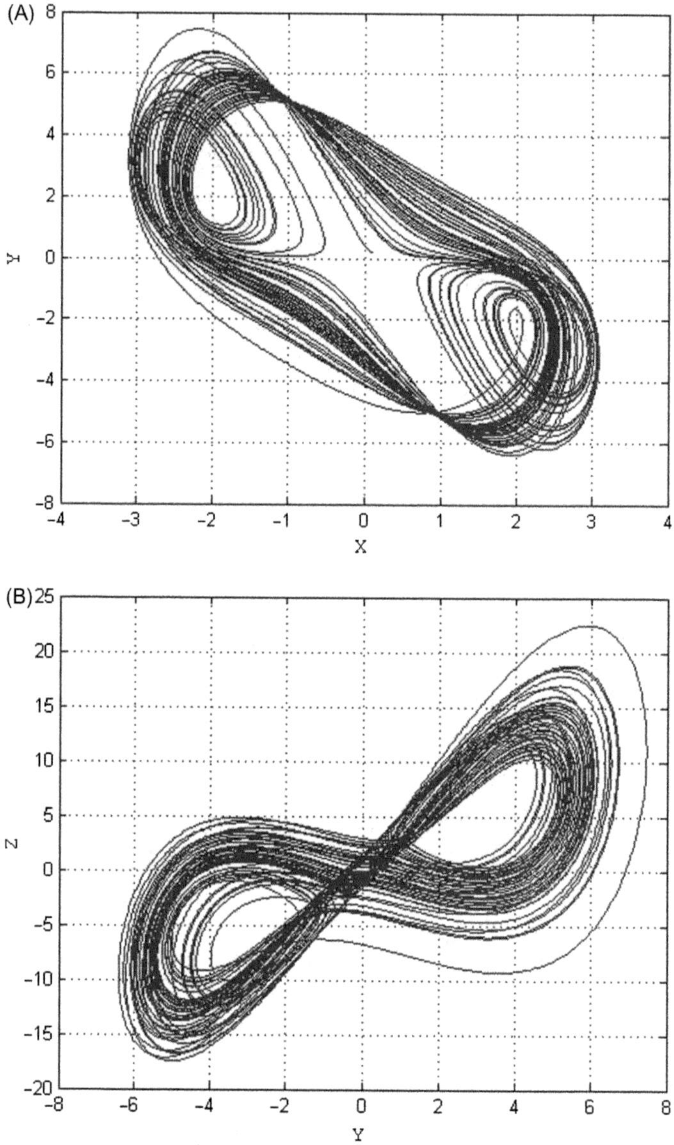

FIGURE 19.2 Phase portrait of fractional order Lorenz–Stenflo system: (A) for states x and y, (B) for states z and y, (C) for states x and w, (D) for states z, y, and x.

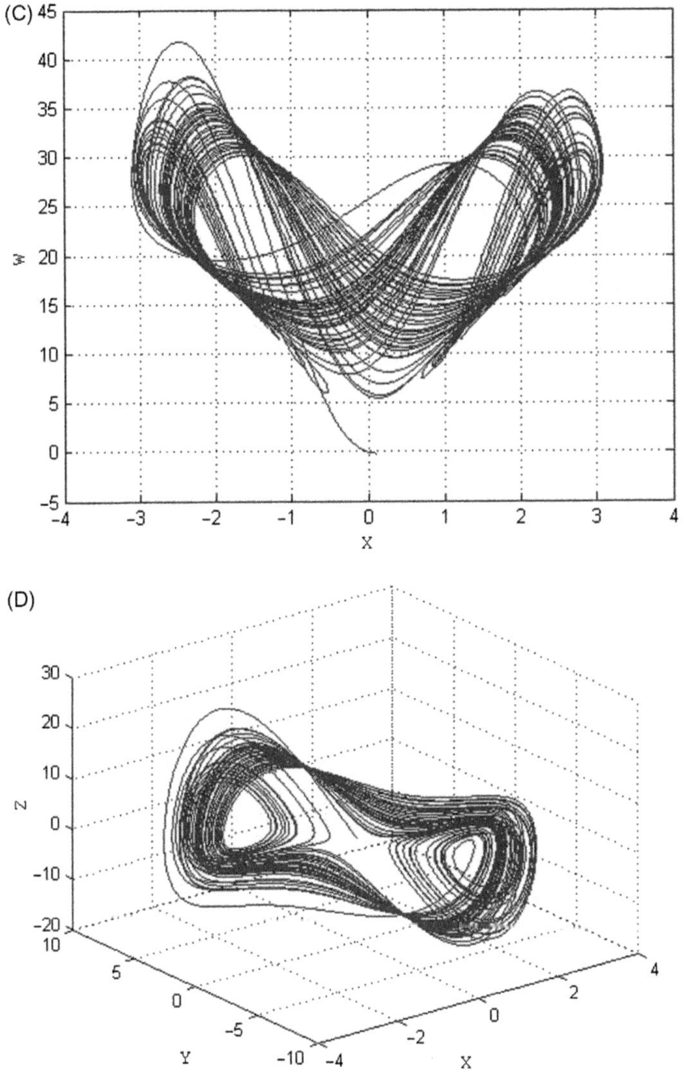

FIGURE 19.2 (Continued).

With the application of Theorem 4, the above expression ensures the asymptotic stability of the system up to this stage. Similarly, for $z_3 = x_3 - \alpha_2$, the derivative of z_2 is expressed as

$$D^q z_2 = a(z_3 + \alpha_2) + dx_1 - ax_2 \tag{19.19}$$

The combined Lyapunov function for the subsystems (19.18) and (19.19) will be:

$$V_2 = V_1 + \frac{1}{2}z_2^2$$

$$\Rightarrow D^q V_2 \leq -az_1^2 - z_1 z_2 + z_2 D^q z_2$$
$$\leq -az_1^2 - az_2^2 - z_1 z_2 + z_2\{a(z_3 + \alpha_2) + dz_1\}$$

Again, virtual controller α_2 is chosen as,

$$\alpha_2 = \frac{1}{a}(1-d)z_1$$

which results into

$$D^q V_2 \leq -az_1^2 - az_2^2 + az_2 z_3$$

Further, stepping back towards the third subsystem in (19.17), one can write,

$$D^q z_3 = cx_2 - x_2 x_4 - x_3 + u - D^q \alpha_2 \tag{19.20}$$

The Lyapunov function for (z_1, z_2, z_3) subsystem defined by (19.18), (19.19) and (19.20), is chosen as:

$$V_3 = V_2 + \frac{1}{2}z_3^2$$

and its fractional order time derivative is given by,

$$D^q V_3 \leq -az_1^2 - az_2^2 + az_2 z_3 + z_3(cx_2 - x_2 x_4 - x_3 + u - D^q \alpha_2)$$

The final control law can be chosen as,

$$u = -kz_3 - cx_2 + x_2 x_4 + x_3 - az_2 + D^q \alpha_2 \tag{19.21}$$

which leads to,

$$D^q V_3 \leq -az_1^2 - az_2^2 - kz_3^2$$

From the above two equations, it's concluded that the system is stable and the transformed states $[z_1 \quad z_2 \quad z_3 \quad z_4]^T$ converge to zero as $t \to \infty$. Using the definition in Theorem 4, the asymptotic stabilization of state vector $[x_1 \quad x_2 \quad x_3 \quad x_4]^T$ can be ensured.

19.4.2 Controller Design for Synchronization

Here, a controller is designed using the backstepping approach for synchronization of fractional order Lorenz–Stenflo system in a master–slave configuration. First, a master and slave dynamics are defined and then error dynamics are written. Further, controller is designed which enables the states of the slave system to follow the states of the master system. The master system can be defined as,

$$\begin{aligned} D_t^q x_1 &= -x_2 - ax_1 \\ D_t^q x_2 &= ax_3 - ax_2 + dx_1 \\ D_t^q x_3 &= cx_2 - x_2x_4 - x_3 \\ D_t^q x_4 &= x_2x_3 - bx_4 \end{aligned} \tag{19.22}$$

Similarly, the slave system can be written as,

$$\begin{aligned} D_t^q y_1 &= -y_2 - ay_1 \\ D_t^q y_2 &= ay_3 - ay_2 + dy_1 \\ D_t^q y_3 &= cy_2 - y_2y_4 - y_3 + u \\ D_t^q y_4 &= y_2y_3 - by_4 \end{aligned} \tag{19.23}$$

For the errors defined as, $e_1 = y_1 - x_1$, $e_2 = y_2 - x_2$, $e_3 = y_3 - x_3$, and $e_4 = y_4 - x_4$, the error dynamics are written as:

$$\begin{aligned} D_t^{q_1} e_1 &= -e_2 - ae_1 \\ D_t^{q_2} e_2 &= ae_3 - ae_2 + de_1 \\ D_t^{q_3} e_3 &= ce_2 - e_2e_4 - e_3 - e_2x_4 - x_2e_4 + u \\ D_t^q e_4 &= e_2e_3 - be_4 + x_3e_2 + x_2e_3 \end{aligned} \tag{19.24}$$

With the transformation, $e_1 = z_1$, $z_2 = e_2 - \alpha_1$, where e_2 as the controller and α_1 as the virtual controller for the first subsystem in (19.24), yields,

$$D^q z_1 = -(z_2 + \alpha_1) - az_1 \tag{19.25}$$

Lyapunov function for the above dynamics can be written as,

$$V_1 = \frac{1}{2} z_1^2$$

$$\Rightarrow D^q V_1 \leq z_1(-(z_2 + \alpha_1) - az_1)$$

For, $\alpha_1 = 0$;

$$D^q V_1 \leq -az_1^2 - z_1z_2$$

Next for the second subsystem, and using the transformation, $z_3 = e_3 - \alpha_2$,

$$D^q z_2 = a(z_3 + \alpha_2) - ae_2 + de_1 \tag{19.26}$$

The new Lyapunov function can be written as,

$$V_2 = V_1 + \frac{1}{2} z_2^2$$

$$\Rightarrow D^q V_2 \leq -az_1^2 - z_1z_2 + z_2\mathrm{a}(z_3 + \alpha_2) - \mathrm{a}e_2 + \mathrm{d}e_1$$

Virtual controller α_2 can be chosen as,

$$\alpha_2 = \frac{1}{a}(1-d)z_1 \tag{19.27}$$

which results in,

$$D^q V_2 \leq -az_1^2 - az_2^2 + az_2 z_3$$

Finally, from the third equation,

$$D^q z_3 = ce_2 - e_2 e_4 - e_3 - e_2 x_4 - x_2 e_4 + u - D^q \alpha_2$$

And the final Lyapunov function is:

$$V_3 = V_2 + \frac{1}{2} z_3^2$$

$$\Rightarrow D^q V_3 \leq -az_1^2 - az_2^2 + az_2 z_3 + z_3\{ce_2 - e_2 e_4 - e_3 - e_2 x_4 - x_2 e_4 + u - D^q \alpha_2\}$$

The final controller is:

$$u = -kz_3 - (c+a)e_2 + e_2 e_4 + e_3 + e_2 x_4 + x_2 e_4 + D^q \alpha_2 \tag{19.28}$$

which gives

$$D^q V_3 \leq -az_1^2 - az_2^2 - kz_3^2$$

The above expression leads to convergence of transformed states $[z_1 \quad z_2 \quad z_3 \quad z_4]^T$ to zero as $t \to \infty$. This further ensures asymptotic convergence of errors $[e_1 \quad e_2 \quad e_3 \quad e_4]^T$ to zero as proved in the previous section. Convergence of state errors leads to asymptotic synchronization of states of slave system in (19.23) to the states of master system in (19.22), i.e., $\lim_{t\to\infty} [x_1 \quad x_2 \quad x_3 \quad x_4]^T = [y_1 \quad y_2 \quad y_3 \quad y_4]^T$.

19.5 CONTROL AND SYNCHRONIZATION BY USING ACTIVE BACKSTEPPING TECHNIQUE

In this section, controller design is proposed for the stabilization and synchronization of a fractional order Lorenz–Stenflo system by employing an active backstepping strategy which is a combination of backstepping technique and active control technique.

19.5.1 Controller Design for Stabilization

Foremost, an appropriate controller is designed to drive the system to stable state (an unstable orbit or an equilibrium point). Controlled system can be expressed as,

$$\begin{aligned}
D_t^q x_1 &= -x_2 - ax_1 + u_1 \\
D_t^q x_2 &= ax_3 - ax_2 + dx_1 + u_2 \\
D_t^q x_3 &= cx_2 - x_2x_4 - x_3 + u_3 \\
D_t^q x_4 &= x_2x_3 - bx_4 + u_4
\end{aligned} \tag{19.29}$$

where, $u_i(i = 1, 2, 3, 4)$, are the controllers to be designed. The target periodic orbit of the system can be defined as, $\tilde{x} = (\tilde{x}_1, \tilde{x}_2, \tilde{x}_3, \tilde{x}_4)^T$, which is a solution of the system dynamics. Hence,

$$\begin{aligned}
D_t^q \tilde{x}_1 &= -\tilde{x}_2 - a\tilde{x}_1 \\
D_t^q \tilde{x}_2 &= a\tilde{x}_3 - a\tilde{x}_2 + d\tilde{x}_1 \\
D_t^q \tilde{x}_3 &= -c\tilde{x}_2 - \tilde{x}_2\tilde{x}_4 - \tilde{x}_3 \\
D_t^q \tilde{x}_4 &= \tilde{x}_2\tilde{x}_3 - b\tilde{x}_4
\end{aligned} \tag{19.30}$$

Now the control objective is to select appropriate control functions such that the trajectory of the controlled system asymptotically approaches the target periodic orbit $\tilde{x}$, in the sense that

$$\lim_{t\to\infty} \|x - \tilde{x}\| = 0$$

where, $x = (x_1, x_2, x_3, x_4)^T$ and $\|\cdot\|$ is the Euclidean norm. For the errors defined as, $e_1 = x_1 - \tilde{x}_1$, $e_2 = x_2 - \tilde{x}_2$, $e_3 = x_3 - \tilde{x}_3$ and $e_4 = x_4 - \tilde{x}_4$, the error dynamics are written as:

$$\begin{aligned}
D_t^{q_1} e_1 &= -e_2 - ae_1 + u_1 \\
D_t^{q_2} e_2 &= ae_3 - ae_2 + de_1 + u_2 \\
D_t^{q_3} e_3 &= ce_2 - e_2e_4 - \tilde{x}_4e_2 - e_4\tilde{x}_2 - e_3 + u_3 \\
D_t^q e_4 &= e_2e_3 + \tilde{x}_3e_2 + \tilde{x}_2e_3 - be_4 + u_4
\end{aligned} \tag{19.31}$$

Backstepping technique can be applied to design active controllers for the above error dynamics. With the transformation, $e_1 = z_1$, $z_2 = e_2 - \alpha_1$, where e_2 as the controller and α_1 as the virtual controller for the first subsystem in (19.31),

$$\begin{aligned}
D^q z_1 &= D^q e_1 = -e_2 - ae_1 + u_1 \\
\Rightarrow D^q z_1 &= -(z_2 + \alpha_1) - az_1 + u_1
\end{aligned} \tag{19.32}$$

Lyapunov function for the above dynamic can be written as:

$$V_1 = \frac{1}{2} z_1^2$$

$$\Rightarrow D^q V_1 \le z_1(-(z_2 + \alpha_1) - az_1 + u_1)$$

For, $u_1 = 0$ and $\alpha_1 = 0$;

$$D^q V_1 \le -az_1^2 - z_1 z_2$$

The above expression ensures the stability of the system up to this stage. Next, for the second subsystem,

$$D^q z_2 = ae_3 - ae_2 + de_1 + u_2$$

Here, $e_3 = \alpha_2$, is the virtual controller

$$D^q z_2 = a\alpha_2 - az_2 + dz_1 + u_2 \tag{19.33}$$

The new Lyapunov function for the above system can be written as,

$$V_2 = V_1 + \frac{1}{2}z_2^2$$

$$\begin{aligned}\Rightarrow D^q V_2 &\leq -az_1^2 - z_1z_2 + z_2 D^q z_2 \\ &\leq -az_1^2 - z_1z_2 + z_2\{a\alpha_2 - az_2 + dz_1 + u_2\}\end{aligned}$$

For $\alpha_2 = 0$ and $u_2 = z_1 - dz_1$,

$$D^q V_2 \leq -az_1^2 - az_2^2$$

The above expression satisfies the stability condition. Next, for $z_3 = e_3 - \alpha_2$,

$$D^q z_3 = ce_2 - e_2e_4 - \tilde{x}_4 e_2 - e_4\tilde{x}_2 - e_3 + u_3$$

After substitution,

$$\Rightarrow D^q z_3 = cz_2 - z_2e_4 - \tilde{x}_4 z_2 - e_4\tilde{x}_2 - z_3 + u_3 \tag{19.34}$$

and the Lyapunov function up to this stage is:

$$V_3 = V_2 + \frac{1}{2}z_3^2$$

$$\Rightarrow D^q V_3 \leq -az_1^2 - az_2^2 + z_3 D^q z_3$$

Now, for $e_4 = \alpha_3$, as the virtual controller,

$$D^q V_3 \leq -az_1^2 - az_2^2 + z_3[cz_2 - z_2e_4 - \tilde{x}_4 z_2 - e_4\tilde{x}_2 - z_3 + u_3]$$

Keeping $\alpha_3 = 0$, the controller can be chosen as:

$$u_3 = -cz_2 + \tilde{x}_4 z_2$$

which leads to

$$D^q V_3 \leq -az_1^2 - az_2^2 - z_3^2$$

Next from the last equation with $z_4 = e_4 - \alpha_3$

$$\begin{aligned}D^q z_4 &= e_2e_3 + \tilde{x}_3 e_2 + \tilde{x}_2 e_3 - be_4 + u_4 \\ \Rightarrow D^q z_4 &= z_2z_3 + \tilde{x}_3 z_2 + \tilde{x}_2 z_3 - be_4 + u_4\end{aligned} \tag{19.35}$$

Lyapunov function will be:

$$V_4 = V_3 + \frac{1}{2}z_4^2$$

$$\Rightarrow D^q V_4 \leq -az_1^2 - az_2^2 - z_3^2 + z_4(z_2z_3 + \tilde{x}_3 z_2 + \tilde{x}_2 z_3 - be_4 + u_4)$$

The controller shall be chosen as,

$$u_4 = -z_2z_3 - \tilde{x}_3z_2 - \tilde{x}_2z_3 \tag{19.36}$$

The stability condition will be satisfied

$$D^qV_4 \leq -az_1^2 - az_2^2 - z_3^2 - bz_4^2$$

As the stability conditions are satisfied, with the application of theorem 4, one can conclude that transformation variables $[z_1 \quad z_2 \quad z_3 \quad z_4]^T$ approach to zero as $t \to \infty$. This further ensures asymptotic convergence of errors $[e_1 \quad e_2 \quad e_3 \quad e_4]^T$ to zero which further implies that the trajectory of the controlled system (19.29) asymptotically approaches the target periodic orbit $\tilde{x}$ which can be an unstable periodic orbit or an equilibrium point.

19.5.2 Synchronization Using Active Backstepping

In this subsection, an active backstepping approach is applied to design a controller to achieve the synchronization of a fractional order Lorenz–Stenflo system in a master–slave configuration. The master system can be defined as,

$$\begin{aligned} D_t^q x_1 &= -x_2 - ax_1 \\ D_t^q x_2 &= ax_3 - ax_2 + dx_1 \\ D_t^q x_3 &= cx_2 - x_2x_4 - x_3 \\ D_t^q x_4 &= x_2x_3 - bx_4 \end{aligned} \tag{19.37}$$

The corresponding slave system can be described as,

$$\begin{aligned} D_t^q y_1 &= -y_2 - ay_1 + u_1 \\ D_t^q y_2 &= ay_3 - ay_2 + dy_1 + u_2 \\ D_t^q y_3 &= cy_2 - y_2y_4 - y_3 + u_3 \\ D_t^q y_4 &= y_2y_3 - by_4 + u_4 \end{aligned} \tag{19.38}$$

Control objective is to design the control functions $u_i (i = 1, 2, 3, 4)$, such that the states of the slave system asymptotically approaches the states of master system, i.e.,

$$\lim_{t \to \infty} \|y - x\| = 0$$

where, $x = (x_1, x_2, x_3, x_4)^T$, $y = (y_1, y_2, y_3, y_4)^T$, and $\|\cdot\|$ is the Euclidean norm. For the errors defined as, $e_1 = y_1 - x_1$, $e_2 = y_2 - x_2$, $e_3 = y_3 - x_3$, and $e_4 = y_4 - x_4$, the error dynamics are written as:

$$\begin{aligned} D_t^{q_1} e_1 &= -e_2 - ae_1 + u_1 \\ D_t^{q_2} e_2 &= ae_3 - ae_2 + de_1 + u_2 \\ D_t^{q_3} e_3 &= ce_2 + e_2e_4 - y_4e_2 - e_4y_2 - e_3 + u_3 \\ D_t^q e_4 &= -e_2e_3 + y_3e_2 + y_2e_3 - be_4 + u_4 \end{aligned} \tag{19.39}$$

Now the objective is to design controllers such that errors converge to zero. With the transformation, $e_1 = z_1$, $z_2 = e_2 - \alpha_1$, where e_2 as the controller and α_1 as the virtual controller for the first subsystem in (19.39),

$$\begin{aligned} D^q z_1 &= D^q e_1 = -e_2 - ae_1 + u_1 \\ \Rightarrow D^q z_1 &= -(z_2 + \alpha_1) - az_1 + u_1 \end{aligned} \tag{19.40}$$

Lyapunov function for the above dynamic can be written as

$$V_1 = \frac{1}{2} z_1^2$$

$$\Rightarrow D^q V_1 \leq z_1(-(z_2 + \alpha_1) - az_1 + u_1)$$

For, $u_1 = 0$ and $\alpha_1 = 0$;

$$D^q V_1 \leq -az_1^2 - z_1 z_2$$

Next, for the second subsystem,

$$D^q z_2 = ae_3 - ae_2 + de_1 + u_2$$

For, $e_3 = \alpha_2$, as the virtual controller, the above dynamics changes to,

$$D^q z_2 = a\alpha_2 - az_2 + dz_1 + u_2 \tag{19.41}$$

The new Lyapunov function for the system up to this stage can be written as:

$$V_2 = V_1 + \frac{1}{2} z_2^2$$

$$\begin{aligned} \Rightarrow D^q V_2 &\leq -az_1^2 - z_1 z_2 + z_2 D^q z_2 \\ &\leq -az_1^2 - z_1 z_2 + z_2\{a\alpha_2 - az_2 + dz_1 + u_2\} \end{aligned}$$

For $\alpha_2 = 0$, $u_2 = z_1 - dz_1$, which results in,

$$D^q V_2 \leq -az_1^2 - az_2^2$$

Next, for $z_3 = e_3 - \alpha_2$,

$$\begin{aligned} D^q z_3 &= ce_2 + e_2 e_4 - y_4 e_2 - e_4 y_2 - e_3 + u_3 \\ \Rightarrow D^q z_3 &= cz_2 + z_2 e_4 - y_4 z_2 - e_4 y_2 - z_3 + u_3 \end{aligned} \tag{19.42}$$

and the combined Lyapunov function for z_1, z_2, z_3 system can be written as,

$$V_3 = V_2 + \frac{1}{2} z_3^2$$

$$\Rightarrow D^q V_3 \leq -az_1^2 - az_2^2 + z_3 D^q z_3$$

for $e_4 = \alpha_3$, as the virtual controller,

$$\Rightarrow D^q V_3 \leq -az_1^2 - az_2^2 + z_3[cz_2 + z_2 e_4 - y_4 z_2 - e_4 y_2 - z_3 + u_3]$$

Keeping $\alpha_3 = 0$, the controller can be chosen as:

$$u_3 = -cz_2 + y_4 z_2$$

which leads to,

$$D^q V_3 \leq -az_1^2 - az_2^2 - z_3^2$$

Next from the last equation in (19.39), with $z_4 = e_4 - \alpha_3$

$$\begin{aligned} D^q z_4 &= -e_2 e_3 + y_3 e_2 + y_2 e_3 - b e_4 + u_4 \\ \Rightarrow D^q z_4 &= -z_2 z_3 + y_3 z_2 + y_2 z_3 - b e_4 + u_4 \end{aligned} \quad (19.43)$$

Overall Lyapunov function for the whole z_1, z_2, z_3, and z_4 system will be,

$$V_4 = V_3 + \frac{1}{2} z_4^2$$

Taking the fractional order derivative with respect to time,

$$\Rightarrow D^q V_4 \leq -az_1^2 - az_2^2 - z_3^2 + z_4(-z_2 z_3 + y_3 z_2 + y_2 z_3 - b e_4 + u_4) \quad (19.44)$$

The final controller shall be chosen as,

$$u_4 = z_2 z_3 - y_3 z_2 - y_2 z_3 \quad (19.45)$$

The stability condition will be satisfied as form (19.44) and (19.45),

$$D^q V_4 \leq -az_1^2 - az_2^2 - z_3^2 - bz_4^2$$

The above expression leads to convergence of transformed states $[z_1 \quad z_2 \quad z_3 \quad z_4]^T$ to zero as $t \to \infty$. This further ensures asymptotic convergence of errors $[e_1 \quad e_2 \quad e_3 \quad e_4]^T$ to zero as proved in the previous section. Convergence of state errors leads to asymptotic synchronization of states of slave system in (19.38) to the states of master system in (19.37), i.e., $\lim_{t\to\infty} [x_1 \quad x_2 \quad x_3 \quad x_4]^T = [y_1 \quad y_2 \quad y_3 \quad y_4]^T$

19.6 SIMULATION RESULTS

The simulations for showing the performance of proposed control approaches have been performed using MATLAB. The FODE have been solved by using the numerical approximation given in Section 19.2. The parameters values and the initial conditions for the simulation purpose have already been mentioned in Section 19.3. Fig. 19.3A and B depict the state stabilization and synchronization of fractional order Lorenz–Stenflo system achieved via backstepping technique, respectively. It is evident from the Fig. 19.3A that states of the systems converge to zero in finite time, which conveys that the states of the system are stabilized. The convergence of synchronization errors to zero ensures that the state of slave system approaches the states of master system in finite time. The synchronization of corresponding states of master and slave system is shown in Fig. 19.4. Figs. 19.5 and 19.6 show the corresponding simulation results for the implementation of active backstepping approach.

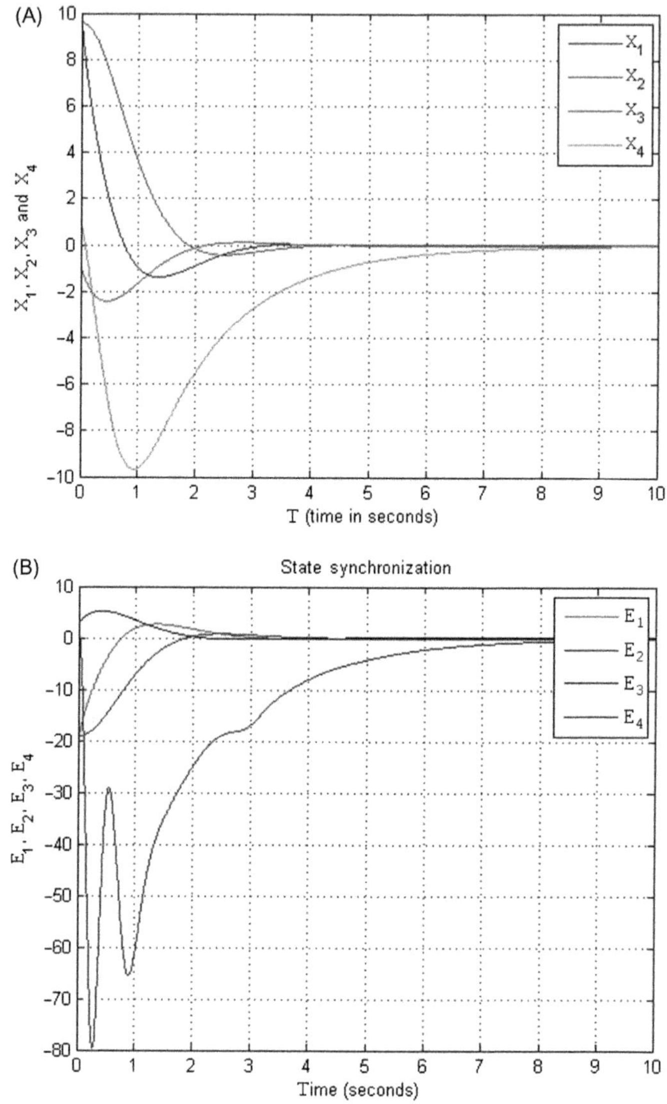

FIGURE 19.3 Results of backstepping approach. (A) State Stabilization. (B) Convergence of synchronization errors.

19.7 DISCUSSIONS

The simulation results obtained above ascertain the efficacy of the proposed scheme for control and synchronization of hyperchaotic fractional order Lorenz–Stenflo system. The convergence of the states to zero confirms that the system is stable with the controller obtained with the application of two

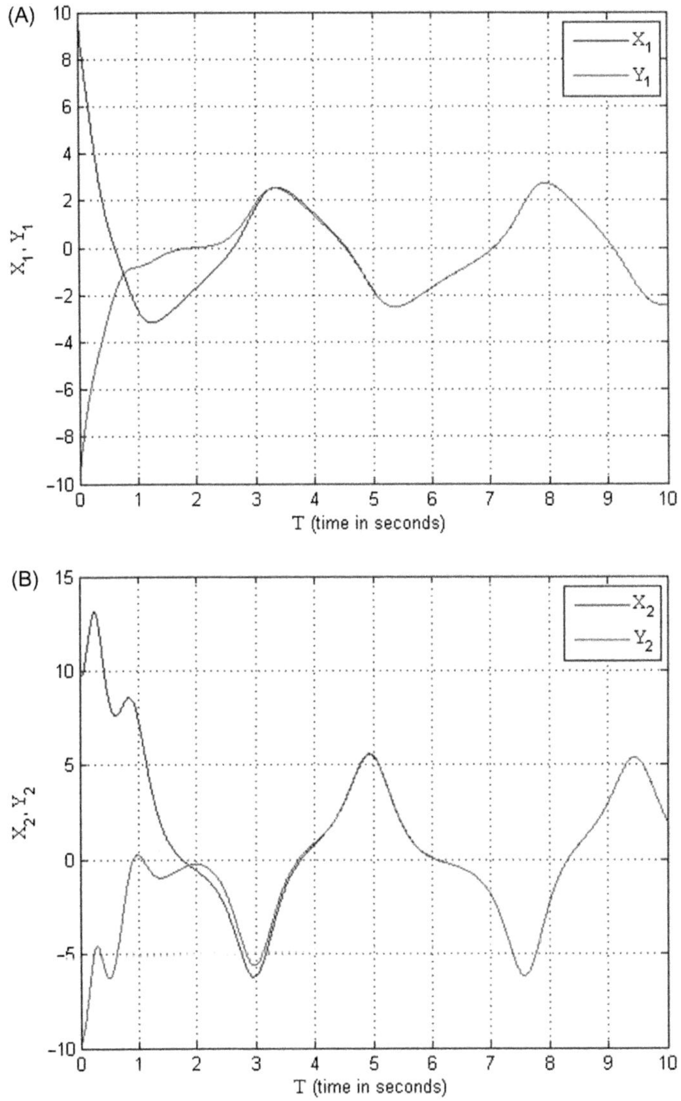

FIGURE 19.4 Synchronization via backstepping approach. (A) Synchronization between states x_1 and y_1. (B) Synchronization between states x_2 and y_2. (C) Synchronization between states x_3 and y_3. (D) synchronization between states x_4 and y_4.

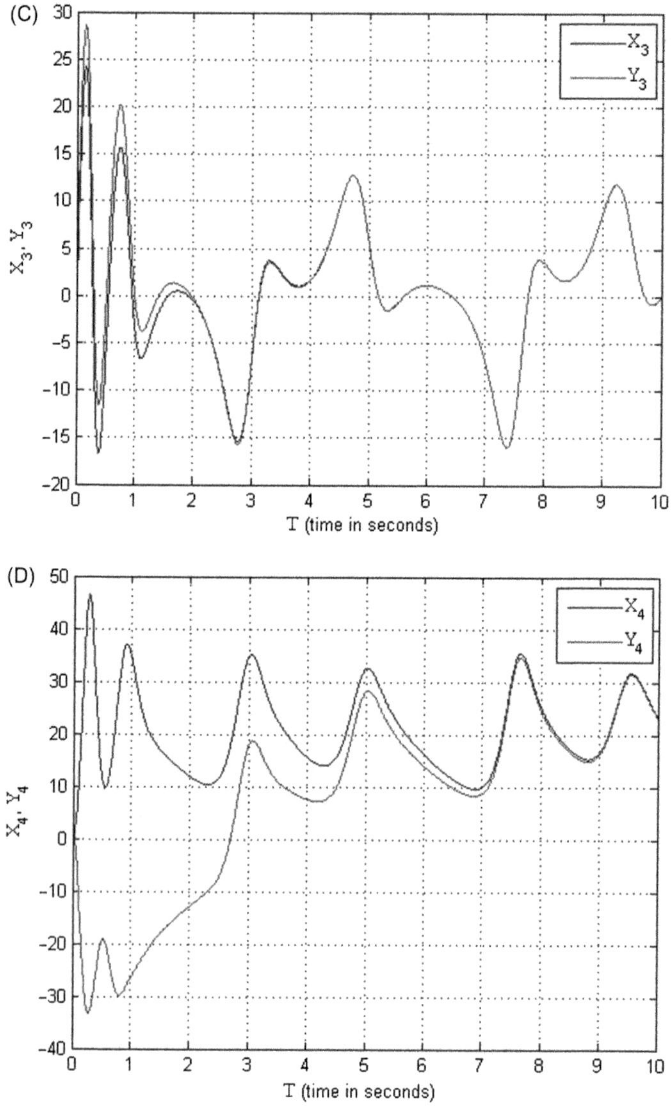

FIGURE 19.4 (Continued).

different schemes. The synchronization between the master and slave system is confirmed as the corresponding states converge on each other in finite time. A comparison can be drawn between the simulation results of the two approaches. One can conclude that the results are nearly same for the two approaches but the ease of deriving the controller is different for both cases. In the case of a backstepping approach one single controller has to be designed but the complexity is greater. Whereas, in the case of an active

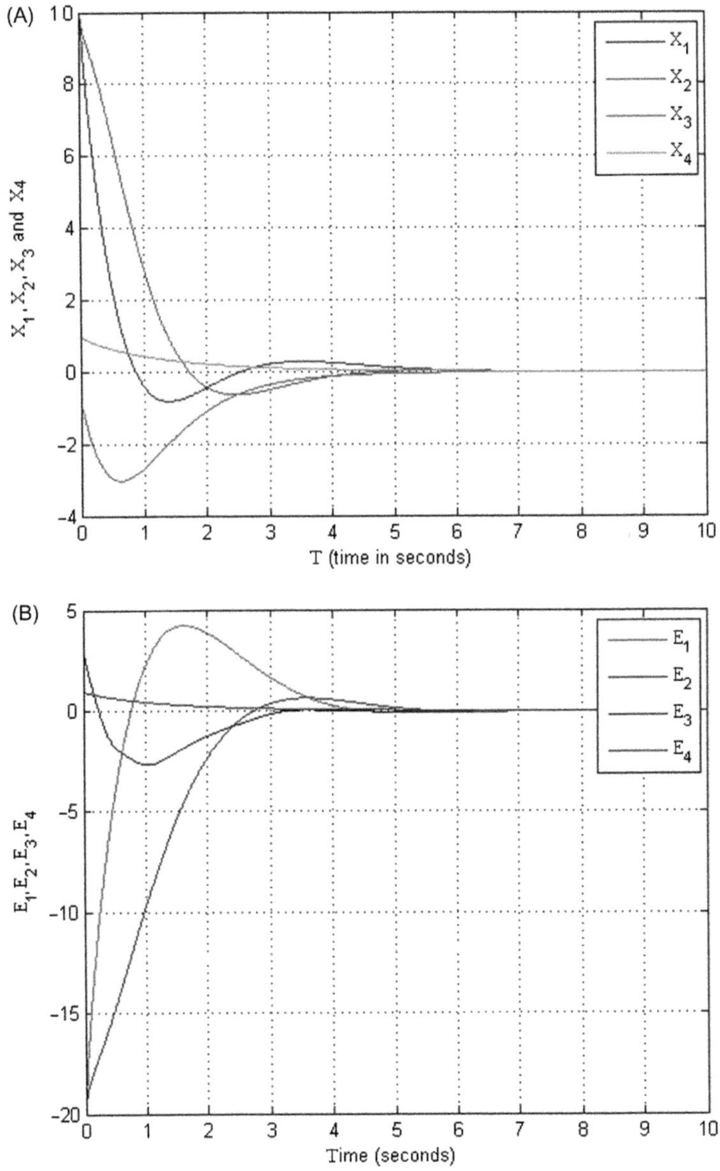

FIGURE 19.5 Results of active backstepping approach. (A) State stabilization. (B) Convergence of synchronization errors.

backstepping multiple controllers are to be designed but the technique is simple in analysis and implementation. The results prove the efficacy of proposed approach which utilizes the combination of Mittag–Leffler stability, Lyapunov stability and backstepping technique.

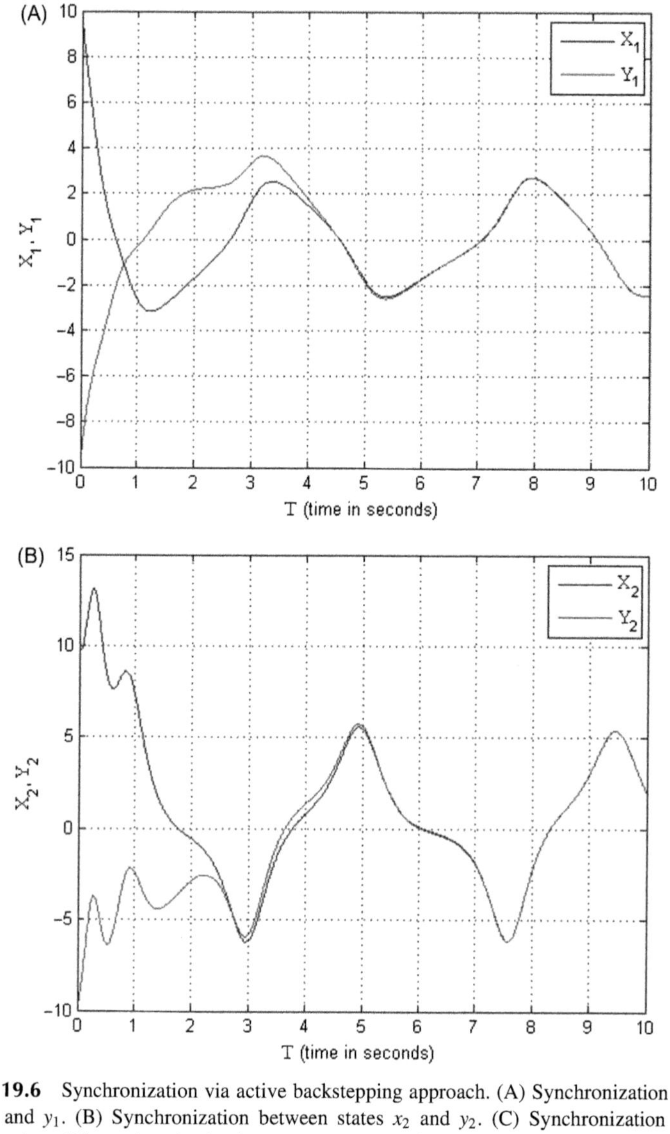

FIGURE 19.6 Synchronization via active backstepping approach. (A) Synchronization between states x_1 and y_1. (B) Synchronization between states x_2 and y_2. (C) Synchronization between states x_3 and y_3. (D) synchronization between states x_4 and y_4.

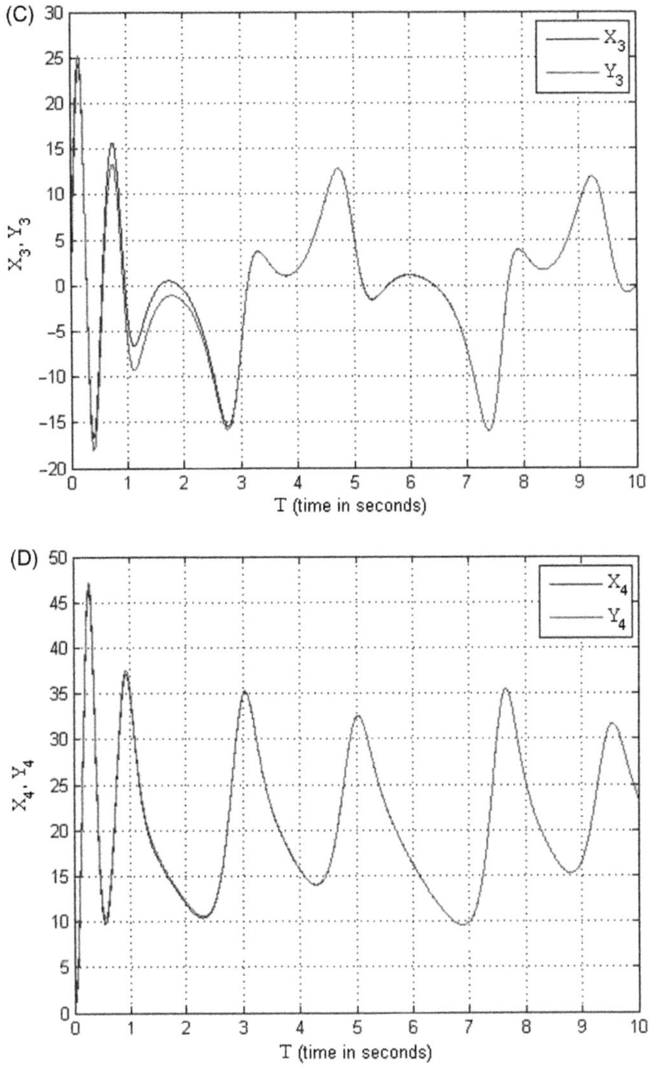

FIGURE 19.6 (Continued).

19.8 CONCLUSION

The control and synchronization of a fractional order Lorenz–Stenflo system was the main focus of this chapter. Two different approaches derived from backstepping approach have been employed for this purpose. The backstepping approach which uses the fractional order extension of Lyapunov stability criterion is initially used for control and synchronization. Later, an active backstepping approach which uses the same stability concepts is utilized for the same purpose. Analytical results for both stabilizing and synchronizing controllers are derived while using a systematic backstepping procedure along with Mittag–Leffler and Lyapunov stability results. The proposed approach ensures global stability and asymptotic synchronization which is evident from the simulation results presented at the end. The results obtained here can also be used to address the problem of secure communication and image encryption. The future research can target the control and synchronization of fractional order systems with uncertain parameters. The implementation of these outcomes on software and finally on hardware is the challenge which can act as the driving force for upcoming researchers in the area of fractional order chaotic systems. Moreover, stability of fractional order systems and its interpretation is still an issue and has to be answered.

REFERENCES

Aghababa, M.P., 2012. Robust stabilization and synchronization of a class of fractional-order chaotic systems via a novel fractional sliding mode controller. Commun. Nonlinear Sci. Numer. Simul. 17 (6), 2670–2681.

Agrawal, S.K., Srivastava, M., Das, S., 2012. Synchronization of fractional order chaotic systems using active control method. Chaos Solitons Fractals 45 (6), 737–752.

Aguila-Camacho, N., Duarte-Mermoud, M., Gallegos, J., 2014. Lyapunov functions for fractional order systems. Commun. Nonlinear Sci. Numer. Simul. 19 (9), 2951–2957.

Angulo, J.M., Ruiz-Medina, M.D., Anh, V.V., Grecksch, W., 2000. Robust adaptive synchronization of different uncertain chaotic systems subject to input nonlinearity. Commun. Nonlinear Sci. Numer. Simul. 15 (2), 430–441.

Asheghan, M.M., Beheshti, M.T.H., Tavazoei, M.S., 2011. Robust synchronization of perturbed Chen's fractional-order chaotic systems. Commun. Nonlinear Sci. Numer. Simul. 16 (2), 1044–1051.

Astakhov, V.V., Anishchenko, V.S., Kapitaniak, T., Shabunin, A.V., 1997. Synchronization of chaotic oscillators by periodic parametric perturbations. Phys. D Nonlinear Phenom. 109 (1), 11–16.

Atangana, A., Bildik, N., 2013. The use of fractional order derivative to predict the groundwater flow. Math. Probl. Eng. 2013. Available from: https://doi.org/10.1155/2013/543026, Article ID 543026, 9 pages.

Atangana, A., Vermeulen, P.D., 2014. Analytical solutions of a space-time fractional derivative of groundwater flow equation. Abstr. Appl. Anal. 2014. Available from: https://doi.org/10.1155/2014/381753, Article ID 381753, 11 pages.

Azar, A.T., Vaidyanathan, S., 2015a. Handbook of Research on Advanced Intelligent Control Engineering and Automation. Advances in Computational Intelligence and Robotics (ACIR) Book Series. IGI Global, USA, ISBN 9781466672482.

Azar, A.T., Vaidyanathan, S., 2015b. Computational Intelligence Applications in Modeling and Control. Studies in Computational Intelligence, Vol. 575. Springer-Verlag, Germany, ISBN 978-3-319-11016-5.

Azar, A.T., Vaidyanathan, S., 2015c. Chaos Modeling and Control Systems Design, Studies in Computational Intelligence, Vol. 581. Springer-Verlag, Germany, ISBN 978-3-319-13131-3.

Azar, A.T., Vaidyanathan, S., 2016. Advances in Chaos Theory and Intelligent Control. Studies in Fuzziness and Soft Computing, Vol. 337. Springer-Verlag, Germany, ISBN 978-3-319-30338-3.

Azar, A.T., Zhu, Q., 2015. Advances and Applications in Sliding Mode Control systems. Studies in Computational Intelligence, Vol. 576. Springer-Verlag, Germany, ISBN 978-3-319-11172-8.

Azar, A.T., Vaidyanathan, S., Ouannas, A., 2017a. Fractional Order Control and Synchronization of Chaotic Systems. Studies in Computational Intelligence, Vol. 688. Springer-Verlag, Germany, ISBN 978-3-319-50248-9.

Azar, A.T., Volos, C., Gerodimos, N.A., Tombras, G.S., Pham, V.T., Radwan, A.G., et al., 2017b. A novel chaotic system without equilibrium: dynamics, synchronization and circuit realization. Complexity 2017. Available from: https://doi.org/10.1155/2017/7871467, Article ID 7871467, 11 pages.

Azar, A.T., Ouannas, A., Singh, S., 2017c. Control of new type of fractional chaos synchronization, Proceedings of the International Conference on Advanced Intelligent Systems and Informatics 2017, Advances in Intelligent Systems and Computing series, Vol. 639. Springer-Verlag, Germany, pp. 47–56.

Borah, M., Singh, P.P., Roy, B.K., 2016. Improved chaotic dynamics of a fractional-order system, its chaos-suppressed synchronisation and circuit implementation. Circuit. Syst. Signal Proc. 35 (6), 1871–1907.

Boulkroune, A., Bouzeriba, A., Bouden, T., Azar, A.T., 2016a. Fuzzy adaptive synchronization of uncertain fractional-order chaotic systems. In: Azar, A.T., Vaidyanathan, S. (Eds.), Advances in Chaos Theory and Intelligent Control. Studies in Fuzziness and Soft Computing, Vol. 337. Springer-Verlag, Germany.

Boulkroune, A., Hamel, S., Azar, A.T., 2016b. Fuzzy control-based function synchronization of unknown chaotic systems with dead-zone input, Advances in Chaos Theory and Intelligent Control. Studies in Fuzziness and Soft Computing, Vol. 337. Springer-Verlag, Germany.

Carlson, G.E., Halijak, C.A., 1964. Approximation of fractional capacitors (1/s)^1/n by a regular newton process. IEEE Trans. Circuit Theory 11 (2), 210–213.

Chen, D.Y., Liu, Y.X., Ma, X.Y., Zhang, R., 2012. Control of a class of fractional-order chaotic systems via sliding mode. Nonlinear Dynam. 67 (1), 893–901.

Chialvo, D.R., Gilmour, R.F., Jalife, J., 1990. Low dimensional chaos in cardiac tissue. Nature 343 (6259), 653–657.

Ding, D., Qi, D., Wang, Q., 2015. Non-linear Mittag – Leffler stabilisation of commensurate fractional-order non-linear systems. IET Control Theory Applicat. 9 (5), 681–690.

Dorcak, L., 1994. Numerical models for the simulation of the fractional-order control systems. UEF-04-94, The Academy of Sciences, Inst. of Experimental Physic, Košice, Slovakia.

Feeny, B.F., Yuan, C.-M., Cusumano, J.P., 2001. Parametric identification of an experimental magneto-elastic oscillator. J. Sound Vibration 247 (5), 785–806.

Field, R.J., Schneider, F.W., 1989. Oscillating chemical reactions and nonlinear dynamics. J. Chem. Educ 66 (3), 195.

da Graça Marcos, M., Duarte, F.B.M., Machado, J.A.T., 2008. Fractional dynamics in the trajectory control of redundant manipulators. Commun. Nonlinear Sci. Numer. Simul. 13 (9), 1836–1844.

Grassi G., Ouannas A., Azar A.T., Radwan A.G., Volos C., Pham V.T., et al., 2017. Chaos Synchronisation Of Continuous Systems Via Scalar Signal. The 6th International Conference on Modern Circuits and Systems Technologies (MOCAST), 4-6 May 2017, Thessaloniki Greece.

Grigorenko, I., Grigorenko, E., 2003. Chaotic dynamics of the fractional Lorenz system. Phys. Rev. Lett. 91 (3), 34101.

Hardy, H.H., Beier, R.A., 1994. Fractals in Reservoir Engineering. Publishing Company, River Edge, NJ.

Hartley, T.T., Lorenzo, C.F., Qammer, H.K., 1995. Chaos in a fractional order Chua's system. IEEE Trans. Circuits Systems I: Fund. Theory Applicat. 42 (8), 485–490.

Kebriaei, H., Javad Yazdanpanah, M., 2010. Robust adaptive synchronization of different uncertain chaotic systems subject to input nonlinearity. Commun. Nonlinear Sci. Numer. Simul. [Internet] 15 (2), 430–441. Available from: https://doi.org/10.1016/j.cnsns.2009.04.005. Available from: citeulike-article-id:7865896%5Cn.

Krstic, M., Kanellakopoulos, I., Kokotovic, P.V., 1995. Nonlinear and Adaptive Control Design. Wiley, New York.

Lamamra, K., Azar, A.T., Ben Salah, C., 2017. Chaotic system modelling using a neural network with optimized structure, Studies in Computational Intelligence, Vol. 688. Springer-Verlag, Germany, pp. 833–856.

Laskin, N., 2002. Fractional schrödinger equation. Phys. Rev. E 66 (5), 56108.

Lesieur, M., 2012. Turbulence in fluids, Fluid Mechanics and Its Applications, Vol. 84. Springer, Netherlands.

Li, C., Chen, G., 2004. Chaos and hyperchaos in the fractional-order Rössler equations. Phys. A Statist. Mech. Applicat. 341 (2004), 55–61.

Li, C., Peng, G., 2004. Chaos in Chen's system with a fractional order. Chaos, Solitons Fractals 22 (2), 443–450.

Li, Y., Chen, Y., Podlubny, I., 2010. Stability of fractional-order nonlinear dynamic systems: Lyapunov direct method and generalized Mittag – Leffler stability. Comput. Math. Applicat. 59 (5), 1810–1821.

Liu, C., Lu, J., 2010. A novel fractional-order hyperchaotic system and its circuit realization. Int. J. Modern Phys. B 24 (10), 1299–1307.

Lorenz, E.N., 1963. Deterministic nonperiodic flow. J. Atmosph. Sci. 20 (2), 130–141.

Lu, J.G., 2006. Chaotic dynamics of the fractional-order Lü system and its synchronization. Phys. Lett. A 354 (4), 305–311.

Mathai, A.M., Haubold, H.J., 2008. Special Functions for Applied Scientists. Springer, New York, NY.

Matignon, D., 1996. Stability results for fractional differential equations with applications to control processing. Comput. Eng. Syst. Applicat. 2, 963–968.

Meghni, B., Dib, D., Azar, A.T., Ghoudelbourk, S., Saadoun, A., 2017a. Robust adaptive supervisory fractional order controller for optimal energy management in wind turbine with battery storage. Studies in Computational Intelligence, Vol. 688. Springer-Verlag, Germany, pp. 165–202.

Meghni, B., Dib, D., Azar, A.T., Saadoun, A., 2017b. Effective supervisory controller to extend optimal energy management in hybrid wind turbine under energy and reliability constraints. Int. J. Dynam. Control, Springer. Available from: https://doi.org/10.1007/s40435-016-0296-0.

Meghni, B., Dib, D., Azar, A.T., 2017c. A Second-order sliding mode and fuzzy logic control to Optimal Energy Management in PMSG Wind Turbine with Battery Storage. Neural Comput. Applicat. 28 (6), 1417–1434. Available from: https://doi.org/10.1007/s00521-015-2161-z.

Miller, K.S., Ross, B., 1993. An Introduction to the Fractional Calculus and Fractional Differential Equations. John Wiley & Sons, New York, NY.

Mishra, S.K., Chandra, D., 2014. Stabilization and tracking control of inverted pendulum using fractional order PID controllers. J. Eng. 2014. Available from: https://doi.org/10.1155/2014/752918, Article ID 752918, 9 pages.

Mohadeszadeh, M., Delavari, H., 2017. Synchronization of fractional-order hyper-chaotic systems based on a new adaptive sliding mode control. Int. J. Dynam. Control 5 (1), 124–134.

Mormann, F., Lehnertz, K., David, P., Elger, C.E., 2000. Mean phase coherence as a measure for phase synchronization and its application to the EEG of epilepsy patients. Phys. D Nonlinear Phenom. 144 (3-4), 358–369.

Nakamura, Y., Sekiguchi, A., 2001. The chaotic mobile robot. IEEE Trans. Robot. Automat. 17 (6), 898–904.

Oldham, K.B., Spanier, J., 1974. The Fractional Calculus: Theory and Applications of Differentiation and Integration to Arbitrary Order. Dover Publications Inc, United States.

Ott, E., Grebogi, C., Yorke, J.A., 1990. Controlling chaos. Phys. Rev. Lett. 64 (11), 1196–1199.

Ouannas, A., Azar, A.T., Abu-Saris, R., 2016a. A new type of hybrid synchronization between arbitrary hyperchaotic maps. Int. J. Mach. Learning Cybernetics. Available from: https://doi.org/10.1007/s13042-016-0566-3.

Ouannas, A., Azar, A.T., Radwan, A.G., 2016b. On Inverse Problem of Generalized Synchronization Between Different Dimensional Integer-Order and Fractional-Order Chaotic Systems. The 28th International Conference on Microelectronics, IEEE, December 17-20, 2016, Cairo, Egypt.

Ouannas, A., Azar, A.T., Ziar, T., Vaidyanathan, S., 2017a. On new fractional inverse matrix projective synchronization schemes, Studies in Computational Intelligence, Vol. 688. Springer-Verlag, Germany, pp. 497–524.

Ouannas, A., Azar, A.T., Ziar, T., Vaidyanathan, S., 2017b. Fractional inverse generalized chaos synchronization between different dimensional systems, Studies in Computational Intelligence, Vol. 688. Springer-Verlag, Germany, pp. 525–551.

Ouannas, A., Azar, A.T., Ziar, T., Vaidyanathan, S., 2017c. A new method to synchronize fractional chaotic systems with different dimensions, Studies in Computational Intelligence, Vol. 688. Springer-Verlag, Germany, pp. 581–611.

Ouannas, A., Azar, A.T., Ziar, T., Radwan, A.G., 2017d. Study on coexistence of different types of synchronization between different dimensional fractional chaotic systems, Studies in Computational Intelligence, Vol. 688. Springer-Verlag, Germany, pp. 637–669.

Ouannas, A., Azar, A.T., Ziar, T., Radwan, A.G., 2017e. Generalized synchronization of different dimensional integer-order and fractional order chaotic systems, Studies in Computational Intelligence, Vol. 688. Springer-Verlag, Germany, pp. 671–697.

Ouannas, A., Azar, A.T., Vaidyanathan, S., 2017f. On a simple approach for Q-S synchronization of chaotic dynamical systems in continuous-time. Int. J. Comput. Sci. Math. 8 (1), 20–27.

Ouannas, A., Azar, A.T., Vaidyanathan, S., 2017g. New hybrid synchronization schemes based on coexistence of various types of synchronization between master-slave hyperchaotic systems. Int. J. Comput. Applicat. Technol. 55 (2), 112–120.

Ouannas, A., Azar, A.T., Ziar, T., 2017h. On inverse full state hybrid function projective synchronization for continuous-time chaotic dynamical systems with arbitrary dimensions. Differ. Equat. Dynam. Syst. Available from: https://doi.org/10.1007/s12591-017-0362-x.

Ouannas, A., Azar, A.T., Vaidyanathan, S., 2017i. A robust method for new fractional hybrid chaos synchronization. Math. Methods Appl. Sci. 40 (5), 1804–1812. Available from: https://doi.org/10.1002/mma.4099.

Ouannas, A., Grassi, G., Azar, A.T., Radwan, A.G., Volos, C., Pham, V.T., et al., 2017j. Dead-Beat Synchronization Control in Discrete-Time Chaotic Systems. The 6th International Conference on Modern Circuits and Systems Technologies (MOCAST), 4-6 May 2017, Thessaloniki Greece.

Pecora, L.M., Carroll, T.L., Johnson, G.A., Mar, D.J., 1997. Fundamentals of synchronization in chaotic systems, concepts, and applications. Chaos (Woodbury, N.Y.) 7 (4), 520–543.

Petras, I., 2011. Fractional-order nonlinear systems: modeling, analysis and simulation. Nonlinear Physical Science, Springer-Verlag, Berlin Heidelberg.

Pham, V.T., Vaidyanathan, S., Volos, C.K., Azar, A.T., Hoang, T.M., Yem, V.V., 2017. A three-dimensional no-equilibrium chaotic system: analysis, synchronization and its fractional order form, Studies in Computational Intelligence, Vol. 688. Springer-Verlag, Germany, pp. 449–470.

Podlubny, I., 1994. Fractional-order systems and fractional-order controllers. Proc. of the Conference Internationale Francophone d'Automatique, p. 2002.

Podlubny, I., 1998. Fractional Differential Equations: An Introduction to Fractional Derivatives, Fractional Differential Equations, to Methods of Their Solution and Some of Their Applications. Academic press, San Diego, Boston, New York.

Podlubny, I., 1999. Fractional-order systems and PIλDμ controller. IEEE Trans. Automatic Control 44, 208–214.

Podlubny, I., 2002. Geometric and physical interpretation of fractional integration and fractional differentiation. Fract. Calculus Appl. Anal. 5 (4), 367–386.

Samko, S.G., Kilbas, A.A., Marichev, O.I., 1993. Fractional Integral and Derivatives: Theory and Applications. Gordon and Breach, Switzerland, Philadelphia, PA.

Sharma, B.B., Kar, I.N., 2009a. Chaotic synchronization and secure communication using contraction theory. In: Chaudhury, S., Mitra, S., Murthy, C.A., Sastry, P.S., Pal, S.K. (Eds.), Pattern Recognition and Machine Intelligence. PReMI 2009. Lecture Notes in Computer Science, Vol. 5909. Springer, Berlin, Heidelberg.

Sharma, B.B., Kar, I.N., 2009b. Contraction based adaptive control of a class of nonlinear systems. In 2009 American Control Conference, St. Louis, MO, 2009, pp. 808-813. https://doi.org/10.1109/ACC.2009.5160111.

Sharma, B.B., Kar, I.N., 2011. Observer-based synchronization scheme for a class of chaotic systems using contraction theory. Nonlinear Dynam. 63 (3), 429–445.

Shukla, M., Sharma, B.B., 2015. Hybrid Projective Synchronization of Fractional Order Volta's System via Active Control. 2015 2nd International Conference on Recent Advances in Engineering & Computational Sciences (RAECS), Chandigarh, 2015, pp. 1-6. doi: https://doi.org/10.1109/RAECS.2015.7453415.

Shukla, M.K., Sharma, B.B., 2017a. Backstepping based stabilization and synchronization of a class of fractional order chaotic systems. Chaos Solitons Fractals 102 (2017), 274–284.

Shukla, M.K., Sharma, B.B., 2017b. Stabilization of fractional order discrete chaotic systems. In: Azar, A., Vaidyanathan, S., Ouannas, A. (Eds.), Fractional Order Control and Synchronization of Chaotic Systems. Studies in Computational Intelligence, vol 688. Springer, Cham.

Shukla, M.K., Sharma, B.B., 2017c. Stabilization of a class of fractional order chaotic systems via backstepping approach. Chaos Solitons Fractals 98 (2017), 56–62.

Shukla, M.K., Sharma, B.B., 2017d. Stabilization of a class of uncertain fractional order chaotic systems via adaptive backstepping control. 2017 Indian Control Conference (ICC), Guwahati, 2017, pp. 462-467. https://doi.org/10.1109/INDIANCC.2017.7846518

Shukla, M.K., Sharma, B.B., 2017e. Investigation of chaos in fractional order generalized hyperchaotic henon map. AEU-Int. J. Electr. Commun. 78 (2017), 265–273.

Shukla, M.K., Sharma, B.B., 2018. Control and synchronization of a class of uncertain fractional order chaotic systems via adaptive backstepping control. Asian J. Control. Available from: https://doi.org/10.1002/asjc.1593.

Singh, S., Azar, A.T., Ouannas, A., Zhu, Q., Zhang, W., Na, J., 2017. Sliding ModeControl Technique for Multi-switching Synchronization of Chaotic Systems. 9th International Conference on Modelling, Identification and Control (ICMIC 2017), July 10-12, 2017, Kunming, China.

Soliman, N.S., Said, L.A., Azar, A.T., Madian, A.H., Radwan, A.G., Ouannas, A., 2017. Fractional Controllable Multi-Scroll V-Shape Attractor with Parameters Effect. The 6th International Conference on Modern Circuits and Systems Technologies (MOCAST), 4-6 May 2017, Thessaloniki Greece.

Srivastava, H.M., Owa, S., 1989. Univalent Functions, Fractional Calculus, and Their Applications. Halsted Press, Ellis Horwood; New York; Toronto.

Stenflo, L., 1996. Generalized Lorenz equations for acoustic-gravity waves in the atmosphere. Phys. Scripta 53 (1), 83–84.

Strogatz, S.H., 2014. Nonlinear Dynamics and Chaos: With Applications to Physics, Biology, Chemistry, and Engineering. Westview Press, Boulder, CO.

Tabari, M.Y., Kamyad, A.V., 2013. Design optimal fractional PID controller for inverted pendulum with genetic algorithm. Int. J. Sci. Eng. Res. 4 (2), 2–5.

Tavazoei, M.S., Haeri, M., 2007. A necessary condition for double scroll attractor existence in fractional-order systems. Phys. Lett. A 367, 102–113.

Tavazoei, M.S., Haeri, M., 2008. Chaotic attractors in incommensurate fractional order systems. Phys. D Nonlinear Phenom. 237 (20), 2628–2637.

Tenreiro Machado, J.A., Silva, M.F., Barbosa, R.S., Jesus, I.S., Reis, C.M., Marcos, M.G., et al., 2010. Some applications of fractional calculus in engineering. Math. Probl. Eng. 2010. Available from: https://doi.org/10.1155/2010/639801, Article ID 639801, 34 pages.

Tolba, M.F., AbdelAty, A.M., Soliman, N.S., Said, L.A., Madian, A.H., Azar, A.T., et al., 2017. FPGA implementation of two fractional order chaotic systems. Int. J. Electr. Commun. 28 (2017), 162–172.

Tseng, C.-C., 2007. Design of FIR and IIR fractional order Simpson digital integrators. Signal Processing 87 (5), 1045–1057.

Vaidyanathan, S., Azar, A.T., 2015a. Anti-synchronization of identical chaotic systems using sliding mode control and an application to Vaidyanathan-Madhavan Chaotic Systems. In: Azar, A.T., Zhu, Q. (Eds.), Advances and Applications in Sliding Mode Control systems, Studies in Computational Intelligence book Series, Vol. 576. Springer-Verlag GmbH Berlin, Heidelberg, pp. 527–547. Available from: http://dx.doi.org/10.1007/978-3-319-11173-5_19.

Vaidyanathan, S., Azar, A.T., 2015b. Hybrid synchronization of identical chaotic systems using sliding mode control and an application to Vaidyanathan Chaotic Systems. In: Azar, A.T., Zhu, Q. (Eds.), Advances and Applications in Sliding Mode Control systems, Studies in Computational Intelligence book Series, Vol. 576. Springer-Verlag GmbH Berlin, Heidelberg, pp. 549–569. Available from: http://dx.doi.org/10.1007/978-3-319-11173-5_20.

Vaidyanathan, S., Azar, A.T., 2015c. Analysis, control and synchronization of a nine-term 3-D novel chaotic system. In: Azar, A.T., Vaidyanathan, S. (Eds.), Chaos Modeling and Control Systems Design, Studies in Computational Intelligence, Vol. 581. Springer-Verlag GmbH Berlin, Heidelberg, pp. 3–17. Available from: http://dx.doi.org/10.1007/978-3-319-13132-0_1.

Vaidyanathan, S., Azar, A.T., 2015d. Analysis and control of a 4-D novel hyperchaotic system. In: Azar, A.T., Vaidyanathan, S. (Eds.), Chaos Modeling and Control Systems Design, Studies in Computational Intelligence, Vol. 581. Springer-Verlag GmbH Berlin, Heidelberg, pp. 19–38. Available from: http://dx.doi.org/10.1007/978-3-319-13132-0_2.

Vaidyanathan, S., Azar, A.T., 2016a. Takagi-Sugeno Fuzzy Logic Controller for Liu-Chen Four-Scroll Chaotic System. Int. J. Intel. Eng. Inf. 4 (2), 135–150.

Vaidyanathan, S., Azar, A.T., 2016b. Dynamic analysis, adaptive feedback control and synchronization of an eight-term 3-D novel chaotic system with three quadratic nonlinearities, Studies in Fuzziness and Soft Computing, Vol. 337. Springer-Verlag, Germany, pp. 155–178.

Vaidyanathan, S., Azar, A.T., 2016c. Qualitative study and adaptive control of a novel 4-D hyperchaotic system with three quadratic nonlinearities, Studies in Fuzziness and Soft Computing, Vol. 337. Springer-Verlag, Germany, pp. 179–202.

Vaidyanathan, S., Azar, A.T., 2016d. A novel 4-D four-wing chaotic system with four quadratic nonlinearities and its synchronization via adaptive control method, Advances in Chaos Theory and Intelligent Control. Studies in Fuzziness and Soft Computing, Vol. 337. Springer-Verlag, Germany, pp. 203–224.

Vaidyanathan, S., Azar, A.T., 2016e. Adaptive control and synchronization of Halvorsen Circulant Chaotic Systems. Advances in chaos theory and intelligent control, Studies in Fuzziness and Soft Computing, Vol. 337. Springer-Verlag, Germany, pp. 225–247.

Vaidyanathan, S., Azar, A.T., 2016f. Adaptive Backstepping Control and Synchronization of a Novel 3-D Jerk System with an Exponential Nonlinearity. Advances in Chaos Theory and Intelligent Control. Studies in Fuzziness and Soft Computing, Vol. 337. Springer-Verlag, Germany, pp. 249–274.

Vaidyanathan, S., Azar, A.T., 2016g. Generalized Projective Synchronization of a Novel Hyperchaotic Four-Wing System via Adaptive Control Method. Advances in Chaos Theory and Intelligent Control. Studies in Fuzziness and Soft Computing, Vol. 337. Springer-Verlag, Germany, pp. 275–296.

Vaidyanathan, S., Sampath, S., Azar, A.T., 2015a. Global chaos synchronisation of identical chaotic systems via novel sliding mode control method and its application to Zhu system. IJMIC 23 (1), 92–100.

Vaidyanathan, S., Azar, A.T., Rajagopal, K., Alexander, P., 2015b. Design and SPICE implementation of a 12-term novel hyperchaotic system and its synchronization via active control (2015). IJMIC 23 (3), 267–277.

Vaidyanathan, S., Idowu, B.A., Azar, A.T., 2015c. Backstepping controller design for the global chaos synchronization of Sprott's Jerk Systems. In: Azar, A.T., Vaidyanathan, S. (Eds.), Chaos Modeling and Control Systems Design, Studies in Computational Intelligence, Vol. 581. Springer-Verlag GmbH Berlin, Heidelberg, pp. 39–58. Available from: http://dx.doi.org/10.1007/978-3-319-13132-0_3.

Vaidyanathan, S., Zhu, Q., Azar, A.T., 2017b. Adaptive Control of a Novel Nonlinear Double Convection Chaotic System. Studies in Computational Intelligence, Vol. 688. Springer-Verlag, Germany, pp. 357–385.

Vaidyanathan, S., Azar, A.T., Ouannas, A., 2017c. Hyperchaos and Adaptive Control of a Novel Hyperchaotic System with Two Quadratic Nonlinearities. Studies in Computational Intelligence, Vol. 688. Springer-Verlag, Germany, pp. 773–803.

Vinagre, B.M., Chen, Y.Q., Petráš, I., 2003. Two direct Tustin discretization methods for fractional-order differentiator/integrator. J. Franklin Institute 340 (5), 349–362.

Wang, C., Ge, S.S., 2001. Synchronization of two uncertain chaotic systems via adaptive backstepping. Int. J. Bifurcation Chaos 11 (6), 1743–1751. Available at: http://www.worldscientific.com/doi/abs/10.1142/S0218127401002985.

Wang, H.H., Sun, K.H., He, S.B., 2014. Dynamic analysis and implementation of a digital signal processor of a fractional-order Lorenz–Stenflo system based on the Adomian decomposition method. Phys. Scripta 90 (1), 15206.

Wang, J., Zhang, Y., 2006. Designing synchronization schemes for chaotic fractional-order unified systems. Chaos Solitons Fractals 30 (5), 1265–1272.

Wang, J., Xiong, X., Zhang, Y., 2006. Extending synchronization scheme to chaotic fractional-order Chen systems. Phys. A Statist. Mech. Applicat. 370 (2), 279–285.

Wang, Z., Volos, C., Kingni, S.T., Azar, A.T., Pham, V.T., 2017. Four-wing attractors in a novel chaotic system with hyperbolic sine nonlinearity. Optik – Int. J. Light Electr. Opt. 131 (2017), 1071–1078.

Wei, Y., Tse, P.W., Yao, Z., Wang, Y., 2016. Adaptive backstepping output feedback control for a class of nonlinear fractional order systems. Nonlinear Dynam. 86 (2), 1047–1056.

West, B.J., 2007. Fractional calculus in bioengineering. J. Statist. Phys. 126 (6), 1285–1286.

Yang, J., Hu, G., Xiao, J., 1998. Chaos synchronization in coupled chaotic oscillators with multiple positive Lyapunov exponents. Phys. Rev. Lett. 80 (3), 496–499.

Yang, N., Liu, C., 2013. A novel fractional-order hyperchaotic system stabilization via fractional sliding-mode control. Nonlinear Dynam. 74 (3), 721–732.

Yassen, M.T., 2005. Chaos synchronization between two different chaotic systems using active control. Chaos Solitons Fractals 23 (1), 131–140.

Yin, C., Zhong, S., Chen, W., 2012. Design of sliding mode controller for a class of fractional-order. Commun. Nonlinear Sci. Numer. Simul. 17 (1), 356–366.

Yin, C., Dadras, S., Zhong, S., Chen, Y., 2013. Control of a novel class of fractional-order chaotic systems via adaptive sliding mode control approach. Appl. Math. Model. 37 (4), 2469–2483.

Yu, Y., Li, H., Wang, S., Yu, J., 2009. Dynamic analysis of a fractional-order Lorenz chaotic system. Chaos Solitons Fractals 42 (2), 1181–1189.

Yuan, J., Shi, B., Ji, W., 2013. Adaptive sliding mode control of a novel class of fractional chaotic systems. Adv. Math. Phys. 2013. Available from: https://doi.org/10.1155/2013/576709, Article ID 576709, 13 pages.

Zhang, H., Ma, X., Li, M., Zou, J., 2005. Controlling and tracking hyperchaotic Rossler system via active backstepping design. Chaos Solitons Fractals 26 (2), 353–361.

Zhang, H., Huang, W., Wang, Z., Chai, T., 2006. Adaptive synchronization between two different chaotic systems with unknown parameters. Phys. Lett. A 350 (5), 363–366.

Zhao, Y., Wang, Y., Zhang, X., Li, H., 2016. Feedback stabilisation control design for fractional order non-linear systems in the lower triangular form. IET Control Theory & Applications 10 (9), 1061–1068.

Zhou, C., Lai, C.H., Yu, M.Y., 1997. Chaos, bifurcations and periodic orbits of the Lorenz–Stenflo system. Phys. Scripta 55 (4), 394.

Zhu, Q., Azar, A.T., 2015. Complex system modelling and control through intelligent soft computations, Stud. Fuzziness Soft Computing, Vol. 319. Springer-Verlag, Germany, ISBN 978-3-319-12882-5.

FURTHER READING

Vaidyanathan, S., Azar, A.T., Ouannas, A., 2017a. An Eight-Term 3-D Novel Chaotic System with Three Quadratic Nonlinearities, its Adaptive Feedback Control and Synchronization. Studies in Computational Intelligence, Vol. 688. Springer-Verlag, Germany, pp. 719–746.

Chapter 20

Enhanced Fractional Order Indirect Fuzzy Adaptive Synchronization of Uncertain Fractional Chaotic Systems Based on the Variable Structure Control: Robust H_∞ Design Approach

Khatir Khettab[1], Yassine Bensafia[2], Bachir Bourouba[3] and Ahmad Taher Azar[4,5]

[1]Mohamed Boudiaf University of M'sila, M'sila 28000, Algeria, [2]Mohand Oulhadj University of Bouira, Bouira 10000, Algeria, [3]University of Sétif, Sétif 19000, Algeria, [4]Faculty of Computers and Information, Benha University, Benha, Egypt, [5]School of Engineering and Applied Sciences, Nile University, Giza, Egypt

20.1 INTRODUCTION

Recently, there are many important developments related to the design of nonlinear systems for many practical applications. Several approaches have been developed such as optimal control, nonlinear feedback control, adaptive control, sliding mode control, nonlinear dynamics, chaos control, chaos synchronization control, fuzzy logic control, fuzzy adaptive control, fractional order control, and robust control and their integrations (Azar & Vaidyanathan, 2015a,b,c, 2016; Azar & Zhu, 2015; Boulkroune et al, 2016a,b; Azar & Serrano, 2015a,b,c,d, 2016a,b, 2017; Azar et al., 2017a,b,c,d; Azar 2010a,b, 2012; Mekki et al., 2015; Vaidyanathan & Azar, 2015a,b,c,d, 2016a,b,c,d,e,f, g; Vaidyanathan et al., 2017a,b,c; Zhu & Azar, 2015; Grassi et al., 2017; Ouannas et al., 2016a, 2017f,g,h,j; Singh et al., 2017; Vaidyanathan et al, 2015a,b,c; Wang et al., 2017).

In recent years, fractional calculus has been applied in the modeling and control of various kinds of physical systems, as is well known and

Mathematical Techniques of Fractional Order Systems. DOI: https://doi.org/10.1016/B978-0-12-813592-1.00020-9

documented in many control theory or in the application literature (Azar et al., 2017a; Meghni et al, 2017a,b,c; Soliman et al., 2017; Tolba et al., 2017; Ghoudelbourk et al., 2016; Ouannas et al., 2016b, 2017a,b,c,d,e,I; Ladaci and Charef, 2006). Fractional calculus deals with derivatives and integrations of arbitrary order (Lin et al., 2011; Podlubny, 1999) and has found many applications in many fields of physics, applied mathematics, and engineering. Moreover, many real-world physical systems are well characterized by fractional order differential equations, i.e., equations involving both integer and noninteger order derivatives. It is observed that the description of some systems is more accurate when the fractional derivative is used. For instance, electrochemical processes and flexible structures are modeled by fractional order models (Lin et al., 2011; Lin and Kuo, 2011; Ladaci et al., 2008). Nowadays, many fractional order differential systems behave chaotically, such as the fractional order Chua system, the fractional order *Duffing* system (Arena et al., 1997), the fractional order Lu system, the fractional order Chen system (Petráš, 2006), the fractional order cellular neural network (Arena and Caponetto, 1998; Petráš, 2006; Lamamra et al., 2017).

The synchronization problem of fractional order chaotic systems was first investigated by Deng and Li who carried out synchronization in the case of the fractional Lü system. Afterwards, they studied chaos synchronization of the Chen system with a fractional order in a different manner (Hartley et al., 1995; Hilfer, 2001; Pham et al., 2017; Ouannas et al., 2016b, 2017a,b,c,d,e,i).

In this chapter we are interested by the problem of uncertain fractional order chaotic systems synchronization by means of robust adaptive fuzzy control. Variable structure control is a very suitable method for handling such nonlinear systems because of low sensitivity to disturbances and plant parameter variations and its order reduction property, which relaxes the burden of the necessity of exact modeling (Lin et al., 2004; Utkin,1977; Hung et al., 1993).

Based on the universal approximation theorem (Wang and Mendel, 1992; Wang, 1992; Wang, 1994; Lin et al., 2012) (fuzzy logic controllers are general enough to perform any nonlinear control actions), there is rapidly growing interest in systematic design methodologies for a class of nonlinear systems using fuzzy adaptive control schemes. An adaptive fuzzy system is a fuzzy logic system equipped with a training algorithm in which an adaptive controller is synthesized from a collection of fuzzy IF–THEN rules and the parameters of the membership functions characterizing the linguistic terms in the IF–THEN rules change according to some adaptive law for the purpose of controlling a plant to track a reference trajectory.

By incorporating the H_∞ tracking design technique (Chen et al., 1996, Bensafia et al., 2017) and based on the fractional Lyapunov stability theorem (Aguila-Camacho et al., 2014; Duarte-Mermoud et al., 2015; Sastry and Bodson, 1989; Khettab et al., 2017a), an efficient adaptive control algorithm by means of fuzzy logic models is proposed that guarantees the feedback control system stability and that is able to attenuate the effects of additive noises and estimation errors on the tracking performance to any prescribed error level via the sliding mode robust tracking design technique (Bourouba and Ladaci, 2017). However, the important problem of sliding mode techniques from the control perspective is the discontinuity of the control signal required to obtain robustness (Khettab et al., 2017b; Bensafia et al., 2017). This destructive phenomenon, so-called chattering may affect control accuracy or incur an unwanted wear of a mechanical component.

Various solutions to reduce the chattering have been studied in the literature (Ge and Ou, 2008; Lin and Balas, 2011; Ho et al., 2009). Comparing with a similar previous work, an improved synchronization technique is proposed here for a robust sliding mode control of nonlinear systems with fractional order dynamics that is able to eliminate the chattering phenomena for uncertain systems with unknown parameters' variation. The main contribution of this work consists in combining a saturation function with the sliding mode controller in order to improve the control signal quality by eliminating the undesirable chattering. The Grünwald–Letnikov numerical approximation method is used for fractional order differential equation resolution with improved performance result. Stability analysis is performed for the proposed control strategy using Lyapunov theory and numerical simulation results on the synchronization of two fractional order chaotic systems illustrate the effectiveness of the proposed fractional fuzzy adaptive synchronization strategy (Khettab et al., 2015; Khettab et al., 2017b,c; Boulkroune et al., 2016a,b)

This chapter is organized as follows: Section 20.2 presents a basic and brief review on the state of the art for the addressed problem and for fractional calculus, fractional derivatives, and its relation to the approximation solution. In section 20.3 is an introduction to the basics and a description of the T–S fuzzy systems. Section 20.4 and 20.5 generally propose adaptive fuzzy robust H_∞ control of uncertain fractional order systems in the presence of uncertainty and its stability analysis. In section 20.6, application of the proposed method on fractional order expression chaotic systems is investigated. Finally, the simulation results and conclusion will be presented in section 20.7.

20.2 BASIC DEFINITION AND PRELIMINARIES FOR FRACTIONAL CALCULUS

The fractional calculus is relatively an old topic dating back more than 300 years. It is a generalization of integration and differentiation to noninteger order fundamental operators, denoted by ${}_aD_t^\alpha$, where a and t are the limits of the operator.

This operator is a notation for taking both the fractional integral and functional derivative in a single expression defined as (Diethelm and Ford, 2002; Diethelm et al., 2004; Petráš, 2006; Khettab et al., 2017b,c,d)

$$
{}_aD_t^q = \begin{cases} \dfrac{d^q}{dt^q} & q > 0 \\ 1 & q = 0 \\ \int_a^t (d\tau)^{-q} & q < 0 \end{cases} \tag{20.1}
$$

There are some basic definitions of the general fractional integration and differentiation. The commonly used definitions are Riemann–Liouville, Adams–Bashforth–Moulton algorithm and the method of Grünwald–Letnikov: "Numerical evaluation of the fractional derivative of some usual functions."

The simplest and easiest definition is the Riemann–Liouville definition given as:

$$
{}_{t_0}^{RL}D_t^q f(t) = \frac{1}{\Gamma(n-q)} \frac{d^n}{dt^n} \int_{t_0}^{t} (t-\tau)^{n-q-1} f(\tau) d\tau \tag{20.2}
$$

where n is the first integer which is not less than q, i.e., $(n-q) < q < n$, and Γ is the Gamma function.

The numerical simulation of a fractional differential equation is not as simple as that of an ordinary differential equation.

The algorithm which is an improved version of the Adams–Bashforth–Moulton (Diethelm and Ford, 2002; Diethelm et al., 2004) to find an approximation for fractional order systems based on predictor–correctors is given (Diethelm et al., 2003). Consider the following differential equation

$$
{}_a^{GL}D_t^q y(t) = f(y(t), t) \tag{20.3}
$$

where $0 \le t \le \mathrm{T}$ and $y^{(k)}(0) = y_0^{(k)}$ and $k = 0, 1, 2, \ldots, m-1$

Can be expressed as follows

$$
y(t_k) = f(y(t_k), t_k)h^q - \sum_{j=v}^{k} c_j^{(q)} y(t_{k-j}) \tag{20.4}
$$

where ${}_0D_t^q y(t)$ is the Caputo fractional derivative of order q > 0 is defined as:

$$
{}_0D_t^q y(t) = \begin{cases} \dfrac{1}{\Gamma(m-q)} \displaystyle\int_0^t (t-\tau)^{m-q-1} f^{(m)}(\tau) d\tau, & m-1 < q < m \\ \dfrac{d^m}{dt^m} y(t) q = m \end{cases} \tag{20.5}
$$

and m is the first integer larger than q.

The solution of Eq. (20.4) is equivalent to the Volterra integral equation described as (Diethelm and Ford, 2002; Diethelm et al., 2004; Khettab et al., 2015; Ladaci and Khettab, 2012; Ladaci et al., 2009; Rabah et al., 2017):

$$y(t) = \sum_{k=0}^{[q]-1} y_0^{(k)} \frac{t^k}{k!} + \frac{1}{\Gamma(q)} \int_0^t (t-\tau)^{q-1} f(\tau, y(\tau)) d\tau \tag{20.6}$$

Let $h = T/N$, $t_n = nh, n = 0, 1, 2, \ldots, N$. Then Eq. (20.6) can be described as follows:

$$\begin{aligned} y_h(t_{n+1}) = &\sum_{k=0}^{[q]-1} y_0^{(k)} \frac{t_{n+1}^k}{k!} + \frac{h^q}{\Gamma(q+2)} f\left(y_n^p(t_{n+1}), t_{n+1}\right) \\ &+ \frac{h^q}{\Gamma(q+2)} \sum_{j=0}^{n} a_{j,n+1} f\left(y_h(t_j), t_j\right) \end{aligned} \tag{20.7}$$

where the predicted value $y_h^p(t_{n+1})$ is determined by the fractional Adams–Bashforth method:

$$y_h^p(t_{n+1}) = \sum_{k=0}^{[q]-1} y_0^{(k)} \frac{t_{n+1}^k}{k!} + \frac{1}{\Gamma(q)} \sum_{j=0}^{n} b_{j,n+1} f\left(y_h(t_j), t_j\right) \tag{20.8}$$

and

$$a_{j,n+1} = \begin{cases} n^{q+1} - (n-\text{q})(n+1)^q, & \text{if } j = 0 \\ (n-j+2)^{q+1} + (n-j)^{q+1} - 2(n-j+1)^{q+1}, & \text{if } 1 \leq \text{j} \leq n \\ 1 & \text{if } j = n+1 \end{cases} \tag{20.9}$$

$$b_{j,n+1} = \frac{h^q}{q}\left((n-j+1)^q - (n-j)^q\right) \tag{20.10}$$

The approximation error is given as

$$max_{j=0,1,2,\ldots,N} = \left|y\left(t_j\right) - y_h\left(t_j\right)\right| = O(h^p) \tag{20.11}$$

$p = min(2, 1+q)$. Therefore, the numerical solution of a fractional order system can be obtained by applying the abovementioned algorithm.

In this work, the method of numerical evaluation of the fractional derivative of some usual functions (Grünwald–Letnikov method) is given (Petráš, 2006; Khettab et al., 2017a,c,d), the definition is expressed as :

$${}_{t_0}^{GL}D_t^q f(t) = \lim_{n \to 0} \frac{1}{h^n} \sum_{j=0}^{\left[\frac{t-q}{h}\right]} (-1)^j \begin{pmatrix} q \\ j \end{pmatrix} f(t - jh) \tag{20.12}$$

where $\left[\frac{t-q}{h}\right]$ indicates the integer part and $(-1)^j \begin{pmatrix} q \\ j \end{pmatrix}$ are binomial coefficients $c_j^{(q)} (j = 0, 1, \ldots)$.

The calculation of these coefficients is done by formula of following recurrence:

$$c_0^{(q)} = 1,\ c_j^q = \left(1 - \frac{1+q}{j}\right) c_{j-1}^q$$

The general numerical solution of the fractional differential equation

$${}_{a}^{GL}D_{t}^{q}y(t) = f(y(t), t),$$

Can be expressed as follows:

$$y(t_k) = f(y(t_k), t_k)h^q - \sum_{j=v}^{k} c_j^{(q)} y(t_{k-j}) \tag{20.13}$$

This approximation of the fractional derivative within the meaning of Grünwald–Letnikov is on the one hand equivalent to the definition of Riemman–Liouville for a broad class of functions (Petráš, 2006), on the other hand, it is well adapted to the definition of Caputo (Adams method) because it requires only the initial conditions and has a physical direction clearly.

Remark: . As noted in (Petráš, 2008, 2006), both numerical methods in the time domain mentioned (Grünwald–Letnikov and Adams–Bashforth–Moulton) have approximately the same order of accuracy and good digital solutions.

20.3 BASICS AND DESCRIPTION OF THE T–S FUZZY SYSTEMS

Fuzzy logic systems address the imprecision of the input and output variables directly by defining them with fuzzy numbers (and fuzzy sets) that can be expressed in linguistic terms (e.g., small, medium and large). The basic configuration of the T–S system includes a fuzzy rule base, which consists of a collection of fuzzy IF–THEN rules in the following form (Wang, 1997; Zelinka and Youssef, 2013; Tanaka et al., 2001; Takagi and Sugeno, 1985; Azar, 2010a,b, 2012):

$$R^{(l)}\!:\!IF\ x_1\ is\ F_1^l, and \ldots, and\ x_n\ is\ F_n^l\ THEN\ y_l = f_l(x)$$

$$y_l = q_0^l + q_1^l x_1 + \ldots + q_n^l x_n = \underline{\theta}_l^T[1x^T]$$

where $(F_1^l, \ldots, F_i^l, \ldots, F_n^l)$ are input fuzzy sets and $\underline{\theta}_l^T = [q_0^l, q_1^l + \ldots + q_n^l]$ is a vector of the adjustable factors of the consequence part of the fuzzy rule. Also y_l is a crisp value, and a fuzzy inference engine to combine the fuzzy IF–THEN rules in the fuzzy rule base into a mapping from an input linguistic vector $X = \underline{x}^T = [x_1 + x_2 \ldots + x_n] \in \Re^n$ to an output variable $y \in \Re$. Let M be the number of fuzzy IF–THEN rules. The output of the fuzzy logic systems with central average defuzzifier, product inference, and singleton fuzzifier can be expressed as

$$y(X) = \frac{\sum_{l=1}^{M} v^l . y_l}{\sum_{l=1}^{M} v^l} = \frac{\sum_{l=1}^{M} v^l . \underline{\theta}_l^T [1x^T]^T}{\sum_{l=1}^{M} v^l} \tag{20.14}$$

where $v^l = \prod_{i=1}^{n} \mu_{F_i^l(x_i)}$ is the true value of the lth implication and $\mu_{F_i^l(x_i)}$ is the membership function value of the fuzzy variable x_i (Lin et al., 2004; Hartley et al., 1995)

Eq. (20.14) can be rewritten as:

$$y(X) = \underline{\theta}_l^T \xi(X) \tag{20.15}$$

where $\underline{\theta}_l^T = \left[\underline{\theta}_1^T \underline{\theta}_2^T \ldots \underline{\theta}_M^T\right]$ is an adjustable parameter vector and

$\xi^T(X) = [\xi^1(X), \xi^2(X), \ldots, \xi^M(X)]$ is a fuzzy basis function vector defined as:

$$\xi^l(X) = \frac{v^l\left[1X^T\right]}{\sum_{l=1}^{M} v^l}$$

When the inputs are fed into the T−S, the true value $v^l$of the lth implication is computed. Applying the common defuzzification strategy, the output expressed as Eq. (20.14) is pumped out.

Based on the universal approximation theorem (Wang and Mendel, 1992; Wang, 1994; Castro, 1995), the above fuzzy logic system is capable of uniformly approximating any well-defined nonlinear function over a compact set Uc to any degree of accuracy. Also, it is straightforward to show that a multioutput system can always be approximated by a group of single-output approximation systems.

20.4 FUZZY ADAPTIVE ROBUST H_∞ CONTROL: SLIDING MODE APPROACH (VSC)

Consider a fractional order SISO nonlinear dynamic system of the form (Lin and Kuo, 2011, 2012; Khettab et al., 2017c):

$$\begin{cases} x_1^{(q_1)} - x_2 \\ \vdots \\ x_{n-1}^{(q_{n-1})} = x_n \\ x_n^{(q_n)} = f(X,t) + g(X,t)u + d(t) \\ y = x_1 \end{cases} \tag{20.16}$$

where

$X = [x_1,\ x_2,\ \ldots,\ x_n]^T = \left[x,\ x^{(q)},\ x^{(2q)}, \ldots,\ x^{((n-1)q)}\right]^T \in R^n$ is the system's state vector, $u \in R$ is the control input, and $y \in R$ is the output, with the initial conditions:

$u(0) = 0$ and $y(0) = 0$.

The initial conditions are set to zero to avoid unrobustness for Nussbaum type adaptive controller as proved by Georgiou and Smith (1997),

If $q_1 = q_2 = \cdots = q_n = q$ the above system is called a commensurate order system. Then equivalent form of the above system is described as:

$$\begin{aligned} x^{(nq)} &= f(X,t) + g(X,t)u + d(t) \\ y &= x_1 \end{aligned} \tag{20.17}$$

where $f(X,t)$ and $g(X,t)$ are unknown but bounded nonlinear functions which express system dynamics and $d(t)$is the external bounded disturbance. The control objective is to force the system output y to follow a given bounded reference signal y_d, under the constraint that all signals involved must be bounded.

The reference signal vector $\underline{y}_d$ and the tracking error vector $\underline{e}$ are defined as,

$$\begin{aligned} \underline{y}_d &= \left[y_d,\ y_d^{(q)},\ \ldots,\ y_d^{((n-1)q)}\right]^T \in R^n \\ \underline{e}_d &= \underline{y}_d - \underline{y} = \left[e,\ e^{(q)},\ \ldots,\ e^{((n-1)q)}\right]^T \in R^n, \\ e^{(iq)} &= y_d^{(iq)} - y^{(iq)} \end{aligned} \tag{20.18}$$

Let $\underline{k} = [k_1, k_2, \ldots, k_n]^T \in R^n$ be chosen such that the stable condition $\left|\arg(eig(A))\right| > q\pi/2$ is met, where $0 < q < 1$ and $eig(A)$ represents the eigenvalues of the system state matrix given in Eq. (20.19).

By substituting Eq. (20.18) into Eq. (20.17) we obtain the closed-loop control system in the state space domain as follows:

$$\begin{aligned} \underline{x}^{(nq)} &= A\underline{x} + B[f(X) + g(X)u] \\ y &= c^T\underline{x} \end{aligned} \tag{20.19}$$

where

$$A = \begin{bmatrix} 0 & 1 & 0 & 0 & \cdots & 0 & 0 \\ 0 & 0 & 1 & 0 & \cdots & 0 & 0 \\ \vdots & \vdots & \vdots & \vdots & \ddots & \vdots & \vdots \\ 0 & 0 & 0 & 0 & \cdots & 0 & 1 \\ -k_1 & -k_2 & -k_3 & -k_4 & \cdots & -k_{(n-1)} & -k_n \end{bmatrix}, B = \begin{bmatrix} 0 \\ 0 \\ \vdots \\ 0 \\ 1 \end{bmatrix} \text{ and } c = \begin{bmatrix} 1 \\ 0 \\ \vdots \\ 0 \\ 0 \end{bmatrix}$$

By using the relation $y_d^{(q)} = Ay_d + By_d^{(nq)}$, the following equation is obtained:

$$\begin{aligned} \underline{e}^{(q)} &= A\underline{e} + B\left[f(X) + g(X)u^* - y_d^{(nq)}\right] \\ e &= c^T\underline{e} \end{aligned} \tag{20.20}$$

In what follows, a fuzzy adaptive control will be designed to stabilize the system Eq. (20.16) or an equivalent system Eq. (20.19).

i. If the functions $f(X,t)$ and $g(X,t)$are known and the system is free of external disturbance $d(t)$(i.e., $d(t) = 0$),

The following *assumptions* are considered (Liu and Wang, 2009; Boulkroune and M'saad, 2012),

Assumption IV.1. The control gain $g(X,t)$ is not zero and of known sign. It is also strictly positive or strictly negative.
Assumption IV.2. The external disturbance is bounded: $|d(t)| \le D$ with D an unknown positive constant.

Then the control law of the certainty equivalent controller is obtained as:

$$u^* = \frac{1}{g(X,t)}\left[-f(X,t) + y_d^{(nq)} + k^T e\right] \tag{20.21}$$

where $\underline{y}_d = \left[y_d,\, y_d^{(q)},\, \ldots,\, y_d^{((n-1)q)}\right]^T \in R^n$

$$\underline{e}_d = \underline{y}_d - \underline{y} = \left[e,\, e^{(q)},\, \ldots,\, e^{((n-1)q)}\right]^T \in R^n,$$

$e^{(iq)} = y_d^{(iq)} - y^{(iq)}$ is the tracking error vector.
Substituting Eq. (20.21) into Eq. (20.20), we have

$$e^{(nq)} = k_n e^{(n-1)q} + \cdots + k_1 e = 0 \tag{20.22}$$

which is the main objective of control, $\lim_{t\to\infty} e(t) = 0$.
The sliding surface is defined as:

$$\begin{aligned} s(X,t) &= -(K\mathbf{e}) \\ &= -(k_1 e + k_2 e^{(q)} + \ldots + k_{n-1} e^{((n-2)q)} \\ &\quad + e^{((n-1)q)}) \end{aligned} \tag{20.23}$$

when $e(0) = 0$, the tracking problem $x = y_d$ implies that the sliding surface $s(\mathbf{e}) = 0,\ \forall t \ge 0$.
This later classic derivative can be decomposed into a fractional type,

$$\begin{aligned} \dot{s}(X,t) = D^{(1-q)}(D^q(s(X,t))) = 0 \\ \text{then} \quad D^q(s(X,t)) = 0 \end{aligned} \tag{20.24}$$

If the functions $f(X,t)$ and $g(X,t)$ are known and the system is free of external disturbance, i.e., $d(t) = 0$. The control signal in the following equation drives the dynamic to reach to the sliding surface:

$$\begin{aligned} s^{(q)} &= -\left(\sum_{i=1}^{n-1} k_i e^{(iq)} + e^{(nq)}\right) \\ &= -\sum_{i=1}^{n-1} k_i e^{(iq)} + f(X,t) + g(X,t)u_{eq} - \mathbf{y}_d^{(nq)} \\ &= 0 \end{aligned} \tag{20.25}$$

Therefore, the equivalent control law is given by:

$$u_{eq} = \frac{1}{g(X,t)}\left(\sum_{i=1}^{n-1} k_i e^{(iq)} - f(X,t) + \mathbf{y}_d^{(nq)}\right) \tag{20.26}$$

Substituting Eq. (20.26) into Eq. (20.19), we have

$$e^{(nq)} + k_n e^{(n-1)q} + \ \ldots \ + k_1 e = 0 \tag{20.27}$$

which is the main objective of control $\lim_{t\to\infty} e(t) = 0$.

In the reaching phase we get $s(X,t) \neq 0$, and a switching-type control u_{sw} must be added in order satisfy the sufficient condition $s(X,t)\dot{s}(X,t) \leq -\eta|s(X,t)|, \quad \eta > 0$ which implies the global control.

Therefore the global sliding mode control law is given by:

$$u^* = \frac{1}{g(X,t)}\left(\sum_{i=1}^{n-1} k_i e^{(iq)} - f(X,t) + y_d^{(nq)} - u_a - K\text{sgn}(s)\right) \tag{20.28}$$

We can show by taking a Lyapunov function candidate defined as

$$V = \frac{1}{2}s^2(\mathbf{e}) \tag{20.29}$$

And differentiating Eq. (20.29) with respect to time to the fractional order q, $V^{(q)}(t)$ along the system trajectory (Aguila-Camacho et al., 2014), we obtain

$$\begin{aligned} V^{(q)} &= ss^{(q)} = s\left(\sum_{i=1}^{n-1} k_i e^{(iq)} + e^{(nq)}\right) \\ &= -s\left(\sum_{i=1}^{n-1} k_i e^{(iq)} - y_d^{(nq)} + y^{(nq)}\right) \\ &= -s\left(\sum_{i=1}^{n-1} k_i e^{(iq)} + f(X,t) + g(X,t)u_{eq} - y_d^{(nq)}\right) \\ &\leq -\eta|s(X,t)| \end{aligned} \tag{20.30}$$

Hence, the sliding mode control u^* guarantees the sliding condition of $s(X,t)\dot{s}(X,t)$.

However, the functions $f(X,t)$ and $g(X,t)$ are usually unknown in practice and it is difficult to apply the control law Eq. (20.28) for an unknown nonlinear plant. Moreover, the chattering problem appears when adding the switching control term $u_{sw} = K\ \text{sgn}(s)$.

To deal with these problems, we consider the adaptive sliding mode control scheme using a fuzzy logic system and the *saturation function* to avoid chattering problem.

The *Chattering phenomenon* can be reduced by replacing the function sign by a function of adequate saturation which filters the high frequencies.

The modified resulting control law $u(t)$, which includes a fuzzy system to approximate the unknown functions $f(X,t)$ and $g(X,t)$ and a saturation function that attenuates the chattering and improves performance, is as follows:

$$u = \frac{1}{g(\mathrm{X}|\underline{\theta}_g)}\left[-f\left(\mathrm{X}|\underline{\theta}_f\right) + y_d^{(nq)} + \underline{k}^T\underline{e} - u_a - \mathrm{sat(s(X,t))}\right] \tag{20.31}$$

Note that the control law Eq. (20.28) is realizable only while $f(X,t)$ and $g(X,t)$ are well known.

However, $f(X,t)$ and $g(X,t)$ are unknown and external disturbance $d(t) \neq 0$, the ideal control effort Eq. (20.28) cannot be implemented. We replace $f(X,t)$, $g(X,t)$ by the fuzzy logic system $f(X|\theta_f)$, $g(X|\theta_g)$ in a specified form as Eqs. (20.14, 20.15), i.e.,

$$f(X|\theta_f) = \xi^T(X)\theta_f,\ g(X|\theta_g) = \xi^T(X)\theta_g, \tag{20.32}$$

Here the fuzzy basis function $\xi^T(X)$ depends on the fuzzy membership functions and is supposed to be fixed, while θ_f, θ_g and are adjusted by adaptive laws based on a Lyapunov stability criterion (Aguila-Camacho et al., 2014; Duarte-Mermoud et al., 2015; Sastry and Bodson, 1989).

The optimal parameter estimations θ_f^*, $\theta_g^*$and are defined as:

$$\begin{aligned} \theta_f^* &= \arg\min_{\theta_f \in \Omega_f} \left[\sup_{x \in \Omega_x} \left| f(X|\theta_f) - f(X,t) \right| \right] \\ \theta_g^* &= \arg\min_{\theta_g \in \Omega_g} \left[\sup_{x \in \Omega_x} \left| g(X|\theta_g) - g(X,t) \right| \right] \end{aligned} \tag{20.33}$$

where $\Omega_f, \Omega_g, \Omega_p$, and Ω_x are constraint sets of suitable bounds on $\theta_f, \theta_g, \theta_p$, and x respectively, and they are defined as $\Omega_f = \{\theta_f|\ |\theta_f| \le M_f\}$, $\Omega_g = \{\theta_g|\ |\theta_g| \le M_g\}$, $\Omega_p = \{\theta_p|\ |\theta_p| \le M_p\}$, and $\Omega_x = \{x|\ |x| \le M_x\}$ where M_f, M_g, M_p and M_x are positive constants.

Assuming that the fuzzy parameters θ_f, θ_g, and θ_p never reach the boundaries.

Let us define the minimum approximation error,

$$\omega = \left[f(X,t) - f(X|\theta_f^*) \right] + \left[g(X,t) - g(X|\theta_g^*) \right] u \tag{20.34}$$

and define the errors: $\tilde{\theta}_f = \theta_f - \theta_f^*$, $\tilde{\theta}_g = \theta_g - \theta_g^*$, and $\tilde{\theta}_p = \theta_p - \theta_p^*$.

Then, the equation of the sliding surface Eq. (20.17) can be rewritten as:

$$\begin{aligned} s^{(q)} &= \omega + \left[f(X|\theta_f^*) - f(X\theta_f) \right] + \left[g(X|\theta_g^*) - g(X\theta_g) \right] u \\ &\quad - p(\mathbf{s}|\theta_p) + p(\mathbf{s}|\theta_p^*) - p(\mathbf{s}|\theta_p^*) + d(t) \\ &= \omega - \tilde{\theta}_p^T \xi(\mathbf{s}) - \tilde{\theta}_f^T \xi(X) - \tilde{\theta}_g^T \xi(X) u \\ &\quad - p(\mathbf{s}|\theta_p^*) + d(t) \end{aligned} \tag{20.35}$$

20.5 STABILITY ANALYSIS

Consider the commensurate fractional order SISO nonlinear dynamic system Eq. (20.17) with control input Eq. (20.28), if the robust compensator u_a and the fuzzy-based adaptive laws are chosen as:

$$u_a = -\frac{1}{r} B^T P \underline{e} \tag{20.36}$$

$$\underline{\theta}_f^{(q)} = -r_1 \xi(X) B^T P \underline{e} \tag{20.37}$$

$$\underline{\theta}_g^{(q)} = -r_2 \xi(X) B^T P \underline{e} u \tag{20.38}$$

where $r > 0, r_i > 0, i = 1 \sim 2$, and $P = P^T > 0$ is the solution of the following Riccati-like equation.

$$PA + A^T P + Q - PB\left(\frac{2}{r} - \frac{1}{\rho^2}\right)B^T P = 0 \tag{20.39}$$

where $Q = Q^T > 0$ is a prescribed weighting matrix. Therefore, the H_∞ tracking performance can be achieved for a prescribed attenuation level ρ which satisfies $2\rho^2 \geq r$ and all the variables of the closed-loop system are bounded.

In order to analyze the closed-loop stability, the fractional Lyapunov function candidate (Aguila-Camacho et al., 2014; Duarte-Mermoud et al., 2015) is chosen as

$$V = \frac{1}{2}\underline{e}^T(t) P \underline{e}(t) + \frac{1}{2r_1}\left(\underline{\tilde{\theta}}_f^T\right)\left(\underline{\tilde{\theta}}_f\right) + \frac{1}{2r_2}\left(\underline{\tilde{\theta}}_g^T\right)\left(\underline{\tilde{\theta}}_g\right) \tag{20.40}$$

Taking the derivative of (20.40) with respect to time, we get

$$V^{(q)}(t) = \frac{1}{2}\left(\underline{e}^{(q)}(t)\right)^T(t) + \frac{1}{2}\underline{e}^T(t) P \underline{e}(t) + \frac{1}{r_1}\left(\underline{\tilde{\theta}}_f^T\right)\left(\underline{\tilde{\theta}}_f^{(q)}\right) + \frac{1}{r_2}\left(\underline{\tilde{\theta}}_g^T\right)\left(\underline{\tilde{\theta}}_g^{(q)}\right) \tag{20.41}$$

$$\begin{aligned}
&= \frac{1}{2}\left\{A\underline{e} + B\left[\xi(X)^T\underline{\tilde{\theta}}_f + \xi(X)^T\underline{\tilde{\theta}}_g u + u_a + w_1\right]\right\}^T P\underline{e} \\
&\quad + \frac{1}{2}\underline{e}^T(e) P\left\{A\underline{e} + B\left[\begin{array}{c}\xi(X)^T\underline{\tilde{\theta}}_f + \xi(X)^T\underline{\tilde{\theta}}_g u + u_a \\ + w_1\end{array}\right]\right\} \\
&\quad + \frac{1}{r_1}\left(\underline{\tilde{\theta}}_f^T\right)\left(\underline{\tilde{\theta}}_f^{(q)}\right) + \frac{1}{r_2}\left(\underline{\tilde{\theta}}_g^T\right)\left(\underline{\tilde{\theta}}_g^{(q)}\right) \\
&= \frac{1}{2}\underline{e}^T\left(A^T P + PA\right)\underline{e} + \underline{e}^T PBu_a + \underline{e}^T PBw_1 \\
&\quad + \left\{\underline{\tilde{\theta}}_f^T\left[\xi(X) B^T P\underline{e} + \frac{1}{r_1}\left(\underline{\tilde{\theta}}_f^{(q)}\right)\right]\right\} \\
&\quad + \left\{\underline{\tilde{\theta}}_g^T\left[\xi(X) B^T P\underline{e} u + \frac{1}{r_2}\left(\underline{\tilde{\theta}}_g^{(q)}\right)\right]\right\}
\end{aligned} \tag{20.42}$$

From the robust compensator u_a and the fuzzy-based adaptive laws, given in Eqs. (20.36)–(20.38), $V^{(q)}(t)$ in Eq. (20.42) can be rewritten as

$$\begin{aligned} V^{(q)}(t) &= -\frac{1}{2}\underline{e}^T Q\underline{e} - \frac{1}{2\rho^2}\underline{e}^T PBB^T\underline{e} + \underline{e}^T PBw_1 \\ &= -\frac{1}{2}\underline{e}^T Q\underline{e} - \frac{1}{2}\left(\frac{1}{\rho}B^T P\underline{e} - \rho w_1\right)^T \left(\frac{1}{\rho}B^T P\underline{e} - \rho w_1\right) + \frac{1}{2}\rho^2 w_1^T w_1 \\ &\le -\frac{1}{2}\underline{e}^T Q\underline{e} + \frac{1}{2}\rho^2 w_1^T w_1 \end{aligned} \tag{20.43}$$

Integrating (20.43) from $t = 0$ to $t = T$, we have

$$V(T) - V(0) \le -\frac{1}{2}\int_0^T \left(\underline{e}^T Q\underline{e}dt + \frac{1}{2}\rho^2 w_1^T w_1\right)dt \tag{20.44}$$

Since $V(T) \ge 0$, Eq. (20.35) can be rewritten as follows:

$$\int_0^T \underline{e}^T Q\underline{e}dt \le e^T(0)Pe(0) + \theta^T(0)\theta(0) + \rho^2 \int_0^T w_1^T w_1 dt \tag{20.45}$$

Therefore, the robust H_∞ tracking performance can be achieved. The proof is completed.

The diagram of the proposed control is given in Fig. 20.1.

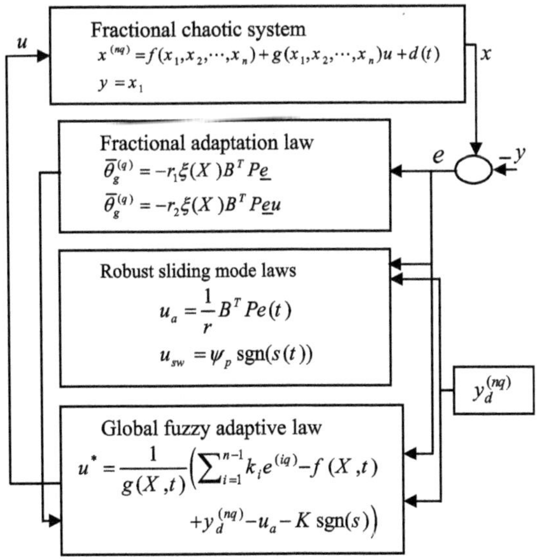

FIGURE 20.1 Global block diagram of the proposed FRAF-VSC controller.

20.6 SIMULATION RESULTS

The chaotic behaviors in a fractional order modified Duffing system (gyro systems) studied numerically by phase portraits are given by Lin et al. (2012) and Hosseinnia et al. (2010). In this section, we will apply our adaptive fuzzy robust H_∞ controller via sliding mode to synchronize two different fractional order chaotic gyro systems.

Consider the following two fractional order chaotic systems (Lin and Balas, 2011, Hosseinnia et al., 2010; Kuo et al., 2011):

- Response system:

$$\begin{aligned} D^q_{x_1} &= x_2 \\ D^q x_2 &= \frac{-100}{4}x_1 + \frac{1}{12}x_1^3 - 0.7x_2 - 0.08x_2^3 + \sin(x_1) + 33\sin(2t)x_1 \\ &\quad - \frac{1}{6}x_1^3 - 0.1\sin(x_1) + d(t) + u(t) \end{aligned} \tag{20.46}$$

- Drive system:

$$\begin{aligned} D^q_{y_1} &= y_2 \\ D^q_{y_2} &= -\frac{100}{4}y_1 + \frac{1}{12}y_1^3 - 0.5y_2 - 0.05y_2^3 + \sin(y_1) + 35.5\sin(2t)y_1 - \frac{1}{6}x_1^3 + d(t) \end{aligned} \tag{20.47}$$

where the external disturbance $d(t) = 0.3\sin(t)$. The main objective is to control the trajectories of the response system to track the reference trajectories obtained from the drive system. The initial conditions of the drive and response systems are chosen as:

$$\begin{bmatrix} x_1(0) \\ x_2(0) \end{bmatrix} = \begin{bmatrix} 0.25 \\ 0.25 \end{bmatrix} and \begin{bmatrix} y_1(0) \\ y_2(0) \end{bmatrix} = \begin{bmatrix} 0.2 \\ 0.2 \end{bmatrix}, \text{ respectively.}$$

For the other constants of design are fixed as follows: $k_1 = k_2 = 1$, $r_1 = 175$, $r_2 = 37$, $r_3 = 75$, $r_4 = 7$, $h = 0.001$, and $Tsim = 40s$.

The simulations results for fractional order $q = 0.98$ are illustrated as follows:

Fig. 20.2 represents the 3D phase portrait of the drive and response systems without control input. It is obvious that the synchronization performance is bad without a control effort supplied to the response system.

The different values of $0 < q < 1$ are considered in order to show the robustness of the proposed adaptive fuzzy H^∞ control with our law.

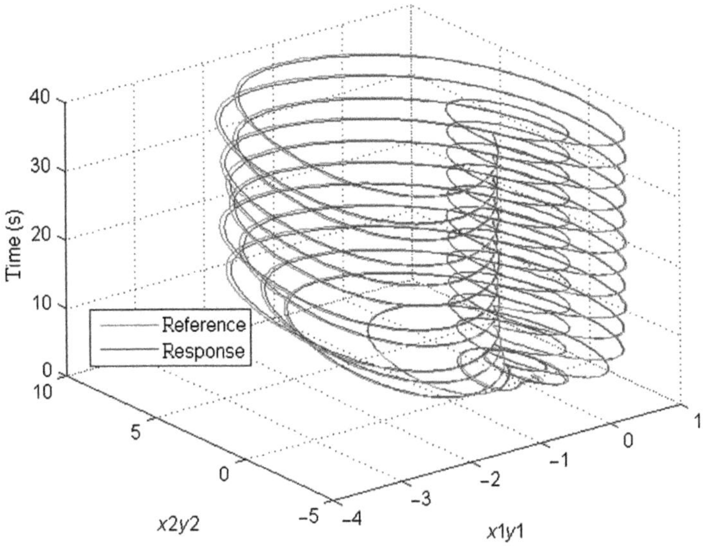

FIGURE 20.2 3D phase portrait of the drive and response systems without control input. (without the control input law).

According to the two state output ranges, the membership functions of x_i, for $f\left(X\middle|\underline{\theta}_f\right)$ and $g\left(X\middle|\underline{\theta}_g\right)$ are selected as follows:

$\mu_{F_i^l}(x_i) = \exp\left[-0.5\left(\frac{x_i-\bar{x}}{0.8}\right)^2\right]$ $i=1,2$ and $l=1,\ldots,7$ where $\bar{x}$ is selected from the interval $[-1,2]$.

From the adaptive laws Eqs. (20.30)–(20.33) and the robust compensator Eq. (20.29), the sliding mode control law of the response system can be obtained as:

$$u = \frac{1}{\xi^T(X)\underline{\theta}_g}\left[\xi^T(X)\underline{\theta}_f + y_d^{(nq)} + \underline{k}^T\underline{e} - u_a - \mathrm{K.sgn(s)}\right] \tag{20.48}$$

According to stability analysis section, the controlled error system can be stabilized, i.e., the master system Eq. (20.46) can synchronize the slave system Eq. (20.47) with the control law Eq. (20.28).

The Figs. 20.3 and 20.4 represent the different simulation results of the drive and response systems with control input Eq. (20.48) for the fractional order $q=0.98$.

It is clearly seen from Fig. 20.4E that the tracking errors $e_1(t)$ and $e_2(t)$ converge both to zero in less than 5s. Synchronization is perfectly achieved as shown by the state trajectories in Fig. 20.4A and B.

The control signal can be observed in the Fig. 20.4D. It indicates that the obtained results are comparable with the solution presented in

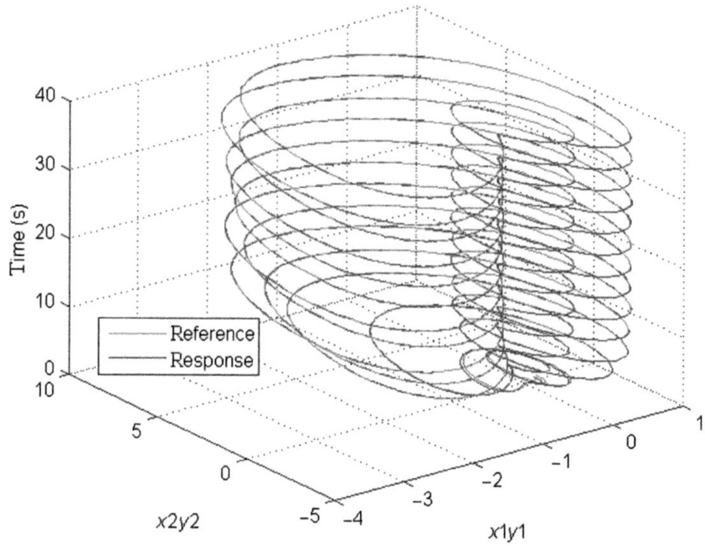

FIGURE 20.3 3D phase portrait, synchronization performance, of the drive and response systems (with control effort).

(Kuo et al., 2011; Lin et al., 2011, 2012), but fluctuations of the control function are much smaller (Fig. 20.5).

- *Step 1: Fractional RAFSM Controller*
- *Step 2 - Chattering elimination: Using the saturation function "sat"*

Let us consider now the problem of eliminating the chattering that appeared in the above results, and introduce the complementary saturation function. The equation of the saturation function is given as follows:

$$\begin{cases} \text{sat}(S) = 1 & \text{if } \ S > \mu \\ \text{sat}(S) = -1 & \text{if } \ S < \mu \\ \text{sat}(S) = S/\mu & \text{if } \ |S| < \mu \end{cases} \tag{20.49}$$

where μ is a small and positive parameter.

The control low as Eq. (20.31):

$$u = \frac{1}{g\left(\text{X}\middle|\underline{\theta}_g\right)\left[-f\left(\text{X}\middle|\underline{\theta}_f\right) + y_d^{(nq)} + \underline{k}^T\underline{e} - u_a - \text{sat}(\text{s}(\text{X},\text{t}))\right]} \tag{20.50}$$

The simulation results are given in Fig. 20.6A for the states trajectories $x_1$and y_1 of the chaotic drive and response systems, Fig. 20.6E for the error signals. It can be seen that the tracking performance is good even in presence of disturbance $d(t)$ and the chattering phenomena is eliminated in the sliding

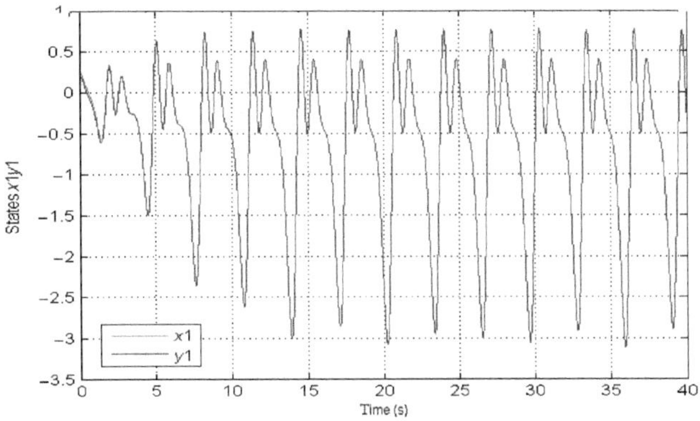

(A) State trajectories: x_1y_1.

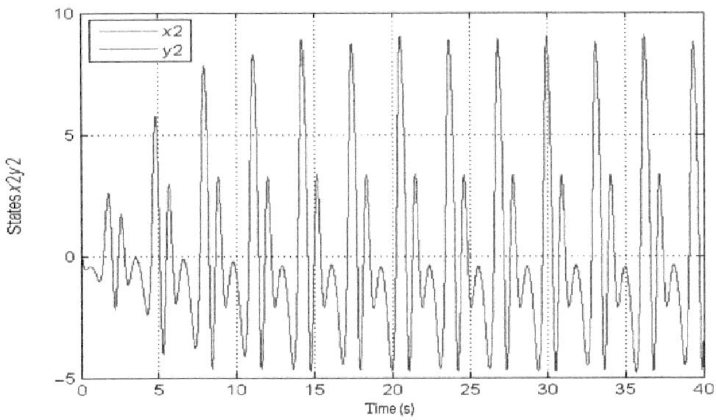

(B) State trajectories: x_2y_2.

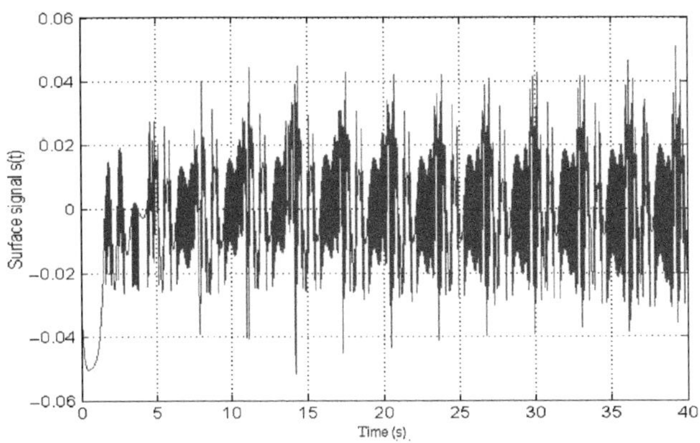

(C) Trajectory of the sliding surface s(t)

FIGURE 20.4 Simulation Results with the RAFSM control law.

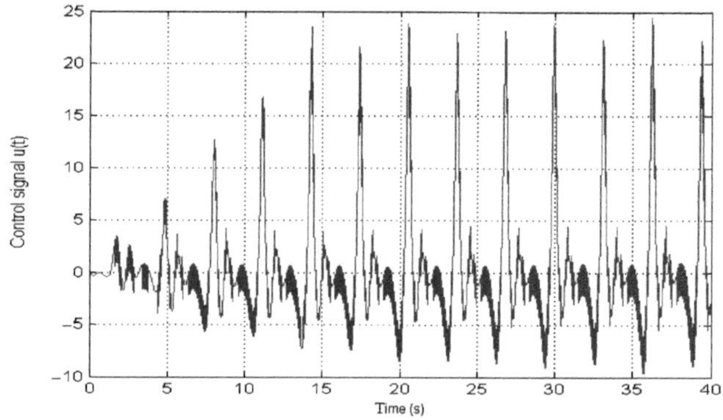

(D) Trajectory of the control effort u(t)

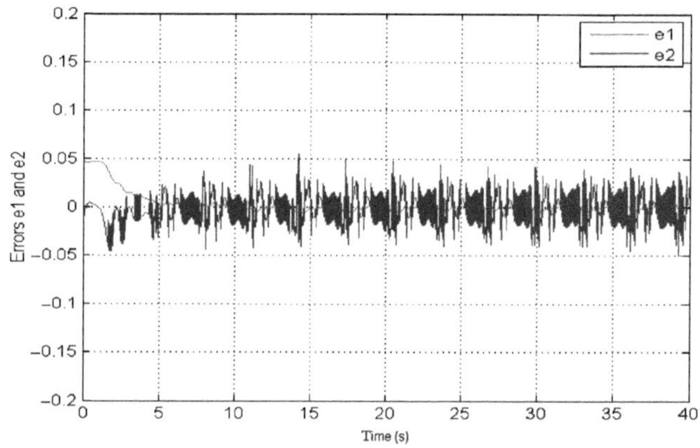

(E) Error signals $e_1(t)$ and $e_2(t)$

FIGURE 20.4 (Continued)

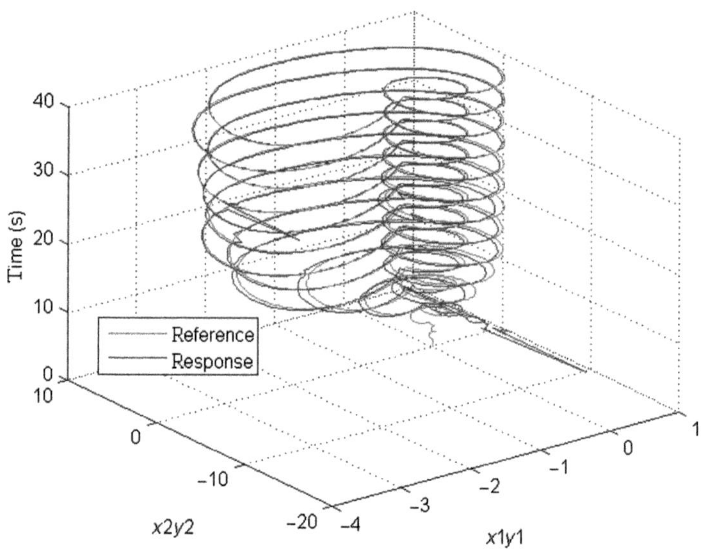

FIGURE 20.5 3D phase portrait, synchronization performance, of the drive and response systems (with the control effort).

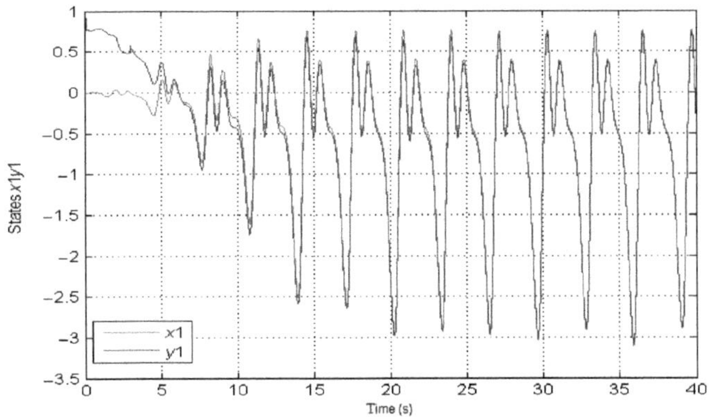

(A) State trajectories: x_1y_1.

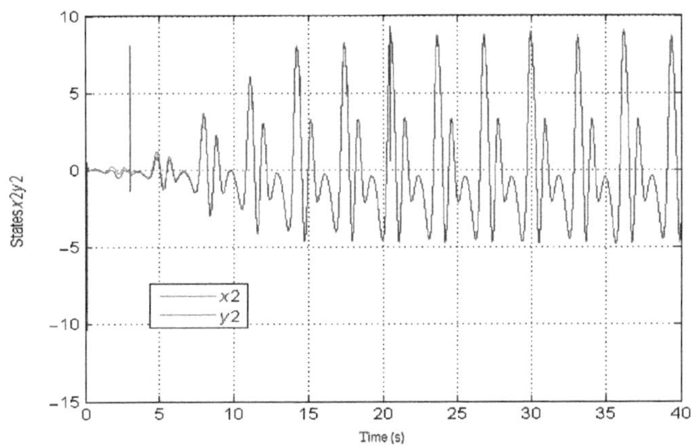

(B) State trajectories: x_2y_2.

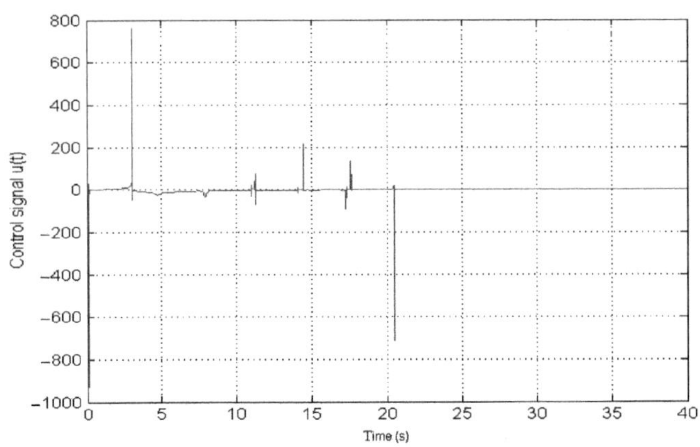

(C) Trajectory of the control effort u(t)

FIGURE 20.6 Simulation Results with the RAFSM-sat control law.

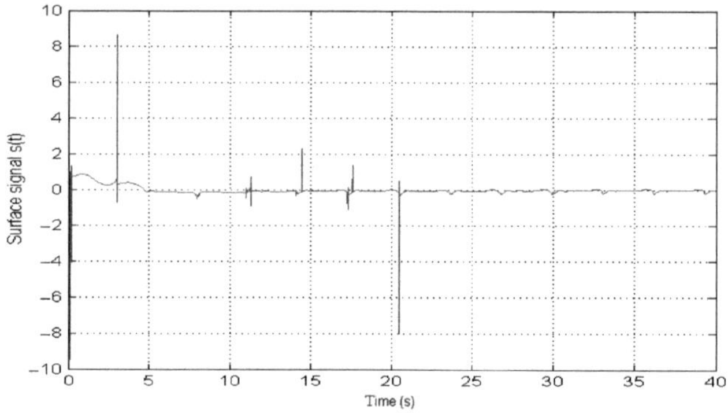

(D) Trajectory of the surface s(t)

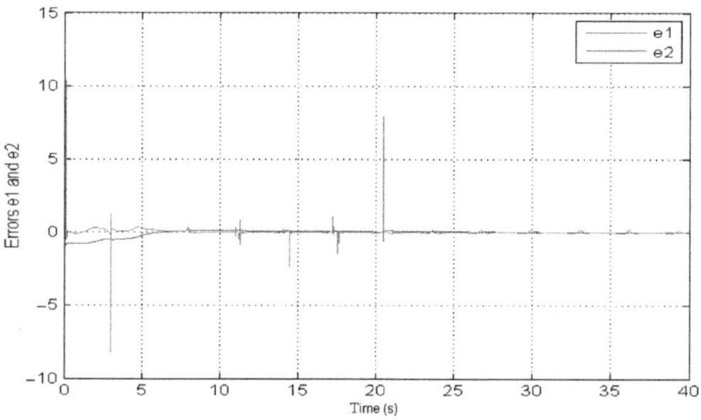

(E) Error signals $e_1(t)$ and $e_2(t)$

FIGURE 20.6 (Continued)

surface trajectory and the control signal $u(t)$ as illustrated from Fig. 20.6D and C respectively.

20.7 CONCLUSION

In this chapter a robust adaptive fuzzy control strategy using robust H_∞ control and VSS-sliding mode theory is proposed to deal with chaos synchronization between two different uncertain fractional order chaotic systems. The use of the fuzzy logic helps to minimize the added computational burden, and hence renders the overall system to be more practically applicable. The use of a saturation function with the sliding mode controller in order to improve the control signal quality by eliminating the undesirable chattering action in the control and surface signals. Based on the fractional Lyapunov synthesis approach, free parameters of the adaptive fuzzy controller can be tuned online by the output feedback control law and adaptive laws. The simulation example, chaos synchronization of two fractional order systems, is given to demonstrate the effectiveness of the proposed methodology. The significance of the proposed control scheme in the simulation for different values of q is manifest. Simulation results show that a fast synchronization of drive and response can be achieved and as q is reduced the chaos is seen reduced, i.e., the synchronization error is reduced, accordingly. The asymptotic stability of the overall control system is established and an illustrative simulation example, chaos synchronization of two fractional order systems, is realized with the Grünwald–Letnikov numerical approximation approach to demonstrate the effectiveness of the proposed methodology.

Future research efforts will concern observer-based nonlinear adaptive control of uncertain or unknown fractional order systems. The problem of online identification and parameters estimation for such systems is also a good challenge. Another topic of interest is the design of new robust adaptive control laws for the class of fractional discrete nonlinear systems based on various control configurations.

REFERENCES

Aguila-Camacho, N., Duarte-Mermoud, M.A., Gallegos, J.A., 2014. Lyapunov functions for fractional order systems. Commun Nonlinear Sci Numer Simulat 19 (9), 2951–2957.

Arena, P., Caponetto, R., 1998. Bifurcation and chaos in non-integer order cellular neural networks. Int J Bifurcat Chaos 8 (7), 527–1539.

Arena, P., Caponetto, R., Fortuna, L. and Porto, D. 1997. Chaos in a fractional order Duffing system. In: Proceedings ECCTD. pp. 1259–1262.

Azar, A.T., 2010a. Fuzzy Systems. IN-TECH, Vienna, Austria. 978-953-7619-92-3.

Azar, A.T., 2010b. Adaptive Neuro-Fuzzy Systems. In: Azar, A.T. (Ed.), Fuzzy Systems. IN-TECH, Vienna, Austria. 978-953-7619-92-3.

Azar, A.T., 2012. Overview of Type-2 Fuzzy logic systems. International Journal of Fuzzy System Applications (IJFSA) 2 (4), 1–28.

Azar, A.T., Serrano, F.E., 2015a. Stabilization and Control of Mechanical Systems with Backlash. In: Azar, A.T., Vaidyanathan, S. (Eds.), Advanced Intelligent Control Engineering and Automation, Advances in Computational Intelligence and Robotics (ACIR) Book Series. IGI-Global, USA.

Azar, A.T., Serrano, F.E., 2015b. Design and Modeling of Anti Wind Up PID Controllers. In: Zhu, Q., Azar, A.T. (Eds.), Complex system modelling and control through intelligent soft computations, Studies in Fuzziness and Soft Computing, Vol. 319. Springer-Verlag, Germany, pp. 1–44. Available from: http://dx.doi.org/10.1007/978-3-319-12883-2_1.1.

Azar, A.T., Serrano, F.E., 2015c. Adaptive Sliding mode control of the Furuta pendulum. In: Azar, A.T., Zhu, Q. (Eds.), Advances and Applications in Sliding Mode Control systems, Studies in Computational Intelligence, Vol. 576. Springer-Verlag GmbH Berlin/Heidelberg, pp. 1–42. Available from: http://dx.doi.org/10.1007/978-3-319-11173-5_1.

Azar, A.T., Serrano, F.E., 2015d. Deadbeat Control for Multivariable Systems with Time Varying Delays. In: Azar, A.T., Vaidyanathan, S. (Eds.), Chaos Modeling and Control Systems Design, Studies in Computational Intelligence, Vol. 581. Springer-Verlag GmbH Berlin/Heidelberg, pp. 97–132. Available from: http://dx.doi.org/10.1007/978-3-319-13132-0_6.

Azar, A.T., Serrano, F.E., 2016a. Robust control for asynchronous switched nonlinear systems with time varying delays. Proceedings of the International Conference on Advanced Intelligent Systems and Informatics 2016, Volume 533, Advances in Intelligent Systems and Computing, pp 891–899.

Azar, A.T., Serrano, F.E., 2016b. Stabilization of Mechanical Systems with Backlash by PI Loop Shaping. International Journal of System Dynamics Applications (IJSDA) 5 (3), 20–47.

Azar, A.T., Serrano, F.E., 2017. Passivity Based Decoupling of Lagrangian Systems. Proceedings of the International Conference on Advanced Intelligent Systems and Informatics 2016, Volume 533, Advances in Intelligent Systems and Computing, pp 891-899.

Azar, A.T., Vaidyanathan, S., 2015a. Handbook of Research on Advanced Intelligent Control Engineering and Automation. Advances in Computational Intelligence and Robotics (ACIR) Book Series. IGI Global, USA. 9781466672482.

Azar, A.T., Vaidyanathan, S., 2015b. Computational Intelligence applications in Modeling and Control. Studies in Computational Intelligence, Vol. 575. Springer-Verlag, Germany. 978-3-319-11016-5.

Azar, A.T., Vaidyanathan, S., 2015c. Chaos Modeling and Control Systems Design, Studies in Computational Intelligence, Vol. 581. Springer-Verlag, Germany. 978-3-319-13131-3.

Azar, A.T., Vaidyanathan, S., 2016. Advances in Chaos Theory and Intelligent Control. Studies in Fuzziness and Soft Computing, Vol. 337. Springer-Verlag, Germany. 978-3-319-30338-3.

Azar, A.T., Zhu, Q., 2015. Advances and Applications in Sliding Mode Control systems. Studies in Computational Intelligence, Vol. 576. Springer-Verlag, Germany. 978-3-319-11172-8.

Azar, A.T., Vaidyanathan, S., Ouannas, A., 2017a. Fractional Order Control and Synchronization of Chaotic Systems. Studies in Computational Intelligence, Vol. 688. Springer-Verlag, Germany. 978-3-319-50248-9.

Azar, A.T., Volos, C., Gerodimos, N.A., Tombras, G.S., Pham, V.T., Radwan, A.G., et al., 2017b. A novel chaotic system without equilibrium: Dynamics, synchronization and circuit realization. Complexity 2017. Available from: https://doi.org/10.1155/2017/7871467. Article ID 7871467, 11pages, 2017.

Azar, A.T., Ouannas, A., Singh, S., 2017c. Control of New Type of Fractional Chaos Synchronization. Proceedings of the International Conference on Advanced Intelligent Systems and Informatics 2017, Advances in Intelligent Systems and Computing series, Vol. 639, pp 47-56. Springer-Verlag, Germany.

Azar, A.T., Kumar, J., Kumar, V., Rana, K.P.S., 2017d. Control of a Two Link Planar Electrically-Driven Rigid Robotic Manipulator Using Fractional Order SOFC. Proceedings of the International Conference on Advanced Intelligent Systems and Informatics 2017, Advances in Intelligent Systems and Computing series, Vol. 639, pp 47-56. Springer-Verlag, Germany.

Bensafia, Y., Khettab, K., Ladaci, S., 2017. Robust Fractionalized PID controller Design Using the Sub-optimal Approximation of FOTF. 6th International Conference on Systems and Control (ICSC, 2017); University of Batna 2, Batna, Algeria.

Boulkroune, A., M'saad, M., 2012. On the design of observer-based fuzzy adaptive controller for nonlinear systems with unknown control gain sign. Fuzzy Sets and Systems 201, 71–85 **(2012)**.

Boulkroune, A., Bouzeriba, A., Bouden, T., Azar, A.T., 2016a. Fuzzy Adaptive Synchronization of Uncertain Fractional-order Chaotic Systems. Studies in Fuzziness and Soft Computing, Vol. 337. Springer-Verlag, Germany, pp. 681–697.

Boulkroune, A., Hamel, S., Azar, A.T., 2016b. Fuzzy control-based function synchronization of unknown chaotic systems with dead-zone input. Studies in Fuzziness and Soft Computing, Vol. 337. Springer-Verlag, Germany, pp. 699–718.

Bourouba, B., Ladaci, S., 2017. Stabilization of class of fractional-order chaotic system via new sliding mode control. 6th International Conference on Systems and Control (ICSC'17), 7-9 May 2017, pp. 470-475, University of Batna 2, Batna, Algeria.

Castro, J.L., 1995. Fuzzy logical controllers are universal approximators. IEEE Trans Syst Man Cybern 25 (4), 629–635.

Chen, B.S., Lee, C.H., Chang, Y.C., 1996. H∞ Tracking design of uncertain nonlinear SISO systems: Adaptive fuzzy approach. IEEE Trans on systems 4 (1), 32–43.

Diethelm, K., Ford, N.J., 2002. Analysis of fractional differential equations. Journal of Mathematical Analysis and Applications 265 (2), 229–248.

Diethelm, K., Ford, N.J., Freed, A.D., 2003. A predictor–corrector approach for the numerical solution of fractional differential equations. Nonlinear Dynamics 29 (1-4), 3–22.

Diethelm, K., Ford, N.J., Freed, A.D., 2004. Detailed error analysis for a fractional Adams method. Numerical Algorithms 36 (1), 31–52.

Duarte-Mermoud, M.A., Aguila-Camacho, N., Gallegos, J.A., Castro-Linares, R., 2015. Using general quadratic Lyapunov functions to prove Lyapunov uniform stability for fractional order systems. Communications in Nonlinear Science and Numerical Simulation 22 (1-3), 650–659 (2015).

Ge, Z.M., Ou, C.Y., 2008. Chaos synchronization of fractional order modified Duffing systems with parameters excited by a chaotic signal. Chaos Solitons Fractals 35 (4), 705–717.

Georgiou, T.T., Smith, M.C., 1997. Robustness Analysis of Nonlinear Feedback Systems: An Input-Output Approach. ransactions on Automatic Control 42 (9), 1200–1221. 1997.

Ghoudelbourk, S., Dib, D., Omeiri, A., Azar, A.T., 2016. MPPT Control in wind energy conversion systems and the application of fractional control (PIα) in pitch wind turbine. International Journal of Modelling, Identification and Control (IJMIC) 26 (2), 140–151.

Grassi, G., Ouannas, A., Azar, A.T., Radwan, A.G., Volos, C., Pham, V.T., et al., 2017. Chaos Synchronisation Of Continuous Systems Via Scalar Signal. The 6th International Conference on Modern Circuits and Systems Technologies (MOCAST), 4–6 May 2017, Thessaloniki Greece.

Hartley, T.T., Lorenzo, C.F., Qammer, H.K., 1995. Chaos on a fractional Chua's system. IEEE Trans Circuits Syst. Theory Appl. 42 (8), 485–490.

Hilfer, R., 2001. Applications of fractional calculus in physics. World Scientific, New Jersey.

Ho, H.F., Wong, Y.K., Rad, A.B., 2009. Adaptive fuzzy sliding mode control with chattering elimination for nonlinear SISO systems. Simulation Modeling Practice and Theory 17 (7), 1199–1210.

Hosseinnia, S.H., Ghaderi, R., Ranjbar, N., Momani, M., 2010. Sliding mode synchronization of an uncertain fractional order chaotic system. Computers and Mathematics with Applications 59 (5), 1637–1643.

Hung, J.Y., Gao, W., Hung, J.C., 1993. Variable structure control: a survey, IEEE, Trans. Indust. Electron. 40 (1), 2–22.

Khettab, K., Bensafia, Y., Ladaci, S., 2015. Fuzzy adaptive control enhancement for non-affine systems with unknown control gain sign, In: Proc. 16th Int. IEEE Conf. on Sciences and Techniques of Automatic control and computer engineering, STA'2015, Monastir, Tunisia, pp. 616–621.

Khettab, K., Ladaci, S., Bensafia, Y., 2017a. Fuzzy adaptive control of fractional order chaotic systems with unknown control gain sign using a fractional order Nussbaum gain. IEEE/CAA Journal of Automatica Sinica 1–8. **PP**(99).

Khettab, K., Bensafia, Y., Ladaci, S., 2017b. Robust Adaptive Fuzzy Control for a Class of Uncertain Nonlinear Fractional Systems. In: Chadli, M., et al., (Eds.), Recent Advances in Electrical Engineering and Control Applications - Lecture Notes in Electrical Engineering 411. Springer International Publishing, pp. 276–294. A.G.

Khettab, K., Bensafia, Y., Ladaci, S., 2017c. Robust Adaptive Interval Type-2 Fuzzy Synchronization for a Class of Fractional Order Chaotic Systems. In: Azar, A.T., et al., (Eds.), Fractional Order Control and Synchronization of Chaotic Systems, Series: Studies in Computational Intelligence 688. Springer- Verlag, Germany, pp. 203–224.

Khettab, K., Bensafia, Y., Ladaci, S., 2017d. Chattering Elimination in Fuzzy Sliding Mode Control of Fractional Chaotic Systems Using a Fractional Adaptive Proportional Integral Controller. International Journal of Intelligent Engineering and Systems 10 (5), 255–265. Available from: https://doi.org/10.22266/ijies2017.1031.28.

Kuo, C.H., Lin, T.C., Valentina, E.B., 2011. Uncertain Fractional Order Chaotic Systems Tracking Design via Adaptive Hybrid Fuzzy Sliding Mode Control. Int. J. of Computers, Communications & Control 1 (3), 418–427.

Ladaci, S., Charef, A., 2006. On fractional adaptive control. Nonlinear Dynamics 43 (4), 365–378.

Ladaci, S., Khettab, K., 2012. Fractional Order Multiple Model Adaptive Control. International Journal of Automation & Systems Engineering 6 (2), 110–122.

Ladaci, S., Loiseau, J.J., Charef, A., 2008. Fractional Order Adaptive High-Gain Controllers for a Class of Linear Systems, Communications in Nonlinear Science and Numerical Simulations, Vol 13. Elsevier, pp. 707–714 (4).

Ladaci, S., Charef, A., Loiseau, J.J., 2009. Robust fractional adaptive control based on the strictly positive realness condition. International Journal of Applied Mathematics and Computer Science 19 (1), 69–76.

Lamamra, K., Azar, A.T., Ben Salah, C., 2017. Chaotic system modelling using a neural network with optimized structure. Studies in Computational Intelligence, Vol. 688. Springer-Verlag, Germany, pp. 833–856.

Lin, T-C., Balas, V.E., 2011. Fractional Order Chaotic System Tracking Design Based on Adaptive Hybrid Intelligent Control. 2011 IEEE International Conference on Fuzzy Systems (FUZZ-IEEE 2011), Taipei, 2011, pp. 2890-2896. doi: 10.1109/FUZZY.2011.6007356.

Lin, T.-C., Kuo, C.-H., 2011. H^{∞} synchronization of uncertain fractional order chaotic systems: Adaptive fuzzy approach. ISA Transactions 50 (4), 548–556.

Lin, T.-C., Chi, W., Han, L.L., 2004. Observer-based indirect adaptive fuzzy-neural tracking control for nonlinear SISO systems using VSS and H^{∞} approaches. ELSEVIER Fuzzy Sets and Systems 143 (2), 211–232.

Lin, T.-C., Lee, T.-Y., Balas, V.E., 2011. Adaptive fuzzy sliding mode control for synchronization of uncertain fractional order chaotic systems. Chaos, Solitons & Fractals 44 (10), 791–801.

Lin, T.-C., Chia-H, K., Tun, Y.L., Valentina, E.B., 2012. Adaptive fuzzy H^{∞} tracking design of SISO uncertain nonlinear fractional order time-delay systems. Nonlinear Dyn. 69 (4), 1639–1650.

Liu, Y.J., Wang, Z.-F., 2009. Adaptive fuzzy controller design of nonlinear systems with unknown gain sign. Nonlinear Dynamics 58, 687–695 **(2009)**.

Meghni, B., Dib, D., Azar, A.T., Ghoudelbourk, S., Saadoun, A., 2017a. Robust Adaptive Supervisory Fractional order Controller For optimal Energy Management in Wind Turbine with Battery Storage. Studies in Computational Intelligence, Vol. 688. Springer-Verlag, Germany, pp. 165–202.

Meghni, B., Dib, D., Azar, A.T., Saadoun, A., 2017b. Effective Supervisory Controller to Extend Optimal Energy Management in Hybrid Wind Turbine under Energy and Reliability Constraints. International Journal of Dynamics and Control, Springer. Available from: https://doi.org/10.1007/s40435-016-0296-0.

Meghni, B., Dib, D., Azar, A.T., 2017c. A Second-order sliding mode and fuzzy logic control to Optimal Energy Management in PMSG Wind Turbine with Battery Storage. Neural Computing and Applications 28 (6), 1417–1434. Available from: https://doi.org/10.1007/s00521-015-2161-z.

Mekki, H., Boukhetala, D., Azar, A.T., 2015. Sliding Modes for Fault Tolerant Control. In: Azar, A.T., Zhu, Q. (Eds.), Advances and Applications in Sliding Mode Control systems, Studies in Computational Intelligence book Series, Vol. 576. Springer-Verlag GmbH Berlin/Heidelberg, pp. 407–433. Available from: http://dx.doi.org/10.1007/978-3-319-11173-5_15.

Ouannas, A., Azar, A.T., Abu-Saris, R., 2016a. A new type of hybrid synchronization between arbitrary hyperchaotic maps. International Journal of Machine Learning and Cybernetics. Available from: https://doi.org/10.1007/s13042-016-0566-3.

Ouannas, A., Azar, A.T., Radwan, A.G., 2016b. On Inverse Problem of Generalized Synchronization Between Different Dimensional Integer-Order and Fractional-Order Chaotic Systems. The 28th International Conference on Microelectronics, IEEE, December 17-20, 2016, Cairo, Egypt.

Ouannas, A., Azar, A.T., Ziar, T., Vaidyanathan, S., 2017a. On New Fractional Inverse Matrix Projective Synchronization Schemes. Studies in Computational Intelligence, Vol. 688. Springer-Verlag, Germany, pp. 497–524.

Ouannas, A., Azar, A.T., Ziar, T., Vaidyanathan, S., 2017b. Fractional Inverse Generalized Chaos Synchronization Between Different Dimensional Systems. Studies in Computational Intelligence, Vol. 688. Springer-Verlag, Germany, pp. 525–551.

Ouannas, A., Azar, A.T., Ziar, T., Vaidyanathan, S., 2017c. A New Method To Synchronize Fractional Chaotic Systems With Different Dimensions. Studies in Computational Intelligence, Vol. 688. Springer-Verlag, Germany, pp. 581–611.

Ouannas, A., Azar, A.T., Ziar, T., Radwan, A.G., 2017d. Study On Coexistence of Different Types of Synchronization Between Different dimensional Fractional Chaotic Systems. Studies in Computational Intelligence, Vol. 688. Springer-Verlag, Germany, pp. 637–669.

Ouannas, A., Azar, A.T., Ziar, T., Radwan, A.G., 2017e. Generalized Synchronization of Different Dimensional Integer-order and Fractional Order Chaotic Systems. Studies in Computational Intelligence, Vol. 688. Springer-Verlag, Germany, pp. 671–697.

Ouannas, A., Azar, A.T., Vaidyanathan, S., 2017f. On A Simple Approach for Q-S Synchronization of Chaotic Dynamical Systems in Continuous-Time. Int. J. Computing Science and Mathematics 8 (1), 20–27.

Ouannas, A., Azar, A.T., Vaidyanathan, S., 2017g. New Hybrid Synchronization Schemes Based on Coexistence of Various Types of Synchronization Between Master-Slave Hyperchaotic Systems. Int. J. Computer Applications in Technology 55 (2), 112–120.

Ouannas, A., Azar, A.T., Ziar, T., 2017h. On Inverse Full State Hybrid Function Projective Synchronization for Continuous-time Chaotic Dynamical Systems with Arbitrary Dimensions. Differential Equations and Dynamical Systems. Available from: https://doi.org/10.1007/s12591-017-0362-x.

Ouannas, A., Azar, A.T., Vaidyanathan, S., 2017i. A Robust Method for New Fractional Hybrid Chaos Synchronization. Mathematical Methods in the Applied Sciences 40 (5), 1804–1812. Available from: https://doi.org/10.1002/mma.4099.

Ouannas A, Grassi G, Azar AT, Radwan AG, Volos C, Pham VT, et al., 2017j. Dead-Beat Synchronization Control in Discrete-Time Chaotic Systems. The 6th International Conference on Modern Circuits and Systems Technologies (MOCAST), 4-6 May 2017, Thessaloniki Greece.

Petráš, I. 2006. A Note on the Fractional-Order Cellular Neural Networks. In: Proceedings of the IEEE International world congress on computational intelligence, international joint conference on neural networks, pp. 16–21.

Petráš, I., 2008. A note on the fractional-order Chua's system. ELSEVIER, Chaos Solitons Fractals 38 (I1).

Pham, V.T., Vaidyanathan, S., Volos, C.K., Azar, A.T., Hoang, T.M., Yem, V.V., et al., 2017. Studies inComputational Intelligence, Vol. 688. Springer-Verlag, Germany, pp. 449–470.

Podlubny, I., 1999. Fractional differential equations. Academic Press, San Diego.

Rabah, K., Ladaci, S., Lashab, M., 2017. A Novel Fractional Sliding Mode Control Configuration for Synchronizing Disturbed Fractional order Chaotic Systems. Pramana, Springer 89 (3), 1–13. Available from: https://doi.org/10.1007/s12043-017-1443-7.

Sastry, S., Bodson, M., 1989. Adaptive Control: Stability, Convergence and Robustness. Prentice-Hall, New York.

Singh, S., Azar, A.T., Ouannas, A., Zhu, Q., Zhang, W., Na, J. 2017. Sliding ModeControl Technique for Multi-switching Synchronization of Chaotic Systems. 9th International Conference on Modelling, Identification and Control (ICMIC 2017), July 10-12, 2017, Kunming, China.

Soliman, N.S., Said, L.A., Azar, A.T., Madian, A.H., Radwan, A.G., Ouannas, A., 2017. Fractional Controllable Multi-Scroll V-Shape Attractor with Parameters Effect. The 6th International Conference on Modern Circuits and Systems Technologies (MOCAST), 4-6 May 2017, Thessaloniki Greece.

Takagi, T., Sugeno, M., 1985. Fuzzy Identification of Systems and Its Applications to Modeling and Control. IEEE Transactions on Systems, Man, and Cybernetics 15 (1), 116–132.

Tanaka, K., Wang, H.O., 2001. Fuzzy Control Systems Design and Analysis: A Linear Matrix Inequality Approach John Wiley & Sons, Inc. ISBNs: 0-471-32324-1 (Hardback); 0-471-22459-6 (Electronic).

Tolba, M.F., AbdelAty, A.M., Soliman, N.S., Said, L.A., Madian, A.H., Azar, A.T., et al., 2017. FPGA implementation of two fractional order chaotic systems. International Journal of Electronics and Communications 28, 162–172. 2017.

Utkin, V.I., 1977. Variable structure systems with sliding mode. IEEE Trans Autom Control 22 (2), 212–222.

Vaidyanathan, S., Azar, A.T., 2015a. Anti-Synchronization of Identical Chaotic Systems using Sliding Mode Control and an Application to Vaidyanathan-Madhavan Chaotic Systems. In: Azar, A.T., Zhu, Q. (Eds.), Advances and Applications in Sliding Mode Control systems, Studies in Computational Intelligence book Series, Vol. 576. Springer-Verlag GmbH Berlin/Heidelberg, pp. 527–547. Available from: http://dx.doi.org/10.1007/978-3-319-11173-5_19.

Vaidyanathan, S., Azar, A.T., 2015b. Hybrid Synchronization of Identical Chaotic Systems using Sliding Mode Control and an Application to Vaidyanathan Chaotic Systems. In: Azar, A.T., Zhu, Q. (Eds.), Advances and Applications in Sliding Mode Control systems, Studies in Computational Intelligence book Series, Vol. 576. Springer-Verlag GmbH Berlin/Heidelberg, pp. 549–569. Available from: http://dx.doi.org/10.1007/978-3-319-11173-5_20.

Vaidyanathan, S., Azar, A.T., 2015c. Analysis, Control and Synchronization of a Nine-Term 3-D Novel Chaotic System. In: Azar, A.T., Vaidyanathan, S. (Eds.), Chaos Modeling and Control Systems Design, Studies in Computational Intelligence, Vol. 581. Springer-Verlag GmbH Berlin/Heidelberg, pp. 3–17. Available from: http://dx.doi.org/10.1007/978-3-319-13132-0_1.

Vaidyanathan, S., Azar, A.T., 2015d. Analysis and Control of a 4-D Novel Hyperchaotic System. In: Azar, A.T., Vaidyanathan, S. (Eds.), Chaos Modeling and Control Systems Design, Studies in Computational Intelligence, Vol. 581. Springer-Verlag GmbH Berlin/Heidelberg, pp. 19–38. Available from: http://dx.doi.org/10.1007/978-3-319-13132-0_2.

Vaidyanathan, S., Azar, A.T., 2016a. Takagi-Sugeno Fuzzy Logic Controller for Liu-Chen Four-Scroll Chaotic System. International Journal of Intelligent Engineering Informatics 4 (2), 135–150.

Vaidyanathan, S., Azar, A.T., 2016b. Dynamic Analysis, Adaptive Feedback Control and Synchronization of an Eight-Term 3-D Novel Chaotic System with Three Quadratic Nonlinearities. Studies in Fuzziness and Soft Computing, Vol. 337. Springer-Verlag, Germany, pp. 155–178.

Vaidyanathan, S., Azar, A.T., 2016c. Qualitative Study and Adaptive Control of a Novel 4-D Hyperchaotic System with Three Quadratic Nonlinearities. Studies in Fuzziness and Soft Computing, Vol. 337. Springer-Verlag, Germany, pp. 179–202.

Vaidyanathan, S., Azar, A.T., 2016d. A Novel 4-D Four-Wing Chaotic System with Four Quadratic Nonlinearities and its Synchronization via Adaptive Control Method. Advances in Chaos Theory and Intelligent Control. Studies in Fuzziness and Soft Computing, Vol. 337. Springer-Verlag, Germany, pp. 203–224.

Vaidyanathan, S., Azar, A.T., 2016e. Adaptive Control and Synchronization of Halvorsen Circulant Chaotic Systems. Advances in Chaos Theory and Intelligent Control. Studies in Fuzziness and Soft Computing, Vol. 337. Springer-Verlag, Germany, pp. 225–247.

Vaidyanathan, S., Azar, A.T., 2016f. Adaptive Backstepping Control and Synchronization of a Novel 3-D Jerk System with an Exponential Nonlinearity. Advances in Chaos Theory and Intelligent Control. Studies in Fuzziness and Soft Computing, Vol. 337. Springer-Verlag, Germany, pp. 249–274.

Vaidyanathan, S., Azar, A.T., 2016g. Generalized Projective Synchronization of a Novel Hyperchaotic Four-Wing System via Adaptive Control Method. Advances in Chaos Theory and Intelligent Control. Studies in Fuzziness and Soft Computing, Vol. 337. Springer-Verlag, Germany, pp. 275–296.

Vaidyanathan, S., Sampath, S., Azar, A.T., 2015a. Global chaos synchronisation of identical chaotic systems via novel sliding mode control method and its application to Zhu system. International Journal of Modelling, Identification and Control (IJMIC) 23 (1), 92–100.

Vaidyanathan, S., Azar, A.T., Rajagopal, K., Alexander, P., 2015b. Design and SPICE implementation of a 12-term novel hyperchaotic system and its synchronization via active control (2015). International Journal of Modelling, Identification and Control (IJMIC) 23 (3), 267–277.

Vaidyanathan, S., Idowu, B.A., Azar, A.T., 2015c. Backstepping Controller Design for the Global Chaos Synchronization of Sprott's Jerk Systems. In: Azar, A.T., Vaidyanathan, S. (Eds.), Chaos Modeling and Control Systems Design, Studies in Computational Intelligence, Vol. 581. Springer-Verlag GmbH Berlin/Heidelberg, pp. 39–58. Available from: http://dx.doi.org/10.1007/978-3-319-13132-0_3.

Vaidyanathan, S., Azar, A.T., Ouannas, A., 2017a. An Eight-Term 3-D Novel Chaotic System with Three Quadratic Nonlinearities, its Adaptive Feedback Control and Synchronization. Studies in Computational Intelligence, Vol. 688. Springer-Verlag, Germany, pp. 719–746.

Vaidyanathan, S., Zhu, Q., Azar, A.T., 2017b. Adaptive Control of a Novel Nonlinear Double Convection Chaotic System. Studies in Computational Intelligence, Vol. 688. Springer-Verlag, Germany, pp. 357–385.

Vaidyanathan, S., Azar, A.T., Ouannas, A., 2017c. Hyperchaos and Adaptive Control of a Novel Hyperchaotic System with Two Quadratic Nonlinearities. Studies in Computational Intelligence, Vol. 688. Springer-Verlag, Germany, pp. 773–803.

Wang, L.X., 1992 Fuzzy systems are universal approximators, IEEE International Conference on Fuzzy Systems, pp. 1163–1170, San Diego.

Wang, L.X., 1994. Adaptive fuzzy systems and control: design and stability analysis. Prentice-Hall, Englewood Cliffs, NJ.

Wang, L.X., 1997. A Course in Fuzzy Systems and Control. Prentice-Hall International, Inc.

Wang, L.X., Mendel, J.M., 1992. Fuzzy basis function, universal approximation, and orthogonal least square learning. IEEE Trans Neural Netw 3 (5), 807–814.

Wang, Z., Volos, C., Kingni, S.T., Azar, A.T., Pham, V.T., 2017. Four-wing attractors in a novel chaotic system with hyperbolic sine nonlinearity. Optik - International Journal for Light and Electron Optics 131 (2017), 1071–1078.

Zelinka, I., Youssef, T., 2013. Unknown inputs observer design for fuzzy systems with application to chaotic system reconstruction. Computers & Mathematics with Applications 66 (2), 147–154.

Zhu, Q., Azar, A.T., 2015. Complex system modelling and control through intelligent soft computations. Studies in Fuzziness and Soft Computing, Vol. 319. Springer-Verlag, Germany. 978-3-319-12882-5.

FURTHER READING

Azar, A.T., Serrano, F.E., 2014. Robust IMC-PID tuning for cascade control systems with gain and phase margin specificationsNeural Computing and Applications 25 (5), 983–995 Springer. Available from: http://dx.doi.org/10.1007/s00521-014-1560-x.

Chapter 21

On the Synchronization and Recurrence of Fractional Order Chaotic Systems

Mohsen Alimi[1], Ahmed Rhif[2] and Abdelwaheb Rebai[3]
[1]*University of Kairouan, Kairouan, Tunisia,* [2]*University of Carthage, La Marsa, Tunisia,* [3]*University of Sfax, Sfax, Tunisia*

21.1 INTRODUCTION

In recent years, fractional calculus has become an excellent tool in the modeling and analysis of many nonlinear phenomena which do not satisfy the principle of superposition and which are governed by certain systems of a fractional nature. One of the exciting areas of fractional calculus research is the theory of chaos. Chaos is one of the most complex dynamics that nonlinear systems can exhibit (Wang et al., 2017). To highlight a complex chaotic system and control its dynamical behavior, many computational intelligent solutions are given (Azar and Vaidyanathan, 2015a,b,c; Zhu and Azar, 2015). Indeed, the theory of chaos that has been intensively studied over the last two decades is found to be useful for many applications in various basic actual fields such as physics, mathematics, signal processing, data encryption, medical, business cycles, financial systems, and various other engineering problems.

However, the fractional order chaotic dynamical systems can be considered as a generalization of integer order chaotic dynamical systems. The conventional fractional order chaotic dynamic systems are numerous, yet interesting. Indeed, in recent years they have started attracting more attention of many researchers in diverse domains, because they exhibit complex, chaotic behavior that needs to be studied. A wide range of this dynamic behavior are manifested via various forms of chaotic behavior such as fractional order Lorenz's system (Li and Yan, 2007), fractional order Rössler's system (Li and Chen, 2004), fractional order Chua's system (Petráš, 2008), fractional order Chen's system (Lu and Chen, 2006), fractional order Lü's system (Deng and Li, 2005), fractional order Liu's system (Liu et al., 2009),

Mathematical Techniques of Fractional Order Systems. DOI: https://doi.org/10.1016/B978-0-12-813592-1.00021-0

fractional order Arneodoe's system (Lu, 2005), and fractional multi-scroll chaotic attractors (Lü et al., 2004; Soliman et al., 2017), etc. It has been demonstrated that all these fractional order systems can produce chaotic attractors with different orders strictly less than three (Silva, 1993; Das et al., 2016). To realize an attractor for any fractional order chaotic system, many innovative numerical simulation methods are applied, such as the Field Programmable Gate Array method (Tolba et al., 2017). Moreover, the problem of conceiving a system that imitates the behavior of another chaotic system is called synchronization. In general, the two chaotic systems are respectively called master and slave systems.

The study of the concept of synchronization goes back to the 17th century. Recently, researchers on chaos control and complex systems synchronization problems have been interested in fractional order chaotic systems. And since the latter are characterized by the very sensitivity to the initial conditions, the synchronization between two fractional order chaotic systems seem impossible. Nevertheless, the work of several scientists in this field has shown the opposite. Thus, many recent studies show that chaotic fractional order systems cannot be only synchronized but also controllable (Azar et al., 2017a; Ouannas et al., 2017d).

In addition, since the seminal contribution of (Pecora and Carroll, 1990), the study of synchronization of fractional order chaotic systems has become an active field. Newly, it has attracted much attention of researchers because of its multiple potential applications, especially in the field of engineering, particularly mechanical and electrical field (Ouannas et al., 2017c). So far, a wide variety of approaches and techniques have been proposed to address the problem of synchronization control of fractional order chaotic and hyperchaotic systems (Azar et al., 2017a,b; Ouannas et al., 2017i), also in discrete-time chaotic systems (Ouannas et al., 2017j). Among the frequently applied methods of synchronization control in continuous-time chaotic systems, there are the sliding mode control method (Tavazoei and Haeri, 2008a; Singh et al., 2017); active and adaptive control methods (Bhalekar and Daftardar-Gejji, 2010; Vaidyanathan et al., 2017b, c); feedback control method (Vaidyanathan et al., 2017a); linear and nonlinear control techniques (Odibat et al., 2010; Chen and Liu, 2012); fuzzy adaptive control method (Boulkroune et al., 2016a,b); scalar signal technique (Grassi et al., 2017) etc. In addition, many different traditional types of synchronization for fractional order chaotic systems have been presented, such as complete synchronization (Li et al., 2008a); antisynchronization (Wedekind and Parlitz, 2001); projective synchronization (Mainieri and Rehacek, 1999; Ouannas et al., 2017a,h); generalized synchronization (Lu, 2008; Ouannas et al., 2017b,e); and Q-S synchronization (Li, 2007; Ouannas et al., 2017f). Although, during this decade, all these types of synchronization of chaos and their methods of control are encompassed only under two coupling modes. The first mode relies on a mutual coupling between two or more chaotic systems. The second is called master–slave or unidirectional coupling mode (Ouannas et al., 2017g). Its principle is to choose a special configuration involving two coupled systems so that the behavior of the second depends on the behavior of the first, but reciprocally, the first is not influenced by the behavior of the second. The first system generating chaos is the emitter (master) system, and the second is the receiver (slave) system. This is described by recurrent equations and characterized by its state variables constituting the state vector.

Nevertheless, in spite of this multitude of research, the synthesis of the fractional order chaotic systems remains an open problem in many fields of research because of the nature of these systems which are, on the one hand, considered as systems with long memory. On the other hand, they present a complex dynamic. Thus, several experimental and theoretical studies show that the problem of the synchronization of the fractional order systems in the presence of chaos is still an open attractive subject to many scientists in many fields of research.

In this chapter, the focus is meant to show that the recurrences associated with fractional order chaotic systems are an efficient way of ensuring the synchronization of continuous-time chaotic attractors. The efficiency for the representation of the synchronization phenomenon of these systems will be evaluated in terms of performance and robustness. The objectives are, first of all, consisting of developing new synchronization schemes based on the notion of recurrence for fractional order chaotic systems (Vladimirsky and Ismailov, 2015a). Then, based on recurrence property (Afraimovich, 1999), new topological synchronization criteria are derived from different dimensional continuous-time in the case of fractional order chaotic systems in different dimensions. Finally, several illustrative numerical applications and computer simulations are used to confirm the theoretical results and to prove the effectiveness of the proposed schemes.

The rest of this chapter is organized in the following way: the first section, describes general definitions and preliminaries of nonlinear dynamic systems, the theory of deterministic chaos, and the methods of Poincaré recurrence visualization and measures. The second section focuses on the basic notions of fractional systems, namely the noninteger order derivation in the Grünwald–Letnikov sense, the fractional integral of Riemann–Liouville, the fractional derivation in Caputo's sense, and the stability condition of fractional order chaotic systems. The third section exposes the topology of fractional order systems. The fourth section presents the problem of synchronization of fractional order chaotic systems and their principal used methods. The fifth section illustrates the content of the work, which consists of a new scheme on the topological synchronization for some concrete examples on fractional order chaotic systems. At the end, the work will be summarized by a general conclusion followed by a bibliography.

21.2 GENERAL INFORMATION ON CHAOTIC DYNAMIC SYSTEMS

21.2.1 Concept of the Deterministic Chaos

The theory of chaos appeared in the early 1960s in meteorology and rapidly spread to several areas of research. There are several mathematical definitions of chaos in the literature, but so far there is no universal mathematical definition of chaos. The habitual definition considers chaos as aperiodic long-term behavior in a deterministic system that exhibits sensitive dependence on initial conditions and some recurrence properties. Mathematically, the most recognized chaos definition is that of (Devaney, 1989), given as:

Definition: *For a consider function F, defined from and to a real interval I, is said to be constituted of a chaotic dynamics if:*

1. *F is topologically transitive;*
2. *The sum of the periodic points of F is dense in I;*
3. *F has sensitive dependence on initial conditions.*

In a nonchaotic deterministic system, neighboring initial conditions lead to neighboring evolutions. Contrary, a chaotic system is a deterministic system fully determined by initial state and subject to a law of evolution which can be extremely simple and regular, but its evolution is extremely complicated and perfectly irregular. This gives it unpredictable behavior over time which is known as the butterfly effect developed by (Poincaré, 1890) as: "It so happens that small differences in the initial state of the system can lead to very large differences in its final state. A small error in the former could then produce an enormous one in the latter. Prediction becomes impossible, and the system appears to behave randomly." Moreover, these systems have a very high sensitivity to the initial conditions. Thus, chaotic behavior is usually defined as a deterministic low-dimensional dynamics with high sensitivity to initial conditions and some recurrence properties. The concept of recurrence goes back to (Poincaré, 1890).

This sensitivity to the initial conditions explains the fact that, for a chaotic system, a minimal modification of the initial conditions can lead to unpredictable results over the long term. The degree of sensitivity to initial conditions quantifies the chaotic nature of the system. The presence of this chaos property is detected on the basis of the Lypunov exponent indicator or on the basis of analysis of the quantification of recurrences. The recurrence analysis methods for the chaotic dynamical system are based on both the recurrence plot and the recurrence quantification analysis.

21.2.1.1 The Recurrence Plot

Definition: *A recurrence plot (RP) is an advanced technique of nonlinear data analysis. It is a visualization (or a graph) of a square matrix, in which* *the matrix elements correspond to those times at which a state of a dynamical system recurs (columns and rows correspond then to a certain pair of times). Techniqually, the RP reveals all the times when the phase space trajectory of the dynamical system visits roughly the same area in the phase space.*

The Recurrence Plot concept has been introduced by (Eckmann et al., 1987) in order to quantify the recurrence properties of chaotic dynamics systems. The aim of this work is to represent whether an attractor possesses, in its temporal evolution, states which are repeated. The RPs denoted R_{ij} are considered as an associated diagram to a square matrix constructed around the indicated assumption as follows. If each point y_i of the phase portrait $\{y_i\}_{i=1,\,2,\ldots,N}$ is near to or not of another point y_j, then a recurrence between point's y_i and y_j of the trajectory will take place. Thus, if the distance between two points is less than a predetermined threshold ε, the points are said to be recurrent and are associated with a black point; otherwise, they are not recurring and are associated with a white point. Mathematically, this is written by an order N square matrix such that (2.1):

$$R_{ij} = \Theta\left(\varepsilon - \left\| y_i - y_j \right\|\right)_{i,j=1,\,2,\ldots,N}, \quad y_i,\ y_j \in \mathfrak{R} \tag{21.1}$$

where N is the number of considered states, $\Theta(.)$ is the Heaviside function,ε is a threshold distance, $\|.\|$ is the norm, and y_i and y_j are two points of the trajectory. The principle of this technique consists in the visual analysis and inspection of the graph matrix of recurrence constructed with vector distances when each portion of the curve is compared to all the others and represented on a recurrence map, as well as the quantitative characteristics are necessary for the evaluation of the distances, between the points in the temporal space and in the phase space.

Several methodological advantages, of which not all are allowed in the analysis of nonlinear dynamic systems by other less adapted tools, justifying the need to use this relevant technique since it does not require some constraints on the stability over time of the statistical or stationary properties of systems, or on the form of the statistical distribution of the associated measures.

Influenced by the characteristic behavior of the trajectory of the phase space, the structure of RPs is composed of elements of minimal sizes, such as single points, diagonal lines, and vertical or horizontal lines, or a mixture of them. The macroscopic structure, known as texture, can be visually dominated by homogeneity, periodicity, drift, or perturbation. The visual appearance of an RP can provide the necessary information about the dynamics of the system. (Marwan and Kurths, 2002) proposed measurements based on vertical and/or horizontal structures capable of highlighting, in particular, chaos–chaos transitions. Thus, RP allows two broad and small-scale topological approaches. For a historical review of recurrence plots, it is appropriate to see Marwan (2008).

As far as the topological interpretation of the graphic matrix of recurrences is concerned, Marwan et al. (2007) distinguish three levels of analysis: first, homogeneity in the RP graphic representation implies a homogeneous distribution of the recurring points in the matrix of recurrences; which characterizes the stationarity of the system, both in the case of a uniformly distributed random phenomenon (white noise), and in the case of a deterministic chaotic structure. On the other hand, the nonstationarity of the system is at the change in the distribution of recurrence points, visualized graphically by brighten areas. Second, the periodicity of a system implies the presence of rhythmicity represented by long parallel diagonal segments at the principal diagonal of the RP matrix. These long diagonal segments are displayed by shadow areas related to the occurrence of the recurring points, in a dense manner in the RP matrix. Third, the drift is characterized by the distance of the diagonal lines, with respect to the main diagonal. The more the diagonal lines are distant from the main diagonal, the higher the drift is. On the contrary, the closer they are to the main diagonal, the lower the drift is. A weak drift of the recurrence matrix is synonymous with nonstationary in a time series and vice versa.

As for the scaled topological interpretation of the recursive graph matrix, four levels of analysis are distinguished: First, the isolated recurring points imply a stochastic behavior of the system. Second, the small diagonal segments (small segments parallel to the main diagonal) signify the presence of some vectors which can be repeated several times during a period of observation. In the context of attractor dynamics, this implies that the system revisits the same region of the attractor at different times. Whereas the small segments perpendicular to the principal diagonal signify the symmetry of the two triangular halves of the matrix of the recurrences, with respect to the principal diagonal, each half triangle reflects the image of the other. But, the small horizontal or vertical segments mean the presence of a sequence of vectors close to each other. Then, the checkered structure is induced by a grouping of segments of parallel lines, at the main diagonal, in small regions of the matrix of recurrences. In the context of attractor dynamics, this means that the system revisits different surfaces of the attractor and switches from one to the other, quite a few times, during a period of observation. Finally, the white space band characterizes the nonrecurring points and indicates a change in structure or a sudden change in the state of the system.

From these topological analyses associated to the matrix of recurrences, various statistical indexes are also introduced to transform this graphical representation into statistical measurements. In this context, a second statistical technique has been developed for measuring complexity, allowing the analysis of the recurrence quantification of the recurrence matrix.

21.2.1.2 The Recurrence Quantification Analysis

The recurrence quantification analysis was introduced by (Zbilut and Webber, 1992) as a method based on statistical measures, able to provide an efficient extension of RPs.

Definition: *The recurrence quantification analysis (RQA) is a method of nonlinear data analysis which quantifies the number and duration of recurrences of a dynamical system presented by its state space trajectory.*

The structures and quantification of RP were developed by Zbilut et al. (1998), Webber and Zbilut (2005) and extended with new measures of complexity by Marwan et al. (2002). If the symbols N ; N_1 ; M ; R_{ij} ; $P(l)$; $P(v)$, and $p(l)$ denote (respectively): the number of points on the phase space trajectory; the number of diagonal lines in the recurrence plot; the maximal number of diagonals parallel to the line of identity which will be considered for the calculation of the trend; the correlation rate; the frequency distribution of the histogram of the line lengths l of the diagonal lines; the frequency distribution of the lengths v of the vertical lines and the Shannon entropy characterizing the probability distribution of the length l of the diagonal line. Then, these statistical measures are based on several indicators which are given in (Table 21.1) as follows:

where,

- *The recurrence rate*: Is the number of recurring points relative to the total number of points. This percentage gives the total plot occupied by points. It's the proportion of black points.
- *The determinism rate*: Is the number of recurring points that form diagonal pieces having a length greater than 1 (off main diagonal), in relation to the total number of points. This percentage makes the possibility to quantify, if information can be repeated over time. That is, the more the diagonal structures are longer, the more the behavior is deterministic (% DET tends to 1) and the more the diagonal structures are shorter, the more the behavior is stochastic (%DET tends to 0). Hence, this measure is related to the predictability of the dynamic system, because the graphic representation of the recurrence of a stochastic system and decorative data is formed only by black dots and very few lines of diagonal pieces, but, a deterministic process is formed only by diagonal lines and very few black spots.
- *The lamirarity rate*: Gives the percentage of blackheads that form vertical lines of minimum lengths, it is a characteristic of the intermittence of the system during the chaos–chaos transition.
- *The trapping time*: Is the average length of vertical lines, which measures the average time for a system to keep certain states.

TABLE 21.1 Some Measures of Recurrence Quantification

Measures	Symbols	Definitions
The Recurrence Rate	RR	$RR = \frac{1}{N^2} \sum_{i,j=1}^{N} R_{i\,j}$
The Determinism Rate	DET	$DET = \frac{\sum_{l=l_{min}}^{N} lP(l)}{\sum_{l=1}^{N} lP(l)}$
The Lamirarity Rate	LAM	$LAM = \frac{\sum_{v=v_{min}}^{N} vP\,(v)}{\sum_{v=1}^{N} vP\,(v)}$
The Trapping Time	TT	$TT = \frac{\sum_{v=v_{min}}^{N} vP(v)}{\sum_{v=v_{min}}^{N} P(v)}$
The Ratio	rENT	$_{r}ENT = N^2 \frac{\sum_{l=l_{min}}^{N} lP(l)}{\left(\sum_{l=1}^{N} lP(l)\right)^2}$
The Shannon Entropy	ENT	$ENT = -\sum_{l=l_{min}}^{N} p(l)\log\left[p(l)\right]$
The Longest Diagonal Lines	max$L$	$_{max}L = \max(\{l_i; i = 1, \ldots, N_l\})$
The Trend	T	$T = \frac{\sum_{i=1}^{M} (i - M/2)\ (RR_i - \langle RR_i \rangle)}{\sum_{i=1}^{M} (i - M/2)^2}$

- *The ratio*: Is the relation between the percentage of recurrence and the percentage of determinism, which gives the ratio of the recurrence points, forming the structure of the diagonal pieces (off main diagonal), and relative to the total number of recurrence points. It is a deterministic measure of the dynamic system.
- *The Shannon entropy*: Is the frequency of the distribution whose probabilities are derived from the percentage of appearance of oblique segments of variable lengths. It was correlated with the inverse of the largest Lyapunov exponent; which is contrary to the usual sense of Shannon entropy which habitually measures the degree of complexity of a dynamical behavior.
- *The longest diagonal lines*: It gives a measure of the sensitivity to the initial conditions; it is inversely proportional to the Lyapunov exponent.
- *The trend*: It quantifies the filling of the recurring points observed when moving away from the principal diagonal and gives a measure of the nonstationarity described by the recurrence diagram.

With these measures, it will be able to study the spatial and temporal variability of the abstract mechanism, behind an endogenous structural change and to assess the stability, over time, of its spatial distribution. That is, to quantify a pattern of recurrences and to identify its structures. Then, these measures of recurrence diagrams can be used for studying the dynamics of fractional order chaotic systems.

21.3 BASICS ON FRACTIONAL ORDER SYSTEMS

21.3.1 Concept of Fractional Order System

The fractional order systems or simply a noninteger order system can be considered as a generalization of integer order systems. The fractional calculus is the generalization of the derivative and the integral to a fundamental operation of noninteger order integro-differential operator denoted by ${}_{\alpha}D_t^p$ where α and t are the limits of the operator and $p \in \Re$. The continuous integro-differential operator is defined as follows:

$$ {}_{\alpha}D_t^p = \begin{cases} \dfrac{d^p}{dt^p} & p>0 \\ 1 & p=0 \\ \int_{\alpha}^{t} (d\nu)^p & p<0 \end{cases} \tag{21.2}$$

In literature of fractional calculus, there exist many definitions for fractional derivatives. The three most frequently used definitions are the Grünwald–Letnikov definition, the Riemann–Liouville definition, and the Caputo definition.

21.3.1.1 Definition of Grünwald–Letnikov

The fractional order derivation in the sense of Grünwald–Letnikov can be obtained by the generalization to the real of the integral and the derivative of integer order, where all the difference with respect to the integer case is located at the extension of the factorial through the gamma function. The pth fractional order derivative in the sense of Grünwald–Letnikov for any real continuous function $f(t)$, is defined by the following relation (21.3):

$$ {}_{\alpha}D_t^p f(t) = \lim_{h \to 0} \frac{1}{h^p} \sum_{j=0}^{\left[\frac{t-\alpha}{h}\right]} (-1)^j C_j^p f(t-jh) \tag{21.3}$$

where [.] denotes the integer part, t and α are the lower and upper limits of the derivative, $(t-\alpha)$ is the simulation time, h is the time step of calculation and $(-1)^j C_j^p$ are the binomial coefficients $c_j^{(p)}$. In this sense, Dorcák (1994) proposed to calculate them using the following formula (21.4):

$$ c_j^{(p)} = \begin{cases} 1 & \forall j=0 \\ \left(\dfrac{j-p-1}{j}\right) c_{j-1}^{(p)} & \forall j \in \aleph^* \end{cases} \tag{21.4}$$

21.3.1.2 Definition of Riemann–Liouville

The derivative in the sense of Riemann–Liouville is based on the first primitive or integral of a function $f(t)$, is defined as Eq. (21.5):

$$ {}_{\alpha}D_t^p f(t) = \frac{1}{\Gamma(n-p)} \frac{d^n}{dt^n} \int_{\alpha}^{t} \frac{f(\nu)}{(t-\nu)^{p-n+1}} d\nu \tag{21.5} $$

where $\Gamma(.)$ the Euler's Gamma function is defined for all $n \in \aleph$ and $p \in [n\text{-}1, n[$ as:

$$ \Gamma(n) = \int_{0}^{+\infty} t^{n-1} e^{-t} dt \tag{21.6} $$

21.3.1.3 Definition of Caputo

The definition of the fractional derivative of $f(t)$ in the sense of Caputo incorporates the initial conditions of the function to be treated, as well as its integer derivatives, the derivative of a function $f(t)$ in the Caputo sense is defined by the following relation (21.7):

$$ {}_{\alpha}^{C}D_t^p f(t) = \frac{1}{\Gamma(p-n)} \int_{\alpha}^{t} \frac{\frac{d^n f(\nu)}{dt^n}}{(t-\nu)^{p-n+1}} d\nu \tag{21.7} $$

where $\Gamma(.)$ the Gamma function, n is the smallest integer larger than the fractional order p and $p \in]0, 1]$.

21.3.2 The *n*-Dimensional Fractional Order Chaotic System

The n-dimensional fractional order chaotic system is defined as following relation (21.8):

$$ {}_{\alpha}D_t^P X_t = f(X_t, X_0, \Theta) \tag{21.8} $$

where $X_t = (x_{1t}, x_{2t}, \ldots, x_{nt})' \in \mathfrak{R}^n$ is the n-dimensional state vector of the original system; X_0 is the initial state of the system, $P = (p_1, p_2, \ldots, p_n)' \in \mathfrak{R}^n$ is a set of fractional order of the original system verifying that for all $i = 1, 2, \ldots, n$ $0 < p_i \leq 1$, f. $\mathfrak{R}^n \to \mathfrak{R}^n$ is the linear or nonlinear function and $\Theta = (\theta_1, \theta_2, \ldots, \theta_m)' \in \mathfrak{R}^m$ is the n-dimensional vector value of original system parameters.

21.3.3 The General Numerical Solution of the Fractional Differential Equation and Stability

A remarkable property that distinguishes a fractional order derivation from an integer derivation is that a noninteger derivation assumes a global character in contrast to an integer derivation. It turns out that the fractional order derivative of the function $f(t)$ requires the knowledge of $f(t)$ expression over an interval, whereas in the integer case only the local knowledge of function f around t is necessary. This property of fractional order systems to be interpreted as long-memory systems and integer systems can be interpreted as short memory systems. So, let's consider the following fractional differential Eq. (21.9):

$$_{\alpha}D_t^p f(t) = f(x_t, t) \tag{21.9}$$

Its general numerical solution of (21.9) using the Grünwald–Letnikov method (21.3) can be formulated as:

$$x_{t_k} = f(x_{t_k}, t_k)\ h^p - \sum_{j=\tau}^{k} c_j^{(p)} x_{t_{k-j}} \tag{21.10}$$

where $k = 1, 2, \ldots, [(t-\alpha)/h]$, $\tau = 1$ for all k and if the short memory principle is used, it is written as $\tau = 1$ for $k < [(t-\alpha)/h]$ and $\tau = k - [(t-\alpha)/h]$ for $k > [(t-\alpha)/h]$.

On the other hand, the stability of the fractional order nonlinear system can be analyzed at its equilibrium points which are calculated by solving the equation $f(X) = 0$. Let's consider a fractional order nonlinear system whose Jacobian matrix at the equilibrium is a square matrix of order n denoted J, then the characteristic equation of the matrix J at equilibrium points is a polynomial of order n which has for all $i = 1, 2, \ldots, n$ the eigenvalues λ_i as roots. These eigenvalues λ_i of the Jacobian matrix J evaluated at the equilibrium points are given by solving the following Eq. (21.11):

$$\det\left(diag\left(\left[\lambda_1^{p_1} \quad \lambda_2^{p_2} \quad \ldots \quad \lambda_n^{p_n}\right]\right) - J\right) = 0 \tag{21.11}$$

- If $p_1 = p_2 = \ldots = p_n = p$, (Tavazoei and Haeri, 2008b) demonstrate that the stability of the equilibrium points of a fractional order nonlinear system must be satisfying the following condition (21.12):

$$\left|\arg(\lambda_i)_{i=1,\, 2,\ldots,n}\right| > p\frac{\pi}{2} \tag{21.12}$$

- If $p_1 \neq p_2 \neq \ldots \neq p_n$, (Tavazoei and Haeri, 2008b) demonstrate that these equilibrium points are locally asymptotically stable. The asymptotic stability of the fractional order nonlinear system at a saddle point must be satisfying the following condition (21.13):

$$\left|\arg(\lambda_i)_{i=1,\, 2,\ldots,n}\right| > \frac{1}{\omega}\frac{\pi}{2} \tag{21.13}$$

where ω is the Least Common Multiple and λ_i, for all $i = 1, 2, \ldots, n$ are roots of the following Eq. (21.14):

$$\det\left(diag\left(\left[\lambda_1^{\omega\, p_1} \quad \lambda_2^{\omega\, p_2} \quad \ldots \quad \lambda_n^{\omega\, p_n}\right]\right) - J\right) = 0 \tag{21.14}$$

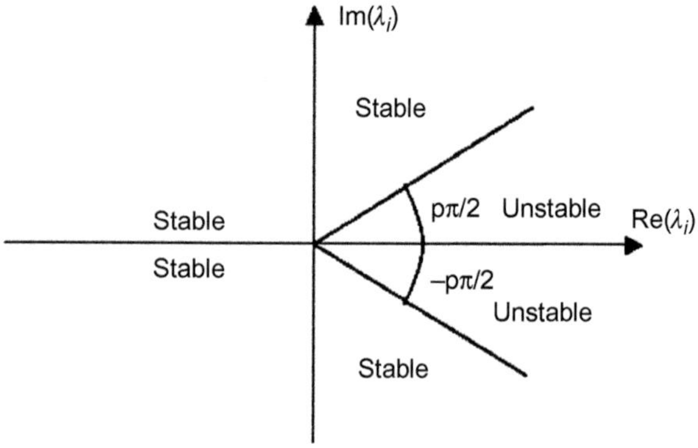

FIGURE 21.1 Stability region of the fractional order system, $0 < p \leq 1$.

If $p_1 = p_2 = \ldots = p_n = p$, the fractional order chaotic system (21.8) is named a commensurate fractional order system. Otherwise, it is named an incommensurate fractional order system. For given parameters, according to Tavazoei and Haeri (2007a,b) the necessary condition for minimal commensurate or incommensurate derivatives order, for which a fractional order system can exhibit a chaotic attractor, is given by the following condition (21.15):

$$p \geq \frac{2}{\pi} \tan^{-1}\left(\frac{|\mathrm{Im}(\lambda_i)|}{\mathrm{Re}(\lambda_i)}\right)_{i=1,\,2,\ldots,n} \tag{21.15}$$

where for $i = 1, 2, \ldots, n$, λ_i are the corresponding all unstable eigenvalues of saddle points of considered fractional order system. Therefore, both stability and instability regions of these complexes fixed points can be represented graphically in Fig. 21.1 as follows:

21.4 TOPOLOGY OF FRACTIONAL ORDER SPACE

When the long-term dynamics of the considered dynamical systems are attracted by certain regions of the phase space, then it is about the attractors studied by Takens (1981). Hence, after a transitory regime, all the trajectories are carried by the attractor, so if under the hypothesis of ergodicity, it is possible to get rid of the study of individual trajectories and to consider that all the dynamics of the system are contained in the attractor. In the phase space, the set of initial conditions leading to an attractor forms a basin of attraction. The type of attractor, and therefore its dimension, fully characterizes a dynamic system.

- If the dimension is 0, the system is stationary.
- If the dimension is 1, the system is periodic.

- If the dimension is greater than or equal to 2 and integer, the system is quasi-periodic.
- If the dimension is not an integer (and greater than 2), the system is chaotic.

The attractors associated with chaotic behaviors are called strange attractors and they have a fractal structure if their individual dimension isn't a natural number. The fractal geometry is a priori notion completely independent of the theory of chaos (due to the fractal character of chaotic attractors) that was developed by Mandelbrot (1982). The starting idea of Mandelbrot is that in nature many objects are self-similar, i.e., a small part resembles to the whole and it is replicated. It is a phenomenon of morphology of endogenous structures by memorization using the relatively homogeneous and synchronized invariance recomposition. Then, it will be possible to account for the invariance measures of dynamic structures by using the topological synchronization of fractional order chaotic systems thanks to this theory of self-similar fractal geometry. The basic assumption implies that the knowledge of attractors and their previous history characteristics determines the dynamic systems in phase space. Thus, the characterization of the topological properties of the chaotic attractor consists in calculating its invariant measures, such as important fractal dimensions and the maximum Lyapunov exponent.

21.4.1 The Hausdorff–Besicovitch Dimension

According to Vladimirsky and Ismailov (2015a) the Hausdorff–Besicovitch dimension of fractional order chaotic system is defined as:

Definition: *For any compact metric space X,*

$$d_H(X) = \inf\left\{\lim_{\varepsilon\to 0}\frac{-\log N_{\varepsilon,\,d}(X)}{\log(\varepsilon)}\right\} \tag{21.16}$$

where ε is the sphere of radius ε, d is a metric of X, (X, d_f) is a compact fractional metric space with dimension d_f, $N_{\varepsilon,\,d}(X) = \min|U|$ and U is a finite open covering of X with mesh less than ε.

21.4.2 The Correlation Dimension

This measure is based on the idea that a chaotic process with dimension m does not fill a space of dimension m + 1. According to Kantz and Schreiber (1997) such indicator is defined as:

Definition: *For two-dimensional square matrices with order N and any compact metric space X,*

$$D_c = \lim_{\varepsilon \to 0} \frac{\log\left(2[N(N-1)]^{-1} \sum_{i=1}^{N} \sum_{j=i+1}^{N} R_{ij}\right)}{\log(\varepsilon)}$$

where ε is the sphere of radius ε, N is the number of considered condition x_i and R_{ij} is the expression as defined in (21.1).

21.4.3 The Maximum Lyapunov Exponent

This measure starts from the idea that two points are initially very close in data space set have trajectories divergent exponentially. This indicator was proposed by Wolf et al. (1985); therefore it makes possible to know whether there is sensitivity to the initial conditions in the data. The exponent of Lyapunov is expressed as follows:

Definition: *The Lyapunov exponent is a measure of divergence rate of two trajectories after an infinitesimal deviation*

$$\lambda_{max} = \lim_{T \to \infty} \frac{1}{T} \log\left|\lambda_i^T\right|$$

where for all i, λ_i are all eigenvalues of the Jacobian matrix J evaluated at the equilibrium points corresponding to the considered fractional order chaotic system.

The maximum Lyapunov exponent estimation value is interpreted as flows:

- If $\lambda_{max} < 0$: it is a converged dynamical system to a stable fixed point.
- If $\lambda_{max} = 0$: it is a limit cycle, the dynamical system is stable in sense of Lyapunov.
- If $\lambda_{max} > 0$: it is an instable dynamical system with a chaotic behavior.
- If $\lambda_{max} = \infty$: it is a noise.

Recently, using the fractional order Lyapunov direct method, according to Aguila-Camacho et al. (2014), the Lyapunov function is given in the quadratic form as:

$$V(X_t) = \frac{1}{2} \Omega_p' \, \Omega_p \tag{21.17}$$

where $X_t = (x_{1t}, x_{2t}, \ldots, x_{nt})'$, p is a rational number between 0 and 1 and Ω_p is the memory function of order p. According to Chen et al. (2014) the fractional order Lyapunov stability using Lyapunov direct method has been developed by the following theorem.

Theorem:: *If there exists a positive definite Lyapunov function for all* $t > 0$, $V(X_t)$ *such that* $D_t^p V(X_t) < 0$, *for all* $t > 0$, *then the trivial solution of system* $D_t^p X_t = F(X_t)$ *is asymptotically stable.*

21.5 GENERALIZED SYNCHRONIZATION OF FRACTIONAL ORDER CHAOTIC SYSTEMS

The synchronization of two dynamic systems means that each system evolves according to the behavior of the other system. Since the pioneering contribution of Zhou and Li (2005), chaos control and synchronization have attracted the attention of several researchers from various scientific fields. Particularly, the research on the synchronization of fractional order chaotic systems consists in designing a process allowing the synchronization of two or more equivalent or not equivalent fractional order chaotic systems by coupling (unidirectional or bidirectional) and/or by forcing (Li and Deng, 2006). In the literature there are several types of synchronization. In this section, the main five types of traditional synchronization are introduced, namely complete synchronization, antisynchronization, projective synchronization, generalized synchronization, and Q-S synchronization. But, the problem of the synchronization of two chaotic systems will bring researchers back to another problem which is that of the stability of the error system in the vicinity of the origin. For this purpose it will take a rather general form of the error system. After that, a new type of synchronization of fractional order chaotic systems based on the notion of Poincaré recurrences will be proposed as indicators of topological synchronization (Afraimovich et al., 2000).

21.5.1 Review of the Traditional Methods of Synchronization of Fractional Order Chaotic Systems

We consider two fractional order chaotic systems. The first is the slave system which can be synchronized with the second which is the master system. Both systems are respectively represented as,

$$\begin{aligned} {}_{\alpha}D_t^p X_t &= f(X_t, X_0, \Theta) \\ {}_{\alpha}D_t^q Y_t &= g(Y_t, Y_0, \Phi) + U_t \end{aligned} \tag{21.18}$$

where ${}_{\alpha}D_t^p$ and ${}_{\alpha}D_t^q$ are the two fractional derivatives of Caputo of respectively order p and q verifying that $0 < p,\ q \leq 1$, $X_t \in \mathfrak{R}^n$ and $Y_t \in \mathfrak{R}^m$ are two multidimensional state vector of respectively the two original fractional order chaotic systems; X_0 and Y_0 are respectively the initial states of the two systems, f: $\mathfrak{R}^n \to \mathfrak{R}^n$ and g: $\mathfrak{R}^m \to \mathfrak{R}^m$ are respectively two linear or nonlinear functions, Θ and Φ are the two multidimensional vectors value of respectively the two original systems parameters, and $U_t \in \mathfrak{R}^m$ is a control vector to be determined.

It is assumed that these two systems can manifest some traditional types of synchronization for a specific time, independent of initial conditions X_0 and Y_0 in a large zone of $\mathfrak{R}^{n+m}$. Different notions of fractional order chaotic systems synchronization are possible. The most useful traditional types of synchronization which can be satisfied by fractional order chaotic systems are: complete synchronization, antisynchronization, projective synchronization, generalized synchronization, and Q-S synchronization. In this first subsection, a definition of each possible synchronization type will be given with exploring the fundamental property of its characteristics of invariant trajectories associated to the synchronous chaotic systems.

- The concept of complete synchronization has been proposed by Li et al. (2008a) in dynamical systems. Zhang et al. (2011) studied the complete synchronization of a coupled fractional order system. More recently, Razminia and Baleanu (2013) has extended this concept for fractional order chaotic systems as a complete coincidence between the state variables of the two synchronized fractional order chaotic systems. The error of complete synchronization is defined as:

$$e_t = Y_t - X_t$$

  Thus, the complete synchronization problem consists to determine the controller U_t so that,

$$\lim_{t \to \infty} \| e_t \| = 0$$

  where $\| . \|$ is the Euclidean norm. If f is equal to g, the relation becomes an identical complete synchronization. If f is not equal to g; then the complete synchronization is not identical.
- The concept of antisynchronization of two different chaotic systems is proposed by Wedekind and Parlitz (2001) as a phenomenon for which the components of the state vectors of the antisynchronized systems have the same amplitudes in absolute values and are of opposite signs. As a result, Li et al. (2008b) prove that the sum of two signals is assumed to converge to zero in the case where the antisynchronization property is satisfied. Bhalekar and Daftardar-Gejji (2011) studied nonidentical fractional order differential systems and confirmed that theoretically two fractional order chaotic systems are antisynchronized if, on the one hand, the two systems have identical state vectors in absolute value but with opposite signs, on the other hand, the sum of the state vectors of the two systems tends towards zero when time tends towards infinity. The antisynchronization error can thus be defined as follows:

$$e_t = Y_t - X_t$$

- The concept of projective synchronization is presented by Mainieri and Rehacek (1999) to justify the state variables of two fractional order chaotic systems master and slave synchronized and which are characterized by a multiple constant of the near state, such as:

$$\exists \beta_i \neq 0, \quad \lim_{t \to \infty} \left\| y_{it} - \beta_i \, x_{it} \right\|_{i=1,\,2,\ldots,n} = 0, \quad \forall x_0, \ \forall y_0$$

  The case where all coupling parameters β_i are equal to 1 represents a case of complete synchronization. The case where all coupling parameters β_i are equal to -1 represents a complete antisynchronization case.
- The concept of generalized synchronization is proposed by Lu (2008) as a new general scheme of complete synchronization, antisynchronization and projective synchronization of discrete-time chaotic and hyperchaotic systems. According to Huang and Xin (2010), this generalized scheme can offer different synchronization of different fractional order chaotic systems of different dimensions and models. More recently, Ouannas et al. (2017e) show that the proposed method can be used for realizing a generalized synchronization of different dimensional integer order and fractional order chaotic systems if it is manifested by a functional relation between the two coupled chaotic systems. Hence, if there is a function $\varphi:\mathfrak{R}^n \to \mathfrak{R}^m$ such as all the trajectories of the two fractional order chaotic systems, with the initial conditions X_0 and Y_0, satisfying the following property:

$$\lim_{t \to \infty} \left\| Y_t - \varphi(X_t) \right\|_{i=1,\,2,\ldots,n} = 0, \quad \forall X_0, \ \forall Y_0$$

Then the two fractional order chaotic systems are synchronized with the generalized sense with respect to the function φ.

- The concept of Q-S synchronization is proposed by (Li, 2007) as a generalization of all previous synchronization types when the response system contains scaling matrix. We say that a master n-dimensional fractional order chaotic system X_t and a slave m-dimensional fractional order chaotic system Y_t are in Q-S synchronization in dimension d, if there is a controller $U_t \in \mathfrak{R}^m$ and two functions $Q:\mathfrak{R}^n \to \mathfrak{R}^d$ and $S:\mathfrak{R}^m \to \mathfrak{R}^d$ such as synchronization error $e_t = Q_t - S_t$ verifying:

$$\lim_{t \to \infty} \left\| e_t \right\| = 0$$

21.5.2 The Topological Synchronization

Unlike traditional methods of synchronization, when the coupled subsystems are nonidentical, then the last fundamental property of manifold invariance cannot be valid and the concept of synchronization must be treated differently. In this chapter, the focus will be on a new form based on the topological synchronization. This work aims to combine two essential phenomena in nonlinear dynamics analysis: synchronization of fractional chaotic systems and systematical quantification of the different structures of recurrences in phase space systems. So, it is not only necessary to study the fundamental conditions under which synchronization of these chaotic systems arises, but also to establish some interesting tests for the detection of systems synchronization. The proposed tests for synchronization are based on the crucial property of recurrences in phase space. The new measures of the different structures of recurrences that will be proposed to find the topological synchronization analysis need to be robust with respect to the RP technique that visualizes the recurrences of chaotic systems. Based on this nonlinear dynamical conception, recently Vladimirsky and Ismailov (2015a) proposed the topological synchronization of fractional orders chaotic systems using the notion of Poincaré recurrences which is defined as:

Definition: *Two fractional orders chaotic systems are topologically synchronized, if Poincaré return times behave a similar way.*

According to Vladimirsky and Ismailov (2015b) the Poincaré return times of fractional orders chaotic systems are calculated using the fractal dimension (see Afraimovich et al., 2000). Hence, the implementation of topological synchronization of two fractional orders chaotic systems must be accomplished with a topological control, defined as:

Definition: *Two fractional order chaotic systems are topologically controllable if and only if they are synchronized topologically.*

Definition: *The fractional order chaotic system is topologically controllable if and only if coincides with master's $\hat{X}_t$ system on the basis of the criterion metrics "proximity" Hausdorff distance.*

According to Vladimirsky and Ismailov (2015a,b) the procedure of topological synchronization and topological control of two fractional orders chaotic systems must be modeled as a network of fractional order models in the structure of topological synchronization with period T, via tracing control and stability with generalized memory. Mathematically, for all $i = 1, 2, \ldots, N$ this procedure was presented as:

$$D_t^p x_i = S(x_i) + C(x_i) + \hat{S}(x_i) + \eta\,(x_i) + GM(x_i) + PL(x_i)$$

where p is the order of derivative, $S(x_i)$ is synchronization of algorithm i^{th} system, $C\,(x_i)$ is control algorithm i^{th} system, $\hat{S}(x_i)$ is stability i^{th} system,

$GM(x_i)$ is generalized i^{th} system with memory, $PL(x_i)$ is percolation lattice i^{th} system, and $\eta\,(x_i)$ is the fractional part map of sawtooth wave, verifying:

$$\eta(x_i){:}D_t^p \vee A \times frac\left(\frac{x}{T} + \phi\right)$$

where A is the amplitude, T is the period of the wave, ϕ is its phase, and $frac\,(x_i)$ is the fractional part, verifying:

$$frac\,(x) = x - [x]$$

The considered procedure needs the implementation of the following iterative algorithm:

Step 1: Initialization of the two fractional orders chaotic systems;

Step 2: Simulation of the considered two fractional orders chaotic systems according to the algorithm;

Step 3: Initialization of the master's $\hat{X}_t = {}_{\alpha}\{x_n\}_{n=0}^{N}$ system related to the considered two fractional orders chaotic systems;

Step 4: Using the proposed method to achieve the behavior of synchronization between two new fractional orders chaotic systems;

Step 5: Determined the corresponding slave system $\hat{Y}_t = {}_{\alpha}\left\{y_n\right\}_{n=0}^{N}$ defined as:

$$\hat{Y}_t = \hat{X} + S(X), \quad \hat{Y} \in U$$

where $S(X)$ is the synchronization of algorithm with period T in its phase ϕ and U is the m-dimensional input vector that will be used.

Step 6: Start the iterative algorithm for topological synchronization using the master–slave schema;

Step 7: Using the iterative procedure to evaluate the effect of the "proximity" capture of the average Poincaré return times as a criterion of chaos control of topological synchronization systems with while tracking control of chaotic nonlinear fractional order systems;

Step 8: Comparing and the Poincaré recurrence diagrams for both $D_t^p\hat{X}$, $D_t^p\hat{Y}$ systems and respectively interpreting their related topological structures. Then, define the joint Poincaré recurrence diagram for these coupled systems which are essentially different and the related topological measures of its quantification analysis. Mathematically, a new index will be defined and which is based on the average probability of their joint recurrence over time. It is considered as criterion for the detection of topological synchronization in master–slave dynamical system, and it is given as:

$$R_{ij}^{x,\,y} = \frac{1}{N}\sum\nolimits_{i=1}^{N}\sum\nolimits_{j=1}^{N}\Theta\left(\varepsilon^{x} - \left\|x_i - x_j\right\|\right)_{i,j=1,\,2,\ldots,N}$$
$$\Theta\left(\varepsilon^{y} - \left\|y_i - y_j\right\|\right)_{i,j=1,\,2,\ldots,N}, \quad x,\, y \in U$$

where N is the number of considered states, $\Theta(.)$ is the Heaviside function, ε is a threshold distance, $\|.\|$ is the norm, and y_i and y_j are two points of the trajectory;

Step 9: Estimation of the effects of the proximity principle;

Step 10: Determined the Lyapunov stability.

21.6 TOPOLOGICAL SYNCHRONIZATION FOR SOME EXAMPLES OF FRACTIONAL ORDER CHAOTIC SYSTEMS

21.6.1 Topological Synchronization of Fractional Order Lorenz's Systems

21.6.1.1 The Fractional Order Lorenz's System

The fractional order Lorenz's system was developed by Li and Yan (2007) following this form:

$$\begin{aligned} {}_0D_t^{p_1}x_t &= a\,(y_t - x_t) \\ {}_0D_t^{p_2}x_t &= b\,x_t - x_t z_t - y_t \\ {}_0D_t^{p_3}x_t &= x_t y_t - c\,z_t \end{aligned} \tag{21.19}$$

where t is the continuous time, p_1, p_2, p_3 are derivative orders, a, b, and c are three positive real parameters and x_t, y_t, and z_t are the three dynamic variables specifying the system status over time t. The system has three equilibrium points $E_1 = (0,0,0)$ and $E_{2,3} = (\pm\ (bc\text{-}c)^{1/2},\ \pm\ (bc\text{-}c)^{1/2},\ b\text{-}1)$. The Jacobian matrix related to the fractional order Lorenz's system at the equilibrium point $E^* = (x^*,y^*,z^*)$ is written as:

$$J = \begin{pmatrix} -a & a & 0 \\ b - z^* & -1 & -x^* \\ y^* & x^* & -c \end{pmatrix}$$

Using the formula (21.10), obtained from Grünwald–Letnikov definition (21.3), for a known initial conditions (x_0, y_0, z_0), the general numerical solution of (21.19) is given by:

$$\begin{aligned} x_{t_k} &= \left(a\left(y_{t_{k-1}} - x_{t_{k-1}}\right)\right)h^{p_1} - \sum_{j=\tau}^{k} c_j^{(p_1)} x_{t_{k-j}} \\ y_{t_k} &= \left(b\,x_{t_k} - x_{t_k} z_{t_{k-1}} - y_{t_{k-1}}\right)h^{p_2} - \sum_{j=\tau}^{k} c_j^{(p_2)} y_{t_{k-j}} \\ z_{t_k} &= \left(x_{t_k} y_{t_k} - c\,z_{t_{k-1}}\right)h^{p_3} - \sum_{j=\tau}^{k} c_j^{(p_3)} z_{t_{k-j}} \end{aligned} \tag{21.20}$$

where $k = 1, 2, \ldots, [(t-\alpha)/h]$ and the binomial coefficients $c_j^{(p_i)}$ for all $i = 1, 2, 3$ are calculated based on the previous formula (2.4). For a given parameters ($a = 10$, $b = 28$, $c = 8/3$) and orders $p_1 = p_2 = p_3 = p$, using the relation (21.15), according to Grigorenko and Grigorenko (2003) this system can be chaotic only when the following condition is satisfied $p > 0.993$.

For first example, with fixed values $p_1 = p_2 = p_3 = p = 0.995$, if parameters are chosen as ($a = 10$, $b = 28$, $c = 8/3$), initial conditions as ($x_0 = 0.4$, $y_0 = 0.1$, $z_0 = 0.1$) and computational time 100s for time step $h = 0.01$, then the fractional order Lorenz's system (21.19) exhibits a chaotic behavior. The strange attractor for the system is shown in Figs. 21.2–21.4. Using the relations (21.11) and (21.12), the three unstable equilibrium points of the system (21.19) are $E_1 = (0, 0, 0)$ and $E_{2,3} = (\pm\ (72)^{1/2}, \pm (72)^{1/2}, 27)$. The 3D state space numerical simulation of the system (21.19) for initial conditions ($x_0 = 0.4$, $y_0 = 0.1$, $z_0 = 0.1$) is illustrated in Fig. 21.4. All model variables are represented versus time in Fig. 21.5. The related recurrence diagram is ported in Fig. 21.6. The recurrent points are plotted with blue color, but the nonrecurrent points are left blank.

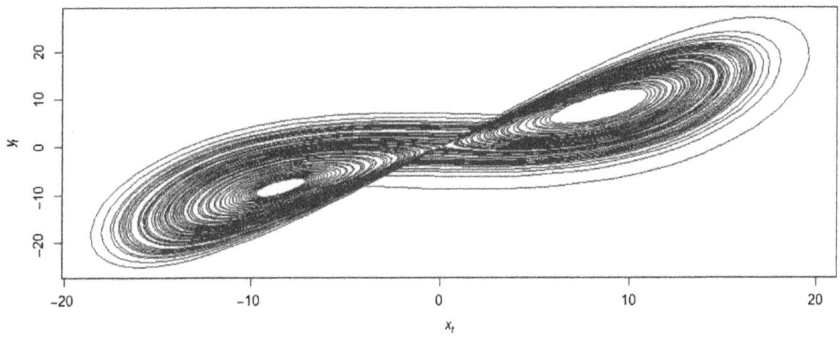

FIGURE 21.2 Simulation result of the fractional order Lorenz's system (21.19) in $x-y$ plane for parameters ($a = 10$, $b = 28$, $c = 8/3$), orders $p_1 = p_2 = p_3 = 0.995$, and initial conditions ($x_0 = 0.4$, $y_0 = 0.1$, $z_0 = 0.1$).

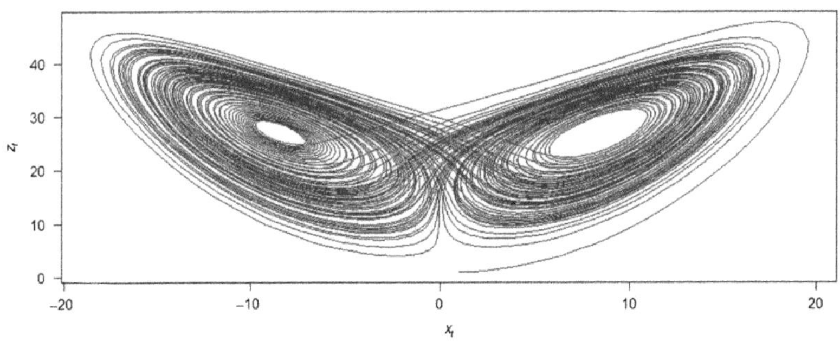

FIGURE 21.3 Simulation result of the fractional order Lorenz's system (21.19) in $x-z$ plane for parameters ($a = 10$, $b = 28$, $c = 8/3$), orders $p_1 = p_2 = p_3 = 0.995$, and initial conditions ($x_0 = 0.4$, $y_0 = 0.1$, $z_0 = 0.1$).

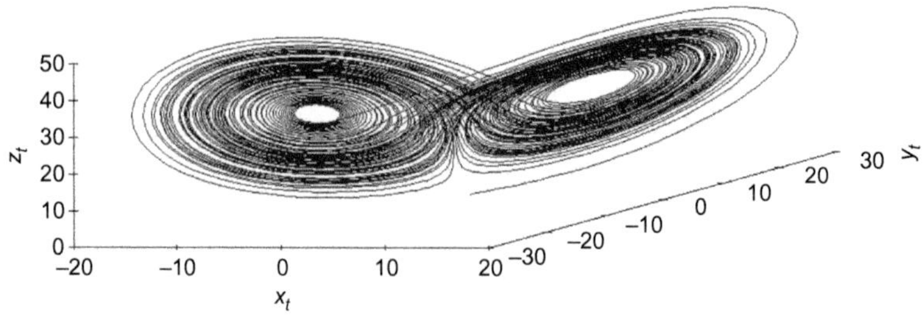

FIGURE 21.4 Simulation result of the fractional order Lorenz's system (21.19) in state space for parameters ($a = 10$, $b = 28$, $c = 8/3$), orders $p_1 = p_2 = p_3 = 0.995$, and initial conditions ($x_0 = 0.4$, $y_0 = 0.1$, $z_0 = 0.1$).

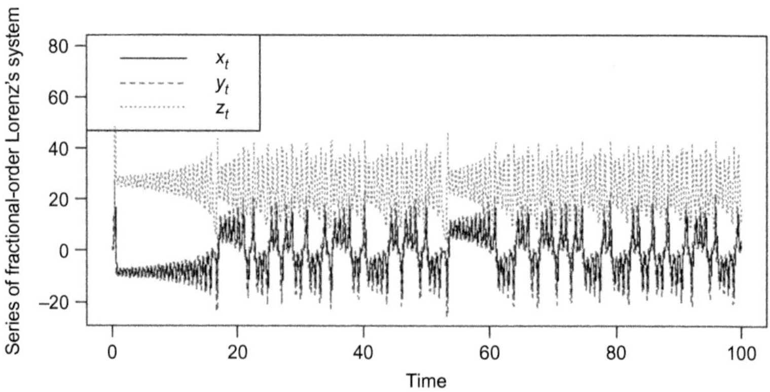

FIGURE 21.5 Simulation result of all fractional order Lorenz's system (21.19) variables versus time for parameters ($a = 10$, $b = 28$, $c = 8/3$), orders $p_1 = p_2 = p_3 = 0.995$, and initial conditions ($x_0 = 0.4$, $y_0 = 0.1$, $z_0 = 0.1$).

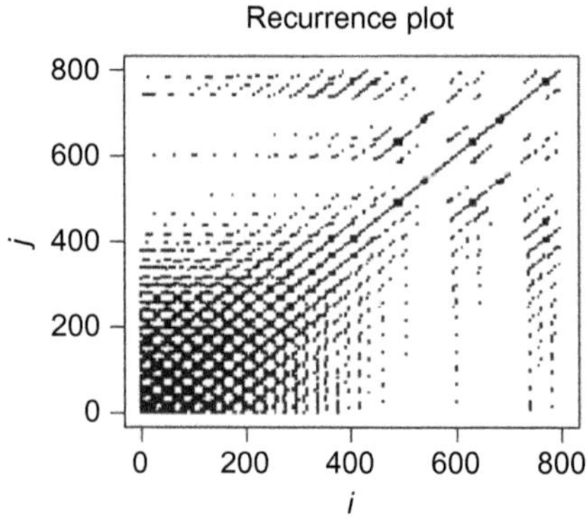

FIGURE 21.6 Simulation result of recurrence diagram of the fractional order Lorenz's system (21.19) in i–j plane for parameters (($a = 10$, $b = 28$, $c = 8/3$), orders $p_1 = p_2 = p_3 = 0.995$, and initial conditions ($x_0 = 0.4$, $y_0 = 0.1$, $z_0 = 0.1$).

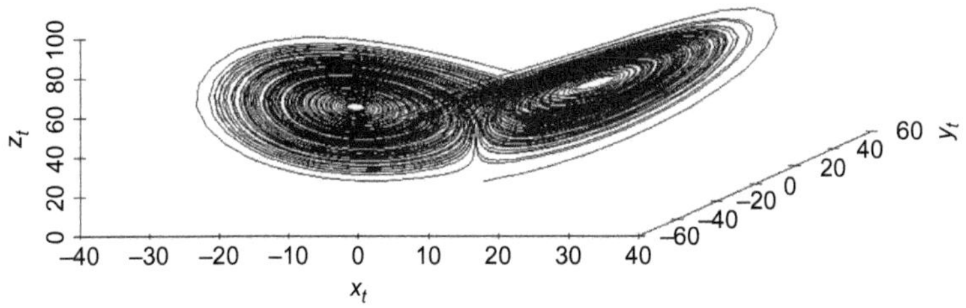

FIGURE 21.7 Simulation result of all fractional order Lorenz's system (21.19) in state space for parameters ($a = 16$, $b = 45.92$, $c = 4$), orders $p_1 = p_2 = p_3 = 0.995$ and initial conditions ($x_0 = 1$, $y_0 = 1$, $z_0 = 1$).

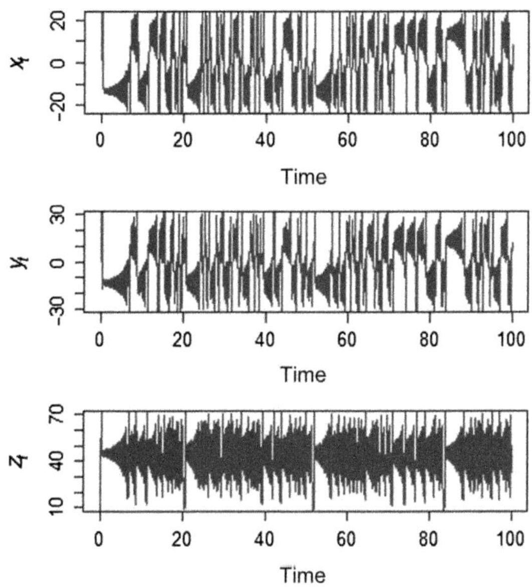

FIGURE 21.8 Simulation result of all fractional order Lorenz's system (21.19) variables versus time for parameters ($a = 16$, $b = 45.92$, $c = 4$), orders $p_1 = p_2 = p_3 = 0.995$ and initial conditions ($x_0 = 1$, $y_0 = 1$, $z_0 = 1$).

In a second case of the same fractional order Lorenz's system (21.19) with even values of the commensurate orders $p_1 = p_2 = p_3 = p = 0.995$, if parameters are fixed as ($a = 16$, $b = 45.92$, $c = 4$), initial conditions as ($x_0 = 1$, $y_0 = 1$, $z_0 = 1$) and computational time 100s for time step $h = 0.01$, then the fractional order Lorenz's system (21.19) exhibits a strange attractor and the chaotic butterfly-effect which are given in Fig. 21.7. All model variables are represented versus time in Fig. 21.8. The related recurrence diagram is depicted in Fig. 21.9. The recurrent points are represented with blue color; however the nonrecurrent points are left blank.

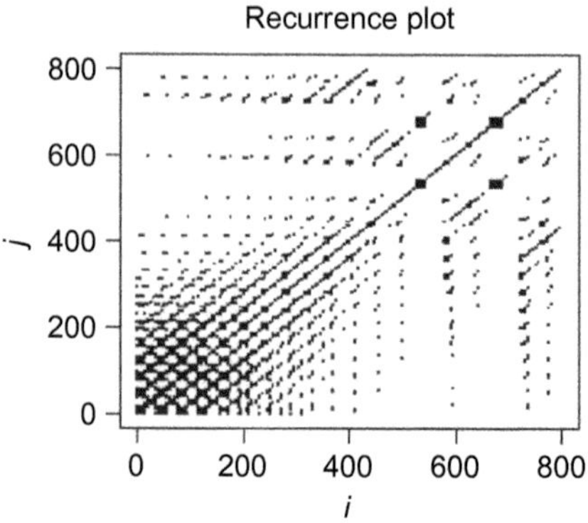

FIGURE 21.9 Simulation result of recurrence diagram of the fractional order Lorenz's system (21.19) in i–j plane for parameters ($a = 16$, $b = 45.92$, $c = 4$), orders $p_1 = p_2 = p_3 = 0.995$ and initial conditions ($x_0 = 1$, $y_0 = 1$, $z_0 = 1$).

First, it is noticeable that both systems are completely different systems. On the other hand, by comparing one condition to another condition in both two experiment simulations, it is remarkable that the recurrence matrices of two considered systems ported respectively in Figs. 21.6 and 21.9 are very similar, but not identical. Then, in order to apply the topological synchronization of fractional order Lorenz's systems, the aim is to build an index that quantifies the degree of similarity between the respective recurrences of both Lorenz's systems.

21.6.1.2 *Algorithm of Topological Synchronization of Fractional Order Lorenz's Systems*

Step 1: Let consider two fractional order Lorenz's systems as (21.19). The first system is given for parameters ($a = 10$, $b = 28$, $c = 8/3$), orders $p_1 = p_2 = p_3 = 0.995$ and initial conditions ($x_0 = 0.4$, $y_0 = 0.1$, $z_0 = 0.1$) which can be topologically synchronized with the second system given for parameters ($a = 16$, $b = 45.92$, $c = 4$), orders $p_1 = p_2 = p_3 = 0.995$ and initial conditions ($x_0 = 1$, $y_0 = 1$, $z_0 = 1$).

Step 2: Both these systems are simulated according to the previous algorithm developed by Vladimirsky and Ismailov (2015a,b) and respectively represented as it was previously visualized.

Step 3: Let $\hat{X}_t = {}_{\alpha}\{x_n\}_{n=0}^{N}$ be associated observable two considered fractional order Lorenz's systems.

Step 4: In order to determine the behavior of synchronization between two new chaotic systems by using the proposed method, we suppose the master is$\hat{X}_t$.

Step 5: Determined the corresponding slave system$\hat{Y}_t = {}_{\alpha}\{y_n\}_{n=0}^{N}$.

Step 6: Execute the iterative algorithm for topological synchronization on master–slave system with while tracking control for chaotic nonlinear fractional order systems.

Step 7: Consider the average Poincaré return times as a criterion for chaotic topological synchronization systems with while tracking control. The first-return map of the master–slave system is ported in Fig. 21.10.

Step 8: The Poincaré recurrence diagram and fractal dimensions for master–slave system are determined. After executing the iterative algorithm for topological synchronization for two iterations, the recurrence diagram for master–slave joint system is in the phase coherent. Then, it may be synchronized as visualized in Fig. 21.11. For these two iterations, their respectively related fractal dimensions are $D_1 = 1.770$ and $D_2 = 1.003$ and their respectively related averages Poincaré return times are $\tau_1 = 3.2566$ and $\tau_2 = 3.2002$.

Step 9: As the averages Poincaré return times are determined, then the effect of the "proximity" is computed as:

$$\varepsilon = \tau_1 - \tau_2 = 0.0564$$

The first-return map of the master–slave system is ported in Fig. 21.10.

Step 10: Define the Lyapunov stability.

We can remark that both the master's system and slave's system are considered separately independent from each other. On the other hand, as given in Fig. 21.12 the autorecurrence diagram of their joint recurrence plot obtained using these two coupled systems show that the recurrent points are plotted with blue color, but the nonrecurrent points are marked with red. As explained in principles, related to the previous subsection 21.2.1.2 above, in order to establish an adequate analysis of topological synchronization by means of recurrences, it is appropriate to focus on this autorecurrence diagram and to consider the following quantification recurrence analysis measures, which are given in Table 21.1. On the diagonal lines, the pattern of

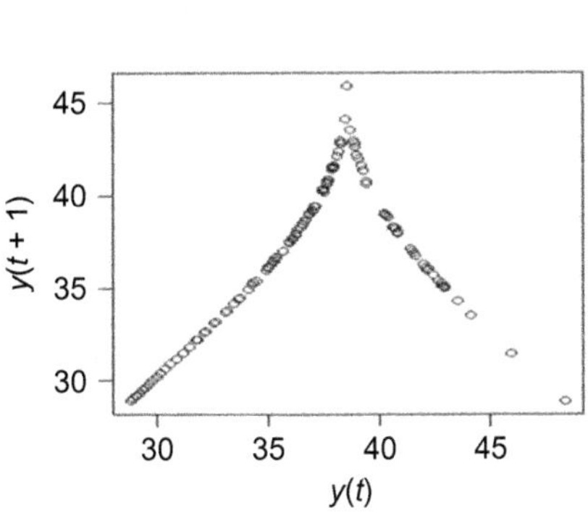

FIGURE 21.10 Simulation result of first-return map of the master–slave related to two considered fractional order Lorenz's systems in plane.

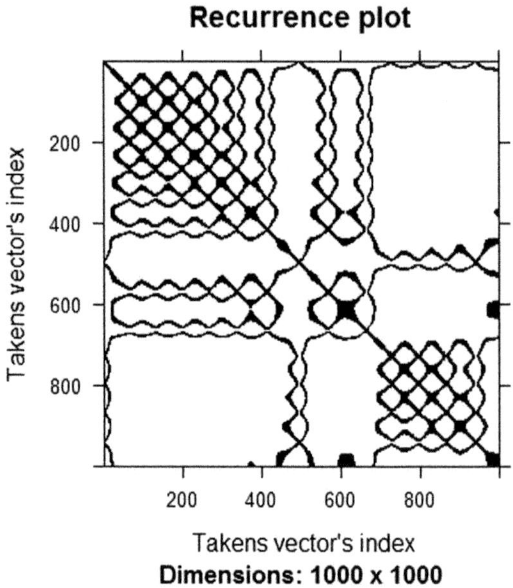

FIGURE 21.11 Simulation result of recurrence diagram of the master–slave related to two considered fractional order Lorenz's systems.

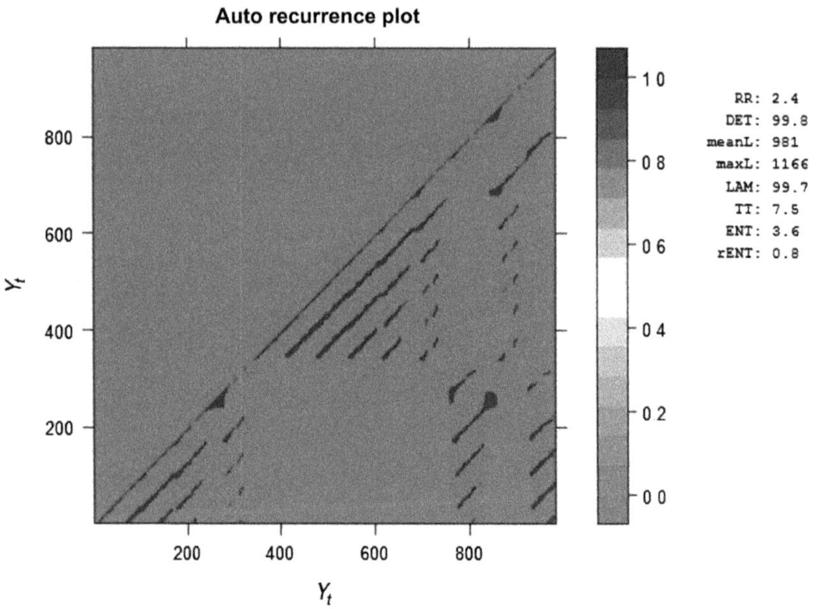

FIGURE 21.12 Simulation result of autorecurrence diagram of the fractional order Lorenz's system (21.19) in i–j plane for parameters ($a = 16$, $b = 45.92$, $c = 4$), orders $p_1 = p_2 = p_3 = 0.995$ and initial conditions ($x_0 = 1$, $y_0 = 1$, $z_0 = 1$).

interaction is observable between the two dynamical systems. The calculated characterizing considered measures are recurrence rate (RR), percentage determinism (DET), average diagonal length (meanL), maximal diagonal length (maxL), and entropy (ENT). On the vertical lines, the stability of the two coupled dynamical systems and relative independence of recurrence over a particular state are detected. The measures characterizing this information are laminarity (LAM) and trapping time (TT).

The values obtained on the quantification recurrence analysis measures related to the diagram given in Fig. 21.12 are: RR = 2.4; DET = 99.8; meanL = 981; maxL = 11.66; LAM = 99.7; TT = 7.5; ENT = 3.6; rENT = 0.8. It's clear, DET is higher than RR, with DET often quite high (90% or higher) and RR normally must be considerably lower (10% or less), so 2% would be considered slightly high. Then the result of topological synchronization is clearly confirmed by these indicators of quantification recurrence analysis as indicated on the Fig. 21.12.

21.6.2 Topological Synchronization of Fractional Order Rössler's Systems

21.6.2.1 The Fractional Order Rössler's System

The fractional order Rössler's system was proposed by Li and Chen (2004) following this form:

$$\begin{aligned} {}_0D_t^{p_1} x_t &= -(y_t + z_t) \\ {}_0D_t^{p_2} x_t &= x_t + a\, y_t \\ {}_0D_t^{p_3} x_t &= b + z_t(x_t - c\,) \end{aligned} \tag{21.21}$$

where t is the continuous time, p_1, p_2, p_3 are derivative orders, a, b, and c are three positive real parameters and x_t, y_t, and z_t are the three dynamic variables specifying the system status over time t. The general numerical solution of (21.21) using the formula (21.10) is obtained from Grünwald–Letnikov definition (21.3), for a known initial conditions (x_0, y_0, z_0), is given by:

$$\begin{aligned} x_{t_k} &= \left(-\left(y_{t_{k-1}} + x_{t_{k-1}}\right)\right)h^{p_1} - \sum_{j=\tau}^{k} c_j^{(p_1)} x_{t_{k-j}} \\ y_{t_k} &= \left(x_{t_k} + a y_{t_{k-1}}\right)h^{p_2} - \sum_{j=\tau}^{k} c_j^{(p_2)} y_{t_{k-j}} \\ z_{t_k} &= \left(b + z_{t_{k-1}}\left(x_{t_k} - c\right)\right)h^{p_3} - \sum_{j=\tau}^{k} c_j^{(p_3)} z_{t_{k-j}} \end{aligned} \tag{21.22}$$

where $k = 1, 2, \ldots, [(t-\alpha)/h]$ and the binomial coefficients $c_j^{(p_i)}$ for all $i = 1, 2, 3$ are calculated based on the previous formula (21.4). For given parameters ($a = 0.5$, $b = 0.2$, $c = 10$) and orders $p_1 = p_2 = p_3 = p$, using the relation (21.15), we can verify that for $0.9 \leq p \leq 1$ the system is chaotic with a positive Lyapunov exponent 0.0368. The minimal commensurate order from which the system (21.21) can exhibit a chaotic behavior is $p > 0.839$.

For first example, if it is assumed that the commensurate order has an arbitrary fixed values $p_1 = p_2 = p_3 = p = 0.95$, parameters are chosen as ($a = 0.5$, $b = 0.2$, $c = 10$), initial conditions are selected as ($x_0 = -0.5$, $y_0 = 0$, $z_0 = 1$), and computational time 100s for time step $h = 0.01$, then the fractional order Rössler's system (21.21) satisfies the condition of presence of chaotic behavior as is visualized in Figs. 21.13–21.15. The 3D state space numerical simulation of the system (21.21) for initial conditions ($x_0 = -0.5$, $y_0 = 0$, $z_0 = 1$) is illustrated in Fig. 21.13. The positively estimating maximal Lyapunov exponent for various value of embedding dimension is visualized

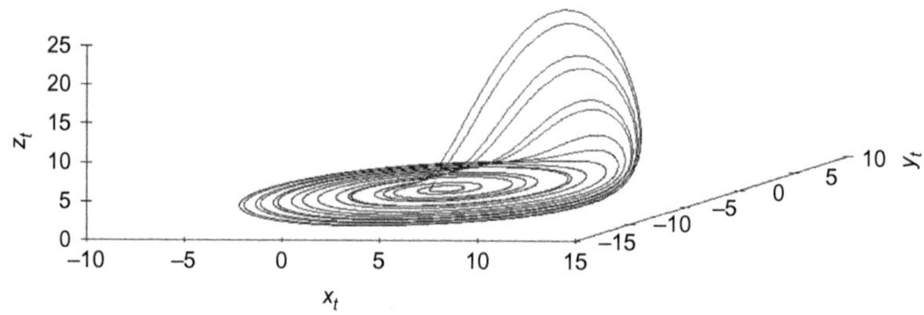

FIGURE 21.13 Simulation result of the fractional order Rössler's system (21.21) in state space for parameters ($a = 0.5$, $b = 0.2$, $c = 10$), orders $p_1 = p_2 = p_3 = 0.95$, and initial conditions ($x_0 = -0.5$, $y_0 = 0$, $z_0 = 1$).

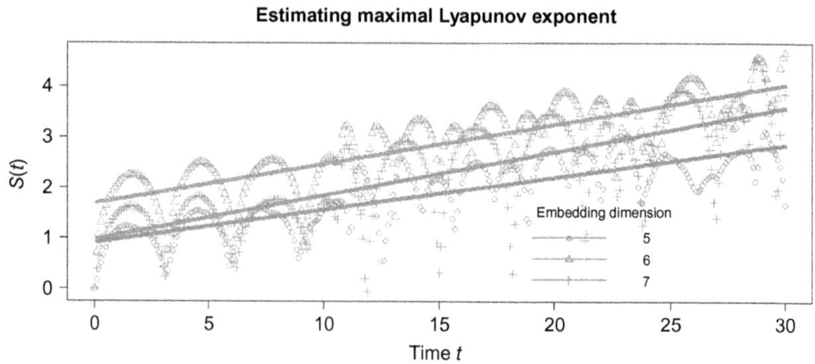

FIGURE 21.14 Simulation result of the maximal Lyapunov exponent of fractional order Rössler's system (21.21) for parameters ($a = 0.5$, $b = 0.2$, $c = 10$), orders $p_1 = p_2 = p_3 = 0.95$, and initial conditions ($x_0 = -0.5$, $y_0 = 0$, $z_0 = 1$).

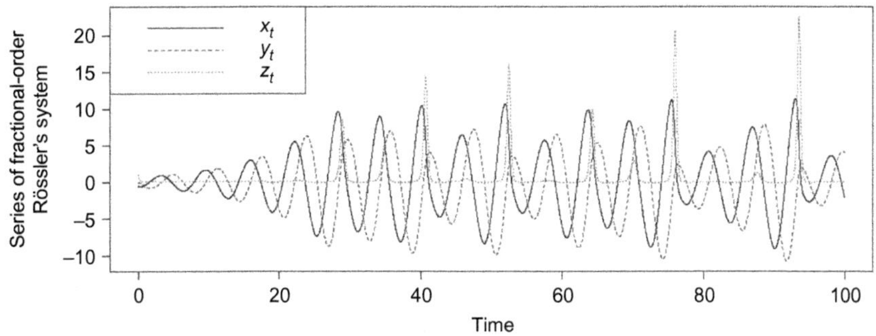

FIGURE 21.15 Simulation result of all fractional order Rössler's system (21.21) variables versus time for parameters ($a = 0.5$, $b = 0.2$, $c = 10$), orders $p_1 = p_2 = p_3 = 0.95$, and initial conditions ($x_0 = -0.5$, $y_0 = 0$, $z_0 = 1$).

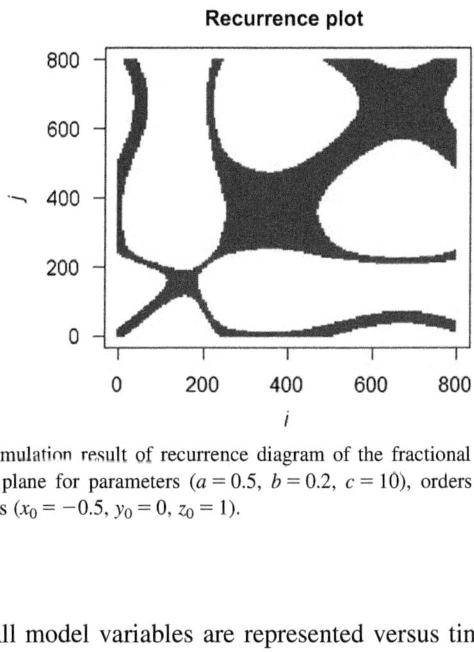

FIGURE 21.16 Simulation result of recurrence diagram of the fractional order Rössler's system (21.21) in i–j plane for parameters ($a = 0.5$, $b = 0.2$, $c = 10$), orders $p_1 = p_2 = p_3 - 0.95$, and initial conditions ($x_0 = -0.5$, $y_0 = 0$, $z_0 = 1$).

in Fig. 21.14. All model variables are represented versus time in Fig. 21.15. The related recurrence diagram is ported in Fig. 21.16.

In a second case of the same fractional order Rössler's system (21.21) with even values of the commensurate orders $p_1 = p_2 = p_3 = p = 0.95$, if parameters are fixed as ($a = 2.5$, $b = 5$, $c = 4$), initial conditions as ($x_0 = 1$, $y_0 = 1$, $z_0 = 1$), and computational time 100s for time step $h = 0.01$, then the fractional order Rössler's system (21.21) exhibits a strange attractor and the chaotic butterfly-effect which are given in Fig. 21.17. All model variables are represented versus time in Fig. 21.18. The related recurrence diagram is depicted in Fig. 21.19.

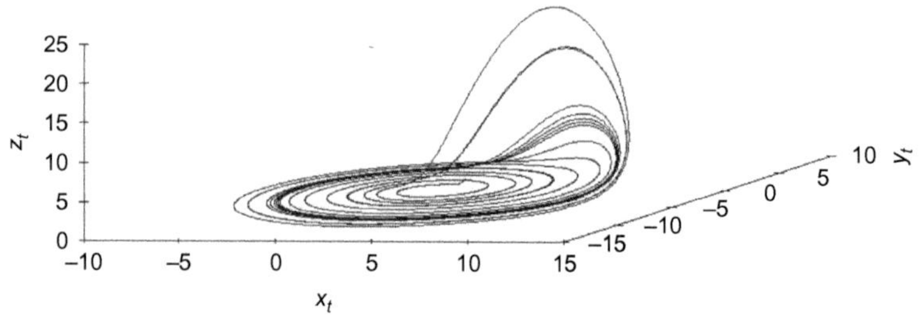

FIGURE 21.17 Simulation result of all fractional order Rössler's system (21.21) in state space for parameters ($a = 2.5$, $b = 5$, $c = 4$), orders $p_1 = p_2 = p_3 = 0.95$ and initial conditions ($x_0 = 1$, $y_0 = 1$, $z_0 = 1$).

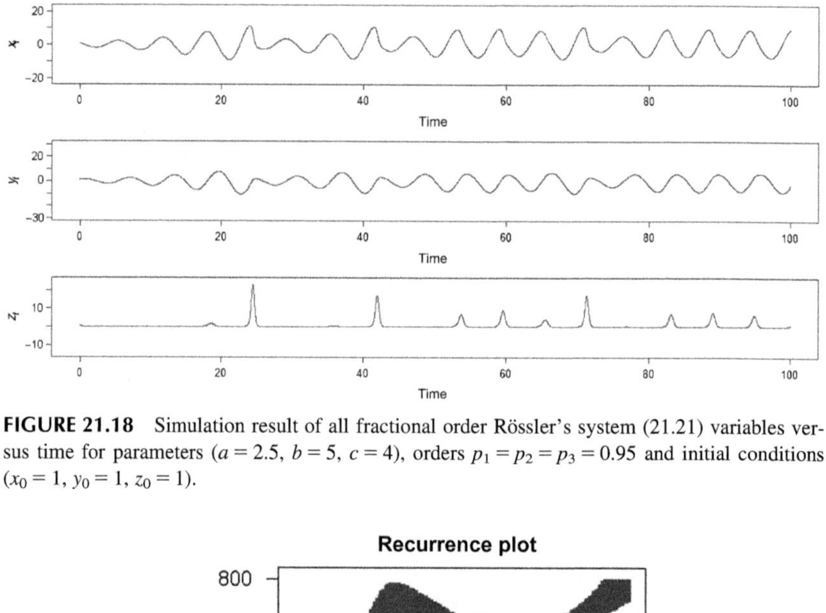

FIGURE 21.18 Simulation result of all fractional order Rössler's system (21.21) variables versus time for parameters ($a = 2.5$, $b = 5$, $c = 4$), orders $p_1 = p_2 = p_3 = 0.95$ and initial conditions ($x_0 = 1$, $y_0 = 1$, $z_0 = 1$).

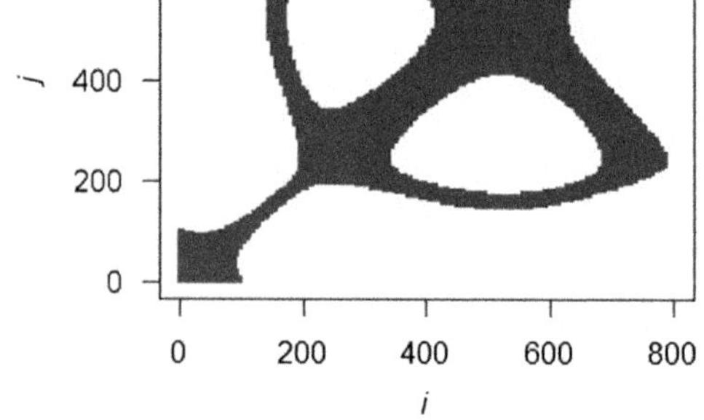

FIGURE 21.19 Simulation result of recurrence diagram of the fractional order Rössler's system (21.21) in i–j plane for parameters ($a = 2.5$, $b = 5$, $c = 4$), orders $p_1 = p_2 = p_3 = 0.95$ and initial conditions ($x_0 = 1$, $y_0 = 1$, $z_0 = 1$).

21.6.2.2 Algorithm of Topological Synchronization of Fractional Order Rössler's Systems

Step 1: Let consider two fractional order Rössler's systems as (21.21). The first system is given for parameters ($a = 0.5$, $b = 0.2$, $c = 10$), orders $p_1 = p_2 = p_3 = 0.95$ and initial conditions ($x_0 = -0.5$, $y_0 = 0$, $z_0 = 1$) which can be topologically synchronized with the second system given for parameters ($a = 2.5$, $b = 5$, $c = 4$), orders $p_1 = p_2 = p_3 = 0.95$ and initial conditions ($x_0 = 1$, $y_0 = 1$, $z_0 = 1$).

Step 2: Both systems are simulated according to the previous algorithm developed by Vladimirsky and Ismailov (2015a,b) and respectively represented as it was previously visualized.

Step 3: Let $\hat{X}_t = {}_{\alpha}\{x_n\}_{n=0}^{N}$ be associated observable two considered fractional order Rössler's systems.

Step 4: In order to determine the behavior of synchronization between two new chaotic systems by using the proposed method, it is supposed that $\hat{X}_t$ is the master.

Step 5: The corresponding slave system $\hat{Y}_t = {}_{\alpha}\{y_n\}_{n=0}^{N}$ is determined.

Step 6: The iterative algorithm for topological synchronization on master–slave system with while tracking control for chaotic nonlinear fractional order systems is executed.

Step 7: We consider the average Poincaré return times as a criterion for chaotic topological synchronization systems with while tracking control. The first-return map of the master–slave system is ported in Fig. 21.20.

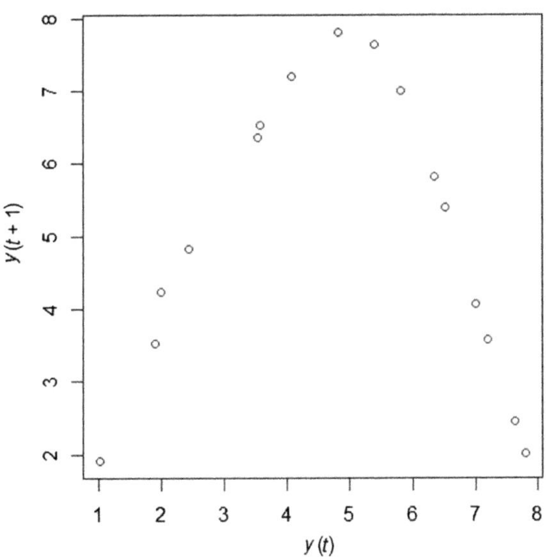

FIGURE 21.20 Simulation result of first-return map of the master–slave related to two considered fractional order Rössler's systems in plane.

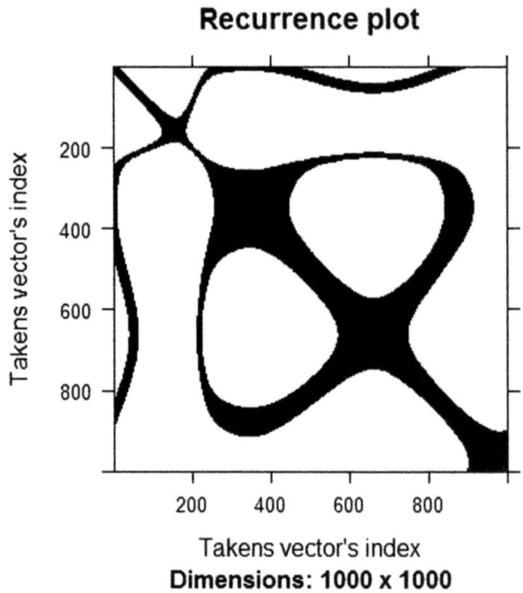

FIGURE 21.21 Simulation result of recurrence diagram of the master–slave related to two considered fractional order Rössler's systems.

Step 8: The Poincaré recurrence diagram and fractal dimensions for master–slave system are determined. After executing the iterative algorithm for topological synchronization for two iterations, the recurrence diagram for master–slave joint system is in the phase coherent. Then, it may be synchronized as visualized in Fig. 21.21. For these two iterations, their respectively related fractal dimensions are $D_1 = 1.619$ and $D_2 = 1.401$ and their respectively related averages Poincaré return times are $\tau_1 = 3.8452$ and $\tau_2 = 3.8192$.

Step 9: As the averages Poincaré return times are determined, then the effect of the "proximity" is computed as:

$$\varepsilon = \tau_1 - \tau_2 = 0.026$$

Step 10: Define the Lyapunov stability.

21.7 CONCLUSION

To conclude, the synchronization of complex systems that has been widely studied in the last two decades has several applications in various fields. Now, synchronization analysis and recurrence in fractional order chaotic systems has become an open and exciting axis of research. In particular, topological synchronization by means of recurrences for some fractional order chaotic systems is discussed in this chapter. The results show that the control of topological synchronization is confirmed via control stability of fractional order chaotic systems. This could be explained based on a given synchronization which is reflected in higher values of Recurrence percent (%RR) and Determinism percent (%DET), caused by the presence of an auto-organization phenomenon. So, Recurrence plot is a powerful tool to determine the fractional order chaotic system periodicity and control its stability.

The next research will be concentrated on the study of topological synchronization of other coupled fractional order hyperchaotic systems. In particular, the focus will be on systems with different classes because of the lack of study of the topological synchronization of these difficult systems with complexes and various dynamical structures.

REFERENCES

Afraimovich, V., 1999. Poincaré recurrences of coupled subsystems in synchronized regimes. Taiwanese J. Math. 3 (2), 139–161.

Afraimovich, V.S., Lin, W.W., Rulkov, N.F., 2000. Fractal dimension for Poincaré recurrences as an indicator of synchronized chaotic regimes. Int. J. Bifurcat. Chaos 10 (10), 2323–2337.

Aguila-Camacho, N., Duarte-Mermoud, M.A., Gallegos, J.A., 2014. Lyapunov functions for fractional order systems. Commun. Nonlinear Sci. Numer. Simul. 19 (10), 2951–2957.

Azar, A.T., Vaidyanathan, S., 2015a. Chaos modeling and control systems design, Studies in Computational Intelligence, 581. Springer-Verlag, Germany, ISBN: 978-3-319-13131-3.

Azar, A.T., Vaidyanathan, S., 2015b. Handbook of research on advanced intelligent control engineering and automation. Advances in Computational Intelligence and Robotics (ACIR) Book Series. IGI Global, USA, ISBN: 9781466672482.

Azar, A.T., Vaidyanathan, S., 2015c. Computational intelligence applications in modeling and control, Studies in Computational Intelligence, 575. Springer-Verlag, Germany, ISBN: 978-3-319-11016-5.

Azar, A.T., Vaidyanathan, S., Ouannas, A., 2017a. Fractional order control and synchronization of chaotic systems, Studies in Computational Intelligence, 688. Springer-Verlag, Germany, ISBN: 978-3-319-50248-9.

Azar, A.T., Ouannas, A., Singh, S., 2017b. Control of new type of fractional chaos synchronization, Proceedings of the International Conference on Advanced Intelligent Systems and Informatics 2017, Advances in Intelligent Systems and Computing series, 639. Springer-Verlag, Germany, pp. 47–56.

Bhalekar, S., Daftardar-Gejji, V., 2010. Synchronization of different fractional order chaotic systems using active control. Commun. Nonlinear Sci. Numer. Simul. 15 (11), 3536–3546.

Bhalekar, S., Daftardar-Gejji, V., 2011. Antisynchronization of nonidentical fractional-order chaotic systems using active control. Int. J. Differ. Eq. 2011, 1–13.

Boulkroune, A., Bouzeriba, A., Bouden, T., Azar, A.T., 2016a. Fuzzy adaptive synchronization of uncertain fractional-order chaotic systems, Studies in Fuzziness and Soft Computing, 337. Springer-Verlag, Germany, pp. 681–697.

Boulkroune, A., Hamel, S., Azar, A.T., 2016b. Fuzzy control-based function synchronization of unknown chaotic systems with dead-zone input, Studies in Fuzziness and Soft Computing, 337. Springer-Verlag, Germany, pp. 699–718.

Chen, D., Zhang, R., Liu, X., Ma, X., 2014. Fractional order Lyapunov stability theorem and its applications in synchronization of complex dynamical networks. Commun. Nonlinear Sci. Numer. Simul. 19 (12), 4105–4121.

Chen, X.R., Liu, C.X., 2012. Chaos synchronization of fractional order unified chaotic system via nonlinear control. Int. J. Mod. Phys. B 25 (3), 407–415.

Das, S., Pan, I., Das, S., 2016. Effect of random parameter switching on commensurate fractional order chaotic systems. Chaos, Solitons Fractals 91 (2016), 157–173.

Deng, W.H., Li, C.P., 2005. Chaos synchronization of the fractional Lü system. Phys. A 353 (1-4), 61–72.

Devaney, R.L., 1989. An Introduction to Chaotic Dynamical System. Addison Wesley Publ. Comp, Redwood City, CA.

Dorcák, L., 1994. Numerical models for the simulation of the fractional-order control systems. UEF-04-94, The Academy of Sciences, Inst. of Experimental Physic, Kosice, Slovakia Republic, 12p.

Eckmann, J.P., Kamphorst, S.O., Ruelle, D., 1987. Recurrence plots of dynamical systems. Europhys. Lett. 4 (9), 973–977.

Grassi, G., Ouannas, A., Azar, A.T., Radwan, A.G., Volos, C., Pham, V.T., et al., 2017. Chaos synchronisation of continuous systems via scalar signal. The 6th International Conference on Modern Circuits and Systems Technologies (MOCAST), Thessaloniki Greece, 4-6 May.

Grigorenko, I., Grigorenko, E., 2003. Chaotic dynamics of the fractional Lorenz system. Phys. Rev. Lett. 91 (3), 034101–034104.

Huang, L., Xin, F., 2010. Generalized synchronization between different fractional-order chaotic systems based on nonlinear feedback. IEEE, Fifth International Conference on Bio-Inspired Computing: Theories and Applications (BIC-TA), 10141–1017.

Kantz, H., Schreiber, T., 1997. Nonlinear time series analysis. Cambridge Nonlinear Science Series 7. Cambridge University Press, Cambridge, UK, ISBN: 0521 551447.

Li, C., Chen, G., 2004. Chaos and hyperchaos in the fractional-order Rössler equations. Phys. A 341 (1-4), 55–61.

Li, C., Deng, W., 2006. Chaos synchronization of fractional-order differential systems. Int. J. Modern Phys. B 20 (7), 791–803.

Li, C., Yan, J., 2007. The synchronization of three fractional differential systems. Chaos Solitons Fractals 32 (2), 751–757.

Li, C.D., Chen, Q., Huang, T., 2008a. Coexistence of anti-phase and complete synchronization in coupled Chen system via a single variable. Chaos Solitons Fractals 38 (2), 461–464.

Li, G.H., 2007. Modified projective synchronization of chaotic system. Chaos Solitons Fractals 32 (5), 1786–1790.

Li, W.L., Chen, X.Q., Shen, Z.P., 2008b. Anti-synchronization of two different chaotic systems. Phys. A 387 (14), 3747–3757.

Liu, C., Liu, L., Liu, T., 2009. A novel three-dimensional autonomous chaos system. Chaos Solitons Fractals 39 (4), 1950–1958.

Lu, J.F., 2008. Generalized (complete, lag, anticipated) synchronization of discrete-time chaotic systems. Commun. Nonlinear Sci. Numer. Simul. 13 (9), 1851–1859.

Lu, J.G., 2005. Chaotic dynamics and synchronization of fractional-order Arneodo's systems. Chaos Solitons Fractals 26 (4), 1125–1133.

Lu, J.G., Chen, G., 2006. A note on the fractional-order Chen system. Chaos Solitons Fractals 27 (3), 685–688.

Lü, J.H., Chen, G.R., Yu, X.H., Leung, H., 2004. Design and analysis of multi-scroll chaotic attractors from saturated function series. IEEE Trans. Circuits Syst. I Regul. Pap. 51 (12), 2476–2490.

Mainieri, R., Rehacek, J., 1999. Projective synchronization in three-dimensional chaotic systems. Phys. Rev. Lett. 82 (15), 3042–3045.

Mandelbrot, B., 1982. The Fractal Geometry of Nature. Flammarion, New York, NY.

Marwan, N., 2008. A historical review of recurrence plots. Eur. Phys. J. Spec. Top. 164 (1), 3–12.

Marwan, N., Kurths, J., 2002. Nonlinear analysis of bivariate data with cross recurrence plots. Phys. Lett. A 302 (5-6), 299–307.

Marwan, N., Wessel, N., Meyerfeldt, U., Schirdewan, A., Kurths, J., 2002. Recurrence plot based measures of complexity and its application to heart rate variability data. Phys. Rev. E 66 (2), 026702.

Marwan, N., Romano, M.C., Thiel, M., Kurths, J., 2007. Recurrence plots for the analysis of complex systems. Phys. Rep. 438 (5-6), 237–329.

Odibat, Z.M., Corson, N., Aziz-Alaoui, M.A., 2010. Synchronization of chaotic fractional-order systems via linear control. Int. J. Bifurc. Chaos 20 (1), 81–97.

Ouannas, A., Azar, A.T., Ziar, T., Vaidyanathan, S., 2017a. On new fractional inverse matrix projective synchronization schemes, Studies in Computational Intelligence, 688. Springer-Verlag, Germany, pp. 497–524.

Ouannas, A., Azar, A.T., Ziar, T., Vaidyanathan, S., 2017b. Fractional inverse generalized chaos synchronization between different dimensional systems, Studies in Computational Intelligence, 688. Springer-Verlag, Germany, pp. 525–551.

Ouannas, A., Azar, A.T., Ziar, T., Vaidyanathan, S., 2017c. A new method to synchronize fractional chaotic systems with different dimensions, Studies in Computational Intelligence, 688. Springer-Verlag, Germany, pp. 581–611.

Ouannas, A., Azar, A.T., Ziar, T., Radwan, A.G., 2017d. Study on coexistence of different types of synchronization between different dimensional fractional chaotic systems, Studies in Computational Intelligence, 688. Springer-Verlag, Germany, pp. 637–669.

Ouannas, A., Azar, A.T., Ziar, T., Radwan, A.G., 2017e. Generalized synchronization of different dimensional integer-order and fractional order chaotic systems, Studies in Computational Intelligence, 688. Springer-Verlag, Germany, pp. 671–697.

Ouannas, A., Azar, A.T., Vaidyanathan, S., 2017f. On a simple approach for Q-S synchronization of chaotic dynamical systems in continuous-time. Int. J. Comput. Sci. Math. 8 (1), 20–27.

Ouannas, A., Azar, A.T., Vaidyanathan, S., 2017g. New hybrid synchronization schemes based on coexistence of various types of synchronization between master-slave hyperchaotic systems. Int. J. Comp. Appl. Technol. 55 (2), 112–120.

Ouannas, A., Azar, A.T., Ziar, T., 2017h. On inverse full state hybrid function projective synchronization for continuous-time chaotic dynamical systems with arbitrary dimensions. Diff. Eq. Dynam. Syst. 21, 1–14.

Ouannas, A., Azar, A.T., Vaidyanathan, S., 2017i. A robust method for new fractional hybrid chaos synchronization. Math. Methods Appl. Sci. 40 (5), 1804–1812.

Ouannas, A., Grassi, G., Azar, A.T., Radwan, A.G., Volos, C., Pham, V.T., et al., 2017j. Dead-beat synchronization control in discrete-time chaotic systems. The 6th International Conference on Modern Circuits and Systems Technologies (MOCAST), Thessaloniki Greece, 4–6 May.

Pecora, L.M., Carroll, T.L., 1990. Synchronization in chaotic systems. Phys. Rev. Lett. 64 (8), 821–824.

Petráš, I., 2008. A note on the fractional-order Chua's system. Chaos Solitons Fractals 38 (1), 140–147.

Poincaré, H., 1890. Sur le problème des trois corps et les équations de la dynamique. Acta. Math. 13, 1–270.

Razminia, A., Baleanu, D., 2013. Complete synchronization of commensurate fractional order chaotic systems using sliding mode control. Mechatronics 23 (7), 873–879.

Silva, C.P., 1993. Shil'nikov's theorem - A tutorial. IEEE Trans. Circuits Systems-I: Fund. Theory Applicat. 40 (10), 675–682.

Singh, S., Azar, A.T., Ouannas, A., Zhu, Q., Zhang, W., Na, J., 2017. Sliding mode control technique for multi-switching synchronization of chaotic systems. 9th International Conference on Modelling, Identification and Control (ICMIC 2017), Kunming China, 10-12 July.

Soliman, N.S., Said, L.A., Azar, A.T., Madian, A.H., Radwan, A.G., Ouannas, A., 2017. Fractional controllable multi-scroll V-shape attractor with parameters effect. The 6th International Conference on Modern Circuits and Systems Technologies (MOCAST), Thessaloniki Greece, 4-6 May.

Takens, F., 1981. Detecting strange attractors in turbulence. Dynamical Systems and Turbulence. In: Rand, D.A., Young, L.S. (Eds.), Lecture Notes in Mathematics, 898. pp. 366–381.

Tavazoei, M.S., Haeri, M., 2007a. Unreliability of frequency-domain approximation in recognizing chaos in fractional-order systems. I.E.T. Signal Proc. 1 (4), 171–181.

Tavazoei, M.S., Haeri, M., 2007b. A necessary condition for double scroll attractor existence in fractional-order systems. Phys. Lett. A 367 (1), 102–113.

Tavazoei, M.S., Haeri, M., 2008a. Synchronization of chaotic fractional-order systems via active sliding mode controller. Phys. A 387 (1), 57–70.

Tavazoei, M.S., Haeri, M., 2008b. Chaotic attractors in incommensurate fractional order systems. Phys. D 237 (20), 2628–2637.

Tolba, M.F., AbdelAty, A.M., Soliman, N.S., Said, L.A., Madian, A.H., Azar, A.T., et al., 2017. FPGA implementation of two fractional order chaotic systems. Int. J. Electr. Commun. 78, 162–172.

Vaidyanathan, S., Azar, A.T., Ouannas, A., 2017a. An eight-term 3-D novel chaotic system with three quadratic nonlinearities, its adaptive feedback control and synchronization, Studies in Computational Intelligence, 688. Springer-Verlag, Germany, pp. 719–746.

Vaidyanathan, S., Zhu, Q., Azar, A.T., 2017b. Adaptive control of a novel nonlinear double convection chaotic system, Studies in Computational Intelligence, 688. Springer-Verlag, Germany, pp. 357–385.

Vaidyanathan, S., Azar, A.T., Ouannas, A., 2017c. Hyperchaos and adaptive control of a novel hyperchaotic system with two quadratic nonlinearities, Studies in Computational Intelligence, 688. Springer-Verlag, Germany, pp. 773–803.

Vladimirsky, E.I., Ismailov, B.I., 2015a. Synchronization, control and stability of fractional order hyperchaotic systems in the context of the generalized memory. Int. J. New Tech. Res. 1 (8), 42–48.

Vladimirsky, E.I., Ismailov, B.I., 2015b. Topological control and stability of dynamic fractional order systems with generalized memory. Inf. J. Eng. Innovat. Res. 4 (3), 468–472.

Wang, Z., Volos, C., Kingni, S.T., Azar, A.T., Pham, V.T., 2017. Four-wing attractors in a novel chaotic system with hyperbolic sine nonlinearity. Optik – Int. J. Light Electr. Opt. 131 (2017), 1071–1078.

Webber, C.L., Zbilut, J.P., 2005. Recurrence quantification analysis of nonlinear dynamical systems. In: Riley, M.A., Van Orden, G.C. (Eds.), Tutorials in Contemporary Nonlinear Methods for the Behavioral Sciences. Digital Publication Available through the National Science Foundation, Arlington, VA, pp. 26–94.

Wedekind, I., Parlitz, U., 2001. Experimental observation of synchronization and anti-synchronization of chaotic low-frequency-fluctuations in external cavity semiconductor lasers. Int. J. Bif. Chaos Appl. Sci. Eng. 11 (4), 1141–1147.

Wolf, A., Swift, J.B., Swinney, H.L., Vastano, J.A., 1985. Determining Lyapunov exponents from a time series. Phys. D. 16 (3), 285–317.

Zbilut, J.P., Webber, C.L., 1992. Embeddings and delays as derived from quantification of recurrence plots. Phys. Lett. A 171 (3-4), 199–203.

Zbilut, J.P., Giuliani, A., Webber, C.L., 1998. Recurrence quantification analysis and principal components in the detection of short complex signals. Phys. Lett. A 237 (3), 131–135.

Zhang, J.F., Pei, Q., Zhang, X.L., 2011. A new kind of nonlinear phenomenon in coupled fractional-order chaotic systems: coexistence of anti-phase and complete synchronization. Chinese Phys. B 20 (8), 162–168.

Zhou, T., Li, C., 2005. Synchronization in fractional-order differential systems. Phys. D 212 (1-2), 111–125.

Zhu, Q., Azar, A.T., 2015. Complex system modelling and control through intelligent soft computations, Studies in Fuzziness and Soft Computing, 319. Springer-Verlag, Germany, ISBN: 978-3-319-12882-5.

Index

Note: Page numbers followed by "*f*" and "*t*" refer to figures and tables, respectively.

D

E

F

G

H

I

Q

R

S

T

CPI Antony Rowe
Chippenham, UK
2018-10-11 19:49